CURRENT CONCEPTS IN BOTANY

Current Concepts in Botany

Editors

K.G. Mukerji
Distinguished Mycologist and Microbial Ecologist
University of Delhi

C. Manoharachary
Osmania University,
Hyderabad, India

I.K. International Publishing House Pvt. Ltd.
NEW DELHI • BANGALORE

Published by

I.K. International Publishing House Pvt. Ltd.
S-25, Green Park Extension, Uphaar Cinema Market
New Delhi - 110 016 (India)
E-mail : ik_in@vsnl.net

ISBN 81-88237-64-7

10 9 8 7 6 5 4 3 2

Published by Krishan Makhijani for I.K. International Publishing House Pvt. Ltd., S-25, Green Park Extension, Uphaar Cinema Market, New Delhi - 110 016 and Printed by Rekha Printers Pvt. Ltd., Okhla Industrial Area, Phase II, New Delhi - 110 020.

Prof. Triloki Nath Lakhanpal

Current Concepts in Botany
is dedicated to
Prof. Triloki Nath Lakhanpal
(Himachal Pradesh University, Shimla, H.P.)
on completing his 60th year. He is internationally recognised for his taxonomic work on myxomycetes and mushrooms. For the last three decades he has extensively surveyed the North West Himalayas for myxomycetes and mushrooms and has worked on their taxonomy, ecology and physiology.

Preface

During recent years enormous advances have been made in all aspects of plant sciences and to bring the widely different aspects under one cover is indeed a herculean but subjective task. *Current Concepts in Botany* is a collection of review articles as well as original research papers from contemporary fellow botanists from all over the world and is dedicated to Professor T.N. Lakhanpal, an internationally known mycologist (fungal taxonomist) and plant pathologist. His work on Indian Myxomycetes is a monumental contribution which not only brought name and fame to him but also brought glory to India. For his significant contributions he will always be remembered with respect.

The volume contains 31 authoritative and thought provoking articles of both applied and fundamental value. These have been written by leading scientists in the area of their specialisation. In developing this volume we decided that the treatment of the subject should be comprehensive but concise enough to enable completion within set time. Further, we wanted to develop a treatise that would offer a detailed overview of applied aspects of Botany in terms of its theoretical, methodological and empirical contributions. By emphasizing the interdisciplinary aspects of the subject we hope that this volume will highlight recent prospective linkages among diverse research paradigms.

Originally we intended to bring out the volume on the date of his superannuation (60 years on 1st July, 2005) but due to some unforeseen circumstances it could not be accomplished. We are extremely sorry for this unintentional delay and tender our sincere apologies to all the contributors of this volume.

Because of the multidisciplinary nature of the subject, judgement had to be made about the relative importance and significance of the topics. Only selected contributions have been included. In a work of this nature, it is just possible that some mistakes might have crept in inadvertently and we own the responsibility. The credit for anything of value goes to the contributors who responded instantaneously to our request for writing comprehensive review articles and for sticking to the time schedule. We thank them all from the core of our heart. It was because of their active cooperation that we completed this uphill task within a short time.

We take great pleasure in presenting this festschrift volume to Professor Triloki Nath Lakhanpal as a token of our appreciation and love. We wish him many more happy and fruitful years of research and a peaceful family life. Hope this book will be useful to those interested in the pursuit of plant sciences.

Editors

Contents

List of Contributors

A.K. Bhatnagar, Department of Botany, University of Delhi, Delhi

A.N. Balakrishna, Department of Agricultural Microbiology, University of Agricultural Sciences, GKVK Campus, Bangalore, India

Ajay Kumar Pandey, Department of Biotechnology, Thapar Institute of Engineering and Technology, Patiala, India

Anupama Shukla, Department of Botany, A.N.D. College, Kalkaji, University of Delhi, New Delhi-110019

Ashok Gaur, Division of Seed Science and Technology, Indian Agricultural Research Institute, New Delhi

B.D. Sharma, National Bureau of Plant Genetic Resources, Regional Station, Shimla (HP), India

B.N. Prasad, Biotechnology Unit, Central Department of Botany, Tribhuvan University, Kritipur, Kathmandu, Nepal.

Bhavneet, Department of Botany, Punjab University, Chandigarh-160 014

Bhoopander Giri, Department of Botany, Hindu College, University of Delhi, Delhi-110007

C. Manoharachary, Department of Botany, Osmania University, Hyderabad – 500 007, A.P., India

C.L. Jandaik, Department of Mycology & Plant Pathology, University of Horticulture & Forestry, Nauni, Solan-173 230, India

C.V. Sarswat, Department of Tree Improvement & Genetic Resources, Dr. YS Parmar University of Horticulture & Forestry, Nauni, Solan (HP) India

D.J. Bagyaraj, Department of Agricultural Microbiology, University of Agricultural Sciences, GKVK Campus, Bangalore, India

D.K. Agarwal, Division of Plant Pathology, Indian Agricultural Research Institute, New Delhi

Davood Eradatmand Asli, Department of Botany, Punjab University, Chandigarh-160 014

Dilfuza Egamberdiyeva, Center of Agro Ecology, University of Agriculture, University str. 1, 700140, Tashkent, Uzbekistan

Firoz Anwar, Department of Environmental Biology, University of Delhi, Delhi-110 007

H. Govindapyari, Department of Botany, University of Delhi, Delhi

Harinder Singh, Department of Botany, Punjabi University, Patiala, Punjab

Ishwar Singh, Applied Microbiology and Biotechnology Laboratory, Department of Botany, University of Delhi, Delhi, India

I. Usha Rao, Department of Botany, University of Delhi, Delhi-110 007

I.K. Kunwar, Botany Department, Osmania University, Hyderabad

I.S. Dua, Department of Botany, Punjab University, Chandigarh-160 014

J.C. Rana, National Bureau of Plant Genetic Resources, Regional Station, Shimla (HP), India

J.R. Sharma, Botanical Survey of India, 192, Kaulagarh Road, Dehra Dun (U.P.)

K.G. Mukerji, Formerly Professor, Department of Botany, University of Delhi, Delhi

K.V.B.R. Tilak, Department of Botany, Osmania University, Hyderabad, A.P.

Kaori Tomita-Yokotani, Graduate School of Life and Environmental Sciences, University of Tsukuba

Lokendra Singh, Applied Microbiology and Biotechnology Laboratory, Department of Botany, University of Delhi, Delhi, India

M. Sharma, Department of Botany, Punjabi University, Patiala, Punjab

M. Sudhakara Reddy, Department of Biotechnology, Thapar Institute of Engineering and Technology, Patiala, India

Manju Tomar-Balhara, Department of Botany, University of Delhi, Delhi

N. Raaman, Centre for Advanced Studies in Botany, University of Madras

Narain Singh Chauhan, Department of Forest Products, Dr. YS Parmar University of Horticulture and Forestry, Nauni-Solan (HP) India

Naresh Arora, International Center for Genetic Engineering and Biotechnology (ICGEB) P.O. Box 10504, Aruna Asaf Ali Marg, New Delhi-110067

Neema Agarwal, International Center for Genetic Engineering and Biotechnology (ICGEB) P.O. Box 10504, Aruna Asaf Ali Marg, New Delhi-110067

Nuray Özer, Department of Plant Protection, Faculty of Agriculture, Trakya University 59030, Tekirdao/Turkey

P. Pardha Saradhi, Department of Environmental Biology, University of Delhi, Delhi-110 007

P. Sharmila, Department of Environmental Biology, University of Delhi, Delhi-110 007

Parvinder Kaur, Department of Microbiology, University of Delhi, South Campus, Benito Juarez Road, New Delhi-110 021

Prem L. Uniyal, Department of Botany, University of Delhi, Delhi

Priyanka Kapoor, Department of Botany, University of Delhi, Delhi-110 007

Rajesh Kumar, Applied Microbiology and Biotechnology Laboratory, Department of Botany, University of Delhi, Delhi, India

R.K. Sharma, Division of Plant Pathology, Indian Agricultural Research Institute, New Delhi

R.V. Gandhe, Postgraduate Research Centre, Department of Botany, Modern College, Pune.

Raj K. Bhatnagar, International Center for Genetic Engineering and Biotechnology (ICGEB) P.O. Box 10504, Aruna Asaf Ali Marg, New Delhi-110067

Rajni Gupta, Shyam Lal College, Shahadara, University of Delhi, Delhi-32

Ritu Sarsoha, Department of Botany, Punjab University, Chandigarh-160 014

Rupam Kapoor, Department of Botany, University of Delhi, Delhi

S. Neupane, Biotechnology Unit, Central Department of Botany, Tribhuvan University, Kritipur, Kathmandu, Nepal.

S. Patharajan, Guindy Campus, Chennai, India

S. Sharath, Department of Agricultural Microbiology, University of Agricultural Sciences, GKVK Campus, Bangalore, India

S. Swarupa Rani, Botany Department, Osmania University, Hyderabad

Shipra Gupta, Department of Environmental Biology, University of Delhi, Delhi-110 007

Surinder Kaur, Sri Guru Tegh Bahadur Khalsa College, Department of Botany, University of Delhi, Delhi

T. Satyanarayana, Department of Microbiology, University of Delhi, South Campus, Benito Juarez Road, New Delhi-110 021

Takako Kato, Graduate School of Life and Environmental Sciences, University of Tsukuba

Toshisada Suzuki, Faculty of Agriculture, Kagawa University

Usha Dev, National Bureau of Plant Genetic Resources, Pusa Campus, New Delhi

Ved Pal Singh, Applied Microbiology and Biotechnology Laboratory, Department of Botany, University of Delhi, Delhi, India

Vidya Thakur, Department of Tree Improvement & Genetic Resources, Dr. YS Parmar University of Horticulture & Forestry, Nauni, Solan (HP) India

Vikas Kumar, Department of Botany, Punjab University, Chandigarh-160 014

Vivek Mitter, National Bureau of Plant Genetic Resources, Pusa Campus, New Delhi

1

Preservation of Biodiversity of Fungi: Herbarium Specimens and Culture Collection

D.K. AGARWAL and R.K. SHARMA
Division of Plant Pathology,
Indian Agricultural Research Institute,
New Delhi

I. INTRODUCTION

Since the time immemorial man has been curious for identifying, classifying and preserving objects present around himself. Formally, it was according to his basic requirements, but as science progressed newer and newer approaches came into being, which also affected the ancient methods of identification, classification and preservation.

The multiple variation in the types of living organisms in any geophysical area is called biodiversity. With reference to large organisms (plants, animals etc.) the distinctive visible anatomical and morphological features enable us to identify the different genera and species and, therefore, we can easily assess the extent of biodiversity. However, most fungi are not visible to the naked eye and they need to be studied under the microscope. The culturable fungi can be identified, provided they sporulate but the biotrophs which are not amenable to culture media go unnoticed. It is, therefore, imperative that there is a gap in knowledge of the biodiversity of all microbes including fungi.

The study of systematic mycology depends on the type, quality and conditions of the material available in preserved state in various repositories. It is often difficult to obtain desirable types of microorganisms due to lack of fresh or preserved material. It, therefore, becomes imperative to study the diseased plant material in herbaria where they can be stored and made easily accessible to any worker. Recent estimates show that more than seventy percent of the active investigators are engaged in systematic study of plant diseases incited by fungi, yeasts or bacteria (Hawksworth, 1974; Hawksworth *et al*, 1983), while the rest are busy in studying ecology, physiology, cytogenetics and molecular biology. According to Martin (1951), there are at least 1,00,000 species of fungi existing in the world and the number of species of Phanerogams is not less than 2,50,000. This estimate seems conservative. In the literature, a large number of new species are being added at the rate of 1500 every year and more than half that rate new combinations are also added. Hence, the above estimate cannot be taken as valid. Therefore, whatever aspect of systematic mycology is pursued, patience, persistence, care and adequate facilities are pre-requisite. To categorize such a vast number of fungi in their taxonomic groups and relative orders, a hierarchical system of classification is used, as in plant nomenclature, based on the affinity of the different ranks.

India is one of the 12 mega-biodiversity countries bestowed with rich flora and fauna diversity. Though, modern biological studies and surveys in India started during the late 18th century, India has a long history and tradition of ecofriendly and sustainable, nature conservation practices deeply rooted in its lifestyle and culture. Botanical Survey of India (BSI) was constituted in 1890 and Zoological Survey of India (ZSI) in 1916, Forest Survey of India (FSI) was created in 1981. These agencies are involved in maintaining biological collections in addition to numerous university departments and colleges and other research institutions. Till date there are about 200 herbariums, 100 zoological museums and more than 30 microorganisms collections spread across the length and breadth of the country. It is estimated that these collections together house more than 100 million specimens. These collections are, in effect, time capsules to analyze conditions from the past and compare them with our present-day state of affairs.

It is necessary to define the term 'specimen' in true mycological terminology. The term is applied to a single fruit body of one of the larger fungi such as agaric but more often to one or more colonies of microorganisms together with the natural substratum on which it occurs (it may be a leaf, stem, root or an insect).

Systematic collection provides the basis of important taxonomic research. The ideal way to learn and collect diseased specimen or material is to join field excursions in small groups led by local experts. Careful screening of limited number of sites is often more fruitful than visits to different places. A good collection should reflect what grows in a particular locality. Correct identification of hosts and substrates is a must. In the absence of identity of the host plant, adequate material for identification, such as flowers, fruits and leaves etc. should also be collected for correct identification.

II. PRESERVATION

A preservative is a substance used for killing or stopping the growth of an organism in or on the substrate on which it grows and the phenomenon is known as preservation.

The preservation of diseased plant material or organism is done in two ways as mentioned below.

Preservation in Dry Form

The necessary four major steps for preserving a diseased specimen in dry form are as follows:

Collection: Diseased material is collected in the field and then it is properly pressed between the blotter sheets. Paper slips or bags are used to indicate locality, host, date and place of collection or other substrate on which it is present. For collecting damp material and fleshy fungi, the collection should be properly wrapped in moisture absorbing sheets to avoid their damage by moulds and insects etc. Dry specimens (e.g., flowering plants) can be kept for long periods (even for hundred years or more), but preservation in natural conditions is difficult.

In short, an experienced collector during collection always pays attention to quality representative material. It should be of good quality preferably should have reproductive parts, and quantity. The material should be collected in sufficient quantity so that it can be made available to any one for investigation. It would be appropriate here to state that "Specimens in the herbarium need to be seen as working tools, like books and journals in a library. However, in a book only the description of the disease and a fungus can be seen, but in the specimen the disease and the fungus themselves are there. Therefore, one good specimen is worth a thousand words."

Drying and accession: Drying is a process in which dehydration of the collected material occurs; this helps in long-term preservation. This is done in hot dry air or in the sunlight for a few days after carefully pressing the specimens between blotter sheets. Best drying is obtained by frequently changing the blotter sheets, particularly during the first few days. Sun-drying is economical and easy but for fleshy fungi, hot air or vacuum-drying is the best. After drying, an entry is made for the specimen collected on the herbarium sheet and then it is accessioned in a large, strongly bound accession register (16″ × 12″). This practice is followed in all important herbaria of the world and also in our national herbarium [Herbarium Cryptogamae Indiae Orientalis (H.C.I.O.), IARI, New Delhi]. The serial accession number is allocated to the specimen entered in a column on the extreme left in the register and the name of the fungus, host and country from which the specimen came along with the name of the collector are also entered. The right-hand column is kept reserved for the remarks of the specialist who may examine the specimen later on.

Labelling and packeting: The accessioned material is kept in a stout, good quality herbarium packet (6″ × 4″). This is attached to a chart paper sheet (16.5″ × 10.5″). The two sides of the packet are folded first, then the bottom flap and finally the top flap; this gives an appearance of an envelop. On the extreme lower left of the chart paper, entries regarding the name of the fungus, host (if applicable), locality, date and identified by the scientist or specialist are mentioned on a printed label (5″ × 3″) which is already pasted there. The entire sheet along with the folder is then placed inside a file cover (16.5″ × 10.5″). On the lower side of cover, accession number, name of the fungus, host and place of collection are given to locate it easily.

Storage and disinfection: These folders along with the specimens are then stored in steel or wooden almirahs. Before putting the folders inside the almirah, these are fumigated in order to protect them from climatic and environmental hazards, e.g., moulds, mites and insects. Fumigation is done by methyl bromide vapours in a fumigation chamber for 24-48 h. Naphthalene balls, para-dichlorobenzene, xylene, thymol or flit is sprayed/placed in a small container in each compartment of the shelf along with folders, whenever facility of methyl bromide fumigation is not available. The method effectively preserves the specimens in their original shape for years. For storing fleshy fungi, dehydration is carried out at 50°C in an electric oven. Periodic fumigation or spraying of specimens in the folders is carried out to avoid any damage by mites and other insects (Miller and Richard, 1974).

Preservation by Slide Collection

Preservation of materials by slide collection can be obtained in two ways.

(i) Microscopic slide preservation
(ii) Photographic slide preservation

Microscopic slide preservation: Preservation of fungi sporulating in pure form in culture can be made by mounting the material in DPX mount or in 0.1% lactophenol. Such slides can be stored for 2-3 years in a good shape. The cover-slip is ringed with the help of nail-polish or Canada balsam to avoid drying of the material.

Photographic slide preservation: For the preservation of dry spore masses of Agarics and Boletus, photographic slides of rare species of Agarics showing spore prints are taken to study the structural features of various species. These slides are then numbered and accessioned in the herbarium. A list is prepared for each genus and species. These can be used as visual aids in teaching as well as in identification.

Dry Preservation of Living Fungal Cultures

This is a semi-freeze-drying method and is useful for mature spore-bearing fungal cultures or specimens. Specimens which are rare and scanty can be transferred into the culture form (if possible) and can be preserved for years to come. In this method, McCartney screw cap vials or bottles are half-filled with self-indicating silica gel. A small piece of aluminium foil is then placed in each bottle so that it forms a diagonal slope at the width of the bottle. The bottles are then autoclaved for one hour at 15 lb ppi (120°C temp.). Inoculum is then placed on the foil in the form of 5 mm diameter agar discs, six to eight per bottle. The discs are conveniently cut using a cork-borer and are lifted with a nichrome wire flattened at the tip and inserted into the bottles from the side. They are then placed on the foil right side up. The bottles are placed in refrigerator immediately after inoculation. The discs dry out in about two hours. In order to subculture from the foil, a piece bearing an agar disc is torn away using sterile forceps. This is then placed on a fresh agar plate. Cultures grow rapidly from this type of inoculum (Bagg, 1967).

Preservation by Polyurethane Coating

An elegant technique for the preservation of fragile or rapidly decomposing diseased plant material has been developed by Bebbington and Burrell (1968). The specimens are dried, either in an electric oven or in a freeze drier and then they are coated with undiluted clear polyurethane by dipping. The specimens are then dried in an oven at 50-60°C. This forms a hard, durable and transparent coating around the specimen which does not obscure its finer details and imparts a natural lustre to it. The coated material can be handled freely without fear of damage. Care is taken to ensure that the polyurethane completely covers the surface of the plant material.

A large number of diseased specimens can be preserved by this method, specially grasses and apple leaves. For the preservation of potato and pea leaves, the polyurethane is diluted (2 : 1) with rectified spirit (ethyl alcohol), resulting in much thinner coating (Hawksworth, 1974).

Direct Preservation of Specimens Collected in Nature

For the collection and preservation of myxomycetes, lichens, aquatic and coprophilous fungi, the following procedures are used.

Myxomycotina: Myxomycetes are found in abundance on woods after two/three days of heavy rains. Usually they are found on rotting wood but a few also occur on dung and damp straw. The specimen should be collected together with the substrate on which it is growing with the help of a sheath knife and it should be immediately pinned to cork in small box of cardboard or wood to reduce the risk of damage. For preserving, the specimen should be dried slowly and glued to box for permanent storage. Members of the Labyrinthulales are isolated by placing algae on 1% serum-sea water agar when they appear after 2-7 days (Carlile, 1971).

Lichens: Lichens are slow in growth and also rare in nature. Therefore, no specimen should be collected unless otherwise needed for scientific purpose or identification. For collections, lichen should be removed with great care so that the basal part of the thallus in fruiticose species is not damaged, and the substratum is removed properly by using a sheath knife if corticolous, or a geological hammer and chisel if saxicolous. The specimen should then be placed in newspaper or tissue paper packet and carefully labeled. It should be dried by leaving it open in warm dry place overnight. This will help in keeping its identity except that the colour of the thallus may change. When dry, it should be kept in storage by pinning or gluing to cards placed in stiff paper packet bearing full details of the collection. These packets are then fixed to herbarium sheets.

Aquatic fungi: Many aquatic fungi occur in fresh and sea water, on fishes, algae, wood and different types of decaying materials in slow flow streams, lakes and ponds. Hyphomycetes are often found associated with decaying leaves. Specimens are placed in polythene bags or tubes together with water from the site. Baiting technique is generally used for isolating aquatic fungi. Hair, seeds, fruits, bread, meat and/or twigs are placed in a small wire mesh basket or suspended in water by nylon cord and examined after 1-5 weeks. This technique is effective and satisfactory, and gives desired results in a short time. A comprehensive method for collecting aquatic and marine fungi is provided by Jones (1971).

Coprophilous fungi: Several interesting fungi, such as members of mucorales, ascomycotina, basidiomycotina and myxomycotina thrive on dungs of various animals and birds. The different stages of decomposition of dung give different mycoflora. For isolation, dung should be placed in damp chambers (petridishes with moist filter paper or bell jars) or agar plate dishes to follow the succession of different species of fungi which cannot be detected in the field, but can be observed easily in the laboratory. Dung having fungi can be dried, fumigated and kept in herbaria in paper packets or glued onto small boxes, as preserved specimens. Lindqvist (1972) provides a useful account of methods for detecting coprophilous pyrenomycetes.

Preservation in Fluid

Different conventional methods of preserving fungi in original form are commonly employed by using various fixatives and preservatives depending upon the type of material. Fixatives are helpful in keeping the material indefinitely in its perfect state. The most commonly used fixatives and preservatives are mentioned below.

Fixatives

Formol-acetic alcohol (F.A.A.): Formalin - 13 ml; Acetic acid (glacial) - 5 ml; Ethyl alcohol (50%) - 200 ml. This fixative helps in restoring the natural shape of the material by checking its dehydration (Ainsworth, 1971).

Flemming's week solution: Chromic acid (1%) - 25 ml; Acetic acid (1%) - 10 ml; Water - 60 ml; Osmic acid (2%) - 5 ml, is added before using this solution.

Preservatives

The general preservative used for museum specimens consists of 5 percent formaldehyde (40%) in water.

Formaldehyde-alcohol solution: Formaldehyde (40%) - 25 ml; Ethyl alcohol (95%) - 150 ml; Water - 1,000 ml. Alcohol helps in dissolving air and formaldehyde maintains the material in its natural form.

Hesler's preservative: This is used for the preservation of coloured vegetables and fruits. It contains the following ingredients: Zinc chloride - 5 g; Formaldehyde (40%) - 25 ml; Glycerol - 25 ml; Water - 1,000 ml.

Preservation of Coloured Fungi

Preservation of coloured fungi depends largely on the solubility of their colour in water and accordingly the following preservatives are used depending on the type of material to be preserved.

Colour not soluble in water: Mercuric acetate - 10 g; Glacial acetic acid - 5 ml; Water - 1,000 ml.

Colour soluble in water: Two main types of preservatives are used for fungi which dissolve their colour in water.

(a) Mercuric acetate - 1 g; Neutral lead acetate - 10 g; Glacial acetic acid - 10 ml; Alcohol (90%) - 1,000 ml.

(b) Zinc sulphate - 25 g; Formaldehyde (40%) - 10 ml; Water - 1,000 ml.

Preservative for Diseased Green Plant Parts

This is accomplished by boiling plant parts in a mixture of one part of glacial acetic acid saturated with copper acetate and four parts of water till the green colour reappears and then keeping it in a 5 percent formalin. Another satisfactory method is to place the material in 5 percent copper sulphate solution for at least six and not more than twenty-four hours. The material is then washed for several hours in running tap water and preserved in sulphurous acid solution (5-6 percent SO_2) (15 ml) and distilled water (1000 ml) in sealed containers (Goncalves, 1931).

III. RECENT TECHNIQUES: PRESERVATION OF FUNGI

The Commonwealth Mycological Institute (now CABIMI) maintains a living collection of fungi to support the taxonomic and information service that it provides. The culture collection also supplies fungi at a charge for teaching, research and testing purposes, as well as services as a depository for all microfungi.

These cultures are kept in a number of different ways which fall into two categories—continuous growth involving sub-culturing, and induced dormancy achieved by dehydration or freezing. The former is carried out by periodical transfer onto suitable substrates and varying storage conditions can lengthen the time between sub-cultures. The latter is achieved by storage at ultra low temperatures or by dehydrating by freeze-drying or silica gel storage.

IV. METHODS OF CULTURE MAINTENANCE

Mushroom Strains

Mainly three methods are used to propagate mushroom strains, these are multispore or single spore culture, tissue culture and mycelial transfer. By periodic subculturing of the mycelium, mushroom cultures can be maintained for many years. However, sometimes frequent subculturing may result in its degeneration (loss of vigour, vitality and yield potentiality) e.g., *Pleurotus eous.* In such cases, time to time tissue culturing from the mushroom fruit body helps in overcoming degeneration problem. In case of homothallic mushrooms like *Agarecus bisporus,* tissue culture and single spore culture methods are employed to maintain the genetic characteristics of the strains. However, tissue cultures are often responsible for giving lower yield than the original culture. In tissue culturing from the mushroom fruit body, after its surface sterilization, small piece of tissue is taken under aseptic condition from the junction of the stipe and pileus and put on the agar slants for its culturing and incubated in B.O.D. incubator at 24-26°C. At this temperature mushroom cultures grow within 8-10 days. But great care should be taken in selecting a fruit body of the best quality, size, colour, shape, aroma and other desirable traits.

In heterothallic mushrooms like *Pleurotus* spp., multispore culture and tissue culture methods are employed to maintain the strain characteristics of mushrooms.

Frequent Transfer

This is the simplest method for maintenance of living fungi and is achieved by frequent or serial transfer from staled media to fresh, suitable solid or liquid media and storage in the most suitable conditions for the individual isolate. Some dermatophytes survive best on hair (Al Doory, 1968), some water moulds are stored in water with the addition of plant material (Goldie-Smith, 1956), and other more sensitive water moulds may require aeration (Clark and Dick, 1974; Webster and Davey, 1976). Such methods are very labour-intensive and time consuming when large collections are involved.

The main disadvantages of frequent transfer are:

(i) Danger of variation, loss of pathogenicity or other physiological or morphological characteristics.

(ii) Danger of contamination by air-borne spores or mite carried infections.

(iii) Requires constant specialist supervision to ensure that the fungus is not replaced by a contaminant or sub-cultured from a deteriorated sector.

The time period between transfers varies from fungus to fungus, for some it may be required once every 2-4 weeks though others may servive for 6-12 months without transfer. The use of cold storage can slow the rate of metabolism and thus increase the intervals between subcultures.

Storage at 4-7°C in a refrigerator or cold room can extend the transfer interval 4-6 months from the average period of 2-3 months. Storage in a deep freeze (–17° to –24°) will allow most fungi survive 4-5 years between transfers though freezing damage may occur in some fungi.

The above-mentioned disadvantages are not overcome by slowing the growth rate though it is thought that the rate at which variation may occur is similarly reduced.

V. STORAGE UNDER MINERAL OIL

Covering cultures on agar slants in 1 oz universal bottles with mineral oil prevents dehydration and slows down the metabolic activity and growth through reduced oxygen tension. This method was first extensively used by Buell and Weston (1947), and subsequent reports have indicated its wide application and success (Dade, 1960; Fennell, 1960; Little & Gordon, 1967; Smith *et al*, 1970; Onion, 1971, 1977).

Mature healthy cultures are covered by 1 cm of mineral oil (liquid paraffin or medicinal paraffin, specific gravity 0.830-0.890). The oil must be sterile and this can be achieved by heating and cooling twice in an autoclave at 121°C for 15 minutes.

Retrieval from an oiled culture is achieved by removal of a small amount of the fungal colony on a mounted needle and draining away as much oil as possible. The fungus is then streaked onto a suitable agar medium. More than one sub-culture may be necessary and after retrievals the growth rate can very often remain reduced due to adhering oil. The fungal mycelium can normally recover when it is re-isolated from the agar plated onto fresh media. There is an added risk of personnel contamination by spattering of oil containing fungus when sterilising inoculating needles in the Bunsen flame (Fennell, 1960).

A wide range of fungi survive this method, Saprolegniaceae and other water moulds survive 12-30 months (Reischer, 1949) and species of *Aspergillus* and *Penicillium* have remained viable for 30 years at CMI (Hawksworth *et al*, 1983). Many cultures have shown that they deteriorate under mineral oil and are regularly subcultured in every two years. Some species of *Cercospora, Arthrobotrys, Colletotrichum, Conidiobolus, Corticium* and *Nodulisporium*, are transferred at two yearly intervals at CMI.

The main disadvantages of oil storage are:

(i) Contamination by air-borne spores.

(ii) Retarded growth on retrieval.

(iii) Continuous growth under adverse conditions that could lead to selection.

Despite these disadvantages preservation under oil is recommended for storage of fungi for the laboratory with limited resources.

VI. SILICA GEL STORAGE

This method has proved to be very successful at CMI where 250 sporulating fungi have been stored for 7 years in silica gel and have revived well remaining morphologically stable. C.F. Roberts of the Department of Botany of Leicester University has successfully stored fungi by this method for long periods and they remained genetically stable.

Method

(i) Universal glass bottles are quarter-filled with non-indicating silica gel and sterilized in an oven (at 180°C) for 3 h.

(ii) The bottles are placed in a water tray and frozen in a deep freezer.

(iii) Spore suspensions are made in cooled 5% skimmed milk.

(iv) The silica gel is three-quarters wet with the spore suspension.

(v) After 10-14 days incubation at 25°C, when the crystals readily separate, the bottles are stored at 4°C in airtight containers with indicator silica gel to absorb moisture.

A wide range of sporulating fungi survive by this method (Perkins, 1962; Ogata, 1962). Thin walled spores, spores with appendages and mycelium tend not to survive by this technique.

VII. SOIL STORAGE

At CMI this method involves inoculation of double autoclaved soil (121°C for 15 min) with 1 ml of spore suspension in sterile distilled water and incubating at room temperature for 5-10 days depending on the rate of growth of the fungus being stored. This initial growth period allows the fungus to use the available moisture and they are then stored in a refrigerator at 4-7°C. This method of storage has proved very successful with *Fusarium* species (Booth, 1971).

Atkinson (1953, 1954) found that *Rhizopus, Alternaria, Aspergillus, Circinella* and *Penicillium* species survived 5 years storage without any morphological or physiological changes. More recently Shearer *et al* (1974) working with *Septoria,* and Reinecke and Fokkima (1979), working with *Pseudocercosporella herpotrichoides,* have reported stable and successful results.

VIII. WATER STORAGE

Storage periods of 20 years have been obtained with species of *Phytophthora* and *Pythium* at CMI before any loss of viability was noted (Boeswinkel, 1976).

Method

(i) 6 mm cubed agar blocks are cut from the growing edge of a fungal colony.

(ii) The blocks are placed in sterile distilled water in McCartney bottles and the lids tightly screwed down, they are stored at room temperature.

This method of storage was originally described by Castellani (1939, 1967) who stored fungi pathogenic to man. Figueiredo (1967) was able to preserve 22 plant pathogens up to three years without any loss in pathogenicity. Figueiredo & Pimentol (1975) subsequently reported 10 years successful storage by this method. Boeswinkel (1976) stored 650 plant pathogens including Phycomycetes, Ascomycetes, Fungi Imperfecti and Basidiomycetes for more than 10 years but they all remained viable and pathogenic for 7 years only. Clark & Dick (1974) with Oomycetes, Ellis (1979) with Entomophthorales, Pyrenomycetes, Hymenomycetes, Gasteromycetes and Hyphomycetes, and Marx & Daniel (1976) with ectomycorrhizal plant pathogens, reported successful results.

IX. FREEZE-DRYING

The preservation of fungal cultures by drying under vacuum from the frozen state by the sublimation of ice was first extensively used by Raper & Alexander (1945). The methods and machinery has been developed over the years to produce a reliable and successful preservation technique for sporulating microfungi. Stability and long storage periods are the main advantages of freeze-drying though the expense of the modern and quite complex machinery can be a deterrent factor.

It is not essential to use expensive equipment to carry out freeze-drying successfully. The basic requirements for a system are, a means of freezing the suspension, of generating and maintaining a vacuum and of absorbing the water vapour evolved. The ampoules containing the spore suspension can be attached to a vacuum pump via a desiccant trap and immersed in an alcoholic freezing mixture or 'Cellusolve bath', to freeze the cultures and then the ampoules can be evacuated. Drying is more or rapid once the ampoules have been removed from the freezing mixture though care must be taken not to allow thawing to occur. A vacuum gauge is very useful to monitor the process giving an indication of its different stages i.e. degree of dryness. Using the correct suspending medium helps in the achievement of optimum residual moisture. The use of skimmed milk protects the culture during freezing and the sugars present help in the retention of some moisture. The recommended final moisture content is between 1-2%. With a 'home made' piece of freeze-drying apparatus careful investigation into the method must be carried out to set the parameters in order to achieve the best results.

At CMI a two-stage process is used, the details of this process are as follows:

Method

(i) A spore suspension is prepared in 10% skimmed milk 5% inositol (sterilized by autoclaving).

(ii) 0.2 ml quantities are placed in sterile 0.5 ml neutral glass ampoules and covered with a lint cap.

(iii) The ampoules are placed in the centrifuge container of the Edwards High Vacuum EF6 primary drier and lowered into the chamber.

(iv) They are centrifuged, evacuated and dried for three hours accelerated by heating the chamber walls.

(v) The chamber is brought to atmospheric pressure, the ampoules removed, plugged with sterile cotton wool and constricted. This stage must be as short as possible as exposure to the atmosphere of the partially dried material can cause deterioration (Rey, 1977).

(vi) They are then placed on a Edwards 30S2 secondary drier and evacuated. The ampoules are sealed after a 17 hour drying process.

(vii) An ampoule is opened after 4 days storage, reconstituted with sterile distilled water and streaked onto an agar plate to test viability.

Controlled freezing at 1°C/min has been found to enable the freeze-drying of cultures that do not normally survive the above method. Using such prefreezing methods some delicate or non-sporulating fungi can be preserved by lyophilisation (Smith, 1982, 1983a).

Work on freeze-drying fungi include attempts to preserve mycorrhizal fungi (Jackson *et al*, 1973; Crush & Pattison, 1975) though good revivals were not reported. Clifford (1973) & Shain (1979) were able to freeze-dry uredospores satisfactorily, Kokko & Elliott (1977) attempted to preserve whole apothecia of *Mollerodiscus lentus.*

Further surveys of the process have been carried out by Heckly (1961, 1978), Nei (1964), Jong (1978), Alexander *et al* (1980) and Smith (1983b).

A list of genera that have failed the CMI centrifugal freeze-drying technique can be seen in Smith (1983a, 1983b). However, it must be borne in mind that success with freeze-drying varies between isolates of the same species. In general those fungi which grow and sporulate well in culture survive the process while the weak or deteriorated isolates tend to fail. It may, therefore, be misleading to state categorically that one particular species will not survive freeze-drying.

Ellis & Robertson (1968) revived most of the original cultures processed by Raper & Alexander (1945) 23 years ago and such a long-term storage speaks highly of this preservation technique for fungi. Results of freeze-drying at CMI are quite good and encouraging. Over 500 isolates of the fungi at CMI have survived 14 years storage and in times to come will hopefully remain viable for many more years with the advent of newer techniques.

X. LIQUID NITROGEN STORAGE

This method employs ultra low temperatures to halt the metabolism of microorganisms. It can be widely applied to sporulating and non-sporulating cultures of fungi which are kept stable throughout their storage period. No morphological or physiological changes have been observed with the 3000 isolates frozen and stored in liquid nitrogen collection comprising 2075 species and 595 genera. This method involves the precooling of 0.5 ml aliquots of fungal suspension in 10% glycerol sealed in 1 ml borosilicate glass ampoules to 7°C and freezing at a rate of 1°C/min at –35°C. The ampules are then plunged into liquid nitrogen and cooled rapidly to –196°C. For thawing and testing the ampoules are removed from the storage refrigerator and immersed in a water bath at 37°C before opening and streaking the contents onto a suitable growth medium.

A similar method has been in use at the American Type Culture Collection (ATCC) since 1965 and has given very good results (Hwang, 1966, 1968; Hwang *et al*, 1976; Butterfield *et al*, 1974; Alexander *et al*, 1980).

This method has been in use at CMI since 1968 (Onion, 1971, 1977) for the preservation of fungi. The only group to give high failure rates belong to the subdivision Mastigomycotina though very little work has been done with the Myxomycota and obligate pathogens. Where failure occurred due to the use of a mixed cryoprotectant (10% dimethyl sulphoxide and 8% glucose) proved successful in some cases (Smith, 1983b).

*** Comparative table showing different methods of preservation**

Methods	*Cost materials*	*Labour*	*Longevity*	*Genetic stability*
Frequent transfer	Low	High	Variable, 1-6 months	Poor
Dried on host	Low	Low	Variable	Good
Living on host	High	High	0-20 years	Good
Oil storage	Low	Medium	2-20 years	Poor
Water storage	Low	Medium	2-5 years	Moderate
Soil storage	Low	Medium	5-7 years	Moderate
Silica gel	Low	Medium	5-7 years	Good
Freeze-drying	High	Initially high	4-40 years	Good. Recent work has shown injury to DNA
Deep freeze	Low	Medium	4-5 years	Moderate
Liquid nitrogen	High	Medium infinite to date 15-17 years	Theoretically	Good

* Adapted from a table in: National Work Conference on Microbial Collections of Major Importance to Agriculture, University of Maryland, Maryland, 1981. American Phytopathological Society.

Fungi difficult or impossible to grow in culture can be stored by the above method. Rust and smut spores (Loegering, 1965; Prescott and Kernkamp, 1971), *Sclerospora* species (Gale *et al*, 1975; Long *et al*, 1978; Smith, 1982) have been successfully preserved by liquid nitrogen storage techniques. Phytopathogenic fungi have survived retaining their pathogenicity (Dahmen *et al*, 1983).

The initial cost of labour for freeze-drying is high due to the processing and the first viability tests but maintenance of a freeze dried collection is not expensive.

XI. CONCLUSIONS

Thus, it can be concluded that the main advantages of preserving fungi are that these can be stored and kept in dry as well as in their original form for long periods. Secondly, the preserved specimen can be studied in its natural form for investigations. In spite of the several methods available for preserving fungi, it has been observed that the number of fungal specimens is dwindling slowly as mycologists are not very serious, and sometimes do not deposit the important specimen in any recognized herbaria. It was rightly pointed out by Clark & Loegering (1967) that "the scientist who discards the biological materials is neglecting part of his responsibility as a scientist". It is, therefore, imperative to deposit these specimens at various important repositories and herbaria of the world so that at least one representative of the type is available for future studies.

Research on the preservation of specimens should be an integral function of herbaria and culture collection, and support of such researches is one of the aims of the World Federation of Culture Collections to facilitate many investigations in years to come. For the information of readers and research workers, a "World Data Centre" has been established and a precise system of recording the data has been developed to help in conserving the taxonomic literature at one place which will be useful in the identification of many taxa of fungi.

REFERENCES

Ainsworth, G.C. 1971. *Ainsworth's Dictionary of Fungi.* Sixth edition, Commonwealth Mycological Institute, Kew, England, 663 pp.

Al Doory, Y. 1968. *Mycologia,* 60: 720-723.

Alexander, M., P.M. Daggett, R. Gherna, S. Jong, F. Simione & H. Hatt 1980. American Type Culture Collection Methods I. *Laboratory Manual on Preservation, Freezing and Freeze-drying.* Rockville, Maryland, American Type Culture Collection, U.S.A.

Atkinson, R.G. 1953. *Canadian Journal of Botany,* 31: 542-547.

Atkinson, R.G. 1954. *Canadian Journal of Botany,* 32: 673-678.

Bagg, H.S. 1967. *Plant Disease Reporter,* 51: 1055-1058.

Bebbington, R.M. & M.M. Burrell 1968. *Bulletin British Mycological Society,* 2: 75.

Boeswinkel, H.J. 1976. *Transactions of the British mycological Society,* 66: 183-185.

Booth, C. 1971. *The Genus Fusarium.* CMI., Kew, England.

Buell, C.B. & W.H. Weston 1947. *Journal of Botany,* 34: 555-561.

Butterfield, W., S.C. Jong & M.J. Alexander 1974. *Canadian Journal of Microbiology,* 20: 1665-1673.

Carlile, M.J. 1971. *In Methods in Microbiology* (ed. Booth, C.) 4: 237-265. Academic Press, London, N.Y.

Castellani, A. 1939. *Journal of Tropical Medicine and Hygiene,* 42: 225-226.

Castellani, A. 1967. *Journal of Tropical Medicine and Hygiene,* 70: 181-184.

Clark, G. & M.W. Dick 1974. *Transactions of the British mycological Society,* 63: 611-612.

Clark, W.A. & W.Q. Loegering 1967. *Annual Review of Phytopathology,* 5: 319-342.

Clifford, B.C. 1973. *Cereal Rusts Bulletin,* 1: 30-34.

Crush, J.R. & A.C. Pattison 1975. Preliminary results on the production of vesicular-arbuscular mycorrhizal inoculum by freeze drying. In : F.E. Sanders, B. Mosse and P.B. Tinker (ed.). *Endomycorrhizas* pp. 485-509. Academic Press, London.

Dahmen, H., T.H. Staub & F.J. Schwinn 1983. *Phytopathology,* 73: 241-246.

Dade, H.A. 1960. Laboratory methods in use in the culture collection, CMI. In : *Herbarium Handbook,* pp. 78-83, CMI., Kew, England.

Ellis, J.J. 1979. *Mycologia,* 71: 1072-1075.

Ellis, J.J. & J.A. Robertson 1968. *Mycologia,* 60: 399-405.

Fennell, D.I. 1960. *The Botanical Review,* 26 : 79-141.

Figueiredo, M.B. 1967. *O Biolgico,* 33: 9-15.

Figueiredo, M.B. & C.P.V. Pimentol 1975. *Summa Phytopathologica,* 1: 299-302.

Gale, A.W., C.G. Schmitt & K.R. Bromfield 1975. *Phytopathology,* 65: 828-829.

Goldie-Smith, E.K. 1956. *Journal of the Elisa Mitchell Scientific Society,* 72: 158-166.

Goncalves, G.W. 1931. *Revista Agric. S. Paulo.,* 6: 315.

Hawksworth, D.L. 1974. *Mycologists Hand Book.* Commonwealth Mycological Institute, Kew, Surrey, England, pp. 1-231.

Hawksworth, D.L., B.C. Sutton & G.C. Ainsworth 1983. *Dictionary of the fungi.* CMI., Kew, England, 7th Ed. 412 pp. + vvi.

Heckly, R.J. 1961. *Advances in Microbiology,* 3: 1-76.

Heckly, R.J. 1978. *Advances in Applied Microbiology,* 24: 1-53.

Hwang, S.W. 1966. *Applied Microbiology,* 14: 784-788.

Hwang, S.W. 1968. *Mycologia,* 60: 613-621.

Hwang, S.W., W.F. Kwolek & W.C. Haynes 1976. *Mycologia,* 68: 205-212.

Jackson, N.E., R.H. Miller & R.E. Franklin 1973. *Soil Biology & Biochemistry,* 5: 205-212.

Jones, E.B.G. 1971. In: *Methods in Microbiology.* Booth, C. (ed.) 4: 336-365, Academic Press, London and N.Y.

Jong, S.C. 1978. Conservation of cultures. In: *The Biology and Cultivation of Edible Mushrooms.* Pp. 119-135, London and New York, Academic Press.

Kokko, E.G. & M.E. Elliott 1977. *Mycologia,* 69: 1206-1209.

Little, G.N. & M.A. Gardon 1967. *Mycologia,* 59: 733-736.

Loegering, W.Q. 1965. *Phytopathology,* 55: 247.

Long, R.A., J.M. Woods & G.G. Schmitt 1978. *Plant Disease Reporter,* 62: 479-481.

Lindqvist, N. 1972. *Iat. Symb. Bot. Upsal.,* 20: 1-374.

Martin, J. 1951. Hard Woods, Alens Key, Ed. 2. (F.P.R. Bull. No. 25), London, H.M.S.O.

Marx, D.H. & W.J. Daniel 1976. *Canadian Journal of Microbiology,* 22: 338-341.

Miller, C.S. & G.M. Richards 1974. *Transactions of the British Mycological Society,* 63: 607-610.

Nei, T. 1964. *Cryobiology,* 1: 87-93.

Ogata, W.N. 1962. *Neurospora Newsletter,* 1: 13.

Onion, A.H.S. 1971. *Methods in Microbiology.* Volume 4, pp. 113-151. London and New York, Academic Press.

Onion, A.H.S. 1977. *Proceedings of the Second International Conference on Culture Collections.* University of Queensland, Brisbane, World Federation of Culture Collection.

Perkins, D.D. 1962. *Canadian Journal of Microbiology,* 8: 591-595.

Prescott, J.N. & M.F. Kernkamp 1971. *Plant Disease Reporter,* 695-696.

Raper, K.B. & D.F. Alexander 1945. *Mycologia,* 37: 499-525.

Reinecke, P. & N.J. Fokkema 1979. *Transaction of the British Mycological Society,* 72: 329-331.

Reischer, H.S. 1949. *Mycologia,* 41: 177-179.

Rey, L.R. 1977. In : *Developments in Biological Standardisation.* International Symposium on Freeze Drying of Biological Products (ed. V.J. Cabasso and R.H. Regamey), pp. 19-27. Basel: S. Karger.

Shain, L. 1979. *Plant Disease Reporter,* 63: 368-369.

Shearer, B.L., R.J. Zeyen & J.J. Ooka 1974. *Phytopathology,* 64: 163-167.

Smith, D. 1982. *Transactions of the British Mycological Society,* 79: 415-421.

Smith, D. 1983a. *Transactions of the British Mycological Society,* 80: 333-337.

Smith, D. 1983b. *Transactions of the British Mycological Society,* 80: 360-363.

Smith, D.H., F.H. Lewis & C.L. Fergus 1970. Long term preservation of *Botryosphaeria ribis* and *Dibotryon morbosum. Plant Disease Reporter,* 54: 217-218.

Webster, J. & R.A. Davey 1976. Simple method for maintaining cultures of *Plastocladiella emersonii. Transactions of the British Mycological Society,* 67: 543-544.

2
Trichomycetes: Forgotten Fungi

C. Manoharachary
Department of Botany,
Osmania University,
Hyderabad – 500 007, A.P., India

I. INTRODUCTION

Insect gut inhabiting fungi live in the digestive tract of arthropods. Leidy (1849 a, b) was the first scientist to discover these fungi. These fungi were studied at length in Europe and other countries and only a moderate information is available from India (Misra, 1998). It is now known that arthropod hosts also benefit from the presence of these fungi, hence considered mutualistic.

II. BIODIVERSITY AND TAXONOMY

Trichomycetes is one of the two classes of phylum zygomycota. Harpellales, Asellariales and Eccriales are three orders. Earlier Amoaebidiales were included in Trichomyces and now it is believed to be protists. At present the class Trichomycetes comprises 53 genera and 213 species (Misra and Lichtwardt, 2000). India is rich in biodiversity and 27,000 fungal species are reported. Out of which, 1% species belong to Trichomycetes. This clearly indicates the sad state of scientific discoveries pertaining to Trichomycetes. Rajagopalan (1967) has reported *Enterobryus cingalobali* from India as new species. Misra (1998) has reviewed status of Trichomyceteous fungal species belonging to *Harpella, Genistellospora, Graminelloides, Pennella, Smittum, Spertiella, Stachylina, Arundinula, Enterobryus, Leidomyces, Passalomyces, Asellaria* gerera.

Trichomycetous fungal thalli may be single or in clusters, branched or unbranched and septate or aseptate. These fungi reproduce asexually by cystospores, arthrospores, trichospores or sporangiospores. However *Harpella* reproduces sexually by zygospores.

The fungi are associated with arthropoda hosts as mutualists. Insect larvae, millipedes, beetles and freshwater arthropod guts or digestive tracts get colonized by these fungi. Fungi of this group can be identified using standard manuals (Lichtwardt, 1986).

Lichtwardt (1986) recognizes four orders and each is characterized by distinctive type of asexual spore; trichospore is produced in the order Harpellales; arthrospores are produced in Asellariales; sporangiospores in Eccrinales and amoeboid cells or sporangiospores in Amoebidiales.

Order: Harpellales

The members of this order have branched or unbranched septate thalli attached to the hindgut lining or peritrophic membrane of immature insects. Trichospores are the characteristic spores produced in this order and possess one or many appendages attached to their basal end. Though these appendages lack the 9+2 structure of flagella but their function is the same as performed by the flagella. The trichospores are produced exogenously on generative cells but the appendages arise from within the generative cells. A trichospore is an exogenous, dehiscent, usually elongate sporangium containing a single uninucleate sporangiospore and has one to several basally attached filamentous appendages continuous with the sporangiospore. Sexual reproduction occurs by biconical zygospores.

Two families, Harpellaceae and Genistellaceae, are recognized in this order. The members of the former family possess unbranched thalli attached to the peritrophic membrane of the midgut while the members of the latter family have branched thalli attached to the cuticle lining the hindgut. 14 genera have been included in this order, 3 genera in the Harpelaceae and 11 genera in the Genistellaceae. Some of the important genera in this order are: *Harpella,* where trichospores have 4 appendages and are cylindrical and slightly curved or coiled. In *Stachylina* and *Smittium*, the trichospores are ellipsoidal with 1 appendage, but in the former the thallus is unbranched while in the latter it is branched. In *Genistellospora*, the holdfast is distinct and the zygospores are formed without conjugation.

Order: Asellariales

In this order, the branched thalli are attached to the hindgut lining of isopods or insects. Asexual reproduction takes place by fragmentation of the thallus to form arthrospores. Sexual reproduction is unknown.

This order consists of a single family, the Asellariaceae with three genera. *Asellaria* is a typical member of this order. This fungus attaches to an isopod and is characterized by the basal cell which is swollen, lobed and has rhizoids or tooth-like outgrowths; the basal cell is unbranched or branched only at the distal end.

Order: Eccrinales

In this order, the thalli are coenocytic and unbranched or branched only at the base and are attached to the chitinous lining of the hindgut and/or foregut. Asexual reproduction takes place by endogenously produced sporangiospores. The sporangiospores are usually produced singly in a series of terminal sporangia that release the spores in a sequence. The sporangiospores may be uni- or multinucleate and sometimes have thick walls. Sexual reproduction is unknown. The members are found in millipedes, isopods, amphipods, decapods and insects.

Three families, Palavasciaceae, Parataeniellaceae and Eccrinaceae, are recognized. The representative genera of these families are *Palavascia, Parataeniella* and *Enterobryus* respectively. In *Palavascia*, thalli produce one type of sporangiospores directly and sporangia sometimes germinate *in situ*. In *Parataeniella*, the thalli produce uninucleate or binucleate sporangiospore; while in *Enterobryus*, thalli produce thin-walled uninucleate and multinucleate sporangiospore that transform into multispored sporangia.

Order: Amoebidiales

In this order, the thalli are aseptate, unbranched and occur in the hindgut or on the external parts of aquatic crustacea and insects. Asexual reproduction takes place by amoeboid cells or sporangiospores. The entire thallus is converted into a mutliamoeboid or multispored sporangium that releases amoeboid motile cells or spores with rigid walls. The amoebae encyst and produce cystospores which give rise to new thalli. Sexual reproduction is unknown.

The order has a single family, the Amoebidiaceae and consists of two genera, *Amoebidium* and *Paramoebidium*. In the former, the thallus is attached to the outer surface of the host and produces spores with either rigid walls or amoeboid cells, while in the latter, the thallus is attached to the hindgut lining of the host and produces only amoeboid cells. *Amoebidium parasiticum* is the first trichomycete to have been grown axenically. This order is considered protists.

Other Details

Most of these fungi have haploid somatic cells. The growth of these fungi when cultured axenically is very slow but for *Smittium culisetae.* Ultrastructural details have been worked out in-depth in Harpellales while the other needs further studies.

Phylogenetically the fungi belonging to Harpellales and kickxellales which have a common ancestor based on morphological and serological evidences. Current molecular studies such as 18 S rDNA were confined to Harpellales and others except Asellariales and Eccrinales. Wall chemistry has been worked out in three orders of Trichomycetes and it is concluded that mostly the cell wall contains chitin in Dimargaritales, Harpellales and kickxellales while cellulose has been reported in the wall of Eccrinales. Dictyosomes are reported as cell organelles. The fungi are Eukaryotes.

III. GUT FUNGI AND HOST RELATIONSHIP

Trichomycetes are associated with arthropods as an obligate mutualists/commensalists. These fungi are associated with chitinous lining of digestive tract of arthoropoda. The commonest hosts are insects aquatic larvae at Diptera, may flynymphs, beetles, crustacea (crabs, isopods, amphipods) and millipedes.

The evaluation of different arthropods must have gone hand in hand with the evolution of Trichomycetes. These fungi are hair-like fungi, hence called trichomycetes. Asexual and sexual spores are produced during intermolting phase and must survive outside the host before reingestion. These fungi may be found in the digested remains of herbivorous Isopoda, Diplopoda, Coleoptera and marine arthropods.

IV. DISTRIBUTION, CULTURING AND MAINTENANCE

Litchtwardt (1986) has reported that these fungi are found distributed in North America, Hawaii, Japan, Australia, New Zealand, Greenland, India, Taiwan, Central and South America, Africa and Asia. Whisler (1960) has reported the pure culturing of *Amoebidium parasiticum.* Litchtwardt (1986) has reported culturing of two species of *Smittium*. Later some more species were cultured axenically and media used are: dilute brain-heart infusion agar, potato dextrose-yeast extract agar, tryptone-glucose medium. 200 isolates of Trichomycetous fungi are currently maintained in the university of Kansas Culture Collection. However majority of these fungi remain uncultured. However it is assumed that their nutritional requirements are similar to other common saprobic fungi.

The members of this group are widely distributed all over the world. Their distribution is limited only by the distribution of their hosts which include marine, freshwater and terrestrial arthropods.

V. TRICHOMYCETES: NEED FOR NEW KINGDOM

The morphology, ultrastructure, cytology, molecular sequence, phylogeny, serological data reproduction and cell wall composition suggest that Trichomycetes cannot be accommodated in fungi. Though cell wall is made up of chitin or cellulose, their inclusion under zygomycota raises many doubts. Zygomycotious fungi have chitin and chitoson while oomycetes have cellulose on cell wall. Recently oomycetes are separated from fungi and are included under chromistra. Further the thallus is like a trichome fungal-host association has been obligate mutualistic/commensalist and this kind of association does not exist in any of the living fungal groups. This group comprises fungi that are intimately associated with arthropods and the hosts occur in aquatic and terrestrial habitats. This group of fungi are of unknown affinity.

Trichomycetes produce either branched or unbranched thalli. These fungi multiply by asexual spores called amoebae, arthrospores, cystospores, sporangiospores or trichospores and are also known to produce zygospores sexually in some members belonging to Harpellales. The molecular aspects Harpellales indicated its relationship with kickxellales. Majority of the criteria and peculiarities indicate that Trichomycetes need to be separated as kingdom. However many more details have to be collected further and an indepth study seems to be essential.

Future Lines for Research

1. Exhaustive surveys are essential. Different habitats need to be investigated.
2. Reproduction, ultra-structure, cell wall composition, biochemistry reproduction, molecular analysis etc. need to be investigated indepth.
3. Metabolites secreted by these fungi need to be elaborated. Utilization of these fungi from the point of biotechnology and other related aspects has to be investigated.

REFERENCES

Leidy, J. 1849 a. *Enterobus*, a new genus of confervaceae. Proc. Acad. Nat. Sci., Philadelphia, 4: 225-233.

Leidy, J. 1849 b. Description (accompanied by drawings) of new genera and species of entophyta. Proc. Acad. Nat. Sci., Philadelphia, 4: 249-250.

Lichtwardt, R.W. 1986. *The Trichomycetes: Fungal associates of Arthropods*. Springer Verlag, New York.

Misra, I.K. 1998. Trichomycetes – Fungi associated with arthoropods: review and world literatures S. Symbiosis, 24: 179-220.

Misra, I.K. and R.W. Lichtwardt, 2000. *Illustrated genera of Trichomycetes: Fungal symbionts of Insects and other Arthropods.* Science publisher, Inc. Enfield, New Hampshire.

Rajagopalan, C. 1967. An *Enterobryus* (Trichomycetes, Eccrinales) in a milliped. Curr. Sci., 36: 20-22.

Whisler, H.C. 1960. Pure culture and nutrition of *Amoebidium parasiticum*. Amer. J. Bot., 49: 193-199.

3

The Laboulbeniales (Ascomycetes): Distribution and Host Parasite Relationships

Surinder Kaur and K.G. Mukerji
Sri Guru Tegh Bahadur Khalsa College
Department of Botany, University of Delhi,
Delhi.

ABSTRACT

The order Laboulbeniales consists of very minute fungi, which are obligate parasites of Arthropods, mainly insects and only a few species of Myriapods and Acarina. They are unusual in not possessing any visual mycelium. Throughout their life cycle, these fungi remain as multicellular thalli. The mature thallus consists of a foot, which attaches it to the host surface; a multilayered, multicellular receptacle; primary and sometimes secondary appendages; male reproductive structure-antheridia, and the female reproductive structure, the carpogonial apparatus, which after fertilization forms the perithecia. The fungus reproduces by means of ascospores, which are produced in asci within the perithecia. Some fungi are dioecious and thus have separate male and female thalli. The species differ in their structures, mainly in the form of receptacles and perithecia.

The apparent restriction of Laboulbeniales to growth only on living arthropods, mostly true insects, and the often remarkable adaptation of a species not only to a given host, but to a given position on the host has long attracted the attention of biologists. The work done on this and other problems, like the nature of the organ of attachment to host and its possible role in the nutrition of the parasite, transmission of the parasite from host to host and various other aspects of the life history of these fungi have been discussed here in detail.

I. INTRODUCTION

The Laboulbeniales are unique among the Ascomycetes because of their habit of growth on the integument of living arthropods, especially insects, and their compact thalli, consisting of a reduced hyphal system organized into a receptacle, bearing one or more perithecia and appendages. Various taxa in this large order, which include many hundreds of species, are characterised by striking morphological features, such as the compact cluster of perithecia and appendages borne at the broad, flattened apex of the massive thallus of *Zodiomyces* and to long, coiled perithecial neck peculiar and secondary feet of *Thaumasiomyces scabellularius* Thaxter. Within a single genus like *Laboulbenia* there may be considerable diversity of form, resulting from variation in size, shape, and colour of cells, as well as the branching of the appendages.

The study of this unique group of fungi began in 1853, when Charles Robin published "Histoire Naturelle des Vegetau Parasites qui croissent sur Phomme et sur les animax vivants". He named a new genus, *Laboulbenia* Mont. et Robin, for two new species, *L. rougetti* Mont. et Robin and *L. guerinii* Mont. et Robin, which were distinct from other fungi in general appearance, though only a little was known about the taxonomic position of these fungi.

After many years following the first report of this fungus group, several fungi with similar habits were found on insects. Peyritsch (1873) first used the name Laboulbeniaceae, in which he recognized only five genera.

In 1896, Roland Thaxter, an American mycologist, published the first volume of his monograph, summarizing his previous work on this fungus group. He reported 28 genera and 152 species mainly collected in North America and discussed in detail the developmental morphology of the Laboulbeniaceae. He also published the following four volumes of the monograph in 1908, 1924, 1926 and 1931. In each volume, he characterized a number of species collected at various localities in the world and most of them were new to science. One hundred and seven genera and 1304 taxa in the total number were recorded through five volumes of his monograph.

Engler (1898) was apparently the first mycologist who treated the group as a separate order, Laboulbeniales. He placed the order under the Laboulbeniomycetes of the Eumycetes, in which he recognized only one family, Laboulbeniaceae.

During the period of Thaxter's publications, many authors started studying this fungus group collected from various areas in the world. The outstanding among them was Spegazzini who described eight new genera and about 270 new species, mainly from Argentina and Europe (1902, 1910a,b, 1912, 1914, 1915a,b, 1917, 1918, 1924). Banhegyi (1940, 1944, 1949, 1950, 1964) described a number of species from Hungry, Poland and other regions in eastern Europe. Lepesme (1941a,b, 1942a,b,c,d, 1943, 1944, 1945, 1946, 1947) added various species as members of the flora of Europe and other localities and made great contributions to the knowledge of the genus *Rhacomyces*. Many new species and genera were described from various regions in the world by Batra (1963), Baumgartner (1923, 1951), Benjamin (1967, 1968a,b), Benjamin and Shanor (1952), Jeekel (1959), Merola (1951), Poelt (1952a,b), Poisson (1954), Scheloske (1969) and Whisler (1968).

As far as the distribution of Laboulbeniales in India is concerned, the first and most extensive work was done by Thaxter, who in his monographs (1896, 1908, 1924, 1926, 1931) described species from south and south east Asia. From the Indian Peninsula, he reported fifteen species of *Laboulbenia* and one species of the genus *Stigmatomyces* i.e., *S. hylemyiae.* His work was extended by Batra (1963), who reported three new species from India. She also found new host and locality records for two species of the fungi, from among those previously described. These two species were *L. vulgaris* Peyr. and *S. hylemyiae* Thaxter. She also described a new species of *Stigmatomyces i.e. S. scatellae* from Ephydrid fly *Scatella* sp. The other two new species were *Laboulbenia olitskyi* Batra from the carabid, *Chlaeminus biguttatus* and *L. camponoti* Batra from male ants *Camponotus* sp.

After the publication of Batra's work, a detailed work was undertaken on the taxonomy, ecology and development of these fungi in India by us. During this work, many species of the Laboulbeniales were reported for the first time from India (Kaur & Mukerji, 1995, 1996a,b, Kaur *et al.*, 1993). Scanning electron microscopy of the insects bearing these fungi was also carried out. Some of these fungi are: *Herpomyces periplanetae* on *Periplaneta americana* (Fig. 3.1 & 3.2), *Dimeromyces anisolabis* on *Labiduara riparia* (Fig. 3.3), *Cantharomyces platystethi* on *Stenus* sp. (Fig. 3.4 & 3.5) and *Laboulbenia egens* on *Bembidion* sp. (Fig. 3.6).

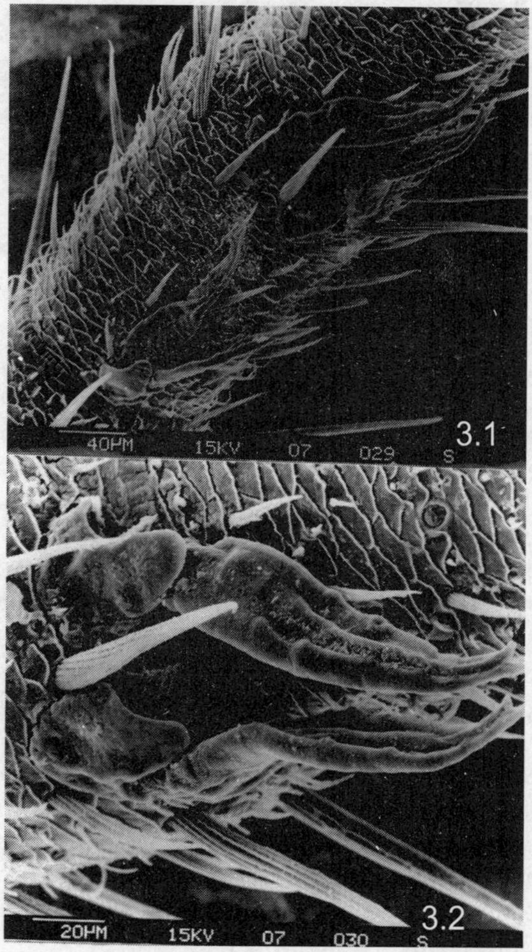

Figs. 3.1 & 3.2: Herpomyces *periplanetae* Thaxter on *Periplaneta americana*

3.1. Part of cockroach antenna showing female thalli with basal shield.

3.2. Part of antenna magnified to show two mature female thalli. The conical shields formed (which are visible below) truncate perithecia.

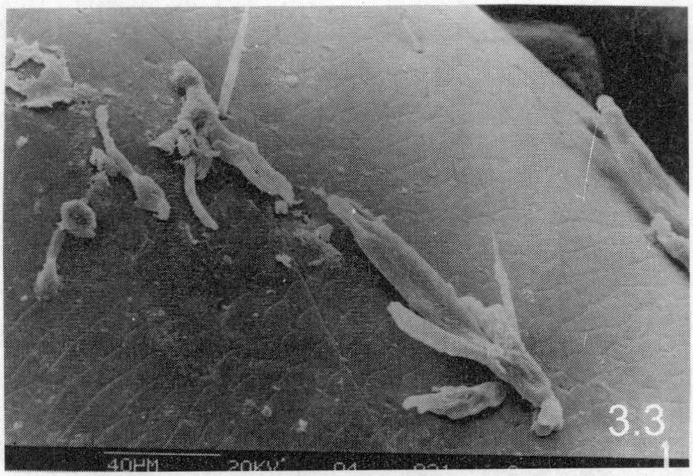

Fig. 3.3: *Dimeromyces anisolabis* Thaxter on earwig, *Labiduara riparia*, part of leg of insect showing thalli.

II. HOST PARASITE RELATIONSHIPS

With few exceptions, the fungus thallus is fixed to the host integument by the lowermost part of the cell. This organelle of attachment commonly is referred to simply as the foot and it usually is derived from what originally was the uppermost cell of the spore relative to its position in the perithecium prior to discharge. In *Herpomyces*, however, one or both of the sexes develop secondary receptacles (Tavares, 1965, 1966; Thaxter, 1908) from a primary receptacle derived directly from the spore after it has become attached to the host. These secondary receptacles then may assume the major role in the attachment of the fungus and presumably, absorption of nutrients. Similarly in *Dipodomyces monostrous* the spore forms a primary foot that soon ceases to function, following the formation by the subbasal cell of the primary receptacle of a rhizoid-like outgrowth, that develops a secondary foot which then assumes the function of the normal foot.

Few species, however, not only penetrate the integument but also send a more or less well-developed haustoria into the body tissues of the host. Rhizoidal haustoria have been demonstrated for all known species of *Arthrorhynchus* (Thaxter, 1896). *Microsomyces* (Thaxter, 1931), *Moschomyces* (Thaxter, 1896), and *Trenomyces* (Thaxter, 1924) and for some species of *Dimeromyces* (Thaxter, 1908), *Gloeandromyces* (Thaxter,1931), *Laboulbenia* (Thaxter, 1908) and *Rhizomyces* (Thaxter, 1908, 1931). Species having such haustoria rarely develop the blackened foot that is so characteristic of species that lack them. However, Kamburov *et al.*, (1966), in a study of *Hesperomyces virescens* (Thaxter, 1891), have shown that this species with a typical blackened foot does indeed perforate the integument and send a fascicle of delicate branched haustoria into the body of its host.

According to Gaumann and Dodge, (1928) and Thaxter (1896, 1908) formation of deeply penetrating haustoria by some species of Laboulbeniales was an adaptation to growth on soft bodied insects and that these structures provided additional support as well as a means of obtaining nourishment. In case of species having the ordinary type of foot there was no physical penetration of any kind into the host, and that in some way these fungi absorbed nourishment without disrupting the insect's cuticle.

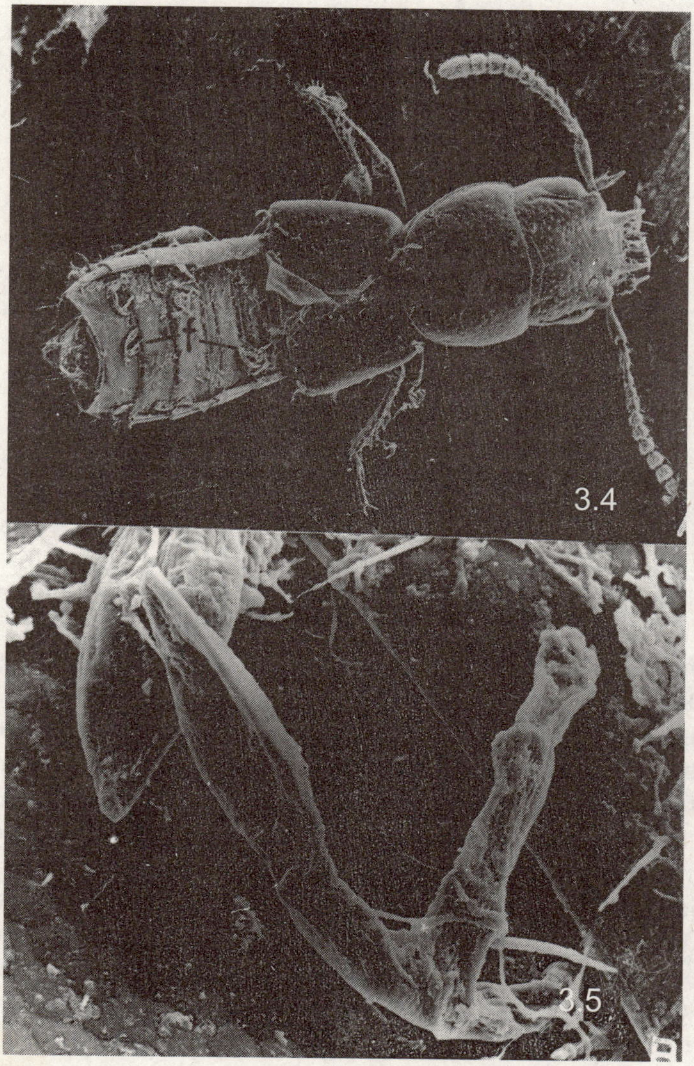

Figs. 3.4 & 3.5: *Cantharomyces platystethi* Thaxter on *Stenus* sp.

3.4. Thallus showing number of thalli at the abdomen.
3.5. One of the thallus attached at the appendage.

Three types of haustoria were observed by Richards & Smith (1956) from a sectioned material of *Herpomyces stylopygae* growing on *Blatta orientalis*:

(i) Minute haustoria measuring about 0.3 μ in diameter that arise from the secondary receptacle of young individuals of the fungus and penetrate the setal shafts of the host.
(ii) Small tubular haustoria about 1.5-2 μ in diameter formed by male individuals. These expand into a small haustorial bulb after reaching the epidermal cell layer of the host.
(iii) Large, less well-defined haustoria 2-5 μ in diameter that arise from female individuals and expand into a relatively large bulb without the epidermal layer.

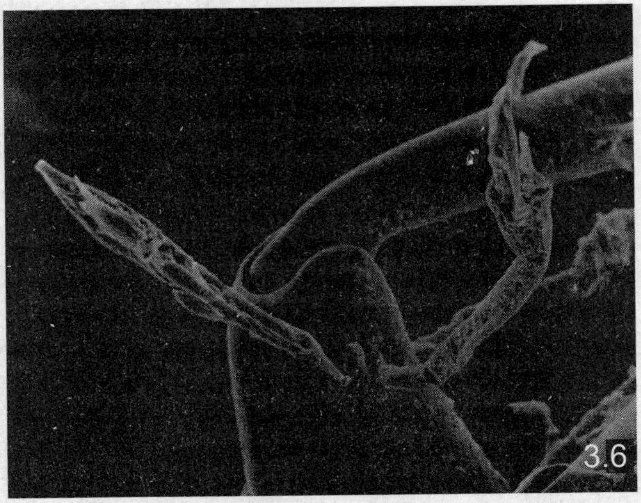

Fig. 3.6: *Laboulbenia egens* Thaxter on *Bembidion* sp. showing thalli attached to the appendage.

The development of the asci is similar to that of majority of Ascomycetes, although ascogenons hyphae do not occur and there is no modification of the apices of the asci. There are no anamorphs. Thalli develop only from ascospores.

The spores in the Laboulbeniales is bicellular throughout the order (Tavares, 1961). The cells of the spore are distinctly unequal in length in most genera with the larger of the two cells being uppermost within the perithecium and assuming the basal position at germination.

The outer wall of the spore usually swells strongly when it emerges from the ascus, especially around the foot. The viscidity of the spore surface makes detachment difficult (Richards and Smith, 1955). It is probably that spore in soil, such as those believed to infect beetles in Lindroth's experiments (Lindroth, 1948), that could become attached to hosts only under humid conditions. Only, the foot is adherent in genera like *Laboulbenia* indicating a greater viscidity here. The outer spore wall (or gelatinous envelope) which protects the spore from desiccation (Thaxter, 1896), persists as a thin, tough membrane surrounding the growing thallus (Benjamin and Shanor, 1950).

Germination normally takes place on the integument of a host, or on a seta protruding from it, although it sometimes occurs on the outside of a thallus or within a perithecium. In a germinating spore remaining inside an old perithecium of *Laboulbenia gyrinidarum*. Tavares (1985) observed a black basal spot immediately adjacent to the inner surface of the perithecial wall, where possibly some absorptive activity had been initiated.

The integument of insects consists of a thin outer layer, the epicuticle, a thick laminated procuticle (composed of protein and chitin), through which pore canals pass outward; an epidermal cell layer and a thin basement membrane, which separates the cuticle from the body cavity, hemocoel. The epicuticle consists of four layers; an outer varnish-like cement layer, which is secreted by the dermal glands, a wax layer, produced by the epidermal cells, and a thin cuticulin and thick underlying homogenous layer, which together make up the protein epicuticle (it contains no chitin) (Locke, 1974, Patton, 1963). There is some variation in the relative thickness of the layer in different insects and in different parts of the body, such as the hard sclerites and soft interconnecting membranes. The exocuticle, which becomes darkened and hardened, may be very thin in larvae, whereas it may occupy almost the entire thickness of the procuticle in adult beetles (Locke, 1974).

Pressure in the perithecium causes spores to emerge through the ostiole. When the spores are deposited horizontally on the surface of the host, which is usual in *Laboulbenia*, the germinating thalli become attached by the side of the foot. Generally the lower side of the foot enlarges considerably, so that the apex of the spore is raised from the surface of the integument (Thaxter, 1896).

In the genera, in which the cells of the spores are unequal in size, the larger cell of the spore (that uppermost in the perithecium) apparently always forms the foot, even when germination begins inside the perithecium. However, in *Herpomyces*, where the cells of the spore are essentially equal in size, the thalli that occur on setae grow with the foot directed toward the base.

The relation of the fungus to the host has been a subject of speculation since the discovery of the Laboulbeniales (Cavara, 1899, Robin, 1853, Scheloske, 1969, Tavares, 1979). Picard (1908a) observed that *Laboulbenia marina* had no visible effect on setae and elytra, both of which he believed to be incapable of regenerating cuticle. He suggested, therefore, that nutrients might be obtained by the fungus from the fluid, waxy surface layer of the cuticle, which he thought was nutritionally comparable to the fat body utilized by *Trenomyces*. Cepede (1914) favoured this view as he did not observe any effect on the integument as a result of infection by *L. blanchardia* Cepede. He cited Robin's observation (1853) that a mucous material favouring fungus growth may appear on the integument of insects attached by destructive parasites. This was presumed to have been produced by the integument.

Wigglesworth (1961) reported that of the upper two layers of the epicuticle if removed by gentle abrasion, the underlying epidermal cells may be induced to secrete more wax to repair the injury. Such a phenomenon would help to support the theory of the use of the surface wax for nutrition.

The ability of spores of *Herpomyces* to divide on waxy surface (Richards & Smith, 1954) and of thalli of *Fanniomyces* to reach the spermatial stage on detached fly wings placed on brain-heart infusion agar (Whisler, 1968) indicates that some growth can occur in the absence of living insect cells or hemolymph. It is possible that there is some utilization of surface secretions and culticular substances during early growth and penetration of the cuticle (Benjamin, 1971; Scheloske, 1969).

Except in *Herpomyces*, in which a small haustorium emerges from the foot, Thaxter (1896, 1908) believed that taxa having a back foot directly absorb nutritive materials through it from the host cuticle with the pale area in the cuticle below the foot of *L. borealis* Spegazzini (Tavares, 1979). Hydrolysis of chitin by the fungi was proposed by Picard (1908b), although chitin actually is present in a relatively small proportion in the procuticle of some hosts (Hackman, 1974; Wigglesworth, 1961).

Conspicuous haustoria occur in several genera of the Laboulbeniales including *Arthrorhynchus* (Blackwell, 1980), *Herpomyces*, *Hydrophilomyces Rhizomyces* & *Trenomyces* (Chatton and Picard, 1909; Meola & Tavares, 1982; Tavares, 1979). The haustorium of *Stigmatomyces* (Daniat and Manier, 1974; Peyritsch, 1871, 1873) and of *Hespermyces* (Tavares, 1979) penetrates the host integument and expands into a small bulb below the culticular layers, whereas the haustorium of *Herpomyces* terminates in a large bulb within the well-developed epidermal cell layer of the host (Richard & Smith, 1956), an extensive slender haustorial system (rhizomycelium) radiates from the small haustorial bulb of *Hesperomyces* into the hemocoel of the host (Kamburov *et al.*, 1967b).

Scheloske (1969) demonstrated that species of *Laboulbenia* parasitizing Coleoptera actually take up material from the hemolymph, when Nile blue sulfate became sufficiently concentrated within an elytron, the dye could be seen to rise into an attached thallus.

Early workers had conflicting opinions about the effect on the host. For instance, Peyritsch (1871) never saw any evidence of damage by the fungus, whereas Karsten (1869) reported resorption of muscle sheaths in the thorax of the host of *Stigmatomyces baeri*. Blackwell (1980) reported that the haustorium of *Arthrorhynchus* extends into the skeletal muscles of the dipteran host. The tergosternal muscles of the abdomen are among the most frequently invaded tissues of the louse hosts of *Trenomyces histophthorus* (Meola & De Vaney, 1976). These authors showed that the muscle striations become diffuse (Meola & Tavares, 1982). Extensive damage to fat body cells was reported by Chatton & Picard (1909). The nuclei of the epidermal cell layer near the haustorial bulbs of *Herpomyces periplanete* lose their normal ability to stain with iron haematoxylin (Tavares, 1979).

There are however, some reports of apparent adverse effects on insects caused by these fungi. Bro Larsen (1952), in her studies of *Bledius spectabilis*, an intertidal species, and *B. diota* and *B. taurus* which inhabit the neighbouring salt marshes only slightly less saline, found an increased tendency to parasites by members of the Laboulbeniales when these insects were inhabiting areas of low salinity. In a somewhat freshwater locality at Skallingen, West Jutland, 40% or more of the individuals in various samples of these beetles examined were infected by the fungus, whereas only one individual among hundreds taken from a neighboring salt water habitat was infested. Bro Larsen (1952) reported that the infected beetles became sluggish. The mortality rate among these was high, especially when the mouth parts and antennae were attached and much of the infected population did not succeed in reaching hibernation localities.

Hesperomyces virescens infects the beetle *Chilocorus bipustalatus* by forming a rhizoidal apparatus that enters the body cavity of the host (Kamburov et al, 1966, 1967). They observed that of the 173 infected beetles taken in late summers of 1966 from an area of coastal citrus groves in Israel, extending from Rosh Hunikra in the north to Ashquelon in the south, 120 were infested with *H. virescens*. These were segregated and provided with ample food, but 103 died within 10 days, only one of the 53 uninfected insects died during the same period. [They reported that the disease reached epizootic proportions by early September in the infected region, when 95% of the insects base collected the fungus. The attacks were accompanied by a sudden decline in the adult beetle population in the groves, and it was suggested that the fungus was associated with this premature mortality of the insects.] Interestingly such decline in the insect population did not occur in the areas where the fungus had not been found. The pathogenic effects on the hosts of the Laboulbeniales have been discussed in detail by Benjamin (1971) and Scheloske (1969).

Distribution

Thalli of a species of the Laboulbeniales may be randomly dispersed on the bodies of individual hosts or they may appear to be limited to a single location (Benjamin, 1971) often thalli that are anatomically similar in many ways may differ in general form, when they occur in different positions on hosts of the same or different sexes. These thalli may be described as different species if their position and the habits of the hosts are not considered. Thalli of *Eusynaptomyces benjaminii* (Sheloske, 1976a) growing on claws of males have short receptacles, whereas those growing on the underside of the pronotum of the female have elongate receptacles and produce perithecial outgrowths. Scheloske (1976b) doubted the existence of sex of host specificity in the Laboulbeniales.

Disposal of thalli on individual hosts is related to present living habits, geographical dispersion and distribution on different host groups reflecting the habitats and host interrelationships of the past. Co-evolution of hosts and parasites has undoubtedly occurred, although Frank (1982) believes that parasite specific reactions by the hosts should be evident.

Distribution on Host Groups

According to Tavares (1985) there are two basic patterns of distribution of Laboulbeniales on the families and orders of hosts:

(i) Unrelated genera of fungi parasitize a particular group of hosts; this results from recurrent waves of infection. The Staphylinidae bear genera from almost every tribe in the Laboullbeniales, with the exception of the Ceratomyceteae, which are limited to aquatic beetles (particularly Hydrophilidae), and the tribes consisting of only one genus, such as Coreomyceteae.

(ii) Many closely related genera parasitize in a single family of hosts indicating diversification of a common ancestral form and suggesting a certain degree of parallel evolution of hosts and parasites on Staphylinidae. There are four genera in the Haplomyceteae, five in the Teratomyceteae and several in the Stigmatomycetinae.

One of the largest families of beetles – the Scarabacidae, with over 12,000 species is parasitized only by *Laboulbenia* and *Rickia*. Another large family, the Tenebrionidae, with over 14,000 species is parasitized by six genera, including *Laboulbenia* and *Dimeromyces,* but *Rickia* has not been reported as a parasite.

Herpomyces occurs only on Blattaria and on no other order of host. The habitats of cockroaches in nature are similar to those frequented by beetles and other arthropods; despite this, the genus apparently has not cross-infected any other group. On the other hand, *Laboulbenia* has been found on a roach in West Africa and probably other collections may be made on roaches living under stones and in other natural habitats. Most collections of *Herpomyces* have probably been obtained from taxa that live in and around buildings. The stable environment of these roaches may be conducive to the retention of archaic parasites. Even among the genera of *Blattaria,* it is unlikely that much cross-infection takes place (Richards & Smith, 1954). Roaches in all stages of development frequently are found living together in large numbers. The heaviest infections are usually found in crowded, old colonies.

On Isoptera, another primitive order of insects, only *Laboulbenia* and *Dimeromyces* have been found. There appears to be no genus of Laboulbeniales that has evolved as with the Isoptera. Isoptera serve as hosts for a number of ectoparasitic anamorphs (Blackwell & Kimbrough, 1976). Two of the species of *Laboulbenia* on termites viz. *L. hagenii* Thaxter and *L. germinata* Buchli, both from East Africa on Termitidae, Macrotermitinae have profusely branched appendages and a cubical cell VI, whereas the third species (*L. felicis-caprae* W. Rossi, on Hodotermitidae from North Africa) has an appendage that is little branched and a flat cell VI. The two species of *Dimeromyces* belong to the typical structural group. Termites in many areas should be examined in order to determine whether evolution has taken place on the termites, whether there are any distinctive distribution patterns and the role that insects play in the dissemination of the fungi.

The species of *Laboulbenia* on different groups of ants (Hymenoptera, Formicidae) are not closely related. The Brazilian species, *L. ecitonis* (Blum 1924), from *Eciton* (Dorylinae), belongs to the *L. galeritae* group, whereas *L. componoti* which was reported on *Camponotus* (Formicidae) in India, by Batra (1963) belongs to the *L. vulgaris* group. It is not known whether closely related species of Laboulbenia occur on ants on a continent or whether only one widespread species such as *L. formicarum* is present in each area.

Arwidsson (1946) reported briefly on some cultural experiments with a species of *Laboulbenia* on *Harpalus* (*Ophonus*) *melleti* Herr (Coleoptera, Carabidae) collected in Sweden. The fungus, which was not identified specifically was found to infest *Harpalus* spp. belonging only to the subgenus *Ophonus*. He further stated that infection of new hosts took place more readily by their contact with soil containing *Laboulbenia* spores than by direct contact of infected insects. Later Lindroth (1948) conducted numerous experiments using uninfected specimens of *Harpalus* (*Ophonus*) *melleti* collected in the wild and also infected earth in which the beetles had lived. Uninfected representatives of three other species of *Harpalus* subgenus *Ophonus*, were readily infected, specially by contact with contaminated soil, less readily by contact with infected hosts. Three species of *Harpalus* belonging to different subgenera as well as *Brachinus crepitans* L. (Carabidae) were not infected by *Laboulbenia ophoni* even after three months of exposure. The fungus required about three weeks or longer to develop to maturity and viability of the spores in soil appeared not to exceed a few weeks. Lindroth (1948) concluded that although contact with infected soil was more important than direct contact for spread of the fungus in local populations of beetles, survival of the fungus from one season to the next and long-range disposal were dependent on the infested adult insect.

Scheloske (1969) found *Euzodiomyces lathraii* not only on several species of *Lathrobuim* (*Staphylindae*) the usual host, but also on species of *Gyrohypnus* as *Xantholinus* and *Othius* (Staphylinidae, Subfam staphylininae), related hosts, and on *Patrobus atrorufus* (Carabidae), undoubtedly a fortuitous host.

Phoretic mites are probably the means of transmission of *Dimeromyces* and *Laboulbenia* as well as *Rickia*, to a number of host groups. These mites generally use the insect hosts merely as a means of transport from one feeding ground (such as dung) to another (Hughes, 1959); however, some mites live more regularly on beetles, such as under the elytra. *Dimeromyces* and *Dichea* which parasitize many host groups show much structural diversity. By contrast, *Trenomyces* which occurs only on Mallophaga and Hippoboscidae (bird and mammal ectoparasites) is a very uniform genus, with little variation among species. The mallophaga on American land birds were found by Kellogg (1896b) to be identical or very similar to those occurring an old world birds that were closely related. Kellogg (1896a) found that although uniformity of living conditions tended to encourage little change, groups of Mallophaga on individual host birds were isolated and close breeding occurred. Perpetuation of slight variations insisted that the members of the community would soon differ from the parental type.

Among aquatic hosts, the greatest diversity of fungi occurs on the Hydrophilidae, on which there is a group of closely related fungi in a single tribe – the Ceratomyceteae. Obviously, the fungi and beetles have evolved together. There have also been several waves of infection on the Hydrophilidae. If the most primitive fungi constituted the first wave, then the Hydrophilidae would have been infected first by the Ceratomycetaceae, then by *Zodiomyces*, with *Chaetorithriomyces*, *Hydrophilomyces* and *Blasticomyces* appearing later.

Position Specificity

The phenomenon of position specificity i.e. precise location of species of Laboulbenials on very restricted area of the host body was observed by Peyritsch (1875). He observed that *Stigmatomyces baeri* usually grows on the upper surface of the female fly and on the lower surface of the male. This can be explained by assuming that the spores of the parasites are transmitted from one sex of the insect to the other at the time of mating. Whisler (1968) observed this in *Stigmatomyces ceratapharus* on fly *Fannia canicularis* where primary infections are on the upper surface of females and the lower surface of males. The parasite may then be spread secondarily to other parts of the body of the fly where it develops normally. Transmission at the time of mating does not, however, provide a hypothesis to explain many examples of position of specificity found in the Laboulbeniales.

In many cases, both sexes of host are equally heavily infested and in the same position on the integument. Location of the primary infection may be very precise, but secondary transfer followed by normal growth and development of the fungus on other parts of the body may take place readily. This demonstrates that some species displaying an apparent position requirement actually are able to grow and develop on any part of the host provided transfer of spores is effected. Among the Laboulbeniales some of the most remarkable examples of position specificity are found in the genus *Chitonomyces*, most species of which occur on beetles of the family Dystiscidae although several are known on Haliplidae and Gyrinidae (Thaxter, 1896, 1908, 1924, 1926). In the case of many species like *C. affinis* and *C. marginatus* (Thaxter, 1896) whose primary position of growth on their host *Laccophilus maculosus* appears to be extremely limited, a careful search of heavily infected insects often reveals isolated individuals of the fungi on other parts of the host body. The location of such secondary infections may be readily correlated with leg movements of the insect itself (Benjamin, 1971).

Sex-of-Host Specificity

Another very unusual type of specificity phenomenon observed in the laboulbeniales was reported for the first time by Benjamin & Shanor (1952) in their work on the parasites of *Bembidion picipes*. In this study six species of *Laboulbenia* were found to be parasitic on this beetle of the 259 specimens of the insects examined, 236 (91%) were parasitized by one or more of the six species of *Laboulbenia*. The infected insects included 96 females and 140 males. He observed that, out of the six species of *Laboulbenia* on this beetle, *L. vulgaris* exhibits no apparent limitations with regard to position of growth; one species, *L. tapirina*, that displays some limitation in that, it grows only on the upper surface, another species, *L. odobena*, displays a high degree of position specificity that differs for the two sexes of host, and three species that not only are position specific but are also the are sex of the specific host. *L. truncata* and *L. bembdio-palpi* grow only on males and *L. perpendicularis* grows only on females. Evidence presented suggests that, it is likely to be due to some regularly occurring movements of the host which aid in transfer of spores from infected to non-infected individuals rather than a physiological phenomenon relating to the nutrition of the parasites. Although, carefully controlled inoculation experiments are needed for better understanding of these phenomenon which have been observed in nature.

There are various factors to take into account when trying to determine the time at which the different host groups probably became infected, the area where the transfer may have been made and the host group that could have been made the source of infection. One is the time that the area became suitable for colonization because of the receding of water and establishment of land connections or because of temperature changes and resultant changes in the vegetation. The amount of phylogenic development that had been achieved by the host and the fungus groups at the time that the initial infection takes place may be reflected in the relationships between the fungi inhabiting a particular host group. The habitats that were occupied by the hosts at different times during their evolution may be indicated by the interrelationships of the parasites.

III. POSITION OF LABOULBENIALES AMONG FUNGI

Although the Laboulbeniales have protunicate ascus walls (Alexopoulos & Mims, 1979), they may be considered to be primitive members of the unitunicate line of Pyrenomycetes (Luttrell, 1955). They have simple, thin walled asci with no apical pore and the ascocarps have neither paraphyses nor pseudoparenchyma in the usual sense. Inner wall cell rows arising from the basal cells of the perithecium can be compared with paraphyses. Those that arise from the upper, inner angles of the outer wall cells are comparable to periphyses.

The close relationship of the Laboulbeniales to filamentous members of the subclass Euascomycetidae was confirmed by Hill (1977), when he demonstrated the presence in *Herpomyces* of flattened spindle pole bodies, woronin bodies associated with simple septal pores and ascus vescicles.

The centrum of the Laboulbeniales is comparable to that of the chaetomiaceae (having the *Xylaria* type of centrum), in which paraphyses arise only from the lateral walls of the perithecium and the thin walls of the asci are deliquescent. As in *Laboulbenia,* there is in the chaetomiaceae a true perithecial wall arising from cells underlying the ascogonium (or from nearby vegetative hyphae) (Luttrell, 1951). In *Laboulbenia*, however, there are no internal cells other than derivatives of the carpogonial cell row (the equivalent of the ascogonium and trichogyne of the chaetomiaceae).

Many other Ascomycetes have two celled spores, they include a large percentage of marine pyrenomycetes, as well as terrestrial genera in the bitunicate and unitunicate series (Denison & Carroll, 1966). Chadefeud (1960) suggested that the two-celled spore is the ancestral form *Spathulospora*, a parasite producing a perithecium on the exterior surface of a red alga, which was proposed as a connecting link between the Laboulbeniales and the Pyrenomycetes (Kohlmeyer, 1973). Although *Spathulospora* produces spermatia from phialides, lacks paraphyses, and has a trichogyne and deliquescing, simple, unitunicate asci, its ascospores are one-celled and bear appendages. The spermatia, at least in some species, bear a filiform, coiled appendage. It was believed that the thallus of *Spathulospora* consisted of a compact surface structure with peglike, penetrating cells. However, Walker *et al.* (1979) have shown that the perithecium is produced by hyphae that emerge from an internal crust reminiscent of parts of lichen thalli.

The thalli of some ancestral Ascomycetes might have been lichen like in form, as suggested by Cain (1972), with the parasitic groups undergoing some reduction in size and becoming more compact. Cain postulated that the earliest Ascomycetes were autotrophic, with a discrete haploid thallus bearing a parasitic dikaryotic stage, he suggested that they colonized soil in a tropical, moist climate and formed the major land flora. Like Cain, Denison & Carroll (1966) proposed that the ancestors of Ascomycetes were autotrophs. It was suggested by Demoulin (1975) that parasitism was the first step in heterotrophic life taken by red algal ancestors of fungi, with saprophytism evolving later from necrotrophic parasitism. He proposed that adelphoparasitic (i.e. parasites that are related to their hosts) red algae were the first step in the development of the Ascomycetes. Kohlmeyer (1973) also thought that the Ascomycetes developed from parasitic ancestors lived on algae.

REFERENCES

Alexopoulos, C.J., & C.W. Mims. 1979. Introductory Mycology. Third edition. John Wiley & Sons, New York, Chichester, Brisbane, Toronto. 632 p.

Arwidsson, T. 1946. Om svenska laboulbeniacefynd. (On Swedish finds of Laboulbeniaceae.) Svensk Bot. Tidskr. 40: 307-309.

Banhegyi, J. 1940. Etudes preliminaries sur les Laboulbeniales de la Hongrie. Index Hort., Bot. Univ. Budapast. 4: 39-59.

Banhegyi, J. 1944. Les Laboulbeniales aux environs du lac de Balaton. Bot. Kozlemenyek 41: 49-61.

Banhegyi, J. 1949. Les Laboulbeniales de la Transylvania. Index. Hort. Bot. Univ. Budapest. 7: 93-101.

Banyegyi, J. 1950. Ritka Laboulbeniak a Karpatmedencebol. Ann. Biol. Univ. Budapest. 1: 189-196.

Banhegyi, J. 1964. Notes sur quelques Laboulbeniales de la Pologne. Ann. Univ. Sci. Budapest. Sect. Biol. 7: 19-27.

Batra, Suzanne W. T. 1963. Some Laboulbeniaceae (Ascomycetes) on insects from India and Indonesia. Amer. Jour. Bot. 50: 986-992.

Baumgartner, R. 1923. Contribution a I'etude des Laboulbeniales de la Suisae. Jahrb. Philos. Fakul. II Univ. Bern 3: 257-265.

Baumgartner, R 1951. Regulates de mes dernieres recherches sur les Laboulbeniales (champignons sur insectes vivants), Mitteil. Naturforsch. Gesellsch. Bern. Neue Folge 8: xxx-xxxiii.

Benjamin. R.K. 1967. Laboulbeniales on semi-aquatic Hemiptera. *Laboulbenia*. Aliso 8(3): 111-136.

Benjamin, R.K. 1968a. *Sandersoniomyces,* a new genus of Laboulbeniales allied to *Diplomyces, Symplectromyces* and *Teratomyces*. Aliso 6: 1-16.

Benjamin, R.K. 1968b. *Balazukia*, a new genus of Laboulbeniales allied to *Cucujomyces* Spegazzini, Aliso 6: 47-56.

Benjamin, R.K. 1971. Introduction and supplement to Roland Thaxter's contribution towards a monograph of the Laboulbeniaceae. Bibliotheca Mycol 30: 1-55, J. Cramer, Lehre.

Benjamin, R.K. & L. Shanor, 1950. The development of male and female individuals in the dioecious species *Laboulbenia formicarum* Thaxter, Amer. J. Bot. 37: 471-476.

Benjamin, R.K. & L. Shanor. 1952. Sex of host specificity and position specificity of certain species of *Laboulbenia* on *Bembidion picipes*. Amer. J. Bot. 39: 125-131.

Blackwell, M. 1980. Developmental Morphology and taxonomic characters of *Arthrorhynchus nycteribige* and *A. aucampsipodae* (Laboulbenomycetes) Mycologia 72: 159-168.

Blackwell, M. & J.W. Kimbrough. 1976. Ultrastructure of the termite associated fungus *Laboulbeniopsis termitarius.* Mycologia 68: 541-550.

Blum, G. 1924. Zwei neue Laboulbenien aus Brasilien. Centralbi. Bakteriol. 2 Abth. 62: 300-302.

Brolarsen, Ellinor 1952. On subsocial beetles from the saltmarsh, their care of progeny and adaptation to salt and tide. Trans. IX Int. Congress Entomol. Amsterdam 1: 502-506.

Cain, R. 1972. Evolution of the fungi. Mycologia 64: 1-14.

Cavara, F. 1899. Di una nuova Laboulbeniacea: *Rickia wasmannii,* nov. gen. e nov. spec. Malpighia 13: 173-188.

Cepede, C. 1914. Etude des Laboulbeniacees europeennes, *Laboulbenia blanchardi* n.sp. et son parasite *Fusarium Laboulbeniae* n. sp. Arch. Parasitol. 16: (1913-1919): 373-403.

Chadefaud, M. 1960. Les Vegetaux non Vascularies (Cryptogamie). (Vol. I, Chadefaud, M. and L. Emberger, Traite de Botanique Systematique). Masson et Cie, Paris. 1018p.

Chatton, E. & F. Picard 1908. Sur une Laboulbeniaciee: *Trenomyces histophtorus* n g., nsp endoparasite des poux *Menopon pallidum* Nitzschet Goniocotes abdominalis P., de la poule domestique compt. Rend. Hebd. Seances Acad. Sci. [Paris] 146: 201-203.

Chatton, E. & Picard, F. 1909. Contribution a' l'etude systematique et biologique des Laboulbeniacees, *Trenomyces histophthorus* Chatton et Picard, endoparasite des poux de la poule domestique, Bull. Soc. Mycol. France 25: 147-170.

Daniet, H., & J.F. Manier. 1974. Haustoria de *Stigmatomyces scaptomyzae* Thaxter (Laboulbeniale) parasite de *Scaptomyza graminum* Fallen (Diptere, Drosophilide). Bull. Soc. Mycol. France 90: 217-221.

Demoulin, V. 1975. The origin of Ascomycetes and Basidiomycetes, the case for a red algal ancestry. Bot. Rev. (Lancaster) 40: (1974): 315-345.

Denison, W.C., & G.C. Carroll. 1966. The primitive Ascomycete: a new look at old problem. Mycologia 58: 249-269.

Engler, A. 1898. Syllabus der Pflanzenfamilien. 2 Auflage. Berlin 214 p.

Frank, J.H. 1982. The parasites of the staphylinidae (Caleoptera). Bull. Florida Agric. Exp. Sta. no. 824 Gebruder Borntraeger 1-118.

Gaumann, E.A. & C.W. Dodge. 1928. Comparative morphology of Fungi. McGraw-Hill Co. New York, London, 701 p.

Hackman, R.H. 1974. Chemistry of the insect cuticle. in: The physiology of insecta. Second Edition, Vol. 6, Ed. M. Rockstein, Academic Press, New York, London. pp. 215-270

Hill, T.W. 1977. Ascocarp ultrastructure of *Herpomyces* sp. (Laboulbeniales) and its phylogenetic implications. Can. J. Bot. 55: 2015-2032.

Hughes, T.E. 1959. Mites: or, the Acari, Athlone Press, University of London, London, 225 p.

Jeekel, C.A.W. 1959. A new carabidicolous species of the genus *Laboulbenia* from Sumatra (Ascomycetes, Laboulbeniales). Acta Bot. Neerl. 8: 257-262.

Kamburov, S.S., D.J. Nadel & R. Kenneth 1966. The fungus *Hesperomyces virescens* Thaxter parasitic on *Chilocorus bipustulatus* L. in Israel. Katavim 16: 135-136.

Kamburov, S.S., D.J. Nadel & R. Kenneth 1967a. Studies on the fungus *Hesperomyces virescens* attacking *Chilicorus bipustulatus.* In: Research Report, 1965,1966, Plant Pathology, Hebrew University, Faculty of Agriculture, Jeruselum, pp. 635-636.

Kamburov, S.S., D.J. Nadel & R. Kenneth 1967b. Observations on *Hesperomyces virescens* Thaxter (Laboulbeniales), a fungus associated with premature mortality of *Chilocorus bipustulatus* L. in Israel. Jour. Agr. Res. 17: 131-134.

Karsten, H. 1869. Chemismus der Pflanzenzelle. Wilhelm Braumiller & Wien. 90 pp.

Kaur, S. & K.G. Mukerji. 1995. Studies on Indian Laboulbeniales IV. Three species of *Laboulbenia*. Mycoscience 36: 311-314.

Kaur, S. & K.G. Mukerji. 1996a. Studies on Indian Laboulbeniales II. Three unrecorded species. Nova Hedwigia 62: 151-156.

Kaur, S. & K.G. Mukerji. 1996b. Studies on Indian Laboulbeniales III. Three unrecorded dioecious genera. Mycoscience 37: 61-64.

Kaur, S., Pathak, A. & K.G. Mukerji. 1993. Studies on Indian Laboulbeniales I. Three unrecorded species of *Laboulbenia* Mont. et Robin. Crypt. Bot. 3: 357-360.

Kellog, V.L. 1896a. New Mallophaga I - With special reference to a collection made from maritime birds of the bay of Monterey, California. Proc. Calif. Acad. Sci. (ser. 2) 6: 31-168.

Kellog, V.L. 1896b. New Mallophaga II,- from land birds; together with an account of the mallophagous mouth-parts. Proc. Calif. Acad. Sci. (ser. 2)6: 431-548.

Kohlmeyer, J. 1973. Spathulosporales, a new order and possible missing link between Laboulbeniales and Pyrenomycetes. Mycologia 65: 614-647.

Lepesme, P. 1941a. Catalogue des Laboulbeniales de la collection Francois Picard. Bull. Mus. Nat. Hist. Nat., Ser. 2. 13: 481-488.

Lepesme, P. 1941b. Deux nouveau *Rhacomyces* de Roumainie. Ibid. 13: 598-600.

Lepesme, P. 1942a. Deux *Laboulbenia exotiques* nouvelles (Laboulbeniaceae). Rev. Mycol., n.s. 7: 105-107.

Lepesme, P. 1942b. Revision des *Rhacomyces palearctiques* (Laboulbeniaceae). Bull. Soc. Mycol. Fr. 58: 57-80.

Lepesme, P. 1942c. Une nouvelle *Laboulbenia* de France. Ibid. 58: 190-91.

Lepesme, P. 1942d. Une nouvelle *Laboulbenia* de Madere (Laboulbeniaceae). Bull. Soc. Entomol. Fr. 1942: 135.

Lepesme, P. 1943. Trois *Laboulbenia* nouvelles parasites de *Chlaenius* malagaches. Bull. Mus. Nat. Hist. Nat., Sier. 2: 15: 244-247.

Lepesme, P. 1944. Note sur quelques Laboulbeniacees de France. Bull. Soc. Entomol. Fr. 1944: 67-68.

Lepesme, P. 1945. Laboulbeniales parasites de Coleopteres. In: Jeannel, et al., Faune des terriers de Rats-Taupes. Mem. Mus. Nat. Hist. Nat., n.s. 19: 149-150.

Lepesme, P. 1946. Quelques mots sur les Laboulbeniales. L'Entomologiste 2: 81-85.

Lepesme, P. 1947. Laboulbeniacees parasites des Coleopteres. Mem. Mus. Nat. Hist nat., Paris , n.s. 20: 111-119.

Lindroth, C.H. 1948. Notes on the ecology of Laboulbeniaceae infesting carabid beetles. Svensk. Bot. Tidskr. 42: 34-41.

Locke, M. 1974. The structure and formation of the integument in insects In: The physiology of insecta, second edition, vol. VI, Ed. M. Roctestein. Academic Press, New York, London. pp. 123-213:

Luttrell, E.S. 1951. Taxonomy of the Pyrenomycetes. Univ. Missouri Studies 243: 1-120.

Luttrell, E.S. 1955. The ascostromatic Ascomycetes. Mycologia 47: 511-532.

Meola, S. & I.I. Tavares, 1982. Ultrastructure of the haustorium of *Trenomyces histophthorus* and adjacent host cells. J. Invert. Pathol. 40: 205-215.

Meola, S. & J. DeVaney. 1976. Parasitism of Mallophaga by *Trenomyces histophorus.* J. Invert. Pathol. 28: 195-201.

Merola, A. 1952. Interessante ritrovamento di laboulbeniologia cavernicola: *Arthrorhynchus acrandros* n. sp. (con considerazioni sul gen. *Arthrorhynchus*). Bol. Soc. Nat. Napoli 60: Suppl. 16 (Studi spelelogici e faunistici sull'Italia meridionale): 1-30.

Patton, R.L. 1963. Introductory Insect Physiology, W.B. Saunders, Philadelphia, 245 p.

Peyritsch, J. 1871. Uber einige pilze aus der Familie der Laboulbenien. Sitzungsber. Kaiserl. Akad. Wissensch. Math.-Naturwissensch. Klasse 64: 441-458.

Peyritsch, J. 1873. Beitrage zur Kenntniss der Laboulbenien. Sitzungsber. Kaiserl. Akad. Wissensch. Math.-Naturwissensch. Klasse 68: 227-254.

Peyritsch, J. 1875. Uber vorkommen und Biologie von Laboulbeniaceae. Sitzungsber. Kaiserl. Akad. Wissensch. Math.-Naturwissensch. Klasse 71: 377-385.

Picard, F. 1908a. Sur une Laboulbeniacees marine (*Laboulbenia marina* n. sp.) parasite d'Aepus *Robini* Laboulbene. Competes rendus hebdom. Seances et Mem. Soc. Biol. Paris 65: 484-486.

Picard, F. 1908b. Les Laboulbeniacees et leur parasitisme chez les insectes. Feuille des Jeunes Naturalistes, Ser. 4. 39: 29-34.

Poelt, J. 1952a. Laboulbenien. Und ihr Vorkommen in Siidbayern. Nachrichtenbl. Bayer. Entomol. 1: 33-36.

Poelt, J. 1952b. Laboulbeniales aus Sudbayern. Mitteil. Bot. Staatssam. Munchen 1: 115-118.

Poisson, R. 1954. Sur une Laboulbeniale ectoparasite de *Vella osborniano* Kirkaldy (Hemipt. Velliidae) *Laboulbenia* (s.g. *Veliomyces* nov.) *titschacki* n. sp. In Titschack, E. Beitrage zur Fauna Perus. 4: 81-82.

Richards, A.G. & M. N. Smith, 1954. Infection of Cockroaches with *Herpomyces* (Laboulbeniales) III. Experimental studies on host specificity Bot. Gaz. [Crawfordsville] 116: 195-198.

Richards, A.G. & M. N. Smith, 1955. Infection of cockroaches with *Herpomyces* (Laboulbeniales) I. Life history studies Biol. Bull. Mar. Biol. Lab. Woods Hale 108: 206-218.

Richards, A.G. & M. N. Smith. 1956. Infection of cockroaches with *Herpomyces* (Laboulbeniales). II. Histology and Histopathology. Ann. Entomol. Soc. Amer. 49: 85-93.

Robin, C. 1853. Histoire Naturelle des vegetaux Parasites qui croissent sur I'Homme et sur les animax vivants. J.B. Bailliere, Paris. 702 pp.

Scheloske, H. 1969. Beitrage zur Biologie, Oekologie und Systematik der Laboulbeniales (Ascomycetes) unter besonderer Berucksichtigung des Parasit-Wirt-Verhaltnisses. Parasitol. Schriftenr. 19: 1-176.

Scheloske, H.W. 1976a. *Eusynaptomyces benjaminii*, spec. nova, (Ascomycetes, Laboulbeniales) und seine Anpassungen an das Fortpflanzungsverhalten seines Wirtes *Enochrus testaceus* (Coleoptera, Hydrophilidae). Plant Syst. Evol. 126: 267-285.

Scheloske, H.W. 1976b. Morphologische Anpassungen eines ektoparasitischen Pilzes (Ascomycetes: Laboulbeniales: *Misgomyces coneglanensis*) an Korperbau und Fortpflanzungrverhalten seines Wirtes (Coleoptera: Hydrophilidae: *Laccobius minutus*). Entomologica Germ, 3: 227-241.

Spegazzini, C. 1902. Mycetes argentinenses (series II). An. Mus. Nac. Hist. Nat. Buenos. Aires 8: 49-89.

Spegazzini, C. 1910a. Laboulbeniaceas nuevas chilenas. Rev. Chilena Hist. Nat. 14: 71-72.

Spegazzini, C. 1910b. Fungi chilensis. Contribucion al etudio de los Hongos chilenos. Buenos Aires, Lib. Nac., J. Lajouane y Cie. 205 p.

Spegazzini, C. 1912. Contribucion al estudio de las Laboulbeniomicetas argentinas. An. Mus. Nac. Hist. Nat. Buenos Aires 23: 167-244.

Spegazzini, C. 1914. Primo contributo alla conoscenza della Laboulbeniali italiani. Redia. 10: 21-75.

Spegazzini, C. 1915a. Laboulbeniali ritrovate nelle collezioni di alcuni musei italiani. An. Mus. Nac. Hist. Nat. Buenos Aires 26: 451-511.

Spegazzini, C. 1915b. Segunda contribuction al conocimiento de las Laboulbeniales italianas. Ibid. 27: 37-74.

Spegazzini, C. 1917. Revision de las Laboulbeniales argentinas. Ibid. 29: 445-688.

Spegazzini, C. 1918. Observaciones microbiologicas. An. Soc. Cient. Argent. Buenos Aires. 85: 311-323.

Spegazzini, C. 1924. *Corethromyces bruchi* nueva Laboulbenial Argentina. Physis 7: 236-238.

Tavares, Isabelle, I. 1961. The rudimentary appendage of *Amorphomyces* (Laboulbeniales) Amer. J. Bot. 48: 539 (abstract).

Tavares 1965. Thallus Development in *Herpomyces paranersis* (Laboulbeniales) Mycologia. 52: 704-721.

Tavares 1966. Structure and development of *Herpomyces stylopygae* (Laboulbeniales). Amer. J. Bot. 53: 311-318.

Tavares, I. I. 1979. The Laboulbeniales and their arthropod hosts. *In*: Insect-fungus Symbiosis, Nutrition, Mutualism, and Commensalism. Ed. L.R. Batra, Allanheld, Osmun & Co., Montclair, N.J. (John Wiley & Sons, New York, Chichester, Brisbane, Toronto. pp. 229-258.

Tavares, I. I. 1985. Laboulbeniales (Fungi, Ascomycetes), Mycologia Memoirs No. 9, J. Cramer Publisher. 627 p.

Thaxter, R. 1896. Contribution towards a monograph of the Laboulbeniaceae. Mem. Amer. Acad. Arts & Sci. 12: 187-429.

Thaxter, R. 1908. Contribution towards a monograph of the Laboulbeniaceae Part II. Mem. Amer. Acad. Arts Sci. 13: 217-469.

Thaxter, R. 1924. Contribution towards a monograph of the Laboulbeniaceae. Part III. Mem. Amer. Acad. Arts & Sci. 14: 309-426.

Thaxter, R. 1926. Contribution towards a monograph of the Laboulbeniaceae. Part IV. Mem. Amer. Acad. Arts & Sci. 15: 427-580.

Thaxter, R. 1931. Contribution towards a monograph of the Laboulbeniaceae Part V. Mem. Amer. Acad. Arts Sci. 16: 1-435.

Walker, D.C., G.C. Hughes, & T. Bisalputra. 1979. A new interpretation of the interfacial zone between *Spathulospora* (Ascomycetes) and *Ballia* (Florideophyceae). Trans. Brit. Mycol. Soc. 73: 193-206.

Whisler, H.C. 1968. Experimental studies with a new species of *Stigmatomyces* (Laboulbeniales). Mycologia 60: 65-75.

Wrigglesworth, V.B. 1961. The principles of Insect Physiology. Fifth edition, Methuen & Co., London, E.P. Dutton & C., New York, 546 p.

4

Arbuscular Mycorrhizal Fungal Taxonomy: An Overview

S. SWARUPA RANI, I.K. KUNWAR and C. MANOHARACHARY
Botany Department, Osmania University, Hyderabad

ABSTRACT

Arbuscular Mycorrhizal (AM) fungi originated more than 460 million years ago, and they, therefore, represent an ancient phylogenetic clade within the fungi. On the basis of comprehensive SSU rRNA analysis, AM fungi can be separated in a monophyletic clade, which is not related to any zygomycetous group but probably shares common ancestry with the Ascomycota and Basidiomycota clade. Based on the comprehensive SSU eRNA analysis and natural relationships of AM fungi and the related fungi, Schler *et al*, (2001) recognized a new fungal phylum Glomeromycota with a single class Glomeromycetes containing 10 genera viz. *Glomus A, Glomus B, Paraglomus, Diversispora, Gigaspora, Scutellospora, Acaulospora, Entrophospora, Archaeospora* and *Geosiphon* and more than 150 described species, some of which are synonyms. Recently one more genus *Pacispora* has been erected in the order Diversisporales.

Key Words: SSU rRNA analysis, monophyletic clade, Glomeromycota, Glomeromycetes, AM fungi

I. INTRODUCTION

Arbuscular Mycorrhizal (AM) fungi originated more than 460 million years ago (Redecker, et al., 2000a, b) and they, therefore, represent an ancient phylogenetic clade within the fungi. Taxonomy of AM fungi is quite old but has received considerable attention during the past 40 years and still remains a puzzle to many of the scientists. Peyronel (1924) was the first to recognize that the AMF were the members of Endogonales rather than Chytrids, *Pythium* spp., or fungi as suggested by earlier workers. Peyronel's discovery followed the revision of the Endogonaceae by Thaxter (1922), who had not realized the mycorrhizal involvement of the family. Gerdemann and Nicolson (1963) then developed procedures for collecting AM fungal spores from soil and described a new species. The family Glomaceae/Endogonaceae was monographed in 1974 with segregation of the genus Endogone into seven genera. Since then rapid and sustained reports of undescribed AM fungal species have occurred during the past 30 years.

Gerdemann and Trappe (1974) reported 30 species of fungi while Trappe (1982) in his synoptic key to the Endogonaceae listed 77 species, excluding *Endogone*. Hall (1984) in his dichotomous keys to the Endogonaceae listed 67 species excluding *Endogone*. Berch (1988) listed 126 species excluding *Endogone*. Approximately 150 species of vesicular arbuscular mycorrhizal fungi have been described based on morphological characters of the spores by Schenck and Perez (1990).

The history of the study of AM fungi was summarized by Butler (1939) by which time the vesicles and arbuscules were clearly illustrated in the 19th century (Janse, 1896) and were already recognized as being produced by a root colonizing fungal symbiont. In the early 1950's Barbara Mosse at East Mallinga (U.K.) first showed experimentally that a fungus, later described as *Glomus mosseae* was responsible for the mycorrhizal colonization of strawberry roots (Mosse, 1953). Morphologically the nearest similar group of fungi with known sexuality belongs to the genus *Endogone* and by analogy the AMF were placed with them in a single family, the Endogonaceae (Zygomycotina).

A comprehensive review of the group was carried out (Gerdemann and Trappe, 1974) during which two genera (*Acaulospora* and *Gigaspora*) were erected within the Endogonaceae. The fungi within this rather unnatural grouping were eventually, formally accommodated in the own order Endogonales, though without further taxonomic classification above genus level (Benjamin, 1979).

A cladistic analysis mainly of morphological features produced a species tree with a new order Glomales containing two suborders and three families (Morton and Benny, 1990). However some of the conclusions of this work were questioned.

II. IDENTIFICATION AND TAXONOMY

The identification of AMF was made chiefly on morphological specimens directly isolated from soil on the maximum observable characters. Morphological criteria used for the classification and identification of AM fungi are viz: hyphal characters, auxiliary cells, subtending hyphae, spores/sporocarp ontogeny, spore germination, germination shield, spore color, spore wall etc. Boichemical, molecular and immunological characters are also important in the identification of AMF.

Taxonomy of Glomalean Fungi continues to change as a result of new information derived from all scales (molecular, organismal, morphological, ecological). Until recently, the order Endogonales (Zygomycotina) has consisted of only one family the Endogonaceae (Benjamin, 1979; Morton, 1988). Several genera *Acaulospora* Gerdemann and Trappe emend, Berch, *Entrophospora* Ames and Schneider, *Gigaspora* (Gerdemann and Trappe, *Glomus* Tulasne and Tulasne, *Sclerocystis* Berkeyley and Broome and *Scutellospora* Walker and Sanders were included whose members formed an arbuscular mutualistic symbiosis with many terrestrial plant families (Trappe, 1987) one genus *endogone link*, Fr whose members were saprobic (Gerdemann and Trappe, 1974) or formed putative ectomycorrhizal associations (Chu-Chou and Grace, 1979; Fassi *et al*, 1969). Pirozynski and Dalpe 1989, described a new family Glomaceae Pirozynski and Dalpe containing two genera, *Glomus* and *Selerocystis*. Morton and Benny (1990) emended the family Glomaceae and erected a new order Glomales and two new families, Acaulosporaceae and Gigasporaceae.

III. CLASSIFICATION

Based on comprehensive SSU rRNA analysis, Schüßler *et al* (2001) showed that the AM fungi can be separated in a monophyletic clade, which is not related to any zygomycetous group but probably shares common ancestry with the Ascomycota, Basidiomycota clade. Further on the basis of natural relationship of Arbuscular mycorrhizal fungi and the related fungi, recognized a new fungal phylum Glomeromycota (Table 4.1) with a single class Glomeromycetes (Cavalier–Smith, 1998) circumscribed as for the Phylum, containing four orders and more than 150 described species, some of which are synonyms (Walker and Trappe, 1993; Walker and Vestberg, 1998). According to this classification 10 valid AMF genera are recognized viz: *Glomus A, Glomus B, Paraglomus, Diversispora, Gigaspora, Scutellospora, Acaulospora, Entrophospora, Archaeospora* and *Geosiphon*. Recently Walker and Schüßler (2004) erected a new family Pacisporaceae in order Diversisporales and placed a new genus *Pacispora* Sieverd and Oehl (Oehl and Sieverding, 2004) in it. Since Pacisporaceae has been described after Schüßler *et al* (2001) proposed their classification of AM fungi, it is not shown in the classification outlined by them.

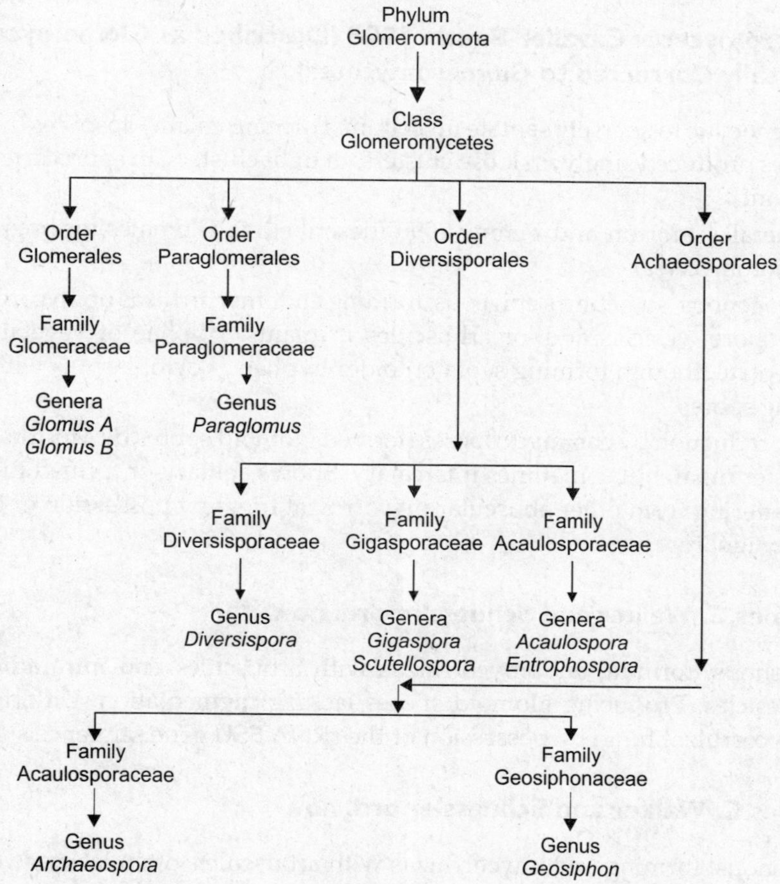

Table 4.1: Classification of Arbuscular Mycorrhizal Fungi (Schüßler *et al*, 2001)

Phylum: Glomeromycota C. Walker and Schuessler, Phylum nov. (Schüßler et *al*, 2001).

Fungi with coenocytic to sparsely septate mycelium, living mostly hypogenously, sometimes epigenously. Forming chlamydospores (in some genera) by blastic development of hyphal tip followed by thickening of structural wall components and occlusion by septum, spore-wall thickening, or deposition of an amorphous plug in the lumen of the subtending (sporogenous) hypha and spore. Complex spores (in some genera) with a rigid, chitinous structural wall component within a blastic terminal saccule, or by extension of a bulbous base, with or without flexible wall components. Spores produced singly, in loose clusters, in tight clusters (without a structured peridium), in sporocarps (with peridial development) or within the roots of the plants.

Habitat: In soil, roots, or on soil surface, vegetation, or decaying segments of substrate. Forming close symbiotic relationships with photoautotrophic organisms. So far known to produce buscular or vesicular-arbuscular mycorrhizas, arbuscular mycorrhiza-like symbiosis with non-vascular plants, and endocytosymbiosis with cyanobacteria.

Class: Glomeromycetes Cavalier-Smith, 1998 (Described as Glomomycetes, Orthographically Corrected to Glomeromycetes)

Fungi with coenocytic to sparsely septate mycelium. Forming chlamydospores by blastic development. Spores produced singly, in loose clusters, in tight clusters, in sporocarps or within the roots of the plant.

Order: Glomerales Morton and Benny, 1990 (described as Glomales, orthographically corrected to Glomeromycetes)

Mostly hypogenous, sometimes epigeous, forming endomycorrhizas or mycorrhiza-like symbiosis within spore, vesicles and/or arbuscules in plants. Hyphae of vegetative mycelium mostly non-septate, though forming septa on older hyphae as cytoplasm is withdrawn or cut off from resting spores.

Asexual reproduction by chlamydospores (termed glomoid spores by Morton and Redecker, 2001), mainly terminal, but sometimes intercalary. Spores solitary or formed in clusters, or in sporocarps, differing from other abuscular mycorrhizal fungi by possession of the rRNA SSU gene sequence signature.

Paraglomerales C. Walker and Schuessler ord. nov.

Fungi hypogenous, forming endomycorrhizas with arbuscules and intraradical mycelium, rarely with vesicles. Producing glomoid spores lacking pigmentation. Differing from other arbuscular mycorrhizal fungi by possession of the rRNA SSU gene sequence signature.

Diversiporales C. Walker and Schuessler ord. nov.

Fungi hypogenous, forming endomycorrhizas with arbuscules, often lacking vesicles with or without hypogenous auxiliary cells. Forming either complex spores produced within a sporiferous saccule (acaulosporoid spore of Morton and Redecker, 2001), complex spores developing from a bulbous base on the soporiferous hypha (termed here gigasporoid spores). They differ from the other arbuscular mycorrhizal fungi by possession of the rRNA SSU gene sequence signature.

Archaeosporales C. Walker and Schuessler ord. nov.

Fungi hypogenous, forming endocytosymbioses with photoautotrophic prokaryotes, or producing mycorrhizas with arbuscules, with or without vesicles. Spores lacking pigmentaton or reaction to Melzer's reagent. Glomoid spores formed singly or in loose clusters on or in the soil, acaulosporoid complex spores formed singly in the soil. Dense spore clusters unknown. They differ from the other arbuscular mycorrhizal fungi by possession of the rRNA SSU gene sequence signature.

Fungi, forming arbuscular endomycorrhizae in mutualistic symbiosis with living plants. Sexual reproduction is rare. Asexual reproduction by spores resembles chlamydospores and azygospores. Spores formed singly, but also in aggregates or in sporocarps.

Family: Glomeraceae (same characters as the order)

Genera: *Glomus A*
Glomus B

Fungi with coenocytic to sparsely septate mycelium, living mostly hypogenously, sometimes epigeously, forming chlamydospores by blastic development of hyphal tip. Spores produced singly, in loose clusters, in tight clusters (without a structured peridium), in sporocarps with peridial development or within the roots of plants. Spores develop on cylindrical, flared or constricted fertile hypha. Spore contents separated from that of the subtending hypha by an amorphous plug, a septum, or wall layers. Maximum number of wall layers amongst the AM fungi.

Family: Paraglomeraceae Morton et Redecker, fam. nov. (Morton and Redecker, 2001).

Studies of other molecular characters, such as monoclonal antibody specificities (Wright *et al*, 1987) and fatty acid profiles (Graham *et al*, 1995) have led to the long-held suspicion that members of this family were phylogenetically distant from species in *Acaulospora* and *Glomus*, but all of these data did not contain enough phylogenetic information to determine its position relative to other lomalean taxa. Even morphological characters were atypical for the *Acaulospora*-like spores (Morton and Benny, 1990). Small subunit (18S) ribosomal DNA sequences finally provided the character set to position this family as ancestral to both Acaulosporaceae and Glomeraceae Redecker *et al*, 2000a; Sawaki et *al*, 1998). The molecular phylogenetic tree suggests that members of this family and Paraglomaceae are more closely related to a non-mycorrhizal fungus, *Geosiphon pyriforme*, a putative Zygomycete fungus, which is the mycobiont in asymbiosis with a *Nostoc* cyanobacterium (the photobiont).

Vegetative structures consist of those inclusive in the suborder Glomineae. There are several characteristics unique to this family.

Molecular characters which appear to be diagnostic for members of this family:

1. Monoclonal antibody cell lines B5 and F8, maintained in the laboratory of Sara Wright (Wright *et al*, 1987).
2. Signature DNA sequence at the 3′ end of the 5.8′S ribosomal gene: GGCATGTCTGTTTGAGGGCACCA separates members of this family from those of Glomeraceae.
3. The primer sequence TGCTAAATAGCCAGGCTGY (ARCH1311) also will selectively amplify species of this family.

Genus: *Paraglomus* Morton and Redecker, 2001, gen. nov.

Development of spores proceeds exactly like that found in *Glomus* (Glomeraceae) by blastic expansion of a hyphal tip. Spores develop terminally on a cylindrical to slightly flared subtending hypha. Spores produced singly, more rarely in loose aggregates of 2-3 spores. Sub-cellular structures identical to *Glomus* species, layers of the spore wall continuous with layers of the subtending hypha. In the species described to date, the spore is occluded by thickening, of the innermost spore wall layer. Germination poorly characterized, but assumed to be like that in *Glomus,* germ tube emergence through the lumen of the subtending hypha. Outer layer of the spore wall often sloughs as the spore ages (in soil or in pot culture storage). Developmentally, this layer is the first component of the spore wall to form in juvenile spores. Usually, organic material will accumulate on the surface; no reaction with Melzer's reagent.

The hypha subtending the spore differentiates at the same rate and synthesizes the same component layer as that was found in the spore wall. In some species the subtending hypha of mature spores is so thin that it is hard to see or separate from the spore.

Only two species have been discovered in this genus, both having been classified first as *Glomus* species (based on spore morphology). Faint to invisible mycorrhizal structures provide morphological clues of possible membership in this genus.

Family: Diversisporaceae Walker and Schüßler, 2004

Spores glomoid, germination not accompanied by the formation of a germination shield and having specific SSU rRNA gene sequence.

Genus: Diversispora Walker and Schüßler gen. nov.

Spores glomoid, spore wall structure consisting of a thin outer wall, a laminated structured wall, and a flexible inner wall not reacting with Melzer's reagent. Germination not preceded by formation of a germination shield, and specific SSU rRNA gene sequence signature. The genus name referring to the diverse nature of spores formed within the order to which the genus lends its name.

Family: Giagasporaceae Bentivenga and Morton (1995).

Spores large (usually >200 µm); spore contents partitioned from that of the bulbous sporogenous cell by a plug, rarely by septum. Germ tubes produced directly through the wall or from a germination shield.

Genus: Gigaspora Gerdemann & Trappe (1974) emend. Walker & Sanders (1986).

Spores without ornamentation; wall bilayered. Germ tubes produced directly through spore wall; inner flexible wall groups absent; thin walled auxiliary cells finely papillate or echinulate. Spores develop blastically from a hypha, which swell and become the "sporogenous cell". After the sporogenous cell reaches its full size (usually about 25-50 μm in most species), the spore begins to develop at the tip. The outer layer and the laminate layer develop simultaneously, and often cannot be distinguished in juvenile spores without the assistance of Melzer's reagent. The laminate wall then thickens and ultimately develops the warty inner layer, from which multiple germ tubes arise.

Genus: **Scutellospora** *Walker & Sanders (1986)*

Spores with or without ornamentation; spore wall bilayered. Germ tubes arise from a persistent, plate like germination shield that is always associated with the innermost flexible wall. Inner flexible wall group always present; auxiliary cells knobby, broadly papillate or smooth. Early stages in spore development mirror those in the sister genus, *Gigaspora*. Spores develop blastically from a hyphal tip, which swell and become the "sporogenous cell". After the sporogenous cell reaches its full size (usually about 25-50 μm in most species), the spore begins to develop at the tip. The outer layer and the laminate layer develop simultaneously, and often cannot be distinguished in juvenile spores without the assistance of Melzer's reagent. The laminate then thickens and the outer ornamentations develop (if present in species). Inner walls develop and the last stage is the formation of a germination shield.

Family: Pacisporaceae Walker & Schüßler (2004) fam. Nov.

Oehl and Sieverding (2004) erected a new genus, *Pacispora* Sieverd. & Oehl. Unfortunately a month later Walker *et al* (2004) erected a new family, Gerdemanniaceae C. Walker *et al*, with a new genus *Gerdemannia* C. Walker *et al* for the same clade of fungi. Hence, the genus *Gerdemannia* is invalid according to the rules of Botanical Code. However, although Oehl's and Sieverding action was pertinent, the location of *Pacispora* in the family Glomeraceae was erroneous (Walker *et al*, 2004). Recently, Walker and Schüßler (2004) transferred the genus *Pacispora* to the family Pacisporaceae C. Walker *et al* renamed from the Gerdemanniaceae.

Genus: Pacispora Sieverd. & Oehl gen. nov. Oehl and Sieverding (2004)

Spores of the new genus are formed terminally on hyphae. An inner, usually three-layered wall is present inside the spores, from which the spores germinate directly through the outer spore wall, which is also usually three-layered.

Family: Acaulosporaceae Stummer & Morton (1999)

Azygospores formed within the neck of a sporiferous saccule that is formed terminally on a fertile hypha. Saccules often cease to expand at the onset of spore formation occurring singly, rarely in loose clusters.

Genus: Acaulospora Gerdemann & Trappe (1974) emend. Berch (1985)

Spores arise laterally from the neck of a sporiferous saccule. In this genus, spores are born, laterally from the neck of a pre-differentiated "sporiferous saccule". The sporiferous saccule develops blastically from a hyphal tip. After the saccule has become fully expanded, a spore begins to develop from the side of the subtending hypha (termed "saccule neck"). As the spore matures, the saccule loses its contents and eventually sloughs off so that, it is often not attached to fully mature spores. Spores usually are sessile (no hyphal attachment). Germ tube emerges from a roughly spherical germination orb.

Genus: *Entrophospora* Ames & Schneider (1979)

Spores formed in the neck of the sporiferous saccule. In this genus, spores are born from within the neck of a predifferentiated "sporiferous saccule". Ontogenesis of spore and inner walls mirror that, found in spores of *Acaulospora*.

Family: Archaeosporaceae Morton and Redecker (2001) fam. nov.

Studies of other molecular characters, such as monoclonal antibody specificities (Wright *et al*, 1987) and the profile of fatty acids (Graham *et al*, 1995) have led to the long-held suspicion that members of this family were phylogenetically distant from species in *Acaulospora* and *Glomus*, but all of these data did not contain enough phylogenetic information to determine its position relative to other glomalean taxa, when morphological characters were atypical for the *Acaulospora*-like spores (Morton and Benny, 1990). Small sub-unit (18 S) ribosomal DNA sequences finally provided the character set to position this family as ancestral to both Acaulosporaceae and Glomaceae (Redecker *et al*, 2000a; Sawaki *et al*, 1998). The molecular phylogenetic tree suggests that members of this family and Paraglomaceae are more closely related to a non-mycorrhizal fungus, *Geosiphon pyriforme*, and a putative Zygomycete fungus that is the mycobiont in a symbiosis with a *Nostoc* cyanobacterium (the photobiont).

Vegetative structures consist of those inclusive in the suborder Glomineae. There are several characteristics unique to this family.

All species in this family can be distinguished from the other glomalean families (including Paraglomaceae) by the signature nucleotide sequence: TCTCKKYTTCGGYSGAGTCC at poison 227 of the 18S rDNA gene (Morton and Redecker, 2001). The de genera in this sequence reflect variation among species in the group and have the positive benefit of providing a more lenient diagnostic tool for the other species.

Genus: *Archaeospora* Morton and Redecker (2001) gen. nov.

All species form spores born laterally from the neck of a pre-differentiated "soporiferous saccule", but mode of formation either is like that of species in *Acaulospora* or from a branch hypha (or "pedicel") to form a *Glomus*-like spore. Some species are dimorphic, forming both acaulosporoid and glomoid spore morphotypes. Three modes of spore formation occur in this family:

1. The spore develops from the side of the subtending hypha of a "sporiferous saccule" similar to that of species in *Acaulospora* (Acaulosporaceae). When the spore detaches, it is sessile (no attached subtending hypha).
2. A hypha ("pedicel") first branches from the subtending hypha of a "sporiferous saccule" and the spore develops more like a species of *Glomus* (Glomaceae), terminally by blastic expansion from its pedicel.

Figs. 1-6: Spores of some of the AM fungal genera.

1. *Acaulospora rehmii* Sieverding and Toro, wih labirinthiform folds in the wall. Bar = 50 µm.
2. *Glomus aggregation* Schenck and Smith. Bar = 50 µm.
3. *Gigaspora margarita* Becker and Hall, spore with bulbous suspensor (arrow). Bar = 80 µm.
4. *Glomus davispora* Trappe, clavate, radially arranged, tightly packed spores in a portion of sporocarp (without peridium). Bar = 50 µm.
5. *Entrophospora schenckii* Sieverding and Toro, spore with portion of sporiferous saccule attached (arrow). Bar = 25 µm.
6. *Scutellospora* sp., spore with germination shield (arrow). Bar = 25 µm.

3. Species with *Glomus*-like spores forming from a "sporiferous saccule" (2 above) also form separate *Glomus*-like spores directly from external hyphae, some of which can also form saccules. Spores formed on the fungal thallus are either monomorphic or dimorphic. When the spores are monomorphic, only acaulosporoid spores are formed. When dimorphic, both acaulosporoid and glomoid are formed.

The spores originate from the neck of a sporiferous saccule that is formed terminally on a fertile hypha. The saccule often ceases to expand at the onset of spore formation, and occurs singly or more rarely in loose aggregates. Spore wall 3-4 layered, of which the outermost layer is continuous with the monolayer wall of the neck of the "sporiferous saccule". Spores do not form thin flexible "germinal walls" like all the species of Acaulosporaceae. One or more layers, even a septum may occlude the opening in the spore to the lumen of a subtending hypha or the neck of the "sporiferous saccule".

Family: Geosiphonaceae

On the basis of molecular analysis, the phylogenetic tree suggests that its members are more closely related to Archaeosporaceae and Paraglomaceae.

Genus: Geosiphon

It is non-mycorrhizal. The single species in the monotypic genus is a putative zygomycete fungus which is the mycobiont in symbiosis with a *Nostoc* cyanobacterium (the photobiont). *Geosiphon pyriformis* (Kütz.) v. Wettstein is the only known fungus forming endosymbiosis with cyanobacteria, *Nostoc punctiforme*. At hyphal tips, the coenocytic soil fungus forms unicellular, multinucleated 'bladders', about 1-2 mm in size, harbouring the cyanobacteria. There have been only five reports of it being found in nature and presently a region in the Spessart mountains (Germany) is the only known natural habitat worldwide.

ACKNOWLEDGEMENTS

The authors are grateful to MoENF, CSIR and UGC for their encouragement and financial assistance.

REFERENCES

Ames, R.N. & R.N. Schneider. 1979. *Entrophospora* a new genus in the Endogonaceae. Mycotaxon. 8: 347-352.

Benjamin, R.K. 1979. Zygomycetes and their spores. In: *The Whole Fungus. Vol II*, (B. Kendriek ed.), National Museums of Canada, Ottawa, Canada. pp. 573–622.

Berch, S.M. 1985. *Acaulospora sporocarpia,* a new species and emendation of the genus *Acaulospora* (Endogonaceae, Zygomycotina). Mycotaxon 23: 409-418.

Berch, S.M. 1988. *Compilation of the Endogonaceae Mycologue publications*, 331 Daleview Place, Waterloo, Ontario, Canada N 2 L 5 M 5.

Butler, E.J. 1939. The occurrence and systematic position of vesicular arbuscular type of mycorrhizal fungi. Trans. Br. mycol. Soc. 22: 274.

Cavalier–Smith, T. 1998. A revised six–kingdom system of life. Biol. Rev. 73: 203-266.

Chu–Chou, M. & L.G. Grace. 1979. *Endogone flammicerona* as a mycorrhiza symbiont of Douglas fir in New Zealand. N.Z.J. For. Sci. 9: 344-347.

Fassi, B., A. Fontana, & J.M. Trappe. 1969. Ectomycorrhizae formed by *Endogone lactiflua* with species of *Pinus* and *Pseudotsuga*. Mycologia. 61: 412-414.

Gerdemann, J.W. & T.H. Nicolson. 1963. Spores of mycorrhizal *Endogone* species extracted from soil by wet-sieving and decanting. Tr. Br. mycol. Soc. 46: 235-244.

Gerdemann, J.W. & J.M. Trappe. 1974. The Endogonaceae in the Pacific North West. Mycol. Mem. 5: 1-197.

Graham, J.H., N.G. Hodge, & J.B. Morton. 1995. Fatty acid methyl ester profiles for characterization of Glomalean fungi and their endomycorrhizae. Appl. Environ. Microbiologia 61: 58-64.

Janse, J.M. 1896-1897. Les endophytes radicaux dequelques Javanousis. Ann. Jard. Bot. Buitrenzorg. 14: 53.

Morton, J.B. 1988. Taxonomy of VA mycorrhizal fungi: Classification, Nomenclature and Identification. Mycotaxon. 32: 267–324.

Morton, J.B., and Benny, G.L. 1990. Revised classification of arbuscular mycorrhizal fungi (Zygomycetes): A new order Glomales, two new suborders, Glominae and Gigasporinae, and two new families, Acaulosporaceae and Gigasporaceae, with an emendation of Glomaceae. Mycotaxon. 37: 473-491.

Morton, J.B. and Redecker, D. 2001. Two new families of Glomales, Archaeosporaceae and Paraglomaceae, with two new genera *Archaeospora* and *Paraglomus*, based on concordant molecular and morphological characters: Mycologia. 93(1): 181-195.

Mosse, B. 1953. Fructifications associated with mycorrhizal strawberry roots. Nature. 171: 974.

Oehl, F., & E. Sieverding. 2004. *Pacispora*, a new vesicular arbuscular mycorrhizal fungal genus in the Glomeromycetes. Journal of Applied Botany. 78: 72-82

Peyronel, B. 1924. Prime ricerche sulle micorrize endotrofiche e sulla microflora radicicola normale delle fanerogame. Riv–Biol. 6: 463-485

Pirozynski, K.A. & Y. Dalpe. 1989. Geological history of the Glomaceae with particular reference to mycorrhizal symbiosis. Symbiosis. 7: 1-36.

Redecker, D., R. Kodner, & L.E. Graham. 2000a. Glomalean fungi from the Ordovician. Science. 289: 1920-1921.

Redecker, D. J.B. Morton, & T.D. Bruns. 2000b. Ancestral lineages of arbuscular mycorrhizal fungi. Mol. Phylogenet. Evol. 14: 276-284.

Sawaki, H., K. Sugawara, & M. Saito. 1998. Phylogenetic position of an arbuscular mycorrhizal fungus. *Acaulospora gerdemannii* and its synanamorph *Glomus leptotichum*, based upon 18 SrRNA gene sequence. Mycoscience. 39: 477-480.

Schenck, N.C. & Y. Perez. 1990. *Manual for the Identification of VA Mycorrhizal Fungi*. Synergistic Publications, Gainsville, Florida, USA. pp. 283.

Schüßler, A., D. Schwarzott & C. Walker. 2001. A new fungal phylum Glomeromycota: Phylogeney and evolution. Mycol. Res. 105: 1413-1421.

Stummer, S.L. & J.B. Morton. 1999. Taxonomic reinterpretation of morphological characters in Acaulosporaceae based on developmental patterns. Mycologia. 91: 849-857.

Thaxter, R. 1922. A revision of the Endogonaceae. Proc. Am. Acad. Arts Sci. 57: 291-351.

Trappe, J.M. 1982. Synoptic Key to the genera and species of zygomycetous vesicular arbuscular mycorrhizal fungi. Phytopathology. 72: 1102-1108.

Trappe, J. M. 1987. Phylogenetic and ecological aspects of mycotrophy in the angiosperms from an evolutionary stand point. In: *Ecophysiology of VAM plants*. (G. Safir ed.). CRC Press, Boca Raton, Florida. pp 5-25

Tulasne, L.R. & C. Tulasne. 1844. Fungi nonnulli hypogaeinovi V. Minus cogniti act. Giornale Botanica Italiano. 2: 55-63.

Walker, C. & F.E. Sanders. 1986. Taxonomic concepts in the Endogonaceae. III. The seperation of *Scutellospora* gen. nov. from *Gigaspora* Gerd. & Trappe. *Mycotaxon*. 27: 169-182.

Walker, C. & J.M. Trappe. 1993. Names and epithets in the Glomales and Endogonales. Mycol. Res. 97: 339–344.

Walker C., & A. Schüßler. 2004. Nomenclatural clarifications and new taxa in the Glomeromycota. Mycol. Res., 108: 981-982

Walker C., J., Blaszkowski D. Schwazott & A. Schüßler. 2004. *Gerdemannia gen. nov.*, a genus separated from *Glomus*, and Gerdemanniaceae fam. nov., a new family in the Diversisporales based on the former *Glomus scintillans*. Mycol. Res. 108(6), 707-718.

Wright S.F., J.B. Morton & J.E. Sworobuk. 1987. Identification of a VAM fungus by using monoclonal antibodies an enzyme-linked immunosorbent assay. Appl. Environ. Microbiol. 53: 2222-2225.

5

Pathological and Taxonomical Aspects of the Genus *Drechslera*

ASHOK GAUR[1], USHA DEV[2] and VIVEK MITTER[2]

[1]*Division of Seed Science and Technology*
Indian Agricultural Research Institute, New Delhi
[2]*National Bureau of Plant Genetic Resources,*
Pusa Campus, New Delhi

ABSTRACT

Drechslera or *Helminthosporium* is a member of the class Deuteromycetes. There is no common agreement among the mycologists regarding the correct identity of the various species of this important genus. Ellis (1971) regarded *Drechslera* and *Helminthosporium* as synonyms. Here an attempt has been made to discuss the various controversies arising for separating or merging the three genera i.e. *Drechslera, Bipolaris* and *Exerohilum*. The type species for anamorph genera *Drechslera, Bipolaris* and *Exerohilum* are connected with *Pyrenophora, Cochliobolus* and *Setosphaeria* respectively. The differentiating criteria and taxonomic variability has been discussed. Major diseases caused by the various species of *Drechslera* have been described in detail. A comprehensive list of all the diseases caused by the pathogen has been prepared to highlight the economic significance of the genus.

I. INTRODUCTION

Link proposed the genus *Helminthosporium* on the basis of the characters of *Helminthosporium velutinum* Link ex. S.F. Gray in 1809. The name *Helminthosporium* was initially proposed by Persoon in 1822 to designate all the forms of *Helminthosporium*. Drechsler (1923) published a monograph on the genus *Helminthosporium* incorporating 21 species occurring in United States and also gave an account of various types of symptoms viz. eye spot, leaf spot, spot blotch, leaf blight, foot rot, stripe etc. produced by various species of the genus.

Luttrell (1963), Ellis (1971) and Hughes (1980) gave a detailed account of the type species *Helminthosporium velutinum* Link ex. S.F. Gray. Here the conidiophores arise from a rudimentary to conspicuous stroma attaining a height of upto 400 μm and the width of 70 μm. Conidiophores occur singly or in tufts and are cylindrical to slightly tapered, dark brown, unbranched, straight or flexuous and septate. The conidia are produced through small pores in walls of distal and intercalary cells.

Production of terminal conidia usually signals cessation of conidiophore growth, although mature conidiophore may resume apical growth, if transferred to a moist atmosphere (Luttrell, 1963). Conidia are solitary, acropleurogenous, developing laterally often through small pores beneath the septa. Conidia usually obclavate, sometimes rostrate, sub-hyaline to brown, pseudoseptate with a prominent dark brown or black scar at the base which is known as hilum. Hilum is nothing but the point of attachment of conidia with the conidiophore. However, there is no corresponding scar on the conidiophore.

Nisikado (1928) made an extensive study of the species of *Helminthosporium* isolated from grasses in Japan, taking into account the morphology, pathogenicity, temperature and pH requirement etc. On the basis of his studies, he divided the genus *Helminthosporium* into two subgenera :

i. **Cylindro-Helminthosporium.** In this genus Nisikado placed all the species with straight, cylindrical conidia that germinate by one or more germ tubes arising from any cell. The conidia, though cylindrical but may be clavate or fusoid, sub-hyaline or pale, pseudopleurogenous with hilum a flat ringed pore.
ii. **Eu-Helminthosporium.** The genus is characterized by typically fusoid conidia germinating only from the end cells. Conidia are usually brown, pseudopleurogenous with flat ringed pore as hilum.

It was Ito (1930) who suggested the creation of a new genus *Drechslera* (after Drechsler) to accommodate all the species belonging to the subgenus Cylindro Helminthosporium of Nisikado. Members of this new genus are characterized by having cylindrical and not curved conidia germinating from every cell. *Drechslera tritici-vulgaris* (Nisik) Ito (=*Drechslera tritici-repentis* [Died.] Shoem.) was designated as the lectotype by Hughes (1958). Later on Shoemaker (1959) established the genus *Bipolaris* to accommodate all the species of the subgenus Eu-Helminthosporium of Nisikado. *Bipolaris maydis* (Nisik & Miyake) Shoem was regarded as lectotype. Leonard and Sugg. (1974) created yet another genus *Exserohilum* to accommodate all the species of *Helminthosporium* whose conidia have strongly protuberant hilum. The new genus was typified as *Exserohilum turcicum* (Pass.) Leonard & Sugg.

Inspite of considerable advanced studies on *Helminthosporium*, there is no common agreement among the mycologists, regarding the correct identity of the various species of this important genus. A perusal of the phytopathological literature reveals that the generic name *Helminthosporium* is being extensively used and the universal acceptance for segregation of graminicolous species into *Drechslera, Bipolaris* and *Exserohilum* is almost non-existent. Thus, mycologists and plant pathologists differ in the usage of *Helminthosporium, Drechslera, Bipolaris* and *Exserohilum*. Ellis (1971) regarded *Helminthosporium* and *Drechslera* as synonyms as he invariably used the two generic names simultaneously, e.g. *Drechslera erythrospila* (Drechsler) and *Helminthosporium erythrospilum* or *Drechslera maydis* (Nisikado) Subram. & Jain and *Helminthosporium maydis* Nisikado. Subramanian and Jain (1966) do not agree with the grouping of the species of *Helminthosporium* in two genera i.e. *Drechslera* and *Bipolaris* and have amended the description of *Drechslera* to include all the species under *Drechslera* and *Bipolaris.* While some workers recognize the generic differences between *Bipolaris* and *Exserohilum*, others stick to the use of *Drechslera* in a wider sense. There are still other hardliners who prefer to use the old generic name *Helminthosporium,* as proposed by Persoon in 1822.

II. DIFFERENTIATING CRITERIA

In *Drechslera,* the conidiogenous cells develop sympodially and have scars where the conidia were born. On the contrary, in *Helminthosporium* the conidiogenous cells are determinate with conidia developing through pores beneath the septa, while conidiophore tip is actively growing. Conidiophore growth ceases when the terminal conidium is formed. The conidiophores are never geniculate and have no dark scars. However, in case of *Drechslera* the first conidium is at the apex of conidiophore (Acropetal). After the development of first conidium, the conidiophore grows out laterally below the scar, often splitting the side walls, pushing the scar to one side and then growing on before forming another conidia at the newly constituted apex (Ellis, 1976). At present, the graminicolous species of *Helminthosporium* are being differentiated on the basis of conidial shape, germination pattern, point of origin, direction of growth of germ tube from the basal cell and progression of septum development (Luttrell, 1963 ; 1964) and hilum morphology (Shoemaker, 1966). Luttrell (1963 ; 1964) compared the two species i.e. *H. avenaceum* Curtis ex Cooke and *H. sativum* Pammel, King & Bakke. In the former, the germ tube from the basal conidial cell is lateral, arising more or less in a median position on the periclinal wall and growing at a wide angle to the longitudinal axis and in the developing conidia the primary septum is formed near the base delimiting the basal cell, while in the latter the germ tube arises close to the hilum and grows approximately in the direction of long axis and the first formed conidial septum is near the mid-point of immature conidia. Shoemaker (1966) highlighted the importance of hilum morphology with *Drechslera* having a flat scar included within the contour of the basal cell and *Bipolaris* having slightly to strongly protuberant hilum. In *Exserohilum,* the hilum is permanently exerted (Leonard & Suggs, 1974). Some of the characters of *Bipolaris, Drechslera,* and *Exserohilum* are being compared in Table 5.1.

Table 5.1: Characteristic features of *Bipolaris, Drechslera* and *Exserohilum.*

Characters	*Bipolaris*	*Drechslera*	*Exserohilum*
Conidial shape	Fusoid, obclavate, rarely truly cylindrical, straight or curved	Commonly cylindrical or obclavate-cylindrical, mostly straight	Fusoid, obclavate fusoid or cylindrical, straight or curved
Hilum	Often slightly protruding and truncate	Non-protruding, rounded, well defined intra-hilar cavity	Strongly protruding, truncate, often with enveloping "bubble"
Germination	Commonly from one or both polar cells, rarely from intermediate cells	From intermediate and/or polar cells	Commonly from one or both polar cells, rarely from intermediate cells
Basal germ tube direction of growth, position of emergence	Semiaxial, close to hilum, rarely lateral	Lateral, midway between hilum and basal septum, rarely semiaxial from near hilum	Semiaxial, close to hilum, rarely lateral
Septum ontogeny			
First	Median to sub-median	Demits basal cell	Sub-median
Second	Demits basal cell	Median	Supramedian
Third	Distal	Distal	Median or variable
Conidiogenous nodes	Smooth or verruculose	Smooth	Smooth or verruculose

According to Kendrick and DiCosmo (1979) and Luttrell (1977; 1978), the genus *Helminthosporium, sensu stricto* generally lacks any known telimorph connections, though Hughes (1953) reported that a 'true *Helminthosporium*' develops in pure culture from ascospores of a Massaria species. The type species for anamorph genera *Drechslera, Bipolaris* and *Exserohilum* are connected with *Pyrenophora, Cochliobolus* and *Setosphaeria* respectively. Another ascomycetous genus *Pseudocochliobolus* has been connected with *Bipolaris* (Tsuda & Ueyama, 1981) but this genus may not be distinguishable from *Cochliobolus* (Alocorn, 1983).

The genus *Pyrenophora* have dark brown ellipsoidal, superficial and solitary ascocarp with a stipitate base covered with numerous dark brown septate setae. Asci are binucleate, cylindrical, slightly curved, rounded at the apex and stipitate at the base containing 4-8 ascospores. The ascospores are uniseriate, pale yellow brown, fusoid, ellipsoid, muriform with up to 8 transverse septa. Young spores develop a primary transverse septum forming a fusoid 4-celled phragmospore. The spore expands and 1 or 2 longitudinal septa are formed in the median cell. Further development varies with the species. Mature spores are constricted at the septa.

Ascocarps of *Cochliobolus* are black, sessile with a globose base and a cylindrical neck which tapers toward the tip. Asci are cylindrical, slightly tapered near the base with an indistinct foot. Ascospores are hyaline, filiform, parallel to slightly coiled, moderately tapered toward the apex with a pronounced tapering base. Some ascospores are bulbous with swollen tips. Occasionally, septa is indistinct as large number of vacuoles are present in some spores.

Ascospores of *Setosphaeria* are superficial on substrate, black, globose or ellipsoid, with or without neck, ostiole surrounded by simple, short, rigid and brown hairs. Similar hairs are scattered on the surface of the upper part of ascocarp. Peridium coriaceous, carbonaceous and pseudoparenchymatous. Asci bitunicate, cylindrical to cylindrical-clavate, with 1-8 spores. Among the asci, there are filamentous pseudoparaphyses. Ascospores phragmosporous, hyaline, fusoid, 2-6 septate (rarely more) enclosed in a thin gelatinous sheath which projects well beyond the end spore.

III. TAXONOMIC VARIABILITY

Considerable variability has been observed in the species of *Helminthosporium* when they grow either in nature or in culture. Variability also occur in carbon and nitrogen requirements as well as the light and temperature. Luttrell (1954) observed greater uniformity of characters when the fungus is grown on cultural media in controlled conditions rather than in diverse natural conditions. He suggested that the fungus may be grown on sterilized barley straw on water agar. For comparative studies of this group, a weak medi (water agar) is preferred as compared to those where sugars and nitrogenous compounds are used (Dennis, 1933; Drechsler, 1923). To study the characters of *Bipolaris* and *Exserohilum* wheat straw or water agar media were used (Sivanesan, 1984; 1987). However, this medium is not suitable for species of *Drechslera, sensu stricto* and V8 juice agar or PDA is recommended (Alcorn, 1983). Sometimes plant material from a seemingly unsuitable host was found useful when incorporated into a media (Sivanesan, 1987). With 3 genera and multiplicity of species involved, it is unlikely that one medium and a set of environmental conditions would be applicable. Nevertheless, detailed comparisons of closely similar taxa within a genus should be concurrent under conditions that are fully documented and as reproducible as possible (Alcorn, 1988).

IV. OTHER TAXONOMICAL APPROACHES

The application of comparative morphological approaches to the systematics of *Helminthosporium* has led to the development of taxonomic concepts and clarity of classifications of this important genera. However, variability in taxonomic characteristics and occurrence of intermediates between currently accepted species warrants the use of expanded taxonomic approaches and techniques like numerical taxonomy, compatibility tests and molecular taxonomy.

Ibrahim and Threfall (1966) studied 28 taxa using over 200 pathological, physiological, cultural and morphological characters. Lam and Chapman (1985) employed 97 pathological, morphological and cultural characters in their numerical taxonomic studies of *Drechsleras* from *Lolium* spp. and *Festuca pratensis.* Compatibility test is another basis of taxonomy, although in all the three species homothallic species have been reported. The use of nucleic acid based molecular approaches is yet to be exploited to its full potential in case of *Helminthosporium.* Polymerase Chain Reaction (PCR) and Restricted Fragment Length Polymorphism (RFLP) based approaches can lead to better resolutions of similarity between taxa of this genus which is a significant plant pathogen. Sivanesan (1987) gave a detailed description of graminicolous species of *Drechslera, Exserohilum* and *Curvularia* as well as their teleomorphs. The largest genus of this group is *Bipolaris* with 52 species followed by *Drechslera* with 25 species and *Exserohilum* with 20 species.

V. CHARACTERISTICS OF IMPORTANT *DRECHSLERA* SPECIES

The genus *Drechslera* is characterized by conidia which are simple, straight or curved with clavate or rounded apical cells, mostly without hilum but protuberant hilum may be present in some species. Conidia straw coloured or pale to dark brown or olivaceous brown, sometimes with cells unequally coloured, the end cells being paler than intermediate ones, mostly smooth, pseudoseptate. Conidiophores simple, straight or branched arising singly or in groups, flexuous, often geniculate, brown to dark brown, smooth walled, bearing solitary conidia or in pairs or clusters and sometimes secondary conidium.

A brief description of colony, conidia and conidiophores of a few species of *Drechslera* is given below.

Drechslera australiensis (Bugnicourt) Subram. & Jain

Colonies effuse, grey to dark blackish brown, conidiophore solitary, flexuous or geniculate, septate. Conidia straight, ellipsoidal or oblong, rounded at the ends, pale brown to mid reddish brown with 3 pseudosepta, rarely with 4 or 5 pseudoseptate, mostly 18-33 × 8-10 μ. (Fig. 5.1).

Drechslera bicolor (Mitra) Subram. & Jain

Conidiophores solitary or in groups, straight or flexuous, dark scars, golden brown. Conidia straight or rarely curved, cylindrical or broader in the middle, tapering toward the rounded apex, often truncated at the base, 3-14 pseudoseptate, measuring 40-8 × 14-18 μ. (Fig. 5.2).

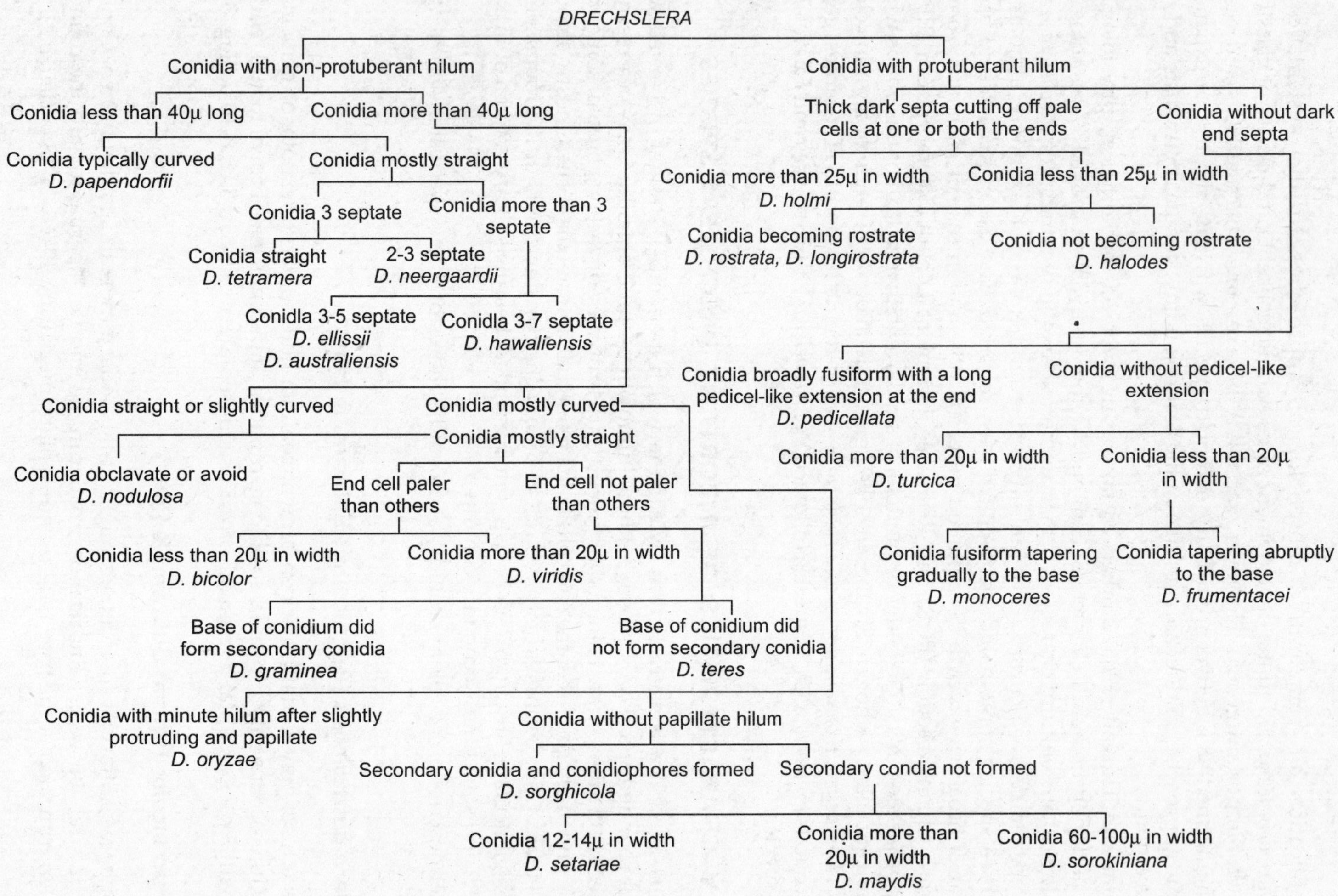
DRECHSLERA
Conidia with non-protuberant hilum
Conidia with protuberant hilum
Conidia less than 40μ long
Conidia more than 40μ long
Thick dark septa cutting off pale cells at one or both the ends
Conidia without dark end septa
Conidia typically curved D. papendorfii
Conidia mostly straight
Conidia more than 25μ in width D. holmi
Conidia less than 25μ in width
Conidia 3 septate
Conidia more than 3 septate
Conidia becoming rostrate D. rostrata, D. longirostrata
Conidia not becoming rostrate D. halodes
Conidia straight D. tetramera
2-3 septate D. neergaardii
Conidla 3-5 septate D. ellissii D. australiensis
Conidla 3-7 septate D. hawaliensis
Conidia broadly fusiform with a long pedicel-like extension at the end D. pedicellata
Conidia without pedicel-like extension
Conidia straight or slightly curved
Conidia mostly curved
Conidia mostly straight
Conidia more than 20μ in width D. turcica
Conidia less than 20μ in width
Conidia obclavate or avoid D. nodulosa
End cell paler than others
End cell not paler than others
Conidia less than 20μ in width D. bicolor
Conidia more than 20μ in width D. viridis
Conidia fusiform tapering gradually to the base D. monoceres
Conidia tapering abruptly to the base D. frumentacei
Base of conidium did form secondary conidia D. graminea
Base of conidium did not form secondary conidia D. teres
Conidia with minute hilum after slightly protruding and papillate D. oryzae
Conidia without papillate hilum
Secondary conidia and conidiophores formed D. sorghicola
Secondary condia not formed
Conidia 12-14μ in width D. setariae
Conidia more than 20μ in width D. maydis
Conidia 60-100μ in width D. sorokiniana

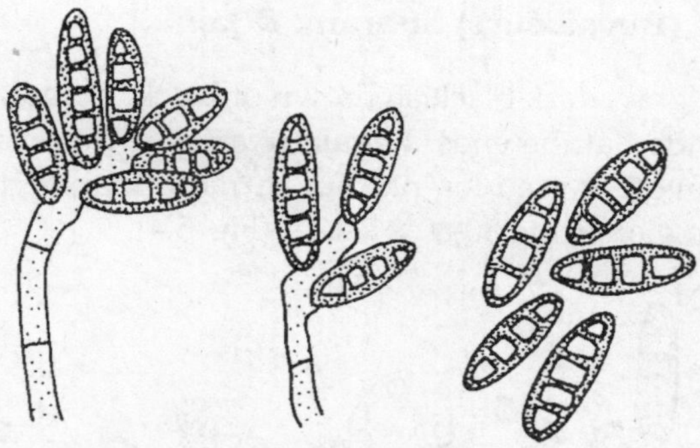

Fig. 5.1: *Drechslera australiensis*

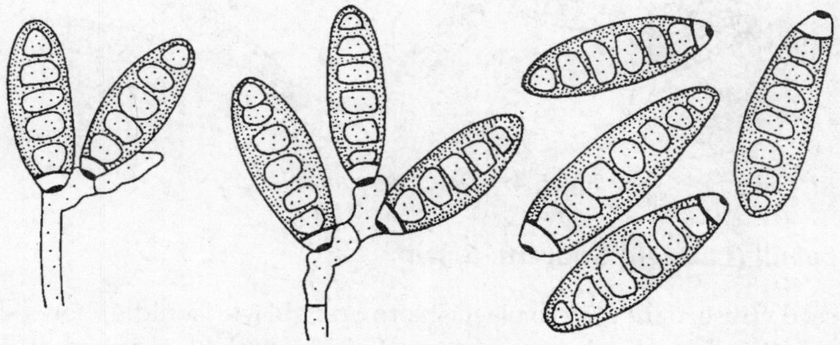

Fig. 5.2: *Drechslera bicolor*

Drechslera halodes (Drechsler) Subram. & Jain

Conidiophores simple, solitary, sometimes in small groups, rarely branched, straight, flexuous, mild to dark brown, olivaceous brown, bearing conidia acropleurogenously. Conidia simple, cylindrical to ellipsoidal, never rostrate, broader in the middle, tapering toward the rounded ends, end cell darker thicker, hilum protuberant, 4-11 pseudoseptate, measuring 48-90 × 15-68 µ (Fig. 5.3).

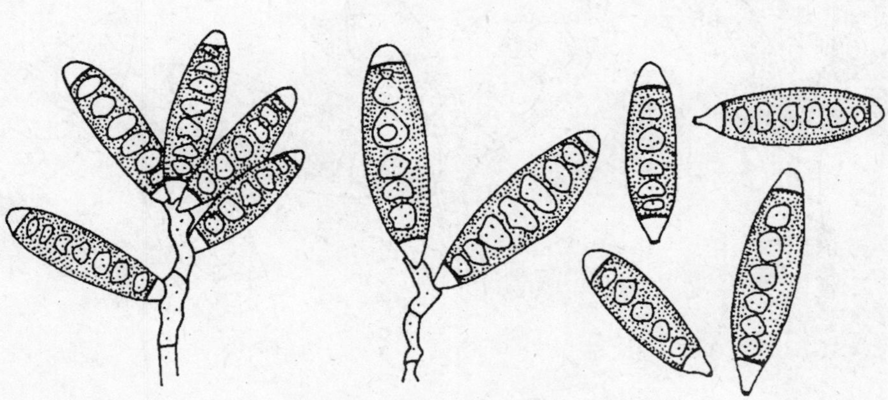

Fig. 5.3: *Drechslera halodes*

Drechslera hawaiiensis (Bugnicourt) Subram. & Jain

Colonies on seed effuse, gray, dark blackish brown or black. Conidia straight, ellipsoidal, oblong or cylindrical, rounded at the ends. Regularly arranged on conidiophores, pale to mid brown. Conidia may have 2-7 pseudosepta but in more than eighty percent cases only 5 pseudosepta are present, measuring 30-80 × 7-12 μ (Fig. 5.4).

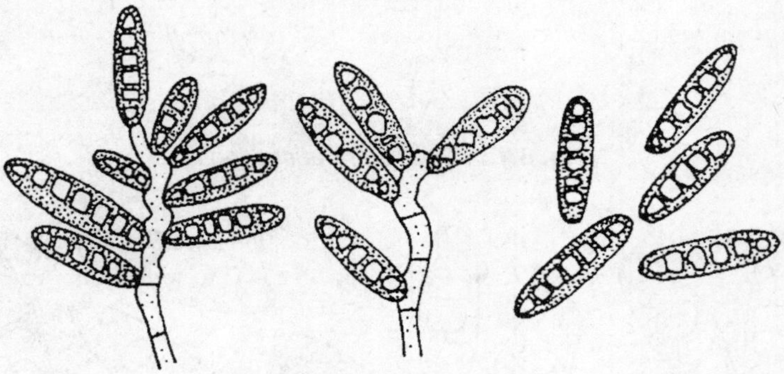

Fig. 5.4: *Drechslera hawaiiensis*

Drechslera holmii (Luttrell) Subram. & Jain

Colonies on seed effuse, dark gray to blackish brown or black. Conidiophores simple, solitary sometimes in groups, septate, pale to mid brown. Conidia straight or curved, obclavate, rostrate with middle portion broader, 6-10 pseudoseptate, end cell pale and cut off by a dark, thick septa, intermediate cells mid dark, golden brown, protuberant hilum measuring 54 × 8.8 μ (Fig. 5.5).

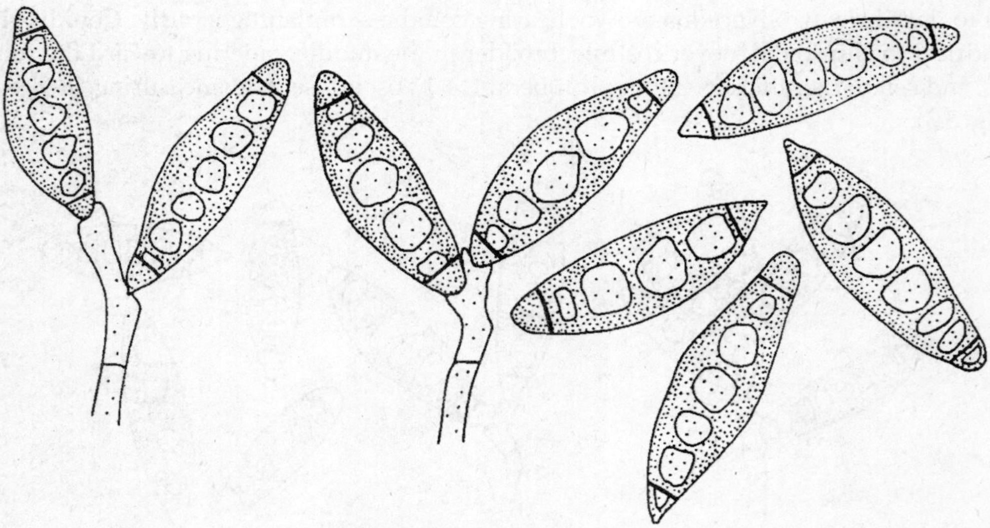

Fig. 5.5: *Drechslera holmii*

Drechslera maydis (Nisikado) Subram. & Jain

Conidiophores simple, solitary or in groups, brown to light black, slender, flexuous, bearing conidia acropleurogenously. Conidia distinctly curved, fusiform, pale to mid dark golden brown, typically curved comparatively more than *D. oryzae*, sharply tapering toward the rounded ends, base clear and flat, hilum within the contour of the basal cell, mostly inconspicuous, 5-10 pseudoseptate measuring 46-76 × 11.76 μ (Fig. 5.6)

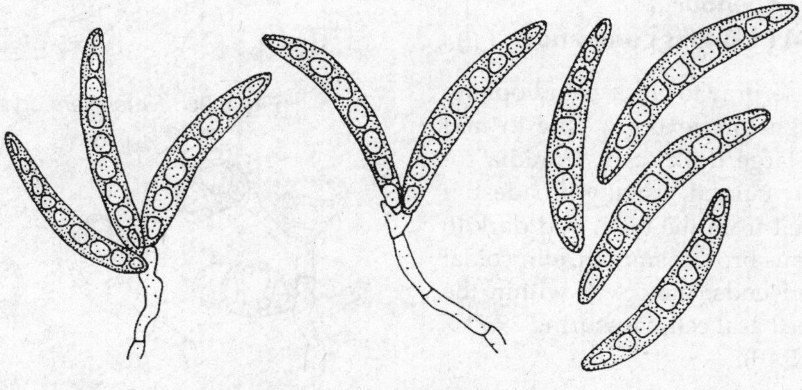

Fig. 5.6: *Drechslera maydis*

Drechslera neergaardii Danquah

Colony effuse, pale to mid brown conidiophore solitary or in groups of 3-4, straight or flexuous, often geniculate toward the apex, pale to mid brown. Conidia straight broadly ellipsoidal, mid to dark golden brown, smooth, mostly 3 pseudoseptate, thick in broadest part, hilum inconspicuous, measuring 20-26 × 13.5-20 μ (Fig. 5.7).

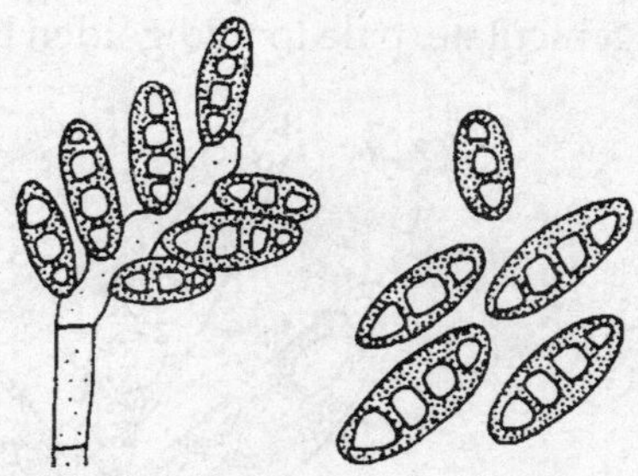

Fig. 5.7: *Drechslera neergaardii*

Drechslera nodulosa (Berk. and Curt.) Subram. & Jain

Conidiophores sometimes solitary but usually in groups, often branched flexuous, mid to dark brown scars. Conidia straight, ovoid to obclavate, pale to mid dark golden brown, 5-7 pseudoseptate measuring 50-65 × 16-19 μ, thick in the broadest part, hilum 3-4 (Fig. 5.8).

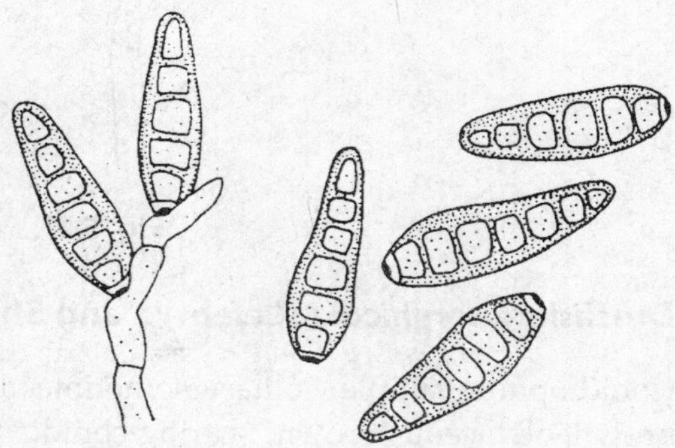

Fig. 5.8: *Drechslera nodulosus*

Drechslera oryzae (Breda de Hann) Subram. & Jain

Conidiophore simple, solitary sometimes in small groups, straight, flexuous, mid to dark brown, olivaceous brown, bearing conidia acropleurogenously. Conidia usually curved, but not as much as *D. maydis*, tapering to rounded end, base rounded with minute but distinct papillate, 6-12 pseudoseptate, measuring 81.60 × 14.8 μ (Fig. 5.9).

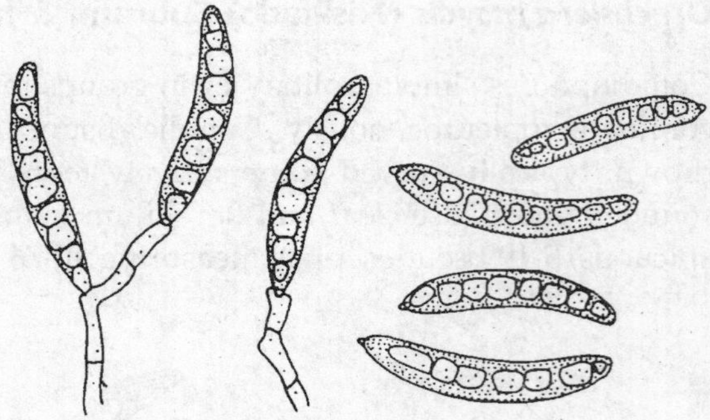

Fig. 5.9: *Drechslera oryzae*

Drechslera papendorfii (Van der Aa) M.B.Ellis comb. nov.

Colonies effuse, gray to black, conidiophore simple, straight or flexuous, pale to mid brown with large dark scars. Conidia robust, typically curved, usually broadest at the second cell from the base, mid dark to dark olivaceous brown, smooth, ellipsoidal with rounded ends, dark scar within the contour of the basal cell, measuring 29.71 × 16.32 μ (Fig. 5.10).

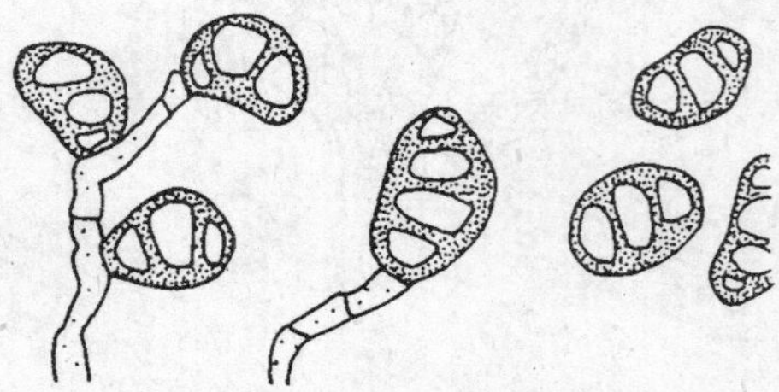

Fig. 5.10: *Drechslera papendorfii*

Drechslera setariae (Sawada) Subram. & Jain

Mycelium scanty, conidiophore simple, solitary or in small groups, straight or flexuous, sometimes geniculate, pale to mid golden brown, smooth, 4-8 pseudoseptate. (Fig. 5.11).

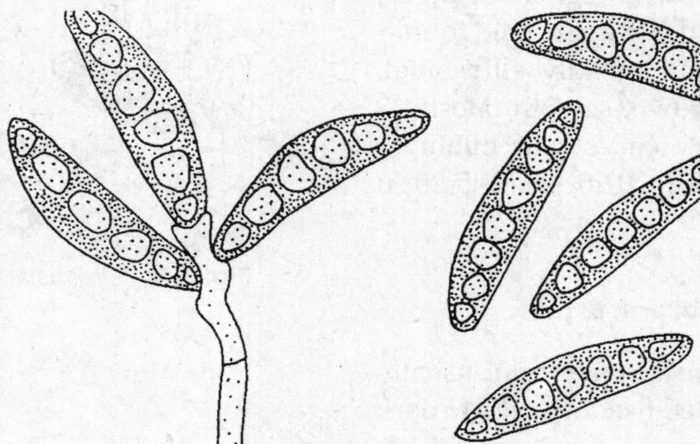

Fig. 5.11: *Drechslera setariae*

Drechslera sorghicola (Lefebvre and Sherwin) Richardson & Fraser

Conidiophore simple, solitary sometimes in small groups, straight or flexuous, mid to dark brown, olivaceous brown, bearing conidia acropleurogenously. Conidia slightly curved, fusiform pale to mid golden brown in the middle, gradually tapering toward the rounded end.

Produce secondary conidiophore with secondary conidia. Conidia 4-7 pseudoseptate measuring 56.72 μ × 12.19 μ (Fig. 5.12).

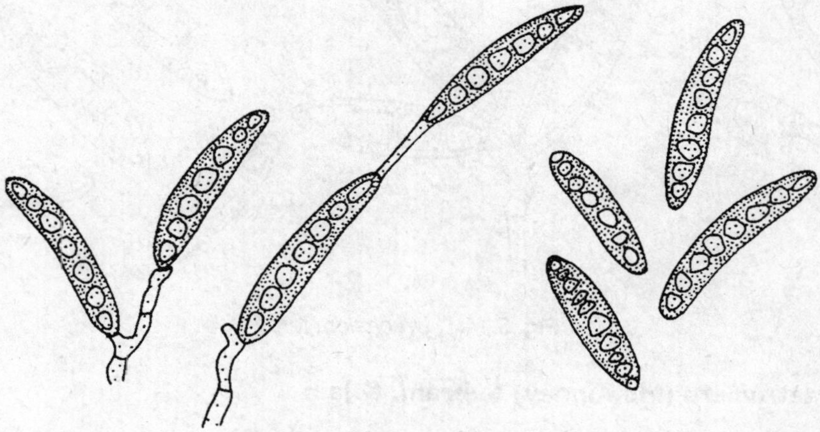

Fig. 5.12: *Drechslera sorghicola*

Drechslera sorokiniana (Sacc.) Subram. & Jain

Conidiophore solitary or in small groups, straight or flexuous, pale to mid dark brown. Conidia slightly curved, sometimes straight, fusiform to broadly ellipsoidal, dark olivaceous brown, smooth, broad in the middle, 6-10 pseudoseptate, measuring 30.60-100 × 18-23 μ (Fig. 5.13).

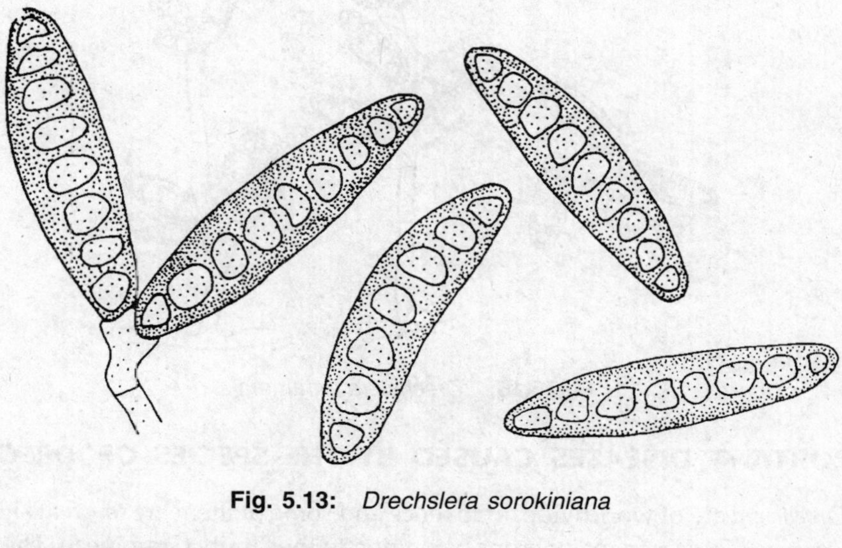

Fig. 5.13: *Drechslera sorokiniana*

Drechslera teres (Sacc.) Shoemaker

Scanty growth of mycelium on seed surface. Conidiophore simple, sometimes in groups of 2 or 3, straight or flexuous, pale to mid brown or olivaceous brown, conidia light, straight, cylindrical, rounded at the ends, smooth, 4-7 pseudoseptate. (Fig. 5.14).

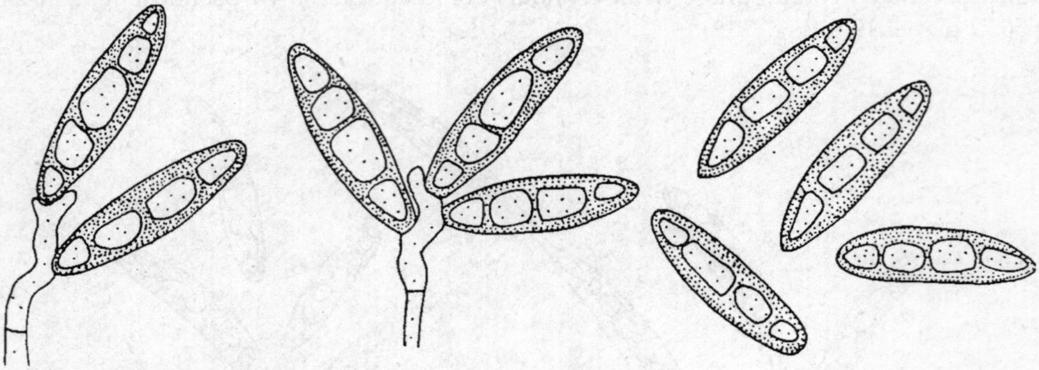

Fig. 5.14: *Drechslera teres*

Drechslera tetramera (Mc Kinney) Subram. & Jain

Colonies on seed effuse, gray, mostly dark, blackish brown to black. Conidiophore simple, solitary rarely more than one septate. Conidia appears acropleurogenously, in groups, mostly at right angle to the conidiophore. Conidia brown to blackish, ellipsoidal, straight with rounded apical ends, lighter toward the terminal cell. Conidia with 3 pseudoseptate, measuring 8-22 × 7.04 μ (Fig. 5.15).

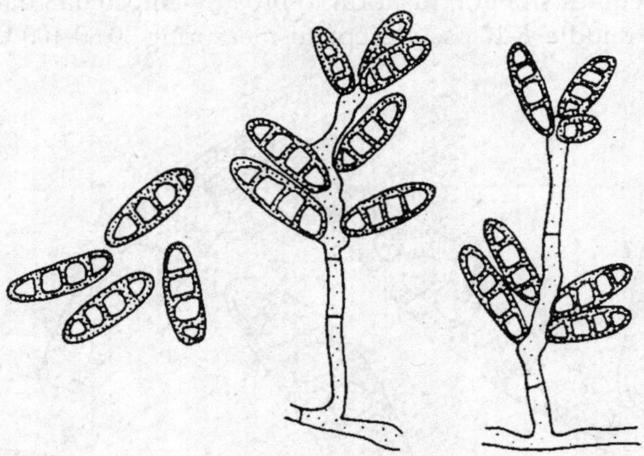

Fig. 5.15: *Drechslera tetramera*

VI. IMPORTANT DISEASES CAUSED BY THE SPECIES OF *DRECHSLERA*

Species of *Drechslera* are of worldwide occurrence and some of them are of great significance as they cause serious diseases in major crops belonging to the family Gramineae. The brown spot disease of rice caused by *Drechslera oryzae* (Breda de Hann) Subram. & Jain (*Helminthosporium oryzae* Breda de Hann) is one of the most serious diseases of rice and was responsible for causing the Great Bengal Famine of 1942. Ghose *et al* (1960) recorded a yield loss of upto 90 percent while Bedi and Gill (1960) reported a loss of up to 20 percent in the weight of grains.

Similarly, leaf blight of maize too, is of great economic importance. The leaf blight is caused by three different species *viz. Drechslera maydis* (Nisikado) Subram. & Jain, *Helminthosporium carbonum* Ullstrap and *Drechslera turcica* (Pass.) Subram. & Jain. The incidence of *D. maydis* was first recorded in India by Munjal and Kapoor (1960), while the incidence of *D. turcica* was reported by Mitra (1923) for the first time in Bihar in 1907. The appearance of race T of *D. turcica* or *H. turcica* caused an epiphytotic in USA in 1970, resulting in a loss of 700 million bushels of maize worth about one million dollar (Moore, 1970; Tatum, 1971). The disease is commonly known as southern corn leaf blight. All the corn hybrids containing Texas cytoplasmic male sterility gene were affected by the T race of this pathogen. Apart from important cereal crops various species of *Cochliobolus* and *Pyrenophora* also attack lawn and golf course grasses. In several infections, the leaves become completely blighted, wither, die and drop off. Moreover, the perennial nature of the turf grasses and the fact that they are frequently mowed and irrigated, creates additional problems. The disease is more destructive in moist weather and may kill the entire turf. A comprehensive account of a few important diseases caused by the pathogen in various crops is given below:

Foliar Blight of Wheat

Spot blotch pathogen *Drechslera sorokiniana* (Sacc.) Subram. & Jain (= *Helminthosporium sativum* Pamm., King & Bakke, perfect stage *Cochliobolus sativus* Ito & Kuribay) appears to be the major pathogen in the foliar blight complex of wheat. The other pathogen associated with the disease is *Drechslera tritici-repentis* (Died.) Shoemaker (*Pyrenophora tritici-repentis*).

The disease is widely distributed and has been reported to cause economic losses from Bihar, West Bengal, eastern Uttar Pradesh and Orissa (Chattopadhyay *et al*, 1968 ; Narain *et al* 1973). Singh *et al* (1993) recorded its presence in the north western region of Indo-Gangetic plains with it assuming epidemic proportions on varieties like HD-2329 and HD-2285 in some western districts of Uttar Pradesh. According to Dublin and Ginkel (1991) *D. sorokiniana* is the predominant species in South East Asia and Indian subcontinent. Losses ranging from 30-100 percent have been reported due to spot blotch from various parts of the world. Yield loss to the tune of 40 percent in Phillippines (Lapis, 1985), 85 percent in Zambia (Raemackers, 1979) and 30-86 percent in Brazil (Mehta, 1985) have been reported. In India, the loss due to spot blotch varies depending upon varieties cultivated and may range from 3-20 percent (Nema & Joshi, 1971).

The disease infects leaf blades and can be seen as small light brown spots of varying shape and size. Fully developed lesions are oval to oblong with dark brown and light brownish center. On mature plants the lesions coalesce forming large necrotic blotches which may cover 50-80 percent leaf area. Severely infected leaves dry and give a burnt look. Symptoms of the disease also appear on leaf sheath, nodes, internodes and glumes. The percentage leaf necrosis caused by *D. tritici-repentis* and lesion length increased with increasing duration of leaf wetness (Sah, 1994). Wolf and Hoffman (1994) reported that frequent rainfall, high RH and temperature more than 10°C enhanced primary infection and production of conidia of *D. tritici-repentis* whereas dry weather for more than 5 days inhibited the process of disease spread.

Breeding for disease resistance is of paramount importance and needs to be supported by studies on pathogen and host variability. Variability among isolates of *D. sorokiniana* has been reported by Gayed (1962), Bidar and Govindu (1975), Misra *et al* (1981) and Mehta (1985). Hetzler *et al* (1991) have reported 15 pathotypes based on their studies with global collection of monoconidial isolates of *Cochliobolus sativus* on a differential set of 11 wheat varieties and one wheat/*Agropyron curviflorum* hybrid line. Bonde *et al* (1993) reports that isozymes analysis can be used as a tool for pathogenic variability in *D. sorokiniana*. Inheritance of resistance to spot blotch is governed by one or two dominant factors (Adlakha *et al*, 1984) or more complex inheritance in varieties showing moderate resistance (de Pauw *et al*, 1984). Resistant sources have been identified against *D. sorokiniana* and efforts are also underway to combine genes for resistance from several sources. Varieties like Horizon, Hadder, BH 146, Suzhae 8, Yang Maif and pat 7219 have been reported to be resistant (Kohli *et al*, 1991; Mehta,1985). It is important to combine resistance from sources other than wheat and in this direction spot blotch resistance from *Agropyron distichm, A. curvifolium* and *Elymus gigantens* is being transferred to wheat at CIMMYT (Kohli *et al*, 1991). In India, Singh *et al* (1995) reported that many varieties like Sonalika, HD 2329, HD 2285, HS 284, OP 238, UP 2003 etc. gave susceptible reaction.

Systemic fungicides procanozole (Tilt 250 EC) and Triadimenol and contact fungicide like Zineb or Maneb have been reported to be effective by various workers (Raemackers, 1979; Mehta & Igarashi, 1985; Singh & Rai, 1986). Ahmad *et al* (1993) reported that mycelial growth of *D. sativus* was effectively reduced by benomyl followed by Panoram (fenfuram).

Various other host plants like *Commelina diffusa, Ekleusine indica, Chloris barbata, Cynodon dactylon, Echinochloa* etc. also harbour the pathogen as reported by Lapis (1985) in Phillippines. In India, Singh *et al* (1995) have reported the presence of the fungus on *Phalaris minor* and *Launia splenifolia.*

Fokemma (1973) observed the role of saprophytic fungi against *D. sosrokiniana.* Lately, Prasad *et al* (1978) reported antagonistic action of *Trichoderma* spp. and *Streptomyces* sp. on the pathogen. Thompson and Wildermuth (1989) have shown that vascular arbuscular mycorrhiza (VAM) can reduce root infection caused by *D. sorokiniana*. Kumar *et al* (1993) reported control of wheat root rot using mixed cultures of *Trichoderma viride* and *Glomus epigenus*. Bello *et al* (1994) also tested the effect of *Trichoderma* spp. on wheat seedling blight on 10 wheat cultivars.

It is important to check the dispersal of the pathogens of foliar blight complex through seed as Schilder and Berstrom (1995) observed that for *P. tritici-repentis*, infected seeds may serve as a source of inolculum and for dispersal of strains of the fungus to new areas.

Cultural practices like wider row spacing and restricted nitrogen application may also help in reducing the disease incidence (Mehta & Gaudencio, 1991; Saunders, 1988).

Stripe Disease of Barley

The disease is caused by *Drechslera graminea* Erikss. (Syn. *Helminthosporium gramineum* Rahb.; perfect stage is *Pyrenophora graminea* (Rahb.) Ito & Kuribay). The disease occurs wherever barley is grown and is responsible for considerable losses in yield. Damage to the crop is from the seedling stage to the maturity. The pathogen has been recorded with incidence of more than 50 percent in seed samples collected from Rajasthan (Agarwal, 1977; Thakkar, 1988). Bhatnagar and Mathur (1991) estimated 60-90 percent loss in grain yield due to stripe disease of barley. According to Pant and Bisht (1983) losses can vary from 70-72 per cent.

Small yellowish brown spots first appear in young seedlings on lower surface of leaf-sheath. Later on, after 6 weeks these spots form long parallel yellow stripes which extend from the base of the leaf blade to its tip. These stripes soon turn brown as the necrosis sets in. There may be 4-6 or more stripes on a single leaf. These stripes finally split and the leaf blade appears shredded. Since the infection is systemic, all tillers arising from an infected plant usually have infected leaves showing characteristic longitudinal brownish stripes. Many a times the brown streaks are separated from the remaining healthy portion by a yellow zone within which chlorophyll has largely disappeared. The discolouration is accompanied by loss of turgidity in the infected leaf and as a result drop down against the stalk. Generally, the lower leaves are attacked first and the infection gradually spreads upward. The leaf sheath and nodes are also attacked. Under humid conditions, the formation of gray to olive gray mass of the conidia and conidiophores on nodes is distinctly visible on infected tissues. In severe cases, the leaves are completely destroyed and the spike fails to emerge and the uppermost inter-node is considerably reduced or sometimes entirely suppressed. The spikes that emerge from infected plant are blighted, twisted and compressed.

The mycelium in the tissues is light yellow to hyaline and is present in abundance. However, in culture the mycelium appears either gray or black in colour and does not produce conidia unless exposed to alternate day and night cycles. Conidiophores occur usually singly or in clustres of 2-3 from the stomata or epidermal cells of infected plants. The basal segment of the conidiophores is enlarged while at the distant end they are slender and grayish in colour. The conidia are 1-7 septate without constriction at the septa and are 40-105μ in size. They are sub-hyaline to yellowish brown, straight to sub-cylindrical to slightly tapering having rounded ends. Conidia germinate by production of 2 polar germ tubes from each end cells and sometimes by intermediate cells.

The disease is transmitted by seed through attached conidia, by mycelium present in the chaff or through mycelium deeply seated in the tissues of the seed. The mycelium remains viable in the seed for an indefinite length of time. Thus, seedborne inoculum is the major source of perennation of the disease from season to season. The infection usually takes place on the meristem of the germinating seedling. Thakkar (1988) reported that during seed germination fungus becomes active and invades coleorrhiza, scutelar node, scutellum, coleoptile and first leaf systemically. The infection reaches up to the new seeds from the mother plant. The mycelium of the fungus may perpetuate in soil and plant debris. The floral infection is more pronounced in area having moist and humid conditions during flowering period. A cool, moist fertile soil favours stripe development in seedling and developing plant.

Seed treatment with Agrosan GN or Ceresan etc. control the seedborne infections. Loughman and Khan (1993) reported that Flutriafol and Traidimenol + Imazalil were the most effective treatment for control of barley leaf stripe in southern belt of Destev, Australia. Singh and Saksena (1971) observed that varieties like C-50, K-19 and K-572-64/6 remained unaffected by the disease. Varieties like K-24, K-125, K-129, K-131, K-133, K-441-7, K-464/11 and K-572/11 were found to be highly resistant against stripe disease infection by Atheya (1974).

Variability studies using isozymeanalysis did not show any species specific banding pattern between *Pyrenophora graminea* and *P. teres* (Johansen *et al*, 1992). Husted (1992) found very little antigenic differences between *P. teres* and *P. graminea* and suggested that monoclonal antibodies may be required for further immunological differentiation of stripe disease and net blotch causal organism. Vale *et al* (1994) studied the activation of genes in barley roots in response to infection by *D. graminea* isolates. Northern blot analysis showed the accumulation of 3 mRNA families in response to inoculations. Within 6 hours of contact the fungus induced a class of transcript homologous to a clone coding for peroxidase. Later a second class of mRNA's induced transcripts homologous to clone coding thaumatin like proteins and third class mRNA's encoding thionin.

Net Blotch of Barley

The disease is caused by *Drechslera teres* (Synonym: *Helminthosporium teres* Sacc.). The perfect stage of the fungus is *Pyrenophora teres* (Died.) Drechsler.

The perithecia of this fungus have been described by Drechsler (1923) and others in USA and by Smith (1929) in England. The perithecia of *Helminthosporium* described by Diedicke probably belonged to this species. Diedicke originally gave the name *Pleospora teres* to a hypothetical perithecial stage of *H. teres*. Perithecia brownish black, 0.5mm, asci sub-cylindrical, 20-230 × 30-36 μ, 8-spored, ascospores biseriate, 3-septate, light brown, 50-60 × 18-22 μ.

Net blotch disease of barley, although common, does not affect the plant so seriously as does the leaf stripe. Hims and Gladdens (1994) reported that net blotch caused 60 percent yield loss in variety GS-77 an UK during 1991-92. The fungus is carried over with the grain and infection on the first leaf may arise as in *P. graminea*, the result being the formation of a pale stripe which may become brown. The symptoms appear first as small brown spots usually near the tip of the leaf blade later increasing in size in longitudinal manner. The spots produce a transverse and longitudinal net-like (reticulate) pattern with internal brown pigmentation. The pigmentation of the interstices within the reticulate pattern is so reduced that the discolouration is present almost exclusively as a sharply defined network of brown lines. The parts of leaf adjacent to discoloured tissue become yellowish and eventually the leaf withers from the tip to the base. Under weather conditions conducive for disease development each successive leaf becomes infected and withers after a certain period.

Net blotch, also like stripe disease, develop from infected seed or plant debris in the soil or straw. But unlike stripe disease, *D. teres* is not systemic in the host plant. The conidia are dispersed by the wind and infect the growing seedling and subsequent infection is also caused by dispersal through wind or water. The pathogen can also infect host plant like *Panicum* spp., *Sorghum vulgare, Pennisetum typhoides* and *Avena sativa* (Misra, 1973).

Toubia-Rahme *et al* (1995) reported that when field grown barley straw was inoculated with *D. teres* and treated with Paraquat, it significantly reduced the sclerotiod structure produced by the pathogen and also modified the morphology and myceliogenesis.

Louw *et al* (1995) characterized *Pyrenophora* isolates associated with spot and net lesions on barley in South Africa using RAPD analysis. They concluded that the isolates occurring in South Africa belong to *P. teres* f spp. *aculata* and not *P. japonica*.

As in case of stripe disease of barley, the disease can be controlled by practising field sanitation and appropriate seed treatment.

Leaf Spot of Rice

Leaf spot of rice is caused by *Drechslera oryzae* (Syn. *Helminthosporium oryzae* Breda de Hann = *Cochliobolus miyabeanus* (Ito & Kuribay) Drech. Ex Dastur).

Leaf spot of rice is widespread disease in almost all the rice growing areas of the world and is also known as brown spot, sesame spot, Helminthosporiosis or seedling and leaf blight of rice. The disease is prevalent in the rice belt of India and the attack is more severe in West Bengal, eastern UP, Assam, Tamil Nadu and part of Kerala where heavy monsoon is recorded. Rice grown in poor soils of low pH and deficient in K_2O tends to be more prone to the disease.

Two major epidemics of the disease reported from India were in 1918 in Godavari delta and the Great Bengal Famine in 1942 in which losses up to 90 percent in yield were reported (Ghose *et al,* 1960; Padmanabhan,1974). Loss in weight of rice grains due to *Drechslera oryzae* ranging from 4.6 to 29 percent were reported by Bedi and Gill (1060) and up to 52 percent by Chattopadhyay *et al* (1975).

The disease affects all the aerial parts of the plant and irregular brown spots with a dark centre can be seen on coleoptile, leaf blade and leaf sheath. However, the symptoms are more pronounced on the leaves. The spots first appear as minute dots, later becoming oval to circular with gray centre and a dark reddish brown margin. In epiphytotic conditions, the spots may coalesce, plants are stunted and leaves wither giving the entire field a characteristic blighted or scorched appearance. At the time of flower emergence typical brown water soaked lesions appear at the principal node below the rachis or neck. Seedlings are also often heavily attacked with numerous spots causing withering of leaves. The fungus has been reported on the grain and the affected seeds may be shriveled and discoloured. The infection or infected plant manifests itself either by causing poor germination of infected seeds, reduction in the effective leaf surface due to spot or on the grain themselves resulting in low yields.

Partially purified toxin(s) from *H. oryzae* inhibited germination and reduced root and shoot length in rice cultivar Giza 180 and Arabi in Cairo (Hegazy *et al,* 1992). They concluded that sensitivity to toxins of *C. miyabeanus* is correlated with susceptibility to pathogen. Vidyasekaran *et al* (1992) reported a decrease in phenolic content and activities of peroxidase and phenylalanine-ammonialyase in infected rice leaves and accumulation of pathogen at later stages of infection.

The fungus grows both intra- and intercellularly within the tissues and conidiophores emerge as tufts through stomata or wounds. The perfect stage of the fungus is *Cochliobolus miyabeanus*. Perithecia in culture are globose, pseudo parenchymatous, black with an ostiolar beak. Asci are cylindrical to long, fusiform, slightly curved and usually contain 4 ascospores.

The disease is both soil and seedborne and the fungus may survive for 2-3 years in infected plant debris and seed. Several grasses such as *Echinocloa colorum* and *Leersia hexandra* growing around rice field have been reported to be natural host of the fungus.

The primary seedborne infection may be controlled by hot water treatment of 10-12 minutes at 52°C (Chaudhuri, 1948). Thirumulachar (1967) suggested a mixture of antibiotic aureofungin (20ppm) and copper sulphate (20ppm) for seed treatment. Antibiotics like Nystatin and Greseofulvin have been recommended by Padmanabhan (1974). Chakrabarthi *et al* (1975) reported seed treatment with Edipenphos to be effective. The secondary spread of the diseases can be controlled by spraying copper fungicides or 0.2 percent dithane Z-78 at regular intervals. Singh *et al* (1975) reported that 0.1 percent hinosan or blitox 0.2 percent controlled the disease when sprayed as foliar spray. Sen and Kaiser (1972) found that herbicide stam F-34 when applied, provides resistance to rice seedlings against the disease. Mukherjee and Bagchi (1965) observed that spraying with trace elements such as boron, copper, manganese and zinc gave significant control of brown spot of rice. Shah and Mariappan (1992) reported that fancotine (guazatine) and panoram (fenfuram) seed treatment was effective in controlling rice seedling mortality. Swaminathan and Selvaraj (1994) found that maneb was most effective in controlling the disease in nursery beds using seed and seedling treatment (180g ai/100Kg seed and 225g ai/100 l water). For secondary air-borne infection in the field, 3 sprays with iprobenfos (48ml ai/100 l water) at 15 days interval after transplanting gave good control.

Padmanabhan *et al* (1966) found Ch-13, Ch-45, T-141, BAM-10, T-2112, T-2118 and T-960 as resistant to brown spot. Singh and Sharma (1975) recommended cultivation of Padma and IR-24 in epidemic areas.

Saifulla (1993) studied 64 genotypes of rice against blast, brown spot and leaf scald disease. IET 11459 was resistant to brown spot as well as blast. Suraksha was found resistant to brown spot but susceptible to blast.

Maydis Leaf Blight

The disease is caused by *Drechslera maydis* (Nishikado) Subram. & Jain[= *Cochliobolus heterostrophies* (Drechsl.) Dreks]. The disease is a problem in Pakistan, India, Nepal, Kampuchea, Phillippines, Indonesia, Vietnam and China (Sharma *et al*, 1993). It has the potential to cause significant yield losses in susceptible cultivars. In India, the disease is reported from Jammu and Kashmir, Himachal Pradesh, Sikkim, Meghalaya, Punjab, Haryana, Rajasthan, Andhra Pradesh, Uttar Pradesh, Bihar, Madhya Pradesh, Karnataka, Gujarat, Maharashtra and Tamil Nadu.

The disease is characterized by the presence of dead cinnamon buff or purplish brown spot surrounded by a darker reddish brown margin. The lesions are elliptical at first, later becoming long and rectangular and are limited by adjacent veins. Lesions may coalesce producing a complete burning of large area of leaves.

Three races of the pathogen have been reported so far viz. race C, race O and race T. Race T which is specific to Texas male-sterile cytoplasm was initially reported in the Phillippines (Mercado & Lantican, 1961) and has also been reported from India (Sharma *et al*, 1978). Race C has been detected in China by Wei *et al* (1988). However, race O remains the most prevalent and widely distributed race.

The resistance to this disease is governed by either qualitative or quantitative nuclear genes and by cytoplasmic factors. Resistance to race T is conditioned by both cytoplasmic factors and nuclear genes, cytoplasmic component being the most important one (Sharma *et al*, 1993). Race T produced HM-T toxin which is specific to plants having Texas male sterility type of cytoplasm (Cms-T). Race C has a specific virulence on Cm-C cytoplasm. Chang and Peterson (1995) reported that resistance to southern leaf blight is controlled by single recessive gene designated rhm. The dominant allele Rhm confers susceptibility. They proposed a hypothesis that there are two linked recessive genes controlling resistance i.e. rhm 1 and rhm 2. They have also postulated that one of the two genes is a copy of the other created by duplication. The model is being currently tested.

Goulant (1993) reported that seed treatment with fungicides like Tolylfluanid or thiram along with iprodione gave best control. Thiram + lindane combination has also been reported to be effective by Ekpo and Banjoko (1994). They have also reported that various ash treatments i.e. 4 wood ashes prepared from *Azadirchita indica, Vernonia amygdalina, Gliricidia sepium* and *Cassia siamea* gave complete control of *C. heterostrophies*. *Pseudomonas fluorescens* strain RJ 0210 has been found to inhibit *D. maydis* both *in vitro* and *in vivo* in infected maize plants by Mohammed and Caenter (1995).

Turcicum Leaf Blight of Maize

The disease is caused by *Helminthosporium turcicum* (*Setosphaeria turcica* (Luttrell) Leon and Suggs. *Exserohilum turcicum* (Pali.) Leon and Suggs.). The disease is widespread in India, Nepal, Pakistan, Bhutan and China (Sharma *et al*, 1993). The disease is prevalent in highland areas where cool temperate conditions prevail during growth. It is also observed in plains during winter months as cool/moderate conditions during that period favours disease development. In India, disease is frequently observed in Jammu and Kashmir, Himachal Pradesh, Sikkim, West Bengal, Meghalya, Tripura, Assam, Rajasthan, Uttar Pradesh, Bihar, Madhya Pradesh, Maharashtra, Andhra Pradesh, Karnataka and Tamil Nadu (Sharma *et al*, 1993). Gowda *et al* (1993) reported incidence of turcicum leaf blight in southern Karnataka in 1987-88 and indicated that the disease intensity depended on location, meterological factors (temperature, RH and rainfall), cultural susceptibility and cultural practices followed by local farmers. On the leaves small chlorotic, oval, water soaked lesions are produced in the early stages of infection which later on spread and become elongated, spindle-shaped necrotic lesions involving 3-4 or more intervascular regions in a single spot. The lesions first appear in the lower leaves and continue increasing in size and number as the plant develops, until a complete necrosis of foliage is conspicuous. This stage is followed by a grayish green emergence of sporophores, usually surrounded by a reddish brown margin. In severe cases the whole leaf gets blighted. When infection occurs at silking stage and conditions are optimum it may cause significant econoimic loss.

Both quantitative and qualitative type of resistance operate against this disease. The quantitative or polygenic resistance controls the number but not the size of the lesions on the leaves. Four single dominant to completely dominant genes (Ht 1, Ht 2, Ht 3 and Ht N) have been identified to confer chlorotic lesion type of resistance, however, monogenic sources of resistance possessing genes Ht 1 and Ht N have been found to be susceptible at adult stage (Sharma *et al*, 1993). Based on their virulence on these genes four races have been identified i.e. race 1, race 2, race 3 and race 2-3-N. The race 2 is the most prevalent strain in Indian sub-continent (Payak *et al*, (1974) and most of the early maturing and local germplasm is susceptible to the disease. Studies carried out in India have shown that the inheritance of disease is polygenic and is an additive character.

Various sources of resistance in different backgrounds have been identified by Payak and Sharma (1979). Durable resistance was demonstrated in two inbred lines CH-104 and CM-105 (Sharma and Payak, (1990). In India, a double-top-cross hybrid Ganga-2 has been released and is resistant to turcicum leaf blight. Spraying of the crop with dithiocarbamates is recommended and has been reported to be effective in reducing the disease incidence.

VII. LIST OF MINOR DISEASES CAUSED BY *DRECHSLERA*

Richardson (1990) has reported 64 species of the fungus to be seed-borne on various crops. *Crop protection Compendium* (CAB *International*, 2004) described a detailed information of their host range and the economic impacts caused. A comprehensive list of *Drechslera* spp. with their synonyms used by different mycologists, disease caused, host range and the economic damage caused are given in Table 5.2.

Table 5.2: List of Minor Diseases caused by *Drechslera* spp.

Genus	*Synonym(s)*	*Disease*	*Host range*	*Remarks*
Cochliobolus geniculatus Nelson	*Curvularia geniculata* (Tracy & Earle) Boedijn	Black kernel of rice, Grain mould of sorghum, black point of wheat	*Oryza sativa, Pennisetum glaucum, Sorghum bicolour, Triticum aestivum, Zea mays*	Grain mould has been classified as a high priority disease of sorghum in Sorghum in Argentina, East Africa and Venezuela (CAB International, 2004).
Cochliobolus lunatus Nelson & Haaris *Drechslera apii* (Gobelez.)	*Curvularia lunata* (Wakker) Boedijin	Seed rot	*Sorghum* spp.	
Richardson & Fraser	*Helminthosporium apii* Gobelez.		*Apium graveolens*	
Drechslera australiensis (Bugn,) Subram. & Jain ex. MBEllis				
Drechslera bicolor (Mitra) Subram. & Jain	*Cochliobolus bicolor* Paul & Perbery		*Sorghum bicolor, Stenotaphrum secundatum*	
Drechslera biseptata (Sacc. & Roum.) Richardson & Fraser:			*Carica papaya, Panicum miliaceum, Triticum aestivum*	
Drechslera buchloes (Lefebure & AGJohnson) Subram. & Jain				
Drechslera campanulata n(Lev.) Sutton	*Drechslera verticillata* (O'Gara) Shoemaker		Sterility disease	*Triticum aestivum*
	Pyrenophora semeniperda			

Contd.

Table 5.2: Contd.

Genus	*Synonym(s)*	*Disease*	*Host range*	*Remarks*
	(Brittebank & Adam.) Shoemaker *Podosporiella vertiillata* O'Gara *Helminthosporium cyclop* Drechsl. *Pleosphaeria semeniperda* Brittlebank & Adam.			
Drechslera catenaria (Drechsl.) Ito	*Helminthosporium catenarium* Drechsl.	Leaf spot	*Agrostis* spp., *Festuca* spp., *Lolium* spp., *Poa* spp.	
Drechslera dictyoides (Drechsl.) Shoemaker	*Helminthosporium dictyoides* Drechsl. *Pyrenophora dictyoides* Paul & Parbery	Net blotch	*Festuca* spp., *Phleum pratense*	
Drechslera ellisii Danquah			*Capsicum* spp.	
Drechslera erythrospila (Drechsl.) Shoemaker	*Helminthosporium erythrospilum* Drechsl. *Pyrenophora erythrospila* Paul	Foot rot, Leaf spot	*Agrostis* spp., *Lolium perenne*	
Drechslera gagantea (Heald & Wolf) Ito			*Musa, Oryza sativa, Poa,*	
Drechslera giganteum (Heald & Wolf.) Ito	*Helminthosporium giganteum* Heald & Wolf.	Zonate eye spot	*Bromus* spp., *Phleum pratense*	
Drechslera graminea (Rabenh. ex Schlecht.) Shoemaker	*Helminthosporium gramineum* Rabenh. ex Schlecht. *Pyrenophora graminea* Ito & Kuribay	Leaf stripe	*Avena* spp., *Hordeum* spp., *Oryza sativa*	Potentially destructive disease. Yield losses upto 12-29% reported.
Drechslera halodes (Drechsl.) Subram.&Jain	*Helminthosporium halodes* Drechsl.	Leaf spot	*Cocos nucifera, Elaeis guineensis, Saccharum officinarum*	Weak parasite
Drechslera hawaiiensis (Bugn.)Subram & Jain ex MSEllis	*Helminthosporium hawaiiensis* Bugn.	From rotting seed	*Cajanus Cajun, Chloris gayana, Cocus nucifera, Elaeis* spp., *Psidium guajava, Oryza sativa,*	Weak parasite

Contd.

Table 5.2: Contd.

Genus	*Synonym(s)*	*Disease*	*Host range*	*Remarks*
			Solanum tuberosum, *Sorghum* spp., *Triticum aestivum*, *Zea mays*	
Drechslera japonica (Ito & Kuribay) Shoemaker	*Drechslera tuberosa* (Atkinson) Shoemaker	Leaf spot	*Hordeum vulgare*	
Drechslera leucostyla (Drechsl.) Subram. & Jain	*Helminthosporium leucostylum* Drechsl. *Helminthosporium nodulosum* Berk. & Curt. *Cochliobolus nodulosus* Luttrell	Head blight, Seedling blight		Weak parasite
Drechslera longirostrata (Subram.) Ramnath, Neergaard & Mathur		From seed	*Oryza* spp., *Sorghum* spp.	
Drechslera maydis (Nisik.) Subram. & Jain	*Helminthosporium maydis* Nisik. *Cochliobolus heterstrophus* (Drechsl.) Drechsl	Southern leaf spot, Blight	*Hordeum* spp., *Lactuca* spp., *Paspulum* spp., *Phaseolus* spp., *Sorghum* spp., *Vigna sinensis*, *Zea mays*	Race T of the fungus caused epidemic in the USA in 1970 resulting in losses upto one billion dollars.
Drechslera neergaardii Danquah			*Oryza* spp.	
Drechslera phlei (Graham) Shoemaker		Leaf streak	*Phleum pratense*, *Trifolium* spp.	
Drechslera poae (Baudyo) Shoemaker	*Drechslera vagans* (Drechsl.) Shoemaker		*Poa* spp.	
Drechslera poaissima Anderson & Broe			*Poa* spp.	
Drechslera rostrata (Drechsl.) Richardson & Fraser	*Helminthosporium rostratum* Drechsler *Setosphaeria rostrata*		Leaf spot.	*Arachis hypogaea*, *Bambusa* spp., *Chloris gayana*, Leonard *Cocos nucifera*, *Curcuma longa*, *Cynodon dactylon*, *Digitaria sanguinalis*, *Eucalyptus* spp.,

Contd.

Table 5.2: Contd.

Genus	*Synonym(s)*	*Disease*	*Host range*	*Remarks*
				Hibiscus cannabinus, Lens culinari, Lycopersicon esculentum, Mangifera indica, Momordica charantia, Oryza sativa, Paspulum, Pennisetum glaucum, Phleum, Psidium guajava, Saccharum officinarum, Solanum melongena, Sorghum bicolor, Triticum aestivum, Vigna spp., *Zea mays*
Drechslera sacchari(butler) Subram. & Jain	*Helminthopsorium sacchari* Butler	Eye spot, Seedling blight	*Pennisetum, Saccharum officinarum, sorghum* spp., *Zea mays*	Widespread in sugarcane growing areas causing upto 50% seedling mortality.
Drechslera siccans (Drechsl.) Shoemaker	*Helminthosporium siccans* Drechsl	Leaf blight	*Festuca* spp., *Hordeum* spp., *Lolium* spp.	
Drechslera sorghicola (Lef. & Sherw.) Richardson &Fraser	*Helminthosporium sorghicola* Lef. & Sherw.	Target spot	*Sorghum* spp.	
Drechslera stenospila (Drechsl.) Subram. & Jain	*Helminthosporium stenospilum* Drechsl.	Brown stripe of sugarcane	*Saccharum oficinarum*	Seedling damage can be controlled by thirum
Drechslera teres (Sacc.) Shoemaker	*Helminthosporium teres* Sacc. *Pyrenoohora teres* Drechsler	Net blotch	*Hordeum vulgare*	
Drechslera triseptata (Drechsl.) Subram. & Jain	*Helminthosporium triseptatum* Drechsl.	*Agrostis* spp., *Dactylis* spp., *Phleum pratense, Poa* spp.		
Drechslera tritici-repentis (Died.) Shoemaker	*Helminthosporium tritici-repentis* Died. *Pyrenophora tritici-repentis* (Died.) Drechsl. *Pyrenophora*	Yellow spot, Leaf blight, Tan spot	*Agropyron* spp., *Agrostis* spp., *Andropogon* spp., *Avena* spp.,*Bromus, Cynodon, dactylis, Elymus repens,*	The epidemic outbreak was reported in North Dacota in USA (Hosford, 1971). Yield reduction upto 28% reported.

Contd.

Table 5.2: Contd.

Genus	*Synonym(s)*	*Disease*	*Host range*	*Remarks*
	trichostoma (Fr.) Fuckel.		*Hordeum vulgare, Leymus, Phalaris, Secale, Setaria, Triticum aestivum*	
Drechslera turcica (Pass.) Subram. & Jain	*Helminthosporium turcicum* Pass *Trichometasphaeria turcica* Luttrell, *Setosphaeria turcica* (Luttrell) Leonard & Suggs.	Northern leaf blight, Black ear rot, Crown rot	*Sorghum* spp., *Zea mays*	Widespread leaf disease, with grain loss upto 50% in sorghum and 46% in maize (CAB International, 2004).
Drechslera verticillata (O'Gara) Shoemaker	*Helminthosporium cyclops* Drechsl. *Pyrenophora semeniperda* (Brittleban&Adam.) Shoemaker	Sterility disease	*Agrostis* spp., *Bromus* spp., *Triticum* spp.	Soil-borne as well as seed-borne causing germination failure.
Drechslera victoriae (Meehan & Murphy) Subram. & Jain	*Helminthosporium victoriae* Meehan & Murphy, *Cochliobolus victoriae* R.R. Nelson	Victoria blight	*Agropyron* spp., *Avena* spp., *Festuca* spp., Millets, *Paspulum* spp., *Phleum* spp.	
Drechslera zeicola (Stout.) Subram. & Jain	*Bipolaris zeicola* (G.L. Stourt) Shoemaker, *Helminthosporium carbonum* Ullstrup *Cochliobolus carbonum* Nelson *Zea mays*	Charred ear mould, Maize leaf spot, Northern leaf spot of corn		Since 1970s, there have been several reports of leaf spot caused by race 3, indicating that it has become widespread in USA and can cause heavy yield losses (CAB International, 2004).
Drechslera spp.	Setosphaeria rostrata	Seedling blight	*Saccharum officinarum*	
Helminthosporium oryzae Breda de Hann	*Drechslera oryzae* (Breda de Hann) Subram. & Jain *Chochliobolus miyabeanus* (Ito & Kurib.) Drechsler ex Dastur *Helminthosporium miyabeanus* (Ito & Kuribay) Drechsl.ex Dastur	Brown leaf spot of rice	Festuca spp., *Oryza* spp., *Poa, Phleum* spp.	It was the main cause of Bengal famine of 1943. Very common disease of rice worldwide causing considerable yield losses ranging from 40-90%.
Helminthosporium sativum Pamm.	*Helminthosporium sorokinianum* Sacc.	Black point, Seedling blight,	*Agropyron* spp., *Avena sativa,*	Diseases caused by the fungus are

Contd.

Table 5.2: Contd.

Genus	*Synonym(s)*	*Disease*	*Host range*	*Remarks*
King. & bake	*Cochliobolus sativus* (Ito & Kuribay) Drechsl. ex Dastur	Crown rot, Head blight	*Bromus, Dactylis glomerata, Elymus repens, Festuca, Lolium, Hordeum vulgare, Phaseolus vulgaris, Oryza sativa, Secale sereale, Sorghum* spp., *Triticum aestivum*	Widespread and serious worldwide, causing heavy yield losses in various crops (CAB International, 2004).
Helminthosporium setariae Saw.	*Cochliobolus setariae* (Ito & Kuribay) Subram. & Jain	Blight	*Eleusine indica, Pennisetum glaucum, Zea mays*	Reduce seed germination resulting in heavy yield losses
Helminthosporium spiciferum (Bainier) Nicot.	*Cochliobolus spicifer* Nelson, *Drechslera spicifera* (Bainier) Von Arx	Foot rot, Leaf blotch	*Arachis hypogaea, Capsicum* spp., *Eucalyptus* sp., *Gossypium* spp., *Hordeum* spp., *Jatropha curcas, Luffa* spp., *Lycopersicon esculentum, Musa* spp., *Piper betel, Sorghum bicolor, Triticum* spp., *Vicia faba, Vigna* spp.	Injurious to germination (Richardson,1990)
Setosphaeria prolata Leonard & Suggs.	*Exserohilum prolatum* Leonard & Suggs.	On seed	*Zea mays*	

VIII. CONCLUSIONS

Drechslera is an important genus causing considerable losses among various members of the family Gramineae. Some of the most important diseases caused by the species of *Drechslera* are leaf blight, brown spot, crown rot, and root rot in wheat, barley, oats, rye, sugarcane, maize, sorghum and turf grasses. The great Bengal famine of 1942 was caused by *Drechslera oryzae* while the epidemic of southern leaf blight in United States of America in 1970 was caused by the race T of *Drechslera maydis*, which resulted in 15 percent loss of the total corn produce.

1. The taxonomy of the genus attracted the attention of mycologists and pathologists in 19th and 20th centuries. Emergence of new techniques and tools have led to the development of taxonomic concepts and more lucid picture of classification of this genus has emerged. The monograph by Sivanesan(1987) is extremely useful and is a significant contribution to our knowledge on graminicolous Helminthsporias. In the present con-

text, a multidisciplinary approach from diverse fields such as genetics, physiology, molecular biology and plant pathology is needed for the construction of species profiles and better resolution of taxonomic difficulties in a genera like *Drechslera* which continues to be an enigma for plant pathologists. Nucleic acid based techniques are being employed for molecular characterization of various species of *Cochliobolus* and *Pyrenophora*. Attempts are also being made to understand the mode of inheritance of resistance as well as molecular elucidation of host-pathogen interaction for development of biotechnological approaches to combat the diseases in an eco-friendly manner.

REFERENCES

Adlakha, K.L., R.D Wilcoxson,. & S.P. Raychaudhuri. 1984. Resistance of wheat to leaf spot caused by *Bipolaris sorokiniana. Plant Disease*, 68: 320-321.

Agarwal, Anshu. 1977. Studies on seed borne mycoflora of barley(*Hordeum vulgare*) with special reference to Rajasthan. Ph.D. Thesis, Univ. of Rajasthan Jaipur, India.

Ahmed, M., M. Din,. A.S. Shakir, R. Ahmed & Z. Iqbal. 1993. Evaluation of some fungicides against *Helminthosporium sativum*-the causal organism of root rot of wheat. *Pakistan J. Phytopathology*, 5: 70-72.

Alcorn, J. L. 1983. Generic concept in *Drechslera, Bipolaris* and *Exserohilum. Mycotaxon*, 17: 1-86.

Alcorn, J. L. 1988. A new species of *Exserohilum. Trans. Br. mycol. Soc.*, 90: 146-148.

Atheya, S.C. 1974. Resistance of barley varieties to stripe disease in Uttar Pradesh. India *J.Mycol. Pl. Pathol.*, 4:83.

Bedi, K.S. & H.S. Gill. 1960. Losses caused by brown spot disease of rice in the Punjab. *Indian Phytopath.* 13: 161-164.

Bello, G. M. , C.I.Manaco Dal & M.N. Sisterna (1994). The effect of *Trichoderma* spp. on control of wheat seedling blight caused by *Bipolaris sorokiniana* in the greenhouse. *Fitopathologia Brasileira*, 19: 394-400.

Bhatnagar, G.C. & A.K. Mathur (1991). Management of barley stripe with mineral elements. *Indian Phytopath.*, 44: 382-383.

Bidari, V.B. & H.C. Govindu (1975). Studies on the pathogenicity of the isolates of *Helminthosporium sativum* Pamm., King and Bakke on wheat at different stages of growth in Karnataka state. *Mysore J. agric. Sci.*, 9: 87-94.

Bonde, M.R., J.A. Micales, & G.L. Peterson (1993). The use of isozyme analysis for identification of plant pathogenic fungi. Plant Disease, 77: 961-968.

CAB *International* (2004) *Crop Protection Compendium*, Wallingford, UK: CAB *International.*

Chakrabarathi, N.K., S.K. Mohanty, A.B. Das & S.B. Lodh (1975). Chemical control of brown spot disease of rice and its effect on the nutritional constituents of rice grains. *Plant Disease*, 59: 1028-1-32.

Chang, R.Y. & P.A.Peterson (1995). Genetic control of resistance to *Bipolaris maydis:* one gene or two genes. *J. Heredity*, 96: 94-97.

Chattopadhyay, S.B. & K.B. Chakrabarti (1968). Role of different pathogens in incititng the *Helminthosporium* disease of wheat in West Bengal. *Indian J. agric. Sci.* 38: 937-940.

Chattopadhyay, S.B., N.K. Chakrabarti & A.K. Ghosh (1975). Estimation of loss in yield of rice due to infection with brown spot incited by *Helminthosporium oryzae* Breda de Hann. *Science & Culture*, 39: 394-397.

Chaudhuri, S.D. (1948). Annual Report of the Economics, Assam for the year 1945-46. *Rep. Dep. Agric. Assam.* 1945-46, pp 85-170.

Dennis, R.W.G. (1933). Studies in the morphology and biology of H*elminthosporium avenae. Trans. Br. mycol. Soc.* 18: 223-238.

de pauw, R.M., R.D. Tinline, H. Harding, T.F. Townley-Smith & E.A. Hurd (1984). Registration of nine wheat germplasm lines with improved resistance to common root rot. *Crop Sci.* 24: 1222-1233.

Drechsler, C. (1923). Some graminicolous species of *Helminthosporium. I.J. agric. Res.* 24: 641-739.

Dubin, J. & M. Ginkel (1991). The status of wheat diseases and disease research in warmer areas. In: *Wheat for the Nontraditional Warm Areas* (eds. Saunders,D.A.) pp 125-145, CIMMYT, Mexico, 542 p.

Ekpo, E.G.A. & K.M. Banjoko (1994). Efficacy of Fernasan D and wood ash in the control of seed-borne fungi, pre-emergence mortality and seedling blight of maize. *Discovery and Innovation*. 6: 84-88.

Ellis, M.B. (1971). *Dematiaceous Hyphomycetes*. Kew, England: Commonwealth Mycological Institute, 608 pp.

Ellis, M.B. (1976). *More Dematiaceous Hyphomycetes*. Kew , England : Commonwealth Mycological Institute. 507 pp.

Fokkema, N.J. (1973). The role of saprophytic fungi in antagonism against Drechslera sorokiniana (*Helminthosporium sativum*) on agar plates and on rye leaves with pollen. *Physiol. Plant Pathol*. 3: 195- 205.

Gayed, S.K. (1962). The pathogenicity of six strains of *Helminthosporium sativum* to three cereals with special reference to barley. *Mycopath. Mycol. Appl*., 18: 271-279.

Ghose, R.L.M., M.B. Ghatge & V. Subramanyam (1960). *Rice in India*. 2nd Edition, ICAR, New Delhi, 474 pp.

Goulant, A.C.P. (1993). Treatment of maize (*Zea Mays*) seeds with fungicides. *Revista Brasileira de Sementes*, 15: 165- 169.

Gowda, A.T.P., H.S. Shetty, B.J. Gowda. & Sangam Lal (1993). Incidence of turcicum leaf blight of maize in southern Karnataka. *Current Research*, Univ. of agric. Sciences, Bangalore, 22: 100- 102.

Hegazy, M.F., D.I. Harfoush, M.H. Mostafa & I.K. Ibrahim (1992). The correlation between sensitivity to brown spot disease. *Annals of Agicultural Science*, 37: 595-601.

Hetzler, J., Z. Eyal, Y.R. Mehta, H. Fehrmann, U. Kvshnir. & J. Zekaria (1991). Interactions between spot blotch (*Cochliobolus sativus*) and wheat cultivars. In: *Wheat for the Nontraditional Warm Areas* (ed. Saunders, D.A.) pp 146-164, CIMMYT, Mexico 549 pp.

Hosford RM (1971) A form of *Pyrenophora trichostoma* pathogenic to wheat and other grasses. *Phytopathology* 61(1): 28-32.

Hims, M.J. & P.L. Gladdens (1994). Use of manipulated fungicides to determine critical period for disease control and yield response in winter barley. In; *Brighton crop protection conference. Pest and Diseases* 1994, vol. 1: 283-288.

Hughes, S.J. (1953). Conidiophores, conidia and classification. *Can. J. Bot*., 31: 577-659.

Hughes, S.J. (1958). Revisiones hyphomycetum aliquot cum appendice de niminibus rejiciendis. *Can. J. Bot*. 36: 727-836.

Hughes, S.J. (1980). *Helminthosporium velutinum*. Fungi Can. No. 163.

Husted, K. (1992). Differentiation of *Pyrenophora graminea* and *Pyrenophora teres* II. Antiserum against soluble mycelial proteins. *Tidsstrupt for planteval*. 96: 399-404.

Ibrahim, F.M. & R.J. Threlfall (1966). The application of numerical taxonomy to some graminicolous species of *Helminthosporium. Proc. Roy. Soc*. London. Ser. B. 165: 662-688.

Ito, S. (1930). On some new ascigerous stages of the species of *Helminthosporium* parasitic on cereals. *Proc. Imp. Acad.* (Tokyo) 6: 352-355.

Johansen, L.H., K. Husted, L. Olson & M. Wand Heide (1992). Differentiation of *Pyrenophora graminea* and *Pyrenophora teres* I. Gel electrophoresis and isozyme analysis of soluble mycelial proteins. *Tidsskrift for Planteaval*. 96: 391-397.

Kendrick, B. & F.Di Cosmo (1979). Teleomorph-anamorph connectionsin Ascomycetes. In: *The Whole Fungus*. Vol. 1 ed. B. Kendrick, pp. 283-410. Ottawa: National Museum of Natural Sciences.

Kohli, M.M., C.E. Mann & S. Rajaram (1991). Global status and recent progress in breeding for the warmer areas. In: *Wheat for Nontraditional Warm Areas* (ed. Saunders, D.A.) pp 96-112, CIMMYT, Mexico 549 pp.

Kumar, C.P.C., L.V. Garibova, L.L. Vellikanov & E.P. Durinina (1993). Biocontrol of wheat root rots using mixed cultures of *Trichoderma viridae* and *Glomus epigaeus. Indian J. Plant Protection*, 21: 145-148.

Lam, A. & P.F. Chapman (1985). Estimation of degree of similarity between species of *Drechslera* on *Lolium* spp. and *Festuca pratensis* using numerical taxonomy. *Trans. Br. mycol. Soc.* 85: 603-614.

Lapis, D.B. (1985). Insect pests and diseases of wheat in the Phillippines. In: *Wheat for more Tropical Environments* (eds. Villareal, R.L. and Klatt, A.R.) Proc. Intl. Symp. CIMMYT, Mexico 354 pp.

Leonard, K.J. & E.G. Suggs (1974). *Setosphaeria prolata*, the assigerous state of *Exserohilum peolatum. Mycologia*. 66 : 281-297.

Loughman, R. & T.N. Khan (1993). Effect of fungicide seed dressing on leaf stripe of barley caused by *Pyrenophora graminea* Ito and Kuribay. *Austr. J. Expt.* Agriculture. 33(4): 465-467.

Louw, L.P.J., D. Victor, P.W.P. Grous, G. Holz, & B.J.H. Jose (1995). Characterization of *Pyrenophora* isolates associated with spot and net type lesions on barley in South Africa. *J. of Phytopathology*, 143:129-134.

Luttrell, E.S, (1954). Approaches to the classification of *Helminthosporium* species. *Plant Dis. Reptr. Suppl.* 228: 111-113.

Luttrell, E.S. (1963). Taxonomic criteria in Helminthosporium. Mycologia, 55: 643-674.

Luttrell, E.S. (1964). Morphology of *Trichmetasphaeria turcica. Am. J. Bot.*, 51: 213-219.

Luttrell, E.S. (1977). Correlations between conidial and ascigerous state characters in *Pyrenophora, Cochliobolus* and *Setosphaera. Rev. Mycol.*, 41: 271-279.

Luttrell, E.S. (1978). Biosystematics of *Helminthosporium:* Impact on agriculture. In: *Biosystematics in Agriculture.* Beltville Symposium in Agricutural Research 2, (eds. J.A. Romberger, R.H. Foote, L.V. Knutson and P.L. Lentz) pp135-209. Mobntclair: Allanheld, Osmon & Co.

Mehta, Y.R. (1985). Breeding wheats for resistance to spot blotch. In: *Wheats for More Tropical Environments* (eds. R.L.Villareal and A.R. Klatt) pp.135 Proceedings of the International Symposium, CIMMYT,Mexico, 354pp.

Mehta, Y.R. & C.A. Gaudencio (1991). The effect of tillage practices -144 and crop rotation on the epidemiology of some major wheat diseases. In: *Wheat for the Nontraditional Warm Areas*. (ed. Saunders) July 29-Aug. 3, 1990. CIMMYT, Mexico, pp 266-283.

Mehta, Y.R. & S. Igarashi (1985). Chemical control measures for the major diseases with special attention on spot blotch. In: *Wheats for More Tropical Environments* (eds. R.L. Villareal and A.R. Klatt) pp. 196-200, Proceedings of the International Symposium, CIMMYT, Mexico, 354 pp.

Mercado, A.D. & R.M. Lantican (1961). The susceptibility of CMS lines of corn to *Helminthosporium maydis* Nis. & Miy. *Philipp. Agric.* 45 : 235-243.

Misra, A.P. (1973). *Helminthosporium* species occurring on cereal and other Gramineae. Final report on PL 480 project. Tirhut college of Agriculture, Dholi, Muzaffarpur (Bihar), India 289 pp.

Misra, A.P., S.P. Pandey, A.K. Misra, & J. Jha (1981). Pathogenic difference in isolates of *Bipolaris sorokiniana*, Shoemaker (H. sativum P.K. & B.) from wheat, barley and triticales from different geographical regions, 3rd Intl. Symp. Plant Pathology, IARI, new Delhi pp. 131.

Mitra, M (1923). *Helminthosporium* species on cereals ans sugarcane in India. Part I Diseases of *Zea mays* and *Sorghum vulgare* by species of *Helminthosporium. Mem. Dep. Agric. India bot. Ser.* 11: 219-242.

Mohammed, S. & I.G. Caenter (1995). Isolation and characterization of a *Pseudomonas fluorescens* strain suppressive to *Bipolaris maydis. J. of Phytopathology*, 143: 111-114.

Moore, W.F. (1970). Origin and spread of southern corn leaf blight in 1970. *Plant Dis. Reptr.* 54: 1104-1108.

Mukherjee, S.K. & B.N. Bagchi (1965). Role of some trace element in controlling brown spot disease of paddy (*Helminthsporium oryzae* Breda de Hann). *West Bengal J. Agric.* 2: 62-65.

Munjal, R.L. & J.N. Kapoor (1960). Some unrecorded diseases of sorghum and maize from India. *Curr. Sci.* 29 : 442-443.

Narain, A., S.K. Sinha, B.K. Mahapatra, D.T. Sechler & J.M. Poehlman (1973). Incidence of *Helminthosporium* blight of wheat in Orissa state, India. *Plant Dis Reptr*. 57: 278-280.

Nema, K.G. & L.M. Joshi (1971). The spot blotch disease of wheat caused by *Helminthosporium sativum*. Proc. 2nd Intl. Symp. Plant Path., IARI, New Delhi, 42 pp.

Nisikado, Y. (1928). Studies on the *Helminthosporium* diseases of gramineae in Japan. *Spec. Report Ohara Inst. Landwirtrch. Foesch. Kurashiki*, 4: 384.

Padmanabhan, S.Y. (1974). Control of rice diseases in India. *Indian Phytopath*. 27: 1-28.

Padmanabhan, S.Y., D. Ganguly & G.H. Chandwani (1966). *Helminthosporium* disease of rice VIII Breeding resistant varieties. Selection of resistant varieties of early duration from genetic stock. *Indian Phytopath*. 19: 72-75.

Pant, S.K. & K.K. S. Bisht (1983). Effect of stripe disease of barley on yield components. *Indian Phytopath*. 36: 103-105.

Payak, M.M. & R.C. Sharma (1979). Resistance to disease in maize, the Indian experience. *Crop Improvement*. 6: 75-85.

Payak, M.M., R.C. Sharma, B.M. Singh & J. Lilaramani (1974). Performance of maize lines carrying single gene resistance to *Helminthosporium turcicum* and *Puccinia sorghi* in India. *Indian J. of Genetics*. 34: 31-35.

Prasad, K.S.K., S. Kulkarni & A.L. Siddaramaiah (1978). Antagonistic action of *Trichoderma* spp. and *Streptomyces* spp. on *D. sorokiniana (Helminthsporium sativum* (Pamm., King and Bakke) Subram. & Jain. *Curr. Res*. 12: 202-203.

Raemackers, R.H. (1979). The efficacy of fungicides in a severe spot blotch epidemic of wheat in Zambia. IX Intl. Cong. Plant Prot., Aug. 5-11, Washington, D.C. (Abstr.).

Raemackers, R.H. (1988). *Helminthosporium:* Disease complex on wheat and sources of resistance in Zambia. In: *Wheat Production Constraints in Tropical Environments* (ed. A.R. Klatt). pp 175-186, CIMMYT, Mexico, 410 pp.

Richardson, M.J. (1990). *An annotated list of seed borne diseases*. 4th edn. International Seed Testing Association, Zurich, Switzerland, 387 p.

Sah, D.N. (1994). Effect of leaf wetness duration and inoculum level on resistance of wheat genotypes to *P. tritici-repentis. J. Phytopathology*, 142: 324-330.

Saifulla, M. (1993). Screenig of rice cultures against blast, brown spot and scald diseases. *Agricultural Science Digest*. 13: 171-172.

Saunders, D.A. (1988). Characterisation of tropical wheat environments: Identification of production constraints and progress achieved- South and South-East Asia. In: *Wheat Production Constraints in Tropical Environments*. (ed. A.R.Klatt) pp. 12-26. CIMMYT, Mexico, 410 pp.

Schilder, A.M.C. & G.C. Bergstrom (1995). Seed transmission of *Pyrenophora tritici-repentis*, causal fungus of tan spot of wheat. *European J. of Plant Pathology*. 101: 81-91.

Sen, C. & S.A. Kaier (1972). The role of herbicides in the control of brown spot of paddy caused by *Helminthosporium oryzae*. *Indian J. Mycol. Pl. Pathol*. 2: 152-155.

Shah, S.C. & V. Mariappan (1992). Efficacy of new fungicides against the diseases of rice and sorghum. *Indian J. Mycol. & Pl. Pathol*. 22: 152-159.

Sharma, R.C. & M.M. Payak (1990). Durable resistance to two leaf blights in two maize inbred lines. *Theoretical and Applied Genetics*. 80: 543-544.

Sharma, R.C., C. Leon De. & M.M. Payak (1993). Diseases of maize in South and South East Asia: Problems and Progress. Crop Protection, 12: 414-422.

Sharma, R.C., J. Lilaramani & M.M. Payak (1978). Outbreak of a new pathotype of *Helminthosporium maydis* on maize in India. *Indian Phytopathology*, 31: 112-113.

Shoemaker, R.A. (1959). Nomenclature of Drechslera and Bipolaris, grass parasites segregated from *Helminthosporium*. *Can. J. Bot*. 37: 879-887.

Shoemaker, R.A. (1966). A pleomorphic parasite of cereal seeds, *Pyrenophora semeniperda. Can. J. Bot.*, 44: 1451-1456.

Singh, A. & R.C.Rai (1986). Chemical control of wheat diseases in India. In: *Problems and Prospects of Wheat Pathology in South Asia* (eds. L.M. Joshi, D.V. Singh, and K.D. Srivastava) pp. 374-401, malhotra Publishing House, New Delhi, 401 pp.

Singh, D.V. & H.K. Saksena (1971). Physiologic specialization in *Helminthosporium gramineum* in India. Abstr. 2nd Intl. Symp. Of Plant Pathology, p. 81.

Singh, D.V., K.D. Srivastava, R. Aggarwal. & P. Bahadur (1993). Wheat Diseases Problems: The Changiing Scenario. In: pests and Pest Management in India. The Changing Scenario

Singh, G., J.S. Chohan & K.S. Sandhu (1975). Fungicidal control of rice blast. *Indian Phytopathology*, 28: 422-423.

Singh, R.A. & V.V. Sharma (1975). Evaluation of rice germplasm and varieties for resistance to brown spot. IL. *RISO*, 24: 383-386.

Singh, R.V., A.K. Singh & S.P. Singh (1995). Progress of work under National Centre of Foliar Diseases during 1994-95. paper presented in the 34th All India Wheat Workers Workshop, AICWIP, ICAR at UAS, Dharwad, August, 1995.

Sivanesan, A. (1984). The bitunicate Ascomycetes and their Anamorphs Vaduz. Liechtenstein: *J. Cramer*, 701 pp.

Sivanesan, A. (1985). New species of *Bipolaris. Tras. Br. mycol. Soc.* 84: 403-421.

Sivanesan, A. (1987). Graminicolous species of *Bipolaris, Curvularia, Drechslera, Exserohilum* and their teleomorph. *CAB Intl. Mycol. Inst., Mycological Paper No.* 158.

Smith, N.J.G. (1929). Observations of the Helminthosporium diseases of cereals in Britain. I. The behaviour of *Helminthosporium gramineum* in a common barley disease. *Ann. Appl. Biol.* 16: 236-260.

Subramanian, C.V. & B.L. Jain (1966). A revision of some graminicolous *Helminthosporia. Curr. Sci.*, 35: 352-355.

Swaminathan, K. & A. Selvaraj (1994). Effect of various fungicides and their mode of application on the incidence of paddy brown spot disease. *Tropical Agriculture*, 71: 204-207.

Tatum, L.A. (1971). The southern corn leaf blight epidemic. *Science N.Y.* 171: 1113-1116.

Thakkar, R. (1988). Studies on some seed born infections of barley (*Hordeum vulgare)* grown in Rajasthan. Ph.D. thesis. Univ. of Rajasthan, Jaipur, India.

Thirumulachar, M.J. (1967). Aureofungin in the control of seed borne *Helminthosporium oryzae* infection and seedling blight. *Indian Phytopath.*, 20: 277-279.

Thompson, J.P. & G.B. Wildermuth (1989). Colonisation of crop and pasture species with vesicular arbuscular mycorrhizal fungi and a negative correlation with root infection by *Bipolaris sorokiniana. Canad. J. Bot.* 69: 687-693.

Toubia-Rahme, H., D.E. Ali-haimoud, G. Barrault. & L. Albertini (1995). Inhibition of *Pyrenophora teres* sclerootioid in barley strain by application of glyphosate or paraquat. *Plant Disease*, 79: 595-598.

Tsuda, M. & A. Ueyama (1981). *Pseudocochliobolus australiensis*. The ascigerous state of *Bipolaris australiensis*. *Mycologia*, 37. 88-96.

Vale, G.P., E: Torrigiani, A. Gatti, G. Delogu, A. Porta-Puglia, G. Vannaii & L. Cattivellie (1994). Activation of genes in barley roots in response to infection by two *Drechslera graminea* isolates. *Physiol. Mol. Pl. Pathol.*, 44: 207-215.

Vidyasekharan, P., Esand Borromeo, & T.W. Mew (1992). *Helminthosporium oryzae* toxin suppresses phenol metabolism in rice plants and aids pathogen colonization. *Physiol. Mol. Pl. Pathol.* 44: 307-315.

Wei, J.K., K.M. Liu, P. Luo, & O.Y. Lee-Stadelmann (1988). Pathological and physiological identification of race C of *Bipolaris maydis* in China. *Phytopathology*, 78: 550-554.

Wolf, P.F.J. & G.M. Hoffman (1994). Epidemiological development of *Pyrenophora tritici-repentis* in wheat crop. *Zeitschrift fur Pflanzenkrankheiten und Pflanzenschutz.*, 101: 22-37.

6

Evaluation of Classificatory Systems in Smut Fungi

R.V. GANDHE
*Postgraduate Research Centre,
Department of Botany, Modern College,
Pune.*
E-mail: rvgandhe@hotmail.com

ABSTRACT

The present paper describes the taxonomy and classification of smut fungi. The various morphological characters used in identification of these fungi has been discussed in detail. Both traditional and modern molecular basis of classification of smut fungi have been discussed.

I. INTRODUCTION

Since last two decades, I have been working on taxonomy, teliospore germination and germination products in smut fungi. During this period, several interesting and new smuts were collected from the Western Ghats, Maharashtra, India. During these collections, necessary field observations of smuts were made and highlighted in the field notebook. Such observations are often useful and support in identification and description of the species. I believe, this is an important and unavoidable criterion for identification of a species in smut fungi. These fresh collections of smuts were tried for their germination on glass slides in sterile distilled water by hanging drop method with minor modifications (Thirumalachar, 1950). Teliospore germinations were also tried on 2% Agar, 0.2% Maltose Agar and 2% Host Extract Agar. Though many smuts showed germination, some did not germinate even after repeated attempts and suggested pretreatments. Therefore, I could not study their germination patterns and their germination products. Germinated fungi were well maintained in pure cultures at 25-28°C in incubator for their progressive growth and to study the culture topography. During last decade, several new taxa were added from different countries with a very wide and considerable morphological variations. Smutologists suggested new genera, species, and families by giving priority to the traditional systems or modern systems of classificatory systems. Therefore, there are number of wrong identifications and misnaming of smut fungi. Basically, there are two methods of classification in smuts and these are supported by two groups of mycologists. Some mycologists like Narsimhan, Thirumalachar, Mundkur, Pavgi, Fischer and Holton, Mac Alpine, Zundel and many others believed in the field observations, spore morphology, spore germination and germination products such as Basidiospores, Ballistospores, Blastospores and Blastoconidia, culture topography, host-specialization and disease development. They critically studied morphometrical characters and gave them priority in the classification. Therefore, classificatory systems proposed by them are considered Traditional Classificatory Systems, which are simple and more convenient for application. During last decade several new and sophisticated techniques were established for detailing of the fungus mainly based on molecular biological data, ultrastructure and biochemical data. These laboratory investigations are more accurate, supplying all the details of the nucleotides and their amino acid sequencing. The species concept is, therefore, discussed and described at the molecular level. The other group of mycologists believed in such molecular data and biochemical analysis for species identification more or less neglecting morphometrical observations. All such classificatory systems are grouped as Modern Classificatory Systems.

During my work on smut fungi, I had an opportunity to discuss some of these problems in smut taxonomy with Professor M.J. Thirumalachar, the well-known mycologist of India. He visited my laboratory in 1995 and expressed his views regarding delimitation of species in smuts. He also believed in field observations and morphometrical observations. During the same period Professor Kalman Vanky, Tubingen University, Germany, world authority on smut fungi, visited the laboratory for some collaborative research on smuts of India. We had several collections of smut specimens from Western Ghats. We had extensive discussions on smut taxonomy and the problems of smut classification. My views in the present paper are based on the above-mentioned discussions with the authorities. However, I still believe in Traditional Classificatory Systems.

II. TRADITIONAL CLASSIFICATORY SYSTEMS

Several smutologists still believe in traditional classification and they include morphological characters and variations, field observations, symptomatology and physiological variations. The traditional systems, therefore, are not much expensive and they do not require much sophistication. These systems are simple to operate with definite and accurate results. Traditional classificatory systems are mainly based on spore morphology, teliospore germination and their germination products, host-parasite interaction and disease development, symptomatology, soral characters and their variations, field observations, culture topography and their variation. All these observations are morphometrical and have prime importance in identification of the fungus. There is wide diversity in all the above-mentioned observations including personal errors. Therefore, sometimes the species described by one has been differently described by the other smutologists which causes confusion in identification.

Teliospore Germination

Since the establishment of L.R. and C. Tulasne's (1847) two family system of classification, many smutologists have given priority to the morphological features of spore germination and their products as taxonomic criteria, particularly at the family level. Accordingly, the order

Ustilaginales is basically classified into Ustilaginaceae, a phragmobasium with lateral and terminal basidiospores and Tilletiaceae, a holobasidium with only apical or terminal basidiospores. A third family, Graphiolaceae, parasiting palms, has been introduced by some mycologists in the Ustilaginales, which others have regarded as imperfect fungi mainly because of atypical pattern of germination and their products. Recently, Obrwinkler *et al.* (1982) showed that *Graphiola phonicis* is a basidiomycete, and described it, as a new order, Graphiolales with a single family.

Liou (1949) also introduced a new family Yeniaceae, with a genus *Yenia,* which produces a rudimentary basidium, which produces progressively 1-4, pedicelled, usually septate "pseudo-promycelia" which in their turn produce either new pseudo-promycelia or basidiospores.

Several smuts have been collected from different countries showing germination different from *Ustilago* and *Tilletia* types. The differences may be so distinct that sometimes it is very difficult to decide to which family the fungus in question belongs; as for instance is the case *Pericladium* or *Schroeteria* species.

Ciferri (1963) proposed four families in smuts, namely: Entylomellaceae, Tilletiaceae. Glomosoriaceae and Ustilaginaceae. Ciferri (1963) introduced family Glomosporiaceae with a single genus, *Glomosporium*, on the basis of a branched promycelium, which partially splits up longitudinally forming two secondary branches in *G. amaranthi*. The splitting of the promycelium, however, was not observed in the type species *G. leptidium.*

During last two decades, (Fischer and Holton 1957, Holton *et al*, 1968) several smuts showed more problematic germinations than the typical germination for the families. The taxonomic status of the smuts has always been uncertain and repeatedly subjected to changes. During the last two decades, several yeasts and other saprophytic basidiomycetons fungi have been isolated and described. These fungi show Heterobasidiomycetons life cycle very similar to that of smut fungi (Bandoni and Johri, 1972, Fell 1970, 1974, Fell *et al,* 1969, 1973, Fell and Tallman, 1980, 1981, Moore, 1980, Nyland, 1950). For a most unusual basidiomycete, *Filobasidium floriforme*, Olive (1968) introduced a new family: Filobasidiaceae, because of its superficial similarities with the smuts. Cox (1976) transferred Filobasidiaceae, to the Homobasidiomycetes, order Aphyllophorales, but Moore (1978) placed it in Order Tremellales, in Heterobasidiomycia.

Several authors Cunningham, 1924; Fischer, 1953, Fischer and Holton, 1957, Linderberg, 1959, and Vanky, 1985 rejected the division in families and considered all smuts in only one family, Ustilaginaceae, because of their random variation in germination with certain limitations.

There is, however, great variation in promycelial morphology within each family. In fact, many species do not germinate in accordance with dictates of either family. Teliospores of species in both families may thus give rise directly to a mycelium without first producing a promycelium or sporidia. During teliospore germination studies, species of both families also exhibited uncertain pattern and showing tendency toward Ustilaginaceae by developing septate promycelia, septate basidia and septate basidiospores.

Teliospore Morphology

Although the taxonomy of the smuts is based largely on teliospore morphology, most species cannot be identified or adequately described solely on the basis of teliospore characters.

Teliospore characters, including size, shape, colour and exospore ornamentation are, nevertheless, criteria of identification, which are indispensable to smut systematics. All the species described thus far have been based on combined study of these characters, along with the host range studies, symptomatology, and to some extent, geographic distribution.

Size of teliospore

There is a considerable variation in spore size in different genera and species of smut fungi. The range of variation in spore size is about 120 percent. The variation in spore size, which often occurs in the same species, can be further demonstrated by comparing spore size in different species. Therefore, spore size has greater importance and taxonomic significance both in small-spored species of *Ustilago, Sorosporium, Sporisorium*, and *Sphacelotheca* as well as relatively large-spored genera like *Cintractia, Tilletia* and *Neovossia*. The smallest spores are observed in *Ustilago minima* in a range of 3.5-5.0µm whereas largest spores are found in *Tilletia paradoxa* 57.0µm, *Neovossia punensis* µm in a range of 30-39µm. Whether spore size alone in some instances can justify creation of new species must, in the final analysis, be left to the judgment of the individual.

Spore colour

Strictly speaking, colour is not a morphological character, but often included in all description of smut fungi. The spores are coloured because of more or less pigmentation. This pigmentation may be evenly or unevenly distributed in the exospores of smuts. In various species of *Ustilago*, uneven pigmentation produces equatorial bands and lighter polar caps in *Plnetella lironis*.

Variation in spore colour in the smut fungi is widely prevalent, expressed primarily in varying shades and tones of the basic colour for the species. For example, brown is the universal colour for spores of the species of Tilletia, which usually exhibit varying shades of this colour in the range of light to dark, including shades on the light side that are yellowish, grayish, and violaceous and the dark side that are reddish, olivaceous, chestnut, and chocolate. Even within the same species variations in spore colour are common, not only between spore masses from different sori but spores of single sori as well. Many species of *Tilletia* and *Neovossia* immature spores are hyaline and characteristically sculptured. There are intergradations in colour ranging from light to heavily pigmented spores. There are several chances of wrong interpretation of colours in spores and depends upon individuality. Spore colour, admittedly, is particularly subject to quantitative and qualitative interpretation. Nevertheless, because it is one of the few phenotypic characters available to ustilaginologist, no doubt it will continue to be extremely useful in smut systematics.

Spore mass

Nearly in all the descriptions of the species the colour of the spore-mass is described. It is generally black, with various shades of brown, fawn or pale yellow.

However, manner in which they are described are not standardized. The colour of the individual spores does not necessarily correspond with that of spore-mass. In several species, spore-mass colour differs as the maturity of the sorus. However, in some smuts spore-mass colour represents characteristic of the genus for example in *Liroa*, the spore-mass is typical chocolate brown.

Spore ornamentation

Exospore ornamentation has special significance other than any other single character in spore morphology. Therefore, in many publications spore ornamentations are studied by scanning electron microscopy. Most of the authorities employed spore ornamentation as an essential criterion for the species delimitation in smut fungi. Basically, exospores are smooth, verrucose, reticulate, echinulate, tuberculate, cerebriform but there is a considerable variation in the species. Echinulations and verruculations are usually described in the form density. As the spore ornamentations are genetically cotrolled characters more or less constant for the species and usually compared with the allied species while describing new species. The species of spore ball forming smuts like *Sorosporium, Sorosporium, Tolyposporium* exhibit variations in spore ornamentation from smooth to verrucose whereas in many *Tilletia* species spores with peripheral rim where the spore ornamentations prominently project. The spore of *Neovossia* often associated with the prominent hyphal appendage and some mycologists considered it an important criterion for separation of *Tilletia* and *Neovossia*. However, this character is confusing and often in the form of rudimentary hyphal structure. Sometimes the ornamentations on the exospores are cleared by bleaching solutions for detailing. *Entyloma nymphae* also show prominent attachment of hyphal fragment at one end and papillate structure at the other end.

Spore shape

The spores of many species of smut fungi are commonly globose to subglobose or occasionally polyhedral. However, these descriptions of the spores are very loosely used, therefore, this character of the spore has minor importance in the identification of species. There is a notable variation in the spore shapes in the same sorus or different sori. It is necessary to observe shapes in several spores so that average spore shape can be described. In *Sorosporium, Tolyposporium* species spore ball configuration is curious where innermost spores of the ball sustain considerable internal pressure, and, hence, tend to assume a polygonal configuration visible when the spores are freed. In *Urocystis hypoxis*, outer few spores are wedge shaped, sometimes hyaline and sterile whereas inner darker spores are fertile and more or less globose. Various combinations of the components such as sterile cells, fertile pores, cortex, pseudoparenchyma and mycelial fragments exhibit different shapes and positions in the spore balls. The spore ball forming genera *Doassansia, Burrillia, Doassansiopsis, Narasimhania, Tracya* showed remarkable compositions.

Spore ball compositions

Nearly half of the genera produce their spores in clumps or aggregate to form semi-permanent-to-permanent spore balls. These spore balls may either consist of only fertile spores few to many in number or of a combination of one or more components such as sterile cells, mycelia, and pseudoparenchymatous cells. The spore balls of *Sorosporium, Sporisorium* and *Tolyposporium* species are composed of only fertile spores. One of the curious spore balls is observed in *Tolyposporium*, wherein all the fertile spores are firmly fitted together by interconnected folds and thickenings on the spores. In the species of *Urocystis* hyaline sterile cells surround one to many fertile spores. Spore balls of *Doassansia, Doassansiopsis* and *Burrillia* were permanently and deeply embedded in the host tissue and composed of various components. There are wide and notable variations in different species of the genera regarding the composition of the spore balls. The species identification, therefore, mostly depends on combination of these components in the spore ball. In Entyloma species spores remain embedded in the common gelatinous matrix whereas in the genus *Georgefischeria* spores are individually encased in gelatinous sheath.

Sterile cells

To a certain extent, the characters of the so-called sterile cells, including their size, shape and colour have been used to characterize certain species or genera in smut fungi. Unfortunately, the term does not refer to structures of common origin. For example, in some *Sphaceloheca* species the cellular components of the sorus peridium, which tend to disintegrate into individual cells or group of cells, are referred to as "sterile cells". In genera in which the spore balls consist of both fertile and sterile elements, the term "sterile cells" has been restricted to the sterile cellular components of the spore ball, as distinguished from the fertile spores, for example, *Urocystis, Thecaphora, Doassansia, Doassansiopsis, Burrillia* etc. However, the sterile cells have greater importance in Tilletia, sometimes supporting character for identification of the species. These sterile cells are intergrated in sculpture, hyaline to slightly tinted and intermixed with the fertile spores. They are few to many in number but do not germinate like fertile spores. However, the exact function of the sterile cells not well understood.

Host Response, Host Specialization, and Symptomatology

Realizing the importance of host-specialization in the taxonomy of the smuts, Fischer (1953), and later Fischer and Shaw (1953), proposed a species concept based principally on morphology and host-specialization at the host-family level. The successive events such as entry of the pathogen, establishment in the host tissue and the disease development resulted into expression of inconspicuous to nonconspicuous symptoms. In smuts these symptoms are mostly systemic but also localized in vegetative or/and reproductive parts. There is a wide variation in the gross morphology of the inflorescence, stem, leaf, anthers, ovaries and the entire inflorescence. Majority of species of *Sorosporium, Sporisorium, Sphacelotheca, Ustilago,* and *Tilletia* are ovaricolous. Thus, they completely or incompletely destroy the inflorescences. They often develop their protruding capsular sori in few to many florets. The species from Ustilaginaceae, therefore, develop sori in the reproductive parts and totally or partially converting them into more or less swollen/hypertrophied sori. A very interesting type of infection pattern is observed in *Urocystis hypoxis* in which floral buds are infected and become hypertrophied and rupture at maturity to expose agglutinated spore mass. However, species of *Doassansia, Doassansiopsis, Burrillia, Entyloma, Melanotaenium* and *Jamesdicksonia* often develop their sori in the stems, leaves and many other vegetative parts of the host. *Enyloma globigena* develop typical button-like galls in the leaves of *Blumea* whereas *Entyloma mundkuri* develop very inconspicuous spots in the leaves of *Centrantherum*. The genus *Pericladium* develops small blister-like structure in the petiole and the rachis of *Grevea*. The species of *Jamesdicksonia* is widespread in the Pune University campus develop typically rust-like pustules in *Dichanthemum* in January. Moreover, elongated, large and often irregular hypertrophy is developed in *Melanopschium pennsylvanicum*. The sorus in this species becomes erumpent to expose agglutinated spore mass at maturity. At this stage, it appears in the form of an irregular ball with thread-like agglutinated structures bursting out at maturity. The species of *Doassansia* and *Burrillia* mostly restricted to the leaves. For example, in *Doassansia khandalensis* and *Burrillia kamatii* the sori are inconspicuous along the upper surface but spore balls are arranged in concentric circle along the lower surface of a leaf. The only species of *Doassansiopsis*, *D. nymphae* develops tumours in the petioles and the pedicels of *Nymphaea*. One of the most interesting and curious infection patterns is observed in the two species of *Georgrfischeria*, *G. riveae* and *G. narasimhanii*. In *G. riveae* the type species there is remarkable shortening of the internodes and the axillary shoots so as to develop rosette-like clustering in the axils. The infected leaves remarkably reduce about ½ size to their normal size. Finally, at maturity, the infected shoot is converted into witches broom-like malformations. The second species *G. narasimhanii* causes yellowing of leaves and blackening of veins in the infected area. Threfore, in my opinion host responses to the pathogens reflected in the form of symptoms are to be taken into consideration while describing the species. Symptoms exhibited by the hosts are very typical and restricted to the genus or species in smut fungi, should not be neglected in the classificatory system.

Field Observations

I feel, it is necessary to study the disease development in nature rather than in the laboratory conditions. Therefore, frequent visits to the disease localities and the relevant collections are to be critically studied for some conclusions. I had an experience during the collection of the species of *Georgefischeria* in which witches broom formation is important and major character of the genus. During the establishment of a new species *G. narasimhanii*, this type of malformation did not observe but the infection is only restricted to the leaves causing blackening of veins in *Argyreia hookeri*. Similarly a very interesting infection pattern was observed in case of the host *Coix lachryma-jobi* in which the entire inflorescence and the anthers are infected by *Sphacelotheca sinhgadanensis* but another smut *Neovossia sumati* is only in the ovaries restricted to female flowers. Whether entire or part of the inflorescence is infected? What is the intensity of infection? What is the percentage of infection at the locality? Whether the disease systemic or localized? Morphological variations in sori of the same species their dehiscence and many other questions can be solved during extensive field work only. I therefore, advise researchers in this field for extensive fieldwork and prepare a field notebook with all minor details. All these field observations are extremely important for laboratory investigations and proper identification of the species.

Cultures of Smuts

Though smuts are obligate parasites but can also be cultured on different on solid or liquid media. However, their growth in culture is comparatively slow to the other fungi. They can be directly dusted on nutrient media such as PDA, HA or 0.2% MA or first germinated on 2% plane Agar and then transferred to the nutrient media. Sometimes fragments of the host are inoculated with the smut spores for their rapid development in the culture. The greatest variation in the smut fungi is displayed by haploid cultures on agar media. Monosporidial cultures of many species usually vary widely in colony characteristics, some of which may represent temporary response to environment; the great majority, however, are inherently permanent. Smut cultures vary in type and rate of growth, topography, colour, pigmentation, and many combinations of these characteristics. The type of growth may be mycelial, sporidial, or a combination of both; the colour may range from white to black, with many intermediate shades. Some lines exhibit widely different rates of growth, especially those that are mycelial.

III. MODERN CLASSIFICATORY SYSTEMS

In the second half of the century, several smuts were added with their complicated and questionable structures from different countries and these smuts cannot be accommodated directly on the basis of established characters of smuts. Numerous smut floras have been published and a number of new smut genera and species were proposed with some ultra structural data, bimolecular data, biochemical analysis and phylogenetic study. Time to time additions of several different smuts become heterogeneous group of smut fungi. For the detailing of such smuts new techniques and protocols were experimented and fixed in the laboratory. The protocols were suggested and successfully applied for DNA extraction from the fungal cultures. The fungal gene markers are also introduced in this procedure. Such DNA is purified and amplified for DNA sequences. In smut fungi DNA sequencing is preferred for mitochondria, therefore, usually referred as LSU rDNA. This advanced technique supplies accurate results and the differences in the gene sequences so that new species can be discussed. On DNA sequencing method several smuts were shifted or merged to their different taxa, more or less neglecting morphometrical characters. On the basis of this technique, many classificatory systems are proposed and phylogenetic study is conducted. Bauer, Oberwinkler and Vanky (1997) suggested a new system of classification by considering ultrastructural characters of the septal pore and that of host parasite interactions, with classical morphological characters. This has been improved, complemented and extended with help of molecular data of nuclear LSU rDNA sequencing (Begerow, Bauer, and Oberwinkler, 1997/1998; Swann, Frieders and McLaughlin, 1999; Pipenbring, Begerow and Oberwinkler, 1999; Begerow, Bauer, and Boeckhout, 2000; Bauer, Begerow, Nagler and Oberwinkler, 2001) and based on classical morphological characters and host plant taxonomy (Denchev, 1997; Ershad, 2000; Vanky, 1999a, b, 2000, 2001a, b, 2002).

IV. CONCLUSION

Traditional Classificatory Systems are simple in application but require skill and very critical field observations, extensive laboratory morphometrical observations, definitely a well established classificatory system. However, in certain problematic taxa Modern Classificatory Systems are advisable though many techniques are expensive and depend upon the purity of cultures.

REFERENCES

Bandoni, R. J. & B. N. Johri. 1972. *Tilletiaria*: a new genus in the Ustilaginales. Canad. J. Bot. 50: 39-43.

Bauer, R., D. Begerow, A. Nagler, & F. Oberwinkler. 2001. The Georgefischeriales, A phylogenetic hypothesis. Mycol. Res. 105: 416-424.

Bauer, R., F. Oberwinkler, & K. Vanky. 1997. Ultrastructural markers and systematics in smut fungi and allied taxa. Canad. J. Bot. 75: 1273-1314.

Begerow, D., R., Bauer, & T. Boekhout. 2000. Phylogenetic placement of ustilaginomycetous anamorphs from nuclear LSU rDNA sequences. Mycol. Res. 104: 53-60.

Ciferri, R., 1924. Seconda contribuzione allo studio degli Ustilaginales. Atti Ist Bot. Univ. Pavia, Ser. 3: 1, 77-97.

Cox, D. E. 1976. a new homobasidiomycete with anomalous basidia. Mycologia, 68: 481-510 .

Cunningham, G. H. 1924. The Ustilagineae, or smuts , of New Zealand. Trans. And Proc. New Zealand Institute, 55: 397-433.

Denchev, C. M. 1997. Anthracoideaceae, a new family in the Ustilaginales. Mycotaxon, 65: 411-417.

Duran, R. & G.W. Fischer. 1961. The genus Tilletia. Washington State University, 138 pp.

Ershad, D. 2000. *Vankya*, a new genus of smut fungi, Rostaniha. 1: 65-72.

Oberwinkler, F. 1998. Phylogenetic studies on nuclear large subunit ribosomal DNA sequences of smut fungi and related taxa. Canad. J. Bot. 75: 2045-2056.

Fell, J. W. 1970. Yeasts with Heterobasidiomycetes life cycles. In Ahearn, D.G., (editor). Recent trends in yeast research. Georgia State University, Atlanta, Georgia, 49-66 pp.

Fell, J. W. 1974. Heterobasidiomycetes yeasts Leucosporidium and Rhodosporidium, Their systematics and sexual incompatibility systems. Trans. Mycol. Soc. Japan. 15: 316-323.

Fell, J.W., A.C. Statzell, I.L. Hunter, & H.J. Phaff. 1969. Leucosporidium gen. n., the Heterobasidiomycetes stage of several yeasts of the genus Candida. Antonae van Leeuwenhoek 35: 433-462.

Fell, J.W., I.L. Hunter, & A.S. Tallman. 1973. Marine basidiomycetous yeasts (Rhodosporidium spp. n.) with tetrapolar and multiple allelic bipolar mating systems. Canad. J Microbiol. 19: 643-657.

Fell, J.W., & A.S. Tallman. 1980. Rhodosporidium paludigenum species nov., a basidiomycetous yeast from interdial waters of South Florida. Internat. J Syst. Bacteriol. 30: 658-659.

Fell, J.W., & A.S. Tallman. 1981. Heterothalism in the basidiomyceous yeast genus Sporidiobolus. Syst. Bacteriol. 5: 77-82.

Fischer G. W. 1953. Manual of the North American smut fungi. New York, Ronald Press Co. 343 pp.

Fischer, G. W. & C. S. Holton. 1957. Biology and control of Smut fungi. New York, Ronald Press Co. 622 pp.

Holton, C.S., J.A. Hoffmann, & R. Duran. 1968. Variations in the smut fungi. Ann. Rev. Phytopathol. 6: 213-242.

Linderberg, B. 1959. Ustilaginales of Sweden. Symb. Bot. Upsal. 16(2): 1-175.

Liou, T.N., 1949. Sur I'existance d'une nouvelle famille des Ustilaginales. Contr. Inst. Bot. Nat. Acad. Peiping 6: 37-47.

Liro, J. I. 1924. Die Ustlagineen Finnlands I. Ann. Acad Sci Fenn Ser A 17: 1-636.

Liro, J. I. 1938. Die Ustlagineen Finnlands II. Ann. Acad Sci Fenn Ser A 42: 1-720.

Moore, R. T. 1978. Taxonomic significance of septal ultrastructure with particular reference to the jelly fungi. Mycologia, 70: 1007-1024.

Moore, R. T. 1980. Taxonomic proposals for the classification of marine yeasts and other yeast like fungi including the smuts. Bot. Mar. 23: 361-373.

Mundkur, B.B. & M. J. Thirumalachar. 1952. Ustilaginales of India. C. M. I., Kew., 84 pp.

Nyland, G. 1950. The genus Tilletiaopsis. Mycologia, 42: 487-496.

Olive, L.S., 1968. An unusual new Heterobasidiomycete with Tilletia –like basidia. J Elisha Mitchell Sci. Soc. 84: 261-266.

Pipenbring, M. D. Begerow, & F. Oberwinkler. 1999. Molecular sequence data assess the value of morphological characteristics for a phylogenetic classification of species of Cintractia. Mycologia, 91: 485-498.

Savile, D. B. O. 1954. Cellular mechanics, taxonomy and evolution in the Uredinales and Ustilaginales. Mycologia, 46: 736-761.

Swann, E. C. E. M. Frieders, & L. McLauglin. 1999. Microbotryum, Kriegeria and the changing paradigm in basidiomycete classification. 91: 51-66.

Tulasne, L.R. & C. Tulasne. 1847. Memoire sur les Ustalaginees comparees Uredinees. Ann. Sci. Nat. Bot. Ser. 3, 7, 12-127.

Thirumalachar, M. J. 1966. Concepts of our knowledge in the genera of Ustilaginales.Indian Phytopahol. 19: 3-13.

Vanky, K. 1985. Carpathian Ustilaginales. Symb. Bot. Upsal. 24 (2) 1-309.

Carpathian Ustilaginales. Symb. Bot. Upsal.24 (2): 1-309.

Vanky, K. 1999a. The new classificatory system for smut fungi and two new genera. Mycotaxon, 70: 35-49.

Vanky, K. 1999b. Three new genera of smut fungi. Mycotaxon, 71: 207-222.

Vanky, K. 2000. New taxa of Ustilaginomycetes. Mycotaxon, 74: 343-356.

Vanky, K. 2001a. The emended Ustilaginaceae of the modern classificatory system for smut fungi. Fungal Diversity, 6: 131-147.

Vanky, K. 2001b. *Lundquistia*, a new genus of Ustilaginomycetes. Mycotaxon, 77: 371-374.

Vanky, K. 2002. Illustrated Genera of Smut Fungi. 2nd ed. APS Press, 245 pp.

7

Pleurotus: A Mushroom of Commerce and Broad Adaptability

C. L. JANDAIK

Department of Mycology & Plant Pathology University of Horticulture & Forestry, Nauni, Solan-173 230, India

I. INTRODUCTION

Species of the genus *Pleurotus* are well known edible fungi and are appreciated for their culinary properties and broad adaptability under varied agro climatic conditions. The genus *Pleurotus* though relatively new to the mushroom industry has gained much popularity world over and presently is believed to be the potential rival of commonly cultivated temperate mushroom *Agaricus bisporus*, which is still the most favourite in production and technology of mushrooms.

Species of oyster mushroom in the genus *Pleurotus* occur commonly as wood-decomposers in forests throughout the world. In many countries, edible *Pleurotus* species are collected and cultivated as choice edibles (Chang and Miles, 1991) and have been subject of many taxonomic and genetic studies. *Pleurotus* has always been attractive to mushroom growers because of the ease with which most species can be cultivated. With over 25 recognized species and varieties cultivated throughout the world, *Pleurotus* is also one of the diverse groups of cultivated edible fungi, which is being exploited industrially as a future mushroom especially of South East Asia region.

Greek Derivation

The *Pleurotus* mushroom, which is generally referred to as 'Oyster mushroom' world over and as 'Dhingri' in India, has its origin from Greek word " Pleuro" which means formed laterally or in a side way position, referring to the lateral position of the stipe (stem) in relation to pileus (cap).

Pleurotus ostreatus (Jack. ex Fr.) Kummer is best known species among Oyster mushrooms and the specific epithet 'Oyster', obviously refers to its Oyster shell appearance of the fruit bodies.

The spiralling food and energy prices witnessed in the recent past warrants research efforts for waste recycling as an alternate energy source. This will not only relieve pressure on environmental pollution but will also help in alleviating health hazards. In this context, *Plerotus* cultivation represents the only current economically profitable biotechnological process for the conversion of waste plant residues (lignocellulogics from agriculture, industry and forestry). Mushroom cultivation is also the only microbial process or product systems which can bio-convert all of the major plant polymers (lignin, cellulose, hemicelluloses) from such residues. The other microbial treatment systems, natural or man-made, utilizing lignocelluloses do not utilize the lignin fraction. The technology of *Pleurotus* cultivation is simple and sound, can operate on many types of wastes and can be carried out in low cost production units of a large range of sizes.

II. CULTIVATION SYSTEMS

Various systems of Oyster mushroom cultivation are in use in different parts of the world. These are:

(i) **Cut wood log cultivation system:** mostly used with hard wood species and generally practised in Europe.

(ii) **Bed cultivation system:** by placing the cultivation substrate on bed frame or the ridge. This system is mostly used with cotton waste/straw substrate.

(iii) **Plastic bag cultivation system:** Cultivation substrate is colonized in plastic bags. This system is mostly used with straw substrate and is common in South East Asian countries.

(iv) **Bottle or jar cultivation system:** In this system, cultivation substrate after treatment is filled in bottles/jars and inoculated with suitable strain. This system is mostly used with sawdust substrate and is used in Japan.

(v) **Modified bed cultivation system:** This system is in practice in mainland China. In this method trench is done which is then roofed with plastic. The trench is about 2 meter deep, 2.5 meter wide and of variable length with 3 meter height from base to plastic roof. In this house either single or double tiers growing system with 3-4 shelves per tier is adopted. The anticipated biological efficiency is 120-150 per cent. It is perhaps due to this very high biological efficiency that during 1993, there was about 430 per cent increase in *Pleurotus* mushroom production in China.

(vi) **Vertical tray system:** Kurtzman (1979) described the vertical tray system for the cultivation of Oyster mushroom but this method was never adopted on large scale and it never got general usage worldwide.

III. SUBSTRATE AVAILABILITY AND TREATMENT

In India, the total quantity of agricultural by-products or wastes, which are cellulosic in nature, account for nearly 25 million tonnes in a year (Ghose & Gosh, 1978). These materials constitute the perennial source of raw material for mushroom cultivation. The important ones among this include different cereal straws, sugarcane bagasse, cotton waste, jute, coir pith (coir waste), cocoa pods different oil cakes including rubber seed cake, sunflower straw, groundnut pod shell, tapioca starch waste, water hyacinth and certain other dry and wet land weeds, banana pseudostems and market wastes.

Various methods of substrate treatment are adopted for *Pleurotus* cultivation. The most commonly adopted substrate treatment methods are:

(i) Non-sterile method: In this method the substrate after soaking in water is used for cultivating Oyster mushroom. This method is based mainly on collection of suitable and chances of success are limited due to contamination.

(ii) Hot water dip method: It consists of mainly dipping of substrate in hot water for variable durations depending upon the nature of substrate used.

(iii) Pasteurisation method: It consists of exposing wet substrate to live stream for different durations (few minutes to few hours). This is the most common and widely adopted method throughout the world especially with cereal straws.

(iv) Fermentation method: It involves application of thermophilic microorganisms in the fermentation of substrate.

(v) Chemical treatment method: This method or substrate treatment though tried by few workers but it never got acceptance worldwide. This is mainly due to increased awareness among mushroom growers and consumers about chemicals as well as due to chances of resistance development among contaminating organisms.

(vi) Sterile Method: In this method, the substrate is sterilized by use of live stream under pressure. This method is suitable under laboratory conditions for experimental purposes but has never been adopted on commercial scale, as sterile substances are more prone to contamination.

IV. SPECIATION IN GENUS *PLEUROTUS*

There are about 38 species described under genus *Pleurotus* from different parts of the world and more than 25 species have been reported from India, of which about a dozen are under cultivation. They mainly include *P. ostreatus, P. sajor-caju, P. eryngii, P. cornucopiae, P. fossulatus, P. flabellatus, P. opuntae, P. citrinapileatus, P. membranaecus, P. platypus, P. eaus, P. sapidus, P. cystidious, P. columbinus* and *P. pulmonarius*.

Oyster mushroom species are of two main types, and they are:

(a) The grey type:
The European strains which fruit at low temperature i.e. below 12°C.

(b) The white type:
Includes those strains, which fruit at temperature up to 30°C.

It is well known that in order to exploit the potentialities of an organism through selection and breeding, its systematic should be well classified. In case of genus *Pleurotus*, there is still a lot of confusion regarding speciation and this is mainly either due to misidentification or isolates or to unreliability of criteria used for identification, as the latter are influenced greatly by environmental parameters. Various researchers are presently occupied in resolving the state of speciation within this genus. However, some controversies on this aspect do exist, particularly concerning isolates having geographically distinct origin. On the other hand, genetic compatibility studies do emphasize about the relatedness of different species/isolate. Now biological molecular approaches coupled with mating compatibility studies are available to correctly identify the strains/species irrespective of its origin and physio-morphological diversity.

V. INTERSTERILITY GROUPS (BIOLOGICAL SPECIES) IN *PLEUROTUS*

Mating compatibility can be assessed by pairing single spore isolates from the same parent (Intra Stock crosses) to determine mating system. Inter stock crosses can be performed among single basidiospore isolates from different parent strains to determine intersterility groups (biological species). Vigalys et al. (1996) have listed fifteen intersterility groups currently known for various species/strains of *Pleurotus.*

Molecular analysis based on ribosomal RNA large sub-unit (LSU) coding region and its associated internal transcribed spacer (ITS) region provides strong support for several taxonomic groupings in *Pleurotus.* Considerable genetic diversity has been detected in *Pleurotus* using molecular and biochemical markers. Several studies of isozymes have demonstrated genetic variation consistent with known species boundaries. Most recent investigations have also demonstrated variation among species using restriction fragment length polymorphism (RFLP) of ribosomal DNA (Iracabi *et al.*, 1995), mitochondrial DNA (Toyomasu *et al.*, 1992), total genomic DNA (Sagaw *et al.*, 1992) and polymerase chain reaction (PCR) studies (Vigalys *et al.*, 1996).

VI. GROWTH PARAMETERS FOR *PLEUROTUS* SPECIES

The growth parameters for *Pleurotus* species (*P. sajor-caju, P. flabellalus, P. sapidus* and *P. ostreatus)* during different cultivation stages (spawn run, pinhead initiation and cropping) are briefly summarized as under.

Table 7.1: Growth parameter

Parameter	*During spawn run*	*During pinhead Initiation*	*During cropping*
Relative humidity	85-90%	90-95%	85-90%
Temperature	75-85°F	55-60° F	60-75° F for white type 55-60° F for grey type
Duration	10-15 days	7-14 days	5-7 weeks
CO_2	20,000 ppm	< 600 ppm	< 600 ppm
Fresh air exchange	0^{-1}hr	3-4-changed^{-1}hr	1500-2500 lux^{-1} hr. for 4-6 hours^{-1} day
Light	Total darkness	1500-2500 lux^{-1}hr for 4-6hr^{-1} day	1500-2500 lux^{-1}hr. for 4-6 hours^{-1} day
Watering	To maintain the desired humidity	Regular misting once or twice daily	To avoid caps from cracking

VII. YIELD POTENTIAL

Average commercial yields are 1-kilogram fresh weight of mushroom^{-1} kilogram of dry substrate. However, with special growing practices and care exercised during pinhead formation and cropping, biological efficiency as high as 150% has also been achieved under commercial conditions in China.

VIII. WORLD PRODUCTION TREND OF OYSTER MUSHROOM

It is apparent that the production of Oyster mushroom in early 80's was negligible but due to simplicity of cultivation procedure and broad adaptability of different *Pleurotus* species, the Oyster mushroom has emerged as a mushroom of significance and it is expected that the production will continue to increase especially in South-East Asian countries. The production trend of oyster mushroom during 1983-84 to 1994-95 is presented as under:

Table 7.2: World production of *Pleurotus* species

Year	*Metric tons (x 1000) Fresh wt.*	*Source*
1983-84	32	Chang and Miles, 1989
1986	169	Chang and Miles, 1991
1990	900	
1994	797.4	Shu-ting Chang, 1996

Table 7.3: Estimated production of Oyster Mushroom species in some countries during 1994.

Country	*Production (Fresh Wt. Tons x 1000)*
China	654.0
Japan	20.8
S. Korea	57.9
Thailand	15.0
Taiwan	4.6
Indonesia	1.0
U.S.A.	0.9
Hungry	4.5
Others	38.7
Total	**797.40**

Source: Shu-ting Change, 1996

In overall view, the world production of Oyster mushroom totalled 170 thousand tons and 900 thousand tons during 1986 and 1990, respectively. In these three years, the Oyster mushroom production increased by 437 per cent, with an annual increase of 145.6 per cent. It should be noted that there is an upward trend in Oyster mushroom production and this trend is expected to continue in the future as well. Due to advances both in basic knowledge and practical technology, the significant impact of Oyster mushroom production can be considered globally but must be implemented according to availability of local substrates, labour and climatic conditions.

Although domestication of Oyster mushroom species dates back to many centuries, research in this field is still in adolescent stage and limited to certain countries. Only in recent years, Oyster mushroom has acquired an international status and presently ranks 3rd in terms of world mushroom production contributing 16.3 % of the total estimated cultivated world mushroom production of 4909.3 thousand tons in 1994. In comparison to 1990 production figures of 900 thousand tons, there has been decrease of around 11.4 per cent in Oyster mushroom during 1994. This decrease may partly be due to increased interest in production of *Agaricus* species in China, which is the major producer of Oyster mushroom in the world.

IX. NUTRITIONAL ATTRIBUTES

Dhingri contains fairly high quantity of ascorbic acid (vitamin C) which helps in the treatment of scurvy. In addition to this it is a good source of calcium and phosphorus which are essential for the formation of bones and teeth. Being low caloric and highly nutritious food, it is also recommended to counter obesity and especially against sugar troubles. The composition of different Dhingri cultivars is given as under:

Table 7.4: Physico-chemical composition of *Pleurotus* species

Species	*Moisture (%)*	*Crude Protein (Nx 4.38)*	*Fat (%)*	*Carbo-hydrate (%)*	*Fibre (%)*	*Ash (%)*
Pleurotus sajor-caju	88.00	33.5	2.26	58.5	-	8.25
P. ostreatus	73.9	10.5	1.6	81.8	7.5	6.1
P. florida	91.5	18.9	1.7	58.0	11.5	9.3
P.flabellatus	91.5	21.6	1.8	57.4	11.9	10.7
Agaricus brunnescens	89.5	26.3	1.8	59.9	10.4	12.0

Table 7.5: Amino acid composition of *Pleurotus* species (g amino acid/100 g protein)

Amino acid	*P. sajor-caju*	*P.ostreatus*		*Agaricus bisporus*
		Grey type	*Florida type*	
Isoleucine*	3.12	2.99	2.97	2.24
Leudine*	4.94	4.38	4.36	3.68
Lysine*	3.99	3.21	2.80	4.48
Phenylalanine*	4.94	2.60	2.41	2.08
Methionine*	1.31	1.08	1.00	0.46
Valine*	3.77	3.64	3.45	3.56
Threonine*	3.53	3.24	2.96	2.72
Tryptophane*	0.88	0.97	0.68	1.02
Alanine*	4.43	4.51	5.04	4.64
Histidine*	1.57	1.20	0.97	1.32
Tyrosine	4.45	2.10	2.02	1.92
Cystine	0.90	0.32	0.32	0.42
Asparatie acid	7.15	6.38	6.32	4.48
Arginine	4.41	3.74	3.42	5.92
Glutamic acid	17.72	12.66	9.96	7.04
Glycine	3.53	3.15	3.05	2.56
Proline	2.80	3.21	3.00	5.12
Serine	3.31	3.45	3.04	3.72
Total aromatic acids	**8.39**	**4.70**	**4.47**	**4.00**
Total sulphur acids	**2.21**	**1.40**	**1.32**	**0.99**
Total essential acids	**31.49**	**28.22**	**26.64**	**25.20**
Total amino acids	**75.76**	**36.25**	**57.68**	**55.48**

*Essential amino acids

Table 7.6: Mineral content of *Pleurotus* species

Species	*Ca*	*P*	*K*	*Fe*	*Cd*	*Zn*	*Cu*	*Pb*
	mg/100g				*ppm*			
P.Sajor-caju	20	760	3200	124.0	0.3	129.0	12.2	3.2
P.florida	24	1850	4660	184.0	0.5	111.4	15.8	1.5
P.ostreatus	33	1348	3799	15.2	-	-	-	-
P.flabellatus	24	1550	3760	12.4	0.5	58.6	21.9	1.4
Agaricus sp.	23	1429	4762	186.0	-	-	12.8	-

*Data presented on dry weight basis

X. MEDICINAL ATTRIBUTES

Throughout history, mushrooms have been used in different cultures and countries for their medicinal and tonic properties. The longest tradition of using edible mushrooms for medicinal purposes is in China and Japan. Primacy among medicinal mushrooms belong to *Lentinus edodes*, a relative of the Oyster mushroom (*Pleurotus* species) also known in oriental medicine for its medicinal properties (Breene, 1990). The beneficial effects of *Pleurotus* were discovered independently on different locations. Knowledge of their medicinal properties originates not only from Asia but from the folklore of Central Europe (Stamets, 1993), South America (Guzman, 1994) and Africa (Oso, 1977). These facts strongly support the notion that *Pleurotus* species indeed contain substances that may help to cure or mitigate certain diseases or may improve the quality of life for the needy.

In Central America and Mexico, *Pleurotus* species have been used in ceremonial as well as traditional medicine. The medicinal properties of Oyster mushroom are supposed to be efficient as blood pressure lowering agent, a diuretic, a cholesterol reducing agent, an adjuvant and an aphrodisiac (Guzman, 1994) besides exhibiting anticancerogenic, antiviral, anti-inflammatory, antibiotic as well as hypochlolesterolemic activities. In Africa, the medicinal traditional with *Pleurotus* is linked to *P. tuberregium*, which is found in the tropical and subtropical regions of the world and is known for its ability to form selerotia. Both the selerotium and fruit bodies are edible and are combined in a variety of specific herbs to cure diseases.

Anticarcinogenic Activity

Most of the research on antitumor effects of mushroom has been conducted in Japan where numerous mushrooms including *Pleurotus* species were tested for their anticarcinogenic potential. Intraperitoneal injection of aqueous extracts of *P. ostreatus* and other edible fungi were highly inhibitory to the growth of tumors by 72 to 92 per cent verses controls arising from Sarcoma 180 implanted in Swiss albino mice (Breene, 1990). The active antitumor component of a water soluble extract of *P. ostreatus* was identified as being as acidic polysaccharide fraction.

Antiviral Activity

Pleurotus mutilis contain the anticarcinogenic and antiviral agent referred to as pleuromutilin which has been reported active against PR8 influenza virus (Jong and Donovick, 1989). Furthermore, pleuran, an another immunmodulatory compound also acts as an antiviral substance when administered prophylactically.

Anti-Inflammatory Activity

It has been reported that a preparation from *Pleurotus* may be used for the treatment of kidney inflammation without reduction in immunity. It is a saccharide and protein extracted with water from the mycelium of *P. ostreatus* and the preparation does not have any side effects.

Antibiotic Activity

Antibiotic substances obtained from *Pleurotus* species may be divided into low molecular weight polysaccharide, which in most cases also exert anticancerogenic and antiviral activity by the stimulation of the immune response of the organism. Water extracts of the fruiting bodies of *P. sapidus* and *P. ulmarius* have antibiotic activity especially against *Staphylococcus aureus*, whereas water extracts of *P. ostreatus* mycelium have been reported active against only *Bacillus subtilis*.

Immune Response

In vitro, glucan isolated from *P. ostreatus* fruiting bodies could enhance the activity of lymphokine activated killer cells (LAK) and natural killer cell (NK) on tumor cells. This immunocompetent glycopeptide enhances LAK and NK cells as well inhibits tumour proliferation (Li et al., 1994).

Hypoeholesterolemic Activity

Increased blood levels of total cholesterol, low density lycoprotein (LDL) and very low density lycoprotein (VLDL) cholesterol, as well as lowered levels of high density lycoprotein (HDL) cholesterol, are identified as risk factors in the development of coronary artery disease (CAD) which is considered the main cause of mobidity in developed countries. The initial steps in the prevention and treatment of CAD and hypercholesterol laemia is the modification of nutritional regime with a diet low infats and saturated fatty acids and rich in crude fibres. Mushrooms in general and *Pleurotus* species in particular, because of high fibre contents, plant sterols proteins micro-elements and low caloric value food are ideal for diets designed to prevent cardiovascular diseases as first suggested by traditional oriental medicine (Breene, 1990).

XI. BIOTECHNOLOGICAL APPLICATIONS

Bioconversion of Lignocellulosic Wastes into Protein

An interesting feature that makes the cultivation of Oyster mushroom so attractive, especially in developing countries, is that this mushroom produce relatively large amounts of high quality protein on a substrate that consists mainly of waste material and the production of mushroom protein is much more efficient.

Spent *Pleurotus* Substrate

The substrate left after *Pleurotus* fruit bodies have been harvested is known as spent *Pleurotus* substrate (SPS). This SPW is present in large amounts and it is certainly not desirable to leave it as a possible source of pollution. It is known that there still remain in SPS a considerable amount of lignocellulosic materials in addition to mushroom mycelium and products found by metabolic activities of the mycelium. Interestingly, in China, in spent compost of *Volvariella volavacea* has been used for production of *P. sajor-caju*. The SPS can be used in many different ways which will help alleviate environmental pollution and avoid health hazards.

As fodder for live stocks

What restricts the use of lignocellulosic agricultural waste as animal feed is the restricted digestibility of the material and this indigestibility is due to lignin, which acts as a physical barrier which prevents the hydrolic enzymes from reaching the polysaccharides. As such digestibility is dependent upon deliginification. The white rot fungi especially *Pleurotus* species have the ability to breakdown lignin. Consequently, *Pleurotus* species are able to increase significantly the digestibility of straw (Calzada *et al.*, 1981).

As soil conditioner and fertilizers

It has been observed that spent *Pleurotus* substrates when added to agriculture soil can act as an effective soil conditioner or even as fertilizer that increases the yield of certain vegetables. It is believed that the SPS helps to provide a balanced nitrogen and carbon source for plant growth and it is also a source of humus formation.

Bioremediation

Bioremediation refers to the process of removing some harmful or potentially harmful situation by biological means. Out concern with bioremediation here, however, involves the use of SPS. Experimentation with SPS in bioremediation is still in very early stages but the results are promising. Regan (1994) has demonstrated in laboratory study that spent mushroom substrates also known as spent compost can be used to remediate solutions containing pesticides by the adsorption of pesticides from aqueous solution. In addition, the SPS may also serve as a source of micro-organisms to detoxify the insecticides. In bioremediation, the white rot fungi have been shown to have a beneficial role in the removal of certain pollutants. In addition to above, in rural areas, the SPS has also been used as a constituent of mud plaster for plastering walls of houses.

XII. BREEDING POTENTIALS

In contrast to the green plants most mushroom species are haploid. The diploid phase is normally transient and restricted to the basidium where meiosis occurs. The meiotic products migrate into the basiospores of which there are characteristically four per basidium.

The breeding of mushrooms has perhaps more parallels with breeding of higher plants then with the breeding of industrial micro-organisms. Both the plant and mushroom breeders are concerned with the simultaneous improvement of a range of characters expressed in a biologically complex structure.

The techniques for breeding Oyster mushroom have been discussed by Eger (1978). Three aspects of particular interest are quantitative biochemical analysis, sporeless strains and response to temperature. Eger (1978) suggested, it should be possible to produce lines of *P.ostreatus* in which spore production is temperature sensitive. A genetic solution to the problem of spore allergy in *Pleurotus* species is a available in the form of a mutant gene which prevent spore production. Eger *et al.* (1966) had obtained a sporeless variants of *P. ostreatus* and the likely monogenic inheritant of this character has since been demonstrated (Leal-Lara, 1979).

The objectives in breeding a new strain is to produce one which is best suited for commercial production and some of the desirable attributes to look for in a new strain are (i) high productivity, (ii) high quality fruit bodies in terms of flush thickness and colour, (iii) favourable cultural characteristics i.e. broad adaptability to temperature and substrates and (iv) stability of strain properties during continuous sub-culture required for spawn production.

With exception of *A. bisporus*, the generation of new strain can be achieved by making crosses between different compatible monkaryons. The isolation and regeneration of protoplasts from a dikaryons provides another approach to the disassociation of the component nuclei and has been applied in a range of species. Protoplasts are used in two areas of genetic manipulation, fusion and transformation. Through protoplast fusion, it is possible to bring together whole genomes of related (isogenic) or non-related strains and even different species and promote recombination leading to the production of novel phenotypes.

Transformation is the approach used to manipulate organisms at the level of individual genes, the introduction of the genes of interest into the cell being mediated by a vector DNA molecule. To-date, the most successful method for the introduction of DNA molecules into the fungal cells has depended on the use of protoplast. However, other approaches are being developed and slowly exploited in edible fungi. DNA mediated transformation related to gene vector advances our understanding of the basic molecular biology of higher basiomycetes. Development of this method also opens up many possibilities for the commercial production of strains having potential for production of medically important substances.

It is suggested that use of multilocus enzyme electrophoresis (MEE), will continue to play a very significant role in genetic study of edible mushrooms especially *Pleurotus* species. This technique has continued value even in the face of growing emphasis on direct examination of DNA variation (RFLP's, RAPD and DNA sequencing). In fact, the greatest power will undoubtedly come from combining allozyme and DNA data. Allozyme permit one of the easiest and cheapest methods of confirming when an actual crosses made between two single spore derived lines, in addition to its application for species identification, variation in germplasm collection, linkage studies as well as patent labeling.

XIII. LIMITATIONS AND FUTURE PROSPECTS

The Oyster mushroom species have a few limitations for which improvement should be made. The main limitations are:

(i) *Pleurotus* species discharge spores in tremendous numbers during cropping, which frequently cause allergies and respiratory problems to workers in mushroom houses.

(ii) The fruit bodies are rather brittle, making it difficult for them to be packed and transported to distant markets without damage.

(iii) The drab colour of fruit bodies is not very appealing and sometimes there is problem in getting population accept this mushroom.

To overcome these drawbacks, research on improvement of quality in terms of:

(i) High yield and broad adaptability to climatic conditions.
(ii) Consistency of stipe and cap with production of more thickness of flesh.
(iii) Firmness of sporocarps and attractiveness of colour.
(iv) Culinary value and grastronomical delight.
(v) Tolerance to biotic and abiotic stresses.
(vi) Production of sporeless types/strains by suitable breeding methods (selection, hybridisation, protoplast fusion etc.) should go a long way in establishing Oyster mushroom as the mushroom of 21st century.

Oyster mushroom cultivation in the country is gaining much importance in the recent past, especially in the southern states as a self-employment avenue for the rural sector. Even though various workers in different areas of the country have standardized the technology for its cultivation, commercialisation of its cultivation is yet to be explored.

REFERENCES

Breene, W.M 1990. Nutritional and medicinal value of specialty mushrooms. J.Food. Prot. 53: 883-894.

Buswell, J.A. 1995 Potential of spent mushroom substrate for bioremediation purposes. Mush. News 43: 28-34.

Calzada, J.F.; E.; dePorres, R.; del con, C. Rolz., & L.F. Franco, 1989. Production of food and feed from wheat straw by *Pleurotus sajor-caju*. Mush. J. for the Tropics 7: 45-46.

Chang, S.T. 1996. Mushroom research and development – Equality and mutual benefit. In: *Mushroom biology and mushroom products* (Ed. D.J. Royse), Penn. State, U.S.A., pp. 1-10.

Chang, S.T. & P.G. Miles, 1989. *Edible Mushrooms and their cultivation*, CRC Press, Boca Raton, Florida, 345 pp.

Chang, S.T. & P.G. Miles, 1991. Recent trends in world production of cultivated edible mushrooms, Mush J. 504: 15-18.

Cimerman, N.G. 1999. Medicinal value of the genus *Pleurotus* (Fr.) P. Karst. (Agaricales s.l., Basidiomycetes). Int. J. Med. Mush. 1: 69-80.

Eger, G. 1976. Rapid method for breeding *Pleurotus ostreatus*, Mush. Sci. 9: 567.

Eger, G. 1978. Biology and breeding of *Pleurotus*. In: The *biology and cultivation of edible mushrooms* (Eds. Chang, S.t. and W.A. Hayes). Academic Press, New York, pp. 497-519.

Ghose, T.K. & P. Gosh, 1978. Bioconversion of cellulosic substances. J. Appl. Chem. Biotechnol. 28: 309-20.

Guzman, G. 1994. The fungi in the traditional medicine in Meso-America and Mexico. Revis lberoamericana Micol. 11: 81-85.

Iracabal, B.; G. Zervakis, & J. Labarere, 1995. Molecular systematics of the genus *Pleurotus*: analysis of restriction polymorphisms in ribosomal DNA. Microbioloty (Reading) 141: 1479-1490.

Jong, S.C. & R. Donovick, 1989. Antitumour and antiviral substances from fungi. *Adv. Appl. Microbiol.* 34: 183-261.

Kurtzman, R.H. Jr. 1979. A vertical tray system for the cultivation of edible fungi. Mush, Sci. 10: 429-436.

Li, H.; Z.; Ling, D. Li & Jicho, C, 1994. Preparation and immunologic competence of glycopetides components from *Pleutotus ostreatus* fungi. Shandong yike Dasue Xuebao 32: 343-346.

Oso, B.A. 1977. *Pleurotus tuber-regium* from Nigeria. Mycologia 69: 271-279.

Regan R.W. Jr. 1994. Use of SMS as a compost matrix to degrade pesticide residuals. Compost Sci. and Utilization 2: 56-62.

Sagawa, I.; M. Tanaka, & Y. Nagata, 1992. Discrimination of mushrooms in genus *Pleurotus* by DNA restriction fragment length polymorphism. J. Gen. Appl. Microbiol 38: 597-603.

Stamets, P. 1993. Growing gourmet and medicinal mushrooms. Ten Speed Press, Berkeley, CA 551 pp.

Toyomasu, T., H. Takazawa, & A. Zennyonji, 1992. Restriction fragment length polymorphisms of mitochondrial DNAs from the basidiomycetes *Pleurotus* species. Biosci. Biotech. Biochem. 56: 359.

Vigalys. R.; J.M.: Moncalvo, S.R. Liou, & M. Volovsek, 1996. Recent advances in molecular systematics of the genus *Pleurotus*. In: *Mushroom biology and mushroom products* (Ed. D.J. Royse), Renn. State, U.S.A., pp. 91-101.

8

Wood-rotting Fungi of Temperate Himalaya

J. R. SHARMA

Botanical Survey of India 192, Kaulagarh Road, Dehra Dun (U.P.)

ABSTRACT

Species of wood-rotting fungi of the order Aphyllophorales form an integral part of the temperate forested Himalayan ecosystems and play the primary role of formation of forest soils by decomposition of substrates. There exists a high diversity in species composition and richness, host preferences and type of wood rots caused by them. The essential but a mammoth task of studying their rich diversity has been limited by taxonomic and logistic difficulties in collecting and defining fungal species. The various difficulties like seasonal fluctuations, succession of substrates and other considerations have led to problems in assessing and documenting their diversity. Dynamics during decomposition of substrate and mechanism of its decay are also emphasized. The various threats to the wood-rotting fungi as a whole and the conservation measures required to be undertaken are also evaluated.

Key words: Diversity, Wood decay, Aphyllophorales, Temperate forests, Conservation.

I. INTRODUCTION

The wood-rotting fungi of the order Aphyllophorales constitute a cosmopolitan group, fruiting mostly on dead woods and wood products throughout the world under different environmental conditions. The sufficient substratum provided both by angiospermic and coniferous tree species, and coupled with a great diversity in ecological habitats provide rich environmental conditions for the growth of these fungi, particularly in the temperate Himalayan zones. "The decades of fungi" by Berkeley (1850-54) was the first systematic information on survey and study of wood-rotting mycoflora of Himalaya and this was further enhanced by Thind (1961); Bakshi (1971); Rattan (1977); Rattan & Khurana (1978) and Sharma (1995, 1999a & 1999b). The other incidental and scattered reports on these fungi from the temperate forests are also found in Bakshi & Bagchee (1950); Bagchee & Bakshi (1954); Bagchee *et al* (1954); Thind & Adlakha

(1956); Reid *et al.* (1958, 59); Rehill and Bakshi (1965, 66); Imazeki *et al.* (1966); Thind and Khara (1968, 1975); Khara (1977a & 1977b); Thind and Rattan (1968-73); Maas Geasteranus (1971); Rattan (1974) and Sharma (1996-1997). Apart from these reports, apparently there is no systematic and consolidated account of the rich and diversified wood-rotting aphyllophoraceous mycoflora of the temperate Himalayan forests.

The temperate forests in Western Himalaya (states of Jammu & Kashmir, Himachal Pradesh and Garhwal and Kumaon regions of Uttar Pradesh) and Eastern Himalaya (states of Sikkim, Arunachal Pradesh and Darjeeling district of West Bengal) occupy a climatic zone with pronounced variations in seasonal temperatures. They extend approximately from 2200 m to 3000 m (-3500 m) and form a major part of the forest vegetation in the Himalayas. The temperate broad leaved forests in the lower parts dominated by species of *Quercus, Platanus, Aesculus, Juglans, Populus, Salix, Alnus, Acer, Pyrus, Lyonia, Viburnum, Elaeocarpus, Lindera, Castanopsis, Ilex* and *Engelhardtia* are followed by rich stands of *Pinus, Cedrus, Abies, Taxus, Cupressus, Larix* and *Tsuga* at higher altitudes. Oak-fir forests occur along the upper limit of temperate forests be-

Table 8.1: Distribution of Fruitbody, Hymenial configuration and Habitat over the families of Aphyllophorales (Donk 1971) in the temperate Himalaya.

Sl.	*Family name*	*Fruitbody*			*Hymenial configuration*							*Habitat*		
		1	*2*	*3*	*4*	*5*	*6*	*7*	*8*	*9*	*10*	*11*		
												C	B	A
1.	Auriscalpiaceae	-	x	x	-	x	-	x	-	-	-	2	1	-
2.	Bankeraceae	-	-	x	-	-	-	x	-	-	4	-	-	-
3.	Bondarzewiaceae	-	-	x	-	-	-	-	x	-	-	1	-	1
4.	Cantharellaceae	-	-	-	x	x	x	-	-	-	6	-	-	-
5.	Clavariaceae	-	-	-	x	x	-	-	-	-	36	1	-	-
6.	Clavullinaceae	-	-	-	x	x	-	-	-	-	15	-	-	-
7.	Coniophoraceae	x	x	x	-	x	x	-	-	-	-	7	1	2
8.	Corticiaceae	x	-	-	-	x	x	x	x	-	-	48	26	27
9.	Echinodontiaceae	-	x	-	-	-	-	x	-	-	-	-	-	1
10.	Fistulinaceae	-	-	x	-	-	-	-	x	-	-	-	-	1
11.	Ganoderrnataceae	-	-	x	-	-	-	-	x	-	-	-	-	3
12.	Gomphaceae	x	-	-	x	x	x	x	-	-	25	3	4	-
13.	Hericiaceae	x	x	-	x	x	-	x	-	-	-	2	3	6
14.	Hydnaceae	-	-	x	-	-	-	x	-	-	3	-	4	-
15.	Hymenochaetaceae	x	x	x	-	-	-	-	x	x	4	4	13	35
16.	Lachnocladiaceae	x	-	-	x	x	-	-	-	-	-	4	6	4
17.	polyporaceae	x	x	x	-	-	-	-	x	x	3	35	16	46
18.	Schizophyllaceae	-	-	x	-	-	-	-	-	x	-	-	-	1
19.	Sparassidaceae	-	-	x	-	x	-	-	-	-	1	-	-	-
20.	Stechherinaceae	x	x	-	-	x	-	x	-	-	-	-	1	3
21.	Stereaceae	x	x	x	-	x	x	-	-	-	-	4	7	8
22.	Thelephoraceae	x	x	x	-	x	x	x	x	-	8	14	4	2
											105	125	86	140

1 = Effused 2 = Reflexed 3 = Pileate 4 = Clavarioid and Cantharelloid 5 = Smooth 6 = Folded 7 = Toothed 8 = Poroid 9 = Lamellate 10 = Humicolous 11 = Lignicolons.

C = Coniferous woods
B = Both woods
A = Angiospermic woods

1

2

tween 2600- 3500 m altitude. The dominant elements of these forests are *Abies spectabilis, A. pindrow, Taxus wallichiana, Quercus semicarpifolia, Rhododendron lepidotum* and *Betula utilis.* The temperate forests ultimately merge into subalpine forests marked by *Quercus-Abies* or *Betula-Abies* or *Rhododendron-Abies-Betula* associations towards the timber line at 3600 m (-4500 in EH).

II. DIVERSITY

A total of 456 aphyllophoraceous fungi belonging to 120 genera and 22 have been recorded from the temperate zones of both the Eastern & Western Himalayas (Table 8.1). Only 18 families have species causing woodrots. Among them six are represented only by one genus; ten by two to eleven genera while two families have more than thirty genera each. In terms of genera, family Corticiaceae (39) is followed by Polyporaceae (34) and Hymenochaetaceae (9) whereas, in terms of species Corticiaceae (101) and Polyporaceae (97) are followed by Hymenochaetaceae (52), Thelephoraceae (20), Stereaceae (19) and Lachnocladiaceae (14). Out of the 120 genera of wood-rotting aphyllophorales found in this zone, 52 genera (43%) have only one species; 48 genera (40%) have two to five species, 15 genera (13%) have six to ten species and only 5 genera (4%) have more than 10 species each.

Phellinus with *ca* 30 species is the largest and widely distributed genus in this zone, followed by *Tomentella* (20), *Inonotus* (14), *Hyphodontia* (11), *Antrodia* (11), *Hymenochaete* (9), *Oligoporus* (9), *Polyporus* (8), *Hyphoderma* (8), *Stereum* (6), *Trechispora* (6), *Scytinostroma* (6), *Vararia* (6), *Phlebia* (6) and *Coniophora* (6).

Though, the order of frequency of occurrence may vary slightly between Eastern and Western Himalayas, but irrespective of the hosts, following are among the most frequently seen species of wood-rotting fungi in the temperate zone.

Trametes versicolor (Fig. 1), *Phellinus xeranticus, Phellinus allardii, Trichaptum abietinum, Xylabolus subpileatus* (Fig. 2), *Hymenachaete tabacina, Stereum hirsutum, Stereum sanguinolentum, Daedalea incana, Fames fomentarius, Antradiella zanata, Inanatus sciurinus, Trametes gibbosa, Amylastereum chailletii, Lenzites betulina* and *Glaeophyllum subferrugineum.*

Nearly 105 species (*ca* 23%) grow on ground probably forming mycorrhizic associations with forest trees, while the remaining, 351 species (*ca* 77%) are lignicolous. About 5 species also share ferns as their hosts. Both angiospermic and coniferous woods provide the major hosts for these fungi. Of 351 lignicolous species, 140 (*ca* 40%) grow on angiosperms, 125 (*ca* 35%) on conifers while the remaining (*ca* 25%) prefer both type of woods.

Out of 140 species which grow exclusively on angiosperms, the following species are more prevalent: *Hymenachaete tabacina; Fomes famentarius; Trametes versicalal; T. gibbasa; Phellinus allardii, P. ferreus; P. igniarius; P. linteus; P. sanfardii; P. xeranticus; Inanatus sciurinus, I. tenuicamis, Bjerkendera adusta, Oxyporus populinus, Polyporus arcularis, P.tricholoma, Rigidoporus ulmarius, R. vinctus, Xylobolus subpileatus* and *Lenzites betulina.* Of the total 125 lignicolous species which prefer coniferous woods, more frequently found are: *Chaetoderma luna, Dacrybolus karstenii, Hyphodontia alutaria, H. arguta.,. Resinicium bicolor; Phlebiopsis ravenelii, P. gigantea, Trechispora farinacea, Gloeocystidiellum sulcatum, Phellinus pini, Fomitopsis rosea, Gloeophyllum subferrugineum,. Heterobasidion insulare, Oligoporus fragilis, Amylostereum chailletii, Stereum sanguinolentum, Tomentella indica, Scytinostromella heterogena, Amylocorticium indicum* and *Vararia rhodospora.* Similarly, the more prevalent of the species (86) preferring both angiospermic and coniferous woods are: *Ganoderma applanatum, Inonotus dryadeus, Phellinus robustus, Antrodia albida, A. gossypina, A. serialis, Daedalea incana, Diplomitoporus lenis, Trichaptum abietinum; Steccherinum ochraceum Stereum hirsutum* and *Xylobolus frustulatus.*

Species like *Fomes famentarius; Ganaderma applanatum; Inanatus dryadeus; Phlebiapsis gigantea; Phlebiapsis ravenelii; Rigidaparus ulmarius; Phellinus igniarius, P. robustus; P. laevigatus; Amylastereum chailletii,. Stereum sanguinalentum, S. hirsutum, S. gausapatum; Phaealus schweinitzii; Inanatus tenuicarnis I. radiatus* and *Glaeacystidiellum sulcatum* are responsible for significant heartrots and weaken the tree trunks, thus making them subject to wind breaks.

Out of the total 120 genera of wood-rotting fungi found in the temperate Himalaya, only 16 (13%) cause brown rots, and the dominant among them are: *Antrodia, Daedalea, Fomitopsis, Gloeophyllum, Oligoporus, Phaeolus, Serpula, Coniophora, Echinodontium,* and *Chaetoderma.* Only 50 (14%) of the known 351 lignicolous species cause brown rots. About 30 species (60%) of them grow exclusively on coniferous woods, 8 species (16 %) grow preferably on hardwoods, while 12 species (24%) prefer both types of woods. It has been found that brown rot fungi are more commonly distributed in the coniferous forests of Himalaya and *Abies, Cedrus, Pinus, Picea* and *Larix* are among the most preferred hosts. It is also evident that among the angiosperms, *Quercus* and Bet*ula* are the major hosts. Some dominant and frequently encountered wood-rotting fungi responsible for more than 80% of brown rot infections in this zone are: *Antrodia albida; Daedalea incana; Chaetoderma luna; Laetiporus sulphureus Gloeophyllum abietinum* G. *subferrugineum; Coniophora arida; Serpula himantioides,* S *lacrimans; Dacrybolus karstenii; Fomitopsis rosea* and *Phaeolus schweinitzii.*

Fomitopsis rosea, Oligoporus fragilis, Piptoporus betulinus and species of *Antrodia* occur preferably towards the upper limit of temperate forests i.e. near the tree line.

The remaining 104 genera cause white rots, and the dominant among them are: *Phellinus, Inonotus, Tomentella, Hymenochaete, Stereum, Phlebia, Phanerochaete, Hyphodontia* and *Hyphodern:ta.* About 301 species (86%) cause white rots and 95 of them (31%) grow on coniferous woods, 132 (45%) prefer angiospermic hosts while only 74 (25%) grow on both types of woods. Among the dominant and frequently seen wood-rotting fungi responsible for more than 80% of white rot infections are: *Phellinus allardii, P. xeranticus, P. pini, P. robustus; P. ferreus; Phylloporia weberiana; Trichaptum abietinum; Trametes gibbosa, T. versicolor, T. hirsuta; Ganoderma applanatum; Hymenochaete tabacina; Antrodiella zonata; Lenzites betulina; Stereum hirsutum; Inonotus sciurinus; Heterobasidion insulare; Amylostereum chailletii; Xylobolus subpileatus; Vararia rhodospora; Phlebiopsis gigantea; Hyphodontia arguta, H. spathulata* and *Phlebiopsis ravenelii.*

Quercus, Betula, Aesculus, Juglans, Alnus, Schima, Viburnum are among the most liked hosts for white rot fungi in this zone.

III. HABITAT

Temperate Himalayan forests are rich in dead trees and provide adequate environments for wood-rotting fungi. When trees are alive, one can hardly expect them because wood-rotting fungi seldom fruit on the surface until their host trees die (Erikson 1958). These fungi usually occur when the forest sites are disturbed to some extent. When trees are broken or damaged, tree-decays are exposed and the fungi fruit.

The secondary growth forest sites formed frequently at edge of the forests consist of vigorously growing young trees. Practically no or very few wood-rotters seem to be growing in such forest sites. Virgin forest sites which mostly consist of healthy mature trees, support only a few fungi like *Climacocystis borealis, Fomitopsis pinicola, Stereum sanguinolentum, Hyphodontia alutaria, Phaeolus schweinitzii* and *Polyporus badius* on living trees.

Forest sites of naturally disturbed or slash dumped areas consisting of dead trees in all stages of decomposition play an important role in ecology of wood-rotting fungi. Wind breaks and wind throws occur where trees are weakened by decay fungi or when they are shallowly rooted on exposed sites. The most favourable condition is achieved when fallen trees and slash are dumped or stacked together on the forest floor. More than 70% of the wood-rotting fungi collection occurred in such type of sites.

When the trees die and defoliate, the overstorey canopy is extensively opened. Increased penetration of direct sunlight down to the forest floor depletes the moisture of the forest. Substrates for fungi are abundant all over but fungal growth is quite limited because of less moisture. In such sites, the fungal fruiting becomes dramatically decreased except when the forest floor remains wet during rainy days. In such sites species like *Antrodia serialis, Amytostereum chailletii; Coniophora arida; Trichaptum fusco-violaceum, T. abietinum; Resinicium bicolor; Gloeopyhyllum sepiarium; Hyphodontia aspera; Trametes versicolor; Stereum hirsutum S. sanguinolentum* and *Ganoderma applanatum* which are adapted to desiccation only survive here.

Species of *Quercus* form the most important and major host for these fungi. Both living and dead fallen trees support a rich aphyllophoraceous fungus flora and a rich species diversity up to the extent that over 120 species grow either exclusively on *Quercus* or share it as one of their liked hosts. *Vararia effuscata Albatrellus dispansus; Antrodiella zonata; Bjerkendera adusta; Daedaleopsis purpurea; Fomes fomentarius; Laetiporus sulphureus, L. percicinus Piloporia indica; Spongipellis unicolor; Tyromyces pelliculosus, Wolfiporia dilatohypha* grow exclusively on *Quercus*. Some other species which also frequently share *Quercus* as one of their hosts are: *Phellinus allardii, P. ferreus, P. xeranticus,. Stereum gausapatum,. Trametes versicolor; T. hirsuta; Xylobolus subpileatus; Inonotus tenuicarnis, I. sciurinus; Aleurodiscus oakesii; Phaeolus schweinitzii, Lenzites betulina; Phlebia radiata; Hericium erinaceus* and *Phylloporia weberiana.*

Abies pindrow and *A, spectabilis* are the most liked hosts among the conifers. Species like *Antrodia serialis; Coniophora arida; Hyphoderma setigerum Fomitopsis pinicola,* F. *rosea; Inonotus radiatus, Phellinus igniarius, Phlebiopsis ravenelii, P. gigantea, Gloeocystidiellum sulcatum, Trichaptum abietinum* and *Stereum sanguinolentum* grow either exclusively on species of *Abies* or share them as one of the most liked hosts.

Near the timberline *Betula utilis* is also a common host. *Phellinus laevigatus,P. nigricans; Piptoporus betulina, Scytinostroma rhizomorpharum* grow exclusively on *Betula,* while species like *Trametes gibbosa, T. versicolor; Inonotus tenuicarnis, I. sciurinus, I. radiatus; Phellinus sanfordii, P. gilvus, P. igniarius; Hyphodontia spathulata; Coniophora betulae; Lenzites betulina; Rigidoporus ulmarius; Spongipellis delectans* and *Polyporus arcularis* also quite frequently grow on *Betula.*

Rhododendron thickets both in exposed slopes or in the shaded areas are very poor sites and *Rhododendron* itself is the poorest host for these fungi. However, the extensive Rhododendron thickets in the shaded areas retain some moisture and in such areas, few species like *Epithele fulva, Hyphodontia crustosa Xenasma subnitens, Phellinus acontextus* grow only on Rhododendrons. The other species which incidentally grow superficially on dead tissue of standing trunks are: *Botryobasidium subcoronatum; Phanerochaete filamentosa; Phlebia radiata; Phellinus sanfordii,' Scytinostroma duriasculum; Hydnum repandum; Antrodia odora* and *A. oleracea.*

IV. DYNAMICS DURING DECOMPOSITION

Pioneer Decayers

The basic trends in the community development of wood-rotting fungi is mainly governed by the physical and chemical properties of a host tree species and microclimate of the growth site (Rayner & Boddy 1988, Boddy 1992). A large number of primary decayers like *Fomitopsis pinicola* and *Heterobasidion annosum* on coniferous trees, *Piptoporus betulina* on oaks and birches do not suffer much due to the death of the host tree as the pathogen usually go on living on the fallen tree trunks for several years or decades.

A few number of species like *Phellinus pini, Inonotus circinatus* and *I. tomentosus* on pines, *I. radiatus* on *Abies* and *Betula* growing as primary decayers seem to suffer indirectly of the death and fall down of the primary host. After the falling of tree, their mycelia give way to more competitive secondary saprophytes and evidently die within two years.

Broken or cut surfaces of dead/living trees provide unique habitats for many wood-rotting fungi and they colonize and fruit frequently by filling tissues around the torn bark and then spreading towards the open space. *Amylostereum chailletii; Trametes versicolor; Antrodia serialis; Coniophora arida; Gloeophyllum sepiarium; Stereum hirsutum; Trichaptum abietinum, T. fusco-violaceum; Hyphodontia arguta* are frequently found on freshly cut or broken surfaces.

When a healthy tree becomes uprooted or broken in a storm, many fungi are more competitive in the primary source capture (Renvall 1995). Species like *Trichaptum abietinum, Stereum sanguinolentum, Antrodia serialis, Hyphoderma setigerum, Phlebiopsis gigantea* and *Chaetoderma luna* rapidly occupy such fallen coniferous trees while abruptly fallen trunks of Oaks, Birch and Aspen are commonly invaded by *Trametes tephroleuca, T. versicolor, Stereum hirsutum, S. gausapatum, Lenzites betulina, Trichaptum abietinum* and *Fomes fomentarius* (Fig. 3). It is also interesting to note that most of the pioneering wood rotting fungi cause white rots.

Decay at Later Stage

The wood-rotting fungi become more complex and the amount of species involved also increase as the decomposition proceeds. However, certain species stand out as more dominant, more extensive and more persistent than the others. Such second stage dominants have been seen in natural forests only because the possibilities of their growth and fruiting are minimal in managed forests in which dead and fallen trees are removed. *Fomitopsis rasea; Glaeaphyllum subferrugineum,* G. *sepiarium; Antradia xantha,* species of *Hyphodantia,. Phlebiella vaga; Oligaparus sericeomollis, Phellinus laevigatus* and other members of the *P. igniarius* group are common as second-stage dominant decayers in the temperate forests. It has also been noted that brown-rot causing species are more dominant at this main decomposition stage. At this stage some initial decayers like *Famitapsis pinicala* (both coniferous and hardwoods), *Fomes famentarius, P. igniarius, Piptaparus betulina* (on hardwoods) may also co-occur regularly and in abundance.

3
d
c
e
b
a

4
b
a

The fallen trunks become decorticated and covered with epiphytes, sink down against the ground and eventually become fragmented towards the end of the stage.

Final Stage Decayers

In temperate boreal zones, the complete decomposition of a fallen big coniferous trunk may even last for decades or even more (Hofgaard 1993). Certain groups of wood-rotting fungi are specialists in utilizing the tree trunks at the final stage of decay. Among such species: *Odonticium romellii; Serpula himantioides;* species of *Tubulicrinis; Antrodia crassa* and *Diplomitoporus lenis* are dominant and inhabit extensively decayed trunks of coniferous trees which are moss covered, soft and about to fall apart. There seems to be a competition for space between fungi and mosses. Decayed trunks are frequently covered with them. Many larger species with great vigour ignore the presence of mosses and override their mechanical barriers, but small corticoid fungi do not fruit well where mosses occur in abundance. Mosses may exclude those fungi by depleting available space for fungal growth on fallen trees. Fungi and mosses often share parts of the trunk with mosses on the upper side for photosynthesis and corticoid fungi on the under side. These wood rotters utilize the last remnants of decaying wood and mostly belong to the family Corticiaceae. They usually occupy small volumes of wood and basidiocarps develop in the shaded spaces on the underside. Most species cause white rots and usually bear annual and small basidiocarps but a few may grow into perennials.

Two types of wood-rotting fungi are found in the final stages of decay. In the first category are a great deal of corticiaceous and polyporaceous fungi which are ephemerals, appearing quickly on suitable pieces of wood, while in the second category are the species which establish slowly and evidently survive only in circumstances staying stable for a longer time. Most of the final stage decayers seem to be independent, appearing or disappearing without evident correlation to other similar species. They may grow side by side on suitable hosts if its volume allows but the co-occurrences seem to take place in a haphazard way and in random combination.

V. WOOD DECAY

A primary function of fungi in any ecosystem is depolymerisation of important plant biopolymers such as cellulose, hemicellulose and lignin. This function is very critical for volatilization of C, H and N as well as elemental release during decomposition and formation of soil and thus wood-rotting fungi are the likely candidate for important function category. The wood inhabiting members of Aphyllophorales form a cosmopolitan group and are capable of utilizing components of wood cell walls for their growth and reproduction. Although, all wood basically is composed of the structural polymers cellulose, lignin and hemicelluloses but there is considerable variation which is particularly evident in the heartwood of living trees in which a wide array of non structural extraneous materials (extractives) are deposited as the maturing cells die (Hillis 1962; Rowe & Conner 1978; Scheffer & Cowling 1966). Cellulose is the most consistent of the structural components varying minimally between wood species. Lignin and hemicelluloses however, vary both in composition and amounts not only between hardwoods and conifers but also among hardwoods (Timell,1967). The conifer wood is far more homogeneous.

Cellulose is a long chain polymer of glucose anhydride units joined by β 1-4 linkages. In general, cellulose components of wood are light in colour, have strong affinity for water and are soft and tough. Hemicelluloses consist of similar polymers of glucose joined by other linkages or polymers of monosaccharides other than glucose. Lignin is quite different from celluloses and hemicelluloses and is also the most resistant to biodegradation. It is a three-dimensional, amorphous, branched polymer of phenylpropane units joined by a variety of interunits linkages (Adler, 1977). It is formed by a random free radical oxidative copolymerization of three different oxy-cinnamyl alcohols which occur in various ratios depending upon the tree species. The greatest differences are between coniferous and hardwoods. It differs from all polymers in being largely nonhydrolizable, forming a protective layer around the wood polysaccharides and limiting the cellulose accessibility within the cell walls. Conifer wood in general has a higher lignin content (27-35%) than wood of angiosperms (19-24%). Unlike cellulose, lignin components are darker, have no affinity for water and are harder and brittle.

The wood-rotting fungi are grouped into two categories i.e. white rot fungi and brown rot fungi depending upon the way in which they decay wood. Since these substances differ markedly in their physical characters, so the effect of wood decaying fungi is different depending upon which substance is removed. The wood that has been acted upon by a lignin dissolving fungus contains a relatively high remainder of cellulose, becoming soft and spongy in texture and is whitened in contrast to normal wood. A dacay of this type is called white rot. A wood from which, the cellulose components have been removed, is darker than the normal wood, will be dry, brittle and of charcoal consistency. The decay of this sort is called brown rot.

In wood products and slash, there is a strong tendency for soft woods to be degraded primarily by brown rot and hardwoods by white rot fungi (Scheffer, 1964). This is probably related to the fact that the lignin in hardwoods is easier to biodegrade than that in coniferous wood (Yang *et al*, 1979) and not due to the difference in the hemicellulose components (Highley, 1976). The white rot fungi in apparent contrast to brown rot fungi must degrade lignin in order to decay wood.

Brown rot fungi utilize the hemicellulose of the cell walls leaving the lignin essentially undigested, but slightly modified (Kirk, 1975; Kirk & Alder, 1970). Evidently the differences between the conditions in culture and decaying wood profoundly affect the lignin-degrading ability of brown rot fungi. The mechanism of hemicelluloses break down by brown rot fungi appears similar to that of white rot fungi (Highley 1976, Keilich *et al*, 1970). But these fungi evidently employ a different mechanism than white rot fungi for attacking the cellulose in the wood (Cowling & Brown, 1969; Highley, 1977; Koenigs, 1974). Hyphae of the brown rot fungi like those of white rot fungi grow inside the lumina in contact with the tertiary wall, into the capillaries of which the secreted enzymes are able to diffuse (Bailey *et at*. 1968; Liese, 1970; Wilcox, 1970). Unlike the white rot fungi, the enzyme attack is not localized near the hyphae but is widespread and deeply diffused. As the decay proceeds, the cellulose and hemicelluloses are gradually destroyed at approximately the same relative rate. Brown rotted wood tends to shrink abnormally when dried giving rise to a characteristic cubical pattern of checking. The brown rot fungi reduce the strength of wood much more than white rot fungi and at the advanced stages, the wood is reduced to a residue of amorphous crumbly brown cubical pieces with excessive vertical and horizontal splitting (Brown cubical rot) composed largely of slightly modified lignin. Brown rot fungi do not produce extracellular phenol oxidases and generally give negative oxidase tests on gallic and tannic acid media and with gum guaic and syringaldazine reagents. Brown rot residues are extremely stable and are important organic components in forest soils (Gilbertson, 1981).

White rot fungi degrade cellulose and hemicelluloses at approximately the same rates relative to the original amounts present (Kirk & Highley, 1973) whereas the lignin is decomposed at a similar rate or usually somewhat at faster rate on a relative basis (Blanchette 1980; Setliff & Eudy 1979).

Hyphae of the white rot fungi are concentrated in the ray cells and vessels although, other cells are invaded very early in decay. The hyphae initially invade other cells from ray cells and vessels via pits or directly by penetration of cell walls (Wilcox, 1970; Liese, 1970). White rot fungi have cellulase and lignase enzyme systems secreted at hyphal tips and on lateral surfaces. These enzymes assist cell wall penetration and enlarge bore holes to perforations. Along the young hyphae, lysis furrows are produced. The degradation products of various cell wall layers are completely absorbed by the hyphae. White rot fungi successively deploymerise cell wall substances only to the extent that the products can be utilized consecutively for metabolism (Cowling, 1961). The action of the enzyme system of white rot fungi is restricted to the cell wall layers in the immediate vicinity of the hyphae.

Wood decayed by white rot fungi tends to lose its strength properties gradually and retains its fibrous structure (white stringy rot) into the advanced stages. The decayed wood becomes spongy, stringy or laminated and is usually bleached and paler in colour than sound wood. White rot fungi produce extracellular phenol oxidases and generally give positive oxidase tests on tannic and gallic acid media and with gum guaiac and syringaldozine reagents. These fungi eventually decay wood completely and the residues are not stable components of forest soils. A rather specialized type of white rot is caused by some species of *Phellinus* where the end result is a series of oval holes (about 1.5 cm × 0.5 cm) distributed evenly throughout the infected zone. This distinctive type of decay is called a white pocket rot.

The wood-rotting fungi that grow and produce basidiocarps on living trees are either restricted to interior, primarily nonliving portions of a living tree (heartwood) or are capable of invading and killing outer living sapwood. In living trees, the sapwood is quite immune to decay fungi unless it is exposed by a wound of some sort. Though, the decay is not likely to progress much beyond the limits of wounded area but these fungi (e.g. *Aurificaria shoreae Phellinus gilvus* etc.) are true pathogens and are capable of invading and killing living sapwood resulting in killing of the tree in many cases. These fungi generally do not attack the heartwood except a limited attack in certain cases.

Of the several thousand wood decaying fungi, only a small number like *Phaeolus schweinitzii* can cause degradation in the wood of the hearts of living trees (Wagner & Davidson, 1954) and are known as necrotrophs or heart rot fungi. They usually invade trees through dead branch stubs (Boyce, 1961) and can tolerate the chemical and physical constraints within the tree trunk. The constraints which are obviously suspect in this regard are levels of O_2, CO_2, variation of moisture content and pH of heart tissue, concentration and nature of volatile organic compounds, interaction with other microbes and above all the host response (Shigo & Hillis, 1973; Shortle & Cowling, 1978; Shain, 1971; Fries, 1973). These factors are probably involved in determining the host specificity of heart-rot-fungi. The heart rot fungi do not normally attack the sapwood. As such the infected tree is not killed but continues to grow and present all outward signs of a healthy tree with vigorous growth. However, the decay is progressive and with increasing years more and more sound heartwood becomes decayed resulting in a considerable volume loss in standing timber particularly in old growth stands. Besides basidiocarps, other symptoms like stem canker (*Phellinus laevigatus, P. nigricans)* and punk knots (*P. caryophyllii* Ryv. and *P. pini*) indicate heart rot decay. In a few cases symptoms like wounds, trunk or butt swells, dead or broken tops, crooked or forked trees, fire scars are also indicative of heart rots.

VI. PROBLEMS IN ESTIMATING DIVERSITY

Following difficulties are faced during the estimation of diversity of wood-rotting fungi in temperate zones.

Seasonability

Seasonability is an important problem when recording macrofungi and trying to estimate their diversity. Sporophore appearance depends on temperature, precipitation, its timings and nutrient availability (C/N ratio of the substrate and the successional patterns in the vegetation). The effects of seasonability create problems. There are often dramatic shifts in the appearance of fungal fruitbodies from the fall to the spring in temperate areas (Bills *et al*, 1986; Fogel, 1976). In the temperate Himalayan forests, there is generally a large flush in the early summer and after prolonged desiccation, the fungus flora developed 26- 30 days after heavy rains. Though, some species appear for a restricted period but some common ones may fruit over a wider period. The diversity estimates taken at different seasons are varied and meaningless. A shift in species composition and diversity can also occur from year to year. It has been observed that following the imposition of several levels of tree cuts in the temperate deciduous and coniferous forests, some years exhibited no or little fruiting. If the study had not been undertaken continuously, the estimates of diversity for wood-rotting fungi during these years would have been inordinately low.

Fungi may occur at certain times of the year and there may be gaps when fruiting does not occur. One either has to be at the right place at the right time or the potential rarity must be recognised by those unfamiliar with it and have it identified by an expert before decomposition takes place. These are the problems which on the whole are not experienced within many other groups of organisms. The only major way of gathering data on wood-rotting fungi by analysis of the species lists prepared from the numerous forays held annually throughout temperate Himalayas, is inadequate as it only covers, limited periods of the year, generally August to late September, and in addition the lists depend on the particular interests of the people attending the forays and these may vary. The seasonal variation of fungi is conditioned by environmental factors. The mycological flora can only be described as characteristic either of time or place by due frequency of individuals. Though, certain conditions must obtain in order that fungus sporophores may be produced in abundance, it may well be that a rapid increase in numbers is determined or influenced by the conditions which were in force prior to such increase.

Precipitation seems to be a major factor in the production of these fungi. Precipitation in May and June is essential for good fruiting in all forest types and the positive influence declines during August. Temperature and precipitation during the growing season explained the annual variation of 19-45% in the biomass of wood-rotting fungi and 24-90% of the species numbers in various forests. Thus a knowledge of the patterns of precipitation over a year and longer periods is essential to study their diversity.

Lack of Knowledge

A multitude of texts have appeared on the topics of biodiversity and the severe consequences, should the diversity of organisms in tropical and other regions not be maintained properly. While many of the principles and discussions of broader issues raised in these texts are relevent to most organisms, most lack any substantive context on fungi (Hawksworth, 1991, 1992). There are no studies on the temperate Himalayan fungal flora comparable to the exhaustive work of Arnolds (1981, 1982). This lack of information on fungi has far reaching implications.

Mycology has had a difficult time keeping pace with other taxonomic disciplines in completing taxonomic circumscription of fungal species. Such works are invaluable in survey and inventorying. According to the estimates of the total number of fungal species found in the world (Hawksworth, 1991, May 1991), only 5-10% of them have been described, which means, mycology is 70-100 years behind the vascular plant taxonomists in the identification and description of taxa.

The fact that there is no group of fungi which has been studied to enable firm diversity calculations to be made, reflects poorly on other disciplines that have neglected to validate the importance of fungi in their studies. Ideally, any description of a plant community should include characterization of its mycoflora as well. In view of their large number of species and ecological functions being fundamentally different from those of green plants and various degree of host specificity, fungi are important bioindicators in the temperate ecosystems. However, the ecology and distribution of most species are so insufficiently known, that the extension of ecological research, taxonomical delimitation and monitoring are very urgent. We need to understand fungal diversity on a local level as well as or regional or global scale to incorporate fungi into conservation management plans (Jaenike 1991, Arnolds & Jansen 1992).

Investigators

At the root of the problem in estimating fungal diversity is the historical and current lack of specialists. The lack of qualified taxonomic mycologists has diminished to a level that qualifies those now working in this speciality as themselves being considered member of an endangered species (Atlas *et al.* 1992). The paucity, often complete lack of information on wood-rotting mycoflora poses a problem in the design of conservation measures. One of the main problems lies in the lack of local expertise. The few mycologists' available work in isolation and are not known to each other. There is a need to establish contacts. The training of mycologists is a contentious issue as the few universities giving courses on mycology concentrate only on the economically important species particularly human and plant pathogens.

Succession

Changes in species composition during succession is another factor affecting measurement of fungal diversity. Though, this successional pattern is not very clearly defined but the fact remains that sampling of wood-rotting fungi at various times during the course of wood degradation, results in differential recovery of species and estimations of fungal diversity from the same substrate. The succession mechanisms are precursors in determining the manner in which environmental factors control species composition and diversity (Holder-Franklin, 1986). The comparative diversity estimates become meaningless until such succession mechanisms are not detected.

Species of wood-rotting fungi succeed each other during the decomposition of wood substrate, though the details vary slightly from tree to tree. Over thirty species of wood- rotting fungi especially polypores, were found to inhabit such trees in the temperate zones which were previously decayed by certain other species. The successor emerges only after the preceding fungus has died and often fruits on dead basidiocarps of the previous one. Little is known about the decay processes on fallen trunks and on the way and order, the fungi are replaced by others during a complete decomposition of a fallen tree trunk (Renvall, 1995). The coexistence may reflect that the species may merely have similar ecological optima (Jahn, 1979) rather than any process of replacement or other invasive interaction and their combat growth side by side is similar to the interspecific occurrence on a single stump (Rayner & Todd, 1982). Further, there can be other reasons for such succession like certain species may parasitize the sporocarps of other species (Helfer, 1991; Jeffries & Young, 1994) or simply using the earlier dead species as a substrate only (Helfer, 1991) or more accidental or neutralistic association because of limited space of growing on a single piece of wood (Cooke & Rayner, 1984). Even within certain genera some species are favoured as preceding fungi e.g. *Fomitopsis pinicola* and *Phellinus ferrugineofuscus* are favourite while others are not. Fungi of this group conquer large volumes of wood for their mycelia, usually several metres along the tree trunks, and the decaying process lasts for long periods of time. Some produce large perennial basidiocarps, while others have large and extensive fructifications. All are strong competitors in the primary resource capture (Cooke & Rayner, 1984). *Fomes fomentarius, Fomitopsis pinicola, Hymenochaete tabacina, Trichaptum abietinum, T. biformis* and species of *Inonotus* and *Phellinus* serve frequently as preceding species. It is interesting to note that members of the family Hymenochaetaceae are especially favourable as predecessors.

Certain genera of wood-rotting fungi include exceptionally many species that have this kind of successional preference. *Antrodiella* is the most famous genus of the successor fungi, followed closely by *Skeletocutis, Junghunia, Piloporia* and *Pycnoporellus.* Among the established successor species (Fig. 4) found in the temperate Himalayan zone are: *Antrodiella semisupina* and *A. zonata* (on species of *Fomitopsis, Fomes, Inonotus, Hymenochaete* & *Phellinus); Junghunia nitida* (on species of *Phellinus* & *Inonotus); Skeletocutis amorpha* (on species of *Trichaptum* & *Phellinus); Pycnoporellus fulgens* (on *Fomitopsis)* and *Piloporia indica* (on *Trichaptum).* Fungi which intimately depend on other species are especially vulnerable and deserve special attention in nature conservation. In most cases they survive only in virgin forests, and seem to be rare there also. Even a slight decrease in the abundancy of the preceding species may drastically suppress the populations of the successor or cause their disappearance.

VII. THREATS TO WOOD-ROTTING FUNGI

In recent years several national and regional lists of threatened fungi have been published. The Red data lists produced for various European countries by Lettau (1982), Winterhoff (1984), Winterhoff & Krieglsteiner (1984), Derbsch & Schmitt (1984, 1987), Krisai (1986), Wojewoda & Lawrynowicz (1986), Rassi & Vaisanen (1987), Hoiland (1988), Hallingback (1988) and Arnolds (1989) consider that 25-30 per cent of the flora of larger fungi has been threatened. The largest number of threatened species are agarics particularly ectomycorrhizae (over 30%) followed by wood-rotting fungi (over 20%). A strong decline in species diversity and sporocarp frequency has also been observed among the wood decomposing fungi in the forests (Arnolds, 1988; Kuyper & De Vries, 1989). Though it is not possible to establish a general decrease but such decline may well cause important changes in decomposition process with unknown consequences for functioning of forest ecosystems. Arnolds (1989) has further identified that after grasslands (44%) the deciduous forests (33%) and temperate coniferous forests (30%) are most vulnerable and have suffered great depletion and significant changes in the fungal flora. Two types of factors are responsible for such decline of wood-rotting mycoflora in temperate Himalaya.

Natural Factors

Succession and natural decline of host tree species are the main natural causes of decline. Spontaneous succession of plant communities to climax communities have led to rarefaction of earlier successional stages and their associated fungi. Similarly, the decreased vitality and subsequent death of host tree species, may also lead to decrease of associated fungi.

Anthropogenic Factors

For centuries, the temperate forest regions have experienced intensive human activity and the dominant effect of which was their conversion to agriculture, urbanization and to large plantations of species, largely represented by few coniferous or hardwoods. Many habitats for wood-rotting fungi have either been destroyed or irreversibly altered. More frequent and selective harvesting of tree species have also exerted a major impact on ecosystem function, biodiversity and precipitation. A serious consequence of clearing activity is the reduction in ecosystem carbon stores and hosts for wood-rotting fungi. This disappearance and destruction of habitats has threatened the existence of wood decaying fungi. Particularly those like *Hymenochaete cruenta, Phellinus acontextus, Fistulina hepatica, Inonotus dryadeus, Piptoporus betulina* which are host specific are more threatened. Moreover, a number of species require continuously humid conditions and are vulnerable to drainage or any form of woodland management which opens up the canopy and causes a reduction in the humidity. This is one of the reasons for the importance of old woodland sites that have escaped heavy disturbance such as the episodes of clear felling. Though, it is difficult to state precisely, how intensive the gathering of fruitbodies can be without harm to further reproduction of the species but excessive collections of fruitbodies has also definitely threatened the existence of many wood-rotting fungi (Arnolds & Jansen, 1991).

Disturbance appears to be an important factor influencing the basidiome fruiting. Thus in a stable habitat within a forest, low but persistent fruiting occurs whereas the fruiting at the forest margin or on forest trails and banks may be extensive but very irregular over long periods.

VIII. CONSERVATION MEASURES

The decline of many macrofungi is not only regrettable from the viewpoint of mycologists, naturalists and nature conservationists but also has far reaching consequences on the functioning of the ecosystems. Fungi are accepted as forming a kingdom separate from animals and plants but in conservation and large scale biodiversity studies, they are usually included within plants and get swamped there. They are treated as curiosities by many conservationists just as their role in nature is frequently given only a lip service by other naturalist disciplines.

Despite their actual and unrealised economic value and vast number of species, many of which are presumably rare and highly endemic, fungi have been ignored in formation of policy for natural resource management and conservation (Fox, 1994). Though interest in fungal diversity has grown in recent years (Hawksworth, 1991; Watling, 1995) but so far, the conservation measures are limited to animals and plants with no policies or measures for the fungi. It is urgently required to start drastic measures for the maintenance of a rich and varied mycoflora and for the conservation of rare species. These measures are not fundamentally different from measures proposed for most other groups of organisms, in particular green plants and mainly based on the principle of habitat conservation. Following steps are aimed at conservation of these fungi at this stage.

(i) The decision to give full legal protection to some rare and beautiful species with splendid and peculiar fruitbodies (Flagship species) will be the first step in this direction of stressing the fact of the threat to fungi. Besides getting protection, these popular and charismatic species will serve as symbols and rallying points for stimulating conservation awareness and action on fungi. In the beginning, a list of following 22 flagship species can be given the legal protection. They are: *Fames famentarius, Famitapsis pinicala, F. rasea, Ganaderma applanatum, Xylabalus subpileatus, Stereum sanguinalentum,* S. *hirsutum, Fistulina hepatica, Inanatus dryadeus, I. tenuicarnis, I. sciurinus, Hericium erinaceus, Piptaparus betulinus, Pilaparia indica, Laetiparus sulphureus, Phaealus schweinitzii, Phellinus rabustus, Pyrrhaderma sendaiense, Sparassis crispa, Daedalea incana, Daedaleapsis purpurea* and *Riagidaparus ulmarius.*

(ii) Due to the difference in the biology of fungi, they differ from the flowering plants as far as the categories of IUCN are concerned. They can be observed only where their fruitbodies are present and the lack of the latter even for a long time does not necessarily mean the extinction of the mycelium. It may happen that a species accepted as extinct in a given region, appears once again, because of the survival of its mycelium in the wood and renoved conditions conducive to the formation of its fruitbodies. There are cases of fungi existing in an invisible form, in timber without producing fruitbodies for a dozen or more years and then producing them on the same substrate. For this reason mycologists can use the concept of indicator species. For example the ancient woodlands are important habitats for wood-rotting fungi and we also know that we can use certain more easily identifiable fungi to indicate whether a wood is ancient and how much disturbance there has been to it in the past. Thus it follows from this that if we protect the more showy and large perennial fungi, we will also go some way, to protecting the species that are less obvious.

(iii) In the face of above mentioned problems and difficulties that may arise within, it will be a daunting task to appreciably increase our knowledge of the diversity and distribution of fungi in endangered habitats before those habitats disappear. Since a species-by-species approach to fungal conservation in a big country like India is unrealistic so the only option is a habitat based approach. Whereby in the absence of knowledge of most species, conservation is achieved by preservation of adequate samples of as many as possible of the whole range of habitats inhabited by wood-rotting species or their communities. It is emphasized that a habitat based approach must be backed up by research on such aspects as the congruence between fungal communities and the vascular plant communities, because it is the latter which is often used to define habitats when conservation reserves are set up. The effectiveness of existing reserves at conserving fungi also needs to be established. Sampling is needed to determine areas of high diversity and to obtain better estimates of the number of taxa which are formally described. A national network of permanent sites in major habitat sites is suggested as a cost effective method of rapidly improving broad scale knowledge and obtaining at least preliminary answers to some of the necessary questions.

The links between habitat preference and species distribution are so significant that the conservation of natural fungal habitats would make a stronger case compared with the efforts at this stage of producing a Red Data List at national level. Such a huge latitudinal range, many distinct contrasting topographies, vegetational communities and varied distribution make it difficult to comprehend a Red Data List which would be meaningful. Perhaps, the better approach under the present financial restrictions would be to produce Red Data Lists for specific areas. These lists will be important conservation tool. Firstly they highlight the important habitats and major threats to the group concerned. The statics they provide can be used by botanists to publicize what needs to be done for wood-rotting fungi. Secondly they are used as one factor in the assessment of which sites are more important for the group under consideration or which species are particularly rare or threatened, so that conservationists can decide which species are in need of most protection. In other words, if we have one site that contains say three Red Data list species and other site which has none, all other factors being equal then we know that former is the priority for protection.

(iv) A broad based mycology curriculum in Indian universities is required to be developed. This could further be developed through conferences and workshops. The few experts available on these fungi need to establish contacts. Owing to the current fragmented approach to taxonomic mycology in India, it is recommended that a review be undertaken of the institutional organisations of taxonomical mycological research and setting up of an institute of fungal diversity in India.

The effective conservation of wood-rotting fungi will depend upon sound taxonomic and ecological research. An important step towards improving the conservation of temperate fungi is to increase the number of workers in taxonomic and ecological mycology. In the event of not having a full range of specialists available, then short cuts are required. One such possibility is to train local people at all levels of interest and this might allow parataxonomists to go some way.

Commendable efforts may be made to sell mycology to younger people at school leve (Kendrick, 1986) which can be intensified, at the higher level and thus exposing more students to this discipline. Short courses on identification of these fungi may be conducted at regular or irregular intervals and the trained mycologists can deliver lectures and lead fungal forays for the general public. This can further be intensified by new ventures like Cryptogamic Sanctuaries or reserves which help general public and enable them to appreciate the full role, fungi play in the ecosystem and various life strategies they display. Likewise the recent idea of Cryptogamic Garden as a new venture can also be of greater educational potential. These funga gardens (Watling, 1993) can act as potential basic tools by school children and public for education and awareness.

ACKNOWLEDGEMENTS

The author is grateful to the Director, Botanical Survey of India, Calcutta and Joint Director Botanical Survey of India, Dehra Dun for providing facilities and encouragement. Thanks are also due to Prof. L. Ryvarden, Oslo University, Norway; Prof. T. Niemla, University of Halsinki Finland and Dr. M. Rajchenberg, University of Buenos Aires, Argentina for constructive suggestions and help in many ways.

REFERENCES

Adler E 1977 Lignin Chemistry -Past Present and Future; Wood Scinece Technology 11 169-218

Arnolds E 1981 Ecology and Oenology of macrofungi in grasslands and moist heathlands in Drenthe, the Netherlands part-I, introduction and Synecology; Bibliotheca Mycologica 30 1-407.

Arnold E 1982 Ecology and Oenology of macrofungi in grasslands and moist heathlands in Drenthe, the Netherlands part-II, Autecology part-III, Taxonomy; Biolotheca Mycologica 31 1- 501.

Arnolds E 1988 The changing macromycete flora in the Netherland; Trans. Brit. Mycol. *Soc.* 90 391-406

Arnolds E 1989 Red Data list of macrofungi in the Netherlands; Persoonia 14 77-125.

Arnold E & A E Jansen 1991 Conclusions of the First Meeting of the European Committee on the protection of Fungi; Soc. Sci. Art. Sci. Tracks 18 114-119.

Arnolds E & A E Jansen 1992 New evidence for changes in the macromycete flora of the Netherlands; *Nova Hedwigia* 55 32-352.

Atlas R M, R R Colwell, D Pramer, J M Tiedje, A K Vidaver & R J Wodzinski 1992 Microbial ecology; Am, Soc. Microbiol. News 58 4-5.

Bagchee K D & B K Bakshi 1954 Studies on Indian Thelephoraceae-I. some species of *Stereum, Peniophora* and *Corticium;* Ind. For Bull. 166 1-11.

Bagchee K D, Y N Puri & B K Bakshi 1954 Principal diseases and decays of Oaks and other hardwoods in India-II; Indian Phytopath. 7 18-42.

Bailey P J, W Liese & R Rosch 1968 Some aspects of cellulose degradation in lignified cell walls; In *Biodeterioration of Materials;* pp 546-557 (Essex, England: Elsevier Publ. Corporation).

Bakshi B K 1971 Indian Polyporaceae; (New Delhi, India: ICAR).

Bakshi B K & K D Bagchee 1950 Principal diseases and decays of Oaks in India; Indian Phytopath. 4-139.

Berkeley M J 1850-1854 Dacades of Fungi 25-30; Hook. J. Bot. 2(1850): Decades of Fungi 31-36; *ibid* 3(1851) : Decades of fungi 37-40 ; *ibid* 4(1852): Decades of Fungi 41-50; *ibid* 4(1854).

Bills G F, G I Holtzmann & O K Jr. Miller 1986 Comparison of ectomycorrhizal basidiomycete communities in red spruce versus northern hardwood forests of West Virginia; Can. J. Bot 64 760-768.

Blanchette RA 1980 Wood decomposition by *Phellinus pini*-a scanning electron microscopy study; Can. J. Bot. 58 1496-1503.

Boddy L 1992 Development and function of fungal Communities in decomposing Wood; In *The fungal community, its organisation and role in the ecosystem;* pp 749-782 ed. D T Wicklow & C G Carrol (New York, USA : Marcel Dekker).

Boyce J S 1961 Forest pathology 3rd ed; (New York, USA : McGraw-Hill).

Cooke R C & A D M Rayner 1984 Ecology of Saprophytic fungi; (London & New York: Longman).

Cowling E B 19671 Comparative biochemistry of the decay of sweetgum by white rot and brown rot fungi; (Washington, USA : USDA).

Cowling E B & W Brown 1969 Structural features of cellulosic materials in relation to enzymatic hy;drolysis; Adv. Chem. Ser. 95 152-187.

Derbsch H & J A Schmitt 1984 Atlas der Pilzes Saarlands Teil I, Verbreitung under Gefahrdung; Aus Natur und Landschaft in Saarland 2 1-535.

Derbsch H & J A Schmitt 1987 Atlas der Pilzes Saarlands Teil 2, Nachweise Okologie, Vorkommen und Beschreibungen; Aus Natur und Landschaft in Saarland 3 1-816.

Donk M A 1971 Progress in the study of the classification of the higher basidiomycetes; In *Evolution in the higher Basidiomycetes;* p 3-25 ed. R H Peterson (Knoxville, USA : The University of Tennesse Press).

Eriksson J 1958 Studies in the Heterobasidiomycetes and Homo-basidiomycetes- Aphyllophorales of Muddus National Park in North Sweden; *Symb. Bot. Upsal.* 16 1.172.

Fogel R 1976 Ecological studies of hypogeous fungi-II, Sporocarp phenology in a Western Oregon Douglas Fir Stand; Can. J. Bot. 54 1153-1162.

Fox J L 1994 Microbial diversity, Low profile, Immense breadth; Am. Soc. Microbiol, News 60 533-536.

Fries N 1973 Effects of volatile organic compounds on the growth and development of fungi; Trans. Brit. Mycol. Soc. 60 1-21.

Gillbertson R L 1981 North American Wood-rotting fungi that cause brown rots; Mycotaxon 12 372-416.

Hallingback T 1988 A Preliminary list of the threatened fungi in Sweden; (Uppsala, Sweden: Department of Ecology and Environmental Research).

Hawksworth D L 1991 The fungal dimension of Biodiversity: Magnitude, Significance and Conservation; : Mycol. Res 95 641-655.

Hawksworth D L 1992 Fungi : a neglected component of Biodiversity crucial to ecosystem function and maIntenance; Can. Biodiversity l 4-10.

Helfer W 1991 Pilze auf Pilzfruchkorpem Untersuchungen zur Okologie, Systematic und Chemie Libri; Botanici 1 1-157.

Highley T L 1976 Hemicellulases of white and brown rot fungi in relation of host preferences; Mater. Or 11 33-46.

Highley T L 1977 Requirements for cellulose degradation by a brown rot fungus; Mater. Org. 12 25-36.

Hillis W E 1962 Wood extractives and their significance to the pulp and paper industry; (New York: Academic Press).

Hofgaard A 1993 50 years of change in a Swedish boreal old-growth *picea-Abies* forest; J. Vegetation Sci. 4 773-782.

Hoiland K 1988 A preliminary list of threatened macrofungi in Norway; (Oslo, Norway : unpublished document).

Holder-Franklin M A 1986 Ecological relationship of microbiota in water and soil as revealed by diversity measurements; In *Microbial autoecology* pp 93-132 ed. R.I. Tate (New York, USA : Academic Press).

Imazeki R, Y Kobayasi & Aoshima 1966 Fungi; In *Flora of Eastern Himalayas:* pp 611-626 ed. H Hara (Tokyo, Japan: Tokyo Press)

Jaenike J 1991 Mass extinction of European fungi; TREE 6 174-175.

Jhan H 1979 Pilze die an Holz wachsen; (Busse, Herford; Germany)

Jeffries P & J W K Young 1994 Interfungal parasitic relationships; (Cambridge, : Cab International)

Keilich G, P Bailey & M Liese 1970 Enzymatic degradation of cellulose, Cellulose derivatives and hemicelluloses in relation to the fungal decay of wood; wood Sci. Technol. 4 273-283

Kendrick W B 1986 A Young person's guide to the fungi; (Waterloo Ont., Canada : Mycologue publications)

Khara H S 1977a Some stipitate Hydnums from North Western Himalayas; Ind. J. Mycol. & Pl. Pathol. 7 127-134

Khara H S 1977b The Hydnaceae of North Western Himalayas; Indian Phytopath 30 94-95

Kirk T K 1975 Effects of a brown rot fungus Lenzites trabea on lignin in spruce wood; Holzforschung 29 99-107.

Kirk T K & E Alder 1970 Methoxyl-deficient structura! elements in lignin of Sweetgun decayed by a brown rot fungus; Acta chem. Scand. 24 3379-3390

Kirk T K & T L Highley 1973 Quantitative changes in structural components of conifer woods during decay by white and brown-rot fungi; Phytopathology 63 1338-1342

Koenigs J W 1974 Hydrogen peroxide and iron: A proposed System for decomposition of wood by brown rot Basidiomycetes; Wood Fiber 6 66-79

Krisai I 1986 Rote Liste gefahrdeter Grosspilze Osterreichs; Grune Reihe Bundesministeriums fur Gesundheit und umweltschutz 5 58-71

Kuyper T W & B D E Vries 1989 Effects of fertilisation on the mycoflora of a pine forest; In: Forest ecosystems and their components; ed. R A A Oldeman, P Schmidt, E J M Arnolds & F Staudt (Wageningen, Netherlands)

Lettau M 1982 Vorlaufige Liste Verschollener und gefahrdeter Grosspilze in Schleswig-Holstein; Landesamtes Naturschutz lnadschaftspfl Schleswig-Holstein Kiel. 5 58-71

Liese W 1970 Ultrastructural aspects of woody tissue disintegration; Annu, Rev. Phytopath. 8 231-258

Maas Geasteranus R A 1971 Hydnaceous fungi of the Eastern Old World; *Verhand*. Kon. Nederl, Akad. Wetensch Natuurk. Ser. 2 1-176

May R 1991 A fondness for fungi; Nature (London) 352 475-476

Rassi P & P Vaisanen 1987 Threatened Animals and Plants in Finland; (Helsinki, Finland).

Rattan S S 1974 *Scytinostroma* in India with notes on extralimital species; Trans. Brit. Mycol. Soc. 63 1-12

Rattan S S 1977 The Resupinate Aphyllophorales of North Western Himalayas; (Vaduz, Germany: J. Cramer)

Rattan S S & I P S Khurana 1978 The Clavarias of Sikkim Himalayas; (Vaduz, Germany : J. Cramer)

Rayner A D M & L Boddy 1988 Fungal decomposition of wood, its biology and ecology; (Chichester, England: John Wiley & Sons)

Rayner A D M & N K Todd 1982 Population structure in Wood decomposing Basidiomycetes; In *Decomposer Basidiomycestes, Their biology and ecology;* pp 109-128 ed. J C Frankland, J N Hedges & M J Swift (Cambridge, England : Cambridge University Press)

Rehill P S & B K Bakshi 1965 Studies on Indian Thelephoraceae Indian species of *Peniophora* and *Corticium;* Ind. For Bull. 242: 1-30

Rehil P S & B K Bakshi 1966 Studies on Indian Thelephoraceae -The genus *Stereum;* Ind. For. Bull. 250 1-20

Reid D A, K S Thind & K L Adlakha 1958 The Thelephoraceae of Mussoorie Hills; Trans. Brit. Mycol. Soc. 41 129-134

Reid D A, K S Thind & M S Chatrath 1959 Polyporaceae of Mussoorie hills; Trans. Brit. Mycol. Soc. 42 40-44

Renvall P 1995 Community structure and dynamics of wood-inhabiting Basidiomycetes on decomposing conifer trunks in northern Finland; Karstenia 35 1-51

Rowe J W & A H Conner 1978 Extractives in eastern hardwoods -A review; *U. S. For Serv. Gen. Tech. Rep. FPL-18.* (Madison, WI : For. Prod. Lab.)

Scheffer T C 1964 Biological observations of significance for improved preservative treatment; Holzforschung 18 88-94

Scheffer T C & E B Cowling 1966 Natural resistence of wood to Microbial deterioration; Ann. Rev. Phytopathol. 4 147-170

Setliff E & W E Eudy 1979 Screening white-rot fungi for their capacity to delignify wood; In *Lignin Biodegradation Microbiology, Chemistry and Practical Applications;* pp 135-149 ed. T K Kirk, T Higuchi & H M Chang (Miami, USA : Chemical Rubber Company Press)

Shain L 1971 The response of sapwood in Norway spruce to infection by *Fomes annosus; phytopathology* 61 301-307

Sharma J R 1995 Hymenochaetaceae of India; (Calcutta, India: Botanical Survey of India)

Sharma J R 1996a Wood-rotting Fungi (Aphyllolphorales) from Sikkim; Bull. Bot. Surv. Ind. 34 89-99

Sharma J R 1996b *Phellinus pachyphloeus* and its related species in India; Bull. Bot. Surv. Ind. 35 16-26

Sharma J R 1997a Fungi; In *Floristic Diversity and Conservation Strategies in India;* pp 91-149 ed. V MudgaI & P K Hajra (Dehra Dun, India: Shiva Offset)

Sharma J R 1997b Wood-rotting fungi of Temperate -Timberline zone; In *Floristic Diversity and Conservation Strategies in India;* pp 151-194 ed. V Mudgal & PK Hajra (Dehra Dun, India: Shiva Offset)

Sharma J R 1999a Genera of Indian Polypores (In press)

Sharma J R 1999b Indian Polyporoid Fungi (In press)

Shigo A L & W E Hillis 1973 Heartwood, discoloured wood and microorganisms in living trees; Annu. Rev. Phytopathol. 11 197-222

Shortle W C & E B Cowling 1978 Interaction of live sapwood and fungi commonly found in discoloured and decayed wood; phytopathology 68 617-623

Thind K S 1961 The Clavariaceae of India (New Delhi, India: ICAR)

Thind K S & K L Adlakha 1956 The Thelephoraceae of Mussoorie hill Indian J. Mycol. Res 2 57-64

Thind K S & H S Khara 1968 The Hydnaceae of North Western Himalayas; Indian phytopath. Soc. Bull. 4 25-33

Thind K S & H S Khara 1975 The Hydnaceae of North Western Himalayas; Indian phytopath. 28 57-75

Thind K S & S S Rattan 1986 The Thelephoraceae of North Western Himalayas; Indian phytopath. Soc. Bull 4 14-24

Thind K S & S S Rattan 1970 The Thelephoraceae of India-the genus *Tubulicrinis* and *Hyphodontia;* Proc. Indian Acad. Sci. 7 118-131

Thind K S & S S Rattan 1971 The Thelephoraceae of India-the genus *Tomentella;* Indian phytopath. 24 32-42

Thind K S & S S Rattan 1972 The Thelephoraceae of India; Trans. Brit. Mycol. Soc. 59 123-128

Thind K S & S S Rattan 1973a The Thelephoraceae of India; Indian phytopath. 26 485-494 & 528-536

Thind K S & S S Rattan 1973b Thelephoraceae of India; Mycologia 65 1250-1258

Timells T 1967 Recent advances in the chemistry of wood hemicelluloses; Wood Sci. Technol. 1 45-70

Wagner W W & R W Davidson 1954 Heartrots in living trees; Bot. Rev. 29 61-134

Watling R 1993 A fungus garden; Mycologist 7 47-48.

Watling R 1995 Assessment of fungal diversity: macromycetes, the problems; Can. J. Bot. 73 (Suppl. 1) 515-524.

Wilcox W W 1970 Anatomical Changes in Wood cell walls attacked by fungi and bacteria; Bot. Rev. 36 1-28

Winterhoff W 1984 Vorlaufige Rote Liste der Grosspilze (Macromycetes); Rote liste der gefahrdeten Tiere und Pflanzen in der Bundersrepublik Deutschland 4 162-184

Winterhoff W & G J Krieglsteiner 1984 Gefahrdeten Pilze in Baden-Wurttemberg; Beihefte Verroff. Naturschutz Deutschland 4 162-184

Wojewoda W & Lawrynowicz M 1986 Red List of threatened macrofungi in Poland; In *List of Threatened Plants in* Poland; pp 47-82 ed. K Zarzyckiego & W Wojewoda (Warsaw, Poland)

Yang H H, M J Effland & T K Kirk 1979 Culture parameters influencing degradation of lignin in thermomechanical pulps by *Phanerochaete chrysosporium;* Biotechnol. Bioeng 21 47-48.

9

Studies on the Growth Forms of Bryophytes in Relation to Habitat Conditions

H. GOVINDAPYARI and PREM L. UNIYAL
Department of Botany, University of Delhi, Delhi, India

ABSTRACT

Bryophytes occupy all possible habitats in all climatic conditions except marine environment. They colonize apparently inhospitable substrata where other group of plants cannot grow. They form a considerable part of vegetation in the cooler zones. The major part of the plant is in direct contact with the substrata and nutrients and other unwanted elements are readily absorbed by all over the surface of the plant. To survive in the variety of habitats and microclimatic conditions, they modify their growth pattern. The shoots or thalli respond to light, gravity, moisture, pH and other stimuli. The characteristic pattern of the shoot growth which makes the harmony with the microclimatic conditions is called "growth form". The main growth forms recognized in bryophytes are cushion, turf, mat, dendroid, weft, tail, fan, feather, bracket and pendents. Under different microclimatic conditions one species may exhibit different types of growth forms. The degree of arrangement of branching may be modified by the direction and intensity of light. In the taxa of dry habitats the shoots remain compact and small due to high light intensity, but the compactness of the shoots conserve water efficiently. On the moist forest floors the turf, weft and dendroid forms are prominent in which the major leaf portion of the shoots is away from the ground to avoid the chemicals leached from the litter. These forms are able to store much water in the capillary system in between the shoots. Pendents are prevalent in the high humid conditions. Some forms are relatively constant in their expression, others show plastic modifications and confer wider ecological amplitude upon the species. In the mat forms the shoots spread over the hard substratum.

The growth form in some cases is relatively invariably specific feature while in others it is subject to variation under the prevailing conditions of the habitat. The use of statistical methods for the structural analysis of natural communities as well as analytical researches into the interrelation between physiological function of life forms and environmental factors are important aspects.

I. INTRODUCTION

Bryophytes with a wide ranging scientific interest are excellent experimental material for studying many fundamental biological problems. There has been a outburst of experimental work on the physiology, ecology and other aspects of bryophytes biology because of their availability in most seasons, manageable size, readily and lending themselves well to culture under axenic conditions, growing fairly fast and completing their life cycles in a short period of time.

Bryophytes are a group of three heterogeneous divisions of the plants kingdom which includes the liverworts (Hepaticopsida), hornworts (Anthoceropsida) and the mosses (Bryopsida). Mosses consist of a species rich group represented by nearly 10,000 species. The bryophytes share certain characteristics with the seed plants, Pteridophytes and green algae. The chloroplast of bryophytes carry a mixture of chlorophyll and carotenoids pigments closely similar to that found in the other green land plants and in the green algae, but different from that found in other groups of photosynthetic organisms. The food reserve consists mainly of starch built of amylase and amylo-pectin units that resemble land plants and green algae. The cell walls comprise of cellulose microfibrils embedded in a matrix of polygalacturonic acid residues and hemicelluloses but lack true lignins (Hebant, 1977).

In general bryophytes are smaller than most other plants and reach average height of 7-10 cm. *Dawsonia superba* is the tallest known moss, may reach a height of about 70 cm (Sainsbury, 1955), and *Polytrichium commune*, when grows in wet conditions, can be nearly as tall as 50 cm. In the pendent moss *Meteoriopsis* and the aquatic moss *Fontalis*, the shoots may grow upto 80 to 100 cm in length. The majority of the bryophytes are quite small ranging down to such tiny plants as species of *Ceplaloziella* and some mosses barely visible to the naked eyes such as species of *Ephemerum* and *Micromitrium*.

Most bryophytes, especially in dry environments or cold climate, grow as turfs or carpets which sometime form a continuous stretch in the forest floors. They usually grow in aggregations. The small stature is a disadvantage for them in competing with vascular plants for space and light. In most climates bryophytes tend to be associated with nutrient poor habitats such as bare rock surfaces, the bark of trees, and unconsolidated alluvium and moronic material on which plants other than lichen and terrestrial algae find it difficult to establish. Some of the bryophytes are epiphllous and grow on the leaves of the living trees and shrubs in the rain forests of the tropics and some humid, warm temperate climates. Fugitive species frequently grow in temporarily available habitats such as patches of bare soil and the muddy edges of lakes and rivers and quickly complete their life cycle. In tropical rain forest, bryophytes are absent on the ground and occur mainly as epiphytes on the trees. The occurrence of a particular species may reflect the microclimate of the locality. Some bryophytes find suitable substrate and microclimate in crevices and caves e.g. *Schistostega pennata*.

Bryophytes have a remarkable ability to survive in the periods of drought, and are able to colonize apparently inhospitable surfaces of rock, sand and gravel in cooler zones of the earth where they form an important part of the vegetation in forest, mires, in oligotrophic and eutrophic conditions and in fresh water, especially in fast-flowing streams. In lowland tropics, bryophytes are scarce or absent in rivers and rocky streams and where a wide range of higher plants species are able to live. The absence of bryophytes in many grassland communities is probably due to competition from higher plants rather than to the physical conditions. Bryophytes are absent in the marine environment. Only a few highly specialized species grow in salt-rich habitats such as *Schistidium maritimum.* Bryophyte life-cycle remains unique as it displays two alternating generations found to be roughly equivalent in importance both highly characteristic in their morphology, yet unlike from one another. The life cycle is far removed from the dominance of the haploid generation found in most green algae. It is also far removed from the overwhelming dominance of the sporophyte generation that occurs throughout pteridophytes. A haploid spore germinates to produce filamentous, ovoid, spherical or plate-like stage which precedes the formation of the leafy shoots or thalloid gametophores. The green gametophores live for few months to several years, and bear antheridia and archegonia. Fusion of motile male gametes from an antheridium and egg in archegonia gives rise to sporophyte generation. The diploid phase develops polarity and differentiates into a basal foot, a stalk-like seta, and a distal spore-producing organ, the capsule. The sporogenous tissue after meiosis leads to the formation of haploid spores. The spores germinate and form the thalloid or filamentous protonema that gives rise to prostrate or erect gametophores.

The importance of bryophytes in modifying their environment and affecting the life of other organisms is quite pronounced, even if it is usually inconspicuous and difficult to assess quantitatively. The vascular plants have characteristically evolved structural adaptations which provide a continuous supply of water, usually from the soil. The adaptation to dry habitats take the form of structural features which minimize water loss and cytoplasm is usually quickly damaged if the water content falls below that at full turgor. On the other hand bryophytes, lichens and some other plants have evolved physiological adaptations which allow suspension of metabolism during periods of drought and its resumption when water is once available. Bryophytes and lower plants show great tolerance to desiccation and dried plants can resume photosynthesis and respiration within minutes or hour after rehydration. Poikylohydric mosses are able to survive complete desiccation for months and even for years without irreversible damage and recover their full activity as soon the water supply is sufficient. In the case of higher plants, the desiccation tolerance is very rare. The use of statistical methods for the structural analysis of natural communities as well as analytical researches into the interrelations between physiological functions of life and environmental factors has made rapid advances in ecology. The correlation between productivity or vitality and environmental factors such as air-temperature, humidity, light intensity and the atmospheric carbon dioxide content are important aspects for the study. The present work deals with the structural modifications in mosses in relation to the environmental factors.

II. GROWTH FORMS

Plants have evolved various adaptive strategies for balancing the benefits and costs of having a high affinity for resources, plasticity of growth allocation and mycorrhizal symbiosis. In bryophytes vegetation, it is difficult to generalize about habitat range due to small size, simple structure and close proximity to the substratum. They are sensitive to minor variations of the habitat, thus enhancing the accuracy with which their distribution responds to micro-habitat factor. The growth in particular is a relatively constant characteristic of some plant species. The general appearance of colonies of mosses is largely governed by the growth forms of their shoots and some species always present a characteristic appearance but others are much more variable in this respect. Using a system of classifying growth forms into types it is possible to demonstrate correlations of certain types and variation in habitat conditions. The moss shoots respond to light, gravity, moisture and other stimuli. They show usual pattern of growth forms modified or transformed by responses to environmental factors. Such studies contribute towards elucidation of the ecological significance of growth form and its influence upon the ecological amplitude of moss species. In bryophytes growth form and ecological amplitude are closely interdependent. The factors of the habitat may select groups of species. They are able to develop certain types of growth-form, and on the other species possessing some degree of plasticity that adopt a particular form under the influence of certain conditions. Gimingham and Robertson (1950) suggested that structural features could be recognized in bryophytic communities and correlated with certain aspects of the habitat. The bryophytes are represented in three of the major life-forms (Gimingham and Birse, 1957). Therophytes; Geophytes and Chamaeophytes.

Therophytes: These are seasonal plants, completing their life cycle in a single favourable season and remain dormant through the unfavourable period of year in the form of seeds.

Geophytes: These are the plants where their buds are completely in the soil as bulbs and rhizomes.

Chameophytes: These are the plants where their buds are situated close to the ground surface. Although the majority of bryophytes fall in the latter group, these distinctions are significant in dealing with habitats.

The general form adopted by colonies of mature plants is determined by shoot morphology and is referred to as "growth-forms" (Du Rietz, 1931). Gimingham and Robertson (1950) described the classification of growth forms based on morphology only and are thus free from any assumptions as to the adaptive significance of structural characteristics which proved a valuable tool in several fields of bryophyte ecology.

As in many higher plants bryophytes are not found as single individuals but in groups of individuals which have characteristic features depending on their family, genus or species. Each individual (moss shoot) has a genetically fixed method of ramification. Assemblage of individuals and growth form, modified by external conditions, together provide the characteristics form called "life forms".

The term "life form" was coined by Warming (1896) and refers to the habit of a plant in harmony with its life conditions i.e. growth form. Giesenhagen (1910) and Herzog (1926) used the term growth form in the sense of life-form. Growth form is defined by Meusel (1935) as the overall character of a plant that can be determined by detailed morphological analysis. The growth-forms are fundamentally established in the organization and are thus a purely morphological term and therefore refer only to homologous structures. To avoid confusion, it is suggested that the growth-form, defined in Meusel's sense considered strictly morphological term based on homologies and life-form suggested by Warming (1896), in the ecological sense based on analogies.

III. CLASSIFICATION OF GROWTH FORMS

Bryophytes with aggregate shoots are termed as gregarious form and those with separated shoots are solitary form. Under different environmental conditions one species may exhibit different types of growth forms, thus a species, an example of one of the classes does not necessarily indicate it invariably develop that particular growth forms. The growth forms are named on the basis of their appearance. Solitary bryophytes are abundant in varied forms in humid tropical forests.

A. Cushion: This is a system with erect shoots radiating in more or less compact dome-shaped groups. The shoot radiates from a central point of origin, branches similar adopting same direction of growth as main shoot. Branching appears from near apex of shoot and adding to the width of the cushion. The acrocarpous mosses of dry sites and some pleurocarpous mosses such as *Orthotrichurn, Leskea, Hypnum* and *Brachythecium* species form cushions. They mainly grow on rocks and bark of trees (photo-epilithic and photo-epiphytic) and also on the ground in extreme conditions (Figs. 1, 2, 23).

(i) Large Cushion: The cushion normally attains a diameter of over 5 cm as found in *Leucobryum glaucum, Schistochila, Andreaea, Arnphidiurn, Orthotrichum.*

(ii) Small Cushions: They normally attain diameter less than 5 cm such as *Grimmia, Ulota, Plagiopus, Octoblepharum, Calymperes* and *Syrrhopodon*.

B. Turfs: System with the individual parallel upright shoot forms the pile of shoots. The main shoot is erect and the direction of growth is vertical. The branches also grow upwards parallel to the main shoots (Fig. 4).

a. Tall Turfs: The upright shoots are generally unbranched and form turfs of considerable height (>5cm). These forms especially grow on temperate forest floors. They are able to hold water and conduct it through capillary system between the shoots (Fig. 3). Some tall turf taxa are *Polytrichum, Dawsonia, Sphagnum, Timmia, Dicranum, Drepanocladus* and *Rhytidiadelphus.*

(i) Tall turfs with erect branches: The shoots are normally over 2 cm high, simple or branched with vertical growth renewed by lateral innovations (Fig. 24) e.g. *Dicranum malus, Polytrichum commune, Tortula ruraliformis, Aulacomnium palustra, Breutelia chrysocoma.*

(ii) Tall turfs with divergent branches: The divergent branches are of limited growth. The shoots are normally over 2 cm high with short laterals scattered or whorled branches. *Campylium stellatum,* and *Sphagnum* sp are examples of this class (Fig. 5).

b. Short Turfs: The short shoots are less than 2 cm tall, stand close together and grow on after ripening of the sporogonia by means of regenerative shoots. Often very spreading turfs are formed and remained for more than a few years preferably growing on mineral soils and on rocks (Figs. 6, 25). *Trichostomum, Didymodon, Barbula, Ceratodon, Gymnomitrium, Bryum argentium,* form shorts turfs.

c. Open Turfs: The shoot is short, distinctly separate from one another, often produced by persistent protonema or rhizoid strand. Open turf mosses are *Diphyscium foliosum, Polytrichum aloide,* and *Trematodon* sp.

C. Canopy Form or Dendroid: Dendroid systems produce a raised leafy canopy and give a miniature tree-like habit. The erect portion of the shoot is unbranched below with scale leaves and normal leaves usually abundant on branches at the tip forming canopy.

(i) Unbranched Dendroid: They are terrestrial forms in which the leafy branches are present only at the tips of main shoot (Fig. 7) such as *Dawsonia, Pogonatum,* and *Climacium.*

(ii) Branched Dendroid: They may be epiphytic or terrestrial have frequently branched, erect, leafy shoot (Fig. 8) e.g. *Hypnodendron, Rhodobryum dendroid* and *Plagiomnium undulatum.*

D. Mats: The shoot system forms a dense and interwoven mat, extending horizontally over the substratum. The main shoots creep over substratum often with rhizoids. There are abundance of lateral branches with limited growth (Fig. 10). They grow on rocks and bark of trees and also on leaves (epiphyllous) in tropics. The mat forms exist in *Brachythecium, Plagiothecium, Homalothecium, Taxithelium* and *Lophocolea.*

(i) Rough Mats: The main shoots adhere to substratum with abundant short erect laterals and the branches are in various directions (Figs. 12, 26). The mat can be readily stripped from substratum e.g. *Brachythecium rivulare, B. albicans, Eurhynchium striatum, Ectropothecium, Sematophyllum* and *Trachypodopsis.*

(ii) Smooth Mats: The branches are closely interwoven, lying chiefly in the same plane as main shoot (Figs. 11, 28). The smooth mat taxa are *Frullania tamarisci, Hypnum cupressiformis, Callicostella, Isopterygium, Radula,* many Hookeriaceae and Lejeuneaceae.

(iii) Thread Forms: The delicate, creeping irregularly and sparingly branched shoots form open mats (Figs. 13, 29). This form is adopted by many species when occurs sparingly among other vegetation. The thread form is displayed by small Lejeuneaceae, including many epiphyllous species, *Eurhynchium praelongium, Hypnum cupressiforme* var. *ericetorum, Lophocolea bidentata.*

(iv) Thalloid Mats: The thalloid taxa form the overlapping dorsiventral thalli (Figs. 14, 27) as found in *Conocephalum conicum, Pellia epiphylla, Symphogynae, Dumortiera,* Marchantiaceae and Metzgeriaceae.

E. Wefts: Systems developed as a result of the loose inter-twining of straggling shoots and branches often ascending and luxuriant. The shoots are long straggling and generally robust and the rhizoids are usually sparse. The lateral branches are often arching or ascending. They occur on forest floor, rotten trunks or trees. A new layer is formed growing every year over previous year shoots (Figs. 15, 30). Weft forming bryophytes are many Hypnaeceae, Brachytheciaceae, Entodontaceae, *Hylocomiace, Thuidium, Rhacopilum, Bazzania, Ptilidium, Trichocolea* and *Lepidozia.*

F. Pendants: The primary stems creep more or less horizontally along twigs to which they are attached by rhizoids. These stems give off secondary stems at intervals which have short tertiary branches of limited growth. The secondary shoots hang down from the branches and twigs of trees like a beard and laterals shoots remain short and stand out horizontally (Figs. 16, 31). Some taxa of pendant forms are Meteriaceae. Phyllogoniaceae, *Meteorium, Meteoriopsis, Cryptoleptodon, Pilotrichopsis, Calyptothecium* and *Bryopteris flaccida.*

G. Tails: These bryophytes are mostly shade loving and of creeping habit and generally grow on the vertical substratum like trunk and rocks. The lateral shoots stands out from the creeping main shoot slightly branched and usually radially leafed (Fig. 17). Tail forms are founds in *Leucodon, Cyathophor Lamprophyllum, Endotrichella schistochila, Tylimanthus saccatus* and *Plagiochila.*

H. Fans: Some creeping mosses growing on a vertical base (trees trunks, rocks) throw branches towards one side in the same plane, project horizontally to obliquely downwards and usually have flattened leaves (Fig. 9). Fan forms are displayed by *Thamnobryum, Ethinodontium, Bryopteris, Archilejeunea thysananthus, Plagioghila* and Neckeraceae.

I. Feather: The secondary stems are more or less pinnately branched and are placed dorsiventrally (Figs. 18, 19, 32). They are epiphytic or saxicolous e.g. *Hypopterygium, Neckeropsis, Callicostabipinnata* and *Bryopteris.*

J. Annuals Forms: The gametophytes stops growing once it has produced gametangia and dies after the production of sporogonium without any production of regenerative shoots (Figs. 20, 33). They are the pioneer bryophytes occur on open mineral soils. The taxa of this category are *Buxbaumia, Phascum, Sphaerocarpos, Calobryum, Riccia* and *Physcomitrium.*

K. Protonemal forms: These taxa have the persistent protonema and leafy shoots are minute and scattered (Figs. 21, 34) found in *Ephemeropsis, Micromitrium, Protocephalozia* and *Schistostegea.*

L. Bracket form: These mosses display horizontal shoots (Fig. 22) and grow mainly as epiphytes e.g. *Leiomela, Spiridens.*

M. Moss Balls (kugel-moose): These are a special form of cushion in Iceland are called joklamys (glacier mice) in the tundra region. The stone contains mosses rolled by winds and the shoots keep on growing from all sides giving appearance of a ball (Fig 35).

Sometimes one and same type can display two different life forms. For example in *Plagiomnium undulatum* the sterile shoots are fans while the fertile shoots are dendroids. *Bazzania trilobata* forms tall turfs or wefts according to environmental conditions. The mosses found in ponds and lakes are mostly found on land and do not change on transfer to water or land as found in *Sphagnum, Drepanoclaus, Calliergon* etc. The life forms of bryophytes are closely connected with light and water. High light intensity inhibits the elongation of the axes. The predominance of short turfs and cushions on brightly lit biotopes is directly connected with this fact. The tails and fans however require a dim light. The influence of environment is reflected in the representation and prevalence of various growth-form types both in the field and under experimental conditions. Some appear to be little modified by varying the intensity of certain habitat factors others have different ranges of plasticity. But light and moisture have a considerable influence in the development of some types of growth form.

IV. DISTRIBUTION OF GROWTH FORMS IN RELATION TO CLIMATE AND SOIL MOISTURE

Growth from spectra of various bryophyte communities are clearly related to physical factors of their environment especially light and humidity. The substratum influences the growth forms. The mats and dendroid types have the tendency to appear on rock while tall turfs with erect branches, thread forms and wefts are generally associated with soil. The exposed conditions bring small cushions and short turfs into prominence while shade and humidity flavour dendroid types, thalloid mats and rhizoidal wefts with tall turfs and mats intermediate. The rough mats have lesser tolerance of low saturation deficits in comparison to smooth mats. Some forms are relatively constant in their expression others show plastic modification thus conferring wider ecological amplitude upon the species concerned by providing for adaptability of specific form in different environment. The gradients in habitat conditions and changes in the

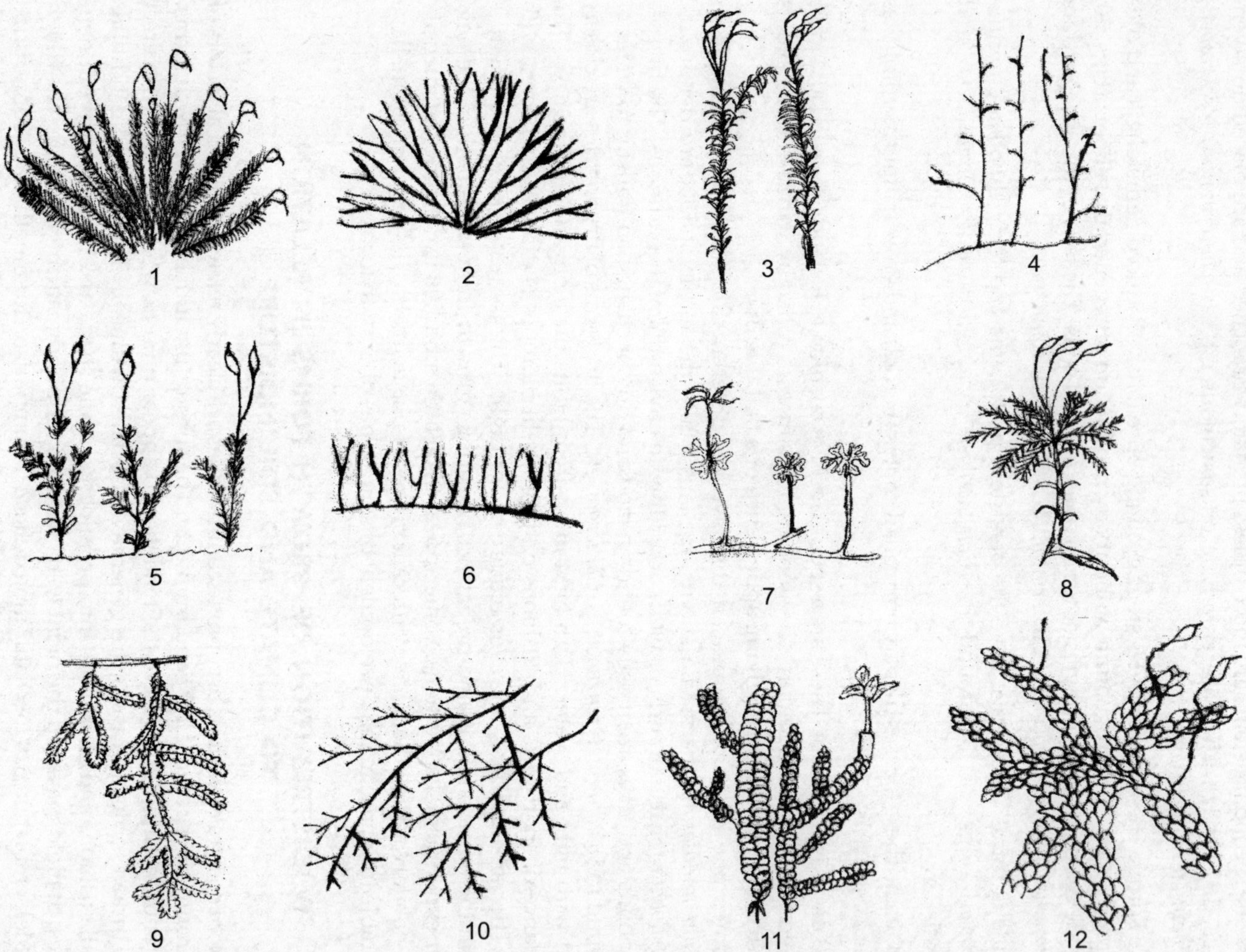

Figs. 1. *Leucobryum*, Cushion with radiating leafy shoots, 2. Line diagram of cushion 3. *Polytrichum,* tall turf with erect parallel shoots, 4. Line diagram of turf, 5. *Sphagnum,* Turf with divergent branches, 6. Line diagram of short turf, 7. Dendroid form in *Frullania*, 8. *Hypnodendron* showing dendroid form, 9. *Neckeropsis,* in fan form, 10. Line diagram of mat form, 11. Smooth mat of *Radula*, 12. *Hookeria* showing rough mat

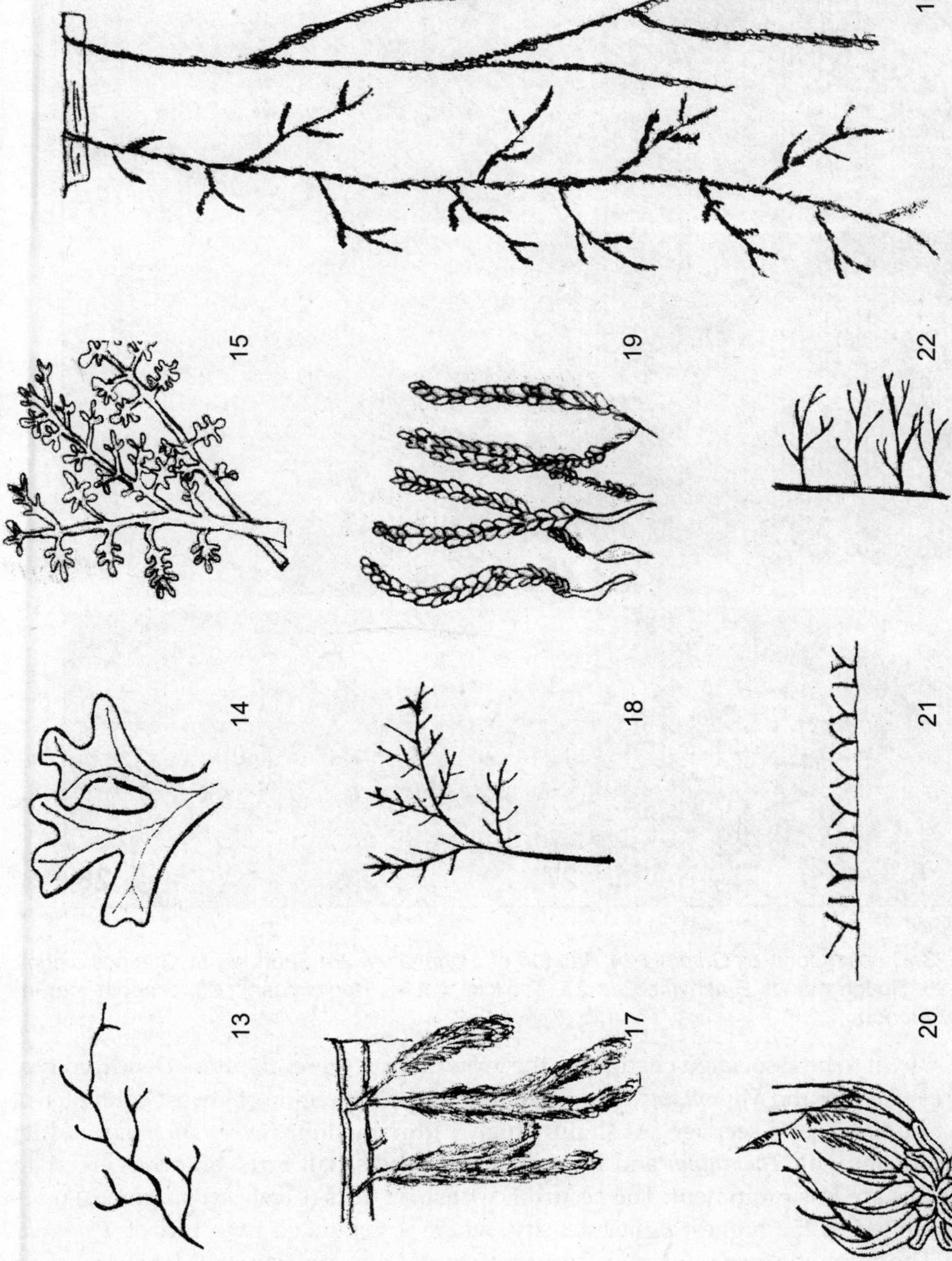

Figs. 13. Thread form in *Isopterygium* 14. Thalloid mat in *Dumortiera* 15. Weft form in *Trichocolea* 16. *Meteorium* showing pendent form 17. Tail form in *Prionodon* 18. Line diagram of feather form 19. Feather form in *Gymnomtrion* 20. *Buxbaumia* showing annual form 21. Line diagram of protonemal forms 22. Bracket form, line diagram

Figs. 23. Cushion form of *Grimmia* 24. Tall turf of *Pogonatum* 25. Short turf in *Gymnostomum* 26. Rough mat in *Brachythecium* 27. Thalloid mat in *Plagiochasma* 28. Smooth mat in *Entodon*

habitat are shown to be associated changes in the growth form representation. Dendroid types (*Thamnium alopecurum* and *Mnium undulatum)* are frequent in the region of lowest light intensity and highest humidity and seepage. At slightly higher illumination, *Isothecium myurum* (dendroid) is dominant with *Thamnium* and *Plagiochila asplenioides* (tall turfs, branches erect) and dendroid forms are less prominent. The occurrence of short turfs (*Ceratodon purpureus*) on the wall top is limited in the highest light intensity, where it is flanked by a turf of *Plagiochila asplenioides*. The epiphytic forms are characteristic of humid montane forests. Meteoriaceae and other hanging forms are often carried about by birds for nest building. Fragmentation is a common method of dispersal in these taxa. Solitary forms especially pendulous mosses are

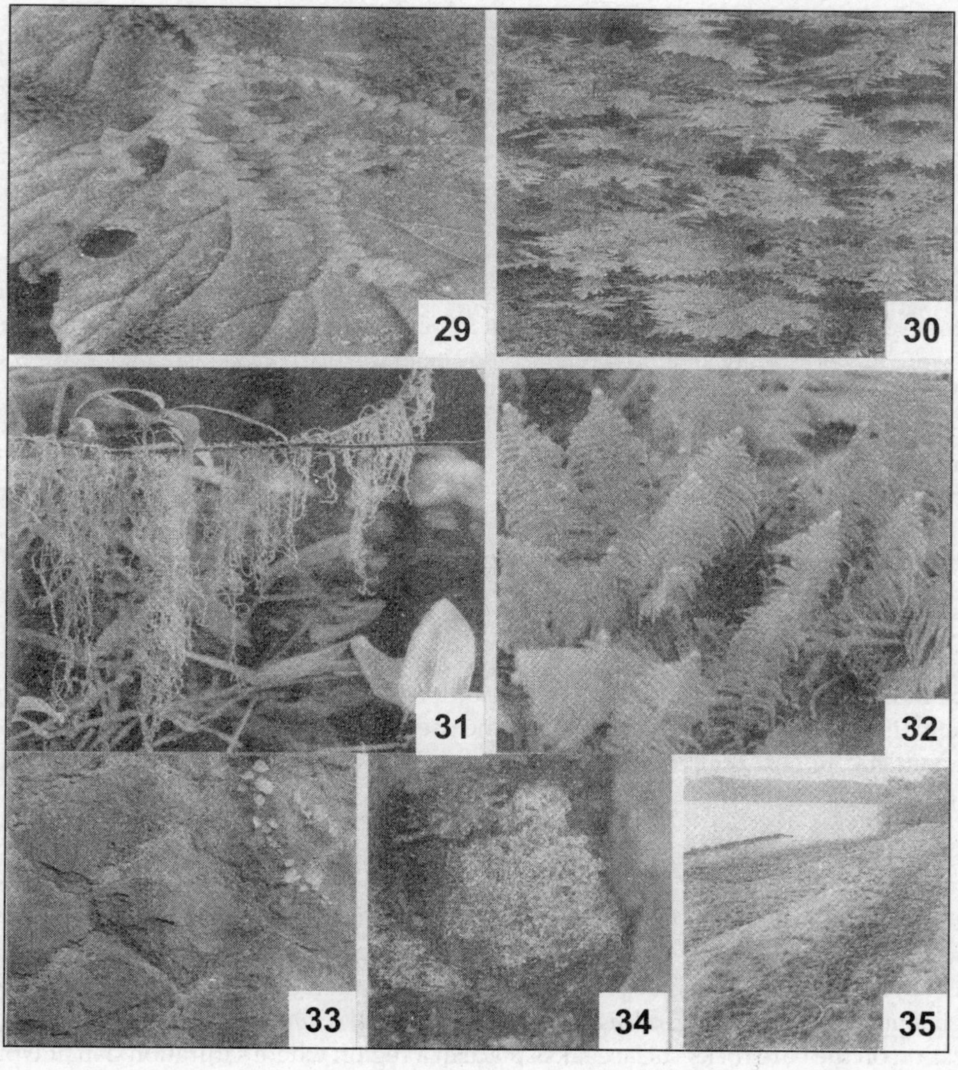

Figs. 29. Thread from of *Cololejeunea* 30. *Thuidium* weft form 31. *Barbella,* pendent form 32. Feather form of *Ptilium* 33. Annual form of *Physcomitrium* 34. Protonemal form of *Schistostega* 35. *Drapanocladus,* moss balls

more exposed to desiccation and extreme of temperature than gregarious bryophytes so they are more common in tropical forests. Pendulous forms are more drought resistant than the terrestrial species but less tolerant than the xerophytic taxa.

V. VARIATIONS IN GROWTH FORMS

The growth form representation of mosses in the ground-stratum reflects the conditions of illumination. The growth forms are found to vary in relation to moisture supply and to the physical nature of the substratum. When water drips constantly on shoots, the primary ascending shoots may be of considerable length and the short laterals are scarcely branched. The shoots hang downwards in the direction of the run-off water, which drips continuously from its apex. Where shoots are periodically subject to spray, they are not all distinctly dendroid and stoloniferous branches may be produced by monopodial or sympodial growth of short laterals from old shoots so that these shoots become more straggling. Plants growing on steep rock faces in summer are dry. The branching in the apical region is very diffuse. The new ascending shoot develops from the stolons. On horizontal rock-surfaces, a markedly dendroid form occurs in the vertical light intensity and in high humidity. Lateral branches grow beyond the apex of the main stem of the ascending shoot. The general appearance is similar to that of lateral branch and they are never radially arranged. The stolons extend from rock on to a clay soil beneath a steep clay bank. The colonizing shoots form merely the rough mat form.

It appears that the primitive bryophytes were erect, radially symmetrical and leafy. The growth is related to a more or less distinct and characteristic range of environmental condition. The range of plasticity varies between growth forms in the dendroid form. The two dendroid forms, for example, *Climacium dendroides* and *Thamnium alopecurum* have a relatively well-defined morphology and relatively narrow ecological amplitude. The alternating stoloniferous and erect growth in *Climacium* may be associated with seasonal fluctuations in the moisture conditions of the habitat. The capacity for plastic variation in these species may be correlated with certain aspects of their ecological amplitude. The bryophyte layer of a forest has been a sensitive indicator of certain environmental conditions (Cantlon, 953). The vegetative response to climate differences shows up well through the use of growth-form spectrum of bryophytes. Gimingham and Robertson (1950) found that bryophyte communities vary with the physical nature of the substratum.

VI. EFFECTS OF HABITAT CHANGE ON THE GROWTH FORMS

When the trees of the ravine or in the forest fall, the conditions formed by the trees may not exist for long, so significant changes takes place in the vertical zonation in the felled area. The alteration in habitat conditions provides the active change in the community structure because of the change in the moisture and light intensity. *Brachythecium rivulare* which showed wide colonization on the bare rocks surface takes place in a region where saturation deficit would be greatest and the growth forms taking part are short turfs (*Grimmia aporcarpa*) and small cushions (*Orthotrichum anomalum*). The belt of *Brachythecium rivulare* is invaded by pioneer species such as members of Pottiaceae (short turf) while below this *Brachythecium rivulares* extends its area downwards overgrowing some of the dead shoots of *Thamnium.* It is evident that when there is a reduction in humidity and shade changes occur in the bryophytic communities the growth forms are shown to be associated with these conditions.

VII. GROWTH FORM PLASTICITY IN RELATION TO CONTROLLED LIGHT INTENSITY AND RELATIVE HUMIDITY

Moisture supply and illumination are the most important factors influencing the development and distribution of bryophytic growth forms. In the high humidity, irrespective of light intensity, the shoots of most mosses become erect or nearly erect. There are exceptions like

Brachythecium rivulare (rough mat), *Hypnum cupressiforme* (thread-like forms) on bark and *Thamnium alopecurum* (dendroids) which produce some growth along the substratum. Shoot growth tends to be vertical, when illumination is fairly even all round. In low humidities which are coupled with light intensity led to frequent production of prostate shoots in mat forms, to a lesser extent in wefts and dendroid types. In low light intensities the regeneration is generally poor, branching is weak and the leaves are small in size. The growth forms, which uniformly produce erect growth, are small cushions, small turfs and tall turfs with erect branches. The species are more plastic in their behaviour; tend to produce erect shoots in high humidity and prostrate or ascending shoots in low humidity with high light intensity. It is associated with a greater amount of regenerative and more richly branched shoots, especially with the wefts and thread-like forms. Mat forms show greater tendency to produce prostrate shoots adhering by rhizoids. Wefts show a smaller capacity to respond in this way. *Thuidium tamariscinum,* however, readily adopts a prostrate form and this type of growth is periodically prostrate in colonizing shoots. Tamm (1953) showed that there is a relationship between the yield of *Hylocomnium splendens* and the light supply where mosses are grown beneath spruce. There is a correlation between the contribution of certain growth-forms to bryophytic communities and the relative intensity of illumination (Gimingham and Robertson, 1950). The degree and arrangement of branching may be modified by the direction and intensity of light supply so in the natural habitat, there is both direct modification of growth morphology and selection of growth form type by the light factor. *Rhytidiadelphus triquetus* branches strongly at irregular interval due to variable light factor. In *Thuidium tamariscinum* the main shoots are arching and ascending, with laterals closely and regularly pinnate while *Eurhynchium praelongum* approaching the weft form tends to be regularly pinnate type.

The continuous moisture condition throughout the year favours the ascending growth in mosses. The growth of robust weft species requires long duration of moderately high light supply. So long as the moisture conditions are suitable the slender thread-like forms grow favourably in low illumination. Certain growth forms of *Thuidium tamariscinum* show resemblance to *Eurhynchium praelongum* particularly in the tendency towards creeping habit production of rhizoids along the stem and pinnate branching.

VIII. GROWTH FORMS IN RELATION TO GROUNDWATER SUPPLY

The bryophyte community shows the predominating influence of groundwater supply in determining the growth form aspect of community structure. The groundwater supply appears to affect the community structure chiefly by its duration and amount, while the floristic composition may be altered by liability to submergence, surface soil pH and humus content or other factors of the habitat. There is differing and characteristic requirements of each growth form and achieves dominance where it find suitable conditions. There is overlapping in the ecological amplitude of growth-forms where the moisture conditions fluctuate varies markedly and a mixed open community is found. The distribution of moss growth forms is correlated with the groundwater supply. Tall turfs with branches erect, short turfs, rough and smooth mats are found in the habitats where the plants are away from the watertable. Wefts and dendroid forms are luxuriant where the surface sand receives an additional supply of water. Tall turfs with divergent branches of limited growth and tall turfs (rhizoidal) are found and in slightly drier localities. Wefts are found in fairly freely drained areas of intermediate moisture.

The distribution of species can be related to the variations in other habitat factors such as surface soil pH, humus content and submergence. *Tortula ruraliformis, Mnium punctatum, Rhacomitrium canescens* and *Barbula cylindrica* are found abundant in the pH range of 5.22 and 6.22. The chief species on the sand with little humus are *Hypnum cupressiformis var lacunosum, Tortula ruraliformis* and *Rhacomitrium canescens. Ceratodon purpureu), Campylium stellatum, Bryum pseudotriquetrum, Fissidens adianthoides, Climatium dendroid* are liable to long period of complete submergence. These species grow in a fairly healthy condition. *Ceratodon dendroids* has a wide tolerance of soil moisture conditions and of submergence. Wefts and tall turfs with erect branches are adversely affected by submergence or high watertables for long periods.

The changes in the moss moisture, at any given site are related to the changing soil moisture, both being determined largely by the watertable. With a constant groundwater supply, a variety of growth forms remain more or less flourishing, especially tall turf(rhizoidal), tall turfs with divergent limited branches, dendroids and others. Whereas in the absence of a groundwater supply, short turfs, rough mats and dendroid form (*Climacium dendroids*) remain prominent.

The morphological features together with the overlap of amplitude of *Thuidium tamariscinium* and thread-like mats of *Eurhynchium praelongum,* which is found in the habitats of similar moisture conditions but of lower light intensity, appears to be rhizoidal weft in the growth-form classification. There may be a lower tolerance of drought with that of *Rhytidialdelphus triquertus,* and greater dependence on the moisture of the substratum. In summer where the intensity of light is below the field-stratum like between 5% and 13%, weft communities are well developed but where the value is 1% or less, slender thread-like forms predominate. The field-stratum by its shading effects is largely responsible for the micro-climatic variations, which determine the prevalence of one or other species of related growth form types. Thus, in the damp deciduous woodland, the wefts form is vigorous and well represented, when the favourable moisture conditions are coupled with relatively little restriction of illumination.

IX. DENDROID GROWTH FORM IN RELATION TO HABITAT CONDITIONS

The growth form in some cases is a relatively invariable specific feature, while in others it is subject to variation under the prevailing conditions of the habitat. *Climacium dendroides* and *Thamnium alopecurum* normally occur in different habitats. In these taxa, initially the growth of side branches begins to exceed and become inactive after establishing an outline resembling a canopy of a tree. Then stolons appear at the base of the trunks of young erect shoots. The tips of the stolons turn up from the substratum. The young ascending shoots become inactive or dead at the tips with the dry weather and growth being renewed in the damper region at the base of the shoots. The young stolons at first are creeping and become ascending with the onset of wetter, cooler weather. The branching of the erect shoots seems to begin where they emerge from the dense shade of the older shoots, so that the new trunks tend to be slightly taller than the older ones.

In *Climacium* a stolon with abundant rhizoid continued to grow along the surface of the stones. Meuse[(1935) observed that once the stolon has turned upwards and the shoot is orthotropic, it responds to the stimuli of light and gravity. The regenerative form 'strong shoots' i.e. portion of the stolons or upright main stems, always give erect shoots when allowing weak lateral branches to regenerate. Fitting (1935) drew attention to the influence of the stimuli of contact with a substratum on bryophytes. The plagiotropic growth of stolons is connected in some way with the proximity of a moist substratum. The upward curve and orthotropic habit with the usual responses to gravity and light are rendered by changes in atmosphere humidity in the direction of saturation. The growth form of *Climacium* bears a relationship to habitat conditions characterized by seasonal fluctuation in the moisture regime.

REFERENCES

Birse, E.M. 1957. Ecological studies on growth form in bryophytes.II. Experimental studies on growth form in mosses. J. Ecol. 45:721-733.

Birse, E.M. I 958. Ecological studies on growth-form in bryophytes. III. The relationships between the growth-form of mosses and ground water supply. J.Ecol. 46:9-27.

Birse,. E.M 1958. Ecological studies on growth-form in bryophytes. IV. Growth-form distribution in deciduous wood. J. Ecol. 46:29-42.

Cantlon, J.E. 1953. Vegetation and microclimate on north and south slopes of Cushetunk Mountain, New Jersey. Ecol. Monog. 23:241-270

Do Reitz, G.E. 1931 Life-forms of terrestrial flowering plants. Acta Phytogeogr.suec.3:1.

Dyer, A.F. & J. G. Duckett, 1984. The Experimental Biology of Bryophytes. Academic Press. London.

Fitting, H. 1935 Untersuchungen uber die Induktion der Dorsiventralitat bei den keimenden Brutkorpen von Marchantia und Lunularia.I. Die 1ndukton und ihre Wirkungen. Jb. wiss.Bot., 82:333.

Giesenhagen, K. 1910. Moostypen der Regenwaider. Ann. Jard. bot. Buitenzorg Suppl. 3:711-790.

Gimmingham, C.H. & E.M. Birse 1957. Ecological studies on growth form in bryophytes. I. Correlations between growth form and habitat. J. Ecol. 45:533-545.

Gimingham, C.H. & E. T. Robertson. 1950. Preliminary investigation on the structure of bryophytic communities. Trans.Brit.Bryol.Soc.1 :330-344.

Hebant, C. 1977 .The conducting tissues of bryophytes. Vaduz. J. Cramer .

Herzog. T. 1926. Geographie der Moose I –xi, 1-440. G Fischer Jena.

Meosel, H. 1935. Wuchsformen und wuchstypen der Eurpaischen laubmoosee Nova Act Leopold N.F.3:123- 277.

Richards, P.W. 1984. The ecology of tropical forest bryophytes. *In* R.M.Schuster (eds) New Manual of Bryology VoI.II. Hattori Botanical Lab. Nichinan, Japan.

Sainsbory, G.O.K. 1955 ..A Handbook of the New Zealand Mosses. W ellington.N. W

Seim, A.L., M.F. Boell, & R.L. Evans. 1955. Bryophytes growth forms and cover in a Jack pine stand Itasca Park, Minnesota. Bryoloist 58:526-529.

Seimmorray, E.I.,, M. F. Buell &R. I. Evans, 1955. Bryophytes growth forms and cover in a Jack pine stand Itasca Park Minnesota. Bryologist 58:327-329.

Smith, A.J.E. 1982. Bryophyte Ecology. (ed). Chapman & Hall. New York.

Tamm, C.O. 1953. Growth, yield and nutrition incarpets of a forest moss (*Hylocomium 5plendense*). Meddel. frdnst. Skogsforskn. Stockholm. 43: 1-140.

Warming, E. 1896. Lehrbuchder okologischen.Pflanzen geographie. Borntraeger, Berlin.

10

An Overview of the Forest Types and Associated Spermatophytic Vegetation of Chamba District (Himachal Pradesh)

M. SHARMA and HARINDER SINGH
Department of Botany, Punjabi University, Patiala, Punjab

ABSTRACT

In all 998 species of spermatophytes belonging to 554 genera and distributed among 132 families have been recorded from Chamba District, HP. The area belongs to N- W Himalayan biogeographic zone. Three main categories of the forest types are recognized, namely, Northern tropical dry deciduous forests, Montane subtropical forests and Montane temperate forests. The Northern tropical dry deciduous forests, characterized by degraded scrub, are distributed between 600-1000 m altitude. *Albizia chinensis, Bombax ceiba, Dalbergia sissoo Phoenix sylvestris, Cassia fistula, Sapindus mukorossi* and *Toona ciliata* are some common trees here. The montane subtropical forests occur between 800-1000 m and are a mixture of temperate and tropical species. The gymnosperms make their appearance through *Pinus roxburghii.* Plant like *Acacia catechu* extends its range from the lower tropical zone. At higher elevation temperate species like *Aesculus indica* and *Rhododendron arboreum* appear. On relatively drier and exposed slopes *Euphorbia royleana* is seen growing individually or in consociations. The montane temperate forests are characterized by the development of large-sized conifers like *Abies pindrow, Cedrus deodara, Picea smithiana* and *Pinus Wallichiana* as dominant species or the constituents of mixed coniferous forests. On southern slopes, the conifers are replaced by oaks (*Quercus leucotrichophora*). Some deciduous species like *Acer cappadocicum, Aesculus indica* and *Celtis australis* are also found distributed among the evergreen forests. A wide variety of herbs, shrubs and climbers are found in all these forest types and have been listed at the proper places. However, it is interesting to note that some herbs are found in all the forest types. Common ones among these are *Arthraxon prionodes, Campanula colorata, Carex filicina, Cerastium glomeratum, Chenopodium album, Cynoglossum zeylanicum, Eragrostis nigra, Medicago lupulina, Poa annua, Sonchus asper, Stellaria media, Taraxacum officinale, Torilis japonica* and *Youngia japonica*.

Key words: Forest types, Chamba district, Himachal Pradesh.

I. INTRODUCTION

Plant collections were made from Chamba district of Himachal Pradesh (N-W Himalayas) for four years (1984–1988). The investigations have brought out the introduction in the Indian flora of a South American plant species *Gnaphalium coarctatum* (Singh & Sharma 1988). Further, in some earlier communications; studies in phytogeography (Singh & Sharma 1989, Sharma & Singh 1996), floristic composition (Sharma & Singh 1990), biological spectrum (Sharma & Singh 1995) and herbaceous vegetation (Sharma & Singh 1997) of Chamba district (excluding Pangi valley) have been published. In continuation, the present paper deals with the forest types of the area.

II. PHYSIOGRAPHY

Chamba district lies between lat, 32° 10′ and 33° 13′ N and long. 74° 45′ and 77° 33′ E. The altitude of study area ranges between 600-2700 m. This hilly and mountainous district is traversed by Chenab and Ravi rivers. Besides, tributaries of rivers Beas also meander through the area. These rivers and streams pass through the valleys named after them. Geologically, in the district the outer or sub-Himalayan zone and the lower Himalayan zone constitute two broad stratigraphical zones. While the former consists predominantly of tertiary formation comprising great thickness of detrital rocks clays and conglomerates; the latter has granites and other crystalline rocks of unfossiliferous sediments in most part of it.

The climate of the area is intensely cold at higher elevations but becomes moderate to hot and dry with the decrease in altitude. At higher altitudes winter months are extremely cold with the temperature touching as low as minus 15°C while summer months are pleasant. At lower elevations, however, summer is harsh and almost dry with maximum temperature reaching to 44°C and winter is quite cold with minimum temperature sometimes dipping to below freezing point. During winter months (November-mid-March) snowfall occurs above 1800 m altitude. The area receives rainfall during July-September. The average annual rainfall varies from 40 cm in Bharmour to 260 cm in Chamba tehsil.

III. MATERIALS AND METHODS

Voucher specimens were gathered from the area after extensive and intensive field trips undertaken in different seasons of the year for four years (1984–88). Usual taxonomic and systematic procedures were followed for identification and confirmation of taxa. The specimens are deposited in the herbarium of Punjabi University, Patiala (Punjab). In all 998 species (992 angiosperms 6 gymnosperms) have been recorded distributed among 554 genera and 132 families of seed plants.

IV. OBSERVATIONS AND DISCUSSION

The West Himalayan biogeographic zone, according to Rao (1994), consists of two separate phytogeographical regions, namely, the Western Himalaya (covering Garhwal and Kumaon) and comparatively drier N-W Himalayas comprising N-W part of Jammu and Kashmir and the entire state of Himachal Pradesh. He has divided the vegetation of this region into categories like tropical (below 1000 m), subtropical (1000–2000 m), temperate (1800–3500 m), subalpine (3500–4000 m) and alpine (4000–5000 m). The mountains here exhibit a wide distributional range and local conditions vary according to the slopes, climate and edaphic situations. This may result in some sort of overlapping in the vegetation types. Since the area under discussion ranges up to 2700 m only, hence it is devoid of subalpine and alpine vegetation. The classification given by Champion and Seth (1968) has been followed here for describing various forest types. However, excessive influence of biotic interference and consequent degradation of vegetation make it rather difficult sometimes to stick strictly to this classification, particularly to the subtypes. Following three main categories of the forest types are recognized:

I. Northern tropical dry deciduous forests.
II. Montane subtropical forests.
III. Montane temperate forests.

Northern Tropical Dry Deciduous Forests

These forests happen to be between 600–1000 m altitude and belong to the northern dry mixed deciduous subtypes. The vegetation is degraded and represented by scrub. The trees and shrubs are deprivated and scattered. The first storey is thin. The trees of the uppermost canopy are *Albizia chinensis, Bombax ceiba, Dalbergia sissoo* and *Phoenix sylvestris.* The middle storey includes the middle-sized trees like *Cassia fistula, Sapindus mukorossi* and *Toona ciliata.* The third storey is constituted by some small-sized trees and medium- to large-sized shrubs. These include *Acacia catechu, Adhatoda zeylanica, Boehmeria macrophylla, Carissa opaca, Colebrookea oppositifoilia, Dodonaea viscosa, Ficus palmata, Flacourtia indica, Jasminum humile, Mallotus philippensis, Murraya koenigii, Oreocnide frutescens, Ougeinia oojeinensis, Wendlandia exserta, Woodfordia fruticosa* and *Zanthoxylum armatum.* At places the area becomes impenetrable because of spiny shrubs like *Lantana camara* and *Mimosa himalayana* which may cover considerable area or may be present in small patches here and there. Among the noteworthy climbers, mention may be made of *Bauhinia vahlii, Cissampelos pareira* var. *hirsuta, Clematis grata., Helinus lanceolatus* and *Pueraria tuberosa.*

The rainy season brings a large number of herbs which grow as undergrowth in these forests. Common ones are *Ageratum conyzoides* subsp. *houstonianum, Anisomeles indica, Artemisia capillaries, Cassia occidentalis, C. tora, Commelina benghalensis, Cyperus cyperoides, Corchorus aestuans., Crotalaria bialata C. prostrata., Eclipta prostrata, Erigeron bonariensis, Indigofera linnaei, Ipomoea eriocarpa, Lindernia crustacea, Polygala arvensis, Portulaca oleracea, pouzolzia hirta, P. zeylanica, Rhynchosia minima, Sida cordata, Sigesbeckia orientalis., Solanum nigrum* and *Tridax procumbens.* The common grasses comprise *Apluda mutica* subsp. *aristata, Dichanthium annulatum, Digitaria ciliaris, D. Stricta Eragrostis unioloides, Heteropogon contortus, Neyraudia arundinacea, Setaria pumila* and *Sporobolus fertilis.*

The common spring season herbs noticed as forest undergrowth are *Amaranthus viridis, Anagallis arvensis, Antirrhinum orontium, Arabidopsis thaliana, Atylosia scarabaeoides, Campanula benthamii, Crotalaria medicaginea, Cyperus niveus, Lepidagathis incurva, Lindenbergia macrostachya, Nepeta hindostana* and *Tridax procumbens.*

Montane Subtropical Forests

Distributed between 800-1800 m, these forests are a mixture of tropical and temperate species. The two categories into which this group can be divided in Chamba district are (a) subtropical pine forests and (b) Himalayan subtropical scrub.

Subtropical pine forests

This class is again split into (i) Lower or Shivalik chir pine forests and (ii) Upper or Himalayan chir pine forests.

(i) Lower or Shivalik Chir Pine Forests

Occurring between 800-1200 m, these forests exhibit Chir pine (*Pinus roxburghii*) as the dominant species. The medium-sized trees like *Acer oblongum, Albizia chinensis, Cassia fistula* and *Grewia disperma* constitute the second storey. A mixture of low-sized tress, tall shrubs and climbers characterize the third storey. Noteworthy among these are *Acacia catechu, Asparagus adscendens, Berberis lycium, Boehmeria macrophylla, Carissa opaca, Clematis grata, Colebrookea oppositifolia, Cryptolepis buchananii, Deeringia amaranthoides, Dodonaea viscosa, Ficus palmata, Helinus lancelolatus, Murraya koenigii, Myrsine africana, Pyrus pashia, Rhus chinensis, Rubus ellipticus, Solanum erianthum, Symplocos paniculata, Wendlandia exserta, Woodfordia fruticosa* and *Zanthoxylum armatum.*

Most of the herbs listed under Northern tropical dry deciduous forests are also found here as undergrowth in forests.

(ii) Upper or Himalayan Chir Pine Forests

In the area, such forests occur from 1000-1800 m altitude. The chir pine forms the top canopy and is associated with the middle-sized trees of the second storey. Prominent among these are *Aesculus indica, Celtis australis* and *Rhododendron arboreum*. The third storey is constituted by the low-sized trees and a large number of shrubs. The important ones among these include *Berberis lycium, Coriaria nepalensis, Desmodium elegans, Deutzia staminea., Grewia disperma., Indigofera atropurpurea, Inula cappa, Leptodermis lanceolata, Lyonia ovelifolia, Myrsine africana, Olea ferruginea, Pyrus pashia, Randia tetrasperma, Rhus chinensis, Rosa Macrophylla, Rubus ellipticus, R. niveus, Symplocos paniculata* and *Zanthoxylum armatum.*

The common ground flora during monsoon season includes *Acalypha ciliata, Achyranthes bidentata, Anisomeles indica, Aristida cyanantha, Artemisia* capillaris, *Barleria cristata, Bidens tripartita, Bupleurum lanceolatum, Cassia mimosoides, C. pumila, Chenopodium ambrosioides, Commelina benghalensis, C. paludosa, Conyza stricta, Cyperus cyperoides, Desmodium microphyllum, Dichanthium, annulatum, Digitaria stricta., Echinops cornigerous, Geranium nepalense, Heteropogon contortus, Pimpinella diversifolia, Plectranthus lophanthoides, Pouzolzia hirta, Roylea cinerea, Rumex nepalensis, Senecio nudicaulis, Setaria pumila., Sida cordata, Sigesbeckia orientalis, Smithia ciliata, Solanum nigrum., Swertia alata, S. Cordata, Thalictrum javanicum, Torenia cordifolia., Tridax procumbens* and *Zornia gibbosa.*

The vernal undergrowth includes *Ajuga bracteosa, Amaranthus blitum., Arabidopsis thaliana, Atylosia scarabaeoides, Campanula benthamii, Cirsium arvense, Cyperus niveus, Digitaria violascens., Geranium rotundiflium., Lepidagathis cuspidata, Mazus pumilus, Medicago polymorpha, Nepeta hindostana, Poa annua, Saussurea heteromalla, Scutellatia repens, Teucrium royleanum* and *Tridax procumbens.*

Himalayan subtropical scrub

Edaphic and biotic factors combine to produce a characteristic type of scrub vegetation which is found on relatively drier soils and on exposed slopes often in Shivalik Chir zone up to 1400 m. The vegetation is sparse and it is rather difficult to categorize it into various storeys. *Euphorbia royleana* may grow individually or forms consociations. *Dodonaea viscosa,* a good colonizer, is the dominant species in considerable large tracts. Other worth-mentioning species are *Carissa opaca,* small-sized trees of *Phonenix sylvestris* and *Zanthoxylum armatum, Caesalphinia decapetala* and *Lantana camara* form thickets. Herbaceous vegetation is rather poor. *Achyranthes aspera, Cassia mimosoides, C. tora, Chenopodium album, Crotalaria prostrata, Dichanthium annulatum, Eragrostis pilosa, Erigeron bonariensis, Portulaca oleracea, Sida coradata* and *Solanum nigrum* are the common herbaceous elements which spring up during rainy period. Common representatives of this class in spring season are *Amaranthus viridis, Campanula benthamii, Crotalaria medicaginea* and *Poa annua.*

Montane Temperate Forests

"Himalayan moist temperate forests" is the subgroup to which the forests of this category belong in Chamba district. It is characterized by the development of conifers which may attain a height of 30-50 m. On southern slopes the conifers are replaced by oaks of low height (10-20 m). Aspect and altitude determine the distribution of species. Champion and Seth (1968) have recognized several subdivision under this category. Out of these four, namely (a) Ban-oak forests (b) Moist deodar forests (c) Mixed coniferous forests and (d) moist temperate deciduous forests are applicable to the present area.

Ban-oak forests

These are distributed between 1600-2100 m altitude. *Quercus leucotrichophora* (Ban-oak) is the dominant species occurring either in pure stands or in association with *Lyonia ovalifolia* and *Rhododendron arboreum.* Considerable part of upper canopy in damp situations is constituted by *Aesculus indica, Neolitsea umbrosa* and *Persea duthiei.* Scattered trees of *Acer cappadocicum, Celtis austalis, Populus ciliata, Pyrus pashia* and *Symplocos paniculata* are also found. Shrubby undergrowth is plentiful and usually consists of *Andrachne cordifolia, Berberis Iycium, Campylotropis eriocarpa, Desmodium concinnum, D. elegans, Deutzia staminea, Goldfussia dalhousiana, Hypericum oblongifolium, Indigofera dosua, I. heterantha, Leptodermis lanceolata, Lespedeza gerardiana, Lonicera quinguelocularis, Myrsine africana, Rubus ellipticus, R. niveus, Sarcococca saligna* and *Vibernum mullaha, Clematis buchananiana, C. grata, C. Montana, Ficus sarmentosa, Hedera nepalensis* and *Rosa brunonii* are the common climbers.

Moist deodar forests

These forests are distributed between 1700-2200 m above msl on cooler aspect. The upper storey consists of nearly pure deodar (*Cedrus deodara*) with sporadic distribution of other gymnospermous trees like *Abies pindrow, Picea smithiana.* and *Pinus wallichiana.* The medium or low sized angiospermous trees that are found here and there, singly or in groups and mostly in depressions are *Lyonia ovalifolia, Neolitsea umbrosa, Pyrus pashia, symplocos paniculata* and *Quercus leucotrichophora.* Shrubby undergrowth is plentiful with the common components being

Andrachne cordifolia, Berberis chitria, B. Iycium, Campylotropis eriocarpa, Coriaria nepalensis, Desmodium concinnum, D. elegans, Deutzia staminea, Goldfussia dalhousiana, Hy;pericum uralum, Indigofera atropurpurea, I. dosua, I. heterantha, Leptodermis lanceolata, Lespedeza gerardiana, L. juncea, Lonicera quinguelocularis, Rosa macrophylla, Sarcococca saligna, Vibernum cotinifolium and *V. mullaha*. The common climbers here are *Clematis buchananiana, C. grata, Ficus sarmentosa, Hedera nepalensis, Parthenocissus semicordata* var. *roylei, Rosa brunonii, Smilax aspera* and *S. glaucophylla*.

Mixed coniferous forests

Distributed between 2200-2700 m altitude, such forests are characterized by a mixture of coniferous species like *Abies pindrow, Cedrus deodara, Picea smithiana, Pinus wallichiana* and *Taxus wallichiana*. Mostly found along water courses or in depressions are some flowering plants such as *Aesculus indica*, Neolitsea umbrosa and *Quercus floribunda*. A large number of shrubs are commonly noticed here. Mention may be made of some common ones that include *Andrachne cordifolia, Berberis Iycium, Desmodium elegans, Indigofera atropurpurea, Leptodermis lanceolata, Lonicera purpurascens, L. quinguelocularis, Myrsine africana, Rubus biflorus, R. macelentus, Sarcococca saligna, Strobilanthes atropurpureus, Vibernum cotinifolium* and *V. mullaha* Climbers frequently noticed in this zone are *Clematis buchananiana, C. connata, C. montana, Euonymus echinatus, Hedera nepalensis* and *Parthenocissus semicordata* var. *roylei*.

Moist temperate deciduous forests

Such forests are not typical of the area. However, a patch exists in the vicinity of village Kathlag near Dalhousie at an altitude of about 1800 m along hill stream and on gentler slopes nearby. There are no large-sized trees and the canopy consists mainly of middle-sized trees like *Acer cappadocicum, Aesculus indica, Celtis australis, Neolitsea umbrosa* and *Rhododendron arboreum*. The common shrubby growth along with small-sized trees and associated climbers recorded from this forests include *Andrachne cordifolia, Berberis lycium, Clematis buchananiana, Desmodium elegans, Deutzia staminea, Ficus sarmentosa, Hedera nepalensis, Jasminum humile, Lyonia ovalifolia, Myrsine africana, Parthenocissus semicordata* var, *roylei, Prinsepia utilis, Pyrus pashia, Smilax aspera, Strobilanthes atropurpureus, Symplocos paniculata, Vibernum cotinifolium* and *V. mullaha*.

Montane temperate forests are the most humid zone in the area which results in the abundance of herbaceous undergrowth. Common herbs of monsoon period are:

Acalypha brachystachya, Achyranthes bidentata, Agrimonia pilosa var. *nepalensis, Agrostis pilosula* var. *royleana, Ainsliaea aptera, Andropogon munroi, Anemone vitifolia, Arisaema jacquemontii, Astragalus chlorostachys., Boenninghausenia albiflora, Bupleurum dalhousianum, B. lanceolatum, Carex nubigena, Carpesium abrotanoides, C. nepalense, Chenopodium umbrosum, Chrysanthemum leucanthemum, Circaea alpina, Craniotome furcata, Digitaria cruciata, Dipsacus inermis, Duchesnea indica, Epilobium cylindricum, Eragrostis minor, Eulalia mollis, Fagopyrum dibotrys, Galium asperifolium, G. elegans, Geranium molle var. glandulosum, G. nepalensis, G. wallichianum, Herminium lanceum, Impatiens amphorata, I. laxiflora, I. racemosa, Lamium album, Leonurus cardiaca, Lychnis fimbriata, Microstegium nudum, Muhlenbergia duthieana, M. himalayensis, Myriactis wallichii, Nepeta laevigata, Ophiopogon intermedius, Oplismenus burmannii, O. compositus, O. undulatifolius, Pilea scripta, P. umbrosa, Pimpinella diversifolia, Plantago erosa, Plectranthus lophanthoides, Polygonum amplexicaule, Prenanthes brunoniana, Plectranthus alatus, Ranunculus diffusus, Rumex nepalansis, Sanicula elata, Scrophularia himalensis, Senecio laetus, Solidago virga-aurea, Stachys sericea, Stelllaria monosperma* var. *paniculata, Swertia cordata, Thalictrum javanicum, T. reniforme, Themeda anathera, Trifolium repens, Trisetum spicatum, Urtica dioica, U. maieri, Verbascum thapsus* and *Vulpia myuros.*

Following herbs are often met here during spring season

Ainsliaea aptera, Androsace rotundifolia, Arabis amplexicaulis, A. pterosperma, Carex foliosa, C. nubigena, C. setigera, Chrysanthemum Leucanthemum, Dactylis glomerata, Duchesnea indica, Fragaria vesca, Gentiana argentea., Geranium lucidum., Geum roylei, Gypsophila cerastioides, Impatiens cristata, Koeleria macrantha, Lepidium apetalum, Mazus pumilus, Plantago lanceolata, Poa alpina, Polygonatum multiflorum, Pseudostellaria heterantha, Rumex angulatus, Senecio nudicaulis, Trifolium repens, Veronica arvensis, and *Viola biflora.*

A special mention is needed here of some herbs that are common in all the forests types of the area. Common ones among these are *Arthraxon prionodes, Atylosia platycarpa, Bupleurum hamiltonii, Campanula colorata, Carex cruciata C. filicina, Chenopodium album, Cyanotis vaga, Cynoglossum zeylanicum, Eragrostis nigra, Erigeron multicaulis, Lecanthus wallichii, Oldenlandia coccinea, Phleum himalaicum, Plectrunthus japonicus, Poa annua, Rubia cordifolia, Sonchus asper, Torilis japonica* and *Viola canescens.*

Among the spring season herbs found in all the forest types, mention may be made of *Campanula colorata, Cardamine impatiens, Cerastium glomeratum Chenopodium album, Erigeron multicaulis, Medicago lupulina, Polygala abyssinica, Stellaria media, Taraxacum officinale, Viola canescens* and *Youngia japonica.*

A comment is needed on the distribution of some shrubs and trees also. It is interesting to note that none of the shrubs and trees is found throughout the climatological and distributional range of these forest types. Species of *Berchemia, Dendrocalamus, Diospyros, Engelahardtia, Feronia, Glochidion, Ougeinia, Sapium, Stephania* and *Terminalia* are either totally absent or scanty from the tropical and subtropical zones of the area under report. Another interesting case is the disappearance of Sal (*Shorea robusta*) from the area. This tree is prevalent almost throughout the tropical zone of the Himalayas. Common temperate element of the Western Himalayas either totally absent from the area or encountered occasionally are species of *Aralia, Berchemia, Betula, Caragana, Cynanchum, Juglans, Meliosma, Ribes, Skimmia, Staphylaea* and *Ulmus.*

ACKNOWLEDGEMENTS

We are thankful to Director, Botanical Survey of India, Calcutta for sanctioning the 'BSI District Flora Project' under which the present studies were carried out. Herbarium and library facilities provided by the authorities of BSD and DD are also gratefully acknowledged.

REFERENCES

Champion H G & S K Seth 1968 *A revised Survey of the Forest types of India*: (Delhi, India: Govt, of India Publication).

Rao R R 1994 *Biodiversity in India (Floristc Aspects)*; (Dehra Dun, India: Bishen Singh Mahendra Pal Singh)

Sharma M & H Singh 1990 Observations on Floristic Composition of Chamba District, Himachal Pradesh; *New Botanist* XVII 273-281.

Sharma M & H Singh 1995 Life-forms and Biological Spectrum of the flora of Chamba District (Himachal Pradesh); In *Taxonomy and Biodiversity*; ed. A K Pandey pp 81-86 (New Delhi, India: CBS Publishers & Distributors).

Sharma M & H Singh 1996 Phytogeographic observations on the flora of Chamba District (Himachal Pradesh) - part II; *New Botanist* 23 103-112.

Sharma M & H Singh 1997 Observations on the herbaceous vegetation of Chamba District; In *Current Researches in Plant Sciences*; pp. 147-153 eds. TA Sarma, S S Saini, M L Trivedi & M Sharma (Dehra Dun, India: Bishen Singh Mahendra Pal Singh).

Singh H & M Sharma 1988 *Gnaphalium coarctatum* willd. - a South American Taxon naturalised in Chamba District (Himachal Pradesh), India: *Bull. Bot. Surv. India* 30 181-184.

Singh H & M Sharma 1989 Phytogeographic observations on the flora of Chamba District (Himachal Pradesh); In *Plant Science Research in India*; pp 217-215 eds. M L Trivedi, B S Gill & S S Saini (New Delhi, India; Today & Tomorrow's Printers & Publishers).

11

Ecological Prospects of Agroforestry Systems in Himachal Pradesh in Western Himalayas of India

VIDYA THAKUR and C. V. SARSWAT
Department of Tree Improvement & Genetic Resources,
Dr. YS Parmar University of Horticulture & Forestry,
Nauni, Solan (HP) India

ABSTRACT

Traditional agroforestry systems are well established in Himachal Pradesh. Farmers undertake cultivation of various tree species on the bunds of the fields along with crops and vegetables on the same piece of land. This is known as 'agri-silviculture' system now. Farm animals and poultry also form a component of the system. The association of ecologically compatible species provides subsistence and commercial products which give additional income to farmers. This paper describes the general scenario of agro-ecological conditions and its relation to existing agroforestry practices and some suggestions for their improvement in Himachal Pradesh in Western Himalayas.

Key Words: Agroforesty system, agri-silviculture ecological compatibility, farming system, multipurpose trees, socio-economy.

I. INTRODUCTION

Himachal Pradesh is an ecologically important Indian State in the Western Himalayan region. It comprises plain areas adjoining to the plains of Punjab and Haryana to peaks under perennial snow cover, adjoining to Tibet. This state is spread over about 55, 67.3 sq km. It is situated between 20° – 20° and 33° – 12 North latitude and 75° – 39° and 69° – 03° East to 79° East longitude. Major area of the tract is mountains, varying from 300 to 7000 m in elevation. Basins of Yamuna, Sutlej, Beas, Ravi, Chenab and their tributaries are situated in this state (Table 11.1). Beas, Yamuna, Sutlet, Baira-Siul and Giri Batta are being harnessed for hydro electric generation and irrigation of vast areas in the neighbouring states. The excessive use of forests and available land mass is leading to a vicious circle of destruction in time. Major part of area is under active erosion and is therefore subject of great concern. Once the top soil cover is washed away and soil fertility is lost, neither agriculture nor animal husbandry or horticulture will be able to sustain the people of hills and plains.

Table 11.1: Distribution of Geographical area (sq. km) in River catchment in Himachal Pradesh (India)

Name of the River	*Area*	*Percentage Distribution*
Sutlej	20,398	36.6
Beas	13,663	24.5
Chenab	7,850	14.2
Yamuna	5,872	10.6
Ravi	5,528	9.9
Indus	1,450	2.6
Markand	360	0.6
Ghaggar	262	0.5

Source: Anonymous (1993).

Himachal Pradesh has human population of 51.7 lacs, 90.7 per cent of which is rural and is directly engaged in agriculture, horticulture and animal husbandry. The land suitable for agriculture approximately may be 30–40 per cent of the total area of the state. It is a well known fact that during the last 20 years the area under agriculture and horticulture has been increased by 1.9 and 3.5 times, respectively (Anonymous, 1994). There has been noted substantial increase in the economic conditions of the state.

Rearing of the livestock is an integral part of the state economy. Census of 1991 reveals that there is approximately 5 lakh of livestock population. This is the intensive population considering the carrying capacity of the pasture in the region. Due to overgrazing, desirable nutritive grasses have depleted considerably. During the last two decades the grazing incidence has decreased with 0.25 ha/unit in 1969 to 0.19 ha/unit in 1982 (Chaudhary *et al.*, 1992). This may be due to bringing more and more area under agriculture, horticulture and closing of existing grazing areas by the State Forest Department as a measure of soil conservation and also under different afforestation programmes. High density of human and livestock populations reflect their effects on the vegetation and indirectly on environment in various ways like soil errosion, increase in temperature, irregular rains and extinction of may species. These are caused mainly by cleaning of forests for agriculture, horticulture, illicit cutting and lopping of forest vegetation for fuel, fodder, timber and charcoal, removal of (litter) leaves from the forest floor for manures, grazing and commercial exploitation of important forest species i.e. *Cedrus deodara, Pinus* species, *Taxus baccata, Abies Pindrow* and *Picea Smithiana,* etc.

II. ECOLOGICAL CONDITIONS AND BIOPHYSICAL ENVIRONMENT

Himachal Pradesh is characterized by variable topography ranging from plain areas at an altitude of 300 metres, to steep mountains with fragile rock systems at 7000 metres, above mean sea level. This, accordingly determines the climate which ranges from sub-tropical, temperate and sub-alpine to alpine. The state has number of valley areas of different sizes at different elevations. The valleys have different lengths varying from 5–6 km to 40–60 km with the breadth ranging from half a kilometre to about 13 km.

Lower and outer hills have a climate similar to plains while higher altitudes are covered with snow most part of the year. Agroclimatic conditions in the state vary from sub-tropical to humid temperate. The state has also the cold desert conditions in the areas bordering Tibet. There is wide variation in rainfall received and temperature that prevails in different areas of the state. The average annual precipitation in the state ranges from about 20 cm in cold desert areas to about 250 cm in Kangra valley. On an average, for the state as a whole, it can be noted that behaviour of rainfall has no specific trend. About 80 per cent of the total annual precipitation is received through, South-West monsoon during July-September. Generally there is an acute shortage of water both for irrigation and drinking during lean periods in many parts of the state.

The soils of the area are light textured the valley soils are medium to high in organic carbon content and low to medium in both available phosphorus and potash. Hill soils are rich in organic matter content (Anonymous, 1993). Soils are well drained but in heavy textured soils of the valley drainage may be a problem. Based on topographical and climatic conditions, the state has been devided into four agroclimatic zones (Fig. 1).

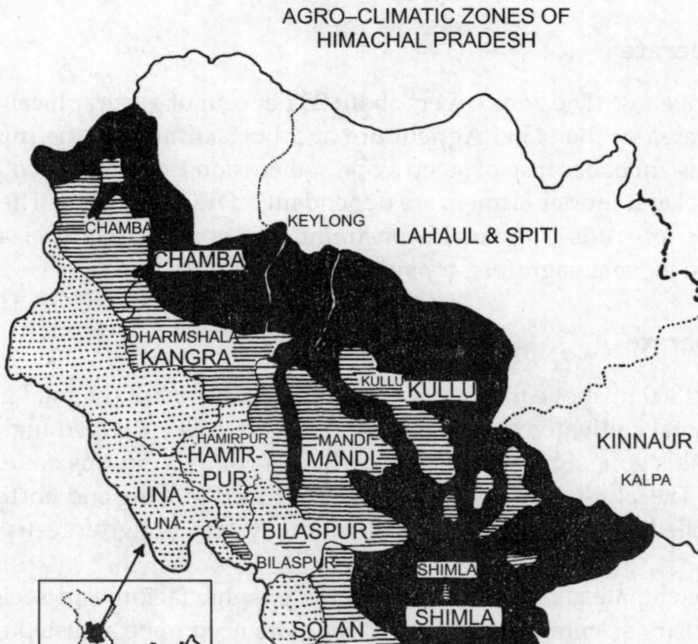

Fig. 11.1

Sub-montane Low Hills Sub-tropical

It consists of the foothills and valley areas upto an elevation of about 8000 m above msl. It occupies 33 per cent of the cultivated area of the state. Agriculture is the main farming system of this zone. Because of heavy demands of fuel, fodder and timber forests here are under pressure. On the basis of farming system, threat to forests and socio-ecological conditions the major agroforestry system are agri-silviculture, agri-horti-silviculture, horti-silviculture and silvi-pastoral. The last two are prevalent only on higher slopes.

Mid Hills Sub-humid

The zone extends from 800-1600 m above msl. It occupies 53 per cent of the cultivated areas. Major farming system of this zone is agriculture. Because of population pressure vis-a-vis conversion of forest land into agriculture, forests are under severe pressure. So the agri-silviculture, silvi-pastoral and horti-silviculture are the three agroforestry systems that require immediate adoption.

High Hills Wet Temperate

It lies above 1600 m above msl. The zone covers about 25 per cent of geographical area and 11 per cent of cultivated area of the state. Agriculture and horticulture are the major farming systems practised in this zone. Because of steep slope soil erosion is the main ecological problem of this zone. For fuel and fodder farmers are dependent on nearby forests. The shortage of wood for packing cases for fruits is the main constraint. The horti-silviculture, silvi-pastoral and agri-silviculture are the main agroforestry systems to be adopted.

High Hills Dry Temperate

This zone lies at about 1800 m above msl occupying about 8 per cent of the total geographical area and 3 per cent of total cultivated area. Soil erosion on these steep sloped high mountains is a major problem of this zone. Sheep and goat rearing is common in this zone. So there is constraint on pastures. The silvi-pastoral system with controlled grazing and horti-silvi-pastoral system with controlled grazing and horti-silviculture system of agroforestry need to be practised in this area.

Depending upon the climate conditions, topography, edaphic factors and socio-ecological conditions, the agroforestry systems and models of the state need to be adjusted according to these factors. The type of forest plants and their association with other components, thus depend on these conditions coupled with socio-economic status of the state.

III. LAND UTILIZATION FARMING SYSTEM AND SOCIO-ECONOMIC CONDITIONS OF THE STATE

The total reported area for land use statistics, in the state is 3349.50 thousand hectares. Forest and permanent pasture land form the major portion of this land area. Area under forest is 1034.60 thousand hectares and this works out to be about 30.5 per cent of the total reported area. The details of land use statistics are presented in Table 11.2.

Table 11.2: Land utilization pattern in Himachal Pradesh (Area in '000 ha) in 1991–92.

Land use category	*Area (in '000 ha)*
Reported area for land use	3390.50
Forest	1034.60
Barren and unculturable land	196.90
Land under non-agricultural use	202.90
Culturabel waste	116.00
Permanent pasture and other grazing lands	1159.60
Land under tree crops	48.10
Current fallow	47.20
Other fallow	20.10
Net area sown	574.10
Total cropped area	980.90

Cultivation is done on varying degrees of slopes and irrespective of soil depth. With increase in population there is always demand for more and more land and sometimes forests are replaced by agriculture and horticulture systems. Degradation of forest areas leads to heavy losses of top soil and water on slopes leaving thin sheet of soil on land surface. The repeated fregmentation and sub-divisions of lands have reduced the size of the holdings as 0.5 to 2 hac. As a result majority are with small and marginal landholdings. Farms are small and unconsolidated.

Main Kharif (summer rainy season) crops are rice, maize, *Amaranthus, Eleucine* etc. *Eleucine* is generally intercropped with black soyabean, Urd and *Setaria etalica*. Chillies, ginger, potato, tomato and peas are the main kharif cash crops. In rabi (winter season) the predominant crop is wheat inter-sown with barley, mustard, black gram and lentil etc. Potato is the major rabi cash crops grown in small irrigated pockets of the lower part of state, crop rotation etc. Potato is the major rabi cash crops grown in small irrigated pockets of the lower part of state. Crop rotation is general practice but with local system. Thus the agricultural status reflects the existence on subsistence economy.

In far rural areas due to small landholdings, lack of transpirations and agricultural developments, men in these areas are forced to seek employment in near by towns to supplement their income, leaving agriculture in the hands of women folk. Thus women play an important role in the socio-economic uplift of rural areas in the state. Generation of additional income through improved agroforestry interactions and cottage industries at farmers door steps will certainly attract more people towards agriculture which is the back bone of rural economy.

Depending on topography and climate, fruit cultivation is generally practised in a systematic way. Orchards of apple (*Pyrus malus*), apricot (*Prunus armeniaca*), plums, peaches (*Prunus persica*) and *Prunus amyodalus* are grown in cool dry and wet temperate zone of the state. Citrus and mango are grown in the sub-tropical and tropical zones. Forests near the villages are often classified as 'Shamlat' which are no man's land and damaged beyond repair. But in some areas they are considered community forests and lopped in a cyclic manner and are well protected by entire village community (Thakur, 1996). Pastures are not properly managed to any level of productivity and are subjected to heavy and uncontrolled grazing.

In Himachal Pradesh traditional forestry systems are well established. Planting and harvesting of trees for wood products, fruits, roots, leaves and fodder have been done here since the earliest days of man's activities. Traditional agroforestry systems of the state are silvi-pastoral and agri-silviculture. Planting around on field boundaries, bunds and homestead aim to diversify and increase the production. However, present-day farmers are trying to adopt suitable and well developed agroforestry system in order to increase the yields. Under traditional agroforestry techniques trees are protected only but they are not planted and managed properly. There is potential for improvement of traditionally managed systems to a great extent in order to realize the real production potential of agroforestry systems. Naturally occurring native agroforestry trees species are i.e. *Quercus leucotroichophora, Morus alba, M. serrata, Celtis australis, Salix alba, Juglans regia, Populus* spp. *Ficus* spp. *Prunus armeniaca* and *Pinus gerardiana* in the higher mountainous temperate zone. *Grewia optiva, Toona ciliata, Bauhinia variegata, Pistacia integermma, Celtis australis, Ficus oxburghii, Acacia catechu*, etc. are common in the existing agroforestry systems in the sub-tropical zone of the state.

Live hedge consists of *Vitex negundo, Woodfordia floribunda, Erythrina suberosa, Rosa moscheta*, and *Agave* mostly around the homesteads and fields etc.

A. Problems of Land Use

Cultivation on slopes without soil and water conservation measures, over-exploitation of community lands, unscientific exploitation of natural resources and over-grazing result in forest degradation of site. These activities lead to slides and avalanches in the high hills of the state.

B. Agroforestry in Retrospect

In the Himalayas, agroforestry system encompasses many well known land use systems practised since long time. Over exploitation for fuel, fodder and packing cases from the forests contribute towards large scale denudation of hills in Western Himalayas. The common agroforestry systems are agri-silviculture, silvo-agriculture, silvi-pastoral, multiple and multistoreyed tree production system. These practices are however, rudimentary and primitive, the economic output from these age old practices are very little to negligible, which is evident from the small holdings by the maximum families. Due to low economic status, the farmers are not able to provide optimum inputs resulting in poor yields. Income of the farmers is mainly through the sale of cash crops like potatoes, chillies, ginger, turmeric, surplus maize, wheat, rice, fruits like apple, peach, plum apricot, almond, Neoja etc. Traditional system of farming in the hills is mixed farming.

IV. CONSTRAINTS IN AGROFORESTRY SYSTEMS: AN ECOLOGICAL PREVIEW

(i) Small size of landholding of the majority of the farmer is the major problem.
(ii) Improper management of land.
(iii) Trees sometimes act as secondary host of insect pest and home for wildlife. i.e. monkeys and birds etc.
(iv) Broad canopy of trees may reduce sunlight to crops.
(v) Shallow rooting behaviour of some tree species affect the growth and yield of associated crops.

(vi) Allelopathic effect of some associated tree crops on agricultural and vegetable crops.
(vii) Lack of ecological niche differentiation/ecological compatibility between and within various components (i.e. trees and crops, fruits crops and vegetable crops etc) of agroforestry systems.
(viii) Thin and immature soil surface due to soil erosion and avalanches on steep slope of hills.

V. ROLE OF ECOLOGY IN AGROFORESTERY

The type of agroforestry system found in a particular area is determined to some extent by agroecological factors. However, sometimes several socio-economic factors such as human and livestock population pressure, constraints on existing flora and fauna, availability of labour and proximity to market also come into play resulting in variations among systems under identical agro-climatic conditions. Agro-ecological factors have a considerable bearing on the functional role of agroforestry practices. The introduction of homegardens and other multispecies system in fertile lowlands and areas with agricultural potential at one end of the ecological scale and extensive silvi-pastoral practices at the other end with other systems in between, reflect that distribution and extent of adoption of specific agro-forestry systems are determined by the ecological potential of an area within the state.

In general depending upon the constraints on various components of the system and demands of local ihabitants, agroforestry system in the Western Himalayas may be classified in three categories:

Need Based System

The growing of trees on farm bunds or on fallow land yielding fuel, fodder, fruits, small timber etc. farm agri-silviculture system. About 50-100 trees are planted over an area of 1 hac. which meets only a part of the fodder and fuel requirements. Agriculture crops include wheat, maize, rice, peas potato, cauliflower, mustard, mash, beans and main tree species include are Grewia optiva, Celtis australis, Bauhinia variegata, Albizia chinensis and Toona ciliata, Acacia catechu, etc.

Economy Based Systems

This management system is the horti-agricultural system. In temperate zone dominant system is horti-agricultural system and in sub-tropical and humid zone agrihorticultural systems predominate. These systems represent an index to optimum land use practices.

Environment Based System

Silvi-pastoral system in the form of natural grasslands and pastures with or without trees as a consequence of ecological succession are common. Himachal Pradesh has an area of 1150.60 thousand hectares under permanent pasture and other grazing lands. The bio-fertility of this system which was once sustainable, is now degraded because of increased number of livestocks. Unrestricted cattle grazing is also highly responsible for soil erosion today and eroded areas bear shrubs or euphorbias. Community land near the villages are known as Ghasni.

In the mountainous region soil erosion is predominant and this can be checked by planting grasses and other native tree species, along or alternate to crop fields. Agroforestry system if modified properly according to the topography and socio-economic conditions of area, play a very important role in reclamation of wastelands and soil conservation.

Agroforestry system envisages integrating perennial trees with fruit plants, agri-crops vegetables grasslands simultaneously on the same land. The interaction between different components is dynamic in nature and may prove complementary or contrary at different stages of growth depending upon the ecological compatibility of different components. In University of Horticulture & Forestry work is being done to stimulate, to implement and to evaluate different agroforestry systems. Models are being developed for horticulture land management system, horti-silvi agriculture system, development of silvi-pastoral system for mid-hill conditions and degraded hill slopes in the high hills (Chaudhary *et al.*, 1992). In addition to this technique of contour plantations on denuded hills (Kumar and Bhardwaj, 1993) i.e. Gradoni method is being practised with the following objectives.

- To assess ecological, physiological and biological compatibility among the component parts of the system.
- Investigation of reciprocal influence of components.
- Economic and ecological viability of the system and influence on the site productivity.

On these basis a common agroforestry model suitable for all types of agroclimatic zones in the Himachal Pradesh has been presented (Fig. 2). It includes timber trees, fruit and fodder trees, grasses and legumes mixture and 3-4 strips of crops and vegetables and again followed by grasses and legumes i.e.

(i) Loss of nutrients due to soil erosion from the barren area with steep slope will be checked by the grasses and legumes.
(ii) The grasses and legumes will be rotated with crops strips successively and sequentially in regular fashion i.e. after four years. Soil fertility and structure will be improved after regular intervals.
(iii) Growing of fruit and fodder trees will fulfil the fruit and fodder requirement of the rural people and also finally check soil erosion.
(iv) Growing timber trees in two-three rows will serve as wind break and also sufice the timber requirement of the farmers.

This can be modified in course of time with the advancement of technology.

On the basis of diagnostic survey and appraisal of existing farming system and agroforestry systems satisfying farmers needs. And to make the systems ecologically and economically feasible, the following are the aspects that need immediate care and attention (Table 11.3).

Table 11.3: Appraisal of agroforestry for satisfying basic need of farmers in rural area of Himachal Pradesh.

Basic Needs	*Desired produce*	*Traditional agroforestry systems and their components (dominating trees, crops, grasses and animals)*	*Adoptable agroforestry systems*
1. Crops Cereals	Maize, wheat, rice, barley, amaranth, millets etc.	Agri-silviculture *Grewia, Bauhinia, Morus Celtis Populus, salix Robinia,*	Crops could be grown with specific association of these trees. The possible sites for

Contd.

Pulses	Black soyabeans, gram, urd, lentil and peas	*Acacia catechu Toona ciliata, Pistacia* etc.	growing trees may be field bund, terraces and main agricultural fields in a scattered fashion. Turmeric, ginger and colocasia have been found best suited crops.
Oil-seeds	Mastard, Toria etc.		
Cash crops	Potato, tomato, chillies, cauliflower seeds, cabbage seed, ginger, turmeric, garlic, onion, sugarcane and peas.		
2. Fruits	Orchards, homegardens, and scattered wild fruits	Apple, pear, peach, plum, apricot, guava, mango, citrus, papaya, banana, strawberries, *Rubus,* fig, *Pyrus pashia,* etc.	Agri-horticulture systems. Turmeric and ginger are suitable crops with peaches and pulses with apple and pears.
3. Vegetables and spices	Kitchen gardens and fields	Potato, tomato, colocasia, beans, bringles, chillies, cucurbits, onion, garlic, coriandrum, black sira, etc.	Agri-horticulture and silvi-agriculture systems
4. Animal products	Animal husbandry, dairying, poultry farm, sheep and goat rearing, etc.	Cows, buffaloes, goats, sheep, piggery paultry, fish and hare etc.	Silvi-pastoral systems or rearing cattle etc. These are the major components of forest farming and other agroforestry system in the state.
5. Non-food needs fuel, fibre, fodder, etc.	Trees, shrubs, live hedges, crop residues, top fodder and grasses, etc.	Minor forest produces from Khair, sal, oaks, etc.	1. Silvi-pastoral system to meet the requirement of fuel and fodder. Planting Poplar, *Robinia, Bauhinia* with grasses.
		Maize, wheat, rice, millets and amaranths as source of crop *Grewia, Celtis, Morus, Bauhinia, Acacia catechu* top fodders and grasses like *Chrysopogan, Heteropogan,* etc.	2. Silvi-agriculture systems
Timber	Forest and private farm	*Shorea robusta, Quercus Toona, Cedrus, Populus, Pinus* and *Robinia,* etc.	For timber requirement grow *Toona, Pinus, Cedrus, Salix, Populus* and *Robinia* in scattered design.
6. Special need supplem-entary income	Cottage industries	Bee keeping, cut flower, canning, dairying etc, farm labour, sale of fire wood, seeds of stone fruits, surplus foodgrains, animals and birds, etc.	Additional income from these sources could be adopted in case of farmer living on small and marginal land holdings and income below poverty lines.
	On farm employment		
	Off farm employment	Employment in towns, cities and army etc.	Supplementary income

- Preservation of genetic resources of the local species mostly exploited by the farmers. In general, conservation of gene and biosphere reserves are needed.

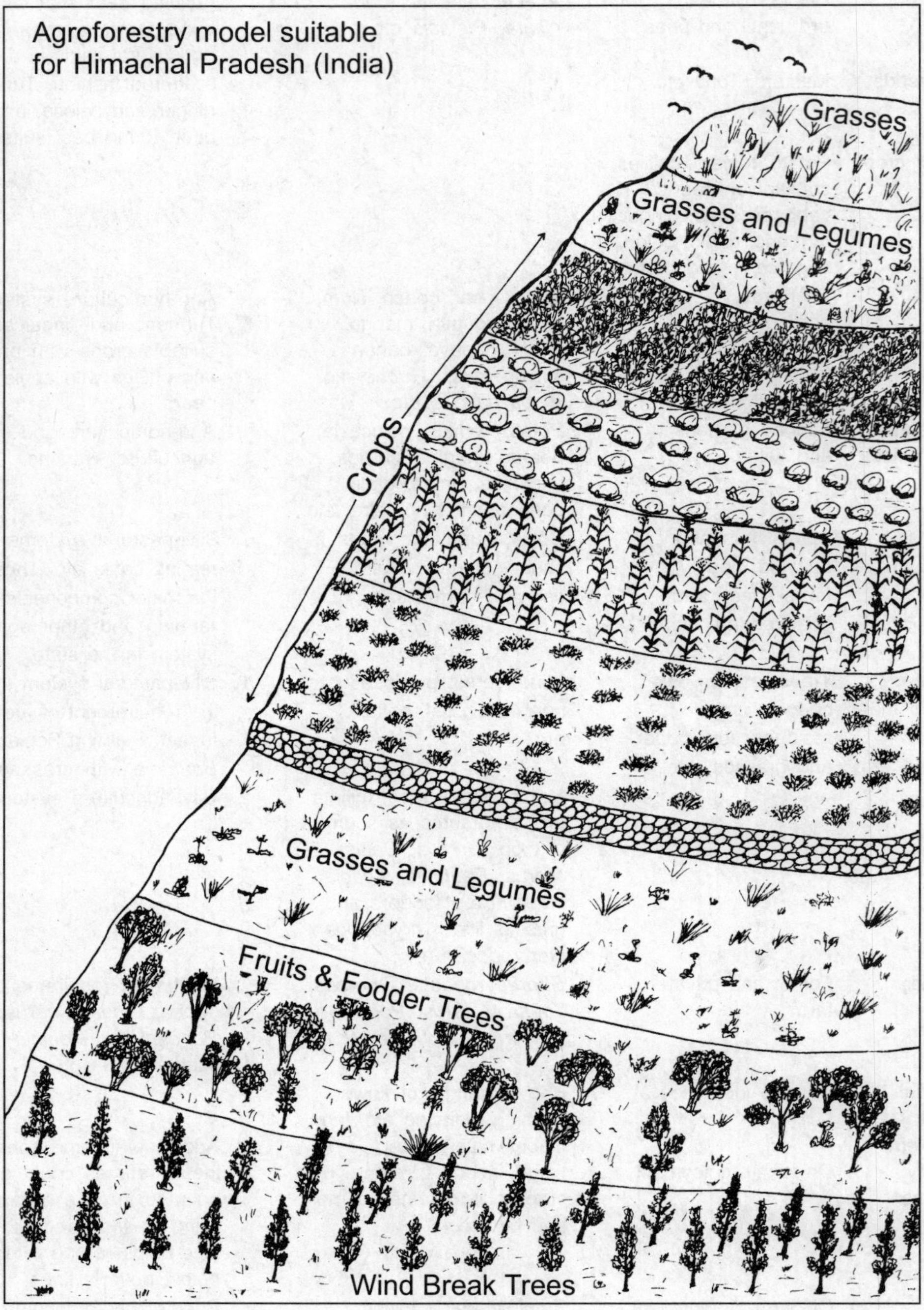
Agroforestry model suitable
for Himachal Pradesh (India)
Grasses
Grasses and Legumes
Crops
Grasses and Legumes
Fruits & Fodder Trees
Wind Break Trees

Fig. 11.2

- Identification of multipurpose woody species.
- Collection, screening, improvement and multiplication of germplasm of multipurpose species.
- Identifying crop associations which can be fitted into different intensities of shade.
- Rooting behaviour and distribution pattern of component tree crops for understanding the optimum nutrients and moisture utilization as available.
- Qualitative and quantitative interactions between plants and soil in the different types of associations.
- Economic evaluation of different systems.
- Awareness among the rural people through training and organizing workshops and seminars etc.
- And last but not the least, involvement and encouragement of rural women in awareness programmes by organizing site and need specific training, workshops and seminars.

REFERENCES

Anonymous 1994 Statistical outline of Himachal Pradesh. Economics and Statistics Department, HP, Shimla.

Anonymous 1993 Forests of Himachal Pradesh. World Environment Day 5th June, 1993.

Chaudhary GC, Y S Negi & M S Pathania 1992 Status of social forestry in Himachal Pradesh. Regional Centre for Wastelands Development (NWDB), Directorate of Extension Education, UHF, Solan (HP) India.

Kumar D & S B Bhardwaj 1993 Techiques of contour plantation on denuded hills. In agroforestry: Indian perspective pp 199-205 ed LK Jha & PK Sen Sharma (Ashish Publishing House, New Delhi)

Thakur V 1996 Gender, environment and sustainable livelihood. Project Report submitted to ICIMOD, - Kathmandu, Nepal pp 1 - 30.

12

Agro Biodiversity in the Indian Himalayas

B.D. SHARMA and J.C. RANA
National Bureau of Plant Genetic Resources, Regional Station, Shimla (HP) India

I. INTRODUCTION

The Himalayas was formed as an enfoldment of earth crust under the Tethys sea resulting from collusion of Indian (peninsular) plate with the Asian continental plate some 15 million years ago. It is 2500 km long occupying a geographical area of about 236000 square kilometres in India. Its width ranges from 240 to 340 km and has a rise of over 8000 m from the Indian plains. It has been a great attraction for spiritual aspirants, trekkers, explorers etc. it is earnestly termed as Devbhumi. Because of its vast diversity in soil, slope aspects, altitudes and ecological conditions, it hosts a large variety of plant life on rocks, rivers and forests as well as in agriculture. This is also true for human ethnic races and wildlife. Agro- biodiversity, which forms the basis of existence of human life on the earth planet, is amply rich resource in the Himalayan Mountains and plain valleys. Agro-biodiversity broadly covers all forms of life that has a link with agricultural production systems that exist among the various farming communities. The various components of agro-biodiversity may include crop plants, weeds, soil micro-organisms, domesticated and wild animals that has bearing on agriculture, pathogens, insect pests, agroforestry plant species etc. This paper however, deals with the diversity in crop plants that are found across the Indian Himalayas.

II. PHYSIOGRAPHY OF THE INDIAN HIMALAYAS

The Indian Himalayas starts in the west from Nanga Parbat and extends to Arunachal Pradesh. The Arunachal Pradesh portion of the Eastern Himalayas further extends in north-easterly direction covering all Norths-eastern states. It includes Indian states such as Jammu & Kashmir, Himachal Pradesh, Uttaranchal, hills of West Bengal, Sikkim and Arunachal Pradesh. Vertically it extends from Indian plains to Tibet highlands and has four parallel ranges from south to north. These are:

(i) Shivalik hills or outer Himalayas are low hills with altitude below 1000 m.

(ii) Middle or lesser Himalayas is about 60 to 80 km width and has an altitude averaging 3500 m above msl.

(iii) Inner or greater Himalaya ranges are high mountain ranges about 120 to 140 km in width.

(iv) Trans-Himalayan region is about 40 km wide and includes arid, temperate areas of Ladakh, Lahual-Spiti, Pangi (Chamba).

The Indian Himalayas mainly comprises

(a) N-W Himalayas comprising Jammu & Kashmir and Himachal Pradesh

(b) Central Himalayas which include the newly formed state of Uttaranchal (Kumaon and Garhwal hills)

(c) Eastern Himalayas comprise state of Sikkirn and Arunachal Pradesh.

(d) North-Eastern Himalayas is the extension of the eastern flank encompassing states of Assam, Tripura, Manipur, Nagaland and Mizoram.

The Western or North-Western ranges are drier, gets snowfall beyond 2000 m during winters and as it advances towards east the elevation for snowfall rises and rainfall increases and the vegetation becomes very rich. The Himalayas are mainly characterised by Boreal and Paleo-tropical or Indo-Malayan type bio-geographical zones. Out of 6000 endemic plant species, 2532 species occur in the Himalayas.

III. AGRO-BIODIVERSITY ACROSS THE INDIAN HIMALAYAS

The Indian Himalayas are a rich reservoir of crop agro-biodiversity. The whole range of species diversity under various crop groups could be classified as:

(i) Cereal crops which include wheat (*Triticum aestivum*), maize (*Zea mays*), rice (*Oryza sativa*), barley (*Hordeum vulgare*), Jau (*H. himalayense* Oujau)

(ii) Millet Crops: Jowar (*Sorghum bicolour*), Bajra (*Pennesetum typhiodes*), Chena (*Panicum miliaceum*), Kodo or Kodra (*Paspalum scrobiculatum*), Kaoni (*Setaria itatica*), Mandira or Jhangera (*Echinochloa frumentacea*), Mandua (*Eleusine coracana*), Sanwa, Kutki (*Panicum sumatrense*).

(iii) Pseudocereals: Buckwheat Ogal (*Fagopyrum esculentum*), Phaphar (*Fagopyrum tataricum*), Pigweed, Bathua (*Chenopodium album*), Chauli (*Amaranthus caudatus, A. hypochondriacus, A. cruentus*)

(iv) Less Known Cereals: Job's tear (*Coix lachryma-jobi*), Small millet, Raishan, (*Digitaria cruciata* var. *esculenta.*).

(v) Pulse or Legume Crops: Black gram, Urd (*Vigna mungo*), Mung or Green gram (*V. radiata*), Rains or Adjuki bean (*V. angularis*), Bharinga or Moth bean (*V. aconitifolia*), Rongi, sonta or Cowpea (*V. unguiculata*), Ricebean or Bhotiya (*V. umbellate*), Matar (*Pisum sativum*), Rajmah (*Phaseolus vulgaris*), Bhatta (*Glycine soja*), Kala Bhatta (*Glycine spp.*), Soybean (*Glycine max*), Chana, Gram (*Cicer arietinum*), Masar, lentil (*Lens esculenta*), Kultha, Gahat or horse gam (*Macrotyloma uniflorum*), Mattari, Khesari (*Lathyrus sativus*), Vakla (*Vicia faba*), Sudu (*V. sativa var. angustifolia*), Tur, Arhar (*Cajanus cajan*).

(vi) Edible Oil Yielding Plant Species: Sarson, Toria (*Brassica campestris*), Banarsi Rai (*Brassica nigra*), Rai Sarson (*B. juncea*), Taramira, Rocket (*Eruca sativa*), Tisi, Alsi, Flax (*Linum usitatissimum*), Sesame, Til (*Sesamum indicum*), Bhanjira, Neileh (*Perilla frutescens*), Chiura

(*Aesandra butyracea = Bassia butyracea*), Chuli (*Prunus armeniaca*), Bhekal (*Princepia utilis*), Erand, Castor (*Ricinus communis*), Akhrot, Walnut (*Juglans regia*).

(viii) Vegetable Crop Plant Species: It is very large group of plant species, which may be cultivated or even collected from natural habitats. The cultivated vegetables include followings;

a. Crucifer Vegetables: Band Gobhi, Cabbage (*Brassica oleracea var. capitata*), Phulgobhi, Cauliflower (*B. oleracea var. botrytis*), Shaljam turnip (*B. campestris var. rapa*), Muli, Radish (*Raphanus sativus*).

b. Cucurbitaceous Vegetables: Pumpkin (*Cucurbita moschata, C. maxima*), cucumber (*Cucumis sativus*), Cho-cho (*Sechium edule*), Bottlegourd, Ghiya, Lauki (*Lagenaria siceraria*), Ridgegourd (*Luffa acutangula*), Spongegourd (*Luffa cylindrica*), Bittergourd (*Momerdica charantia*), Parval (*Trichosanthes dioica*), Snakegourd (*Trichosanthes anguina*), Bhat Karela (*Momordica cochinensis*), Kakral, Kakora (*Momordica dioica*), Kundru (*Cephalandra indica*), Petha, Ashgourd (*Benincasa hispida*).

c. Leafy Vegetables: Palak (*Beta vulgaris var. bengalensis*), Vilayati Palak (*Spinacia oleracea*), Methi (*Trigonella foenum-graecum*), Kasuri Methi (*T. corniculata*), Lai Patta (*Brassica campestris var. cuneifolia*), Lohi patta (*B. campestris var. rugosa*), Karam Sag (*B. oleracea var. acephala*), new Zealand Spinach (*Tetragona expansa*), Bathua (*Chenopodium album*), Khatta Palak (*Rumex vesicarius*), Chaulai (*Amaranthus blitum, A. dubius, A. hypochondriacus, A. caudatus*), Jirag (*Phytolacca acinosa*).

d. Leguminosus Vegetables: Pea, Matar (*Pisum sativum*), Frans bean (*Phaseolus vulgaris*), Bakla (*Vicia faba*), Rongi, Asparagus bean or Yard long bean (*Vigna sinensis var. sesquipedalis*), Lima bean (*Phaseolus lanatus*), *Cabanalia ensiformis*, C. *gladiata, Dolichos lablab, Mucuna pruriens, Psophocarpus tetragonolobus.*

e. Root and Tuber Crops: Alu (*Solanum tuberosum*), Ratalu, Yam (*Dioscorea alata, D. esculenta, D. bulbifera*), Tapioca Yuca or Cassava (*Manihot esculenta*), Suran, Zamikand (*Amorphophallus campanulatus*), Arvi (*Colocasia esculenta*), Sakarkand (*Ipomoea batatas*), Mankand (*Alocasia indica, A. macrorhiza*), Tannia (*Xanthosoma sp.*).

f. Solanaceous Vegetables: Tamatar, Tomato (*Lycopersicon esculentum, L. pimpinellifolium*), Baingan, Brinjal (*Solanum melangena var. esculentum, Serpentum depressum, S. macrocarpon*), Bitter brinjal (*Solanum gilo*), Tree Tomato (*Cyphomandra batacea*), Shimla mirch (*Capsicum annum*), Chilli, Lal (*C. frutescens*), *Solanum torvum.*

g. Others: Bhindi, ladies finger (*Abelmoschus esculentus*), Gajar, Carrot (*Daucus carota*).

(viii) Starchy-tuber Foods: *Moghania vestita, Vigna vexillata, Bunium persicum, Pachyrhizus tuberosus, Maranta arundinacea.*

(ix) Spices and Condiments: *Allium cepa, Carum carvi, Coriandrum sativum, Cinnamomum tamala, Curcuma domestica, Zingiber ojficinale, Amomum subulatum, Allium sativum. Capsicum frutescens* Shallot (*Allium cepa var. aggregatum*), Bara lahsun (*Allium ampeloprasum var. porrum*), Chives (*Allium schoenoprasum*), *Allium fistulosum.*

(x) Fibre Crop Plants: *Boehmeria nivea, Cannabis sativa, Grewia optiva Hibiscus cannabinus, Gossypium hirsutum, G. barbadens, Corchorus oletorius.*

(xi) Fruit Plants: The major temperate fruit plant species that occur in the region represent: *Malus baccata, M. sikkimensis, M. dirangensis; Prunus prostrata, P. wallichi, P. nepaulensis,*

P. persica. P. cornuta, P. armeniaca, P. ceresoides, and *Prunus spp.* (*bhemi*); *Pyrus jacumontiana, P. polycarpa, P. pyrifolia, P. pashia. P. communis; Sorbus lanata. S. cuspidata; Cotoneaster accuminata, C. frigida; Crategus crenulata, C. oxycantha; Cydonia oblonga; Doycinia hookeriana; Vibernum contifolium V. lanata, Rubus elipticus, R. nivus, R. biflorus, R. fruitcosus, R lasiocarpus; Ribes nigrus, R. rubrum; Corylus avellana,* C. *colurna, Cornus capitata, Elaegnus latifolia; Juglans regia* and *Pinus geraradiana.* Besides temperate fruit species diversity also exist for sub-tropical fruits species such as *Punica granatum, Mangifera spp., Psidum guava, Citrus* spp., *Musa* spp., etc. This region is also a treasurer of many less exploited fruit species viz. *Aegle marmelos. Cordia obliqua, Emblica officinalis, Ficus palmata, Ficus roxburghii. Flacourtia sapida, Myrica nagi, Opuntia dillenii, Prunus armeniaca, Punica granatum, Ribes alpestrea, Sorbus lanata, Rubus niveus, Fragaria nubicola, Rubus ellipticus, Vitis lanata, Zizyphus juguba, Spondias mangifera., Buchanania latifolia, Pinus girardiana, Corylus colurna, Ribes rubrum, Hippohae salicifolia,* and *H. rhamnoides.* These fruits have very good potential for processing industries and in local markets as table fruit. These resources may also provide genes to resistance to diseases, insect-pests, cold, drought and a desirable rootstocks. Most of these minor fruit species found in the forests efforts for its collection and conservation have been initiated by NBPGR.

(xii) Grasses and Fodder Trees: In subtropical and temperate region grass cover is *Dichanthium -Cenchrus -Lasiurus, Phragmites -Saccharum -Imperata* and *Themeda - Arundinella.* Alltitude separates *Themeda-Arundinella* cover (restricted to hills) from *Phragmites-Saccharum-Imperata* (in moist to wet habitats) and *Dichanthium- Cenchrus-Lasiurus* (in relatively dry habitats) types which occur in lower elevations (Bhagmal, 1997). The major species of grasses in the Indian hills are *Agropyron cristatum, Agrostis tenuis, Bromus inermis, Dactylis glomerata, Festuca pratensis, F. rubra, Lolium perenne, Phleum pratense, Avena fatua, Elymus* spp., *Poa pratense, Stipa* spp., *Bromus* spp., and *Phalaris tuberosa.* Legume species *Trifolium pratense, T. repens, Lotus corniculatus, Medicago sativa* exist. Diversity for genus like *Andrpogon, Bothrichloa, Cenchrus, Chrysopogon, Echinochloa, Heteropogon, Panicum, Themeda* and *Pennisetum* also occur in sub-tropical parts.

Tree fodder is extensively used particularly in the lean months (winter) as lot of diversity is available. Major tree species occur in the state are: *Celtis australensis,* Bamboo spp., *Grewia* spp., *Debrogeasia hypoleuca , Toona celiata, Acacia catechu, Acer oblongum, Aesculus indica, Albizia* spp., *Ainus nitida, Bauhinia* spp., *Corbus oblonga, Ailonthus excelsa, Artocarpus heterophyllus, Duabonga grandiflora, Erthrina variegata, Exbucklandia populnea, Ficus* spp., *Gleditsia triacenthos, Holoptelea integrifloia, Melia azadarach, Moringa oleifera, Morus* spp., *Prunus* spp., *Kydia calycina, Terminalia myriocarpa, Quercus leucotrichophora, Q. floribunda, Ougeinia oojeinensis, Robinia pseudacaria, Salix babylonica, Pistacia integrimma* and *Myrica nagi.*

The above list of crop plant species under cultivation across the Himalayas includes over 139 crop species. The total crop species in India may be around 166 species (Arora, 1996); which is about 84 per cent of predominant species under cultivation (Table 12.1).

IV. CROP AND VARIETAL DIVERSITY IN N-W HIMALAYAS

In the N- W Himalayas, 135 crop plant species are being cultivated which is about 38 per cent of total crop diversity in Indian agriculture. The crop groups and no of species under each crop groups is given below:

Table 12.1(a): Species biodiversity of cultivated crops of Himalayas in comparison to the country status

Crop group	*No. of species (Indian Himalayas)*			
	Western	*Central*	*Eastern (N-E)*	*Total India*
Cereals, Millets & Pseudocereals	19	16	19	51
Legumes	20	17	10	31
Vegetables	30	28	65	103
Oilseeds	11	9	11	12
Fibre plants	5	7	6	24
Spices & condiments	16	15	11	27
Fruits	34	27	45	109
Total Species	135 (38%)	119 (33%)	167 (47%)	357

Table 12.1(b): Species diversity of crop plants in Western Himalayas

Crop group	*No. of species*	*Varietal diversity under important crop species*
Cereals	5	Rice (60), Wheat (29), Maize (13),
Millets	8	
Pseudocereals	6	
Pulses	20	
Oil yielding	11	
Vegetables	30	
Spices & condiments	16	
Fruit plants	34	Apple (50), Walnut (5), Pear (6),
Fibre plants	5	

The varietal or genetic diversity has been narrowed down to few improved varieties from 60 landraces in rice, 29 landraces in wheat and 13 landraces in maize (Table 12.2).

Table 12.2: Important landraces of cereal crop plants in the Western Himalayas

State	*Crop*	*Landraces*
Himachal Pradesh	Paddy	Madholu, Roda, Rehdi, Phulpatash, Ladheri, Cheena, Japani, Dhan, Ramjawain, Laldhan, Kaladhan, Jhini, Bowru, Katheri, Tapta, Kodri, Sukul, Rangri, Jalpara, Jarati, Rohdu, Jawas, Matalu, Vakhil, Jattoo, Sukhdwas, Jhinjhini, Jaulia, Bathal, Jaldara, Seyartu, Peela dhan, Dewal, Matali, Jiri, Reesa, Bangwai, Nakadu, Begami, Achhain, Till, Dinu, Hansraj, Lal Nakanda, Jhona,
	Wheat	Dharmori, Lalpuri, Kankoo, Shruin, Misri, Gazariya, Mandleu, Bhangroo, Daru, Dharon, Paluwa, Jhuladi, Rundan, Latar, Kihali, Trimundi, Brahdoo, Kalodi, Kiawali, Lal Kanak, Marodu, Chawera, Chiti Kanak, Brada Kanak, Kathi, Rigaliya.
	Maize	Chidkoo, Salhu, Desi Kukari, Temta, Gadda, Rohdu, Doni, Tedi.
Jammu & Kahsmir	Paddy	Baber, Lonazen, Budiji, Mushaka budiji, Niver, Kathwar, Brez, Mazette, Begam, Ital, Lal Mundi, Komu Basmati, Kalini, Perumal, Khuch, Rehman Batli, Pootbear, Mahwan.
	Wheat	Ghoon,Kanak,Kankoo.
	Maize	Chiti Makka, Lal Makka, Sathi.

The cultivation of millets and pseudo-cereals is almost on negligible area. Due to the use of high doses of fertilisers harming natural rhizobium soil cultures, the pulses such as Moong,

Mash, Gram and Kulth has been reduced to negligible acreage. Temperate fruit and nuts diversity is also narrow and because of expansion of area under these fruits has resulted in replacement of traditional crop species such as grain amaranth, chenopods and buckwheat.

Crop and Varietal Diversity in Central Himalayas

The total crop plant diversity in the Central Himalayas is 119, which is about 33 per cent of the total crop agro-biodiversity in the country.

Table 12.3: Species diversity of crop plants in Central Himalayan Hills (Pant, 1998)

Crop group	*No. of species*	*Varieties*
Cereals	4	Rice (49), Wheat (10), Maize (5)
Minor Millets	6	
Pseudocereals	10	
Pulses	17	
Oil yielding	9	
Forage & grasses	22	
Fibre plants	7	
Spices & condiments	15	
Fruits & nuts	27	
Vegetables	28	

The Central Himalayas is constituted of Uttaranchal. The loss of genetic materials has been comparatively slow. But as the communication will improve the replacement of landraces and some traditional crops will increase. These will be replaced by few improved high yielding varieties and major crops will replace some of the famine food crops such as millets and pseudocereals. The genetic diversity in crops like rice, maize and wheat has been to the tune of 49, 5 and 10 respectively (Table 12.4).

Table 12.4: Important landraces of cereal crops in the Central Himalayas

State	*Crop*	*Landraces*
Uttaranchal	Paddy	Andi, Anandi, Bagdia, Bagani, Bank, Binduli, Boora, Chani, Chaunia, Dangya, Dalbadal, Dhania, Dudh, Gadal, Garoo, Geemi, Hansi, Jamali, Jhumari, Kala dhan, Kali Malati, Kafalya, Katuna, Khaja, Lal dhan, Lamadi, pingla, Suriya, Thapachini, Lal Chandan, Anchan, Dhur Basmati, Dahia, Char Dhan, Lekmal, Kajuri, Roti, Sawa, Tolia, Chwatu, Lal Nalendi, Uttali, Jhalla
	Wheat	Kathi, Misri, Rigaliya, Gazariya, Misri Lal, Misri Safed, Misri Moti, Mundriya, Mundriya Lal, Misri Baasar. Kathi & Gazariya are lodging resistant; Rigaliya is weed tolerant
	Maize	Shathi, Safed Makka, Lal Makki, Peeli Mugri, Peeli Makki.

Crop and Varietal Diversity in the Eastern and N-E Himalayas

The Eastern and N-E Himalayas region receives heavy rains and is a rich reservoir of crop plant species as well as genetic diversity. The total crop diversity found in this region is 167 species, which is about 47 per cent of the total agricultural crop diversity in the country. The number of species in each crop group is given in Table 11.5.

Table 12.5: Species diversity of crop plants in Eastern and N-E Himalayas

Crop group	*No. of species*	*Varietal diversity under important crop species*
Cereals	4	Rice (8003), Maize (10 races* and 799 landraces)
Millets	10	
Pseudocereals	5	
Legumes	10	
Oilseeds	11	
Vegetables	65	
Spices & condiments	8	
Fruits & nuts	45	
Fibre plants	6	
Energy rich crops	12	

* Maize races are Poorvi Botapa, Murli (subrace of Poorvi Botapa), Tirap, Nag-Saypung, Arun Tepi, Aloksapa.

There has been large varietal diversity in crops like rice (8003 landraces) and in maize ten races with 799 landraces has been identified by Bhag Singh (1994). A large number of vegetables (65) are grown and consumed in the region. The important are given in Table 11.6.

Table 12.6: Important landraces of cereal food crops in the Eastern & N-E Himalaya

State	*Crop*	*Landraces*
Sikkim and North Bengal Hills	Paddy	Attey, Juari, Krishnabhog, Nuniya, Phouril, Botua, Lama Dhan Lal bathi, Chimti, Champhey, Jalongay, Tharpasing, Jirasali, Kalsati, Khemti, Tulasi, Kalami, Dudhey, Bhuidhan, Champasari, Budheymarsi, Sikreymarsi.
	Maize	Kukkuri, Sweti, Sweti-Chapti.
Arunachal Pradesh	Paddy	Yamuk, Sarali, Yambar, Tape, Lite, Mimre, Mipun, Lemuk, Leli, Ankar, Bali, Kimin, Myu, Piyong, Kya, Ke-Kalti, Ahumang, Khojohalai, Khownamtu, Amlka, Holpeena, Tufum, Dangki, Jajum, Nipun, Pelu, Param, Katchia Am, Amdang, Ampang, Dangum, Khongkhat, Khonu, Khocho, Longkhap, Komi, Champo, Tuisa, Chautim, Bidguti, Shimoi, Deodhan, Badsho Am, Kerjok, Am tee, Tangra, Jou, Nyu, Drukpanu, Grimicki, Yewang, Noh, Remi, Piyti, Nammi, Naga Am, Yarte, Mikka, Ralo, Mikar, Paki, Ningpung, Mesapala, Patkogo, Ampor, Micha, Emme, Patpakhe.
	Maize	Oghum, Oyusum, Fengtang, Ashum, Top puli, Tapio, Tami, Puiritchi, Topp.
Mizoram	Paddy	Total landraces of paddy are about 94 and important are Farel, Mualkawi, Buhban, Liamrau, (Buhsen (red rice), Mangbuh (upland rice), Changkawl (upland, unirrigated), Biamran (sticking var.), Bohpui, Fangsin, Fazu (Black husk), Fazai, Ido, Shan, Zaitlai, Savite, Maw-buh, Tailuain (bold, small grain), Namthu, Idaw, Ralleng (husk deep red), Tialpui (bold grain), Biruchuk (fine grained), Fangsin-hinoi, Mangbuh, Hinar-buh, Buhsakei (fine grained), Ngaijual (early var.).
	Maize	Mimban, Mimium (can be grown twice a year), Puakzo, Mimpui (white), Mimbanbar (white), Miria Liamdiak, Lenliam, Riakzo
Meghalaya	Paddy	Minel, Minel Makhre, Sapha bhog, Mima, Bora, Sarang, Miphisa, Mukhidep, Sphang, Mibisa, Misiken
	Maize	Poorvi Botapa, Khasi Riewhadem, Tsungrhu.
Assam	Paddy	Kalajira, Manohar Sali, Joha, Champli, Mulukana, Jowsabalam, Kalajoha, Balam, Moynagiri, Malsira, Manohar, Kartik Sali, Bornidhan, Parama, Bhola, Nepalibhog
	Maize	Cachar Gomdhan, Maidani Makka, Teesta, Mendi, Silken Tipang.

Contd.

Tripura	Paddy	Total land races are about 32 and important are Khalisatiya, Chanmouri, Binni, Sonamukhi, Gheegoj, Billrong, Hailchinal, Chailmuni, Mainasal, Jhinger, Beti, Pakhang, Seroy, Garcha, Lalmoti, Katikalam, Nagarsail, Jhingot, Gorasal, Maluti, Chikansal. Jhum varieties: Badia, Mai Sai-Ningha, Marao, Maima, Bhadoia, Lankapora, Kamranga, Lumwur, Jhum Maloti, Koprok, Maiasaw, Kanchali, Misini, Bindi.
Nagaland	Paddy	Total land races are about 65 and important are Sirima, Kuki Malan, Supentha, Apachu, Musarie, Y anthemguv , Lazatsok, Thumpaktsok, Konpemo, Moro, Motiro, Ajinme, Medi, Hepet, Kiwame, Ruo, Shamtha, Atunga, Phratori, Asu, Majhi, Matse, Younyu, Kongyoushou.
	Maize	Pochikhuzhi, Mukhutheri, Omum, Tuphrukhuzi, Suphakhouzhi, Khyosobo, Tsungharo, Yalum, Khonoma, Epok-Tsungharo.
Manipur	Paddy	Changhahsan, Kolchang, Changngatpal, Makhommeill, Charhor, Maonap, Teathabi, Changpalman, Lagmanbi, Napdai, Chao, Tangnar, Kahinghih, Langnei, Changman, Satiralbu, Nikhunkathi, Dziikero, Khhangman, Maophoh, Taingod, Napodai hangmeh, Changsan, Manipur-1, Ramlekmei, Kohn, Phaichang, Hate (black hull), Hnapdai, Changhahsám, Ate, Eyroya, Nhpgai.

Another important feature is the cultivation of 12 plant species of energy rich crop plants. These are *Manihot esculenta, Saccharum officnarum, S. sinense, Pachyrhizus erosus. Maranta arundinacea, Dioscorea alata, Dioscorea esculenta, Helianthus tuberosus, Ipomoea batas, Maghania vestita, Vestita, Vigna vexillata, Alocassia indica, Dioscorea bulbifera.*

V. DIVERSITY OF AGRO-ECOSYSTEMS IN THE HIMALAYAS

An agro-ecosystem is essentially constituted of land crops, agroforestry , forests, domestic animals and water resources. It is a man and nature made ecosystem which is aimed at meeting basic human food, fodder, fuel, manure, fibre, timber, commodity crops and also income generation crops or plants. Out of 21 agro-ecosystems identified, five agro- ecosystems fall in the Himalayas (Sehgal *et al.*, 1990). The state of Himachal Pradesh has been divided into five major physio-graphic regions, namely, the Greater Himalayas, the Lesser Himalayas, the outer Himalayas or Siwaliks, Piedmont plains and Floodplains. These are further subdivided into ten agro-ecological zones on the basis of rainfall, potential evapo-transpiration actual evapo-transpiration, length of growing period, soils and physiography (Sidhu *et al.*, 1997). This illustrates the diversity in agro-ecological zones or agro-ecosystems that occurs in the length and breadth of the Himalayas. Implicit in this is the proper choice of crop species so as to develop suitable cropping patterns (Khushoo, 1992) and provides scope for on-farm conservation of traditional crops and landraces. In each of the agro-ecosystem traditional farmers generally grow ten to twelve or even more crop species to meet all their food requirements all through the year (Maikhuri *et al.*, 1996).

VI. CONSERVATION OF HIMALAYAN AGRO-BIODIVERSITY OF TRADITIONAL AGRO-ECOSYSTEMS

The crop diversity in traditional agro-ecosystems is basic to stability in food production as it minimises crop losses due to diseases, insect pests and weather odds, conserves soil fertility, protects from soil erosion and increases productivity per unit area due to maximum utilisation of production resources. In order to conserve traditional crops and to ensure long term stability of agro-ecosystems there is need to lay emphasis on traditional uses of crops and new value added consumption uses should be found out. Moreover maintenance and conservation of traditional agro-ecosystems in their entirety is the only sensible way to preserve *in situ* repositories of crop germplasm (Nabhan, 1985; Alteri and Merrick, 1987). The issue of on-farm conservation has recently attracted the attention of many scientists but the question as to how to promote it still remains obscure. The mechanisms of compensating the traditional growers for losses accruing on not adoption of modern agricultural techniques are still a debatable issue.

VII. AGRO-BIODIVERSITY AS THE BASIS OF PROSPERITY OF HIMALAYAN FOLK

The agricultural prosperity of Himalayan people solely depends upon the wider base of agro-biodiversity. This biodiversity may be at the crop species level or at varietal (Genetic) level. This is because it uses the available resources in an optimal way and ensures the supply of food inspite of weather odds, epidemics of diseases and insect pests. It is therefore pertinent to maintain wide base of agro-biodiversity in order to enhance prosperity of people living in the Himalayan hills.

The potential agro-biodiversity plant components in the Himalayan hills and mountains, which can provide high income to the people, are off-season vegetables and cut flowers, temperate fruits and nuts and cultivation of medicinal and aromatic plants. However in these cases also a wide diversity base must be maintained.

Change in the Food Habits

It is a well-known fact that when we don't put any thing into use it loses its existence or is bound to be extinct. The same has happened with the agricultural biodiversity i.e. all the age-old varieties, crops and technologies, which were not in use since long, have been dropped from our system and got extinct. The study conducted by Rana *et al.* 1998 in Himachal Pradesh showed that people in general and youngsters in particular were nor found very enthusiastic for growing traditional crops and even in using many of the indigenous technologies even though they are eco-friendly. Moreover, they don't want to deprive themselves from the economic benefits that they earn from cash crops. The data given in the Table 8 revealed a major shift from traditional foods to common foods viz. wheat and rice in both Kangra and Kinnaur districts. It was therefore realised during the investigation that changing food habits is also one of the factors responsible for depletion of agri-diversity. The changes in the food habits of the people are given below.

Genetic Erosion of Agro-Biodiversity

The state and central governments have undertaken massive efforts to modernise agriculture laying much emphasis on introduction of high yielding varieties. These varieties have replaced the landraces of wheat, maize and rice at a very fast rate. The landraces could only be found only in remote, landlocked areas, which had difficult means of communication. This is especially true in Himachal Pradesh. However, situation is not so bad in Jammu and Kashmir hills,

Consumption of different food crops by a family for number of days (average) per year

Food crops	*Kangra*		*% ±*	*Kinnaur*		*% ±*
	1977-78	*1997-98*		*1977-78*	*1997-98*	
Rice	205	350	+ 70	144	304	+ 111
Wheat	121	290	+ 140	258	103	+ 103
Maize	145	75	-48	21 85	+ 304	
Millets	30	10	-66	164	38	-76
Amaranth	26	20	-46	246	160	-35
Buckwheat	21	05	-76	320	201	-37
Chenopod	18	04	-77	102	17	-83
Traditional vegetables	142	65	-54	301	118	-34
Pulses (own produce)	241	112	-53	276	98	-64

Uttaranchal, Sikkim, Arunachal Pradesh and other N-E states. In respect of crop diversity also some of the species either have become rare or completely lost from the cultivation. In Meghalaya, a local cereal *Digitaria cruciata var. esculenta* has become rare. Similarly in Himachal Pradesh, crops like black gram (*Vigna mungo*), green gram (*Vigna radiata*), Kultha (*Macrotyloma uniflorum*), Gram (*Cicer arietinum*), velvet bean (*Mucuna utilis*), Mandua (*Eleusine coracana*), Cheena (*Panicum miliaceum*), Kodo (*Paspalum scrobiculatum*), have become either rare or extinct. In Tripura, landraces of rice had been reduced to 32 from 105 in 1975 (Sharma & Hore, 1990). The cultivation of millets and pseudo-cereals is almost on negligible area. Due to the use of high doses of fertilisers harming natural rhizobium soil cultures, the pulses such as Moong, Mash, Gram and Kultha has been reduced to negligible acreage. Temperate fruit and nuts diversity is also narrow and because of expansion of area under these fruits and vegetables has resulted in replacement of traditional crop species such as grain amaranth, chenopods and buckwheat (Rana *et al.*, 1998; Rana & Sharma, 2001).

VIII. REASONS FOR THE DEPLETION OF AGRO-BIODIVERSITY

(i) Better access to the modern means of livelihood e.g. better transport facilities, communication facilities, availability of better technologies etc.

(ii) Indiscriminate and excessive use of chemicals especially fertilisers (e.g. urea) resulting in unsuitable growth conditions for pulse crops.

(iii) Shift in agriculture i.e. from traditional crops to cash crop. Farmers don't want to deprive themselves from the lucrative benefits they receive from high value cash crops.

(iv) Absence of economic valuation of agro-biodiversity—how much can it contribute to the economy of the community, nation, global community and on what terms?

(v) Migration from rural to urban areas. Majority of the rural people has shifted form villages to cities in search of jobs, better education facilities and prospects for their children.

(vi) Agriculture is considered low social status profession as compared to other professions. In the hills any young chap would prefer to join as daily wages employee rather than to opt agriculture as occupation.

(vii) Agriculture occupation considered highly risk oriented i.e. Most of the agriculture holdings are small, and entirely rain-fed, thereby subjected to natural calamities.

(viii) Commercialisation of agriculture leading has led to the cultivation of only high valued crops and in the process many traditional crops and varieties have been ignored.

(ix) Shrinking of the dietary base of the people. Among cereals only rice and wheat are being consumed by majority of the people and all the traditional crops even maize has been dropped from the normal diet. Majority of the young generation farmers does not relish traditional foods e.g. amaranth, buckwheat, chenopod, other coarse cereals etc. and also does not want to waste time in the adoption of traditional technologies as better substitutes are available.

(x) Disintegration of joint families into nucleus families has resulted in the fragmentation of land holdings and reduction of working hands in the family, thereby some portion of land in the villages is left as barren and uncultivated. It is worth mentioning that in each village at least 4-5 families have left their land uncultivated and 15-20 families have given their land on basis to landless families.

(xi) Protection of community forests has caused fodder scarcity and thus leading to erosion in cattle population. The people used to keep a few goat and sheep herds in the village, which were entirely dependent on community forests. After the protection of village forests by the forest department, people had no other alternative but to sell out their animals. Many rural people were of the opinion that is has affected their socio-economic status as they were not able to fetch their basic needs viz. Fuel, small timber and fodder from the forests and now has to opt for LPG, kerosene and other costly means of living.

REFERENCES

Arora, RK. 1996. Indian region provides treasures house of wild plant genetic resources. Diversity 12: 22.

Arora, RK & E R Nayar. 1984. Wild relatives of crop plants in India. In Science Monograph, NBPGR, New Dehli.

Alteri, MA & LC Merrick. 1987. *Insitu* conservation of crop genetic resource trough maintenance of traditional farming systems. Econ. Botany 4186-96.

Bhag Mal. 1997. Management, sustainable use and value addition in Grass covers of India. Paper presented at workshop on "National concern for conservation, management and use of Agro-biodiversity". Shimla, October 15-16, 1997.

Khushoo, TN 1992. Pandit Govind Ballabh pant Memorial Lecture II Delive Institute of Himalayan Environment and Development, Almora.

Maikhuri, RK, KS Rao & KG Saxena. 1996. Traditional crop diversity for sustainable development of central Himalyan agro-ecosystems. Int. J. Sustainable Dev. World Ecol. 3:8-31.

Nabhan, GP. 1985. Native crop diversity in aridoamerica' conservation of regional genepools Econ. Botany 39 : 387-394

Rana, J. C., B. D. Sharma & PL Gautam. 1998. Diversity and Conservation strategies of Medicinal Plants Diversity in the North-Western Himalayas. Presented in the National Symposium on Spices, Medicinal and aromatic plants -Biodiversity, Conservation and utilization held at IISR Calicut on 10-12 August, 1998.

Rana, JC & BD Sharma. 2001. Shrinking crops base–challenge for food security and nutrition. Paper presented in the 88th session of Indian Sceinece Congress held at IARI, New Delhi on 3-7 January 2001.

Rana, JC, BD Sharma & PL Gautam. 2000. Agri-biodiversity Erosion in the North-West Indian Himalayas- Some Case Studies. Indian J. Pl. Genet. Resources. 13 : 58-63.

Seghal, JL, DK Mandal, C Mandal & S. Vadirelu. 1990. Agro-ecological regions of India. NBSS Publ. 24. National Bureau of Soil Survey & and Use Planning, (IACR), Nagpur

Shanna BD & DK Hore. 1990. Rice gennplasm collection in Tripura State. Indian J. Pl. genet. Resources. 3 :71-74.

Shanna, B. D., J. C. Rana & B. D. Joshi. 1997 Agri-diversity in North-Westen Himalayas. Indian J. PI. Genet. Resources 10 : 63-6.

Sidhu GS, KPC Rana, J Sehgal & M Velayutham. 1997. Soils of HimachalPradesh., their kinds, distribution, characterization and interpretation for optimising land use. NBSS Publ. 57. National Bureau of Soil Survey & Land Use Planning, (ICAR), Nagpur.

Singh Bhag. 1977. Races of Maize in India. ICAR, New Delhi. p. 106.

13

Ethno-medicobotany of Tribal Areas in Himachal Pradesh, India

NARAIN SINGH CHAUHAN
Department of Forest Products
Dr. YS Parmar University of Horticulture and Forestry,
Nauni-Solan (HP) India

ABSTRACT

Himachal Pradesh, a mountainous State situated in the lap of the lofty West-Himalayan ranges has a rich and diversified flora in its different geographical regions ranging from 247 m to 7000 m above mean sea level. It is situated between 30° 22′ N to 33° 13′ N latitude and 75° 23′ 24″ E to 79° 00′ 50″ E longitudes having a maximum East-West stretch of 315 km and North- South stretch of 310 km covering an area of 55673 sq. km. Majority of the area is designated as tribal and backward constituting chiefly the districts of Kinnaur, Lahau1-Spiti along with Pangi- Bharmour of Chamba; Chhota and Bara Bhangal areas of Kangra and Dodra-Kawar of Shimla. Many of these areas are highly inaccessible and have many high ranging passes for graziers and trekkers and are the storehouse of medicinal and aromatic plants rarely found elsewhere. Pooh in Kinnaur, Lahau1-Spiti and parts of Pangi-Bharmour are the typical cold deserts in the region.

The tribal way of life, people's daily encounters with their herbal wealth growing around them; existing lores and mores, myths and legends, songs and proverbs about plants coupled with the tradition of herbal lore having been passed down from generation to generation, is the heritage of these localities. Propitiating deities, jantras and mantras and use of herbs is still practised to cure various diseases like snakebite, scorpion sting, rabies, fractured bones, herpes, measles, eczema, eye diseases, malignant ulcers, diarrhoea and dysentery, piles, alopasia etc. The herbs are either administrated singly in the form of juice, infusion, decoction, powder or paste or in combination with other ingredients. More than 300 such plants have been identified through surveys conducted in these areas for over 2 decades. Most important and efficacious species of medicinal herbs have been selected for the present paper. These include many such species the use of which is recorded for the first time.

Key Words: Tribal, Herbal wealth, traditional lores, cold desert

I. INTRODUCTION

Himachal Pradesh, one of the pioneer Himalayan states, is a rich repository of medicinal wealth and occupies an important place in the Vedic treatise In ancient times, it has been the abode of *'rishies'* and *'munies'*, who pursued their meditational and scholarly endeavours here. In this land of Gods, Goddesses and religious resorts, lived Pandavas, demons, Yakshas, Kinners and various Maharishies. Rishi Vyasa, Agnivesha, Bhardwaja, Punarvasu Atreya and Kali Dass wrote about this hill abode including treatises on herbs and herbalores. The first ever seminar to cure the sufferings and ailments of the people is reported to be held in some parts of Himachal Pradesh, at place situated somewhere in Shivalik range (Charkapani, 1941). Punarvasu Atreya, a great scholar in Ayurveda, is believed to have lived in Chandra-Bhaga river catchment in Lahaul and Spiti. Indra, the son of Kashyapa, is reported to have Soma in order to immortalise and rejuvenate himself here. References to Kailash, Kinner Pradesh (Kinnaur) and Gandharva-desha are found in ancient literature. In this very sacred land, under the nature's all embracing paradise; calm, peaceful and tranquil atmosphere; these Rishies propagated the physical, spiritual as well as the social values that form an integral part of the religion and culture prevalent in this state.

The researches and practices of *'rishies'* and *'munies'* on herbs proved a panacea for the people of the region. It created awareness and curiosity among the people to seek real knowledge in close association with these ascetics and scholars. The knowledge acquired by a few dedicated local inhabitants, descended down to generations and still forms the traditional practice of herbal cure. Jantras, mantras and the uses of herbs are still practised to cure various diseases including snakebite, scorpion sting, rabies, bone fractures, eczema, wounds, gangrene and birth control. There are many verses, songs, folk-tales, myths and legends in existence in different parts of the state about plants. The evaluation of this information is likely to yield amazing results and will add many new potent sources to the drug industry.

Despite a rich heritage and enormous herbal-wealth existing in the state, it is disheartening to point out that the knowledge acquired and disseminated by the great saints and scholars, is facing extinction on the very land of its origin. Owing to illiteracy and lack of proper written records, the practice of herbal-cure which mostly. often literally depended on the phenomenally good memory of the practitioner, had to suffer a lot during the course of time. The person who knew the art of healing, usually did not reveal his knowledge to any one for fear of being misused. As a result, much of the valuable knowledge of the practitioners used to be burnt in the funeral pyre with the death of such a person.

Some attempts have been made by various workers for highlighting/documenting the traditional medicinal knowledge and local folklores about the plants of Himachal Pradesh. Notable among these are Uniyal and Chauhan (1971, 1973); Chauhan, 1974, 1984, 1988, 1989, 1997, 1998; Shabnam, 1964; Gupta, 1981, Sindhi, 1996; Gupta, 1998 and Sharma, 1998.

II. ETHNO-BOTANY

The area included in the present study is notified as tribal by the Govt. of India. It is entirely a mountainous terrain and is interspersed with river valleys, like Chandra and Bhaga, Sutlej, Spiti, Pabbar. Rupin Lambadag and Uhal valley in Chhota alld Bara Bhangal of Kangra district; high mountain peaks, many of which are still unnamed and unscaled. Snowfall is heavy to very heavy and persistent glaciers on high mountains are a common scene. Means of transport communications are inadequate and most of the areas remain cut off for 4-8 months due to snow. The main occupation of the people in these areas is rearing of sheep and goats, yaks and churus and agriculture. Only one crop is grown in a year for wheat, barley, potato, and off season vegetables as the growing period is very less. Horticulture activities have made Kilmaur famous for providing best quality apples, almonds, walnut, grapes and drying type apricot. Among the medicinal plants, *Crocus sativus* (Saffron), *Bunium persicum* (Black cumin), *Humulus lupulus* (Hops), *lnula recemosa* (Pushkarmool) and *Saussurea costus* (Kuth) are grown in Kinnaur, Lahaul-Spiti, Bara Bhangal area and Pangi-Bharmour. People still adhere to old traditions and rituals, custom and beliefs, fairs and festivities and adorn traditional costumes and ornaments during such celebrations. The belief in gods and deities, jantras and mantras is still prevalent. Mainly ancient and monumental temples and Gompas can be seen in the region which reflect the strong bondage between the traditions of the remote past. Though the modem healthcare facilities are spreading with the sincere efforts of the government, yet people largely rely upon the traditional herbal lore. Every village has 2-3 local vaids, herb healers or Tantriks to provide free services. Being very near to nature and having daily encounters with the plant life growing around them, people of old generation have a remarkable knowledge about the diverse uses of various plants. Such practitioners do not reveal their knowledge to others and it is only through persuasion, motivation and exchange of experience, the information can be gathered worth documenting. Otherwise, an amateur seeker may be misled by fake and unreliable information given by the informants.

The present information has been generated during the course of surveys and explorations carried out by the author over a period of 3 decades and the voucher specimens are housed in the Herbaria of IRISM, Jogindernagar and UHF, Nauni-Solan. It has been found that vast knowledge still lies with old people living near remote forests, shepherds, gaddies, gujjars and local vaids/lamas. Some of the herbs used by them are so interesting that proper investigations for screening and evaluation were needed. Many such herbs are likely to provide new cures, new materials to the world of medicine. The selected plants have been arranged in alphabetical order giving local names followed by local uses and mode of application.

Achillea millefolium Linn. (Amritdhara ghas, Barighas)

Inflorescence yields an essential oil (0.09-0.45%), used as haemostatic. Leaves yield rutin. Herb is better, pungent and aromatic; used as flavouring agent and can be substituted for Hops in the preparation of beer. Its decoction is given in colic; whereas the essential oil is used as astringent, tonic and diphoretic. It is an ingredient of *Liv.-52*. It is also a useful remedy in stomach pains and the leaves are chewed to relieve toothache.

Achyranthes bidentata BI. (Umblokando, Umblakanata)

The root of the plant is given in snake bite. About three *mashas* of root powder is mixed and powdered with equal quantity of smoked tobacco residue from hooka and is applied in the form of a paste to the snake bitten/swollen organ after giving a proper cross cut. The patient is not allowed to sleep for 12 hours at least. Sufficient ghee is given to the patient orally.

Aconitum deinorrhizum Stapf. (Mohra, Mouro, Bish)

Roots are highly poisonous and instances of suicides and deaths are accredited to it. Roots are kept hidden and out of reach of ladies and children and taken out very cautiously by the old/

matured persons for medicinal purposes in small pieces. The roots are smoked in hooka (humble-bubble) in acute gastric pain, toothache and severe pain in rheumatic joints. Root power mixed with mustard oil is massaged on paralytic organs as an effective remedy. It is never applied on abraised skin for fear of poisoning.

Aconitum heterophyllum Wallich ex Royle (Atis, Patis)
The roots are used as an antiperiodic to combat weakness after fever, as an aphrodisiac, astringent and in the form of tonic. It is given to cure diarrhoea, dyspepsia and cough. The roots are also chewed to subside toothache.

Aconogonum rumicifohum (Royle ex Dab.) Hara (Goronthu, Choarh, Dlvyakanda)
The leaves are cooked and squeezed. The water is thrown away and the cooked leaves are prepared as vegetable by the shepherds in Dodra-Kwar and Jakha-Pandar area. The taste is said to be acidic. The root is regarded as an antidote in Aconite poisoning and is chewed to satiate the thirst. It is reported to be an antidote to snakebites and scorpion stings. The paste of the leaves is applied locally in abscesses and boils.

Acorus calamus L. (Bach, Bare)
Rhizomes are effective in curing dyspepsia, colic, fever, bronchitis, dysentery, snakebites and is also used as a nerve tonic and insectifuge. A small piece of the rhizome is chewed empty stomach for a week in asthma.

Adiantum venustum L. (Dhumtolo, Bai-re-barun)
The paste of the aerial parts of the fern is made into six tablets of the size of the almond nut and one tablet, two times a day is given for three days in cases of leucorrhoea. Its juice can be extracted and two spoons each in the morning and evening are given for three days. Only butter and chapati should be taken as food during this period. Chillies, potatoes, urad, rajmah, oilyand spicy items should be avoided at all costs for the drug to be more effective.

Aesculus indica Co.ebr. Ex Camb. (Kanor, Khanor)
The fruits are dried and beaten into flour, washed several times after delivery; in water to remove the bitter taste. It is then dried and kept for use as tonic for ladies; considered useful in colic and stomach problems. The dried fruits are beaten into coarse particles along with shell cover and mixed with fresh or fried soyabean and are given to young sheep and goats as tonic during winters. Leaves are used as dry fodder. The oil expressed from the fruits is used in healing of wounds and the bark is applied in the from of a paste in dislocated joints.

Ainstiaea aptera DC. (Karvi-Duti)
The powder of the dried roots in the dose of two *mashas* (Approx. 2 g) with lukewarm water is an effective cure for acute gastric pain. Thc root is extremely bitter and may be useful as a vemicide. To avoid the bitter taste, a piece of jaggery is taken if required. Shepherds have been using it effectively in hill diarrhoea also. Fresh roots can be chewed in emergency and its effect in acute stomach pain and diarrhoea is quite encouraging.

Ajuga bracteosa Wallich ex Benth. (Ratpacho, Nilkanthe)
The plant is considered a bitter tonic. The leaves are used to erase deposition on tongues of children suffering from stomach complaints and fever. The infusion or the juice is used in

fevers and gastric pains. Leaf juice is dropped into the ear to treat earache and otorrhoea. The pounded leaves in the form of pea size pills are given in pneumonia and typhoid fever with satisfactory results. Three tablets a day are given in fever, bronchitis and in the cases where blood clots appear in phlegm while coughing.

Allium carolinianum DC. (Laot, Jangli lahsun)

Leaves are cooked as vegetables and flower heads are used as condiment. It is a beautiful plant for ornamental purpose. The dried, pink flower heads are sold in the market.

Allium przewalskianum Regel (Koche, Pharna)

Leaves form a flavoured condiment. Bulbs and flower heads are also used as vegetable and spice. Dried, cut parts of leaves are marketed from the area.

Allium rubellum Beib (Neulagu)

Pseudobulbous are taken as vegetable and the aerial parts are used as a prized spice. Useful as anti-inflammatory and anti-bacterial.

Androsace rotundifolia Hardw. (Golpatte-phulnu, Nirodhak-but)

The leaves of the herb along with a few needles of *Cedrus deodara* are powdered and formed into tablets of the size of apricot kernel with the addition of gur (jaggery). As per one informant, six such tablets are prepared and given in the dose of one tablet three time a day along with about 25 gm of ghee and same quantity of gur for two days trom the first day of the onset of menses for affecting birth control. As per the practice, one tablet a day is given for four days from the first day of menses for similar effect.

Angelica glauca Edgew. (Chora)

Used as condiment in cookery especially while cooking black gram pulse. Also used in dyspepsia and stomach-ache. It is taken internally for bringing out phlegm deposited in lungs.

Aquilegia fragrans Benth. (Lunded darje, Lamo)

The herb is diuretic, diaphoretic, antiscorbutic and tranquilizing. In homoeopathy it is employed for hysteria, somnolency, dysmenorrhoea, and chronic skin troubles. Roots are used for removal of kidney stones. Seeds are used as an oxytoxic and for jaundice.

Arctium lappa Linn. (Jangli kuth, Burdock)

Roots are used as diuretic and diaphoretic in gout and skin infections. Tincture of seeds is used for psoriasis and toothache.

Arnebia euchroma (Royle) Johnston (Khamet, Ratanjot)

Used as an antiseptic and antibiotic in healing of the cuts and wounds when poultice. Also used for subsiding swelling of the organs. Red dye from the roots is in silk and wool. The roots are used against toothache, ear-ache and has anti-cancer properties and found to have antibacterial properties as well. The paste made is applied for effective healing.

Artemisia brevifolia Wall. ex DC. (Bhurchhe, Bhurse)

Aromatic leaves are used as astringent, stomachic, stimulant, ferbrifuge, and also as spice. It can be used as an insecticide. The plant is being investigated for anti-cancer properties.

Artemisia dracunculus Linn. (Barma, Dragon Mugworty)
Aromatic leaves are used as stomachic, stimulant, and ferbrifuge and also as spice. The herb yields an essential oil, used for flavouring vinegar, sauces, salads, canned soups and liquours and is also employed as cheap perfume.

Artemisia vulgaris L. (Chhamber)
The leaves of the plant are crushed and the paste is applied on the cuts and wounds to check bleeding. The wound is wrapped with a cloth and after few minutes, bleeding stops.

Berberis aristata DC. (Kashmal)
The roots are used as fuel wood while the extract of the root bark, roots and stem is used to cure skin diseases, diarrhoea, jaundice, eye diseases, piles and malaria. Berries are eaten as laxative and anti-scorbutic.

Bergenia ciliata (Haw.) Sternb. (Dhoklambu, Patharchat)
The root is used in fever, diarrhoea, pulmonary affections and in swollen joints. Bruised leaves are applied as diuretic to boils and opthalmia. It is credited with expulsion of kidney and bladder stones.

Bergenia stracheyi (Hook. f & Th.) D. Don (Khilche, Gatikpa)
Rhizome is regarded as diuretic and credited with the properties of expulsion of urinary bladder stones. Also found to possess analgesic and anti-inflammatory properties.

Betula utilis D. Don (Takpa, Tajpa, Bhojpatra)
Bark was used in the ancient times as writing sheets of manuscripts. Infusion of bark is aromatic, antiseptic and carminative. Betulin from bark shows anti-fertility activity. Fungal growth on the tree known as 'Bhurja granthi' is useful in muscular pains and the water extract is used as a vaginal douche to avoid conception.

Bidens pilosa Linn. (Kando, Kumbro, Chorpushpi)
The leaves and young shoots are used as a vegetable curry in skin diseases including leprosy. The leaf juice is used in ulcers and wounds. The achenes are used as anthelmintic for domestic cattle.

Boehmeria platyphylla Don (Samrala)
One matured leaf or two to three small leaves are ground into powder with 2-3 grains of black pepper and then applied on the eczema affected parts of the organs as a paste. It is found as a magic cure for any type of eczema, old or new.

Boenninghausenia albiflora (Hook.) Reichb. Ex Meissn. (Pissumar Buti)
The entire aerial parts are used to repel lice, fleas and other insects. Aerial parts and the shoot is usually spread beneath the bed sheet and allowed to remain for few days.

Bupleurum falcatum Linn. (Kaligewar)
Herb is used as diaphoretic and for stomach and liver complaints.

Calotropis procera (Wild.) Dryand ex W. Ait. (Aak)
The leaves are ground to a paste, which is applied to the burnt skin or swellings. The milky juice is used as a purgative. The bark is used to check dysentery.

Cannabis sativa L. (Bhang)
The extract, locally called 'Sulpha', is smoked in stomach pain, colic, common colds and toothache. The seeds form the culinary article and are used in chutneys and for stuffings of the bread. The warm, roasted seeds are inhaled in flue, and eaten as such in coughs and colds These are also mixed in 'muri' prepared out of rice, wheat and *Panicum miliaceum* (Checna) grains and is distributed among relatives and friends during fairs and festivals. The cultivated plants yield a good quality fibre, which is used for preparing ropes and cordages and fibre shoes, "Pula'. The leaves are rubbed to nullify the harmful effects of stings of bees and wasps. The juice of the roots is given as an antidote to check the narcotic effect of *bhang/charas/ganja.*

Capparis spinosa Linn. (Rutokpa {Plant}, Martokpa {Friut}, Kabra)
Flower buds contain glucoside, rutin and pentosans (4%) on dry weight basis. Seeds yields 34-36 per cent fatty oils, used in Spiti as massage in rheumatic joints, Pickled flower buds are useful in scurvy. Bark is diuretic and tonic used in infection of liver, rheumatism and paralysis. It is an ingredient of *Liv.-52.*

Carduus nutans Linn. (Rucharg, Ruchoa, Musk Thistle)
Flowers are used as febrifuge and are edible along with thick pith of the stem. Roots are often eaten raw and the root extracts can be used to kill roundworms.

Carum carvi Linn. (Dru, Shinguzira, Mawo, Kalazira, Caraway)
The fruits are used as condiments. Regarded aromatic, pungent, stomachic, carminative and useful in flatulence. Young plants along with roots are cooked as vegetable.

Cassiope fastigiata D. Don (Hieunshelo, Hieunsheli)
The twigs are ground into a paste and applied to fireburns. It exerts immediate cooling effect and is effective in healing the wound also. It is kept in houses as an emergency medicine for this purpose.

Cedrus deodara (Roxb. Ex Lamb.) G. Don (Diar, Kewti, Keto)
The oil extracted from the wood is applied on rheumatic joints as a massage. The oil is also applied to ulcers and wounds. It acts as a good insect repellent. In cattle wounds, it helps in healing the wound and killing the mites etc. The oil is massaged in inner parts of the thighs to induce sterility in men. Two to three drops of the oil, if given to drink in tea for one week to 15 days is also said to induce sterility in the males.

Celtis australis Linn. (Khouruk, Kharik, Khirak)
The bark of the tree is made into a fine paste and is applied to the hurt (as a result of physical injury) muscular part, where the part of the body has become blue-black with the skin intact and is quite painful. The application of the paste is said to declot the blood and subsides the pain and helps in healing the tissues. Fruits are sweet and eaten in amenorrhoea and colic. Bear enjoys the fruits during July-September.

Chenopodium album L. (Bathu)
A variable plant both wild as well as cultivated in thc field borders for seeds. The young leaves are used as vegetable and given in leucoderma. The juice of the leaves is also applied on the depigmented skin spots upto one month. It is said to be effective in early stages of the disease.

The decoction of the seeds is given in large doses to induce abortion in women. In small doses, it is supposed to be effective in cough and coryza. The flour prepared out of its seeds is utilised as a catalyst for preparing *soor,* a local alcoholic breverage. In far flung areas; the seeds are cooked along with cheena an shathi rice as staple food occasionally/ceremoniously.

Cirsium arvense (L.) Scop. (Bhungchee, Bhungcho)
The root is eaten fresh for stomach complaints, especially useful in pains due to roundworm infestation. Its taste is agreeable with a faint peculiar aroma. The juice extracted from roots is given to relieve the burning sensation in the liver in snakebites. It exercieses a cooling effect.

Clematis buchananiana Benth. (Laujewarh, Kaunibel)
The infusion or decoction of the leaves is given in fever. The stem and branches are intertwined in a ring for keeping on it the kitchen utensils to save the floor staining black.

Codonopsis ovata Benth. (Ruchoukpa, Kampadugdug, Sardandi, Khiri) and C. *rotundifolia* (Sardanda, Khirawlo)
The milky roots are cut into slices and added to goat milk to make it viscous and gel type. The roots can be eaten raw, are sweet in taste, considered a good physical and sexual tonic. The unbranched radish like root of C. *rotundifolia* acts as aphrodisiac and stimulant and is known as sardanda. Sardandi C. *ovata* has thick robust, long, branched roots and is attributed with feminine properties and its use is claimed to be of cooling nature and lowers down the stimulating effect of sardanda.

Colocasia antiquorum (L.) Schott (Arbi, Kuchawari, Ghandiali)
A cultivated plant for its corns, used as staple food and frequently as vegetable. The cooked and peeled corns are cut into pieces in the evening and kept on roof or open places uncovered overnight to allow dew drops to fall on it. In the morning the cut pieces are given with honey to the patient suffering from jaundice.

Colutea nepalensis Sims. (Braa)
Leaves are purgative and also used as an adulterant of senna leaves (Cassia senna). Seeds are emetic.

Corydalis govaniana Wall. (Chingti, Bhutjata, Indrajata)
The decoction of fue whole plant is give in chronic fevers and liver complaints. Root is used as tonic, diuretic and in eye diseases.

Cotoneaster microphyllus Wall. ex Lindl. (Chhaato, Mango, Ruinsh)
The bright red fruits are eaten. The pulp is used to prepare chutney and jams.

Cuscuta reflexa Roxb. (Sutrawlo, Akashbel)
The plant is used in piles. 20-35 gm of crushed herb, 5-7 seeds of 'illaichi' (lesser cardamom) and about 25 gm of kuje-misri are all put in a glass of water in the evening and the decanted solution is given for drinking regularly for eight days in the morning to cure piles. The paste of the plant with purified butter is applied effectively in sinus. In the area, legend goes that whosoever finds the thread-like white root of this plant and keeps it in milk in his house and changes in fresh milk pot daily, it is reputed to bring prosperity. As a parasite, there is no scientific basis of its bearing the roots, but even then this lore is much talked about.

Cynodon dactylon Pers. (Joob, Doob)
The grass shoots are used for sprinkling holy water during poojah. The shoots with honey is offered to the parents of new born male issue as a token of good will and blessings called 'Jubadali'. The parents concerned happily return double the amount to the offerer. The juice of the plant is used to check nose bleeding, called 'nathar' or 'nakseer'. A few drops are dropped into the nostrils and it acts as styptic. The whitish roots, cleaned and washed are used as memory sharpener.

Cynoglossum glochidiatum Wallich ex Benth. (Kumbro, Khoudre, Kuri)
The juice from the properly washed leaves is applied like eye-drops in conjunctivitis and reddening of the eyes. The juice is purplish in colour. The flowers are collected and added in a small bottle containing distilled water and tightly corked. After a few days, the solution so obtained is filtered and used as eye drops with efficacy.

Dactylorhiza hatagirea (D. Don) Soo' (Angbolaghpo, Panja, Salampanja)
Roots are used as farinaceous food, nervine tonic and aphrodisiac. Mucilaginous jelly is nutritious and is useful in diarrhoea, dysentery and chronic fevers. Its decoction with sugar and spices is a useful drink for the sick and the debilitated person. It is useful in seminal debility, used as tonic for women after delivery and also in paralytic affections. The palmate tuber is also used in oracle rites during shanti yagya in erstwhile Bushair State.

Delphinium-vesiitum Wallich ex Royle (Salyan)
Locally the flowers are powdered and mixed with oil and applied to the hairs by the village women to kill lice. The powder of the roots is applied on the cuts and wounds in case of cattle to repel the insects during summer and help heal the wound soon.

Dioscorea bupleuroides Nees. (Pipri, Ludra-buti, Samp-ke-buti)
A fine thick paste of the herb (leaves and top shoot) is applied on the wounds in snakebite. After a few minutes of its applications, yellowish secretion is reported to ooze out from the wound bringing out the venom. The juice of leaves is given to children suffering from fever and stomach troubles in the dose of half to one tea spoon. In bone fractures, the pounded leaves are given in the form of a tablet and is claimed to be effective in cases of cattle, sheep and goats.

Dicliptera kumaonensis Kunth. (Magyainyu, Magaiyun)
The tubers are eaten as vegetable by the women suffering from leucorrhoea. The hard, black shining circular bulbils are pierced and fitted in a thread to prepare a beaded necklace by the young girls in hilly villages.

Dioscorea deltoidea Wall. ex Kunth. (Singli-Mingli, Ganj, Kins)
Locally the tubers are used to kill lice and to poison fish. It is also used for washing the woollen clothes. Cortisone of which Diosogcnin is a content, is used in the treatment of rheumatic and eye diseases in modem medicines and also in birth control pills.

Eleusine coracana (L.) Gaertn. (Koda, Kodra, Mandua)
An annual crop plant, cultivated in all parts of the Pabbar valley as foodgrain. The pudding prepared of its flour with gur is given in colds, coughs and coryza. The flour is also utilised in preparing local alcoholic beverage called as *soor* and good quality product is used as medicine in acute stomach-ache, in whooping cough, common colds etc.

Equisetum arvense L. (Harjori)
The infusion of the herb in the form of a raw tablets or juice is given orally in bone cracks and fractures for 7-8 days. The paste is also applied externally on the affected part after fixing the bone properly. It helps in early unification of broken bony pieces provided, the part is properly bandaged. The juice extracted from the stem and branches is given for 2-3 days with milk and sugar in leucorrhoea.

Euphorbia pilosa Linn. (Shadhan)
The plant has got both emetlc as well as purgative properties. Shoots are emetic whereas the root is purgative. The milky latex has also the similar properties in respective parts. If a part of branch/stem is plucked, the latex oozing out from the separated part is considered purgative. It is mixed in 'sattu' or 'gur' and stored properly for use. The latex from the broken top of the plant which oozes upwards is collected in the same way and used as emetic at the time of need. The root part, when ground into a paste or powder by moving the grinding stone only in one direction repeatedly, causes purgation and if as a normal process, the stone is moved backward and forward while grinding, the use of this material causes both purgation and vomiting (detailed investigation are suggested).

Fraxinus xanthoxyloides (G. Don) DC. (Pipjul, Thuma)
The bark is an esteemed cattle medicine. The decoction of the stems/twigs is applied to wounds and bone fractures. The green paste obtained as a result of extract is locally applied to wounds caused by horse bites, wherein part of skin and flesh is removed as a result of individual fights.

Galinsoga parvijlora Cav. (Pipli-ghas)
Leaf juice is dropped into the ear to cure earache.

Gentiana tianschanica Rupr. Ex Kusn. (Karu, Kour, Nilakant, Tikta)
Roots are used as Indian Gentian for *Gentiana kurroo* as bitter tonic, stomachic, liver stimulant, blood purifier, antipuretic and usefull in urinary infections. Also used as a masala for fattening horses in Spiti and Ladakh.

Gentianella moorcroftiana (Wall. ex G. Don) Airy Shaw (Tikta)
The decoction of whole plant is used as febrifuge, antipyretic, blood purifier, in cough as expectorant and also to cure bile complaints in Lahaul-Spiti area.

Gentianella paludosa (Hook.) Harry Smith (Tikta)
The whole plant is used in fevers and as blood purifier, either singly or in combination with other herbs.

Geranium nepalense Sweet (Laljari)
Generally used in tooth-ache, to check bleeding and in ulcers and jaundice. The power of the roots when mixed with hot butter oil is applied to the cuts and wounds and cures them rapidly.

Geum elatum Wall. (Masreen)
The fine powder of the dried roots with ghee is applied on deep cuts and wounds. It binds the wound together and effects quick healing even by a single application. The best procedure adopted is to wash the wound with hot water properly and dry it by the use of cotton or cloth

piece. Apply lukewarm butter oil all over the wound and the fine root powder is applied on it in a thick layer. Proper bandaging of the wound is done and the healing results within three to seven days time. It is very effective home remedy even for deep cuts.

Girardinia diversifolia (Link) Friis (Bara Bichhua, Karal)
Leaves are used in headache and swollen joints. Its decoction is given in fever. The forced contact of the stinging shoots is made with swollen joints to activate blood circulation. The blisters produced as a result there of give out watery juice and ultimately result in subsiding of the swellings. The paste of the root is applied to boils to enhance suppuration. The fibre from the plant forms a good quality material for ropes, cords and 'PAULA' (fibre shoes).

Hedychium acuminatum Wall. (Banhaldi, Kapur-kachari)
The roots are used in stomach disease as carminative, stimulant, expectorant, useful in vomiting, diarrhoea, pains and inflammations. Also said to be used in snakebites.

Hippophae rhamnoides Hook. f. (Chharma, Tirku, Seabuckthorn).
Plant contains steroids. Fruits are rich in Vitamin-C. Seeds yield fatty oil. Fruits are used for preparation of jelly, syrup and chutney useful against pulmonary complaints.

Hippophae salicifolia D. Don (Chhuchha, Chuk)
The paste of the bark is reported to be useful in healing of the wounds and ulcers. The acidic fruits are used to prepare chutney which is useful in respiratory problems. Fruit juice with powdered common salt is sprinkled on snow to make it granular and solid, which is eaten at leisure by the natives during the sunny days in spring.

Hyoscyanus niger Linn. (Khurasani ajwain, Thuglang)
Plant is regarded as poisonous. The seeds are put on red hot iron and the smoke is inhaled by month to expel-worm from decaying/painful tooth. Mantras are also enchanted in local dialect.

Hyssopus officinalis Linn. (Tengu, Tyangu, Juffa)
It is a highly scented herb yielding upto 1 % essential oil. Flowering tops are used by local lamas in fevers and blood related diseases. Tibetan Medical Centre, Dharamsala sends trained collectors to Spiti every year to collect the flowering tops and it is said to be used as a substitute for saffron in their drug fonnulations.

Juniperus communis Linn. (Hauber, Chhershi)
It is used as incense. Plant is used in rheumatism by the local vaids. Junipers berries are also used in dry gin blending by the breweries.

Juniperus squamata Buch.- Ham. Ex D. Don. (Theleru, Dhoop)
The leafy twigs are used in 'Havans' as an aromatic ingredient for dhoop. The twigs are also mythically used to dispel the effects of bad eye/evil spirit. The stems and branches are used as fuel wood at the alpine pastures above snow line.

Lepidium latifolium Lion. (Chulti)
Infusion of the plant is given in liver and kidney related problems. Also used against skin affections.

Leucas lanata Benth. (Dhurlughas)
The decoction of the leaves and apical shoots with jaggery (gur) acts as a good abortifacient upto or under three months pregnancy. The roasted leaves in ghee have been successfully tried as a remedy to expel the placenta after delivery in buffaloes and cows. The plant is also reported to be a good remedy in diarrhoea and dysentery in cattle and is given with bamboo leaves. If given with buttermilk (lassi), it is reported to treat constipation also.

Lilium giganteum Wall. (Bhook)
Gaddies and grazers line the inner side of their caps with its leaves and it is reported to exert a cooling effect in hot sun and also gives relief in fevers. The hollow stem is used as a musical, instrument by the young village kids.

Lonicera angustijolia Wall. ex DC, (Pidoo, Bandakh)
The fruits are quite sweet and make a delicious desert. The paste of the leaves is said to be usefully applied to cure boils and sores.

Malva rotundifolia Linn. (Chhambasa).
Seeds are used against cough and inflammation of bladder. Mucilaginous extract of the plants is given in leucorrhoea.

Malva verticillata L. (Shorshotle, Shochlo, Shochle)
The juice extracted from the roots is said to be useful in leucorrhoea. The crushed root in water is given to cows and buffaloes to facilitate detachment and expulsion of placenta after delivery. The infusion of the root is used in washing woollen garments. It gives a thick foam and helps in making the garments thick and stuffy, and aids in cleaning.

Melilotus o/ficinalis (Linn.) Pall (Bhuksup)
Dried leaves and flowering tops are used in the preparation of cigarettes which are smoked for relief in asthma.

Mentha longifolia (L.) Hudson (Safed Pudina Chachri, Chachar)
The thin fine paste of the leaves is applied to rheumatic and painful swollen joints. If a thick paste is applied, blisters may be produced on the skin. The watery juice oozing out from these blisters helps in subsiding the swellings and gives permanent relief from the painful problem. Thin paste is advisable to be on the safer side as excessive blisters may lead to wounds.

Micromeria biflora (Buch-Ham.ex D. Don) Benth. (Jwane)
It is used for disinfecting worm infected wounds of cattle. Its paste is applied on eczematous and skin rashes of unknown etiology.

Morina longifolia Wall. ex DC. (Bishkandara, Dhoop)
Root powder is applied as a poultice in boils for sucking the puss out of it and facilitate healing of the wounds. The root is used as incense, whereas the leaves have disagreeable odour.

Nardostachys grandiflora (D. Don) DC) (Machhi, Mansi)
Local shepherds and old people keep the roots in their possession. A small piece of the root is powered and mixed with tobacco and smoked in cases of palpitation of heart and mental

tension. The roots are kept in oil and applied by woman as an aromatic oil to promote the luxuriant growth of hairs.

Nasturtium officinale R. Br. (Chala, Chhuchha)

The young shoots are used as leafy vegetables and its use is said to act as an appetiser, laxative and diuretic. It is believed to increase blood circulation.

Nepeta ciliaris Benth. (Pissu-buti)

The crushed leaves are spread on the floor of the room or on the cot, under bedcover to act as repellent for fleas, bed bugs and other harmful insect pests. The plant has a strong odour, which helps as repellent and keeps the pests away.

Nepeta podostachys Benth. (Ribuksu, Ribhuksu)

The herb is used in cough and fevers.

Onosma bracieatum Wall. (Pasa, Ratanjot, Gaojabam)

The herb is used as tonic, demulcent, diuretic and refrigerant. Useful in spasmolytic troubles.

Onosma hispidum Wall. ex G. Don (Ratanjot, Pasa)

Source of Ratanjot. The red dye obtained from the roots is used for colouring food stufts, oils and medicinal preparations. Dried root is used against skin related diseases.

Origanum vulgare Linn. (Basloghas, Jakham-buti)

The plant is regarded as an important household remedy for various purposes. The parts of leaves and terminals shoot are applied to boils, ulcers, wounds, cuts and weeping eczema. Two to three fruits of black pepper are mixed while making the paste. The paste of the leaves is reported to be highly useful in healing the wounds caused by fire-burns and abscess of the breast. A tablet made of its leaves along with the leaves of *Plectranthus coetsa* is given with flour or folded bread to animals in cases of bone fractures. The root piece of the plant is bound in a cloth piece and tied to the necks of infants as a protective measures against conjunctivitis.

Oxalis corniculata Linn. (Umlori, Alori)

A fine paste of the aerial parts of the herb along with a few seeds of black pepper is applied to boils, abscesses, wounds and weeping eczema. This application helps in drying and healing of the wounds. This paste is reported quite useful in abscesses of the mammary glands. Leaf juice is regarded useful in curing cataract of thc eyes, and also used externally to remove warts.

Oxyria digyna (Linn.) Hill (Suma, Chuchh)

The leaves are considered carminative and digestive. Useful in abdominal problems. Leaves are also used as vegetable and for making chutney.

Paeonia emodi Wall. ex Royle (Udsalib)

Internally the powder of the roots is given in mental diseases and as a blood purifier. The decoction of the roots is externally applied for healing the cuts, wounds and ulcers.

Phlomis bracteosa Royle (Kotha)

Nectar of the flowers is put in the eyes and has soothing effect against irritation.

Picris hieracioides Linn. (Chava)
Leaves are bitter and used as febrifuge.

Pinus wallichiana A. B. Jackson (Cheel Kail)
The resin from the young sapling is used as a lotion for healing of the cuts and wounds. The bark taken out in the from of long circular cylinders is used to bandage the dislocated/fractured organs both in cattle, sheep and goats and even on human beings after setting the dislocated or broken bone in proper place. This helps as a plaster in keeping the organ in resting position and facilitates calcification.

Plantago depressa Willd. (Musal-Ninai)
A paste of the plant, including leaves and roots is applied to the muscular abscess (where the skin has cracked and watery secretion and shooting pain occurs). The paste is applied two to three times a day. The paste of leaves is also applied to skin eruptions, boils and rashes. The root of the plant is tied to the button hole of the infants as a protective to ensure against infection of stomach diseases.

Plantago major Linn. (Yakram, Jharam Naram)
Seeds are used in gastric complaints, burning sensation in stomach and dysentery. Leaves are applied to bruises.

Pleurospermum brunonsis Benth. (Nesar, Losar, Garam, Masala)
A highly scented pleasant smelling, small, tufted herb, 6-20 c tall and is found growing on rock boulders and rock crevices in alpine areas between 3800-4500 m. The herb, including leaves and flowering stems is made into garlands with intermediate fixing of *Saussuria obvallata. Bistorta macrophylla* and *Senecio jaquemontianus* flowers These are distributed to distinguished persons on fairs and festivities held during the months of August and September every year by a special party of the shepherds, who in turn receive Re. 1/- Rs. 5/- or 10/- per garland as per one's own sweet will. Its aroma spreads a scintillating effect on the entire congregation. The powder of the flowering shoot is mixed with cow's butter and massaged over the entire body to allay fever The dried herb is used as a preserving agent against the attack of moth and silver fish to protect woollen garments kept in boxes. It can be of great value in perfume industry. The plant is used as spice and flavouring agent. The shoots are dried, powdered and used in cooking.

Polygonatum verticillatum Aelioni (Salam-Misri)
The rhizomes being sweet in taste are eaten by the local people, as an aphrodisiac and general tonic.

Polygonum hydropiper Linn. (Mirchighas, Ghaneri)
The crushed plant is thrown into the pond and streams as a fish poison to catch fish. The juice of the leaves is reported to be useful as a cure for pussing ear.

Po*lygonum macrophylla* D. Don (Chhota ninayin, Kande-re-ninai)
The purplish coloured, recurved, rhizomatic rootstock is sweetish in taste and the paste of it is given to children and infants as a cure for stomach problems, Older patients chew the roots as such for similar problems. The dried roots are arranged in a string of thread and offered by

Kinners to their favourites as a token of regards and good wishes. The roots are eaten in dry state also. The red flowering spikes are arranged in scented *Pleurospermum* garlands to make it more showy.

Polygonum recumbens Royle (Vishalayakarni)
Powder of the whole plant is applied in abscesses, boils and other skin diseases. The decoction is given internally as a blood purifier. Its paste is reported efficaceous in healing the sore skin between fingers of foot which is common in hills during monsoons.

Potentilla eriocarpa Wall. ex Lehm (Penma)
Used in the form of tea in diarrhoea, kidney stones; arthritis and intestinal troubles.

Potentilla nepalensi Hook. (Laljari)
The root powder or the paste is applied in cuts, burns and wounds as a healing agent. The root is also substituted for ratanjot as a red colouring matter for hair oils.

Pimula nivalis D. Don (Nili-ka-phool)
The beautiful flowers are given preference over all flowers of the alpine regions by the local people. As a legendary tale, it is said that he, who has this flower in possession cannot be affected by any magic, spirit (evil) and tantric tricks. He can even detect the tricks played by the magicians which otherwise none can visualise under ordinary conditions. Under side of the leaves has white powder bloom, which gives a fine impression when pressed against paper, cloth or skill.

Prinsepia utilis Royle (Bhekhali, Bhekhal)
The oil extracted from the seeds is taken internally as tonic and is considered useful in general debility and arthritis. The oil is also massaged on rheumatic joints for relief. The extract of Bhekhal root is given in bloody dysentery. The hollow branches are made into flutes and toy pistols by the children. The long hollow tubes of its branches/stems are also used *for* making pipes for smoking tobacco by fitting it in the outlet of Hubbell-bubble (*Hukka*).

Punus armeniaca L. (Chuli, Sharhi, Wild apricot)
The fleshy part of the fruit is eaten as fruit. The ripe fruits as a whole are dried and kept in storage for use during winters. The fresh fruit is beaten into pulp and preserved in the form of slabs or circular cakes and later used as sour ingredient in chutneys. It is also considered very useful for washing the silver and gold ornaments. The edible part of the fruits is also used in higher altitude areas for preparing local alcoholic beverage 'soor' or 'soro'. The oil from it is highly prized for domestic use i.e. cooking, eating in lieu of ghee and for external application like hair oil. It is regarded equivalent in properties to almond oil. The hard shell of the kernel is used as a coal. The charred oily seeds are rubbed into a paste (black). which is a highly efficacious ointment in healing of the cuts, wounds and sores.

Prunus persica (L.) Batsch (Arua, Aru)
Dried and powdered flowers are given as a cure in menstrual disturbances suffered by the teenaged girls. Fruits are edible. Young leaves are rubbed over the skin to cure the white irritating spots often infecting the neck region.

Quercus dilatata Lindl. (Mohru Oak)
The seeds (kernels) are used in urine troubles and the decoction of the wood is used in scabies and other skin diseases.

Rheum australe D. Don (Moti ninayin, Chunkhri, Archo, Thukla, Tukshu)
The dried roots are kept in every house as a panacea for several diseases, The paste of the root is prepared by rubbing one end of the root on a flat surfaced stone with the addition of a few drops of water periodically, until sufficient quantity of the fine gel like paste is produced. This paste is given orally in stomach ache, constipation and swelling in the throat or in tonsils. Externally, it is applied to forehead for headache, in mumps, on cuts and wounds, muscular pains etc. In skin rashes and allergic conditions, it is given orally. As an anticoagulant, it is applied as a paste on tlle muscular part which has swollen and turned blue as a result of injury, blow or a fall. Its lotion is given in bacillary dysentery and dropped in ears in earache. Its other species, *R. moorcroftianum* is also used for similar purposes.

Rhododendron anthopogon D. Don (Simrish, Talsar)
The powder of the dried flowers is mixed in any bland oil and massaged over the entire body of a patient in complications like fevers. coughs and colds. The aroma of the plant is reported to cause giddiness, nausea and vomiting in its natural habitat.

Rhododendron arboreum Smith (Buras, Burans)
The fresh petals are used in chutneys. The flowers of this tree are used by the local people for various purposes. Petals are used to prepare jams and local drinks. The juice of the flower is used to cure dysentery and diarrhoea. Petals are dried and stored for the use in the later part of the year, when the fresh flowers are not available. The flowers are ground with sugar and the paste is fed to the person suffering from nose bleeding. The flowers and leaves are tied in long ropes made of munja grass and tied around the houses including temples as decorations during Baisakhi (BISHOO) festival. The powder of the dried flowers is used as an efficacious drug to check bloody dysentery. The *Rhododendron* flowers have gained a wide popularity for syrup or sharbat known as 'Rhodos uash'

Rhus wallichii Hook. f. (Arkhol, Rikhal, Devil's trees)
The tree has a highly corrosive and vesicarious juice, which causes, blisters on the skin. The juice turns dark brown on exposure. Because of this, people fear to tread near the tree. The dry logs of the tree are used for preparing drum, trumpets, bee houses and feeding trays for cattle, sheep and goats. The dry plank of its wood is also used for agriculture implements for levelling the soil and in preparing musical instruments.

Rubia cordifolia Linn. (Manjistha, Tususho)
The leaf juice or the crushed leaves are applied on cuts and wounds to stop bleeding and healing of the wounds. The fruits are sweetish in taste and are eaten as blood purifiers.

Rubia nepalensis Spreng: (Jangli Palak, Khourasho)
The leaves are comnonly used to cure constipation. Also used as vegetable. The root paste is applied for removal of warts and to cure the pain caused by stinging nettle, its leaves are rubbed on the affected parts to bring relief.

Saussurea gossypiphora D. Don (Ghuggi badshah, Bhergaodha)
The plant is regarded as sacred. The wool of the herb is applied to the cuts, where it sticks compactly, seals the wound, stops bleeding and facilitates healing. In Jakha, the shepherds rub the powder of this plant on the body of newly born sheep and goats, so that the young mother takes more care to nurse them by milk feeding. (It inculcates in the otherwise reluctant mother a sense of we feeling). The plant is kept in houses and its fumigation after burning on coal is credited with the properties of driving away the evil spirits.

Saussurea obvallata (DC.) Edgew. (Brahmakarnal, Kaunl-doda, Dode-or-phool)
Root paste is applied to wounds and cuts. The floral bracts boiled in water are used as fomentation in hydrosele. Flowers are kept between woollen clothes to protect these against insect attact (Moth or Silver fish)

Sedum rosulatum Edgew. (Mooshe-ke-nonaye)
The paste of the leaves is said to be effective in subsiding and removal of warts (*masha*) on the hands and toes.

Selinum tenuifolium Wall. ex DC. (Mathosal)
Roots are used as a catalyst in the preparation of local alcoholic beverage, Dheli. The smoke produced from the roots is used for killing and repelling the insects and for purifying the atmospheric air. Roots are also used as a substitute for Bhutkesi and used as an ingredient in the preparation of Dhoop.

Senecio chrysanthemoides DC. (Sersan mendo)
Essential oil obtained from the plant is used as perfumery material. The herb is toxic to animals.

Skimmia laureola (DC.) Sieb. & Zucc. ex Walp. (Shimshar, Patti-dhoop, Kasturi patra)
The dried leaves when burnt with desi ghee on red hot coal emit highly scented fumes which purify the atmosphere and are used in worshipping of local deities. The orange or purple berries are sweet in taste and are administered to the fever stricken patients to allay the thirst. The infusion of the leaves is also used in Jalodar (excessive accumulation of watery fluid in the stomach).

Strobilanthus auriculatus Nees. (Kapuriminjri)
Locally used as a magic cure for eye diseases like cataract, trachoma etc. The oil droplets are mixed with collyrium (Kajal) and is then applied in eyes as a collyrium with the help of a blunt needle. Within 15 days of its application, the eyes are cured from the disease—is the claim made by the practitioners in Kangra district.

Swertia chirayita Buch.-Ham. (Chirata)
The plant is a bitter tonic, stomachic, febrifuge and laxative. Locally used in fevers and toning up of liver. Its decoction is given internally to the ladies after delivery to protect them from post natal complications.

Tagetes minuta L. (Gaindaghas, Gaindri)
Raw paste of the bruised leaves is applied on the naval of the children to expel roundworms. The bruised leaves are applied to the cuts to stop bleeding.

Tanacetum longifolium Wall. (Amritdhara-ghas)
Aroma of flowers especially when found bruised and crushed by sheep is said to cause giddiness. The leaves are made into a round pill of the size of almond nut and swallowed with water in stomach pain and indigestion. The expressed juice is useful as antispasmodic, carminative and antipyretic.

Tanacetum nubigenum Wall. ex DC. (Bhurse khamb)
Roots are used as incense to purify houses and to ward off harmful insects. Also yields essential oil known as oil of tansy.

Taraxacum officinale Wigg. (Kanphooti, Dudhia, Khirawalo)
Roots are used as diuretic, stomachic, in liver complaints and in the preparation of beer, wine etc. The sheep and goats browse it as a potent fodder The whole plant is crushed into a mesh and given internally in snakebites. The paste is applied externally on the wound. Leaves are effectively used for fomentation in swollen parts, boils and sprains. In young stage, it is also cooked as vegetable 'Sag'.

Taxus wallichiana (Zucc.) Pilger (Thoon, Birmi)
The branches are lopped for fodder in winters in cases of scarcity. The wood chips are used as tea by Kinners (shepherds) and is reported to be refreshing and invigorating. The wood is used as timber for preparing specially designed water channels (*nalas* or *piayus*) fitted at roadsides in memory of the dead person in a family to provide drinking water facilities to the passersby as a noble cause. The wood is also useful for making engravings and drawings of wooden architectural designs.

Thalictrumfoliolosum DC. (Pili-Jarhi)
The roots are used internally in abdominal pains and as a blood purifier. The paste of the roots is applied on the eyelids to cure eye diseases. The poultice of the root is applied to cure the boils.

Thymus serpyllum Linn. (Lepto, Lepte, Padumba, Khishikpasa)
The decoction of the plant is an effective home remedy for colds, cough, coryza, fever and stomach problems. Infusion of leaves with gur acts as a very good diuretic, increases urine flow and relieves pain. The paste of the herb including a few seeds of black pepper (kali mirch) is said to be efficacious in alopacia (falling of hair in circles from head, chin (beard) etc.). Powdered flowers mixed with gur are given as vermicide in appropriate doses. lt is believed that he, who drinks tea medicated with this herb regularly, never suffers from common colds and general stomach ailments. The Tibetans use it as a condiment.

Typhonium diversifolium. Schott. (Chamush)
The tubers are collected during August-September. These are washed, the central hard core removed and then ground into a thick paste of which 2.5-7.5 cm long x 1-2 cm in diameter pieces are moulded. These pieces are macerated with a mixture of garlic and salt and swallowed as such without chewing. Chewing causes severe itching in thc throat because of the presence of certain chemicals. This is regarded as an efficacious remedy for chronic dysentery. complaints of gastrointestinal tract and piles. In addition, it forms a staple diet for the poor tribal people and supplements their cereal requirements during the lean months.

Urtica dioico Linn. (Kimchhi, Kugsha, Bichhu-buti)
The young leaves are cooked as vegetable and regarded as blood purifier and is said to be a cure for skin diseases. The fine paste of the plant, especially of the root part, is applied in hidden muscular pains/hurt. The shocks of the stings of its twigs two to three times are applied in swollen, rheumatic joints for relief from pain and swelling.

Valeriana jotomonsii Jones (Mushkbala, Mansi, Jatamansi)
Used locally as antispasmodic, carminative and in acute stomach aches. It is added to hair oils as aromatic hair tonic, blended in tobacco to stimulate heart and brain and also added to incense. A sizeable quantity of its rhizomes are also collected and exported.

Verboscum thopsus Linn. (Bhoot-lankra, Bhoot-lankuro, Gidar-tambakhu)
The crushed leaves are made into a pill and given in constipation and stomach pains. Decoction of the leaves along with onion is useful for stomach problems The infusion of plant is given in snakebite. A paste of leaves is also applied lacally on the bitten part (wound). The crushed leaves are warmed in mustard oil and used as massage for curing inflammations of the body. The smoke (dhuni) of the plant is utilised to drive away the ghostly instincts influencing the children. The plant is used by the brahmins and tantriks in all such worships and oracles, where bad spirit is involved and is to be driven out of the place so as to give relief to the inmates.

Viola serpens Wall. (Banfasha, Banaksha)
Locally the leaves and flowers are used as expectorant and emollient. The poultice of it is used in throat pains, cough and cold. It is taken as decoction to relax throat muscles and remove congestion of the lungs and chest.

Zehnerio umbelloto Thwaites (Mokri, Gwalkakru)
The ripe fruits are edible for their sweet taste. Seeds are brittle shelled and break easily inside the mouth. It is considered stomachic and tonic. The root fibres bear spindle shaped tubers at the ends of their tips. This tuberous part of the root is washed, dried, powdered and taken with milk in the dose of about two grams twice a day as a cure for seminal debility, spermatorrhoea and its use improves vitality and vigour.

III. CONCLUSIONS

The modern system of medical treatment has become a sophisticated and costly affair beyond the reach of even the middle class and poor have no access to the facilities. Because this system of medical treatment being cost prohibitive, it has got restricted only to the upper class of the society to provide healthcare. In contrast, the traditional systems of medicine have played a pivotal role in providing healthcare to the people living in the far-flung and remote areas, where the former has failed to reach and flourish. These herbal medicines being safer even after prolonged use have not only cured the sufferers of synthetic drugs, but have also proved their remarkable curative properties even after vigorous testing by modern scientific techniques.

Our traditional system of Ayurveda handed down from generation to gencration uses 80 per of medicinal plants in its formulations. In addition, our home remedies called as grandmother's recipes ofter a wide range of curative and preventive herbal lores. China, Russia and Japan have kept their old systems alive and the results have amazed the world. Since, India, particularly the Western Himalayas, has a great diversity in medicinal plants and a rich heritage of traditional knowledge coupled with scientiflc backup, we can expect to achieve the objectives of Alma-Ata declaration of Health for all by 2000 AD and beyond through the use of herbs. It is expected that with extensive surveys an screening of vegetal wealth, many of newer resources of alkaloids, saponins, tannins, anti-malarial, antineuplastic, anti-AIDS drugs etc. are available for the service of humanity.

The increasing gap between the demand and supply of the synthetic drugs, coupled with high costs will not sustain the healthcare to the ever increasing demand of populace. So, it becomes necessary to explore the alternatives to provide healthcare for all and that lies with the medicinal and aromatic plants which grow wild in the forests, banks, bunds of agricultural farms, meadows, cultivated fields, valley areas and alpine pastures. The role of ethno-medicobotany in meeting such demands will be of immense value in the years to come.

REFERENCES

Anonymous 1977-Himachal Mein Ayurved; Dept. of Ayurved, Govt. of Himachal Pradesh.

Chakrapani 1941 Ed. Charakasamhita Nimaysagar press, Bombay.

ChauhanNS 1984 Herbal Wealth of Pabbar Yalley. HP. PG Thesis. HPKV. Palampur.

Chauhan NS 1974 *Ainslaea aptera* DC a hitherto unknown herb used in acute gastritis. Sachitra Ayurveda, Patna: February, pp. 497-99.

Chauhan NS 1987 Himachal Pradcsh Kc Jan Jatiya Kshctra Chhota Bhangal Ki Audhyaniki Evam Vaniki 2: 9 -19.

Chauhan NS 1988 Ethnobotanical study of medicinal plants of Himachal Pradesh. In P. Kaushik (Ed.) Indigenous medicinal plants: M/s Today and Tomorrow Printers & Publishers. New Delhi. pp. 187-19 8.

Chauhan NS 1988 Exploitation and Utilisation of Hcrbal Resourccs of H.P. Invitged lecture delivered "Workshop on Opportunities identification in Agro-based and Food- Products. pp 23-32.

Chauhan NS 1989 Herbal Wealth of Cold Desert area of Himachal Pradesh. proceedings. National Seminar-cum-Workshop on Cold Desert Area in India, (Govt. Press H.P., Shimla). pp. 65-68.

Chauhan NS 1997 Ecological and Ethnobotanical studies on the Flora of Spiti, Himachal Pradesh. Ph.D. Thesis, Department of Bio-Sciences, HP University, Shimla-5.

Chauhan NS 1997 —(I) Some rare and interesting medicinal and aromatic plants of Western Himalaya. Manphar V aidya Patrika, : 36-41.

Chauhan NS 1998 —(II) Some rare and interesting medicinal and aromatic plants of Western Himalaya. Manphar V aidya Patrika, : 16-19.

Chauhan NS 1998 —(III) Some rare and interesting medicinal and aromatic plants of Western Himalaya. Manphar V aidya Patrika, : 23-25.

Chauhan NS & G K Sharma 1988 Banoushadhiyon Ka Himachal Pradesh Ki Samajik Evam Arthik Sthiti Mein Mahatva. Audhyaniki Evam Vaniki 1 : 32-37.

Chauhan NS, MR, Uniyal & BN Sannd 1976 A preliminary study of the Indigenous drugs used at the Tibetan medical Centre. Dharamshala (Himachal Pradesh) Nagarjun, April, pp. 190-193.

Chauhan V & Chauhan NS 1988 Ethnobotany of Trans Giri area of Sirmour district, H.P. Bulletin of medico-Ethnobotanical Research, CCRAS, New-De1hi.

Gupta Harish 1998 Comparative studies on the medicinal and aromatic flora of Churdhar and Rohtang areas ofHP. P.G. Thesis, Dr. YSP. UHI:, Nauni-Solan.

Gupta Rajinder 1981 Plants in Folk Medicines of the Himalayas. ln S.K. Jain (ED) Glimpses of Indian Ethnobotany 83-90.

Panday G. S. & K. C. Chunekar 1969 Bhav Prak Prakash Nighantu; Chowkhamha Vidya Bhawan, Varanasi-1.

Sharma P K 1998 Ethnobotanical studies of *Gaddie*- A tribal community in District Kangra, HP. M.Sc. Thesis, Dr. YSP, UHI:. Nauni-Solan.

Sindhi S 1996 Diversity and economic uti]ity of Flora of Nauni, District M.Sc. Thesis Dr. YSP, UHF, Nauni-So1an.

Uniyal MR & Chauhan NS 1971 Medicinal Plants of Uhal valley in Kangra Forest Division, Himacha1 Pradesh Jour. Res. Ind. Med. 6 : 287-299.

Uniyal MR & Chauhan NS 1973 Traditionally important medicinal plants of Kangra Valley in Dharamshala Forest Circle, Himacha1 Pradesh Jour. Res. Ind. Med. 8: 76-85.

14

Soil Microbial Diversity in Relation to Chemical Transformations and Biotic Interactions

[1]BHOOPANDER GIRI, *ANUPAMA SHUKLA and #K.G. MUKERJI

[1]*Department of Botany, Hindu College, University of Delhi, Delhi 110007*

**Department of Botany, A.N.D. College, Kalkaji, University of Delhi, New Delhi 110019, India*

E-mail: bhoopg@yahoo.com, Ph. 9312242289 #237 DDA/SFS Flats, Mukerji Apartments, Mukerji Nagar, Delhi 110009, India, E-mail: kgmukerji@rediffmail.com, Ph. 011-27654874

ABSTRACT

Microbial diversity is a general term used to include genetic diversity, that is, the amount and distribution of genetic information, within microbial species; diversity of bacterial and fungal species in microbial communities; and ecological diversity, that is variation in community structure, complexity of interactions, number of tropic levels, and number of guilds. The diversity of microorganisms, has a much longer evolutionary history than plants or animals and thus have had more time to evolve into diverse forms. Most probably, the first living creatures appeared about 4 billion years ago, shortly after the earth's surface had cooled enough to allow liquid water to form. Two radically different kinds of organisms co-exist in the contemporary living world i.e. prokaryotes and eukaryotes. The more complex cells constitute eukaryotes, which include algae, fungi and protists. Evolutionary studies revealed a great diversity of eukaryotic organisms as compared to prokaryotic microorganisms. Microbes play a significant role in biogeochemical cycles through the specific chemical transformations (nitrogen phosphorus, sulphur and iron). Their effects on biogeochemical transformations occur through both direct and indirect means. Due to their vast diversity, large populations and long evolutionary history, microorganisms have contributed greatly to the rich and complex interactions among soil organisms. These interactions range from highly specific symbiosis to diffused mutualisms.

I. INTRODUCTION

Soil is a complex and dynamic biological system. It governs plant productivity of terrestrial ecosystems and it maintains biogeochemical cycles because microorganisms in the soil degrade, sooner or later, virtually all organic compounds including persistent xenobiotic and naturally occurring polyphenolic compounds (Nannipieri *et al*, 2003). Soil contains a wide range of living population including macrofauna, mesofauna, microfauna and microflora, is descried as a 'black box' (Paul & Clark, 1989).

Soil is a natural medium in which microbes live, multiply and die. Nowadays, microbial diversity is a critical environmental topic that concerns people from all walks of life are rallying around. Interest in microbial diversity has grown rapidly in scientific community (Wilson, 1988; Franklin, 1993). Increasingly attention is being drawn to microorganisms because the fertility of soil depends not only on its chemical composition, but also on the qualitative and quantitative nature of microorganisms inhabiting it. Maintenance of viable, diverse populations and functioning microbial communities in the soil is essential for sustainable agriculture (di Castri and Younes, 1990; Beare *et al*, 1995). Microorganisms are generally divided into five major taxonomic categories: algae, bacteria, fungi, protists and viruses. Indeed, bacteria and fungi are highly versatile. They can carry out almost all known biological reactions. In soil, they are closely associated with soil particles, mainly clay-organic matter complexes (Foster, 1988).

II. SOIL AS A MICROHABITAT

The microbial population in soil is very diverse hence it is considered a complex microhabitat (Nannipieri and Badalucco, 2003). There is the presence of about 6000 different bacterial genomes per gram of soil by taking the genome size of *E. coli* as a unit. Microbial biomass is large in a temperate grassland soil the bacterial and fungal biomass amounted to 1-2 and 2-5 t ha-1, respectively (Killham, 1994). Soil is structured, heterogenous and discontinuous system, generally poor in nutrients and energy sources, with microorganisms living in discrete microhabitats (Stotzky, 1997). The chemical, physical and biological characteristics of these microhabitats differ in both time and space. Scales of the habitat depends mainly on the size of the organism. Another distinctive characteristic of soil as a microhabitat is property of the solid phase to adsorb important biological molecules such as proteins and nucleic acids. In this way, some enzymes adsorbed by clay minerals or entrapped by humic molecules can maintain their activity, being protected against proteolysis, and thermal and pH denaturation (Nannipieri *et al*, 2002). The surface of soil mineral components can themselves catalyses many reactions. Clay minerals and Mn and Fe oxides catalyse electron transfer reactions, such as the oxidation of phenols and polyphenols with formation of humic substances.

III. ORIGIN OF MICROBIAL DIVERSITY

The diversity of microorganisms, have a much longer evolutionary history than plants or animals and thus have had more time to evolve into diverse forms. Microorganisms have been exposed to and survived cataclysmic conditions unknown by higher animals and plants. Plants and animals are relative newcomers and have only had to prove their adaptive capacity for several hundred million years, a fairly short span of years in evolutionary time. During this time, conditions on the earth's surface have been conducive to survival of plants and animals. Certainly, there have been many examples of species extinction, but, by and large, the temperature has remained fairly stable, there have been few collisions with really large meteroes, volcanic activity has been moderate, and the oceans have remained well mixed and oxygenated.

Microorganisms have proven their ability to face challenges unimaginable to us today. Moreover, microorganisms did not simply occupy various niches offered by earth. Through their chemical activities, they transformed the earth and the atmosphere in a number of ways. Some of these changes actually contributed to making the earth habitable for the plants and animals that appeared much later.

The earth is about 4.5 billion years old. Scientists estimate that the first living creatures appeared about 4 billion years ago, shortly after the earth's surface had cooled enough to allow liquid water to form. These creatures were most similar to modern-day prokaryotes—bacteria and archaea. Because some microorganisms living on earth today are capable of growing in boiling water, life could clearly have begun while the earth's surface was still very hot. Too, because the sun was only about two-thirds as bright as it is today, the earth's surface would have become habitable faster than if the sun has been brighter.

Life during the high-impact period would not have been easy. Some impacts were powerful enough to vaporize oceans, creating clouds of steam that would have sterilized the earth's surface. These events may not have completely obliterated emerging life forms. Microorganisms could have survived this period deep underground. Some may had the capacity of modern microorganisms to produce tough survival forms called spores. Although direct exposure to steam would have killed them, some spores could have survived slightly cooler conditions that would still have been hot enough to kill an actively growing microorganism.

Some microorganisms may actually have been able to live on the earth's surface. One bacterium, *Deinococcus radiodurans,* can survive doses of radiation 3000 times greater than the lethal dose for humans. Most organisms, however, probably developed in the subsurface of landmasses or beneath the ocean surface, where they were proctected to some degree from the UV radiation.

Oxygen Revolution

The revolutionary development occurred between 2.5 and 2 billion years ago, changing the earth and its atmosphere completely. Oxygen gas began to appear in significant amounts in the earth's atmosphere as a result of a microbial metabolic process called oxygenic photosynthesis. Although many compounds such as water contained bound oxygen, there had been no oxygen gas in the atmosphere. Oxygen photosynthesis differed from earlier forms of photosynthesis, in that it split water and released oxygen. The bacteria responsible to this new type of photosynthesis are called cyanobacteria. The first appearance of oxygen left a tangible geological record: banded iron formations in rock. Iron in the earth's crust created with oxygen to form back iron oxides, producing dark bands. Cyanobacteria left a fossil record, too. Some cyanobacteria accumulated to form large mounds called stromatolites. Geologists have found fossilized stromatolites dating back 3 billion years and microfossils of individual cyanobacteria cells that date to 3.5 billion years ago. Cyanobacteria brought the oxygen level of the earth's atmosphere up to about 10% of today's level, high enough to create conditions that favoured evolution of oxygen-utilizing organisms.

Origin of First Eukaryotes

Because many eukaryotes are oxygen dependent, scientists had theorized that protozoa first appeared about 2 billion years ago. However, there are modern protozoa that live anoxic environments, so protozoa could have emerged before the appearance of oxygen in the atmosphere. It is estimated that the appearance period of first protozoa back to about 3 billion years ago. Algae presumably appeared after cyanobacteria, because their chloroplasts were derived from cyanobacteria. They probably evolved within the last 2 billion years. The fungi appeared only comparatively recently, during the last several hundred million years. It is of opinion that terrestrial fungi might have co-evolved with plants, because they are closely associated with them. Until recently, fungi were thought to be exclusively terrestrial. Now fungi have been reported in marine and other locations far from land (Salyers & Whitt, 2001).

IV. TYPES OF SOIL MICROORGANISMS

Microscopic studies led to the recognition of a profoundly important dichotomy among the various groups of organisms with respect to the internal architecture of the cell: two radically different kinds of organisms co-exist in the contemporary living world. The more complex cells constitute eukaryotes (organisms with true nucleus), which include algae, fungi and protists (Fig. 14.1). Evolutionary studies revealed a great diversity of eukaryotic organisms as compared

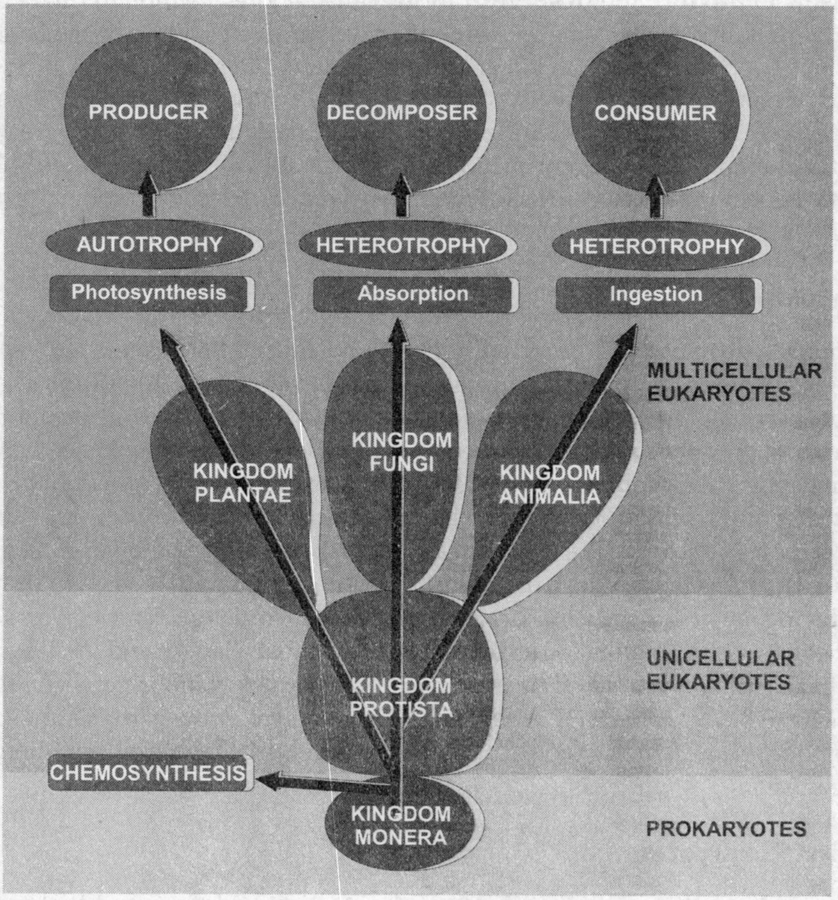

Fig. 14.1: The five-kingdom system showing diversity of organisms (Modified from http://www.npc.edu/bio105/media htm/M1-L7.01htm).

to prokaryotic microorganisms (Fig. 14.2). The less complex cell constitutes prokaryotes, comprising two microbial groups: the eubacteria (including cyanobacteria, this group once known as blue-green algae) and the archaebacteria, a heterogeneous group of microorganisms with prokaryotic structure. These organisms show characteristics and play beneficial roles to mankind. Considering cell structure and function as criteria; there are three groups of cellular organisms: eukaryotes, eubacteria and the archaebacteria. The eukaryotes can be subdivided into three further groups: the plants, animals and protists. The eubacteria can be subdivided into purple, green, gram-positive and gram-negative eubacteria on the basis of cell wall. On the basis of their nutritional requirements, prokaryotes have been categoried as photoautotroph,

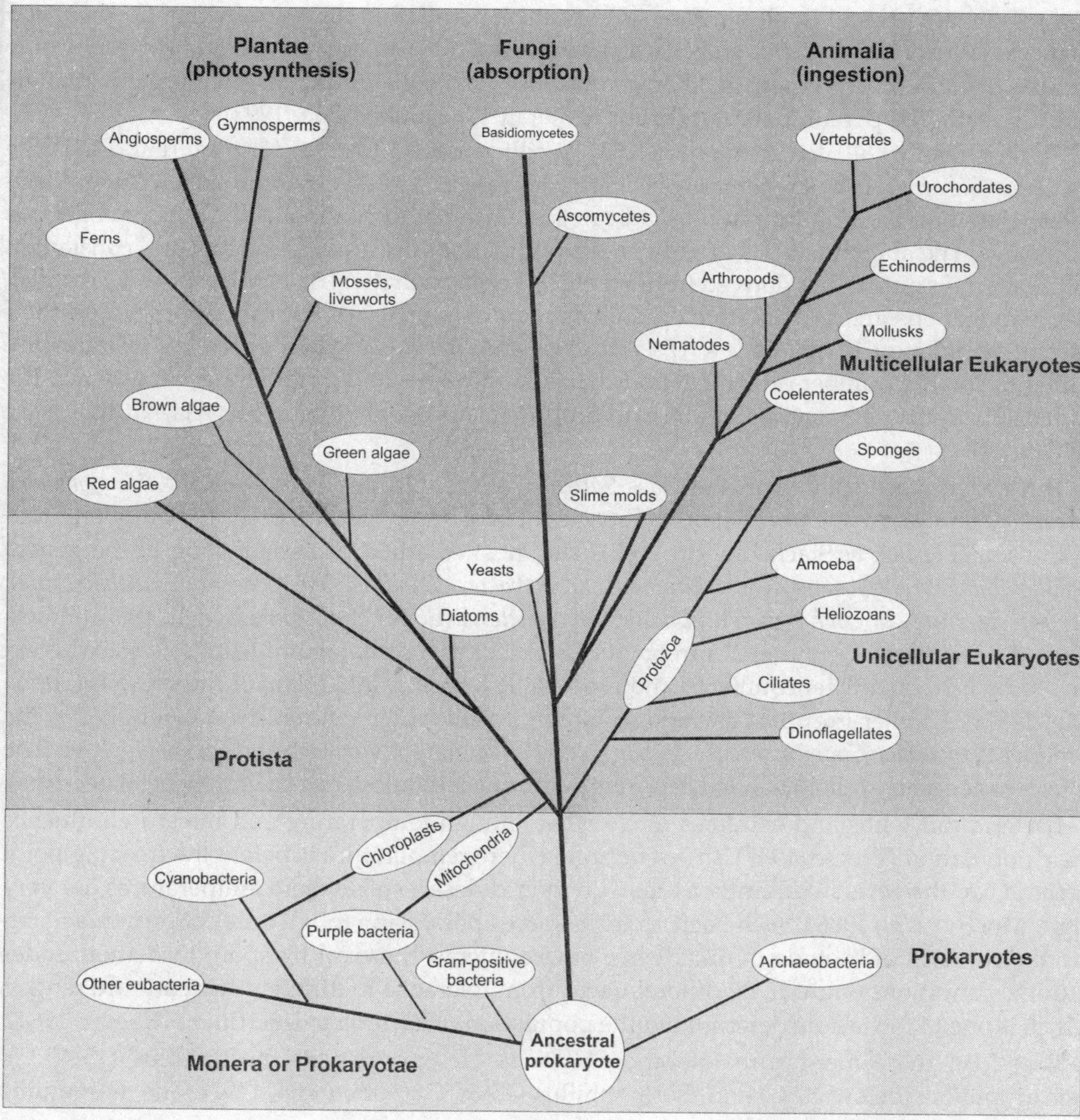

Fig. 14.2: Diagram showing the diversity of organisms (modified from Prescott *et al*, 1996)

Photoheterotroph, Chemolithoautotroph, and Chemolithoheterotroph. Bacteria have also classified as Oxybiont and Anoxybiont on the basis of their oxygen requirement. Prokaryote diversity is not, however, restricted only to relationships to molecular oxygen or to the ability to utilize radiant energy to capture energy. Their optimal diversity also depends on soil pH, temperatures (cold, ambient, hot), inorganic salts (Hurst, 2002).

Eubacteria

Eubacteria are oxybiont prokaryotic microorganisms. They are recognized as most dominant group of microorganisms among the various kinds of soil (Visscher *et al*, 1992; Borneman *et al*, 1996). They are present in all types of soil but their population decreases as the depth of soil increases (Wieland *et al*, 2001). In general, horizon A (soil with organic matter) of a soil profile consists of more microorganisms than horizon B (silicate clay minerals plus organic matter) and C (weathered parent material) (Bruns & Slatar, 1982; Subba Rao, 1997).

Bacteria live in soil as cocci (sphere 0.5μ), bacilli (rods, 0.5 to 0.3μ) or spiral (Fig. 14.3). The bacilli are common in soil whereas spirilli are very rare in natural environments. Winogradsky (1925) classified bacteria into two broad categories, the autochthonous and the zymogenous organisms. The autochthonous or indigenous populations are always uniform and constant in soil, since their nutrition is derived from native soil organic matter (*Arthrobacteria* and *Nocardia*). The zymogenous are fermentative organisms require an external source of energy and their normal population in soil is low (*Pseudomonas* and *Bacillus*). When specific substrates are added to soil the number of zymogenous bacteria increases and gradually declines when the added substrate is exhausted (cellulose decomposers, nitrogen utilizing bacteria, *Nitrosomonas*, *Nitrobacter*).

Ten orders are included in the class Schizomycetes. Of these three orders–*Pseudomonas*, *Eubacteria* and *Actinomycetes* contain the species of bacteria, which are predominantly reported in the soil (Liesack & Stackebrandt, 1992). The most common bacteria belong to the genera *Pseudomonas*, *Arthrobacter*, *Clostridium*, *Achromobacter*, *Bacillus*, *Micrococcus*, *Flavobacterium*, *Corynebacterium*, *Sarcina*, *Azosprillium* and *Mycobacteria* (Loper *et al*, 1985; Lynch, 1987a; Brock, 1987). *Escherichia* is encountered rarely in soils except as a contaminant from sewage whereas *Aerobacter* is frequently encountered and is probably a normal inhabitant of certain soils (Subba Rao, 1997). Another group of bacteria, common in soil, is the *Myxobacteria* belonging to the genera *Myxococcus*, *Chondrococcus*, *Archangium*, *Polyangium*, *Cytophaga* and *Sporocytophaga*. The latter two genera are cellulolytic and hence are dominant in cellulose-rich environment (Slater, 1988).

Bacteria can withstand extremes of climate although temperature and moisture influence their population (Woese, 1987). In Arctic zone where temperature is below the freezing point bacteria can thrive as luxuriantly as they do in arid desert soils where temperatures are very high (Moreno *et al*, 1986). Such bacteria form spores possessing tough outer covering facilitating the survival of bacteria in all adverse environments. Survival by spore formation under extreme conditions ought to be differentiated from tolerance to different temperature ranges, which is one of the factors determining the population of bacteria in soil (Burr & Caesar, 1984).

Based on the temperature tolerance, bacteria are grouped as mesophillus (15-45°C), psychrophillus (below 20°C) and thermophillus (45-65°C) (Subba Rao, 1997). The mesophillic

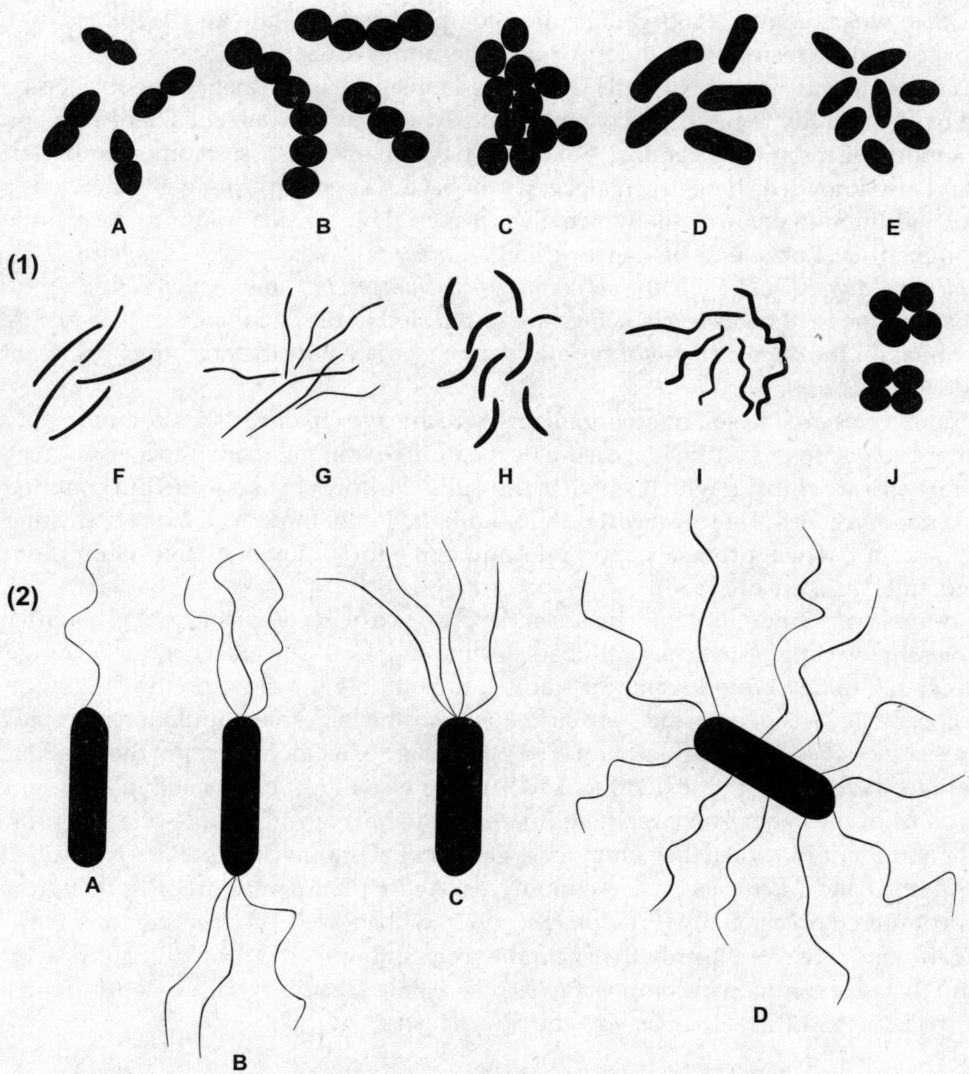

Fig. 14.3: Diversity of major form of soil bacteria (Giri *et al*, 2005)

bacteria, however, constitute the bulk of soil bacteria (Barber & Lynch, 1997). Other factors affecting bacterial population in soil are pH, farm practices, fertilizers and pesticide applications and organic matter amendments (Tate, 1987).

Autotrophic and heterotrophic bacteria are present in a wide range of soils (Tate, 1995). Autotrophic bacteria (purple and green bacteria) synthesize their own foods whereas heterotrophic bacteria depend on pre-formed food for nutrition. Photoautotrophs derive food energy through mediation of sunlight because they have bacteriochlorophyll pigment. Chemoautotrophs oxidize inorganic materials to derive energy and at the same time utilize the carbon from CO_2 for growth (Tate, 1995). There are a group of bacteria known as obligate chemoautotrophs like *Nitrobacter*, which utilizes nitrite, *Nitrosomonas* that utilizes ammonium,

Thiobacillus, which converts inorganic sulfur compounds to sulfate and *Ferrobacillus*, are capable of converting ferrous ions to ferric ions (Alexander & Clark, 1965).

The cyanobacteria are a structurally diverse assemblage of Gram-negative eubacteria characterized by their ability to perform oxygenic photosynthesis. They are considered true procaryotic microorganisms (Stainer *et al*, 1986). They have characteristics common to bacteria and algae and also known as blue-green algae. Cyanobacteria contain a pigment known as phycocyanin in addition to chlorophyll, which gives a special blue-green colour to these organisms. The dominant cyanobacteria belong to genera *Chrooccus, Aphanocapsa, Lyngbya, Oscillatoria, Phormidium, Microcoleus, Cylindrospermum, Anabaena, Nostoc* and *Scytonema.* Some of the cyanobacteria possess heterocysts, which are implicated in nitrogen fixation. The rice field is a good habitat for the development of certain cyanobacteria where they fix atmospheric nitrogen (Prescott *et al*, 1996).

Actinomycetes are the soil microorganisms, which have characteristics common to bacteria and fungi and yet possess sufficient distinctive features to delimit them into a distinct category. Actinomycetes are clubbed with bacteria in the same class of Schizomycetes but confined to the order Actinomycetales. They bear certain similarities to Fungi Imperfecti in the branching of the aerial mycelium, which profusely sporulate, and in the formation of distinct clumps or pellets in liquid cultures (Benson, 1988).

The number of actinomycetes increases in the presence of decomposing organic matter. They are intolerant to acidicity and their numbers decline at pH 5.0. The most conducive range of pH is between 6.5 to 8.0. Waterlogging of soil is unfavourable for the growth of actinomycetes, whereas desertic soils of arid and semi-arid zones sustain sizeable populations, probably due to the resistance of spores to desiccation. The percentage of actinomycetes in the total microbial populations increases with the depth of soil. Actinomycetes can be isolated in sufficient numbers even from the C horizon (weathered parent material) of soil sample of a soil profiles. In order of abundance in soils, the commonest genera of actinomycetes are *Streptomyces* (nearly 70%), *Nocardia* and *Micromonspora*: Actinomyces, Actinoplanes and Streptosporangium have been encountered occasionally (Prescott *et al*, 1996; Subba Rao, 1997). Temperature between 25 to 30°C are conducive for the growth of actinomycetes although thermophillic cultures growing at 55 and 65°C are common in compost heaps where they are numerically extensive and mostly belong to the genera *Thermoactinomyces* and *Streptomyces*.

Archaebacteria

Archaebacteria is a group of primitive prokaryotes, which were the earliest organisms to have appeared on the earth. Therefore, they are called the ancient bacteria. They even live in the extreme hostile environments, like saltpans, salt marshes, hot sulfur springs etc. The archaebacteria are a heterogeneous group that is phylogenetically very distant from the eubacteria and perform very distinct characteristics. They are characterized by the absence of peptidoglycan in their wall. Instead, their wall contains protein and non-cellulosic polysaccharides. Their cell membrane contains branched chain lipids. It enables them to bear extremes of heat and pH. Their rRNA nucleotides are quite different from those of other organisms.

Archaebacteria comprise two sub-groups: obligate and facultative anoxybiont. Obligate anoxybionts live in the absence of oxygen only and get killed in the presence of oxygen. These archaebacteria are of two types: Methanogens and Halophiles. Facultative anoxybinds are found in the presence of oxygen and can live under anaerobic conditions comfortably. They are represented by thermoacidophiles (Tate, 1995).

Methanogens

They are strict anoxybionts occurring in marshy areas. They are characterized by their habit of producing CH_4 (methenogenesis) from CO_2 or fumaric acid. They are, therefore, called methanogens. Methanogens are ubiquitous in highly reducing habitats. Some of the methanogens live as symbionts in the rumen or first chamber of stomach of ruminant animals. The most common species among methanogens are *Methanobacterium, Methanobrevibacter, Methanococcus Methanispirillum* and *Methanosarcina*. Methanogenesis is now attributed to more than 50 known species of bacteria (Jones, 1991). Their growth and survival depends directly on the activities of associated microflora, which enhance methanogenesis through the release of C substrates and the maintenance of reducing conditions.

Halophiles

Highly saline environment harbour large population of a small and distinctive group of halophiles (*Halococcus* and *Halobacterium*). These archaebacteria live in extremely strong brine or salt solutions, salt beds and salt marshes, Some halophiles occur in deep sea volcanic vents at 100°C, a temperature at which water remains liquid because of extreme hydrostatic pressure. In strong light, halophiles develop a purple-pigmented membrane, which can absorb solar radiation. The absorbed light is utilized in the synthesis of ATP. These archaebacteria are unique because they carry out their metabolic processes directly by the ATP produced by their pigmented membrane. They cannot convert CO_2 to sugar as in photosynthesis. Halophiles growing in salt beds give offensive smell and undesirable pigmentation to the salt (Stianer *et al*, 1986).

Thermoacidophiles

The thermoacidophiles occur in the high temperature environments because they are able to tolerate high temperature as well as high acidity. They occur in hot sulfur springs, where temperature may be as high as 80°C and pH as low as 2. These archaebacteria are chemoautotrophic which obtain energy for synthesis of food by oxidizing sulfur. Under aerobic conditions they oxidize sulfur to sulfuric acid. Some form of archaebacteria can reduce sulfur to hydrogen sulfide in the absence of oxygen.

V. FUNGI

Fungi dominate all types of soil and represent greater diversity among soil microorganisms. Fungi possess filamentous mycelium composed of individual hyphae. The hyphae may be uni-, bi- or multinucleate and non-septate or septate (Hawksworth, 1991a,b). All the environmental factors, which influence the distribution of bacteria too, influence the fungal flora of soil. The quality and quantity of organic matter present in the soil have a direct bearing on fungal numbers in soil since most fungi are heterotrophic in nutrition. Fungi are dominant in acid soils because acidic environment is not conducive for the existence of either bacteria or actinomycetes, resulting in the monopoly of fungi for utilization of native substrates in soil (Bolton *et al*, 1993). They are also present in neutral or alkaline soils and some can tolerate pH beyond 9.0. Arable soils contain abundant fungi since they are strictly aerobic and excess of soil moisture decreases their numbers. Fungi exhibit selective preference for various depths of soil. The uses of fungi, which are common in lower depths, are rarely found on the surface of soils. This is due to the availability of organic matter and the ratio between oxygen and carbon dioxide in soil atmosphere at various depths. Farm practices including crop rotation and fertilizer or pesticide applications influence the nature and dominance of fungal species (Hawksworth, 1991a,b).

Fungi are classified into Phycomycetes, Ascomycetes, Basidiomycetes and Fungi Imperfecti (Alexander, 1977). Many fungi, which are commonly isolated from soil, come under the class Fungi Imperfecti by virtue of the fact that they produce abundant asexual spores and lack sexual stages (Lynch, 1987a,b). Members of this class are distinguished by their septate mycelium and a structure called conidiophores from which conidia or spores are continuously produced. The other three classes of fungi have both sexual and asexual means of reproduction. Phycomycetes members possess non-septate and unicellular mycelia and have sacs called sporangia containing spores. A specialized sporangium known as ascus is characteristic of Ascomycetes, which contains a definite number of ascospores usually numbering 4 or 8. Many of the wood-rotting fungi such as *Polyporus* and ectotrophic mycorrhizal fungi (*Boletus*) which inhibit the root region of forest trees come under soils Basidiomycetes (Lynch & Hobbie, 1988). Nevertheless, Basidiomycetes are usually present in soil in a mycelial state and can be recognized by the fructification and fruit bodies, which they produce, on the surface of soil or decaying wood. Many of the Basidiomycetes are capable of utilizing cellulose and in this respect they appear to be good colonizers of forest soils.

The following genera of fungi are most commonly encountered in soils: *Acrostalagmus, Aspergillus, Botrytis, Cephalosporium, Gliocladium, Monilia, Penicillium, Scopulariopsis, Spicaria, trichoderma, Trichothecium, Verticillium, Alternaria, Cladosporium, Pillularia, Cylindrocarpon* and *Fusarium, Absidia, Cunninghamella, Mortierella, Mucor, Rhizopus, Zygorynchus, Pythium, Chaetomium* and *Rhizoctonia* (Mukerji, 1996; Newman, 1985; Hawksworth *et al*, 1991a; Subba Rao, 1997). Many yeasts belonging to true Ascomycetes such as *Saccharomyces* and those belonging to Fungi Imperfecti such as *Candida* have been isolated from soils. However, their numbers in soil are low. Filamentous fungi in soil degrade organic matter and help in soil aggregation. Certain fungi like *Alternaria, Aspergillus, Cladosporium, Dematium, Gliocladium, Helminthosporium, Humicola* and *Metarhizium* produce substances similar to humic substance in soil and, hence, may be important in the maintenance of soil organic matter (Hawksworth *et al*, 1991b).

VI. ALGAE

Soil algae are ubiquitous in nature where moisture and sunlight are available. These microorganisms are visible to unaided eye in the form of green scum on the surface of soils whereas some algae are microscopic. In the soil, algae are not as many as fungi. They may be unicellular (*Chlamydomonas*) or filamentous (*Spirogyra, Ulothrix*). Algae belong to different classes; most of them belong to class Chlorophyceae and Cyanophyceae. Algae are photoautotrophic organisms by virtue of presence of chlorophyll in their cells. They use CO_2 from atmosphere and produce O_2. Algae have been found below the surface of the soil and beyond the reach of sunlight but their number is very less as compared to algae occurring on the surface of soil (Subba Rao, 1997). Some of the common green algae occurring in most of the soils belong to the genera *Chlorella, Chlamydomonas, Chlrococcum, Oedogonium, Chlorochytrium* and *Protosiphone*.

VII. MICROBIAL ACTIVITY IN SOIL

Microbial activity is a term used to indicate the vast range of activities carried out by microorganisms in soil whereas, term biological activity reflects not only microbial activities but also the activities of other organisms in the soil, including plant roots (Nannipieri *et al*, 1990; Mukerji, 2002; Mukerji *et al*, 2005). Various methods have been used to determine microbiological activity (Table 14.1). Some of them measure the rate of entire metabolic processes; for example, the evolution of CO_2 reflects the catabolic degradation under aerobic conditions. Nitrification activity represents the rate of ammonia oxidation to nitrate; thymidine incorporation represents the rate of DNA synthesis in bacteria; and dehydrogenase activity represents the intracellular flux of electron to O_2 and is due to the activity of several intracellular enzymes catalyzing the transfer of hydrogen and electron from one compound to another (Nannipieri *et al*, 1990). The hydrolysis of fluorescein diacetate, a colourless compound, to flourescein, which is coloured, is also a measurement of the contribution of several enzymes such as non-specific estrasew, proteases and lipases, all of which are involved in the decomposition of organic matter in soil.

Table 14.1: Some parameters used to determine microbiological activity (Alef and Nannipieri, 1995)

Basal respiration	*Dehydrogenase activity*
Substrate induced respiration	Fluorescein diacetate hydrolysis
Nitrogen mineralization	Heat output
Nitrification rate	Thymidine incorporation
Potential denitrification activity	Leucine incorporation
Nitrogen fixation	Specific enzyme activity
Adenylate energy change	Arginine ammonification
ATP content	Dimethyl sulphoxide reduction

VIII. CHEMICAL TRANSFORMATIONS BY MICROBES

The importance of microbial diversity to biogeochemical transformations can be viewed most directly through the specific chemical transformations that organisms perform. Their effects on biogeochemical transformations occur through both direct and indirect means. In this section, we review the direct effects of microorganisms (particularly bacteria, fungi) on biogeochemical transformations in soils (Beare *et al*, 1995).

Nitrogen Transformation

Nitrogen availability is a key factor regulating the biological productivity of many aquatic ecosystems (Capone, 2000; Herbert, 1999). Soil microorganisms have long been recognized as important agents affecting N pools through various transformations. The assimilation into organic form and subsequent release of inorganic N, as performed by a broad array of prokaryotic and eukaryotic organisms, comprise the inner core of the N cycle in nature (Alexander, 1977; Paul & Clark, 1989). However, it is the uniquely bacterial processes of N_2 fixation, nitrification, and denitrification that define the broader cycle and can affect directly the availability and form of N within particular ecosystems (Postgate, 1987). In the nitrogen cycle, many bacteria (eubacteria and *Archaea*) are involved in ammonification, but other N transformations are carried out by taxonomically narrow groups of microorganisms. Chemoautotrophic nitrification, is accomplished by relatively few obligate aerobic soil bacteria (ammonium oxidizers and nitrite oxidizers) which oxidize NH_3 to NO_2 (*Nitrosomonas, Nitrococcus*) and NO_2 to NO_3 (*Nitrobacter*) (Kaplan 1983). Heterotrophic nutrition is also known for several bacteria (*Arthrobacter*) and actinomycetes, but probably accounts for relatively low levels of NO_3 production. Other steps in the N cycle, such as dissimilatory NO_3 and NO_2 reduction (*Mycobacterium, Clostridium*) and denitrification (*Pseudomonas, Bacillus, Thiobacillus*), are carried out by a few, widely distributed genera (Payne 1981). Asymbiotic N_2-fixation is carried out by aerobic (*Azotobacter, Beijerinckia*), microaerophilic (*Clostridium*) organotriphic bacteria as well as free-living cyanobacteria that are sometimes abundant in soils. Symbiotic N_2-fixation is best known for bacterial (*Rhizobium, Bradyrhizobium*) associations with numerous genera of non-leguminous angiosperms (*Alnus, Casuarina, Ceanothus, Myrica*).

Fungi are major components of the soil biomass (Hawksworh, 1991a,b) and are of considerable importance in regulating ecosystem processes (Cromack & Cadwell, 1992; Dighton & Boddy, 1989; Wainwright, 1992). Though often grouped by their specific enzymatic capabilities, most fungi have broad versatility in their chemohetrotrophic metabolisms. Despite this versatility and their prominent role in plant litter decomposition (Cromack & Cadwell, 1992; Kjoller & Struwe, 1982), many fungi maintain more specialized mechanisms for obtaining energy and nutrients (Wainwright, 1992).

The important role of many fungi, including ectotrophic mycorrhizal species (Reid *et al*, 1989; Wainwright, 1992), in the ammonification of organic N is well established, but their contribution in other areas of the N cycle has received little attention. Nitrification has long been known for *Aspergillus flavus*, but the broader range of fungal involvement has only recently been described (Killham, 1987). Though autotrophic nitrification by bacteria is often assumed to dominate, the heterotrophic activities of fungi may account for a significant proportion of the nitrification in acid forest soils (Schimel *et al*, 1984). The extent of fungal nitrification in other soil systems remains poorly known. In contrast, several genera of fungi are known to play a role in nitrate reduction (*Fusarium, Acremonium* and *Aspergillus* spp.) though few studies have demonstrated significant levels of complete denitrification in fungi.

Phosphorus Transformation

Phosphorus is considered to be a major growth-limiting nutrient and unlike the case for nitrogen, there is no large atmospheric source that can be made biologically available. It is essential in both cellular energetics (ATP) and cellular structure (DNA, RNA, and phospholipids). Therefore, phosphate-dissolving soil microorganisms play profound role in P deficiency of crop plants. The role of bacteria in the P cycle appears somewhat less specialized. Although there are no microbial mediated gaseous flux of P, however, *Pseudomonas, Bacillus* are involved in the solubilization of inorganic phosphorus. Although bacteria have been used in the growth of plants, fungi seem to be better agents in the dissolution of phosphates. Phosphate-dissolving bacteria are known to reduce the pH of the substrate by secretion of a number of organic acids such as formaic, acetic, propionic, lactic, glycolic, fumaric and succinic acids. As a group, soil bacteria are important to the short-term immobilization of P and the mineralization of organic phosphorus (Subba Rao, 1997).

Somewhat more specialized groups of bacteria are involved in the transformation of metals in soils. Examples of these transformations include the reduction (*Bacillus*) and predipitation (*Chladobacteriaceae*) of iron as well as the chemolithotrophic oxidation of Fe^{2+} under acid conditions (*Thiobacillus ferroxidans*). Some free-living fungi (*Aspergillus* and *Penicillium*) also excrete organic acids and Fe siderophores that solubilize insoluble form of phosphate and contribute to the weathering of soil minerals (Mehta *et al*, 1979; Sollins *et al*, 1981).

Sulfur Transformation

Sulfur is an important element biochemically and geochemically. It constitutes approximately 1% of the dry mass of organisms where it serves many structural and enzymatic functions. Sulfur also acts as a significant electron donor and acceptor during many bacterial metabolisms (Hurst, 2002). Sulfur can be found in a range of valence states from the highly reduced sulfide to the most oxidized form in sulfate. Microbial S transformations are closely linked with the carbon cycle in which S reduction coupled with organic matter utilization is a major mineralization pathway in anoxic habitats, while S oxidations can occur aerobically and anaerobically, some of which are autotrophic and/or phototrophic (Jorgensen, 1982; Jorgensen, 1994).

Microorganisms of the S cycle are extremely diverse. They could be either oxybiont or anoxybiont. The anoxybiont sulfate-reducing bacteria (SRB), which are unique physiologically and genetically, are represented by several genera (Devereux & Stahl, 1993). Sulfate-reducing bacteria are capable of utilizing iron and manganese as electron acceptors (Lovley & Phillips, 1994). Oxygen-reduction has been demonstrated but O_2-dependent growth has not been confirmed. SRB are rather nutritionally diverse with the known suite of electron donors increasing each year (van Niel & Gottschal, 1998). Chemolithotrophic sulfur oxidation is mediated aerobically by colourless sulfur bacteria, some purple sulfur bacteria and SRB. Anaerobically, nitrate respiring chemolithotrophs oxidize sulfide, and both oxygenic and anoxygenic phototrophic bacteria use sulfide as an electron donor for photosynthesis (Visscher & Taylor, 1993). Sulfate-reducing bacteria may diminish the availability of sulfur for plant nutrition and thus influence agricultural production. *Desulfovibrio desulfuricans* is a species belonging to this class of bacteria. Bacteria capable of oxidizing inorganic sulfur compounds varies morphologically from non-filamentous (*Thiobacillus*) to filamentous forms (*Beggiatoa, Thiothrix* and *Thioploca*). Among these bacteria, *Thiobacillus* deserves special mention as it produces sulfuric acid when elemental sulfur is added to soil with the result that the pH of soil may fall as low as 2.0 after prolonged incubation with the bacterium. Several fungi and Actinomycetes have also been reported to be sulfur oxidizers (*Aspergillus, Penicillum, Microsporeum*). *Thiobacilli* can also be used in the manufacture of 'Biosuper' a form of organic fertilizer once favoured in Australia. In Biosuper, a mixture of rock phosphate and sulfur is inoculated with *Thiobacillus thiooxidans.*

The H_2SO_4 produced in the mixture dissolves the phosphate and thereby enhances phosphorus nutrition of plants (Widdel & Hansen, 1991).

Iron Transformation

Certain bacteria oxidize ferrous iron to ferric state, which precipitates as ferric hydroxide around cells (Quastel, 1955). These bacteria commonly known as iron bacteria are usually non-filamentous and spherical or rod-shaped (*Gallionella, Siderophacus, Siderocapsa, Siderophaera, Ferribacterium, Naumanniela, Ochrobium, Sideromanas, Sideronema, Ferro-bacillus, Siderobacter* and *Siderococcus*). Filamentous form resembling algae are also encountered (*Leptothrix, Sphaerotilus, Toxothrix, Crenothrix* and *Colnothrix*). In addition to these bacteria, certain algae belonging to Cyanophyceae, also transform ferrous salts to ferric state and deposit the precipitation around the filaments. The ferric hydroxide deposits give a brown or rust-red colour to these organisms.

The iron bacteria can be grouped into: (1) Obligate chemoautotrophs, capable of utilizing energy released in the process of ferric hydroxide formation (*Gallionella ferruginea, Thiobacillus ferroxidans* and *Ferrobacillus ferroxidans*), (2) facultative chemoautotrophs, utilizing energy derived in the process of ferric hydroxide formation or alternatively, form organic matter (*Leptothrix ochraceae*) and (3) heterotrophs represented by most other iron bacteria which do not derive energy for iron oxidation but depend upon organic matter for their nutrition.

IX. MICROBIAL DIVERSITY AND BIOTIC INTERACTIONS

Due to their vast diversity, large populations and long evolutionary history, microorganisms have contributed greatly to the rich and complex interactions among soil organisms. These interactions range from highly specific symbiosis to diffused mutualisms.

Mycorrhizal Interaction

Mycorrhizal symbiosis is among the best-known examples of plant-microbe interactions and plays a key role in regulating plant productivity and nutrient cycling (Srivastava *et al*, 1996; Mukerji *et al*, 2000; 2002). Mycorrhizal fungi are found in 75-80% of all vascular plants species. Although these associations are often assumed to have weak specificity; many are highly specific, emphasizing the importance of diversity to ecosystem function. The root-microbe interactions are the key to understanding ecosystem function, and places mycorrhizas in perspective with the many other complex interactions taking place in the rhizosphere.

Mycorrhizal fungi interact with a wide range of other microorganisms in the rhizosphere (Bowen & Rovira, 1999). These interactions may be stimulatory or inhibitory; some may be competitive, while others may be mutualistic. Mycorrhizal fungi are found in the endorhizosphere, in the rhizosphere and in the bulk soil. In all these zones, they interact with the soil microbiota. The internal mycelium interacts mainly with the host root and other microorganisms inhabiting the area. The external mycelium interacts with many more organisms, including bacteria, fungi, protozoa, nematodes, arthropods and large animals (Fitter & Garbaye, 1994). Some interactions may be mutualistic while others may be difficult to define. Some bacteria such as fluorescent Pseudomonads may proliferate in the hyposphere of mycorrhizal fungi.

Geosiphon-nostoc Interaction

Geosiphon pyriforme, a coenocytic soil fungus lives in endocytobiotic association with a cyanobacterium, *Nostoc punctiforme*. The symbiotic nature of the system was first recognized by Wettstein (1915), who described it as a symbiosis between a heterotrophic siphonal chlorophycean alga and *Nostoc*. The fungus lives together with the cyanobacterium on the surface and in the upper layer of wet soils poor in inorganic nutrients, particularly in phosphate (Schubler & Kluge, 2001). When fungal hyphae comes into contact with free-living *Nostac* cells, the latter are incorporated by the fungus at the hyphal tip, which, thereafter, swells and forms a unicellular "bladder", about 1-2 mm in size and appearing on the soil surface. Inside this bladder the cyanobacteria are physiologically active and dividing. Due to the physiological activities of the endosymbiont, the consortium is capable of C- and N-autotrophic life. *Geosiphon* can be considered a primitive endocytobiotic system, because the photobiont can be experimentally separated and cultured without the fungal partner, which is obligate symbiotic. It has been suggested that Geosiphon could provide an important model system for another symbiosis, the arbuscular mycorrhiza. It bears a great potential for the study of many fundamental mechanosms and evolutionary questions concerning arbuscular mycorrhiza. (Kluge *et al*, 1997)

Geosiphon pyriforme representing a symbiotic association between a glomalean fungus and a photoautotrophic prokaryotic alga could reflect an ancestral partnership. Thus, it is very plausible to assume that in the beginning of terrestrial plant life also other associations between glomalean fungi and photoautotrophic organisms existed (Redecker *et al*, 2000).

Mollenhauer *et al* (1996) studied the development of the symbiotic association *Geosiphon pyriforme*. Initially, the cells of the cyanobacterium *Nostoc pyriforme* live freely together with the future fungal partner in and on the soil. There, the partners come into contact, but a successful interaction of the fungus with *Nostoc* to form the symbiosis depends on the appropriate developmental stage of the cyanobacterium (Schubler & Kluge, 2001).

X. CONCLUSIONS

Soil microorganic population is very varied consisting of different groups of microbes. They are obligatory for the healthy growth and development of plants. Natural environments are extremely diverse and the majority contains a wide range of microorganisms which reflect the nature of the habitat and the ability of individual members to compete successfully and coexist within the given ecosystem. In general terms, the greater the heterogenicity of the environment, the more diverse and complex will be the microflora. These microbes include bacteria, algae, fungi, actinomycetes, protozoa and nematodes etc. Some survive and function symbiotically (algae, fungi).

ACKNOWLEDGEMENTS

Authors thank the Department of Science and Technology and Council of Scientific and Industrial research for providing financial support.

REFERENCES

Alexander, M. 1977. Introduction to Soil Microbiology, 2nd edn. Academic Press, New York.

Alexander, M. & F.E. Clark. 1965. Nitrifying bacteria. In: Methods of Soil Analysis, Part 2. Chemical and Microbiological Properties, Black CA (ed.) American Society of Agronomy, Inc Madison, Wisconsin, USA, pp. 1477-1483.

Barber, D.A. & J.M. Lunch. 1997. Microbial growth in the rhizosphere. Soil Biol Biochem 9: 305-308.

Beare, M.H., D.C., Coleman, D.A., Crossley, P.F. Hendrix & E.P. Odum. 1995. A hierarchical approach to evaluating the significance of soil biodiversity to biogeochemical cycling. Plant Soil 170: 5-22.

Benson, D.R. 1988. The genus Frankia: actinimycetes symbionts of plants. Microbial Sci 5: 9-12.

Bolton, H. Jr. J.K. Fredrikson & L.E. Elliot. 1993. Microbiology of the rhizosphere. In: Soil Microbial Ecology, Metting FB Jr (ed.) Marcel Dekker, Inc New York, pp. 27-63

Borneman, J., .PW. Skroach, E.W. O'Sullivan, J.A. Palus, N.G. Rumjanek, J.L. Jansen, J. Nienhuis & E.W. Triplett. 1996. Molecular microbial diversity of an agricultural soil in Wisconsin. Appl Environ Microbiol 62: 1935-1943.

Bowen, G. D. & A. D. Rovira. 1999. The rhizosphere, the hidden half. In: Plant roots: the hidden half, Waisel Y, A. Eshel and U. Kafkafi (eds.) Marcel Dekker, New York, pp. 641-669.

Bruck, T.D. 1987. The study of microorganisms *in situ:* progress and problems. In: Ecology of Microbial Communities, Fletcher M., T.R.G. Gray and J.G. Jones (eds.) SGM Symposium 41, Cambridge University Press, Cambridge, pp. 1-17.

Bruns, R.G. & J.H. Slatar. 1982. Experimental microbial ecology. Blackwell Scientific Publications, Oxford, pp. 683.

Burr, J.J. & A. Caesar. 1984. Beneficial plant bacteria. CRC Crit Rev Pl Sci 21: 1-20.

Capone, D.G. 2000. The marine nitrogen cycle. In: Microbial Ecology of the Ocean Kirchman D (ed) Wiley-Liss, New York, pp. 455-493.

Christensen, M. 1989. A view of fungal ecology. Mycologia 81: 1-19.

Cromack, K. & B. A. Caldwell. 1992. The role of fungi in litter decomposition and nutrient cycling. In: The Fungal Community, its Organization and Role in the Ecosystem Carroll, G.C. and D.T. Wicklow (eds.) Marcel Dekker, Inc., New York, pp. 601-618.

Devereux, R. & D.A. Stahl. 1993. Phylogeny of sulfate-reducing bacteria and a perspective foranalysing their natural communities. In: Sulfate-Reducing Bacteria: Contemporary Perspectives: Odom, J.M. and Jr. R. Singleton (eds.) Sringer-Varleg, New York, pp. 131-160.

di Castri, F. & T. Younes. 1990. Ecosystem function of biological diversity. Biol Int Special Issue NO 22.

Dighton, J. & L. Boddy. 1989. Role of fungi in nitrogen, phosphorous and sulphur cycling in temperate forest ecosystems. In: Nitrogen, Phosphorus and Sulphur Utilization by Fungi, Boddy, L., R. Marchent and D.J. Read (eds.) Cambridge University Press, Cambridge, pp. 269-298.

Fitter, A. H. & J. Garbays. 1994. Interactions between mycorrhizal fungi and other soil organisms. Plant and Soil 159: 123-132.

Foster, R.C. 1988. Microenvironment of soil microorganisms. Biol. Fertil Soils 6: 189-203.

Franklin, J.F. 1993. Preserving biodiversity: species, ecosystems, or landscapes? Ecol Appl. 3: 200-205.

Hawksworth, D.L. 1991a. The biodiversity of microorganisms and invertibrates: its role in sustainable agriculture. CAB International, Redwood Press Ltd, Melksham, UK pp. 302.

Hawksworth, D.L. 1991b. The fungal dimension of diversity: magnitude, significance, and conservation. Mycol Res 95: 641-655.

Herbert, R. 1999. Nitrogen cycling in coastal marine ecosystems. FEMS Microbial Rev 23: 563-590.

Hurst, C.J. 2002. Introduction to Environmental microbiology. In: Manual of Environmental Microbiology, Hurst CJ, Crawford RL Knudsen GR, McInerney MJ and Stetzenbach LD (eds.) ASM Press Washington, DC.

Jones, J.W. 1991. Diversity and Physiology of methanogens. In: Microbial Production and Consumption of Greenhouse Gases: Methane, Nitrigen Oxides, and Halomethanes, Rogers JE and Whitman WB (eds.) American Society of Microbiology, Washington DC, pp. 39-35.

Jorgensen, B.B. 1982. Ecology of the bacteria of the sulfur cycle with special reference to anoxic-oxic interface environments. Phil Trans R Soc Lond 298: 543-561.

Jorgensen, B.B. 1994 Sulfate reduction and thiosulfate transformations in a cynobacterial mat during a diel oxygen cycle. FEMS Microbiol. Ecol 13: 303-312.

Kaplan, W.A. 1983. Nitrification. In: Nitrogen in the marine environment, Carpenter EJ and Capone DG (eds.) Academic Press, New York, pp. 139-190.

Killham, K. 1987. Heterotrophic nitrification. In: Nitrification, Prosser (ed) Society of General Microbiology, Spec Public IRL Press, Oxford, pp. 117-126.

Killham, K. 1994. Soil ecology. Cambridge university press, Cambridge.

Kjoller, A. & S. Struwe. 1982. Microfungi in ecosystems: fungal occurrence and activity in litter and soil. Oikos 39: 391-422.

Kluge, M., H. Gehrig, D. Mollenhauer, E.Schnepf, & A. Schubler. 1997. News on Geosiphon pyriforme, an endocytobiotic consortium of a fungus with a cyanobacterium. In: Eukaryotism and symbiosis, Schenk H.E.A., R. Herrmann, K.W. Jeon, N.E. Muller, and W. Schwemmler (eds.) Springer-Verlag Berlin Heidelberg, New York, pp. 469-476.

Liesack, W. & E. Stackebrandt. 1992. Occurrence of novel groups of the domain bacteria as revealed by analysis of genetic material isolated from an Australian terrestrial environment, J Bacteriol 174: 5072-5078.

Loper, J.E., C. Haack & M.N. Schroth. 1985. Population dynamics of soil *Pseudomonads* in the rhizosphere of potato (*Solanum tuberosum* L.). Appl Environ Microbiol 49: 416-422.

Lovley, D.R. & P. Phillips. 1994. Novel processes for anaerobic sulfate production from elemental sulfate by sulfur-reducing bacteria. Appl Environ Microbiol 60: 2394-2399.

Lynch, J.M. & J.B. Hobbie. 1988. Microorganisms in Action: concepts and application in Microbial Ecology. Blackwell Scientific Publications, Oxford, pp. 363.

Lynch, J.M. 1987a. Microbial interactions in the rhizosphere. Soil Microorg. 30: 33-41.

Lynch, J.M. 1987b. Soil biology-accomplishments and potential. Soil Sci Soc Am J 51: 1409-1412.

Mayer, J.R. & R.G. Linderman. 1986. Selective influence on population of rhizosphere or rhizoplane bacteria and actinomycetes by mycorrhizas formed by *Glomus fasciculatum*. Soil Biol Biochem 18: 191-196.

Mehta, A.P., A.E Torma & L.E Murr. 1979. Effect of environmental parameters on the efficiency of biodegradation of basalt rock by fungi. Biotechnol Bioeng 21: 875-885.

Mollenhauer, D., R. Mollenhauer & M. Kluge. 1996. Studies on initiation and development of the partner association in *Geosiphon pyreforme* (Kuiz.) v. Wettstein, a unique endocytobiotic system of a fungus (Glomales) and the cyanobacterium *Nostoc punctiforme* (Kuiz.) Hariot. Protoplasma 139: 3-9.

Moreno, J., J. Gonsalez Loper & G.R. Vela. 1986. Survival of *Azotobacter* spp. in dry soils. Appl Environ Microbial 51: 123-125.

Mukerji, K.G. 1996. Ecological studies on the microorganic population of user soils. Mycopath. Mycol. Appl. 29: 339-349.

Mukerji, K.G. 2002. Soil microbes. In: Techniques in Mycorrhizal Studies, Mukerji K.G., C. Manoharachary & B.P. Chamola (eds.) Kluwer Academic Publishers, Dordrecht, Boston, London. pp. 7-13.

Mukerji, K.G., B.P. Chamola, & J. Singh (eds.) 2000. Mycorrhizal Biology. Kluwer Academic/Plenum Publishers, New York, Dordrecht, London. pp. 336.

Mukerji K.G., C. Manoharachary & B.P. Chamola (eds.) 2002. Techniques in Mycorrhizal Studies, Mukerji K.G., C. Manoharachary & B.P. Chamola (eds.) Kluwer Academic Publishers, Dordrecht, Boston, London. pp. 554.

Mukerji, K.G., C. Manoharachary & J. Singh (eds.) 2006. Microbial Activity in the Rhizosphere. Springer-Verlag, Heidelberg, Germany. pp. 430.

Nanipieri, P & L.Badalucco 2003. Biological processes. In: Processes in the Soil-Plant System: Modelling concepts and applications, Bembi, D. K. & Nieder R. (eds.). The Haworth Press, Binghamton, NY.

Nannipieri, P., J. Ascher, M.T. Ceccherini, L. Landi, G. Pietramellara & G. Renella 2003. Microbial diversity and soil functions. European J Soil Sci. 54: 655-670.

Nannipieri, P., S. Grego & B.Ceccanti. 1990. Ecological significance of the biological activity in soil. In: Soil Biochemistry, Bollag, J.M and G. Stotzky, (eds.) Marcel dekker, New York, pp. 293-355.

Nannipieri, P., E. Kandeler & P. Ruggiero. 2002. Enzymes activities and biochemical processes in soil. In: Enzymes in the environment, Bruns, R.G. & Dick, R. (eds.). Marcel dekker, New York, pp. 1-33.

Newman, E.I. 1985. The Rhizosphere: carbon sources and microbial populations. In: Ecological Interactions in Soil, Plants, Microbes and Animals, Fitter AH, Atkinson D, Read DJ and Usher MB (eds.) Blackwell Scientific publications, oxford. pp. 107-121.

Paul, E.A. & F.E.Clark. 1989. Soil Microbiology and Biochemistry. Academic Press, Inc. San Diego, USA.

Payne, J.W. 1981. Denitrification. Wiley Press, New York.

Postgate, J.R. 1987. Nitrogen fixation, 2nd edn. Edward Arnold, London.

Prescott, L.M., J.P. Harley & D.A. Klein. 1996. The diversity of the microbial world. In: Microbiology, Prescott LM, Harley JP and Klein DA (eds.) WCB Publishers USA.

Quastel, J.H.O 1995. Soil metabolism. Proc R Soc 143: 159-179.

Redecker, D., J.B. Morton & T.D. Bruns. 2000. Ancestral lineages of arbuscular mycorrhizal fungi (Glomales). Mol Phylogenet Evol 14: 276-284.

Reid, D.J., J.R.Leake & A.R. Langdale. 1989. The nitrogen nutrition of mycorrhizal fungi and their host plants. In: Nitrogen, phosphorus and sulfur Utilization by Fungi, Boddy L, R. Marchant and D.J. Read (eds.) Cambridge University Press, Cambridge, pp. 269-298.

Salyers, A.A. & D.D. Whitt. 2001. Diversity and history of microorganisms. In: Microbiology: Diversity, Diseases and the Environment, Salyers AA and Whitt DD (eds.) Fitzgerald Science Press, Inc, Bethesda, Maryland, pp. 19-32.

Schimel, J.P., M.K. Firestone & K. Killham. 1984. Identification of heterotrophic nitrification in a Sieerrs forest soil. Appl Environ Microbiol 48: 802-806.

Schubler, A. & M. Kluge. 2001. Geosiphon pyriforme, an endocytosymbiosis between fungus and cyanobacteria, and its meaning as a model system for arbuscular mycorrhizal research. In: The Mycota Vol IX, Hock B (ed) Springer-Verlag Berlin Heidelberg, pp. 151-161.

Slater, J.H. 1988. Microbial population and community dynamics. In: Microorganisms in Action: Concepts and Application in Microbial Ecology Lunch JM and Hobbie JB (eds.) Blackwell Scientific Publications, Oxford, pp. 51-74.

Smiles, D.E. 1988. Aspects of the physical environment of soil organisms. Biol. Fertil. Soils 6: 204-215.

Sollins, P., K. Jr. Cromack, C.Y Li & R. Fogel. 1981. Role of low-molecular weight organic acids in the inorganic nutrition of fungi and higher plants. In: The Fungal Community, Its Organization and Role in Ecosystem, Carroll G.C. and D.T. Wicklow (eds.) Marcel Dekker, Inc, New York.

Srivastava, D., R. Kapoor, S.K. Srivastava & K.G. Mukerji. 1996. Vesicular arbuscular mycorrhiza: an overview. In: Concepts in Mycorrhizal Research, Mukerji KG (ed) Kluwer Academic Publishers, The Netherlands, pp. 1-39.

Stanier, R.Y., J.L. Ingraham, M.L. Wheelis & P.R. Painter. 1986. The Microbial World. Prentice-Hall, Englewood Cliffs, New Jersey.

Stotzky, G. 1997. Soil as an environment for microbial life. In: Modern Soil Microbiology, van Elsas, J.D., J.T. Trevors and E.M.H. Wellington (eds.) Marcel dekker, New York, pp. 1-20.

Subba Roa, N.S. 1997. Soil Microbiology. Oxford and IBH Publishing Co. Pvt. Ltd.

Tate, R.L III. 1995. Soil Microbiology. John Wiley and Sons, Inc New York.

Tate, R.L. II. 1987. Soil organic matter: biological and ecological effects. John Wiley and Sons, New York, pp. 291.

van Niel, E.W.J. & J.C. Gottschal. 1998. Oxygen consumption by Desulfovibrio strains with and without polyglucose. Appl Environ Microbiol 64: 1034-1039.

Vannier, G. 1987. The porosphere as an ecological medium emphasized in Professor Gilarov's work on soil animal adaptations. Biol Fertil Soil 3: 39-44.

Vishniac, W. & M. Santer. 1957. The thiobacilli. Bac Rev 11: 195-213.

Visscher, P.T., F.P. Vandenede, B.E.M. Schaub & H. van Gemerden. 1992. Competition between anoxygenic phototrophic bacteria and colorless sulfur bacteria in a microbial mat. FEMS Microbial Ecol 101: 51-58.

von Wettstein, F. 1915. Geosiphon Fr. v. Wettst., eine neue, interessante Siphone. Osterr Bot Z 65: 145-156.

Wainwright, M. 1992. The impact of fungi on environmental biogeochemistry. In: The Fungal Community, Its Organization and Role, in the Ecosystem, Carroll GC and Wicklow DT (eds.) Marcel Dekker, Inc, New York pp. 601-618.

Widdel, F. & T.A. Hansen. 1991. The dissimilatory sulfate and sulfur-reducing bacteria. In: The Prokaryotes, Balows A, Truper HG, Dworkin M, Harder W and Schleifer KH (eds.) 2nd edn. Springer Verlag, New York, pp. 583-634.

Wieland, G., R. Neumann & H. Backhaus. 2001. Variation of microbial communities in soil, rhizosphere, and rhizosphere in response to crop species, soil type, and crop development. Appl Envirom Microbiol 67: 5849-5854.

Williams, C.N. 1975. The agronomy of the major tropical crops. Oxford University Press, London.

Wilson, E.O. 1988. Biodiversity. National Academy Press, Washington, DC.

Woese, C.R. 1987. Bacterial evolution. Microbiol Rev 51: 221-271.

15

Role of Endophytic Nitrogen-Fixing Bacteria in Crop Production

K.V.B.R. TILAK
Department of Botany, Osmania University, Hyderabad, A.P.

I. INTRODUCTION

The term endophytes was designated earlier to refer to root interior colonization of plant by microorganism (bacteria and fungi) that usually do not cause damage to the host and live most of their life inside plant tissue without eliciting any pathogenic symptoms (Pereira, 1995). Later, the definition was extended to bacteria and fungi that colonize the root interior that promote benefits to the plant (Kloepper and Beauchamp, 1992). Due to the ability of some diazotrophs to colonize primarily the root interior of graminaceous plants, the term endophyte was also introduced into this area by Dobereiner (1992) and become widespread among scientists interested in nitrogen fixing bacteria. Splitting the term endophyte into facultative and obligate was suggested to distinguish the strains that were able to colonize both the surface and interior and survive well in soil from those that do not survive well in soil but colonize the root interior and aerial parts, respectively (Baldani *et al*, 1997).

A variety of endophytic nitrogen fixing bacteria have already been found that colonize the interior roots of several plants of economic importane (Boddey, 1995; Boddey *et al*, 1995; Barraqui *et al*, 1997; Cocking *et al*, 1990; Hurek *et al*, 1991; Ladha and Reddy, 1995; Webster *et al*, 1997). Although they have been widely reported in grasses, diazotrophic endophytes such as *Gluconacetobacter diazotrophicus* (Cavalcante and Dobereiner, 1988, Muthukumaraswamy *et al*, 1999, Sahai, 2000), *Herbaspirillum* spp. (Baldani *et al*, 1986), *Azoarcus* spp. (Reinhold-Hurek *et al*, 1993). Two species of a new nitrogen fixing bacterium, *Burkholderia* (*B. brasilensis* (strain 130) and *B. tropicalis* (strain ppe8) have been reported by Baldani (1996). Two other strains (STM 678 and STM 815) of the genus have also been described as possessing nitrogen fixation ability and establish a symbiosis with leguminous plants (Moulin *et al*, 2001).

II. FACULTATIVE ENDOPHYTES

Azospirillum Spp.

The genus *Azospirullum* can be considered the starting point of the most ongoing biological nitrogen fixation programme with non-legume plants worldwide. It has now been almost 45

Diazotrophic bacteria associated in the rhizosphere of Grasses

Diazotroph	*Plant species*
Azotobacer paspali	Bahai grass (*Paspalum notatum cv. batatais*)
Azospirillum halopraeferans	Kallar grass (*Leptochloba fusca* L.)
Beijerinckia indica	Sugarcane
Beijerinckia fluminensis	Sugarcane
Azospirillum lipoferm and other *Azospirillum* spp.	*Digitaria decumbens*, rice, wheat
Kklebsiella spp. (K. oxytoca and *K. pneumoniae)*	Rice and wheat
Bacillus sp.	Wheat
Paenibacillus azotofixans	Wheat, sugarcane, rice, and sorghum
Enterobacter spp.	Rice, wheat, kallar grass and others
Pseudomonas spp.	Rice, wheat, sorghum
Clostridium spp.	Rice
Azotobacter spp.	Rice, wheat

years since Dobereiner and Ruschel (1958) isolated the nitrogen-fixing bacterium named *Beijerinkia fluminensis* from the rhizosphere of the sugarcane plants grown in the tropical soils and demonstrated the potential of diazotrophs to associate with graminaceous plants. However, the genus *Azospirillum* was discovered by Dobereiner and Day in 1975. Initially two species of *Azospirillum, Azospirillum lipoferum* and *A. brasilense* were characterized at physiological and molecular level (Tarrand *et al*, 1979).

Colonization of root interior of many cereals and forage grasses showing the presence of *Azospirillum* in cells of the cortex, intercellular spaces between cells of the cortex and endodermis and in the xylem cells of the plant roots applying the tetrazolium reduction staining technique was reported (Lakshmi *et al*, 1977; Patriquin and Dobereiner, 1978; Agarwala and Tilak, 1988a,b; Agarwala-Dutt *et al*, 1991). *Azospirillum* spp. was repeatedly isolated from the inside of roots and shoots of many plants (Dobereiner *et al*, 1995). Using strain-specific monoclonal antibodies against *Azospirillum brasilense* strains Sp7 and Sp245 respectively, it was clearly demonstrated that the isolate Sp245 could efficiently enter and colonize roots and the vascular system of wheat plants (Scholter *et al*, 1994). Sly and Stackbrandt, (1999) reclassified the *Conglomeromonas largomobilis* subsp. parooensis be transferred to the genus *Skermanella* gen. nov. as the type species *Skermanella parooensis* gen. nov. sp. nov. This taxon belongs to an isolated subline of descent in the *Azospirillum* branch of the Alpha-Proteobacteria. The spelling of the specific epithet of *Azospirillium largomobile* is corrected to *Azospirillium largimobile.* A new group of nitrogen-fixing *Azospirillum* spp. bacteria was isolated from the roots of the C4-gramineous plant *Miscanthus*, subsequently proposed a new species *Azospirillum doebereinerae* (Eckert *et al*, 2001).

Many inoculation experiments with *Azospirillum* spp. have been carried out all over the world showing a positive effect of the bacteria on yield (Boddey and populations of diazotrophs. This may be due to the inoculated organism not being able to compete with the indigenous microbial population (Kamura *et al*, 1998). It has already been shown that 60 to 80% of N in the aerial parts of Kallar grass was derived from fixation by rhizobacteria (Malik *et al*, 1997). In rice plant, inoculation with *Azospirillum* contributed to 66% of the total N in the plants (Malik *et al*, 1997).

Facultative diazotrophs isolated from non-legumes

Diazotroph	*Plant*	*Plant parts*
Azospirillum amazonense	Rice, wheat, sugarcane and Palm trees	Roots, stems, seeds, roots, stems, roots, stems, fruits
Azospirillum brasilense	Rice, wheat, barley sugarcane Palm trees, sorghum, millets	Roots, stems, seeds, roots, stems, roots, stems, leaves, Roots, stems, leaves
Azospirillum irakense	Rice	Roots
Azospirillum lipoferum	Cereals, forage grasses, sugarcane palm trees, tubers	Roots, stems, seeds, roots, leaves, roots, stems, leaves roots, stems, fruits, tubers, roots
Klebsiella, spp.	Maize	Stem

III. OBLIGATE ENDOPHYTES

Gluconacetobacter

The discovery in 1988 of *Acetobacter diazotrophicus* (Syn. *Gluconacetobater diazotrophicus*) a nitrogen fixing bacterium inhabiting the interior of roots, stems and leaves of sugarcane opened a new avenue of research in nitrogen fixation and plant-microbe interactions. The endophytic colonization of sugarcane by *G. diazotrophicus* represents a model system for analysis of beneficial association between a monocot and a diazotrophic bacteria species. *G. diazotrophicus*, first reported from Brazil (Cavalcante and Doberiner 1988), Australia (Li and Mac Rae 1991), Mexico (Fuentes – Ramirez *et al*, 1993), Uruguay, (Ureta *et al*, 1995) Canada and Cuba (Dong *et al*, 1995) and India (Muthukumaraswamy, 1996, Indira and Bagyaraj 1997, Suman *et al*, 2000 Sahai,). *G. diazotrophicus* colonizes other sugar rich plants like *Pennisetum purpureum*, *Ipomoea batatas* (Dobereiner *et al*, 1993) and low-sugar rich plants like *Coffea arabica* (Jimenez-salgado *et al*, 1997), *Eleusine coracana* (Loganathan *et al*, 1999) and *Ananus comosus* (Tapia-Hermandez *et al*, 2000). Two new species in the genus Gluconacetobacter, namely, *G. johannae* and *G. azotocaptans* have been reported by Fuentez-Ramirez *et al*, (2001) from coffee plants in Mexico.

G. diazotrophicus capable of colonizing stem and roots of sugarcane plants. It was confirmed by using immunogold labelling in which the bacteria are first seen to colonize the epidermis of stem and roots followed by distribution in xylem vessels and was transported via., the vascular system (James *et al*, 1994). Dong *et al*, (1995) observed the tissues under bright field and electron microscopy but failed to find any bacteria within xylem vessels. They concluded that it is most unlikely that the endophyte inhabits xylem vessels. Li and Mac Rae, (1992) estimated *G. diazotrophicus* population by MPN count and concluded that the strain inhabited tissues of sugarcane roots, stems and leaves.

G. diazotrophicus does not heavily colonize rice; however, the portion of the roots just emerging from the seeds was almost always heavily colonized just after 2 days of inoculation (Sevilla and Kennedy, 2000). This suggests that crack entry is one way by which *G. diazotrophicus* can get into the plant. It was also common to find *G. diazotrophicus* lined up along the groves of the epidermal cells of the roots. Similar observations have also been reported by Rolfe *et al*, (1997), who examined DAP1-stained roots of rice grown with 2,4-D and inoculated with

Obligate endophytic diazotrophs in non-legumes

Diazotroph	*Plant*	*Plant parts*
Gluconacetobacter diazotrophicus	Sugarcane	Roots, stems, leaves
	Cameroon grass	Roots, stems
	Sweet Potato	Roots, stems, tubers
	Coffee	Roots, stems
	Tea	Roots
Alcaligenes faecalis	Rice	Roots
Azoarcus spp.	Kallar gras	Roots, base of stem
Burkholderia spp.	Cereals	Roots, stems, leaves
	Cassava	Roots, stems, tubers
	Banana	Roots
	Sweet Potato	Roots, stems, tubers
Herbaspirillum rubrisubalbicans	Rice	Roots, stems, tubers
	Sugarcane	Roots, stems, tubers
Herbaspirillium seropedicae	Cereals	Roots, stems, leaves, seeds
	Sugarcane	Roots, stems
	Forage grasses	Roots, stems, leaves
	Palm trees	Roots, stems

G. diazotrophicus cells were observed when sections of Gus-positive root tissues were examined under the microscope.

Herbaspirillum

The genus *Herbaspirillum* was first reported by Baldani *et al,* (1986) as a nitrogen-fixing bacterium associated with wide host range specificity such as maize, sugarcane, sorghum, forage grasses grown in Brazil. It was also isolated from other non-leguminous plants including oil palm trees (Ferreria *et al,* 1995). Initially *Herbaspirillum* genus only consisted one species, *H. seropedicae*. A mild pathogen of sugarcane and sorghum called *"Pseudomonas" rubrisubalbicans* (Pimental *et al,* 1991) that fixes N_2 (Baldani *et al,* 1992) was observed to be very closely related by phenotypic and genotypic characteristic to *H. seropedicae* and later it was reclassified as *H. rubrisubalbicans* (Baldani *et al,* 1996). A third species, *Herbaspirillum* species 3, was classified on the basis of DNA-rRNA and DNA-DNA hybridization experiments (Gillis *et al,* 1991; Baldani *et al,* 1996). These isolates were entirely different and were of clinical origin and unable to fix nitrogen. The fourth species *H. fisingense* as a nitrogen-fixing bacterium isolated from C-4 grasses like *Miscanthus* spp., *Pennisetum purpureum* and *Spartina pectina* (Kirchhof *et al,* 2001). The fifth species *H. chlorophenolicum*, a 4-chlorophenol degrading bacterial strain, formerly designated as a strain of *Comamonas testosteroni*, was reclassified as a member of the genus *Herbaspirillum* based on its phenotypic and chemotaxonomic characteristics, as well as phylogenetic analysis using 16S rDNA sequences (Taek Im *et al,* 2004). The sixth species *H. lucitanum* was isolated from the root nodules of *Phaseolus vulgaris* grown in a soil from Portugal (Valverde *et al,* 2004).

Artificial inoculation of sugarcane leaves with *H. seropedicae* showed that the bacteria do not spread into the xylem vessels of the leaves but remains localized at the point of inoculation. The pattern of sugarcane root colonization by *H. seropedicae* is quite similar to that observed for *G. diazotrophicus*, except that the attachment of the cells occurs mostly at junction of secondary roots through which intercellular space colonization takes place. On the other hand, the colonization of rice plants by a selected strain of *H. seropediceae* showed that they occur on root surface and enter through the intercellular spaces derived from cells or secondary root emergence (Baldani *et al*, 1995). Inoculation of *H. rubrisubalbicans* by injecting into leaves of sugarcane variety B-4362, susceptible to mottle striped disease produce typical symptoms of the disease. No symptoms, however, were observed in the commercial Brazilian sugarcane variety B45-3; NA 56-79; when included with this bacteria (Olivares *et al*, 1993).

Azoarcus

Originally isolated from Kallar grass (*Leptochloa fusa* kunth) growing in saline sodic soils of Pakistan, a strain of the gram-negative nitrogen-fixing bacterium *Azoarcus* spp. BH72 has been described by Reinhold-Hurek *et al* (1993). The organism colonizes grasses and rice in laboratory experiments (Hurek *et al*, 1994). These bacteria have a strictly aerobic type of metabolism, fix nitrogen micro aerobically, and grow well on salts of organic acids but not on carbohydrates. Indirect evidence for the colonization of the root interior by these organisms was confirmed by fluorescent antibodies and immunogold electron microscopy (Hurek *et al*, 1991). Based on the morphological, nutritional, biochemical and molecular characteristics a new genus two species of *Azoarcus* (*A. communis* and *A. indigenous*) were propsed (Reinhold – Hurek *et al*, 1993).

As these species have never been isolated from soil itself, fungus resisting stages might resent a colonization site that serves for survival in soil in the absence of the plant (Reinhold-Hurek and Hurek, 1998). Hurek *et al*, (1994) reported that *Azoarcus* does not cause pathogenic reaction when inoculated into Kallar grass grown axenically. They showed that the strain BH72 is capable of invading roots of the original host as well as rice plants, infecting the cortex region and penetrating the stele of the roots. Compared with the non-inoculated controls, inoculation with strain BH72 significantly promoted growth of rice seedlings.

Burkholderia

The genus *Burkholderia* is an important component of the soil microbial community (Dalmastri *et al*, 1999). *Pseudomonas cepacia* was originally described by Burkholder in 1950 as the causative agent of bacterial rot of onion bulbs (Burkholder, 1950). Stanier *et al*, (1966) noted an extraordinary metabolic versatility of another novel pseudomonad for which they proposed the name *P. multivorans*. This novel species was mainly isolated from soil and water samples. A few years later Ballard *et al*, (1970) reported on the synonymy between *P. cepacia* and *P. multivorans*. Jonsson (1970) proposed the name *Pseudomonas kingii* for CDC group EO-1 (Eugonic oxidizer group 1), an opportunistic human pathogen. Subsequent taxonomic analysis (Snell *et al*, 1972; Sinsabaugh and Howard, 1975) again revealed that this novel organism was the same as *P. cepacia*.

Pseudomonas cepacia was not included in the bacterial nomenclature earlier (Skerman *et al*, 1980) until 1981, when it was revived by Palleroni and Holmes (1981). In 1992, *P. cepacia* and six other species belonging to rRNA group II of the genus *Pseudomonas* (*P. solanacearum, P. pickettii, P. gladioli, P. mallei, P. pseudomallei* and *P. caryophilli* (Palleroni *et al*, 1973) were transferred to the new genus *Burkholderia* (Yabuuchi *et al*, 1993). The genus *Burkholderia* now includes 36 validly described species.

IV. PLANT GROWTH PROMOTING ACTIVITY

Agriculturally important grasses such as sugarcane, rice, wheat, sorghum and maize contain numerous diazotrophic bacteria, such as *G. diazotrophicus, Herbaspirillum* spp. *Azosprillum* spp. and *Burkholderia* spp. These bacteria do not usually cause disease symptoms in the plants. They are a key factor in nutrient cycling and the maintenance of soil fertility. The inoculation of cereals with diazotrophic bacteria, so far has mainly been performed with *Azospirillum* spp. with variable responses from the plants to inoculation (Okon and Labandera – Gonazalez 1994; Bashan and Holguin 1997).

Burkholderia vietnamiensis as a promising bacterial candidate for inoculation assays with its high rhizospheric population, and high nitrogenase activity, its ability to inhibit fungal plant pathogens and its ability to produce siderophores and indole acetic acid (IAA) production. Tran Van *et al*, (2000) reported that inoculation of *B. vietnamiensis* in pot experiment resulted in 20% yield increasein rice. Four field and three pot experiments conducted in acid sulphate soil in Vietnam using *Burkholderia vietnamiensis* showed the final increase in rice yield ranging from 13% to 22% due to increased tiller number and grain weight (Tran Van *et al*, 2000). Not much information is available on agricultural uses of this species so far.

Thus, there is some concern about the use of potentially pathogenic strains of *Burkholderia* for biotechnological application such as bio-remediation of polluted sites and as biocontrol agents in agriculture. Among the microorganisms occurring in the rhizosphere of maize, *Burkholderia cepacia* probably represents one of the predominant bacterial species (Nacamulli *et al*, 1997). *B. cepacia* seemed to be strictly associated with maize roots, probably the most representative bacterial species in the maize rhizosphere (Di Cello *et al*, 1997). *B. cepacia* complex (Genomovar III) is a common species occurring in the rhizosphere. It has been identified as a potential biocontrol agent in the control of fungal pathogens and nematodes (Balandreau *et al*, 2001).

Baldani *et al*, (2000) observed that *Burkholderia* spp. contributed relatively high levels of N to rice via BNF ranging from 28% to 31% in contrast to the 19% recorded with the *B. vietnamiensis* strain TVV75, isolated from the rhizosphere of rice grown in Vietnam. The best strains were those originating from washed and surface – sterilized roots of rice, suggesting that the endophytic colonization of plant tissues by these endophytic diazotrophs allows them to express their potential better as compared to rhizosphere bacteria. Rice stems are more suitable niches for N_2-fixing endophytes than roots because more photosynthates are available to the bacteria (Barraquio *et al*, 1997).

The endophytic habitat of *Burkholderia* spp. strain BM16, has been confirmed using MLEE (Multilocus enzyme electrophoresis) profiles. Large numbers of bacterial colonies were recovered from surface sterilized plant tissues (10^3 to and up to 10^5 cfu/G of fresh root (Estrade *et al*, 2002).

REFERENCES

Agarwala R & KVBR Tilak (1984 a) Colonization of *Azospirillum* in leaves of *Eleusine coracana* and *Setaria italica*. Zentralbl. Mikrobiol. 143: 533-537.

Agarwala R & KVBR Tilak (1988 b) Establishment of *Azospirillum brasilense* in roots of fingermillet (*Eleusine coracana*), Zentralbl. Mikrobiol. 143: 539-544.

Agarwala-Dutt R, KVBR Tilak & JPS Rana 1991. Isolation of *Azospirillum* from the interior of various parts of some graminaceous plants. Zentralbl. Mikrobiol. 146: 217-219.

Balandreau J, V, Viallard B, Cournoyer T, Coenye S Laevens & P Vandamme (2001) *Burkholderia cepacia* genomovar III is a common plant-associated bacteriaum. Appl. Environ. Microbiol. 67: 982-985.

Baldani VLD (1996) Efeito da Inoculaqao de *Herbaspirillum* ssp. No processo de colonizaquo e infecquo de plantas de Arroz e ocorrencia e caracterizagao parcial de Uma Nova Bacteria Diazotrofica Tese de Doutorado, Universidade Federal Rural do Rio de Janeiro. RJ. Brazil.

Baldani J I, VLD Baldani, L Seldin & Dobereiner J (1986) Characterisation of *Herbaspirillum seropedicae* gen. nov. sp. novz., a root associated nitrogen fixing bacterium. Int. J. Syst. Bacteriol. 36: 86-93.

Baldani VLD, JI Baldani & J Dobereiner (2000) Inoculation of rice plants with the endophtic diazotrophs *Herbaspirillum seropedicae* and *Burkholderia* spp. Biol Fertil. Soils. 30: 485-491.

Baldani JI, L Caruso, VLD Baldani, SR Goi & J Dobereiner (1997) Recent advances in BNF with non-legume plants. Soil. Biol Biochem. 29: 911-922.

Baldani JI, TB Pot, G Kirchhof, E Falsen, VLD Baldani, FL Olivares, B Hoste, K Kersters, A Hartmann M Gillis & J Dobereiner (1996) Emended description of *Herbaspirillum* a mild plant pathogen, as *Herbaspirillum* rubrisubalbicans comb. Nov., and classification of a group of clinical isolates (EF Group 1) as *Herbaspirillum* species 3. Int. J. Syst. Bacteriol. 46: 802-810.

Ballard RW, NJ Palleroni, M Doudoroff & RY Stanier (1970) Taxonomy of the aerobic Pseudomonads : *Pseudomonas cepacia*. *P. marginata*. *P. allicola* and *P. caryophylli*. J. Gen. Microbiol. 60: 199-214.

Barraquio WL, L Revilla & JK Ladha (1997) Isolation of endophytic diazotrophic bacteria from wetland rice. Plant and Soil. 194: 15-25.

Bashan Y & G Holguin (1997) *Azospirillum*-plant relationships: environmental and physiological advances (1990-1997). Can. J. Microbiol. 43: 103-121.

Boddey RM (1995) Biological nitrogen fixation in sugarcane: a key to energetically viable biofuel production. Crit. Rev. Plant sci. 14(3): 263-279.

Boddey RM. & J Dobereiner (1988) Nitrogen fixation associated with grasses and cereals : Recent results and perspectives for future research. Plant and Soil, 108: 53-65.

Boddey RM, CC-de Oliveira, S Urquiaga, VM Reis, FLde Olivares, VLD Baldani, J Dobereiner, OC De-Oliveira, FL De-Olivares & JK Ladha (1995) Biological nitrogen fixation associated with sugarcane and rice: contributions and prospects for improvement. Plant and Soil, 174: 195-209.

Burkholder WH (1950) Sour skin, a bacterial rot of onion bulbs. Phytopathology, 40: 115-117.

Cavalcante VA & J Dobereiner (1988) A new acid tolerant bacterium associated with sugarcane. Plant Soil, 108: 23-31.

Cocking EC, MK Al-mallah, E Benson & MR Darvey (1990) Nodulation of non-legumes by *Rhizobium*. In: Gresshoff PM, Roth, LE, Stacey G, Newton WE (eds.) Nitrogen fixation achievements and objectives. Chapman and Hall, London and Newyork. pp. 813-823.

Dalmastri C, L Chiarini, C Cantale A Bevivino & S Tabacchioni (1999) Soil type and maize cultivar affect the genetic diversity of maize root associated *Burkholdera cepacia* populations. Microbiol. Ecol. 38: 273-284.

Di cello F, A Bevivino, L Chiarini, R Fani, D Paffetti, S Tabachioni & C Dalmastri (1997) Biodiversity of *Burkholderia cepacia* population isolated from the Maize rhizosphere at different growth stages. Appl. Environ. Microbiol. 63: 4485-4493.

Dobereiner J. (1992) History and new perspectives of diazotrophs in association with non-leguminous plants. Symbiosis. 13: 1-13.

Dobereiner J, VLD Baldani & VM Reis (1995) Endophytic occurrence of diazotrophic bacteria in non-leguminous crops. In: *Azospirillum* VI and related microorganisms. Fendrik I., del Gallo M. Vanderleyden J and de Zamaroczy M). pp. 3-14. Springer-Verlag, Berlin.

Dobereiner J & JM Day (1975) Associative symbiosis in tropical grasses: Characterisation of microorganisms and dinitrogen fixing sites, In: Nitrogen fixation (eds. Newton, WE and Nyman CJN). pp. 518-538. Washington State University Press, Pullman, Washington.

Dobereiner J. VM Reis, MA Paula & F Olivares (1993) Endophytic diazotrophs in sugarcane, cereals and tuber plants. In: New horizons in Nitrogen Fixation (eds. Palacios R, Mora J and Newton WE), pp. 671-676, Kluwer Academic Press, Dordrecht, Netherlands.

Dobereiner J & AP Ruschel (1958) Uma nova especie de *Beijerinkia*, Revista de Bilogia, Rio de Janeiro, 1: 261-272.

Dong Z, M Haydrich, K Bernard & ME Mc Cully (1995) Further evidence that N_2 fixing endophytic bacteria from the intercellular spaces of sugarcane stems is *Acetobacter diazotrophicus*. Appl. Environ. Microbial. 61: 1843-1846.

Eckert B, OB Weber, G Kirchhof, A Halbriter, M Stoffels, & A Hartman (2001) *Azospirillum doebereinerae* sp. nov., a nitrogen-fixing bacterium associated with the C4-grass *Miscanthus*. Int J. Syst Evol Microbiol. 51: 17-26.

Estrada P, P Mavingui, B Cournoyer, F Fontaine, J Balndreau & J Cabellero Mellado (2002) A N_2-fixing endophytic *Burkholderia* sp. associated with maize plants cultivated in Mexico. Can J. Microbiol. 48: 285-294.

Ferreira AC, K Cozzolino, ARV Carvalho & J Dobereiner (1995) Isolation and characterization of diazotrophs in oil palm trees. In: International Symposium on sustainable agriculture for the tropics, the role of Biological nitrogen fixation pp. 210, Angra dos Res, Rio de Janeiro.

Fuentes-Ramirez LE, R Bustillos-Cristales, A Tapia-Harnandez, T Jimenez-Salgado, ET Wang, E Martinez-Romero & J Cabellero-Mellado (2001) Novel nitrogen-fixing acetic acid bacteria, *Gluconacetobacter johannae* sp. nov., and *Gluconacetobacter azotocaptans* sp. nov., associated with coffee plants. Int. J. Syst. Evol. Microbiol. 51: 1305-1314.

Fuentez-Ramirez LE, T Jimenez-Salgado, IR Abarea-Ocampo & J Cabellero-Mellado (1993) *Acetobacter diazotrophicus* an indole-acetic acid producing bacterium isolated from sugarcane cultivars of Mexico. Plant Soil. 154: 145-150.

Gillis M, J Dobereiner, B Pot, M Goor, E Falson, B Hoste, B Reinhold, & K Kersters (1991) Taxonomy relationship between *"Pseudomonas" rubrisubalbcans*, some clinical isolates (EF group 1), *Herbaspirillium seropedicae* and *"Aquaspirillum" autotropicum*. In: Nitrogen Fixation (eds., Polsinelli M, Materassi R and Vicenzini M), pp. 292-294. Kluwer Academic Press, Dordrect.

Hurek T, B Reinhold-Hurek, M Van montage & E Kellenberger (1991) Infection of intact roots kallar grass and seedlings by *Azoarcus*. In: Nitrogen fixation. pp. 235-242. Polsinelli M, Materassi R and MV (eds) Kluwer Academic Publisher, Dordrecht, Netherlands.

Indira BN & DJ Bagyaraj (1997) Associative diazotrophicus of sugarcane (*Saccharum officinarum* L) cultivars. Indian J. Microbiol. 37: 95-98.

James EK, VM Reis, FL Olivares, JI baldani & J Dobereiner (1994) Infection of sugarcane by the nitrogen fixing bacterium, *Acetobacter diazotrophicus*. J. Exp Bot. 45: 757-776.

Jimenez-salgado T, LE Fuentes-Ramirez, A Tapia-Harnandez, MA Mascarua, E Martinez-Romero & J Cabellero- Mellado (1997) *Coffea arabica* L., a new host plant for *Acetobacter diazotrophicus* and isolation of other nitrogen fixing acetobacteria. Appl. Envirn. Microbiol. 63: 3676-3683.

Jonsson V (19970) Proposal of a new species *Pseudomonas kingii*. Int. J. Syst. Bacteriol. 20: 255-257.

Kamura F SL Albrecht, LH Allen & KT Shanmugan (1998) Dry matter nitrogen accumulation in rice inoculation with a depressed mutant of *Anabaena variabilis*. Agron. J. 90: 529-535.

Kirchhof G, B Eckert, M Stoffels, JI Baldani, VM Reis & A Hartmann (2001) *Herbaspirillum frisingense* sp. nov. a new nitrogen-fixing bacterial species that occurs in C_4-fibre plants. Int. J. Syst. Evol. Microbiol. 1: 157-168.

Kloepper JW & CJ Beauchamp (1992) A review of issues related to measuring colonization of plants by bacteria. Can. J. Microbiol. 38: 1219-1232.

Lakshmi V., ASN. Rao, K Vijaya Lakshmi, M Lakshmi-Kumari, KVBR Tilak & NS Subba Rao 1977. Establishment and survival of *Spirillum lpoferum*. Proc.Indian Acad. Sci. 86 B, 6: 397-404.

Ladha JK & G. Reddy (1995) Extension of nitrogen fixation to rice: necessity and possibilities. Geojournal. 35: 363-372.

Li RP & IC Mac Rae (1991) Specific association of diazotrophicus acetobacters with sugarcane. Soil. Biol. Biochem 23: 999-1002.

Li RP & IC Mac Rae (1992) Specific identification and enumeration of *Acetobacter diazotrophicus* in sugarcane. Soil. Biol. Biochem. 24: 413-419.

Loganathan P, R, Sunitha, AK Parida & S Nair (1999) Isolation and characterization of genetically two distant of Acetobacter diazotrophicus from new host plant (*Eleusine coracana* L). J. Appl. Microbiol. 86: 1053-1058.

Malik KA, B Rakhshanda, S Mehnaz, G Rasul, MS Mirza & S Ali (1997) Association of nitrogen-fixing plant growth promoting rhizobacteria (PGPR) with kallar grass and rice. Plant and Soil. 37-44.

Moulin L, A Munive, B Dreyfus & C Boivin-Masson (2001) Nodulation of legumes by members of β-subclass of proteobacteria. Nature. 411: 948-950.

Muthukumarasamy R (1996) Studies on diazotrophic association in sugarcane and their role in cane cultivation. Ph.D. thesis, Bharathidasan University, Trichy, India.

Muthukumarasamy R, G Revathi & C Lakshminarasimhan (1999) Influence of N-fertilisation on the isolation of *Acetobacter diazotrophicus* and *Herbaspirillum* spp. from Indian sugarcane varieties. Biol. Fertil. Soils. 29: 171-178.

Nacamulli C, A, Bevivino, C, Dalmastri S Tabachioni & L Chiarini (1997) Pertubaction of maize rhizosphere microflora following seed bacterization with *Burkholderia cepacia* MCI 7. FEMS Microbiol. Ecol. 23: 183-193.

Okon Y & C Labandera-Gonzalez (1994) Agronomic application of *Azospirillum* : an evolution of 20 years of worldwide field inoculation. Soil. Biol. Biochem. 26: 1591-1601.

Olivares FL, EK James, VM Reis, VLD Baldani, JI Baldani & J Dobereiner (1993) Colonisation of vascular tissue by *Herbaspirillum* spp. in sorghum and sugarcane. In: Proceedings of the 23rd Brazilian Congress of Plant Pathology.

Palleroni NJ & B Holmes (1981) *Pseudomonas cepacia* sp. nov., nom.rev. Int.J. Syst. Bacteriol. 31: 479-481.

Palleroni NJ, R Kunisawa, R Contopoulo & M Doudoroff (1973) Nucleic acid homologies in the genus *Pseudomonas*. Int. J.Syst. Bacteriol. 23: 333-339.

Partriquin DG & J Dobereiner (1978) Light Microscopy observation of Tetrazolium-reducing bacteria in the endorhozosphere of maize and other grasses in Brazil. Can. J. Microbiol. 24: 734-742.

Pereira JO (1995) Microorganisms endofiticos em species tropicais. In: Anais 20a, Reuniao Annual de Genetica de microorganismos. Vol-20, p-3, Sociedade Brasileria de Genetica, piracicaba, Sao Paulo.

Pimental JP, Fl Olivares, RM Pitardl, SC Urquiga, F Akiba & J Dobereiner (1991) Dinitrogen fixation and interaction of grass leaves by *Pseudomonas rubrisubalbicans* and *Herbaspirillum seropedicae*. Plant and Soil. 137: 61-65.

Reinhold-Hurek BR, T Hurek, M Gillis, B Hoste, M Vancanneyt, K Kersters & JD Ley (1993) *Azoarcus* gen.nov. Nitrogen-fixing proteobacteria associated with roots of kallar grass (*Leptochoa fusca* (L.)

Kunth) and description of two species, *Azoarcus indigens* sp. nov. and *Azoarcus communis* sp. nov. Int. J. Syst. Bacteriol. 43: 573-584.

Rolfe BG, MA Djordjeric, JJ Weinman, U Mathesius, C Pittock, E gartner, KM Ride, Z Dong, M Mc Cally & J Melver (1997) Root morphogenesis in legumes and cereals and the effects in legumes and cereals and the effects of bacterial inoculation on root developments. Plant and Soil. 194: 131-144.

Ruvkun GB & FM Ausubel (1980) Interspecies homology of nitrogenase genes. Proc. Natl. Acad. Sci. USA. 77: 191-195.

Sahai R. (2000) association of acetobacter diazotrophicus with sweet sorghum (*Sorghum bicolor* L. Moench). Ph. D. thesis Division of Microbiology, Indian Agricultural Research Institute, New Delhi, India.

Schloter M, G Kirchhof, U Heinzmann, J Dobereiner & A Hartman (1994) Immunological studies of the wheat root colonization by the *Azospillium brasilense* strains Sp 7 and Sp 245 using strain specific monoclonal antibodies. In: Nitrogen Fixation with non-legumes. (eds., Hegaz N, Fayez M and Monib M) pp. 290-295. The American University in Cairo Press, Cairo, Egypt.

Servilla M & C Kennedy (2000) Genetic analysis of N_2 fixation and plant growth stimulating properties of *Acetobacter diazotrophicus*, an endophyte of sugarcane. In: Prokaryotic Nitrogen Fixation, (eds. Triplet E), pp. 737-760, Horizon Scientific Press, Wyndham, UK.

Sinsabaugh HA & GW Howard (1975) Emendation of the description of *Pseudomonas cepacia* Burkholder (synonyms: *Pseudomonas multivorans* Stanier *et al, Pseudomonas kingae* Jonsson: Eo-1 Group). Int J. Syst. Bacteriol. 25: 187-201).

Skerman VBD, V Mc Gowan & PAH Sneath (1980) Approved list of bacterial names. Int. J. Syst. Bacteriol. 30: 225-420.

Sly LI and E Stackebrandt (1999) Description of *Skermanella parooensis* gen. nov., sp. nov. to accommodate *Conglomeromonas largomobilis* subsp. *parooensis* following the transfer of *Conglomeromonas largomobilis* subsp. *largomobilis* to the genus *Azospirillum*. Int. J. Syst. Bacteriol. 49: 541-544.

Snell JJS, LR Hill, SP Lapage and MA Curtis (1972) Identification of *Pseudomonas cepacia* Burkholder and its synonomy with *Pseudomonas kingi* jonsson. Int. J. Syst. Bacteril. 24: 127-138.

Stanier RY, NJ Palleroni & M Doudoroff (1966) The aerobic pseudomonads: a taxonomic study. J. Gen. Microbiol. 43: 159-271.

Suman A, Shasang AK, Singh M, Shahi HN, Gaur A and Khanuja SPS (2000) Molecular assessment of diversity among endophytic diazotrophs isolated from subtropical Indian sugarcane. World. J. Microbial. Biotech. 17: 39-45.

Taek Im W, H Sung Bae, A Yokota & S Taik Lee (2004) *Herbaspirillum chlorophenolicum* sp. nov., a 4-chlorophenol-degrading bacterium. Int J. Syst. Evol. Microbiol. 54: 851-855.

Tapia-Hernandez A, MR Bustillos-Cristalles, T Jimenez-Salgado, J Cabellero-Mellado & LE Fuentes-Ramirez (2000) Natural endophytic occurance of *Acetobacter diazotrophicus* in pineapple plants. Microbial. Ecol. 39: 49-55.

Tarrand JJ, NR Kreig & J Dobereiner (1979) A taxonomy study of the *Spirillum lipoferum* group, with description of a new genus, *Azospirillum lipoferum* (Beijerinkck) comb. nov. and *Azospirillum brasilense* sp. nov. Can J. Microbiol. 24: 967-980.

Tran Van V, O Berge, S Ngo Ke, J Balandreau & T Heulin (2000) Repeated beneficial effects of rice inoculation with a strain of *Burkholderia vietnamiensis* on early and late yield components in low fertility sulphate acid soils of Vietnam. Plant and Soil. 218: 273-284.

Ureta A, A Alvarez, A Ramon, MA Vera & G Martinez-Drets (1995) Identification of *Acetobacter diazotrophicus, Herbaspirillum rubrisubalbicans* using biochemical and genetic criteria. Plant Soil. 172: 271-277.

Valverde A, E Velazquez, C Gutierrez, C Cervantes, A Ventosa & J Mariano Igual (2004) *Herbaspirillum lusitanum* sp. nov., a novel nitrogen-fixing bacterium associated with root nodules of *Phaseolus vulgaris*. Int. J. Syst. Evol. Microbiol. 53: 1979-1983.

Webster G, C Gough, CA Vasse, KJ O'callaghan, SL Kothari, MR Darvey, J Denarie & Cocking EC (1997) Interaction of *Rhizobium* with rice and wheat. Plant and Soil. 194: 115-112.

Yabuuchi E, Y Kawamura, T Ezaki, M Ikedo, S Dejsirilert, N Fujiwara, T Naka & N Kobayashi (2000) *Burkholderia uboniae* sp. nov., L-arabinose assimilating but different from *Burkholderia thailandensis* and *Burkholderia vietnamiensis*. Microbiol. Immunol. 36L 1251-1275.

16

Arbuscular Mycorrhizal Associations in Agricultural Plants Including Weeds

[1]RAJNI GUPTA and [#]K.G. MUKERJI
[1]*Shyam Lal College, Shahadara, University of Delhi, Delhi-32*
[#]*Department of Botany, University of Delhi, Delhi*

I. INTRODUCTION

Weeds cause enormous losses in crop yields both in quantity and quality besides they create health and aesthetic problems. The losses are of subtle nature and are not recognized immediately but if not controlled timely and properly, the yield losses in different cereals, pulses, fibre and oilseeds crops have been reported varying from 16 to 58% under different agro-climate conditions. Mehta *et al* (1989) estimated the losses to a tune of Rs. 4,200 millions per annum by weeds in India in principal agricultural crops. Rao (1983) worked out these losses as Rs. 5,000 to 50,000 millions per annum, when calculated at a conservative estimate of 10-15% losses in yield. Besides the losses in field crops, the weeds also cause considerable losses in plantation, horticultural and vegetable crops and interferes in production of fauna and flora in aquatic system and impede water flow in irrigation canals. They also cause problems in highways, rail tracks and industrial sites.

Weed as Defined

Tull (1931) defined weed as, "a plant growing where it is not desired". The definition is simple as one may note, it may be a crop plant but is not desired where it is growing. Even a absolute weed e.g., *Chicorium intybus* may not be weed when it is grown for specific purpose. Likewise though, *Cynodon dactylon* is among the world's most difficult weeds but when put in lawns it is not a weed. *Imperata cylindrica* is a tough and hard to control weed of uncropped and plantation crops but when used as a soil binder it may not be called a weed. While a gram plant may be weed if it is growing in wheat field where it is not desired. Weeds in general are characterized as a plant having better competitive ability and grow profusely under marginal conditions and escapes adverse conditions by shortening their life cycles. Mortimer (1989) defined weeds as natural hazards to the activities of man as any plant or vegetation interfering with the objectives of people. Weeds are those plants growing in places and at times when man wants either some other plants to grow or no plants to grow at all (Gupta, 1984). So long as a plant is growing at a place and time without interfering with the man's affairs, it is not looked as a weed. In other words, while all weeds are unwanted plants, all unwanted plants may not be weeds. Alstrom (1990) defined agriculture weeds as the non-crop vegetation that reduces the quality and quantity of the existing crop to such an extent that any net benefit of the weed plants themselves cannot compensate for their unfavourable effects on crops.

History of Weed

The history of weed is as old as the human civilization when the man started cultivating crops and then he soon realized that crop yields were more when grown in weed free environment. Thus monoculture, contrary to natural way was identified and concept of weeds as unwanted plant was developed. A lot of exercise was made to check the spread of undesirable plants through seed cleaning, preparatory cultivation and various interculture practices and efforts are continuing to eliminate undesirable plants from crops. The evolution of weed may occur at three levels viz. (i) genetic changes in the existing species due to unpredictable disturbance or in the habitat; (ii) the race formation in which locally adopted races of a weed species occur and (iii) crop mimicry. Where the crop and weed are so intimately associated with man that it has led to the evolution of mimetic weeds. The alien immigration in the import of weed seed through crop seed or any other commodities which settles in the new environment and later becomes problem weeds. The most burning examples are recent introduction of *Parthenium hysterophorus* in India from USA through the import of wheat grain around 1956. *Phalaris minor* and *A. ludoviciana* are some of the other examples. Similarly *Eicchornia crassipes* which has come in India through sea transport long back around 1890 from South America has created havock in aquatic systems in India.

II. MICROBIAL COMMUNITY

Soil microorganisms constitute the world's largest reservoir of biological diversity and are crucial to the functioning of terrestrial ecosystems. Soil provides a spatially and temporally heterogeneous environment, especially at the scale encountered by microbes. Heterotrophic microbial communities inhabiting soil mediate key processes that control ecosystem carbon (C) and nitrogen (N) cycling, and they potentially represent a mechanistic link between plant diversity and ecosystem function. It is expected that plant diversity could affect microbial communities and the ecosystem processes they mediate (Bargett and Shine, 1999; Stephan *et al*, 2000). The availability of growth-limiting resources shapes the composition of biotic communities (Tilman, 1987) and resource availability for microbial communities as constrained by organic compounds in dead leaves and roots that can be used to generate cellular energy (Smith and Paul, 1990). Plant species differ in their biochemical composition, therefore, changes in plant diversity could alter production as well as the range of organic compounds in detritus that limit, and thus control, the composition and function of heterotrophic microbial communities. The present microbial community is responsible for litter decomposition and nutrient cycling in the forest soils. The rhizosphere is continually supplying readily available forms of carbon and nitrogen to the soil system (Kapoor & Mukerji, 2005). Thereby, the growth of the microbial biomass is promoted. The quantity and the composition of soil microbial biomass are particularly sensitive to changes in soil environment. Microbial biomass parameters have frequently been used as indicators of ecosystem stress and disturbance (Mukerji, 2002). Among forest types, different limitations were found acting upon the growth of the microbial biomass (Hackel *et al*, 2000). In more acidic soils, microbial biomers are significantly related to soil pH. Positive correlations between microbial biomass and soil moisture and total soil N are found in soils of all forest types. Low molecular weight compounds enter the soil through plant litter decomposition, rhizodeposition and microbial metabolism as well as from decaying soil organisms. The concentration of sugars and amino acids not only depend on the amount, quality and decomposition rate of litter and on rhizodeposition but also on turnover times of these substances. High concentration of sugars and amino acids may indicate high substrate abundance as well as low turn out rate. The endomycorrhizal associations produce minor changes in the root morphology. The mantle of the root is absent. Orchids have endomycorrhiza which is called glomerate endomycorrhiza because of the presence of glomerules. AM fungi are found associated with most crop plants, forest plants, horticultural plants, ornamental plants and other plants. These fungi are not host specific but differ in their characteristic association with host rhizosphere (Mukerji *et al*, 2000). Most AM fungi form a symbiotic association with a wide range of plants including angiosperms, gymnosperms, pteridophytes and bryophytes (Smith and Read, 1997). In colonizing the plant root system the fungus receives a carbon source, which permits it to fulfill its life cycle. The majority of host plants benefit through an improvement of their mineral nutrition (Smith and Smith, 1990) and protection against biotic and abiotic stresses (Newsham *et al*, 1995; Subramanian *et al*, 1995). Ectomycorrhizal formations are confined to forest and ornamental tree species belonging to Pinaceae, Salicaceae, Betulaceae, some members of Rosaceae, Leguminosae, Ericaceae, Juglandaceae and Dipterocarpaceae (Manoharachary & Mukerji, 2002).

III. VAM IN WEEDS

The AM fungi are ubiquitous in distribution and occur abundantly. 90% plants develop AM fungal association (Remy *et al*, 1994; Taylor *et al*, 1995). The arbuscular mycorrhizal fungi are the most complex group of mycorrhizae which form intra-radical structures i) vesicles/ arbuscules, inter- and intra-cellular hyphae inside the root tissue. ii) spores/sporocarps and branched hyphae (extrametrical hyphae) outside the root and in soil.

Some AM fungi are more effective than others in stimulating plant growth (Menge, 1983). The capacity of AM fungi to supply P to the host, varies with the species of fungus (Pearson & Jacobson, 1993) and host (Ranskov & Jakobson, 1993). The degree to which AM fungi enhance nutrient uptake and growth of plants is frequently proportional to the rate and the extent of colonization in root (Wilson & Tommerup, 1992), though exceptions occur (Viehel & Ocampo, 1991). During colonization hyphae penetrate within the root extending each colonization unit and extra-radical runner hyphae colonize other parts of the root system (Schenck *et al*, 1975) the later is referred to as secondary spread. Initiation of primary colonization units continues along with secondary spread (Wang *et al*, 1993). Decline in the population of AM fungi in soil can lead to decreased colonization and nutrient uptake in plants. Reduced colonization by AM fungi might follow weed-free fallows or rotation with non-mycotrophic hosts (Abbott & Robson, 1991). The extent of decline of the population of fungi during fallows might depend on interaction with other processes. Soil disturbance and cycles of wetting and drying contribute to propagule attrition (Braumberger *et al*, 1991), especially when soil temperatures are high (Pattinson & McGee, 1997). Research on floristics and endomycorrhizal occurrence in natural ecosystems is especially limited in the coastal regions of tropical countries (Mahadevan *et al*, 1989; Raghupathy & Mahadevan, 1991). In the present paper, the significance of mycorrhizal associations in weeds has been discussed. It is now amply known that AM colonization can significantly improve the phosphorus nutrition and yield of plants grown in soils of low fertility (Mosse, 1973). AM association occurs in many types of vegetations including tropical rain forests, semi-arid grassland (Jagpal & Mukerji, 1988) sand dune systems (Nicolson, 1955) and temperate deciduous woodland (Mosse & Hayman, 1980). AM influences soil fertility and thus the growth and development of plants (Mukerji & Dixon, 1992) in social forestry and land reclamation. Gupta & Mukerji (1996) characterized and quantified the VAM association in roots of grass weeds. In *Eragrostis ciliaris* and *Oplisemenus* 100% colonization was observed. External hyphae were observed forming distinct appresoria, at the entry points in the root tissues. The spores of *Glomus fasiculatum* were present mainly in the cells of mycorrhizal roots of *Chloris*.

Out of 14 grass species examined, only in 2 species i.e. *Bracharia reptans* and *Setaria tomentosa*. AM association was not found. AM colonization and morphology were observed over a period of two years. The measurement of mycorrhizal infection indicated that in some species the extent of fungal infection varied greatly both in one year and then in successive years, but the cause of this variation is not clear. Mycorrhizal fungi from hyphal connections between plants of the same and different species and nutrients can be transferred from one plant to another along these connections. If these connections are abundant in the field then this could have important implications for mycorrhizal functioning and plant performance. *Rumex acetosa* has previously been reported as being non-mycorrhizal (Harley & Harley, 1987) but Sanders & Fitter (1992) found it otherwise.

Plants with mycorrhizal associations are known to predominate in most natural eco-systems but little is known about phenological variations in mycorrhizal activity in many habitats (Brundrett, 1991; Bergeso *et al*, 2003). Seasonal fluctuations in numbers of mycorrhizal roots and spores have been examined in deciduous forests (Mayer & Godoy, 1989) grasslands (Sanders & Fitter, 1992), salt marsh (Van Duin *et al*, 1989), sand dune, tropical forests and arid communities (Allen, 1989). Seasonal fluctuations in numbers of spores of AM fungi have been observed and there may also be seasonal variations in their capacity to form mycorrhizas because newly formed spores require periods of dormancy (Mukerji, 1996).

A significant aspect of parasitic plant biology is their association with microorganisms (fungi and bacteria). Atsatt (1973) suggested that haustoria were first produced in response to microbial parasitism and were modified for water and nutrient uptake which act as a bridge between host and parasite. Holoparasite *Conopholis americana* (Scrophulariaceae) are only found attached to mycorrhizal roots of their oak host (Baird & Riopel, 1980). Because essentially all oak roots found in nature are mycorrhizal, it is difficult to determine whether this association is obligate. Experimental studies using exudates from intact oak mycorrhizal roots yielded a 3% germination rate, comparable to rates seen from soil samples (Baird & Riopel,

1986). Sanders *et al.* (1993) observed that *Cascuta* preferentially parasitizes host plant that have mycorrhizal associations. Nickrent (2002) reported that parasitic plants alternate source of food is from another plant through AM fungal hyphae.

A total of 737 plant species from 121 families of angiosperm and four species of pteridophytes were examined for mycorrhizal colonization (Raghupathy & Mahadevan, 1993). Of these mycorrhizal root colonization was recorded in 372 species. The quantum of colonization was in the range of 10-90%. Of the four pteridophytes only *Isoetes coromandelina* displayed mycorrhizal association. Out of 549 dicotyledonous species surveyed, about 301 displayed mycorrhizal association, whereas in monocotyledons, 71 species out of 188 were mycorrhizal. In fabaceae, the largest family, 94 species were examined of which 62 species were mycorrhizal. Of the 85 species of Poaceae, 39 were mycorrhizal. About 51 species of Cyperaceae were screened for VAM colonization but only 12 species were mycorrhizal. 50% members of the Asteraceae, Euphorbiaceae, Convolvulaceae and Rubiaceae showed mycorrhizal association. Other families such as Cyperaceae, Asteraceae, Scrophulariaceae, Amaranthaceae and Poaceae displayed less than 40% mycorrhizal colonization.

Gupta & Mukerji (1996b) characterized and quantified the AM association in roots of Compositae weeds. Most of the weeds belonging to Compositae were arbuscular mycorrhizal. The amount of vesicles and arbuscules formed in roots varied in all plants. The shape of vesicles were oval to rounded. In plants of *Galinsoga parviflora* arbuscules were densely present. They were dichotomously branched and were of dominant type. Percent arbuscules formation was 40% in most of these plants. Dominant AM fungal spores in the rhizosphere soil consisted of *Glomus macrocarpum* and *Glomus fasciculatum.* A comparative study of diseased and healthy plants of *Sonchus arvensis* was made. Severally infected *Sonchus arvensis* plants with downy mildew *Bremia lactuca* showed presence of dense extramatrical hyphae and appresoria formation was quite distinct, while in the healthy plants number of vesicles and arbuscules per one cm was quite high. AM increased the foliar infection in this plant (Zaidi & Mukerji, 1983). Meney *et al* (1993) reported the presence of AM fungi in Cyperaceae and Rostionaceae.

Harley & Harley (1987) compiled the mycorrhizal association of plants in the British Isles and concluded that most plants in Britain were mycorrhizal. Mycorrhizal colonization was more frequent in forest areas, where there was a higher diversity of plant species than in the cultivated fields. It is likely that fertilizer application to cultivated fields reduced AM fungal species (Mosse & Hayman, 1980).

Raghupathy & Mahadevan (1993) found 47% of the acquatic plants to be mycorrhizal. Chaubal *et al* (1982) earlier reported the occurrence of AM association in aquatic plants. Singh & Tyagi (1989) reported the root colonizations of *Aeschynomene indica* by indigenous AM fungi grown under normal moisture conditions. The non-mycotrophic plant species make up only a small proportion of plant species and appear to be confined to Brassicaceae, Chenopodiaceae and the Amaranthaceae (Tester *et al*, 1987). Gupta and Mukerji (1999) reported about the incompatibility of various weeds belonging to the family Cyperaceae and Aizoaceae.

Mycorrhizal fungi are known to improve plant growth under water stress conditions. Soil water and aeration have considerable impact on the distribution and effectiveness of VAM fungi due to their aerobic nature. Levels of O_2 and CO_2 were found to influence VAM response on plants although root colonization with VAM fungi have been observed in aquatic plants

(Raghupathy & Mahadevan, 1993). Low and high soil moisture content of soil directly effects root colonization by mycorrhizal fungi (Braunberger *et al*, 1996).

Microorganisms in agricultural soils are known to exert profound influence on the soil's fertility status in particular with respect to the availability of plant nutrients (Kennedy & Smith, 1995). This key life support functions of soil are likely to be dependent on crucial constituents of the soil microbial community. The health of the soil can be defined in terms of its microbiological capacity to counteract the activity of plant, pathogenic or plant–deleterious microorganisms (VanBruggen & Somenov, 2000). The extent and degree of microbial activity in soils are clearly affecting the soil microbial communities and thereby diseases (Smalla *et al*, 2001). A major driving force in these effects is the influx of sources of carbon and energy into the grossly oligotrophic soil system (Bago, 2000; Yang & Crowley, 2000). Differences in these inputs by different crop plant induce different microbial community structures (Smalla *et al*, 2001).

Arbuscular mycorrhizal fungi which form mycorrhizas with the roots of nearly all major crop and native grassland and many tree species, were found to host obligate nitrogen-fixing endosymbiotic bacteria at intracellular locations in fungal spores and hyphae (Minerdi *et al*, 2002). AM fungi also helps in the establishment of exotic plants in baren areas. Haskins & Gehring (2004) found that exotic plants from burned and non-burned areas exhibited levels of AM colonization that were 50% greater than native weeds. Fire did not reduce AM inoculum potential, or that AM fungal populations recovered in the 5 years after the slash was burned. Changes in under storey plant communities and increases in exotic species with burning could result from soil changes due to presence of AM fungi.

IV. MICROBIAL DYNAMICS IN TERRESTRIAL ECOSYSTEM

Biological activities essentially control soil productivity. Soil invertebrates, fungi and bacteria recycle all the carbon, nitrogen, and other mineral nutrients in plant and animal residues into forms that can be used by plants. Soil microbes regulate the destruction of toxic environmental pollutants like nitrous oxides and methane. Increasing the diversity of microorganisms in the soil will increase the ability of the soil to adopt and tolerate a variety of environmental conditions.

In undisturbed soil, most of the nutrient cycling, roots and biological activity are found in the top 20 to 30 cm, sometimes called the rooting zone, and known as the rhizosphere. More specifically the rhizosphere is the root and adjacent soil which is influenced by the root. It is characterized as a zone of intense microbial activity. The rhizosphere is bathed in energy-rich carbon compounds like sugars, amino acids and organic acids that leak from the roots. These compounds are called root exudates. Bacteria and fungi use root exudates and the dead sloughed cells from the root to grow and reproduce, but competition for a space on or near the root is stiff. In the battle for carbon, bacteria often produce antibiotics that remove the competition, and plant growth promoting substances which increase root growth. Scavenging and predator fungi, protozoa, nematodes and mites are also looking to increase their carbon and mineral nutrient intake. The sticky secretions and webs of fungal hyphae bind even smaller soil particles into aggregates further preventing erosion.

Every plant species leaks a unique signature of compounds from its roots. The quantity and quality of these compounds depend to a certain extent on the soil chemical and physical properties, but in all cases determine the microbial community of the rhizosphere. Symbionts like *Rhizobium* and diseases causing pathogens may be particularly well tuned to the composition and quantity of root exudates (Van Rhijin *et al*, 1997). Vesicular arbuscular mycorrhizal fungi form a mutually beneficial or symbiotic relationship with 95% of all land plants, including cereals, pulses and some oilseeds. They appear to be essential to the establishment, growth and survival of many plant species. VAM fungi are known to increase drought and root disease resistance of the plant host. However, they are most well known for their ability to increase the uptake of the contents less available i.e., mineral nutrients such as phosphorus (P), calcium (Ca), Zinc (Zn) and Copper (Cu). These fungi penetrate the cells of the root without harming the plant. From the root, the hyphae extend like a network of silk threads through the bulk soil where they absorb water and mineral nutrients and glue and tie soil particles into more erosion-resistant aggregates (Rillig *et al*, 2001). VAM/AM fungi can be considered a highly effective transport system like a pipeline between the soil and the plant, moving water and nutrients to the plant in exchange for direct access to the carbon-rich products of photosynthesis. The degree to which a particular plant relies on VAM/AM for access to nutrients is termed its level of dependency. High dependent crops often have limited root systems, with thick roots and few root hairs. Less dependent plants will have larger fibrous root systems that are well adapted to competing for nutrients. Even less dependent plant species may rely on VAM fungi. VAM hyphae will tie and glomalin secreted by the hyphae glue soil particles into more erosion-resistant aggregates. Once plant roots are colonized by VAM fungi, their physiology and biochemistry change. These structures require changes in the cell wall flexibility and cytoskeletal re-arrangements. Mycorrhizal fungi activate plant genes coding for cell-wall associated proteins like ENOD11, possibly to modulate cell wall plasticity (Journet *et al*, 2001; Blancaflor *et al*, 2001). AM fungi usually cause differentiation regulated by plant hormones especially auxins and cytokinins (Hirsch *et al*, 1997). Mycorrhizal fungi produces phytohormones including cytokinins that indirectly affect plant hormone system (Barker & Tagu 2000). Sometimes weeds-AM association helps transferring nutrients to adjoining crop plants (Ryan, 1998; Mukerji & Kochar, 1983).

Mycorrhizal fungi alter the root exudation of plants and also produce their own exudates, which has prompted development of the mycorrhizosphere concept (Linderman, 1992). The potential impacts of rhizobacteria on mycorrhizal fungi include changes in root fungus signalling recognition and receptivity, as well as effects on fungal growth and germination (Johansson *et al*, 2004). Rhizobacteria can either increase or inhibit plant growth indirectly through effects on the competency of arbuscular mycorrhizas (Hetrick *et al*, 1988) and ectomycorrhizas (Garbaye, 1994). Some interaction between rhizobacteria and mycorrhizal fungi probably occurs by default, through joint occupation of the rhizosphere, but specialized associations are also evident. Free-living diazotrophic bacteria have the potential to act in synergism with arbuscular mycorrhizal fungi to increase plant growth (Subba Rao *et al*, 1985; Paula *et al*, 1992). The same applies for interaction between mycorrhizas and the symbiotic diazotrophs such as the *Rhizobia* and *Frankia.* Biological nitrogen fixation expends large amounts of energy, requiring the synthesis of around 16 molecules of adenosine 5' triphosphate for every molecule of N_2 converted to ammonia (Gottschalk, 1986). Phosphorus supply, therefore, is a limiting factor to symbiotic nitrogen fixation. The capacity for mycorrhizas to enhance phosphorus uptake by plants enables synergistic interaction between Frankia and both endo-

and ectomycorrhizal fungi (Tian *et al*, 2002). Similarly synergistic interaction has been observed between arbuscular mycorrhizal fungi and rhizobia (Vazquez *et al*, 2001; Mathesius, 2003). Rhizobial symbiosis in turn can also be influenced by rhizobacteria.

V. CONCLUSION

AM fungi can influence plant community composition by differentially affecting the growth of different plant species. AM fungal symbiosis is highly dynamic interaction that affects nearly all physiological aspects of the host. Restoration of disturbed land has prompted the urgent need for understanding mycorrhizas. Soils which are disturbed and turned into wastelands/ unproductive land were found to have reduced mycorrhizas and that many plant species dependent of mycorrhizas failed to establish or survive in these regions. Therefore, greening of such areas is possible only with AM fungal symbionts.

Desert, arid and semi-arid soils suffer most and pose technical problems with discrete patches of vegetation. Therefore, mycorrhizal fungi can be important in the primary production and nutrient flow in these specialized ecosystems mycorrhizas are of immense utility in the established of forest seedlings, commercial crops, economically viable plants, medicinal and aromatic plants, horticultural and ornamental plants. P uptake which is an essential process gets meticulously monitored by AM fungi. This process is of great help in increasing N, P and K status of host tissues. Mycorrhizal fungal networks enhance soil capacity and soil fertility.

REFERENCES

Abbott, L.K. & A.D. Robson. 1991. Factors influencing the occurrence of vesicular arbuscular mycorrhizas, Agriculture Ecosystems and Environment. 35: 121-150.

Allen, M.F. 1981. Formation of vesicular arbuscular mycorrhizae in *Atriplex gardneri* (Chenopodiaceae): A seasonal response in a cold desert. Mycologia. 75: 773-776.

Alstrom, S. 1990. Fundamentals of weed management in hot climate peasant agriculture. Swedish Univ. Agric. Sci., Uppasala, Sweden.

Anderson, T.H. & K.H. Domesch. 1993. The metabolic quotient for CO_2 (qCO_2) as a specific activity parameter to assess the effects of environmental conditions, such as pH, on the microbial biomass of forest soil. Soil Biol. Biochemists 25: 393-395.

Atsatt, P.R. 1973. Parasitic flowering plants: how did they evolve? American Naturalist. 107: 02-510.

Baird, W.W. & J.L. Riopel. 1980. Studies on the life history of *Conopholis americana* (Orobanchaceae). Bot. Soc. Am. Misc. Ser. Publ. 159: 9-11

Baird, W.W. and J.L. Riopel. 1986. Life history studies of *Conopholis americana* (Orobanchaceae). Amer. Midl. Nat. 116: 140-151.

Bargett, R.D. & A. Shine. 1999. Linkages between plant litter diversity, soil microbial biomass and ecosystem function in temperate grasslands. Soil Biol. and Biochemist. 31: 317-321.

Barker, S.J. & D. Tagu. 2000. The roles of auxins and cytokinins in mycorrhizal symbiosies. J. Plant Growth Regal. 19: 144-154.

Bergero, R., L.A. Harrier & P. Franken. 2003. Reporter genes: applications to the study of arbuscular mycorrhizal (AM) fungi and their symbiotic interactions with the plant roots. Plant and Soil, 255: 143-155.

Blancaflor, E.B., L.M. Yhao & M.J. Harrison. 2001. Microtubule organization in root cells of *Medicago truncatula* during development an arbuscular mycorrhizal symbiosis with *Glomus versiforme*. Protoplasma. 217: 154-165.

Braunberger, P.G., L.K. Abbott & A.D. Robson. 1996. Infectivity of arbuscular mycorrhizal fungi after wetting and drying. New Phytologist. 134: 673-684.

Braunberger, P.G., M.H. Miller & R.L. Peaterson. 1991. Effect of phosphorus nutrition on morphological characteristics of vesicular arbuscular colonization of maize. New Phytologist, 119: 107-113.

Brundrett, M. C. 1991. Mycorrhizas in natural ecosystem. In: Advances in Ecological Research, Vol. 21: A.A. caefaydin, M. Begon, A.H. Fitter, eds. Academic Press, London, pp. 171-313.

Chaubal, R., G. D. Sharma & R.R. Misra. 1982. Vesicular arbuscular mycorrhizal in subtropical acquatic and marsh plant communities. Proc. Indian Acad. Sci. 91B: 69-77.

Garbaye, J. 1994. Helper bacteria: A new dimension to the mycorrhizal symbiosis. New Phytol. 128: 197-210.

Gemma, J.N. & R.E. Koske. 1988. Seasonal variation in spore abundance and dormancy of *Gigaspora gigantea* and in mycorrhizal inoculum potential of a dune soil. Mycolgia. 80: 211-216.

Gottschalk, G. 1986. Bacterial metabolism. 2nd Edition, Springer-Verlag Berlin.

Gupta, O.P. 1984. Scientific weed management. Today and Tomorrow's Printer and Publishers, New Delhi.

Gupta R. & K.G. Mukerji. 1996(a). Studies on vesicular arbuscular mycorrhizal association in the root of common Delhi Grasses. Phytomorphology, 46: 117-121.

Gupta R. & K.G. Mukerji. 1996(b). Vesicular arbuscular mycorrhizae in plants in compositae, occurrence, identity and significance, Comp. News Letter. 29: 40-46.

Gupta R. & K.G. Mukerji. (1999). Incompatibility of various weeds towards Vam fungi; Plant Cell Incompatibility News Letter. 29: 6-9.

Hackel, E., G. Bachmann & S. Zechmeister-Boltenstern. 2000. Soil microbial biomass and rhizosphere effects in natural forest stands. Phyton, 40: 83-90.

Harley, J.L. & E.L. Harley. 1987. A check-list of mycorrhiza in the British Flora. New Phytol., (Supplement). 105: 1-102.

Haskins, K.E. & C.A. Gehring. 2004. Long-term effects of burning slash on plant communities and arbuscular mycorrhizae in a semi-arid woodland. Journal of Applied Ecology. 41:379-388.

Hetrick, B.A.D., G.T. Nilson, D.G. Kitt & A.P. Schwab. 1988. Effects of soil microorganisms on mycorrhizal contribution to growth of big blue stem grass in non-sterile soil. Soil. Biol. Bioch. 20: 501-507.

Hirsch, A.M., Y. Fang, S. Asad & Y. Kapulnik. 1997. The role of phytohormones in plant microbe symbioses. Plant Soil. 194: 171-184.

Jagpal, R. & K.G. Mukerji. 1998. VAM fungi–Tool for reforestation. Physol. Res. 1: 31-35.

Johansson, J.F., L.R. Paul & R.D. Finlay. 2004. Microbial interactions in the mycorrhizosphere and their significance for sustainable agriculture. FEMS. Microbiol. Ecol. 48: 1-13.

Journet, E.P., N. El-Gachtonli, V. Vernout, F. deBilly, M. Pichon, A. Dedieu, C. Arnould, D. Morandi, D.G. Barker. & V. Gianninazi-Pearson. 2001. *Medicago truncatula* ENOD11: A novel RPRP-encoding early nodulin gene expressed during mycorrhization in arbuscule containing cells. Maloc. Plant Microbe Interact. 14: 737-748.

Kapoor, R. & K.G. Mukerji. 2005. Rhizosphere Microbial Community Dynamics. In: Microbial activity in the Rhizosphere. (Mukerji, K.G., Manoharacharya and Singh eds.). Springer Verlag Heidelberg Germany, 450 p. (in press).

Kennedy, A.C. & K.L. Smith. 1995. Soil microbial diversity and sustainability of agricultural soils. Plant. Soil, 170: 75-86.

Linderman, R.G. 1992. Vesicular arbuscular mycorrhizae and soil microbial interactions. In : Bethlenfalvay G.J., and Linderman, R.G. (eds.) Mycorrhizae in Sustainable Agriculture. American Society of Agronomy, Madison, pp. 45-70.

Mahadevan, A., N. Raman & K. Natarajan, eds. 1989. Mycorrhiza for green Asia, Univ. of Madras, Madras.

Mathesius, U. 2003. Conservation and divergence of signaling pathways between roots and soil microbes–The *Rhizobium* legume symbiosis compared to the development of lateral roots, mycorrhizal interactions and nematode induced galls. Plant and Soil, 255: 105-119.

Mayer, R. & R. Godoy. 1989. Seasonal patterns in vesicular arbuscular mycorrhizas in melick beech forest–Agric. Ecosys. & Environ. 29: 281-288.

McGongitle, T.P. 1987. Vesicular arbuscular mycorrhizas and plant performance in a semi-natural grasslands, D. Phil. Thesis, Univ. of New York.

Mehta, S.P., R.S. Kanwar & R.S. Bhatia 1989. Herbicidal control of winter weeds in Autumn cane. Indian mustard intercropping system. Ind. J. Weed Sci. 21: 37-41.

Meney, K.A., K.W. Dixon, M. Scheltema & J.S. Pate. 1993. Occurrence of vesicular arbuscular mycorrhizal fungi in dryland species of Restionaceae and Cyperaceae from South West Western Australia, Australia, J. Bot. 41: 733-737.

Menge, J.A. 1983. Utilization of vesicular arbuscular mycorrhizal fungi in agriculture. Canadian Journal of Botany, 61: 1015-1025.

Minerdi, D., V. Bianciotto & Bonfante. 2002. Endosymbiotic bacteria in mycorrhizal fungi from their morpohology to genomic sequences. Plant and Soil, 244: 211-219.

Manoharachary, C. & K.G. Mukerji. 2002. Introduction: Mycorrhizal Studies. In: Techniques in Mycorrhizal Studies, Kluwer Academic Publishers, Netherlands, pp. 1-6.

Moretimer, A.M. 1989. The biology of weeds. In: Weed Control Hand Book–Principles. R.J. Hense, and K. Hally, eds. British Crop Protection Coun., Black Sci. Pub., Oxford.

Mosse, B. 1973. Advances in the study of vesicular arbuscular mycorrhizas. A. Rev. Phytopath. 11: 171-196.

Mosse, B. & D.S. Hayman. 1970, Mycorrhizae in agricultural plants. In: Tropical Mycorrhizal Research, P. Mikola, ed., Clarendon Press, Oxford, pp. 213-230.

Mukerji, K.G. ed. 1996. Concepts in mycorrhizal research. Kluwer Academic, Publishers, Dordrecht, Boston, London, p. 374.

Mukerji, K.G. 2002. Rhizosphere Biology. In Techniques in Mycorrhizal Studies. Mukerji, K.G., C. Manoharachary and B.P Chamola, eds. Kluwer Academic Publishers. pp. 87-101.

Mukerji, K.G., B.P. Chamola & J. Singh (eds.). 2000. Mycorrhizal Biology. Kluwer Academic/Plenum Publishers, New York, p. 336.

Mukerji, K.G. & R.K. Dixon. 1992. Mycorrhizae in Reforestation. In: Research and Development Rain forest Ecosystem: Research and Development Priorities, N.M. Majia, A.K. Malik, Md. Z. Hamzah and K. Tsuoff, eds., Salanagar, Malaysia, University Pertanian Malaysia, pp. 62-82.

Mukerji, K.G. & B. Kochar, 1983, Vesicular asbuscular mycorrhizae in rape seed plant. 6th cong. Int. Soc. Colza, Paris, pp. 945-950.

Newsham, K.K., A.H. Fitter & A.R. Watterson. 1995. Arbuscular mycorrhiza protectant annual grass from root pathogenic fungi in the field. J. Ecol. 83: 991-1000.

Nickrent, D.L. 2002. Parasitic plants of the world. In. Parasitic Plants of the Iberian Peninsula and Balearic Islands. J.A. Lopezsaez, P. Catalan, and L. Saez (eds.). Mundi Prensa, Madrid. pp. 7-27.

Nicolson, T.H. 1955. The mycotrophic habit in grass. Ph.D. Thesis. University of Nottingham. Nottingham, U.K.

Pattinson, G.S. & P.A. McGee. 1997. High densities of arbuscular mycorrhizal fungi maintained during long fallows in soils used to grow cotton except when soil is wetted periodically. New Phytol. 136: 571-580.

Paula, M.A., S. Urquiaga, J.O. Siquiera, & J. Dobereiner. 1992. Synergistic effects of vesicular arbuscular mycorrhizal fungi and diazotrophic bacteria on nutrition and growth of sweet potato. Biol. Fert. Soil. 14: 61-66.

Pearson, J.N. & I. Jakobson. 1993. The relative contribution of hyphae and roots to phosphorus uptake by arbuscular mycorrhizal plants measured by dual labelling with p^{32} and p^{33}. New Phytol. 124: 489-494.

Raghupathy, S. & A. Mahadevan. 1993. Distribution of vesicular arbuscular mycorrhizal fungi in the plants and rhizosphere soils of the tropical plains. Tamil Nadu India, Mycorrhiza, 3: 123-126.

Raghupathy, S. & P. Mahadevan. 1991. Ecology of vesicular arbuscular mycorrhizal fungi in a coastal tropical. Forest, Ind. J. Microbiol. Ecol. 2: 1-9.

Ranksov, S. & I. Jakobson. 1995. Functional compatibility in arbuscular mycorrhizas measured as hyphal P transport to the plant. New Phytol. 129: 611-618.

Rao, V.S. 1983. Principles of weed science. Oxford & IBH Publications Co., New Delhi.

Remy, W., T.N. Taylor, Hass, H. & H. Kerp. 1994. Four hundred million years old vesicular arbuscular mycorrhizae. Proc. Natl. Acad. Sci. U.S.A. 91: 11841-11843.

Rillig, M.C., S.F. Wright, K.A. Nicholas, W.F. Schmidt & M.S. Torn. 2001. Large contributions of arbuscular mycorrhizal fungi to soil carbon pool in tropical forest soils. Plant and Soil, 233: 167-177.

Ryan, M.H. 1998. The ecology of VAM fungi in contrasting Australian Agricultural Systems. Ph.D. thesis. Australian National University, Canberra, Australia.

Sanders, E.R. & A.H. Fitter. 1992. The ecology and functioning of vesicular arbuscular mycorrhiza in co-existing grassland species, New Phytol. 120: 517-524.

Sanders, I.R., R.T. Koide & D.L. Shumway 1993. Mycorrhizal stimulation of plant parasitism. Can. J. Bot. 71: 1143-1146.

Schenck, N.C., S.O. Graham & N.E. Green. 1975. Temperature and light effect on contamination and spore germination of vesicular arbuscular mycorrhizal fungi. Mycologia, 67: 1189-1192.

Singh, C.S. & S.P. Tyagi. 1988. Study on the occurrence of vesicular arbuscular mycorrhizal fungi in the root of *Aeschynomone indica* under the influence of various ecological factors. Zentralblott fur Microbiology, 144: 241-248.

Smalla, K., G. Wieland, A. Buchner, A. Zock, J. Parzy, S. Kaiser, N. Roskot, H. Henre & G. Berg. 2001. Bulk and rhizosphere soil bacterial communities studied by denaturing gradient gel electrophoresis plant-dependent enrichment and seasonal shifts revealed. Appl. Environ. Microbiol 67: 4742-4751.

Smith, J.L. & E.A. Paul. 1990. The significance of soil biomass estimates. In: Soil biochemistry. Bollag J.M., Stozky, G. eds. Marcel Dekker, New York, pp. 357-396.

Smith, S.E. & D.I. Read. 1997. Mycorrhizal symbiosis. Academic Press. New York, p. 790.

Smith, S.E. & F.A. Smith. 1990. Structure and function of the interfaces in biotrophic symbiosis as they relate to nutrient transport. New Phytol. 114: 1-38.

Sparling, G.P. & P.B. Tinker. 1975. Mycorrhizas in Pennive grasslands. In: Endomycorrhiza, F.E. Sanders, B. Mossee and P.B. Tinker Eds., Academic Press, London, pp. 545-560.

Stephan, A., A.H. Mayer & B. Schmid. 2000. Plant diversity affects culturable soil bacteria in experimental grassland communities. J. Ecol. 88: L988-999.

Subba Rao, N.S., K.V.B.R. Tilak & C.S. Singh. 1985. Synergistic effect of vesicular arbuscular mycorrhizae and *Azospirillium brasilense* on the growth of barley in roots. Biol. Fert. Soil. 17: 119-121.

Subramanian, K.S., C. Charest, L.M. Dwyer & R.I. Hamilton. 1995. Arbuscular mycorrhizas and water relations in maize under drought stress at Tasseling. New Phytol. 129: 643-650.

Taylor, T.N., W. Remy, H. Hass & H. Kerp. 1995. Fossil arbuscular mycorrhiae from the early Devonian. Mycologia. 87: 560-573.

Tester, M., S.E. Smith & F.A. Smith. 1987. The phenomenon of non-mycorrhizal plants. Canadian Journal of Botany, 65: 419-431.

Tian, C.J., X.Y. He, Y. Zhang & J.K. Chen. 2002. Effects of VA mycorrhizae and *Frankia* dual inoculation on growth and nitrogen fixation of Hippophae tibetane. For Ecol. Manag. 170: 307-312.

Tilman, D. 1987. Secondary succession and the pattern of plant dominance along experimental nitrogen gradients. Ecol. monographs. 57: 189-214.

Tull, Jethro. 1931. Horse hoeing husbandry, Baskshire, MDCC 33.

Van Bruggen, A.H.C. & A.M. Semenov. 2000. In search of biological indicators for soil health and disease suppression. Appl. Soil. Ecol. 15: 13-24.

Van Duin, W.E., J. Rozema & W.H.O. Ernst. 1989. Seasonal and spatial variations in the occurrence of vesicular arbuscular mycorrhiza in salt marsh plants, Agric. Ecosys. Environ. 29: 107-110.

Van Rijn, P., X. Pang, S. Galili, D. Shaul, N. Atzmon, S. Winiger, Eshaad, Y. Kapulnik, G. Lum, Y. Li, V. To, Jujishige & A. M. Hirsch. 1997. Expression of early nodulin genes in alfalfa mycorrhizae indicates that signal transduction pathways used in forming arbuscular mycorrhizae and Rhizobium–induced nodules may be conserved. Proc. Natl. Acad. Sci. USA. 94: 5467-5742.

Vazquez, M., J. Barea & R.A.D. Azen. 2001. Impact of soil nitrogen concentration on *Glomus* sp. in *Rhizobium* interactions as affecting growth nitrate reductase activity and protion content of *Medicago sativa.* Biol. Fert. Soil. 34: 57-63.

Viehelig, H. & J.A. Ocampo. 1991. Susceptibility and effectiveness of vesicular arbuscular mycorrhizae in wheat cultivars under different growing conditions. Biol. Fert, Soil, 11: 290-294.

Wang, G.M., D.P. Stribley, P.B. Tinker & C. Walker. 1993. Effects of pH on arbuscular mycorrhiza I. Field observations on the long-term timing experiments at Rothamstead and Wobura. New Phytol., 124: 465-472.

Wilson, J.M. & I.C. Tommerup. 1992. Interactions between fungal symbiosts: VA mycorrhizae. In Mycorrhizal functioning an integrative plant–fungal process (ed. Allen, M.F.), Chapman and Hall, New York, U.S.A. pp. 199-248.

Yang, C.H. & D.E. Crowley. 2000. Rhizosphere microbial community structure in relation to root location and plant iron nutritional status. Appl. Environ. Microbiol. 66: 345-351.

Zaidi, R. & K.G. Mukerji. 1983. Incidence of vesicular arbuscular mycorrhizae in diseased and healthy plants. Indian J. Plant. Pathol. 1: 24-26.

17

Integration of Arbuscular Mycorrhizal Fungi with Micropropagated Plants

[1]N. RAAMAN and [#]S. PATHARAJAN
[1]Centre for Advanced Studies in Botany University of Madras,
[#]Guindy Campus Chennai. India.
E-mail: raaman55@gmail.com

ABSTRACT

One of the important applications of modern biotechnology in agriculture is its use in tissue culture. *In vitro* micropropagation techniques are increasingly being applied to large scale production of quality planting materials especially in medicinal plants, fruit crops and woody timber trees. There are four important stages in the micropropagation of plants irrespective of the various techniques employed for it. They are the regeneration in *in vitro*, proliferation of the regenerates, rooting in *in vitro* and for transplantation of the plantlets in the soil or fields. Of these, the last stage is very crucial and important with respect to the establishment of the *in vitro* derived plantlets. It has been established that tissue culture plantlets have very divergent leaf anatomy and physiology and hence require an acclimatization period during the transition from culture to green house, from a total unnatural system to the very natural environment. In the *in vitro* condition, plantlets are heterotrophic and get very high favourable conditions for their growth. But in the *in vivo* situation, the plantlets have to switch over to autotrophic nutrition, involving normal photosynthetic activity and water relations. The plantlets may not be able to withstand such a sudden shock of the environmental changes, mostly due to some aberrant characteristic features of *in vitro* derived plantlets. Inoculation with arbuscular mycorrhizal (AM) fungi on micropropagated plantlets to overcome this problem and improves the establishment and growth in the field condition. Unfortunately, not much work has been conducted in this line. But the few convincing reports decisively prove that successful hardening and *in vivo* establishments of plantlets could be achieved by inoculation with AM fungi at the time of planting out.

I. INTRODUCTION

Soil quality is an integral indicator of sustainable agricultural ecosystems (Herrick, 2000) and one of the key components of soil quality is the biota, in particularly the microbial component.

One of the most important groups of soil microorganisms is the mycorrhizal fungi. Mycorrhizal associations vary widely in structure and function, but the most common interaction is the arbuscular mycorrhizal association. The AM symbiosis represents an ancient symbiosis (Pirozynski and Malloch, 1975; Pirozynski and Dalpe, 1989; Remy et al., 1994; Taylor et al., 1995; Redecker et al., 2000) with over 80% of all terrestrial flowering plants forming this type of relationship (Smith and Read, 1997). The symbiosis is formed when fungi belonging to the unique Phylum Glomeromycota (Schussler *et al.*, 2001) grow within and around plant roots. It is well established that the growth response of plants to mycorrhizal infection is influenced by the amount of phosphorus supplied to the soil. Mycorrhizae are not only beneficial in absorption of nutrients from soil but also associated with biological weathering of geologic substrate and decomposition of soil organic matter. There is evidence that mycorrhizal plants mobilize nutrient ions from soil minerals that are relatively insoluble in extracting agents typically used in soil analysis. Direct supply of nutrients from litter to living roots through mycorrhizae has been indicated. They play an important role in energy flow and nutrient cycling in undisturbed ecosystems. Mycorrhizae are not only more efficient in utilization of available nutrients from soil but also involved in transfer of nutrients from components of soil minerals and organic residues to solution and in nutrient cycling in an ecosystem. Daft and Nicolson (1972) found that inoculation with *Endogone macrocarpa* var. *geospora* increased P content in tomato plants. Inflow (rate of uptake per unit length of root) of P from soil into the roots of mycorrhizal plants was faster than the non-mycorrhizal roots (Sanders *et al.*, 1977; Graham, 1982). The amount of P in o-phosphate fraction in both shoots and roots increased in mycorrhizal plants (Gianinazzi-Pearson, 1984; Gianinazzi-Pearson and Gianinazzi, 1985). Davis and Menge (1980) reported that leaf P level was higher in *Glomus fasciculatum* infected *Citrus sinensis* than in the non-mycorrhizal plants. The plants, which were inoculated with, AM fungi had increased level of P in leaves than in the pathogen inoculated or control plants. Evidences clearly indicated the beneficial role of AM fungi to higher plants (Hirrel and Gerdemann, 1980; Mahadevan *et al.*, 1988; Khan, 1988; Sambandan *et al.*, 1991; Raman and Mahadevan, 1996).

AM fungi play a very important role in the improvement of plant growth and survival in most natural communities and contaminated soils (Harley and Harley, 1987; Smith and Read, 1997). Enhanced absorption of trace metals (Zn, Cu and Co) from deficient soils by AM plants has been documented (Tinker and Gildon 1983; Killham 1985; Rogers and Williams 1986; Faber *et al.*, 1990; Kothari *et al.*, 1991; Li *et al.*, 1991). Under such conditions, Killham and Firestone (1983) observed an adverse effect of AM on plant growth, which they attributed to the enhanced uptake of Cu, Ni, Pb and Zn with decreasing pH. The fungi serve as a highly efficient extension of the host root system (Bolan, 1991). Minerals like N, P, K, Ca, S, Zn, Cu and Sr are absorbed from the soil by AM fungi and are translocated to the host plant (Mosse, 1957).

II. AM ASSOCIATION IN EXPERIMENTAL PLANTS

In India, Butler (1939) first reported the occurrence of AM fungi in groundnut. Godse *et al*, (1976) found the occurrence of AM fungi on crop plants of Karnataka such as *Capsicum annum*, *Vigna sinensis*, *Eleusine coracana*, *Zea mays*, *Pennisetum typhoides*, *P. purpureum* and *Glycine max*. Sukhada (1988) studied the effect of *Glomus fasciculatum* on *Lycopersicon esculentum* under the field condition and showed that the AM fungi enhanced the growth and yield of tomato.

Mukerji and Kapoor (1990) and Rekha Rani and Mukerji (1990) found that the *Glomus* as the most dominant genus in Indian soils. Raman *et al*, (1992) reported that 19 tree species out of 25 tree species in Mamundur forest of Tamil Nadu. Raman *et al*, (1993) reported the AM association in plants growing in magnesite mine spoils of Salem, Tamil Nadu. They found that *Gl. fasciculatum* and *Gigaspora gigantea* occurred commonly in the mine spoil. Ragupathy and Mahadevan (1993) examined 737 plants in Thanjavur district, Tamil Nadu for AM infection and found 372 plants were mycorrhizal ones. Venkatraman *et al*, (1985) studied the endogonaceae of Jorhat (Assam) in North-Eastern India and found *Gig. nigra* as the common AM fungus followed by *Gl. mosseae, Gl. fasciculatum* and *Sclerocystis rubiformis*.

Rachel *et al*, (1991) surveyed arbuscular mycorrhizae in Angiosperms in semi-arid soil of Andhra Pradesh and they found *Gl. mosseae* as the dominant species. Bilgrami and Prasad (1993) observed that *Gl. fasciculatum* inoculated sugarcane cultivar B.O. 109 under field condition increased the total dry weight, stem thickness and number of leaves. Ragupathy and Mahadevan (1993) observed AM association in *Saccharum officinarum* in the tropical plains of Tamil Nadu.

Kothari *et al*, (1990), Sylvia *et al*, (1993), Ho (1993) and Shen *et al*, (1994) found AM infection in *Z. mays*. In *Sorghum*, AM infection was found by Raju *et al*, (1990). Diem *et al*, (1981) reported the colonization by species of *Glomus* and *Gigaspora* in neem. Bala *et al*, (1989) observed the maximum mycorrhizal association of neem. Ganesan *et al*, (1991) reported that neem was colonized by AM fungi and observed infection up to 40%. Raman *et al*, (1993) reported AM association of neem plants in magnesite mine spoil exhibited 64% AM infection. Harley (1970) reported the importance of occurrence of dual symbiosis in primary colonizers like *Casuarina equisetifolia* which contained root nodules infected with actinomycetes and roots with AM association. Rose (1980) reported the presence of arbuscular mycorrhizal structures in the roots of *C. cunninghamiana* obtained from sand woodlands of Florida and Japan and in the roots of *C. equisetifolia* from marine sand dunes of Florida. Diem and Gauthier (1982) observed that double inoculation of *C. equisetifolia* with *Gl. mosseae* and *Frankia* significantly improved plant growth. Torrey (1983) observed both ectotrophic and endotrophic mycorrhizal association in *Casuarina* species in south Florida in both wet and dry sites. *Gl. fasciculatum* is common inhabitant in *C. equisetifolia* grown in alkaline soils (Raman and Elumalai, 1991). Narayana Bhat *et al*, (1994) reported AM infection of *C. equisetifolia* at Nilgiris. Osundina (1998) reported that *Gl. clarum* inoculation influenced the nodulation of growth of flooded *C. equisetifolia*.

The occurrence of AM infection in *Bacopa monnieri* and *Hybanthus enneaspermus* was not reported earlier. The occurrence of arbuscular mycorrhizal colonization in medicinal plants has been reported (Govinda Rao *et al*, 1989). Malathi *et al*, (1996) reported the occurrence of AM association in root, rhizome and scale leaves of a medicinal plant, *Kaempferia galanga*.

III. MICROPROPAGATION

In 1938, Schwann and Sachleiden expressed that plant cells are 'totipotent', that is a complete plant could be regenerated from a single cell. Totipotency of a plant cell is the basis of plant tissue culture (Sharp *et al*, 1995). In this technique, different plant organs or 'differentiated' tissue or 'dedifferentiated' cells were developed from a single cell. Depending on the medium constituents, the dedifferentiated cells either continue to grow as an unorganized mass of tissue or redifferentiated into organs.

In recent years, plant biotechnology has contributed to the broadening of genetic base of the existing varieties of crop plants and medicinal plants. In India, it has to prove not only more food, but also sustained employment and income generation. The historic development of *in vitro* plant cell and tissue culture has undoubtedly been a major factor in the advancement of our knowledge of cell biology, physiology, biochemistry and more recently in molecular biology. However, its exploitation for applied purposes could be argued as being of even greater consequence. Plant tissue culture techniques have gained greater momentum on commercial application in the field of plant propagation. With the advent of latest techniques for the isolation and purification of the compounds from the medicinal herbs, it has gained the limelight. The current research in plant sciences involves new technologies, the main thrust being the exploitation of the knowledge gained to develop new plants. The prominence gained by various works on medicinal plants over the last decade has renewed the interest of man in the *in vitro* growth and metabolism of medicinal plants.

In the development of plant tissue culture since its pioneering experiments by Laibach (1925, 1929), White (1934, 1939), Skoog (1944), Loo (1945), Van Overbeck *et al*, (1947) and Murashige and Skoog (1962) have contributed more in establishing a strong foundation for the application of the tissue culture technology. Regeneration had been achieved in many medicinal plants from different plant parts. There are several reports on multiple or clonal propagation of different genera of the family Zingeberaceae such as *Costus* (Roy and Pal, 1991), *Alpine* (Chang and Criely, 1993, Agretious *et al.*, 1996) and *Kaempforia rotunda* (Mustafa Anand *et al*, 1997). Young vegetative buds of turmeric (*Curcuma longa* Linn.) cultured on MS medium supplemented with coconut water, kinetin and 6-Benzyl Amino Purine (BAP) or on Smith's medium supplemented with coconut water (CW), Kinetin (KIN) or BAP and inositol produced shoots.

IV. AM FUNGI WITH MICROPROPAGATED PLANTS

Interest in plant tissue culture so far has been greater in the area of rapid clonal propagation of plants. A number of commercial laboratories now routinely use micropropagation for different foliage and ornamental plants. This spurt of interest in micropropagation is partly due to the greatly increased rate of multiplication and more profit than conventional vegetative multiplication. Increased attention is now focused on *in vitro* techniques for the mass production of planting stock from elite trees and also tree improvement by selection at *in vitro* level (Bonga, 1995). Tissue culture technology is being widely used for mass propagation of crop plants (Green and Philips, 1975), medicinal plants (Catapan *et al*, 2002) and trees (Mascarenhas and Muralidharan, 1989). The transfer of plants cultivated *in vitro* to the green house is one of the most important step in the structural and physiological adaptation during preparation of plantlets. This phase known, as acclimatization is the beginning of the autotrophic existence of the plant, with the initiation of the physiological processes necessary for survival. During this time, plantlets must increase the adsorption of water and minerals and the photosynthetic rate (Grattapaglia and Machado, 1990). Micropropagated plantlets are free of diseases, but they also lack arbuscular mycorrhizal (AM) fungi. AM fungi are known to increase the vigour of plants by increasing absorption of water and mineral nutrients, especially phosphorus. Moreover, AM can protect host plants from root pathogens and mitigate the effect of extreme variation in temperature, pH and water stress (Dixon and Marx, 1987; Siqueira, 1994). Successful inoculation of AM fungi at the beginning of the acclimatization period (Granger *et al.*, 1983; Branzati *et al*, 1992; Giullemin *et al*, 1995; Estrada and Davies, 2003) or even during *in vitro* cultivation (Mathur and Vyas, 1995) has been demonstrated.

Micropropagated plantlets inoculated with AM spores increased the survival rate and growth in potted condition. The growth of AM inoculated banana plantlets in sterilized growth media showed significantly higher total dry weight, diameter of pseudostem and plant height than those in non-autoclaved culture media and control plants (Lin and Chang, 1987). Kuo *et al*, (1987) reported that tissue cultured *Vitis vinifera* plantlets inoculated with *Gl. epigeous* significantly increased percent infection, biomass, photosynthesis of host and increased resistance to downy mildew disease, compared to control. Cheema and Adholeya (1987) demonstrated that micropropagated *Populus deltoids* plantlets inoculated with AM spores showed increased survival rate and growth in potted condition. Rojanasiriwong *et al*, (1987) reported that tissue cultured *Zea mays* and *Arachis hypogea* using MS medium, having auxin and cytokinin to produce long roots and inoculated with *Gl. intraradices* and *Gl. mosseae* were highly infected in liquid aerated culture than on agar medium.

Chien and Chang (1988) found that tissue cultured plantlets of *Paulownia albiphrea* clones inoculated with *Gl. epigaeum* in containers containing two different substrates after 30 days of inoculation, showed significant increase in shoot height, root collar diameter and foliage area compared with uninoculated controls. The distribution of several nutrient elements in mycorrhizae was detected with X-ray microanalysis, which showed that arbuscules, vesicles and hyphae had more concentration of P, S and Ca than uninoculated plantlets. Strulla *et al*, (1989) observed the regeneration of AM fungi in the somatic embryos of alfalfa (*Medicago sativa*) encased in a nutrient coating, inoculated with *Gl. fasciculatum* and *Gl. mosseae* in *in vitro* condition. Varma and Schuppe (1994) reported the improved survival and establishment of tissue cultured plantlets when associated with AM fungi. The beneficial effects of the symbiosis formed with the root system of plants from tissue culture manifests itself in the development of vigorous plants with high photosynthetic and transpiration rates, improved absorption of nutrients and water and increased stress tolerance (Jaizme-Vega *et al*, 1991; Azcon-Aguilar *et al*, 1997).

AM association has been reported in more than 80% of terrestrial plant species (Trappe, 1987). Muthukumar and Udaiyan (2001) have reported AM associations in 58 of 60 medicinal plants studied in Maruthamali Hills, Western Ghats, Southern India. The colonization level ranged from 12% to 90% of the root length and spore numbers in the soil ranged between 12 and 72 spores/10 g soil. *Allium* spp., including garlic, are extremely responsive to AM fungal symbiosis and stunted to onion growth in the absence of AM fungi had been well documented (Gerdemann 1968). Garlic plants inoculated with AM fungi appeared with darker green leaves, increased plant height, fresh and dry weight, bulb filling and maturation than uninoculated plants (Koch *et al*, 1997). Pacovsky and Fuller (1986) had reported increased level of growth, biomass production and survivability of AM propagated plants due to enhancement of nutrition. Chang (1990) reported increased survival rate of micropropagated plantlets of *Fragaria ananassa* by the inoculation of *Glomus* spp. during transfer from the test tubes to pots and from pots to field soil. Sbrana *et al*, (1994) reported the enhanced growth and survival of apple, peach and plum rootstocks inoculated with AM fungus *Glomus* spp. on transplanting from *in vitro* to *in vivo* condition. Micropropagated plantlets of banana inoculated with AM spores had greater height, leaf area and fresh weight of shoots as well as higher rates of photosynthesis and transpiration than controls (Adriana *et al*, 1999). Taylor and Harrier (2000) compared 9 species of AM fungi on the growth, development and nutrient status of micropropagated *Rubus idaeus* L.cv. Glen Prosen (Red Raspberry) and have observed that *Glomus* spp. showed higher rate of colonization among the AM species studied whereas *Scutellospora* spp. increase the dry weight of shoots. Root growth had been enhanced in micropropagated plantlets of *Fragaria* x *ananassa* cv. Elvira (Strawberry) inoculated with *Gl. intraradices* and *Gig. margarita* (Taylor and Harrier, 2001). Yao *et al*, (2002) had reported significant increase in shoot fresh weight, root dry weight and the number of tubers produced per plant in micropropagated potato plantlets inoculated with *Gl. etunicatum*. Enhanced growth and survival of *Tylophora indica* inoculated with AM fungi on transplanting from *in vitro* to *in vivo* condition (Shunmuga Sundaram, 2004).

Lister *et al.* (1986) reported that mycorrhizae of conifers increased translocation of photosynthates to root and increase photosynthesis. In *Gl. intraradix* inoculated plants of *Citrus aurantium*, increased level of reducing sugars, sucrose, starch and total non-structural carbohydrates in the leaves and roots were reported (Nemec and Vu, 1990). Nair *et al*, (1991) have reported increased amounts of formononentin found in roots of *Trifolium repen* inoculated with *Glomus* spp. Krishna and Bagyaraj (1983) reported higher protein content in *Gl. fasciculatum* inoculated *Arachis hypogea* roots. Higher acid phosphatase activity had been observed in *Phaseolus vulgaris* plants inoculated with *Gl. etunicatum*. Alkaline phosphatase activity has been reported in *Glomus* spp. infected roots of *Allium porum* and *Plantanus acerifolia* by histochemical studies (Tisserant *et al*, 1993). Wright and Milliner (1994) had documented the role of AM fungi in enhancing the plant growth as well as improving host resistance to certain plant diseases. AM inoculated plants have more resistant to soil-borne pathogens, which are causing diseases (Schenck, 1981; Dehne, 1982; Raman, 1996).

In the last decade, micropropagation has been a most powerful tool for cloning several plant species, especially the ornamentals (Ammirato *et al*, 1989). Despite many successful applications of this technique, there are still several problems limiting its widespread use. The transfer of *in vitro* plantlets to *in vivo* conditions is one of the most critical factors of the micropropagation process and a cause of higher production costs. High mortality is often observed upon transfer to *in vivo* conditions as the cultured plants have a poorly developed cuticle (Wetzstein and Sommer, 1982), non-functional stomata (Lee and Wetzstein, 1988) and a weak root system (Pierik, 1987). In order to increase growth rate and reduction of mortality of plantlets at the state of acclimatization, recent research has been focused on control of the environmental conditions. Pearl millet inoculated with AM fungus *Gig. calospora* alone and in association with phosphorus fertilizer in sterile and unsterile soil showed significant increase in dry weight of shoot and root than non-mycorrhizal control plants (Champwat and Pathak, 1993).

AM inoculation enhanced the growth and survival of the micropropagated sugarcane plantlets (Sbrana *et al*, 1994). AM inoculation during transplantation from *in vitro* to *in vivo* culture enhanced both growth and survival rate of plants. Infectivity of root inocula increased with increasing percentage of root length (All-Radded, 1995). *S. officinarum, Zea mays, Sorghum vulgare, Bacopa monnieri, Hybanthus enneaspermus, Azadiracta indica, casuarinas equisetifolia* were micropropagated and inoculated with 3 species of AM fungi (*Gl. fasciculatum, Gl. mosseae and Gig. gigantea*) the maximum survival and growth were observed by inoculated with *Gl. mosseae* (Sahadevan, 1998). Sylvia *et al*, 2003 reported that the AM fungal host combinations that produced greater plant growth and survivability of micropropagated Sea Oat. The micropropagated plants of garlic and tomato inoculated with *Glomus fasciculatum*, *Gl. mosseae* and *Scu. erythropa* were observed to be in association with mycorrhizal fungi. Among the AM fungi, *Gl. fasciculatum* was highly suitable to the experimental condition (Patharajan, 2004). The genus *Glomus* is the most common AM fungi throughout the world (Gerdemann and Trappe, 1974; Raman *et al*, 1993). Venkateshwarrao (2000) has reported predominance of *Gl. fasciculatum* species among all the plants studied for AM fungal species association. Raghupathy and Mahadevan (1993) have reported *Gl. fasciculatum* as a successful colonizer in cultivated plants and presence of higher frequency of *Gl. mosseae*. *Gig. margarita* was found to be the frequent colonizer in natural forests (Bhatia *et al*, 1996). Quatrini *et al*, 2003 have studied the comparison with *Gl. mosseae* (BEG 116) and a mixture of native Glomus spp. inoculated micropropagated lemon plant (*Citrus limon* (L.) Burm. 'Zagara Bianca') and they found more survivability and growth in the inoculation of native *Glomus* spp. mixed inoculum.

AM colonization of horticultural plant roots can improve growth by increased uptake of phosphorus, zinc and other minerals (Menge *et al*, 1978; Ames *et al*, 1983) and may reduce the incidence of disease (Dehne, 1982). Moreover, colonization with AM fungi may decrease transplant mortality (Biermann and Linderman, 1983) and injury (Menge *et al*, 1978). Recent work has also shown improvement in water relations of the host plant using AM. These fungi may also improve drought tolerance by decreasing leaf water potential (Koide, 1985; Auge *et al*, 1986), by reducing stomatal and root hydraulic resistances (Allen and Boosalis, 1983; Nelson, 1987 a) and by increasing transpiration rate (Koide, 1985). For these reasons, several studies have focused on arbuscular mycorrhizal formation during acclimatization of cultured plantlets (Chavez and Ferrara-Cerato, 1990; Ponton *et al*, 1990; Chabot *et al*, 1992; Vestberg, 1992; Williams *et al*, 1992; Wang *et al*, 1993). It has been shown that AM colonization of micropropagated plantlets can improve rooting, enhance root function and increase survival rates (Strullu, 1985). The benefits of AM have also been demonstrated by higher survival rates and better overall growth (Wang *et al*, 1993). In most cases, the beneficial effects on growth have been observed after *in vitro* plantlets was fully acclimatization if arbuscular mycorrhizal fungi were introduced during the *in vitro* rooting stage. The increase in plant growth observed in these experiment following the inoculation of plants with AM fungi confirms results from *in vivo* conditions (Vestberg, 1992; Williams *et al*, 1992; Wang *et al*, 1993) and demonstrates that *in vitro* mycorrhizal formation has effects on the growth of micropropagated plants similar to those observed for seedlings (Hayman, 1983). *In vitro* AM colonization significantly improves shoot and root growth by decreasing the osmotic potential.

Root exudates stimulated the growth of AM fungi before they penetrated the root (Hepper, 1981; Ferguson and Menge, 1982; Azcon and Ocampo, 1984). According to Elmeskaoui *et al*, (1995), micropropagated plantlets forming a successful association with AM *in vitro*, recovered faster and grew better than non-inoculated plantlets during tripartite culture. AM fungi improved the nutritional state of fruit plants by enhancing mineral absorption through the extrametrical mycelium (Harley and Smith, 1983). Mycorrhizal infection enhanced the water conductivity in *Citrus* plants (Graham and Syversten, 1984) and reduced the root pathogen spread in the rhizosphere and in the root tissues of *Citrus* and tobacco (Davis and Menge, 1981; Graham and Egel, 1988; Giovanetti *et al*, 1991). AM fungi inoculated micropropagated plants, such as strawberry, apple and pear grew faster and higher than control plants during transplantation from *in vitro* to *in vivo* (Morandi *et al*, 1979; Gianinazzi *et al*, 1983; Gianinazzi *et al*, 1985).

V. EFFECT OF AM ON POTTED OR FIELD PLANTS

The absence of mycorrhizal observation was attributed to the secondary metabolites of the host plants (Mohankumar and Mahadevan, 1984). AM fungi have been shown as a tool for improving the growth of agricultural and horticultural plants (Gianinazzi *et al*, 1989). AM colonization greatly improves growth and nutrients of host plants (Mosse, 1973; Abbott and Robson, 1982; Sulochana and Manoharachary, 1989; Padmavathi *et al*, 1990), and induces biological suppression of soil-borne pathogens (Jalali and Thareja, 1981). Safir *et al*, (1972) reported that AM fungi facilitate water transport and thereby increase CO_2 conduction. Gerdemann and Trappe (1974) reported *Gl. mosseae* was more adopted to crop species in more mesic sites or microhabitats. *Gl. fasciculatus* is a dominant arbuscular mycorrhizal fungus in several nutrient and drought stress habitats (Allen and Allen, 1980).

VI. HOST RESISTANCE

Many reports indicated that plants previously inoculated with fungal symbionts exhibit increased resistance to fungal root diseases like wilt and root rot (Schenck and Kellam, 1978; Dehne, 1982; Jalali and Chand, 1988). Zambolim and Schenck (1980) while studying the interaction between *Gl. mosseae* and *Macrophomina phaseolina* in soybean observed that colonization of host root by the mycorrhizal symbionts compensated for the effect of the pathogen, without affecting their incidence in the host. Smith and Gianinazzi-Pearson (1988) reported that by increased phosphate nutrition, AM fungi enhanced root growth, increased the absorptive capacity and affected cellular processes in roots. The increased root growth and phosphate status of the plant may be responsible for increased tolerance of mycorrhizal plants toward the pathogen (Wallace, 1973). AM fungi may increase host tolerance to pathogens by increasing uptake of essential nutrients other than P that would be deficient in a non-mycorrhizal plant (Pacovsky *et al*, 1986). Improved nutritional status of the plant by AM fungal colonization may affect qualitative and quantitative changes in root exudates to alter rhizosphere or rhizoplane microbial populations and population densities (Reid, 1984; Linderman, 1985).

AM fungi not only control the effect of fungal pathogen and nematode but also bacterial pathogens. Halos and Zorilla (1979) found that there was a reduction in *Pseudomonas syringae* colony in rhizosphere of tomato plants when inoculated with *Gl. mosseae*. Nemec and Vu (1990) found that *Gl. intraradix* increased chlorophyll content in *Citrus aurantium*. Increased level of carbohydrate in both roots and leaves in AM fungi infecnted plants was recorded (Nemec and Guy, 1982; Nemec and Vu, 1990). Volpin *et al*, (1994) found that *Gl. intraradix* enhanced the synthesis of flavonoids and glucosamine in *Medicago sativa*. Raman and Gnanaguru (1996) found *Gl. fasciculatum* inoculated *Lycopersicon esculentum* plants showed higher chlorophyll content in leaves, higher dry weight, sugar, phenol, lipid and protein content in both leaf and roots than control plants. Mycorrhizae protected the rootlets from the pathogens of tomato plants (Raman and Gnanaguru, 1996).

ACKNOWLEDGEMENTS

The authors are thankful to Prof. N. Anand, Director, Centre for Advanced Studies in Botany, University of Madras for providing facilities and encouragement.

REFERENCES

Abbott, L.K. & A.D. Robson. 1982. The role of vesicular-arbuscular mycorrhizal fungi in agriculture and the selection of fungi for inoculation. Aust. J. Agric. Res. 33: 389-408.

Abo, M.M. 1977. Organogenesis and embryogenesis in callus cultures of garlic (*Allium sativum* L.). Plant Sci. Lett. 9: 259-264.

Agretious, T.K., K.P. Martin & M. Hariharan. 1996. *In vitro* clonal multiplication of *Alpinia Calcarata* Rosc. Phytomorphology 46: 133-138.

Ali, A. & E.E. Metwally. 1992. Somatic embryogenesis and plant regeneration as a tool for garlic improvement. Egypt J. Appl. Sci. 7: 727-735.

Allen, E.B. & M.F. Allen. 1980. Natural re-establishment of vesicular-arbuscular mycorrhizae following stripmine reclamation in Wyoming. J. Appl. Ecol. 17: 139-147.

Allen, M.F. & M.G. Boosalis. 1983. Effect of two species of VAM fungi on drought tolerance of winter wheat. New Phytol. 93: 67-76.

All-Radded, A. 1995. Mass production of *Glomus mosseae* spores. Mycorrhiza 5: 229-231.

Ames, R.E., C.P.P. Reid, L.K. Porter & C. Cambardella. 1983. Hyphal uptake and transport of nitrogen from two ^{15}N-labelled cource by *Glomus mosseae*, a vesicular-arbuscular mycorrhizal fungus. New Phytol. 95: 381-396.

Ammirato, P.V., D.A. Evans, W.R. Sharp & Y.P.S. Bajaj. 1989. Handbook of plant cell culture, vol. 5, Ornamental species, McGraw-Hill, New York.

Auge, R.M., K.A. Schekel & R.L. Wample. 1986. Osmotic adjustment in leaves of VA mycorrhizal and non-mycorrhizal rose plants in response to drought stress. Plant Physiol. 82: 765-770.

Ayabe, M. & S. Sumi. 1998. Establishment of a novel tissue culture method, stem-disc culture, and its practical application to micropropagation of garlic (*Allium sativum* L.). Plant Cell Rep. 17: 773-779.

——— & ———. 2001. A novel and efficient tissue culture method- 'stem-disc dome culture'- for producing virus free garlic (*Allium sativum* L.). Plant Cell Rep. 20: 503-507.

Azcon, R. & Ocampo. 1984. Effect of root exudation on VA mycorrhizal infection at early stages of plant growth. Plant Soil 82: 133-138.

Azcon-Aguilar C, M. Centos, A. Troncoso & J.M. Barea. 1997. Beneficial effect of arbuscular mycorrhizas on acclimatization of micropropagated Cassava plantlets. Sci. Hortic. 72: 63-71.

Bala, K., A.V. Rao & J.C. Tarafdar. 1989. Occurrence of VAM associations in different plant species of the Indian desert. Arid Soil Research and Rehabitation 3: 391-396.

Barandiaran, X., N. Martin, M.F. Rodriguez-Conde, A.D. Pietro & J. Martin. 1999. An efficient method for callus culture and shoot regeneration of garlic (*Allium sativum* L.). Hort. Science 34: 348-378.

Barg, E., D.E. Lesemann, H.J. Vettren & S.K. Green. 1994. Identification, partial characterization, and distribution of viruses infecting *Allium* crops in South and Southease Asia. Acta. Hortic. 358: 251-258.

Bhatia, N.P., K. Sundari & A. Adholeya. 1996. Diversity and selective dominance of vesicular-arbuscular mycorrhizal fungi. In Concepts in Mycorrhizal Research, (ed. K.G. Mukerji), Kluwer Academic Publishers, the Netherlands, pp. 133-178.

Bhojwani, S.S. 1980. *In vitro* propagation of garlic by shoot proliferation. Sci. Hort. 13: 47-52.

Biermann, B. & R.G. Linderman. 1983. Use of vesicular-arbuscular mycorrhizal roots, intraradicle vesicles and extraradicle vesicles as inoculum. New Phytol. 95: 97-105.

Bolan, N.S. 1991. A critical review on the role of mycorrhizal fungi in the uptake of phosphorus by plants. Plant and soil 134: 189-207.

Bonga, J.M. 1995. Applications of tissue culture in forestry. In: Reinert, J and Y.P.S. Bajaj (eds.), Applied and fundamental aspects of plant cell, tissue, and organ culture. Narosa publishing house, New Delhi. pp. 93-108.

Branzati, B.V., Gianinazzi-Pearson & S. Gianinazzi. 1992. Influence of phosphate fertilization on the growth and nutrient status of micropropagated apple infected with endomycorrhizal fungi during the weaning stage. Agronomie. 12: 841-846.

Butler, E.J. 1939. The occurrence and systemic position of the vesicular arbuscular type of mycorrhizal fungi. Trans. Brit. Mycol. Soc. 22: 274-301. Catapan, E., M. Luis, B. Silva, F.N. Moreno & A.M. Viana. 2002. Micropropagation, callus and root cultures of *Phyllanthus urinaria* (Euphorbiaceae). Plant cell, Tissue and Organ culture 70 (3): 301-309.

Chabot, S., G. Becard & Y. Piche. 1992. Life cycle of *Glomus intraradix* in root organ culture. Mycologia 84: 315-321.

Champwat, R.S. & V.W. Pathak. 1993. Effect of vesicular arbuscular mycorrhizal fungi on growth and nutrition uptake of pearl millet. Indian J. Mycol. Pl. Pathol. 23: 30-34.

Chang, B.K. & A. Criley. 1993. Clonal propagation of pink ginger *in vitro*. Hort. Sci. 28: 1203.

Chang, D.C. 1990. Responses of micropropagated Haurnoka strawberry plantlets to three *Glomus* spp. Vesicular arbuscular mycorrhizal fungi. Journal of the Chinese Society for Horti. Sci. 36: 265-273.

Chavez, M.C. & R. Ferrara-Cerato. 1990. Effect of vesicular-arbuscular mycorrhizae at tissue cultured plantlets of strawberry. Hort. Science 25: 903-905.

Cheema, G.S. & A. Adholeya. 1987. Evalution of mycorrhizal association in micropropagated popular clones. In: Sylvia, D.M., L.L. Hung and J.H. Graham (eds.), Mycorrhizae next decade, Proc 7th North American Conf. Mycorrhiza, May 3-8, 1987, Gainseville, Florida, U.S.A. p. 89.

Chien, K.S. & B.K. Chang. 1988. A study of VAM formed on the tissue cultured plantlets of *Paulownia*. In: Mahadevan, A., N. Raman and K. Natarajan (eds.), Mycorrhiza for green Asia, Proc. First Asian Conf. Mycorrhiza, Jan 29-31, 1988, Univ. Madras, Madras. pp. 311-314.

Conci, V., S.F. Nome & R.G. Milne. 1992. Filamentous viruses of garlic in Argentina. Plant Dis. 76: 594-596.

Daft, M.J. & T.H. Nicolson. 1972. Effect of *Endogone* mycorrhiza on plant growth. IV. Quantitative relations between the growth of the host and the development of the endophyte in tomato and maize. New Phytol. 71: 287-295.

Davis, R.M. & J.A. Menge. 1981. *Phytophthora parasitica* inoculation and intensity of vesicular-arbuscular mycorrhizae in *Citrus.* New Phytol. 87: 705-715.

Dehne, H.W. 1982. Interaction between vesicular-arbuscular mycorrhizal fungi and plant pathogens. Phytopathology 72: 1115-1119.

Diem, H.G. & D. Gauthier. 1982. Effect de I' interactin endomycorrhizienne (*Glomus mosscae*) sur la nodulation et la croissance de *Casuarina equisetibolia*. C.R. Acad. Sci. Paris. 294: 215-217.

———, I. Goyeve, V. Gianinazzi-Pearson, J.A. Fortin & Y.R. Dommergues. 1981. Ecology of VA-mycorrhizae in the tropics; the semi-arid zone of Senegal. Acta. Ecologica. Ecol. Plant 2: 53-62.

Dixon, R.K. & D.H. Marx. 1987. Mycorrhizae. In: Bonga JM, Durzan DJ (eds.), Cell and tissue culture in forestry. (Forestry Science, Vol 2) Nijhoff, Dordrecht.

Elmeskaoui, A., J.P. Damond, M.J. Poulin, Y. Piche & Y. Desjardins. 1995. A tripartite culture system for endomycorrhizal inoculation of micropropagated strawberry plantlets *in vitro*. Mycorrhiza 5: 313-319.

Estrada-Luna A.A., & F.T. Jr Davies. 2003. Arbuscular mycorrhizal fungi influence water relations, gas exchange, abscesic acid and growth of micropropagated chile ancho pepper (*Capsicum annuum*) plantlets during acclimatization and post-acclimatization. J. Plant Physiol. Sep; 160(9): 1073-83.

Faber, B.A., R. J. Zasoski, R.G. Burau & K. Uriu. 1990. Zink uptake by corn affected by vesicular arbuscular mycorrhizae. Plant Soil 129: 121-130.

Fergusen, J.J. & J.A. Menge. 1982. The influence of light intensity and artificially extended photoperiod upon infection and sporulation of *Glomus fasciculatum* on sudangrass and on root exudation of sudangrass. New Phytol. 92: 183-191.

Ganesan, V., S. Ragupathy, B. Parthipan, D.B. Rajini Rani & A. Mahadevan. 1991. Occurrence of vesicular-arbuscular mycorrhizae in coal, lignite and calcite mine spoils of India. Biol. Fertil. Soils. 12: 131-136.

Gerdemann, J. 1968. Vescicular-arbuscular mycorrhiza and plant growth. Annu Rev Phytopathol. 6: 397-418.

Gerdemann, J.W. & J.M. Trappe. 1974. The endogonaceae in the pacific northwest. Mycologia Memoir. No.5, 76 p.

Gianinazzi Pearson, A., & S. Gianinazzi. 1983. The physiology of vesicular arbuscular mycorrhizal roots. Plant soil 71: 197-209.

Gianinazzi Pearson, A., S. Gianinazzi & A. Trouvelot. 1985. Evaluation of the infectivity and effectiveness of indigenous vesicular arbuscular mycorrhizal fungi populations in some agricultural soils in Burgundy. Can.J. Bot. 63: 1521-1524.

Gianinazzi, S., A. Trouvelot & V. Gianinazzi Pearson. 1989. Coceptual approaches in agriculture for the rational use VA endomycorrhizae in agriculture possibilities and limitations. Agriculture, Ecosystems and Environments 29: 153-161.

Giovannucci, E., A. Ascherio, E.B. Rimm, M.J. Stampfer, G.A. Colditz & W.C. Willet. 1991. Intake of carotenoids and retinol in relation to risk of prostate cancer. J. Natl. Cancer Inst. 87: 1767-1776.

Giullemin, J.P., S. Gianinazzi, V. Gianinazzi-Pearson & J. Marchal. 1995. Influence des endomycorhizes a arbuscules sur la croissance et la nutrition minerale de vitroplants d'ananas dans unsol a forte salinite. Fruits 50: 333-341.

Godse, D.S., T. Madhusudan, D.J. Bagyaraj & R.E. Patil. 1976. occurrence of VA mycorrhizae on crop plants in Karnataka. Trans. Brit. Mycol. Soc. 67: 169-171.

Govinda Rao, Y.S., C.K. Suresh, N.S. Suresh, R.R. Mallikarjunaiah & D.J. Bagyaraj. 1989. Vesicular-arbuscular mycorrhizae in Medicinal plants. Indian Phytopath. 42: 476-478.

Graham, J.H. 1982. Effect of citrus root exudates on germination of chlamydospores of the vesicular arbuscular mycorrhizal fungus, *Glomus epigaeum*. Mycologia. 74: 831-835.

——— & D.S. Egel. 1988. *Phytophthora* root rot development on mycorrhizal and phosphorus-fertilized nonmycorrhizal sweet orange seedlings. Plant Dis. 72: 611-614.

——— & J.P. Syversten. 1984. Influence of vesicular arbuscular mycorrhizae on the hydralic conductivity of roots of two *Citrus* rootstocks. New Phytol. 97: 277-284.

Granger, R.L., C. Plenchette & J.A. Fortin. 1983. Effect of a vesicular arbuscular (VA) endomycorrhizal fungus (*Glomus epigaeum*) on the growth and leaf mineral content of two apple clones propagated *in vitro*. Can. J. Plant Sci. 63: 551-555.

Grattapaglia, D. & M.A. Machado. 1990. Micropropagacao. In: Torres AC, Caldas LS (eds.), Tecnicas e aplicacoes da cultura de tecidos de plantas. ABCTP/Embrapa-CNPH, Brasilia, 99-170 pp.

Green, C.E. & R.L.Phillips. 1975. Plant regeneration from tissue cultures of maize. Crop Sci. 15: 417-421.

Halos, P.M. & R.A. Zorilla. 1979. Vesicular-arbuscular mycorrhizae increase growth and yield of tomatoes and reduce infection by *Psedomonas solanacearum.* Philippines Agriculture. 62: 309-315.

Harley, J.L. 1970. The importance of microorganisms to colonizing plants. Trans. Bot. Soc. (Edinburgh) 41: 65-70.

——— & E.L. Harley. 1987. A check list of mycorrhiza in the British flora. New Phytol. (Suppl). 105: 1-102.

——— & S.E. Smith. 1983. Mycorrhizal symbiosis. Academic Press, London and New York. 483 p.

Hayman, D.S. 1983. The physiology of VA endomycorrhizal symbiosis. Can. J. Bot. 61: 944-963.

Hepper, C.M. 1981. Techniques for studying the infection of the plants by vesicular arbuscular fungi under axenic conditions. New Phytol. 88: 641-647.

Herrick, J.E. 2000. Soil quality an indicator of sustainable land management. Appl. Soil Ecol. 15: 75-83.

Hirral, M.C. & J.W. Gerdemann. 980. Improved growth of corn and bell pepper in saline soils by two vesicular arbuscular mycorrhizal fungi. J. Soil Sci. Soc. Am. 44: 654-655.

Ho, I. 1993. Analysis of nutrient uptake by six species of vesicular-arbuscular mycorrhizal fungi in *Zea mays.* Indian J. Mycol. Pl. Pathol. 23: 64-69.

Jaizme-Vega, M.C., V. Galan-Savcoa & J. Cabrera Cabrera. 1991. Preliminary results of VAM effects on banana under field conditions. Fruits 46: 19-22.

Jalali, B.L. & M.L. Thareja. 1981. Management of soil-borne plant pathogens by vesicular arbuscular mycorrhizal model system. Indian Phytopath. 34: 115-116.

Khan, A.G. 1988. Inoculum density of *Glomus mosseae* and growth of onion plants in unsterilized bituminous coal spoil. Soil Biol. Biochem. 20: 749-753.

Killham K. 1985. Vesicular arbuscular mycorrhizal mediation of trace and minor element uptake in perennial grasses: relation to livestock herbage. In: Fitter AH, Atkinson D, Read DJ, Usher MB (eds.), Ecological interactions in soil: plants, microbes and animals. Blackwell Palo Alto Calif pp. 225-232.

——— & M.F. Firestone. 1983. Vesicular arbuscular mycorrhizal mediation of grass response to acidic and heavy metal deposition. Plant Soil 72: 39-48.

Koch, M., Z. Tanami & R. Salomon. 1995. Improved regeneration of shoots from garlic callus. Hort. Science 30: 378.

———, Z. Tanami, H. Bodani & S. Wininger. 1997. Field application of vesicular-arbuscular mycorrhizal fungi improved garlic yield in disinfected soil. Mycorrhiza. 7: 47-50.

Koide, R. 1985. The effect of VAM infection and phosphorus status on sunflower hydraulic and stomatal properties. J. Exp. Biol. 36: 1087-1098.

Kothari, S.K., H. Marschner & E. George. 1990. Effect of AM fungi and rhizosphere organisms root and shoot morphology, growth and water relations in maize. New Phytol. 116: 303-311.

Krishna, K.R. & D.J. Bayyaraj 1983. Changes in the free amino nitrogen and protein fractions of groundnut caused by inoculation with VA-mycorrhiza, Ann. Bot. 51: 399-401.

Kuo, S.C., K.C. Bhai & Z.S. Li. 1987. Effect of VA mycorrhiza on the growth of tissue cultured grape vine shoots. In: Sylvia, D.M., L.L. Hung and J.H. Graham (eds.), Mycorrhizae in the next decade, Proc. 7th North American Conf. on mycorrhizae, May 3-8. Gainsville, Florida, USA. p. 29.

Laibach, F. 1925. Das Taubverden von Bastardswen and die kunstiliche Anfzucht frnch absterbender Bastard embryos. Z. Bot., 17: 417-459.

———. 1929. Ectogenesis in plants. Methods and genetic possibilities of propagating embryos otherwise dying in the seed. J. Herad. 20: 201-208.

Lee, N. & H.Y. Wetzstein. 1988. Quantum flux density effects on the anatomy and surface morphology of *in vitro* and *in vivo* developed sweetgum leaves. J. Am. Soc. Hortic. Sci. 113: 167-171.

Li, X.L., H. Marschner & V. Romheld. 1991. Acquisition of phosphorous and copper by VA mycorrhizal hyphae and root-to-shoot transport in white clover. Plant Soil 136: 49-57.

Lin, C. & D.C. Chang. 1987. Effect of three *Glomus* endomycorrhizal fungi on the growth of micropropagated banana plantlets. Trans. Mycol. Soc. R.O.C. 2: 37-45.

Linderman, R.G. 1985. Microbial interactions in the mycorrhizosphere. In "Proc. 6th N. Am. Conf. on Mycorrhizae" Molina, R. (ed.), pp. 117-120.

Lister, G.R., V. Slankis, G. Krotkov & C.D. Nelson. 1986. The growth and Physiology of *Pinus strobus* L. Seedlings as affected by various nutritional levels of nitrogen and phosphorus. Ann. Bot. (NS). 32: 33-434.

Loo, S.W. 1945. Cultivation of excised stem tips of *Asparagus in vitro*. AM. J. Bot. 32: 13-17.

Mahadvan, A., N. Raman & K. Natarajan. 1988. Mycorrhizae for Green Asia, University of Madras, Madras, India, pp. 351.

Malathi, M.R., K. Ravikumar & P. Sivaprasad. 1996. Occurrence of vesicular-arbuscular mycorrhizae in *Kaempferia galanga* Linn. Ancient Science of Life 16: 82-85.

Mascarenhas, A.F. & E.M. Muralidharan. 1989. Tissue culture of Forest trees in India. Curr. Science 58(11): 606-613.

Mathur, N & A. Vyas. 1995. *In vitro* production of *Glomus deserticola* in association with *Ziziphus nummularia*. Plant Cell Rep. 14: 735-737.

Mayer, J.M & P.W. Simon. 1998. Continuous callus production and regeneration of garlic (*Allium sativum* L.) using root segments from shoot tip-derived plantlets. Plant Cell Reports 17: 726-730.

Menge, J.A., R.M. Davis, E.L.V. Johnson & G.A. Zentmeyer. 1978. Mycorrhizal fungi increase growth and reduce transplant injury in Avacado. Calif. Agric. 32: 6.

Mohankumar, V. and A. Mahadevan. 1984. Do secondary substances inhibit mycorrhizal association? Curr. Sci. 53: 377-378.

Morandi, D., S. Gianinazzi & V. Gianinazzi-Pearson. 1979. Interet de I'endomycorrhization dans la reprise et al croissance du Framboisier issu de multiplication vegetative *in vitro*. Ann. Amerl. Plantes 29: 623-630.

Mosse, B. 1957. Growth and chemical composition of mycorrhizal and non-mycorrhizal apples. Nature 179: 922-924.

Mukerji, K.G. & A. Kapoor. 1990. Taxonomy of VAM fungi with special reference to Indian Taxa. In: G.P. Agarwal (Ed.). Perspectives in mycological research. Today and Tomorrow publishers, New Delhi, pp. 7-16.

Murashige, T. & F. Skoog. 1962. A revised medium for rapid growth and bio assays for tobacco tissue cultures. Physiol. Plant. 15: 431-491.

Mustafa Anand, P.H., K.N. Harikrishnan, M.P. Martin & M. Hariharan 1997. *In vitro* propagation of *Kaempferia rotunda* Linn. "Indian corcus" – A medicinal plant. Phytomorphology 47: 281-286.

Muthukumar, T. & K. Udaiyan. 2001. Vesiculer Arbuscular mychorrhizal Association in medicinal plants of Maruthamalai Hills. Western Ghats, Southern India. J. Mycol. Pl. Pathol. 31: 180-184.

Nagakubo, T., A. Nagasawa & H. Ohkawa. 1993. Micropropagation of garlic through *in vitro* bulblet formation. Plant Cell Organ Culture. 32: 175-183.

Nair, M.G., G.R. Safir & J.Q. Siqueira. 1991. Isolation and identification of vesicular-arbuscular mycorrhiza stimulatory compounds from clover (*Trifolium repens*) roots. Appl. Environ. Microbiol. 57: 434-439.

Narayana Bhat, M., R. Jeyarajan & B. Ramaraj. 1994. Occurrence of vesicular-arbuscular mycorrhizae in forest trees at Nilgiris. Ind. J. Fores. 17: 175-177.

Nelsen, C.E. 1987. The water relations of vesicular arbuscular mycorrhizal system. In: Sai, G.R. (ed.), Ecophysiology of VA mycorrhizal plants, CRC Press, Boca Raton, Fla. pp. 71-91.

Nemec, S. & C. Guy. 1982. Carbohydrate status of mycorrhizal and non-mycorrhizal citrus root stocks. J. Am. Soc. Hortic. Sci. 107: 177-180.

——— & J.C.V. Vu. 1990. Effect of soil phosphorus and *Glomus intraradices* on growth, non-structural carbohydrates and photosynthetic activity on *Citrus aurantium*. Plant Soil 128: 257-263.

Novak, F.J. 1990. *Allium* tissue culture. In: Rabinowitsch, H.D., Brewster, J.L. (Eds.), Onion and Allied Crops. CRC Press, Boca Raton, FL, pp. 233-250.

———, L. Havel & J. Dolezel. 1986a. *Allium.* In: Evans, D., Sharp, W.R., Ammirato, P.V. (eds.), Handbook of Plant Cell Culture. Macmillan, New York, pp. 419-456.

———, ——— & ——— . 1986b. Onion, garlic and leek (*Allium* species). In: Bajaj, Y.P.S. (ed.), Biotechnology in Agriculture and Forestry. Springer, Berlin, pp. 387-404.

Osundina, M.A. 1998. Nodulation and growth of mycorrhizal *Casuarina equisetifolia.* Biology and Fertility of Soils 26: 95-99.

Pacovsky, R.S. & Fuller. 1986. Development of two endomycorrhizal symbiosis on soybean and comparison with phosphorus fertilization. Plant Soil 95: 361-377.

———, G.J. Bethlenfalvay & E.A. Paul. 1986. Comparisons between P-fertilized and mycorrhizal plants. Crop Sci. 26: 151-156.

Padmavathi, T., J. Veeraswamy & K. Venkateswarlu. 1990. Occurrence of *Sclerocystis* species in semiarid soils of India. Proc. Indian. Acad. Sci. (Plant Sci.) 100: 259-262.

Patharajan, S. 2004. Investigations on selenium enrichment by AM fungi for anticancer activity and micropropagation of garlic and tomato. Ph.D., Thesis, University of Madras.

Pierik, R.L.M. 1987. Handicaps for the large-scale commercial application of micropropagation. Acta. Hortic. 230: 63-71.

Pirozynski, K.A. & D.W. Malloch. 1975. The origin of land plants: a matter of mycotrophism. Biosystems. 6: 153-164.

——— & Y. Dalpe. 1989. Geological history of the Glomceae with particular reference to mycorrhizal symbiosis. Symbiosis. 7: 1-36.

Ponton, F., Y. Pichie, S. Parent & M. Caron. 1990. The use of vesicular-arbuscular mycorrhizal in boston fern production. I. Effect of peat based mixed. Hort Science 25: 183-189.

Quatrini, P., M. Gentile, F. Carimi, F. de Pasquale & A.M. Puglia. 2003. Effect of native arbuscular mycorrhizal fungi and *Glomus mosseae* on acclimatization and development of micropropagated *Citrus limon* (L.) Burm. The Journal of Horticultural Sciences and Biotechnology, 78(1): 39-45(7).

Rachel, E.K., S.M. Reddy & S.R. Reddy. 1991. VA mycorrhizae in semiarid soil (sandy loam) of Andhra Pradesh. Ind. J. Mycol. Pl. Pathol. 21: 174-178.

Ragupathy, S. & A. Mahadevan. 1993. Distribution of vesicular arbuscular mycorrhizae in the plants and rhizosphere soils of the tropical plains, Tamil Nadu, India. Mycorrhiza 3: 123-136.

Raju, P.S., R.B. Clark, J.R. Ellis & J.W. Maranville. 1990. Effect of species of VA-mycorrhizal fungi on growth and mineral uptake of Sorghum at different temperatures. Plant Soil 121: 165-170.

Raman, N. 1996. Role of Mycorrhizae in biocontrol of plant pathogens. In: Current trends in Life Sciences Vol. 21- Recent Developments in biocontrol of plant pathogens. K. Manibhushan Rao and A. Mahdevan (eds.), Today and Tommorrow's Printers and Publishers, New Delhi, pp. 11-20.

——— & S. Elumalai. 1991. Studies on mycorrhizal and actinorhizal association in *Casuarina equisetifolia* in Coramendal coastal region. J. Trop. For. 7: 138-150.

——— & A. Mahadevan. 1996. Mycorrhizae research – a priority in agriculture. In: *Concepts in mycorrhizal research.* Mukerji, K.G. (ed.), pp. 41-75. Kluwer Academic Publishers, The Netherlands.

———, N. Nagarajan & K. Sambandan. 1992. Vesicular arbuscular mycorrhizae of tree species of Mamandur forest of Tamil Nadu. J. Tree Sci. 11: 135-139.

———, N. Nagarajan, S. Gopinathan & K. Sambandan. 1993. Mycorrhizal status of plants species colonizing a magnesite mine spoil in India. Biol. Fertil. Soil 16: 76-78.

Redecker, D., R. Kodner & L.E. Graham. 2000. Glomalean fungi from ordovacian. Science 289: 1920-1921.

Reid, C.P.P. 1984. Mycorrhizae: A root-soil interface in plant nutrition. In: Microbial plant interactions Todd, RL. And J.E. Giddens (eds.), ASA special Pub. 47: pp. 29-50.

Rekha Rani & K.G. Mukerji. 1990. Distribution of vesicular arbuscular mycorrhizal fungi in India. Acta. Mycol. Hungarica. 37: 347-350.

Remy, W., T.N. Taylor, H. Hass & H. Kerp. 1994. Four hundred million year old vesicular arbuscular mycorrhizae. Proc. Natl Acad. Sci. USA. 91: 11841-11843.

Rogers RD & S.E. Williams. 1986. Vesicular arbuscular mycorrhizal influence on plant uptake of calcium and cobalt. Soil Biol Biochem 18: 371-376

Rojanasiriwong, W., O. Napamornobodi & S. Thamsurakul. 1987. Production of VA mycorrhizal fungi in tissue culture. In: Sylvia, D.M., L.L. Hung and J.H. Graham (Eds.). Mycorrhizae in the next decade, Proc. North. Amer. Con. Mycorrhizae, May 3-8, Gainesville, Florida. p. 217.

Rose, S.L. 1980. VA mycorrhizal association of some desert plants of Bajacalifornia. Can. J. Bot. 59: 1056-1060.

Roy, A & A. Pal. 1991. Propagation of costus speciosus (Koen) sm. through *in vitro* rhizome production. Pl. Cell Rep. 10: 525-528.

Safir, G.R., J.S. Boyer & J.W. Gerdemann. 1972. Nutrient status and mycorrhizal enhancement water transport in soybeans. Plant Physiol. 49: 700-703.

Sahadevan, C. 1998. Investigations on effect of vesicular and arbuscular mycorrhizal fungi on some micropropagated plants. Ph. D., Thesis, University of madras, India.

Sambandan, K., K. Kannan & N. Raman. 1991. Effect of VAM fungus, *Glomus geosporum* on effluent tolerance of Casuarina equisetifolia. Poll. Res. 10: 55-67.

Sanders, F.E., B.P. Tinker, R.L.B. Black & S.M. Palmerley. 1977. The development of endomycorrhizal rott systems; I. Spread of infection and growth-promoting effects with four species of vesicular-arbuscular endophyte. New Phytol. 78: 257-268.

Sbrana, C., M. Giovannetti & C. Vitagliano. 1994. The effect of mycorrhizal infection on survival and growth renewal of micropropagated fruit root stocks. Mycorrhiza 5: 153-156.

Schenck, N.C. 1981. Can mycorrhizal control root disease? Plant Disease 65: 230-234.

Schussler, A., D. Schwarzott & C. Walker. 2001. A new fungal phylum, the Glomeromycota: phylogeny and evolution. Mycol. Res. 105: 1413-1421.

Sharp, J.M., R. Brian, & K.H.K. Crowely. 1995. Plant Cell Culture. TAFE Publications. Collingwood, Victoria, Australia.

Shen, L.U., P.G. Braunberger & M.H. Miller. 1994. Response of vesicular arbuscular mycorrhizas of maize to various rates of P addition to different rooting zones. Plant Soil. 158: 119-128.

Shunmuga Sundaram, M. 2004. Investigations on micropropagation, mycorrhization, pharmacognosy and antimicrobial activity of *Tylophora indica* (burm. f) Merrill. Ph.D. thesis, University of Madras.

Shuto, H. & T. Abe. 1993. *In vitro* propagation of plants from root apex-derived calli in Chinese chive (*Allium tuberosum* Rottler) and garlic (*Allium sativum* L.). Japan J Bread 43: 349-354.

Siqueira, J.O. 1994. Micorrhizas. In: Aranjo RS Hungria M (eds.), Microorganisms de importancia agricola. Embrapa – CNPAF. Embrapa – CNPSO, Embrapa – SPI, Brasilia, pp. 151-194.

Skoog, F. 1944. Growth and organ formation in tobacco tissue cultures. Am. J. Bot. 31: 19-24.

Smith, S. E. & D.J. Read. 1997. *In* "Mycorrhizal symbiosis" 2nd edition. Academic Press, California.

——— & V. Gianinazzi-Pearson. 1988. Physiological interactions between symbiont in vesicular arbuscular mycorrhizal plants. Annu. Rev. Plant Physiol. Mol. Biol. 39: 221-244.

Strullu, D.G. 1985. Les mycorrhizas, Handbuch der Pflazen Anatomie. Portraeger, Berlin.

———, C. Romand, P. Callac, T. Teoule & Y. Demarly. 1989. Mycorrhizal synthesis of in vitro between *Glomus* spp. ad artificial seeds of alfalfa. New Phytol. 113: 545-548.

Sukhada, M. 1988. Response of papaya (*Carica papaya*) to VAM fungal inoculation. In: A. Mahadevan, N. Raman and K. Natarajan (Eds.). Mycorrhizae for green Asia: Proc. First Asian Conference on Mycorrhizae, Univ. Madras, Madras, pp. 260-261.

Sulochana, T. & C. Manoharachary. 1989. Vesicular arbuscular mycorrhizal associations of castor and safflower. Curr. Sci. 58: 459-461.

Sumi, S., T. Tsuneyoshi & H. Furutani. 1993. Novel rod-shaped viruses isolated from garlic, *Allium sativum*, possessing a unique genome organization. J. Gen. Virol. 74: 1879-1885.

Sylvia, D.M., L.C. Hammond, J.M. Bennet, J.H. Hass & S.B. Linda. 1993. Field response of maize to VAM fungus and water management. Agron. I. 85: 193-198.

Sylvia, M., Abid K. Alagely, Michael E. Kane & Nancy L. Philman. 2003. Compatible host/mycorrhizal fungus combinations for micropropagated sea oats. Mycorrhiza 13:177–183

Taylor, J. & L. Harrier. 2000 A Comparison of nine species of arbuscular mycorrhizal fungi on the development and nutrition of micropropagated *Rubus idacus* L.Cv. Glen Prosen (Red Raspberry). Plant and Soil 225: 53-61.

Tinker, P.B. & A. Gildon. 1983. Mycorrhizal fungi and ion uptake. In: Robb DA, Pierpoint WS (eds.), Metals and micronutrients uptake and utilization of metals by plants. Academic Press London Pp 21-32.

Tisserant B., V. Gianinazzi- Pearson, S. Gianinazzi & A. Gollotte 1993. In plant histochemical staining of fungal alkaline phospatase activity for analysis of efficient arbuscular mycorrhizal infections. Mycol. Res. 97: 245-250.

Torrey, J.G. 1983. Root development and root nodulation in Casuarina: In Casuarina ecology, management and utilization. Midgley, S.J., Turnbull, J.W. and Johnson, R.D. (Eds.), CSIRO, Melbourne. pp. 180-197.

Trappe, J.M. 1987. Phylogenetic and Ecological aspects of Mycotropy the angiosperms from an evolutionary stand point. In: G.R. Safir (ed.), Ecophysiology of VA Mycorrhizal plants. CRC Press. Boca Rators. Florida. pp. 5-25.

Tsuneyoshi, T. & S. Sumi. 1996. Differentiation among garlic viruses in mixed infections based on RT–PCR procedures and direct tissue blotting immunoassay. Phytopathology. 86: 253-259.

Van Overbeek, J., M.E.Conklin & A.F. Blaksler. 1947. Factors in coconut milk essential for growth and development of very young *Datura* embryos. Science 94: 350-351.

Varma, A. & H. Schuppe. 1994. Positive influence of arburculer mycorrhizal fungus *in vitro* raised *Hortensia* plantlets. Angew Botanik. 68: 108-115.

Venkataraman, M.N., P. Borthakur & H.D. Singh. 1995. Endogonaceae of North Eastern India with special refernce to *Gigaspora nigra*. Acta Bot. Indica 13: 73-76.

Venkateshwar Rao, G., C. Manoharachary I.K. Kunwari & B.R. Rajeshwar Rao. 2000. Arbuscular Mycorrhizal fungi Associated with some economicaly important spices and aromatic plants. Plant Journal of Science 129: 1-5.

Vestberg, M. 1992. Arbuscular mycorrhizal inoculation of micropropagted strawberry and field observations in Finland. Agronomie 12: 865-867.

Vin Dijk, P. 1991. Mite-borne viruses isolates from cultivated *Allium* spp. and their classification into two new rymoviruses in the family *Potyviridae*. Neth. J. Plant Path. 97: 381-399.

——— & R. Sutarya. 1991. Virus survey of garlic, shallot and Welsh onion streak of rakkyo (*Allium chinense* G. Don), a newly recognized disease caused by garlic latent virus and onion yellow dwarf virus. Ann. Phytopathol. Soc. Jpn. 57: 65-69.

———. 1993a. Survey and characterization of potyviruses and their strains of *Allium* species. Neth. J. Plant Path. 99: 1-48.

———. 1993b. Carlavirus isolates from cultivated *Allium* species represent three viruses. Neth. J.Plant Path. 99: 233-257.

Volpin, H., Y. Elkind, Y. Okan & Y. Kapulnik. 1994. Embryogenesis and Plantlet formation in tissue culture of Kaempferia galanga L. a medicinal plant. Phytomorphology 42: 253-256.

Walkey, D.G.A. 1990. Virus Disease. In: Rabinowitch, H.D., Brewster, J.L (Eds.) Onion and Allied crops, vol. 2 CRC Press, Boca Ratol, pp. 191-211.

Wallace, H.R.1973. Nematode ecology and plant disease. Alden Press, London and Oxford.

Wang, H., S. Parent, A. Gosselin & Y. Desjardins. 1993. Study of vesicular-arbuscular mycorrhizal peat-based substrates on symbiosis establishment, acclimatization and growth of three micropropagated species. J. Am. Soc. Hortic. Sci. 118: 896-901.

Wetzstein, H.Y. & H.E. Sommer. 1982. Leaf anatomy of tissue cultured *Liquidambar styraciflua* planlets during acelimatization. J. Am. Sco. Hortic. Sci. 108: 475-480.

White, P.R. 1934. Potentially unlimited growth of excised tomato root tips in a liquid medium. Plant Physiol. 9: 585-600.

———. 1939. Potentially unlimited growth of excised plant callus in an artificial medium. American Journal of Botany 26: 59-64.

Williams, S.C.K., M. Vestber, M. Uosukain, J.C. Dodd & P. Jeffries. 1992. Effect of fertilizers and arbuscular mycorrhizal fungi on the post vitro growth of micropropagated strawberry. Agronomie 12: 851-857.

Wright, S.F., & P.D. Millner. 1994. Dynamic process of vesicular arbuscular mycorrhizal: a mycorrhiza system within the agro ecosystem, In: J.A. Hatfield and B.A. Stewart (eds.) *Soil Biology: Effects on Soil Quality*. Lewis Publishers, Boca Raton, USA. 29-59 p.

Yao, M.K., R.J. Tweddell & H. Desilets. 2002. Effect of two vesicular-arbuscular mycorrhizal fungi on the growth of micropropagated potato plants and on the extent of disease caused by *Rhizcotonia solani*. Mycorrhiza 12: 235-242.

18

Influence of Mycorrhizal Fungi on the Growth and Survival of Micropropagated Forest Trees

M. SUDHAKARA REDDY* and AJAY KUMAR PANDEY

*Department of Biotechnology, Thapar Institute of Engineering and Technology, Patiala, India

ABSTRACT

Mycorrhizal inoculation has been proposed as a way of protecting plants against environmental stress and increasing the sustainability of plant production. Micropropagation of forest trees showing attributes of superior growth provide the means of increasing productivity in reforestation programmes, but unfortunately, micropropagated plants are adversely affected by different factors. Water stress, poorly developed cuticles and non- or partially functional stomata appears to be the major causes for the failure of acclimation of micropropagated plants. As a consequence of vitrification micropropagated plants have poorly developed constitutive disease resistance, with hypoligification and reduced phytoalexine content. Micropropagated plants also require a long period of transition to become adapted to *ex vitro* conditions, which is called transplantation shock. To maximize the benefits of micropropagation it will be necessary to ensure that these plants form effective mycorrhizas. The presence of mycorrhizas could increase the availability of limiting nutrients such as phosphorus and nitrogen by facilitating their absorption and also enable them to acclimatize to *ex vitro* conditions. In this chapter, an attempt has been made to review the role of mycorrhizal fungi in improving the growth and survival of micropropagated plants.

I. INTRODUCTION

The survival of mankind requires that we modify deeply our relationship to natural resources, and especially those with living organisms. In particular, new management roles have to be defined to exploit what for us is an unavoidable resource: the plant. We can postulate from present-day knowledge that plant survival and fitness in a given ecosystem are not conceivable without the establishment of beneficial relationships between plants and micro-organisms, in particular those living in the soil. Roots, the "hidden half' of plants, serve a multitude of functions. They act as anchoring of whole plant system and supply the plants with water and nutrients, and helps in exchange of various growth substances with the shoots.

Forest tree species, which are the boon of any country, in terms of providing feed, fodder and timber play a major role in the advancement of country. But, due to extra exploitation of such rich resources, led to the diminishing of forest cover. India is having a true forest cover of about 11 %, mainly concentrated towards the Himalayan region to northeast and expanding towards some parts of southern and middle region of India. The main forest tree species that are commonly found are deodar, spruce, pines, oak, and eucalypt. These species are rich sources of high quality timber; efforts are made to increase the number of such plantation by multiplying such important forestry tree species using tissue culture techniques.

Micropropagation is a set of procedures that multiply plants in tissue culture with minimal genetic and epigenetic variability. The advantages of micropropagation are high multiplicity rates, propagation of some woody plants that are difficult to propagate by cuttings, layers or grafts, easy to produce pathogen free clones, rejuvenate or reinvigorate clones, making it easier to propagate by cuttings or other conventional techniques and also tissue culture methods are the foundation of genetic engineering.

What Lacks in Micropropagated Plants?

Micropropagation of forest trees showing attributes of superior growth provide the means of increasing productivity in reforestation programs (Libby & Rauter, 1984), but unfortunately, micropropagated plants are adversely affected by different factors. Water stress appears to be one of the major causes for the failure of acclimation of micropropagated plants. This may be either due to a low nutrient absorption capacity of their roots or due to deficient stomatal regulation of water loss (Bonga, 1977; Flick *et al*, 1983). Micropropagated plants often show symptoms of vitrification i.e., they have poorly developed cuticles and non- or partially functional stomata, which result in drought, stress at acclimatization in *in vitro* (Casselle *et al*, 1996; Diettrich, 1992). As a consequence of vitrification micropropagated plants have poorly developed constitutive disease resistance, with hypoligification and reduced phytoalexine content (Sutter, 1988). Aseptically micropropagated plants represent "biological vaccum" available for promiscuous facultative pathogens (Varma & Schuepp, 1996). Micropropagated plants also require a long period of transition to become adapted to *ex vitro* conditions, which is called transplantation shock. A semi-sterile substrate such as peat/sand mixture is generally used in the transition period of micropropagated plantlets. Hence, the subsequent growth of plantlets may be slow because of the absence of compatible mycorrhizal fungi or plant growth promoting microorganisms.

Micropropagation techniques do not permit a spontaneous mycorrhization; besides, the high nutrient contents of the cultural medium ensure a good growth of the plants. Nevertheless, many micropropagated species are mycorrhiza-dependent and many problems can arise from the absence of symbiotic fungi during the acclimatization phases in the green house. To maximize the benefits of micropropagation it will be necessary to ensure that these plants form effective mycorrhizas. The presence of mycorrhizas could increase the availability of limiting nutrients such as phosphorus and nitrogen by facilitating their absorption and also enable them to acclimatize to *ex vitro* conditions.

II. MYCORRHIZA

Mycorrhiza is a symbiotic association between some soil fungi and the roots of higher plants, where carbon fluxes from plant root to mycobiont and in turn, the fungal hyphae make available the essential nutrients to plant (Mukerji; 1996; Mukerji *et al*, 2000). The mycorrhizal relationship is essential to the healthy growth of trees, especially where soil nutrients are limiting or environment conditions are harsh (Smith & Read, 1997). The colonized plant is better nourished and better adapted to its environment. It obtains increased protection against environment stresses (Sylvia & Williams, 1992), including drought (Subramanian *et al*, 1995), cold (Charest *et al*, 1993; Paradis *et al*, 1995), salinity (Davis & Young, 1985) and pollution (Leyval *et al*, 1994; Shetty *et al*, 1995). In addition, symbiosis tends to reduce the incidence of root diseases and minimizes the harmful effect of certain pathogenic agents (Dehne, 1982). By and large, the growth and health of colonized plant is improved. At the same time they obtain increased protection against environmental conditions detrimental to their survival.

The major types of mycorrhizas are ectomycorrhiza and endomycorrhiza (Fig. 18.1). The diagnostic feature of ectomycorrhiza is the presence of hyphae between root cortical cells producing a net like structures called as Hartig net. Many ectomycorrhizas also have a sheath or mantel, of fungal tissue that may completely cover the absorbing root usually the feeder roots. Ectomycorhiza are found on woody plants ranging from shrubs to forest trees. Many of these hosts belong to the family Pinaceae, Fagaceae, Betulaceae and Myrtaceae. Over 4,000 fungal species, belonging primarily to Basidiomycetes and fewer to Ascomycotina, are known to form ectomycorrhizas. The diagnostic feature of arbuscular mycorrhiza (AM) is the development of a highly branched arbuscule within the root cortical cells. The fungus initially grows between the cortical cells, but soon penetrate the host cell wall and grow within the cell. The general term for all the mycorrhizal type where the fungus grows within cortical cells is endomycorrhiza. Other structures produced by such AM fungi include vesicles, auxiliary cells and asexual spores (Fig. 18.1). The AM type of symbionts is common as the fungi involved can colonize a vast taxonomic range of both herbaceous and woody plants, indicating a general lack of host specificity among this type.

Tree species differ in their dependency on mycorrhizae. Some tree species like pines fail to establish and grow healthy in the absence of mycorrhizal colonization on their roots. Some tree species may be partly dependent on mycorrhiza, but there is hardly any tree species, which do not have mycorrhizae of one type or other on their roots. Tree species especially exotic, fail to establish on aforestation sites in the absence of their fungal symbionts. For example repeated attempts to establish pine plantations in Puerto Rico failed due to lack of ectomycorrhizae on their roots. Transfer of ectomycorrhizal soil from successful plantations in the mainland to Puerto Rico nurseries helped in the successful establishment of pine plantations on this island (Hacskaylo, 1967).

A. Mycorrhizas and Micropropagated Plant

The integrated use of mycorrhiza management tools with other cultural practices and the potential use of selected beneficial fungi for mycorrhizal inoculation of seedlings will help ensure

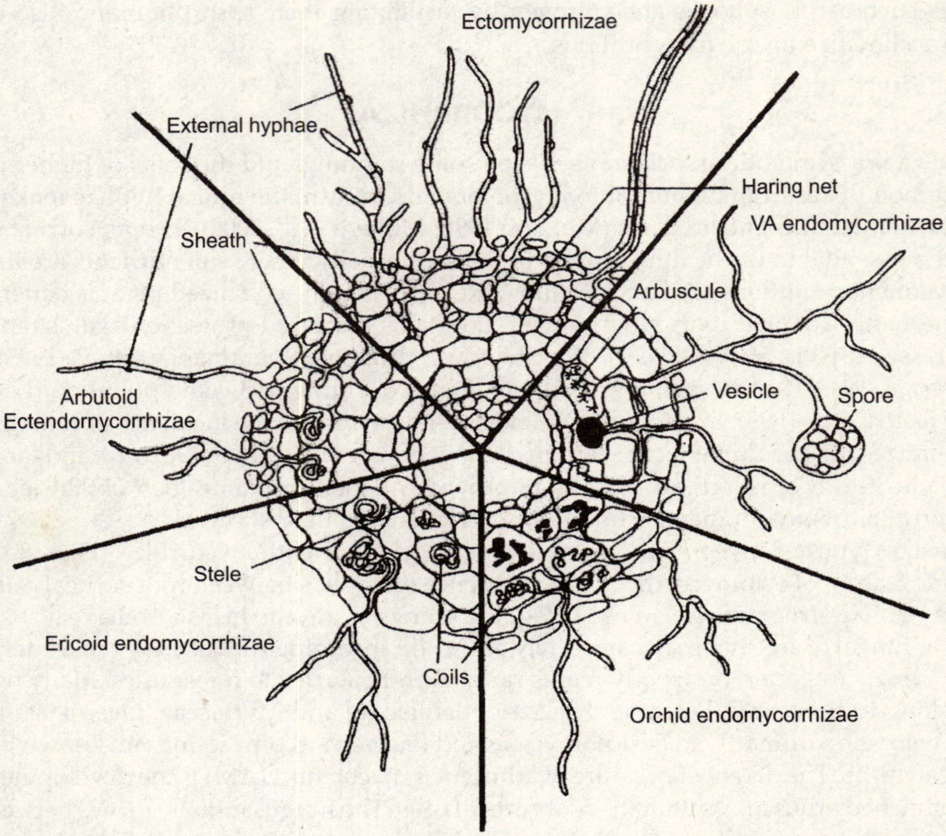

Fig. 18.1: Characterization of mycorrhizal types (from Selosse & Le Tacon, 1998)

the successful production of vigorous planting stock. Plant micropropagation has an outstanding place in plant biotechnology and can be applied to the production of many plants from herbaceous to hardwood species, including those of interest in mountain areas. Understanding of the mechanisms by which these biota interact and influence root quality, especially the ways in which they promote quality, is a critical factor in the development of improved micropropagation systems. The use of microbial inoculants raises a series of technical issues related to their effective use with micropropagated and other plants, and a number of key questions linked to safety and environmental impact. Mycorrhizal fungi can act as biofertilizers, bioprotectors and bioregulators. The fitness of micropropagated plants could be greatly increased by exploiting these symbiotic fungi, so ensuring their successful introduction into mountain production areas of low chemical input.

Ectomycorrhizal fungi have been used to enhance rooting of hardwood cuttings of non-ectomycorrhizal plants such as fruit trees (Branzanti *et al*, 1992; Cristoferi *et al*, 1986) Mycelia showing high IAA-synthesizing activity could be particularly suitable to enhance rooting of cuttings. Such mycelia could be selected by chromosomal genetics (Gay & Debaud, 1987). These mycorrhizas presumably improve the resistance of the hypocotyls to transplantation shock.

AM Fungi and Micropropagated Plantlets

The beneficial effects of AM fungal inoculation of micropropagated plants are widely documented in the literature. Positive effects on survival and growth were seen after AM inoculation both in micropropagated horticultural plants and fruit species, as in Gerbera (Wang *et al*, 1993), avocodo (Vidal *et al*, 1992), apple (Granger *et al*, 1983; Branzanti *et al*, 1992), cassava (Azcon-Aguilar *et al*, 1997), grapevine (Schubert *et al*, 1988; Lovato *et al*, 1992), kiwi fruit (Schubert *et al*, 1992), peach and pear (Rapparini *et al*, 1994, 1996), pineapple (Guillemin *et al*, 1992; Lovato *et al*, 1992), plum (Fortuna et. al., 1992), *Prunus cerasifera* (Berta *et al*, 1995), persimmon (Martin *et al*, 2003), straw berry (Vestberg, 1992; Sharma & Adholeya 2004; Vestberg *et al*, 2004) and walnut (Dolcet- Sanjuan *et al*, 1996). Improved performance of endomycorrhizal plants is usually due to enhanced uptake of nutrients, resistance to pathogens (Dehne, 1982, Cordier *et al*, 1996) and protect against nematode parasitism, as reported for banana (Pinochet *et al*, 1997) and plums (Pinochet *et al*, 1998). The technique to inoculate at the start of the acclimatization stage is normally employed, due to its easier feasibility (Ravolanirina *et al*, 1989). In some cases a better plant response is produced when the inoculation is carried out of the transplantation phase (Vidal *et al*, 1992). Pre-colonization of the roots takes place only in young secondary roots and these can be produced later during the *ex vitro* development (Azcon-Aguilar & Barea, 1997). The time course for development of the infection and the entity of the infection can vary depending the fungal species (Berta *et al*, 1995) and the mycorrhizal effects may become evident in later stages (de la Vina *et al*, 1996). Another aspect to consider is the effect of the amount of inoculum, which can affect root-infection (Schubert *et al*, 1992). The volume ratio of soil/inoculum used could be insufficient for a massive mycorrhizal establishment. Moreover the characteristic of the substrate (pH, phosphorus content) used in the acclimatization can affect the success of the mycorrhization: they must ensure the spore germination and the mycelial growth as well as the plant development (Azcon-Aguilar & Barea, 1997).

Arbuscular mycorrhizal fungi have been shown to increase resistance of plants to various environmental stresses. Investigations were carried out to overcome transplantation shock in three multipurpose tree legumes viz. *Leucaena leucocephala, Sesbania seban* and *Sesbania grandiflora.* Subsequent to nodulating roots of plantlets with *Rhizobium* strains PRGL 1 and S-7, they were transferred to garden soil with and without spores of VAM fungal species viz. *Glomus fasciculatum* and *G. macrocarpum.* Only 25 to 30% of the plantlets transferred to soil without VAM fungi, survived (six weeks after transplantation). In contrast, over 95% of the plantlets survived from those transferred to soil with VAM fungi inoculum. Moreover, mycorrhizal plants were stronger than non-mycorrhizal, indicating that mycorrhizal association helps in overcoming transplantation shock (Puthur *et al*, 1998).

Multimicrobial Influence

Multimicrobial inoculation has been proposed as a way of protecting plants against environmental stress and increasing the sustainability of plant protection. Inoculation of *G. fasciculutum* and *Rhizobium* during *ex vitro* hardening stage increased the growth and survival of micropropagated *L. leucocephala.* The dual inoculation has resulted in higher growth rate of root and shoot, number of leaves, number of nodules, mycorrhizal colonization and 100% survival of plants (Naqvi & Mukerji, 1998). Monticelli *et al* (2000) reported that inoculation of

Glomus mossae and *Scutellospora calospora* to micropropagated *Prunus* plants at the transplantation phase increased the height and dry weight of the plants when compared to control plants (Table 18.1). Mark *et al* (1999) reported the establishment of growth and nutritional conditions, which support *in vitro* of the *Casuarina equisitifolia* with *Frankia* and VAM fungi. To facilitate dual colonization of roots of micropropagated C. *equisitifolia* with *Frankia* and VAM fungi *in vitro* it is necessary to select the appropriate media. The colonization of callus by VAM and *Frankia* is not only of scientific interest but also often the possibility of using callus material as a carrier material for inoculation of *in vitro* systems. Srinath *et al* (2003) reported the enhanced growth and nutrition of micropropagated *Ficus benjamina* to *Glomus mossae* co-inoculated with *Trichoderma harzianum* and *Bacillus coagulans.* These results suggest that dual inoculation can increase the growth and survival of micropropagated plants much better when compared to the single inoculation, especially in case of legume trees and also in case of *Casurina* where these plants are associated with *Frankia.*

Table 18.1: Growth parameters observed in micropropagated *Prunus* plants after inoculation of AM fungi at transplantation phase. (Monticelli *et al*, 2000)

	Percent mycorrhizas in roots	*Root mean Diameter (mm)*	*Root dry wt (mg)*	*Plant Height (mm)*	*Shoot dry wt (mg)*
Control	0.0	0.17	46.6	18.8	161.8
G. mossae	142	0.3	105.7	100.0	431.8
S.calospora	23.9	0.66	126.5	100.0	411.7

Ectomycorrhizal Fungi and Micropropagated Plantlets

Ectomycorrhizas offer several advantages to plants including an increased root absorbing area, enhanced nutrient uptake and increased host resistance to plant pathogens and to drought. Several reports are available about the inoculation of ectomycorrhizal fungi for micropropagated plants of different tree species. Some of the ectomycorrhizal fungi produce edible mushrooms, which are of great economic value in the world market. Inoculation of such ectomycorrhizal fungi to plants can increase the growth of the seedlings and also produce edible mushrooms in the plantations. The common edible ectomycorrhizal fungi are *Cantharellus cibarius, Boletus* spp., Truffels, *Tricholoma* and Morels. Malajczuk & Hartney (1986) have suggested that inoculation of micropropagated plantlets with mycorrhizal fungi at the root initiation stage helps the fungus to colonize the root surface without competition or antagonism from other soil microflora prior to outplanting. Tonkin *et al* (1989) reported major difference in the rate of ectomycorrhizal development on eucalypt roots when inoculated with different isolates of *Pisolithus tinctorius* in aseptic conditions. Although reaction between various hosts and fungal species can well be regarded as compatible, the greater rapidity of mycorrhizal development between the isolates and host implies that there are varied degrees of host compatibility among isolates. As suggested by Malajczuk *et al* (1990) it is obvious that the events associated with the process of recognition between the symbionts allow for discrimination between strains of the same fungus. Oak and pine seedlings mycorrhizal with *P. tinctorius* showed differences in survival related to the extent of mycorrhization of their roots, and seedlings with abundant *P. tinctorius* ectomycorrhizas tolerated environmental stresses, such as soil water deficit and high temperatures, better than seedlings without *P. tinctorius* ectomycorrhizas (Marx & Cordell, 1989). Different ectomycorrhizal fungi were tested on micropropagated plantlets of *Eucalyptus* (Malajczuk & Hartney, 1986; Tonkin *et al*, 1989; Poissonier, 1986; Reddy & Satyanarayana, 1998a) Poplar (Heslin & Douglas, 1986; Navratil & Rochan, 1981; Reddy & Satyanarayana, 1998b), birch (Grellier *et al*, 1984) and chestnut (Martins *et al*, 1996; 1997).

Species of *Eucalyptus* have many desirable characteristics, including fast growth, and have been planted extensively both in dry and irrigated areas in India. *Eucalyptus tereticornis* has been introduced on a large scale in all parts of country because of its good growth in most of the localities and it has been identified suitable for dry zone afforestation. Ectomycorrhizas are characteristic of all eucalypts species and the response of various species of eucalypts to inoculation with different species of ectomycorrhizal fungi suggests that there is a potential for increasing productivity from eucalypts grown as short rotation crops in plantations (Garbye *et al*, 1988; Bougher & Malajczuk, 1990). Extensive colonization of roots was observed by Reddy Satyanarayana (1998a) using the local isolates of *Pisolithus* from India with micropropagated *E. tertecornis* plantlets, when compared with other isolates. Nursery experiments showed, that, mycorrhizal plants have significantly increased in height, dry weight of shoot, root and root collar diameter compared to non-mycorrhizal plants (Table 18.2). Their results suggested that in selecting compatible isolates for a particular host species, host specificity should be considered along with the colonization ability of the fungus.

Table 18.2: Performance of *Pisolithus* sp. inoculated micropropagated mycorrhizal plants

Material	*Shoot Height (cms)*	*Shoot dry wt (rng)*	*Root dry wt (mg)*	*Shoot/ Root ratio*	*Root collar (mm)*	*Leaves (Nos)*	*% mycorrhizas*
			Eucalvptus tereticornis (Reddy and Satyanarayan, 1998)				
Mycorrhizal	59.1	2230	340	7.0	5.0	-	51
Control	42.6	1179	158	8.6	3.4	-	ND
			Castenea sativa (Martins et al, 1996)				
Mycorrhizal	-	233.70	114.46	2.01	-	96.93	-
Control	-	175.90	51.56	3.4	-	51.08	-

ND: Not detected, -: Not reported

Malajczuk & Hartney (1986) developed a procedure for inoculation of micropropagated plantlets of *E. camaldulensis* with ectomycorrhizal fungi and compared with seedlings inoculated using inoculum contained in peat/vermiculite carrier. Their findings showed that inoculation of ectomycorrhizal fungi to *E. camaldulensis* seedlings at the root initiation stage improved the growth and colonization when compared to the inoculation of peat/vermiculite carrier suggesting that root initiation stage helps the fungus to colonize the root surface without competition or antagonism from other soil microflora prior to outplanting. Tonkin *et al* (1989) studied the ectomycorrhiza formation by micropropagated clones of *E. marginata* inoculated with isolates of *P. tinctorius.* They formed the differences in colonization by different isolates of *P. tinctorius* with *E. marginata.* There are other reports, which also indicate the improvement of growth of micropropagated *Eucalyptus* plants by inoculation with different ectomycorrhizal fungi (Poissonier, 1986).

Poplar grows satisfactorily on soil types of varying fertility (Harris & Jurgensen, 1977). Different tree species of poplar were introduced in India to meet the growing demand for wood and because of their rapid growth rate and ability to coppice, which are valuable characteristics of their utilization in short rotation forestry. Ectomycorrhizas, endomycorrhizas and ectendomycorrhizas have been recorded in poplar species (Vozzo & Hacskaylo, 1974). The most common ectomycorrhizal fungi associated with poplar were *Amantia muscaria, Boletus edulis, Hebeloma crustiliniforme, Russula delica, Paxillus involutus* and *Coenococcum qraniforme* (Trappe, 1962). Inoculation of different ectomycorrhizal fungi to micropropagated plantlets of *Populus deltoids* to synthesize mycorrhizas was reported by Reddy & Satynarayana (1998b). Their results showed that inoculation of *P. involutus* at the root initiation stage increased the growth of the plantlets when compared to other fungi. The other fungi such as *Cenocoocum geophilum, Laccaria laccata, P. tinctorius* isolates did not form mycorrhizal association with *P. deltoides.* The formation of mycorrhiza of *P. deltoides* with *P. involutus* support the view that this fungus has a broad host range at the species level and at the isolate level. The broad host range of *P. involutus* was previously reported by Molina & Trappe (1982). In addition, this species has been reported to form mycorrhizas *in vitro* with seedlings of *Populus tremula* (Laiho 1970), *P. euramericana* (Fontana, 1961) and *Populus* hybrid TT32 (*P. trichocarpa* and *P. tacamahaca)* (Heslin & Douglas, 1986). Ectomycorrhizal fungi stimulate growth in conventionally propagated plants of several species (Rifle & Tinus, 1982) and in cuttings of several poplar hybrids. With poplar cuttings, an increase in the number and dry weight of roots has been reported for mycorrhizal plants (Navratil & Rochan, 1981). Mycorrhizal fungi may affect growth by production of auxins (Ulrich, 1960) or cytokinins (Miller, 1971) or by altering either the uptake of minerals or other metabolic processes (Turner, 1962).

Ectomycorrhizal fungi rapidly metabolize exogenous supplied tryptophan to IAA and serve as a suitable tool to enhance *in vitro* rooting of micropropagated gymnosperms (Gay, 1990). Enhanced rooting of woody cuttings of *Vaccinium ovatum* and *Arctostaphylos uva-ursi* was reported by inoculation of ectomycorrhizal fungi (Lindermann & Call, 1977). Supriyanto & Rohr (1994) showed that *Hebeloma cylindrosporum* forms mycorrhiza with the tissue cultured plantlets of *Pinus sylvestris in vitro* under fully controlled conditions. The percent survival rates have increased from 20% to 70% in *P. sylvestris* when inoculated with *H. cylindrosporum* (Supriyanto & Rohr, 1994). Their work showed that *P. sylvestris* could be multiplied continuously *in vitro* using micropropagation pathways that are new for conifers. The plantlets produced are vigorous and their *ex vivo* survival is improved by a mycorrhizal association developed *in vitro.* Ectomycorrhizas also helps in improving the growth and survival of different other trees species. Martins *et al* (1996) reported the improvement of micropropagated *Castanea sativa* (Chestnut) growth by inoculating with ectomycorrhizal fungi. *C. sativa* is a forest tree of great economic interest for wood and it is difficult to propagate through cuttings and also because of high heterosis of seeds. Their results showed that *P. tinctorius* mycorrhization improved the general conditions of chestnut plants and also increased survival and growth (Table 18.2). Diez *et al* (2000) reported the mycorrhization of *in vitro* plants raised from somatic embryos of *Quercus suber* (Cork oak). They showed that, *in vitro* mycorrhization increased the formation of secondary roots and the survival after acclimatization of *in vitro* plants rose from somatic embryos. This approach is likely to leave behind the bottleneck so far encountered by Q. *suber* micropropagated by somatic embryogenesis.

Ericoid mycorrhiza belongs, together with orchidaceous and arbuscular mycorrhiza types, to the group of endotrophic mycorrhizal associations (Fig. 18.l). Ericoid mycorrhizas are associated between ascomycetous fungi and plant species belonging to the families *Ericaceae, Epacridaceae* and *Empetraceae* (Smith & Read, 1997). The common genera where ericoid mycorrhizal association reported were *Calluna, Erica, Rhododendron, Vaccinium* and *Empetrum*. Jansa & Vosatka (2000) studied *in vitro* and *post vitro* inoculation of micropropagated *Rhododendron* with ericoidmycorrhizal fungi. They found that the most efficient ericoid mycorrhizal fungal strains were found beneficial for the growth of micropropagated *Rhododendron* plants when inoculated *post vitro* transplantations to peat based substrate.

Mycorrhization also influence the physiological parameters of micropropagated plantlets. Martins *et al* (1997) showed that *in vitro* mycorrhization of *C. sativa* with *P. tinctorius* increased protein content, photosynthetic rates; decrease the respiratory rates and CO_2 compensation point of seedlings. Martins *et al* (2001) also studied the macro and micronutrient content of mycorrhizal and non-mycorrhizal micropropagated *C. sativa* plants. They reported that N, P, K, Ca and Mg content and also micronutrient content are higher in mycorrhizal plants when compared to non-mycorrhizal plants. Their results suggest that mycorrhization of micropropagated plants helps in increase the nutritional status of the plant.

Micropropagated plants become associated with different symbiotie/pathogenic organisms when they are outplanted. However, an early inoculation of these plants with appropriate symbiotic organisms promotes to improve plant survival and performance. *In vitro* coculture of plant tissue explants with beneficial microorganisms induces developmental and metabolic changes in the derived plantlets, which enhance their tolerance to abiotic and biotic stresses. The induced resistance response caused by the inoculants is referred to as 'biotization' (Nowak, 1988). There is enough experimental evidence with bacteria (bacterization) and ecto- and endomycorrhiza (mycorrhization) inoculations to recommend utilization of this technology in commercial micropropagation. Micropropagation of *in vitro* propagated plantlets is having a positive impact on their post transplanting performance.

III. CONCLUSIONS

Growth, development and nutrient status of micropropagated plants can be enhanced with the inoculation of mycorrhizal fungi. As mycorrhizal fungi play an important role in mineral nutrition, the presence of mycorrhizas could increase the availability of limiting nutrients such as phosphorus, nitrogen and other minerals by facilitating their absorption and also enable them to acclimatize to *ex vitro* condition. Apart from these mycorrhization also influence the physiological parameters of micropropagated plants such as increased protein content, photosynthetic rates, decrease in respiratory rates.

REFERENCES

Azcon-Aguilar, C. & Barea, J. M. 1997. Applying mycorrhiza biotechnology to horticulture: significance and potentials. Sci. Hort. 68: 1-24.

Azcon-Aguilar, C., Cantos, M. & Troncoso, A. Barea, J. M. 1997. Beneficial effect of arbuscular mycorrhizas on acclimatization of micropropagated cassava plantlets. Sci. Hot. 72:63-71.

Berta,G., .Totta, A., Fusconi, A., Hooker, J. E., Munro, M., Atkinson, D., Giovanetti, M., Morini, S., Fortuna, P., Tisserant, B., Gianinazzi-Pearson, V. & Gianinazzi, S. 1995. Arbusular mycorrhizal induced changes to plant growth and root system morphology in *Prunus cerasifera.* Tree Physiol. 15: 281-293.

Bonga, J. M. 1977. Application of tissue culture in forestry, pp. 93-107. In: Reinert, J. and Bajaj, Y.P.S (Eds) Plant Cell Tissue and Organ Culture. Springer, New York/Berlin/Heidelberg.

Bougher, N. L. & Malajczuk, N. 1990. Effects of high soil moisture on formation of ectomycorrhizas and growth of karri (*Eucalyptus diversicolor*) seedlings inoculated with *Descolea maculata, Pisolithus tinctorius* and *Laccaria laccata.* New Phytol. 114: 87-91.

Branzanti, B., Gianinazzi-Pearson, V. & Gianinazzi, S. 1992. Influence of phosphate fertilization on the growth and nutrient status of micropropagated apple infected with endomycorrhizal fungi during the weaning stage. Agronomie 12: 841-845.

Casselle, A. C., Mark, G. L. & Periappuram, C. 1996. Agronomie 16: 625-632.

Charest, C., Dalpe, Y. & Brown, A. 1993. The Effect of Vesicular-Arbuscular Mycorrhizae and Chilling on two hybrids of *Zea mays* L. Mycorrhiza 4: 89-92.

Cordier, C., Trouvelot, A. Gianinazzi, S. & Gianinazzi-Pearson, V. 1996. Arbuscular mycorrhiza technology applied to micropropagated *Prunus avium* and to protection against *Phytophthora cinnamomi.* Agronomie 16: 679-688.

Cristoferi, G., Branzanti, B. & Zocca, A. 1986. Role of mycorrhizae on fruit root stock rooting. In Physiological and genetical aspects of mycorrhiza, In: V. Gianinazzi- Pearson and S. Gianinazzi (eds), INRA, Paris. Pp. 483-488.

de la Vina, G., Azcon-Aguilar C. Azcon, R. Barea, J. M. Barcelo A. & Pliego-Alfaro, F. 1996. Effects of arbuscular mycorrhizas in the acclimatization of micropropagated avocado plants. In: C. Azcon-Aguila., J.M. Barea (eds), Mycorrhiza in Integrated Systems. From Genes to Plant Development EC-DGXII, Brussells, pp. 491-494.

Dehne, H. W .1982. Interaction between vesicular-arbuscular mycorrhizal fungi and plant pathogens. Phytopathology 72:1115.

Diettrich, B., Mertinant, H. & Luckner, M. 1992. Biochem. Physiol. Pflanz. 188: 23-31.

Diez, J., Manjon, J. L. Kovacs, G. M. Kovacs, Celestino, C. & Toribio, M. 2000. Mycorrhization of vitroplants raised from somatic embryos of cork oak (*Quercus suber* L.). Applied Soil Ecology 15: 119-123.

Dolcet-Sanjuan, R., Claveria, E. Camprubi, A. Estaun, V. & Calvet, C. 1996. Micropropagated of walnut trees (*Juglans regis* L.) and response to arbuscular mycorrhizal inoculation. Agronomie 16: 639-645.

Flick, C. E., Evans, D. A. & Sharp, W. R. 1983. Organogenesis. In: D.A. Evans, W.R. Sharp, P.V. Ammirato and Y.Yamada (eds), Handbook of Plant Cell Culture, Vol I. McMillan, New York. pp. 13-81.

Fontana, A. 1961. Primo contributo allo studio delle micorrize dei pioppi in Piedmonte. Allionia 7: 87-129.

Fortuna, P., Citernesi, S. Morini, S. Giovanetti, M. & Loreti, F. 992. Infectivity and effectiveness of different species of arbuscular mycorrhizal fungi in micropropagated plants of Mr S 2/5 plum rootstock. Agronomie 12: 825-829.

Garbaye, J., Delwaulle, J. C. & Diangana, D. 1988. Growth response of *Eucalyptus* to ectomycorrhizal inoculation in the Congo. For. Ecol. Manag. 24: 151-157.

Gay, G. & Debaud, J. C. 1987. Genetic study on indole –3- acetic acid production by ectomycorrhizal *Hebeloma* species: Inter- and intra-specific variability in homo- and di-karyotic mycelia. Appl. Microbiol. Biotechnol. 26: 141-146.

Gay, G. 1990. Effect of ectomycorrhizal fungus *Hebeloma hiemale* on adventitious root formation in derooted *Pinus halepensis* shoot hypocotyls. Can. J. Bot. 68:1265- 1270.

Granger, R. L., Planchette, C. & Fortin, J. A. 1983. Effect of a vesicular-arbuscular (VA) endomycorrhizal fungus (*Glomus epigaeum*) on the growth and leaf mineral content of two apple clones propagated *in vitro.* Can. J. Plant. Sci. 63:551-555.

Grellier B., Letouze, R. & Strullu, D. G. 1984. Microporpagation of birch and mycorrhizal formation in vitro.New Phytol. 94: 591-599.

Guillemin, J.P., Gianinazzi, S. & Trouvelot, A. 1992. Screening of arbuscular endomycorrhizal fungi for establishment of micropropagated pineapple plants. Agronomie 12: 831-836.

Hacskaylo, E. 1967. Mycorrhizae. Indispensable invasions by fungi. *Agriculture Science Review* 5: 13-20.

Harris, M. M. & Jurgensen, M. F. 1977. Development of *Salix* and *Populus* mycorrhizae in metallic mine tailings. Plant and Soil 47: 509-517.

Heslin, M. C. & Douglas, G. C. 1986. Effects of ectomycorrhizal fungi on growth and development of poplar plants derived from tissue culture. Sci Hort. 30: 143-149.

Jansa, J. & Vosatka, M. 2000. *In vitro* and *post vitro* inoculation of micropropagated *Rhododendrons* with ericoid mycorrhizal fungi. Applied Soil Ecology 15: 125-136.

Laiho, O. 1970. *Pa.xillus involutus* as a mycorrhizal symbiont of forest trees. Acta Forest alia Fennica 106: 1- 72.

Lata, H., De Andrade, Z. Schaneberg, B. Bedir, E. Khan, I. & Moraes, R. 2003. Arbuscular mycorrhizal inoculation enhances survival rates and growth of micropropagated plantlets of Echinacea paliida. Palnta Medica. 69:679-782.

Leyval, C., Weissenhorn, I. L. Glashoff A. & Berthelin, J.1994. Influence of heavy metal on germination of arbuscular-mycorrhizal fungal spores in soils. Acta Bot. Gallica 141: 523-528.

Libby, W. J. & Rauter, R. M. 1984. Advantages of clonal forestry. For. Chron. 60: 145-149.

Lindermann, R. C. & Call, C. A. 1977. Enhanced rooting of woody plant cuttings by mycorrhizal fungi. J. Am. Soc. Hot. Sci. 102:629-632.

Lovato, P., Guillemin J. P. & Gianinazzi, S. 1992. Application of commercial arbuscular endomycorrhizal fungal inoculants to the establishment of micropropagated grapevine rootstock and pineapple plants. Agronomie 12: 873-880.

Malajczuk, N. & Hartney, V. J. 1986. Procedure for inoculation of micropropagated plantlets of *Eucalyptus camaldulensis* with ectomycorrhizal fungi, and comparison with seedling inoculation using inoculum contained in a peat/vermiculite carrier. Aust. For. Res. 16: 199-206.

Malajczuk, N., Lapeyrie F. & Garbaye, J. 1990. Infectivity of pine and eucalypt isolates of *Pisolithus tinctorius* on roots of *Eucalyptus urophylla in vitro.* New Phytol. 114: 627-633.

Mark, G. L., Hooker, J. E. Hahn, A. & Wheeler, C. T. 1999. *In vitro* culture of arbuscular mycorrhizal fungus and *Frankia* for inoculation of micropropagated *Casuarina equistifolia.* Can. J. Bot. 77:1391-1397.

Martin, M., Mari., A. Ibarra, M. & Garcia-Ferriz, L. 2003. Arbuscular mycorrhizal inoculation of micropropagated persimmon platlets. *J Hort. Sci. Biotechno[* 78:734-738.

Martins, A., Bapista, P. Sousa, J. & Pais M. S. 2001. Macro (N, P, K, Ca and Mg) and micronutrient (Fe, Mn, Cu and Zn) content of mycorrhizal and non-mycorrhizal micropropagated *Castanea sativa* Mill. Plants along three months of *in vitro* growth. ICOM-3, held at Adelaid, Australia.

Martins, A., Barroso, J. & Pais, M. S. 1996. Effect of ectomycorrhizal fungi on survival and growth of micropropagated plants and seedlings of *Castanea sativa* mill. Mycorrhiza 6:265-270.

Martins, A.., Casimiro, A. & Pais, M. S. 1997. Influence of mycorrhization on physiological parameters of micropropagated *Castanea sativa* Mill. Plants. Mycorrhiza 7: 161-165.

Marx, D. H. & Cordell, C. E. 1989. The use of specific ectomycorrhizas to improve artificial forestation practices. In: J.M. Whipps and R.D. Lumsden (eds), Biotechnology of Fungi for improving plant growth. British Mycological Society, Cambridge University Press. Cambridge, pp,1-25.

Miller, C.O. 1971. Cytokinin production by mycorrhizal fungi. Pages 168-174 in: E. Hacskaylo (Ed), Mycorrhizae, U.S. Dep. Agric. For. Serv. Misc. Publ. No.1189.

Molina, R. & Trappe, J. M. 1982. Patterns of ectomycorrhizal host specificity and potential among Pacific Northwest conifers and fungi. For. Sci. 28:423-458.

Monticelli, S., Puppi G. & Damiano, C. 2000. Effects of *in vivo* mycorrhization on micropropagated fruit tree root stocks. Applied Soil Ecology. 15: 105-111.

Mukerji, K. G. (ed.) 1996. Concepts in Mycorrhizal Research. Kluwere Academic Publishers, Dordrecht. Boston, London, p. 374.

Mukerji, K. G., Chamola, B. P. & Singh, J. (eds.) 2000. Mycorrhizal Biology. Kluwer Academic/Plenum Publishers, New York, Dordrecht, London, p. 336.

Naqvi, N.S. & Mukerji, K. G. 1998. Mycorrhization of micropropagated *Leucaena leucocephala* (Lam.) de Wit. Symbiosis 24: 103-113.

Navratil, S. & Rochan, G. C. 1981. Enhanced root and shoot development of poplar cuttings induced by *Pisolithus* inoculum. Can. J. For. Res. 11: 844-848.

Nowak, J. 1988. Benefits of *in vitro* biotization of plant tissue cultures with microbial inoculants. In vitro cell Dev. Biol-Plant 39:122-130.

Paradis, R., Dalpe, Y. & Charest, C. 1995. The combined effect of arbuscular mycorrhizas and short-term cold exposure on wheat. New Phytol. 129: 637-642.

Pinochet, J., Camprubi, A. Calvet, C. Fernandez C. & Kabana, R. R. 1998. Inducing tolerance to the root-lesion nematode *Pratylenchus vulnus* by early mycorrhizal inoculation of micropropagated myrobalan 29 C plum rootstock. J. Am. Soc. Hort. Sci. 123: 342-347.

Pinochet, J., Fernandez, C. Jaizme M. D. & Tenoury, P. 1997. Micropropagated banana infected with *Meloidogyne javanica* responds to *Glomus intraradices* and phosphorus. Hor Sci. 32: 101-103.

Poissonier, M. 1986. Mycorrhization in vitro de clones d'Eucalyptus. Note de Laboratoire. Ann. AFOCEL, 80-93.

Puthur, J. T., Prasad, K. V. S. K. Sharmila P. & Saradhi, P. P. 1998. Vesicular arbuscular mycorrhizal fungi improves establishment of micropropagated *Leucaena leucocephala* plantlets. Plant Cell tissue Organ Cult. 53: 41-47.

Rapparini, F., Baraldi, R. & Bertazza, G. 1996. Growth and carbohydrates status of *Pyrus communis* L. plantlets inoculated with *Glomus* sp. Agronomie 16: 653-661.

Rapparini, F., Baraldi, R. Bertazza, G. Branzanti, B. & Predieri, S. 1994. Vesicular-arbuscular mycorrhizal inoculation of micropropagated fruit trees. J. Horti Sci. 69: 1101-1109.

Ravolanirina, F., Blal, B. Gianinazzi, S. & Gianinazzi-Pearson, V. 1989. Mise au point d'une methode rapide d' endomycorhization the vitroplants. Fruit 44: 165-170.

Reddy, M. S & Satyanarayana, T. 1998a. Inoculation of micropropagated plantlets of *Eucalyptus tereticornis* with ectomycorrhizal fungi. New Forests 16:273-279.

Reddy, M. S & Satyanarayan, T. 1998b. Ectomycorrhizal formation in Micropropagated Plantlets of *Populus deltoids.* Symbiosis 25:343-348.

Riffle, J. w. & Tinus, R. W. 1982. Ectomycorrhizal characteristics, growth, and survival of artifically inoculated Ponderosa and Scots pine in a greenhouse and plantation. For. Sci. 3:646-660.

Schubert, A., Bodrino, C. & Gribaudo, I. 1992. Vesicular-arbuscular mycorrhizal inoculation of kiwifruit (*Actinidia deliciosa*) micropropagated plants. Agronomie 12: 847-850.

Schubert, A., Cammarata, S. & Eynard, I. 1988. Growth and root colonization of grapevines inoculated with different mycorrhizal endophytes. Hort Sci. 23: 302- 303.

Selosse, M.A., Le Tacon, F. 1998. The land flora: a phototroph-fungus partnershiop? Trends Ecol. 13: 15-20.

Sharma, M. P. & Adholeya, A. 2004. Effect of arbuscular mycorrhizal fungi and phosphorus fertilization on the post vitro growth and yield of micropropagated strawberry grown in a sandy loam soil. Can. J. Bot. 82: 322-328.

Shetty, K. G., Herick, B. A. D. & Schwab, A .P. 1995. Effects of mycorrhizae fertilizer amendments on zinc tolerance of plants. Env. Poll. 88:307-314.

Smith, S. E. & Read, D. J. 1997. Mycorrhizal symbiosis, 2nd Ed, Academic Press, London.

Srinath, J., Bagyaraj, D. J. & Satyanarayan, B. N. 2003 .Enhanced growth and nutrition of micropropagated Ficus benjamina to Glomus mosseae co-inoculated with Trichoderma harzianum and Bacillus coagulans. World J. Microbiol. Biotech. 19: 69-72.

Subramanian, K. S., Charest, C. Dwyer, L. M. & Hamilton, R. I. 1995. Arbuscular mycorrhizae and water relations in maize under drought stress at tasseling. New Phytol. 129: 643-650.

Supriyanto & Rohr, R. 1994. *In vitro* regeneration of plantlets of Scot pine (*Pinus sylvestris*) with mycorrhizal roots from subcultured callus initiated from needle adventitious buds. Can. J. Bot. 72: 1144-1150.

Sutler, E.G. 1988. J. Am. Soc. Hort. Sci. 113: 234-238.

Sylvia, D. M. & Williams, S. E. 1992. Vesicular-arbuscular mycorrhizae and environmental stress. Pages 101-124 in: G.J. Bethlenfalvay and R.G. Linderman (eds), Mycorrhizae in Sustainable Agriculture, ASA Special Publication Number 54, Madison Wisconsin.

Tonkin, C. M., Malajczuk, N. & McComb, J. A. 1989. Ectomycorrhizal formation by micro propagated clones of *Eucalyptus marginata* inoculated with isolates of *Pisolithus tinctorius.* New Phytol. 111: 209-214.

Trappe, J. M. 1962. Fungus associates of ectotrophic mycodhizae. The Botanical Review 28: 538-606.

Turner, P. D. 1962. Morphological influence of exudates of mycorrhizal and non- mycorrhizal fungi on excised root culture *Pinus sylvestris* L. Nature (London), 194:551-552.

Ulrich, J. M. 1960. Auxin production by mycorrhizal fungi. Physiologia Plantarum 13: 429-443.

Varma, A.& Schuepp., H. 1996. In K. S. Mukerji (ed), Concepts in Mycorrhizal Research, Kluwer Academic, London. 113-132.

Vestberg, M. 1992. Arbuscular mycorrhizal inoculation of micropropagated strawberry and field observations in Finland. Agronomie 12: 865-867.

Vestberg, M., Kukkonen, S. Saari, K. Parikka, P. Huttunen, J. Tainio, L. Devos, N. Weekers, F. Kevers, C. Thonarts, P. Lemoine, M. C. Cordier, C. Alabouvette C. & Gianinazzi., S. 2004. Microbial inoculation for improving the growth and health of micropropagated stawbwrry. Appl. Soil Ecol. 27: 243-258.

Vidal, M. T., GAzcon-Aguirar C. Barea, J. M. & Pliego-Alfaro, F. 1992. Mycorrhizal inoculation enhances growth and development of micropropagated plants of avocado. Hort Sci. 27: 785- 787.

Vozzo, J. A. & Hacskaylo, E. 1974. Endo and ectomycorrhizal associations in five *Populus* species. Bulletin of the Torrey Botanical Club 101: 182-186.

Wang, H., Parent, S. Gosselin, A. Desjardins, Y. 1993. Vesicular-arbuscular mycorrhizal peat-based substrates enhance symbiosis establishment and growth of three micropropagated species. J. Am. Soc. Hort. Sci. 118: 896-901.

19

Effect of Dual Inoculation of AM Fungi and *Bradyrhizobium japonicum* on Nodulation and Nitrogen Content in Soybean

S. NEUPANE and B.N. PRASAD

Biotechnology Unit Central Department of Botany
Tribhuvan University, Kritipur, Kathmandu, Nepal.
E-mail: prasadbn@tubraj.wlink.com.np

ABSTRACT

The experiment in pot culture was conducted to study the effect of AM fungi and *Bradyrhizobium japonicum* on nodulation and nitrogen content in nodule and shoot in soybean. In this investigation double symbiosis of nitrogen fixing bacteria (*B. japonicum*) inoculated with AM fungi in soybean increased the nodule number, nodule dry weight and nitrogen content in nodule as well as shoot found highest. The biochemical analysis showed the increment of nitrogen content in nodule was 35.09% and 39.95% on 35th and 60th day after seed sowing respectively over the control plant. Similarly, in shoot the increment of nitrogen content was 38.24% and 39.90% on 35th and 60th day after seed sowing respectively. These microorganisms stimulated the nodulation and also increased the total nitrogen accumulation in plant.

Key Words: AM fungi, *Bradyrhizobium japonicum*, nodulation, N content.

I. INTRODUCTION

Soybean has the capacity to fix large amounts of atmospheric nitrogen by symbiotic association with biological nitrogen fixer. The leguminous plants can benefit from the symbiotic association with *B. japonicum* as well as AM fungi. Rhizobial symbiosis provides the nitrogen to the plant necessary for the growth. Likewise, AM symbiosis enhances the plant growth through the ability of extraradial fungal hyphae to take up the low diffusing nutrients such as phosphorus from soil (Varma and Hock, 1998). Thus, it is used to improve the soil fertility and reduce the production cost for sustainable soil fertility.

The tripartite association of plant, AM fungi and *B. japonicum* significantly improves P nutrition, which requries for nodulation and nitrogen fixation. Legume requires high p levels for their growth and effective nodulation (Mclachlan *et al.*, 1975; Bergersen, 1971). However, the mycorrhiza acted both ways by p mediated nitrogen fixation mechanism and by directly enhancing N uptake from soil to improve nitrogen status of the plant (Barea *et al.*, 1988). Nitrogen plays a vital role in plant metabolism since it is the main component of amino acids and proteins.

The present paper reports the effect of AM fungi and *B. japonicum* on nodule number, nodule dry weight and nitrogen content in nodule and shoot.

II. MATERIALS AND METHODS

Seed of soybean cultivar "Sathiya" was collected from Agronomy Division of Nepal Agricultural Research Council (NARC), Khumaltar, Nepal. The inoculum of *B. japonicum* was provided by Soil Science Division of NARC. The mycorrhizal inoculum (resting spores) was isolated from the maize rhizosphere via wet sieving (Gerdmann and Nicolson, 1963) and sucrose centrifugation method (Jenkins, 1964; Daniels and Skipper, 1982; Gupta and Mukerji, 2002).

The pot culture experiment was conducted during August 2002 to October 2002 with four replications to study the response of microbial treatments. There were four inoculant treatments viz., plant inoculated with *B. japonicum* (+R), plants inoculated with AM fungi (+M), plants co-inoculated with *B. japonicum* and AM fungi (+R+M) and control (-R-M). The pots were placed under natural condition of light and temperature. The observations were taken on 35th and 60th day after seed sowing.

The percentage of mycorrhizal infection was determined after clearing roots in 10% KOH and staining with 0.05% trypan blue (Philips and Hayman 1970).

The nodule number and nodule dry weight were recorded. Nitrogen and content in nodule and plant tissue were estimated by Micro-Kjeldahl method (PCARR, 1980).

III. RESULT AND DISCUSSION

In soybean cultivar "Sathiya" the microbial treatments increased the nodule number, nodule dry weight, nitrogen content in nodule and plant tissue. The nodule number and nodule dry weight of plant inoculated with only AM fungi or *B. japonicum* was found more than uninoculated plant, where dual inoculation of both AM fungi and *B. japonicum* was found highest (Figs 1 and 2). This is because presence of suffient bradyrhizobial cells in the soil to nodulate the soybean. The precise exchange of molecular signals between bradyrhizobia and host plant is essential to the development of an effective nodulation (Varma, 1992). Several legumes grow poorly and fail to effective nodulation in autoclaved soil unless they were mycorrhizal (Asai, 1944). But in the case of uninoculated plant only ineffective nodulation occurs. The numbers and dry weight o these ineffctive nodules are less.

Nitrogen content on nodule and plant tissue in co-inoculated plant with AM fungi and *B. japonicum* found highest than the single inoculation of the AM fungi or *B.japonicum* and control plant. The increment percent of nitrogen in nodule of co-inoculated plant was 35.09% and 39.95% on day 35th and 60th after seed sowing respectively (Fig. 3). Similarly the increment percent of nitrogen in plant tissue of co-inoculated plant was found 38.24% and 39.90 on day 35th and 60th after seed sowing respectively (Fig. 4), over theuninoculated plant. This is because of AM fungi increase the effectiveness of *Rhizobium* on soybean (Ross and Harper, 1970) where AM fungi stimulate nitrogen fixation by *legume-Rhizobium* symbiosis (Daft and EI-Gialmi,

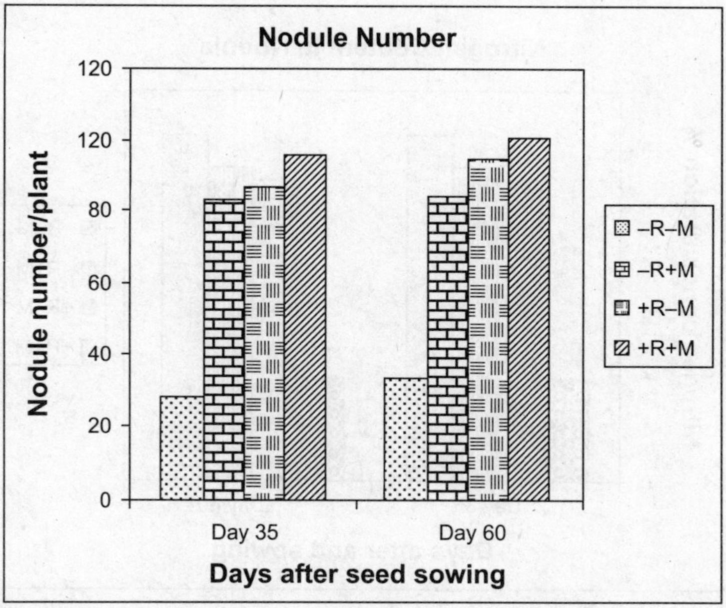

Fig. 19.1: Effect of microbial treatment on nodule number

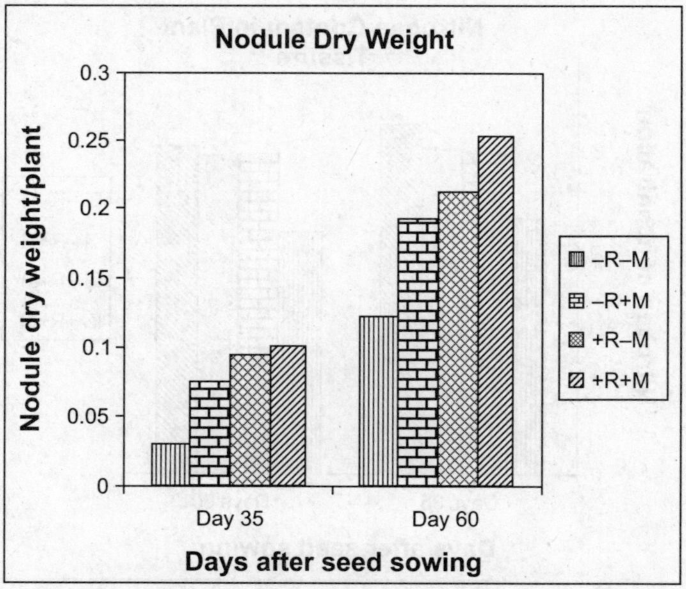

Fig. 19.2: Effect of microbial treatment on nodule dry weight

1974). AM fungi produce the some growth hormones like auxins, gibberellins, cytokinins. Gibberellins are found to stimulate the nodulation and *in vivo* nitrogenase activity (Prasad and Mathur, 1992). The additional phosphate uptake through fungal hyphae also stimulated the nitrogen fixation and accumulation of nitrogen content in nodule and plant tissue occurred.

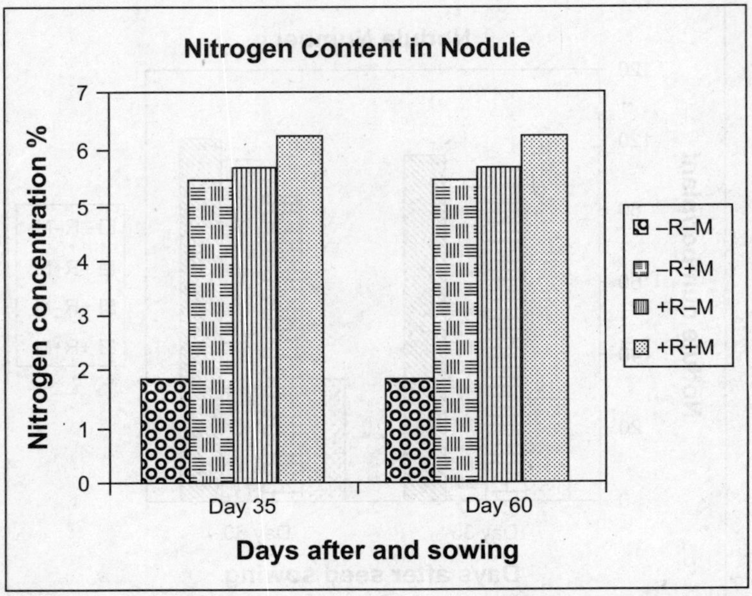

Fig. 19.3: Effect of microbial treatment on nitrogen content in nodule

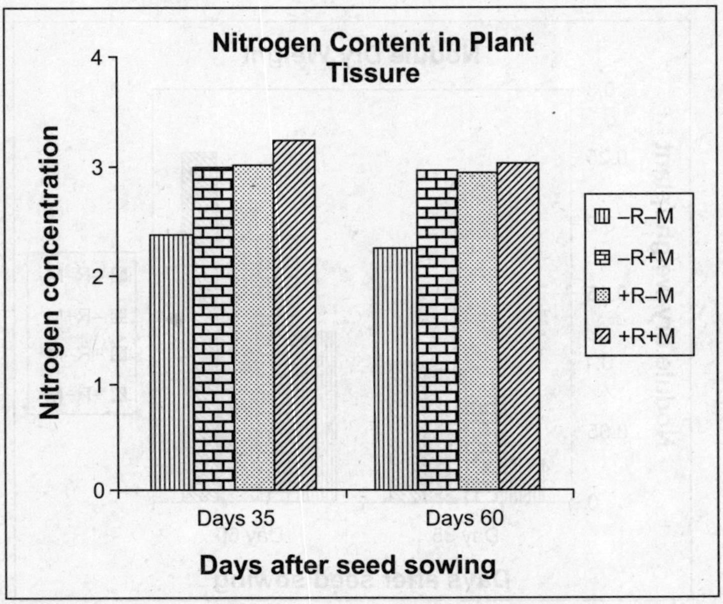

Fig.19.4: Effect of microbial treatment on nitrogen content in plant tissue

IV. CONCLUSION

AM fungi markedly improve the nodulation and nitrogen fixation by interaction with *B. japonicum,* a legume bacterium, mainly by providing the high phosphorus requirement for the fixation process. Thus, these microorganisms can be used to improve the soil fertility and reduce the production cost of soybean crop.

ACKNOWLEDGEMENTS

The first author want to express sincere thanks to Dr. B.R. Khadage, Senior Scientist, Plant Pathology division, Nepal Agricultural Research Council, Khumaltar, Nepal, for his kind cooperation to complete this work.

REFERENCES

Bergersen, F.T. 1971. Biochemistry of symbiotic nitrogen fixation in legume. Ann. Rev. Plant Physiology. pp 121-130.

Daniels, B.A. & H.D. Skipper. 1982. Methods for the recovery and quantitative estimation of propagules from soil. In: Methods and Principles of Mycorrhizal Research, N. C. Schenck, ed. The Amer. Phytopathol, Soc, St. Paul, Minnessota, U.S.A. pp. 29-35.

Gerdmann, J.W. & T.H. Nicolson 1963. Spores of mycorrhizal Endogone species extracted from soil by wet sieving and decanting. Trans. Bri. Mycol. Soc. 46: 235-244.

Gupta, R., & K.G. Mukerji. 2002. Techniques for the isolation of VAM/AM fungi is soil. In: Techniques in Mycorrhizal studies, K.G. Mukerji, C. Manoharachary & B.P. Chamola eds. Kluwer Academic Publishers, Dordrecht, pp. 201-216.

Jenkins, W.R. 1964. A. rapid centrifugal-flotation technique for separating nematodes from soil. Plant Dis. Rep. 48: 692-694.

Philips, J.M. & D.S. Hayman 1970. Improved procedure for clearing rootts and staining parasitic and vesicular-arbuscular mycorrhizal fungi for rapid assessment of infection. Trans. Briti. Mycol. **55**: 158-160.

Prasad, B. N. & S. N. Mathur 1992. Effect of GA_3 on nitrogen fixation in forage legumes in Nepal. In: *Role of Biotechnology and Agriculture* Prasad, Ghimire and Agrawal eds. Oxford and IBH Publication. pp. 85-90.

Ross, J.P. & J.A. Harper 1970. Effects of endogone mycorrhiza on soybean yields. Phytopathology **60**: 1552-1557.

Smith, N. C. & D. J. Read 1997. Mycorrhizal symbiosis (second edn.) Academic Press. p 605.

Varma, A & B. Hock (eds.) 1998. Mycorrhiza; strucure, functon, moledular biology and Biotechnology. Spring-Verlag, Germany.

Varma, D. P. S. 1992. Signals in root nodule organogenesis and endocytosis of *Rhizobium.* Plant and Soil 47: 373-382.

20

Microbial Degradation of Hazardous Dyes

VED PAL SINGH, LOKENDRA SINGH, ISHWAR SINGH and RAJESH KUMAR

Applied Microbiology and Biotechnology Laboratory,
Department of Botany, University of Delhi,
Delhi, INDIA
E-mail: vpsingh_biology@rediffmail.com, E-mail: lokendra121@rediffmail.com,
E-mail: ishwarsdedha@rediffmail.com, E-mail: rajesh_bot@yahoomail.com

ABSTRACT

Microorganisms are known to degrade various environmental pollutants by introducing either a minor change in their complex chemical structure or by fragmenting them into two or more simpler molecules. Dyes constitute one of the major groups of pollutants released into the environment as textile industrial effluent, which pose potential threat to both human and animal health. There are at least three categories of microbes, namely, fungi, bacteria and actinomycetes which have been reported to degrade an array of hazardous dyes. Mechanisms employed by microorganisms to degrade these compounds have not yet been understood completely. However, investigations suggest that most of the microorganisms degrade or remove various dyes from the environment either by adsorbing in biomass or by converting them enzymatically into non-toxic molecules or by both. In the light of all these parameters, the role of fungi, bacteria and actinomycetes in the degradation of hazardous dyes has been discussed in details in this review article.

I. INTRODUCTION

Microbial degradation refers to the breakdown of complex molecules by microorganisms to smaller and simpler ones, the original complex molecules being often environmentally objectionable. It is a biological process by which environmental pollutants are eliminated or converted into less toxic (or even useful) substances. Natural biodegradation is often largely catalyzed by indigenous microbial or plant populations in soil or aquatic ecosystems.

Microbial degradation is of three types: (i) a minor change in an organic molecule leaving the main structure still intact, (ii) fragmentation of a complex organic molecule in such a way that the fragments could be reassembled to yield the original structure and (iii) complete mineralization, which refers to transformation of organic molecules to mineral forms, including carbon dioxide or methane, plus inorganic forms of other elements that might have been contained in the original structures. All these three types of microbial degradation are depicted in Figure 20.1.

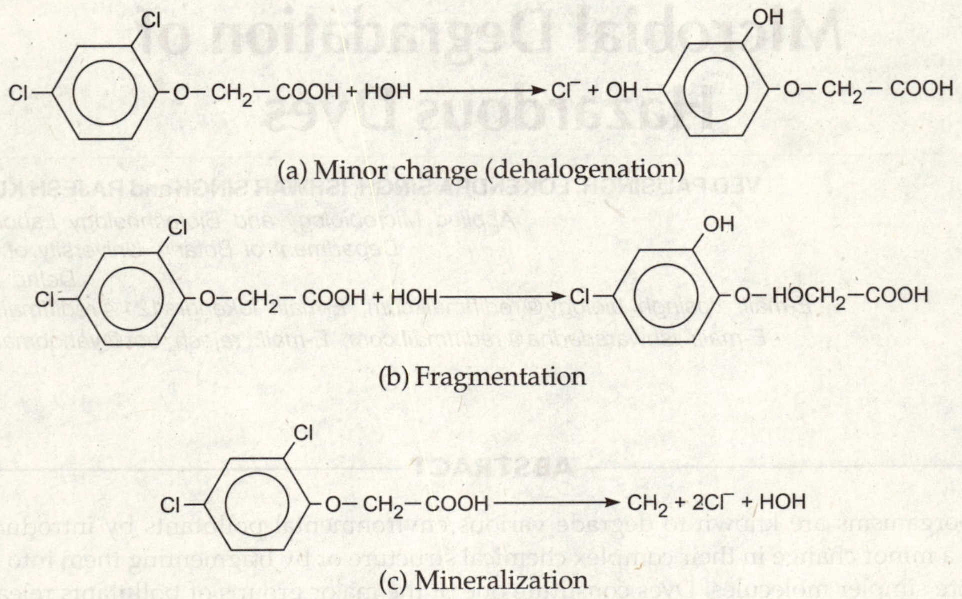

Fig. 20.1: Types of microbial degradation

Degradation of a complex compound takes place in several stages. In the case of halogenated compounds, dehalogenation often occurs early in the overall process. Dehalogenation of many compounds containing chlorine, bromine or fluorine occurs faster under anaerobic than under aerobic conditions. The study of reductive dehalogenation, especially its commercial application, is expanding rapidly. Once the anaerobic dehalogenation steps are completed, degradation of the main structure of many xenobiotics often proceeds more rapidly in the presence of O_2.

Observations of natural and synthetic organic compound accumulation in natural environment, however, raises questions about the ability of microorganisms to degrade these varied substances and the role of environment in protecting some chemicals. The chemical recalcitrance (resisting authority or control) resulted from the apparent fallibility of microorganisms, or their inability to degrade some industrially synthesized chemical compounds.

White-rot fungi are unique among eukaryotes in their ability to cleave carbon-carbon bonds in polycyclic aromatic hydrocarbons (PAHs). The biological degradation of these compounds formerly having been considered an exclusively bacterial process. The breakdown of most organo-pollutants by ligninolytic fungi is closely linked with ligninolytic metabolism, in that the process is stimulated by nutrient limitation, and it is generally thought that enzymes, whose normal function is lignin degradation, also catalyse the highly non-specific xenobiotic oxidation that is the characteristic of these organisms. The present article deals with microbial degradation of hazardous dyes, as these form an important source of pollutants of textile industrial effluents which pose potential threat to both human and animal health and environment.

II. MICROBIAL DEGRADATION OF DYES

Dye industries are the major source of pollution among the industrial effluents, which are directly poured into the water bodies. In India, dye industries play a major role to destroy the water quality. The wastage of dye industry and dye process is not only responsible for water contamination but also responsible for many diseases in living beings. In dye industry, the azo dye and many synthetic dyes are used, which are highly toxic and carcinogenic in nature due to their persistent nature in the environment. Azo dyes, one of the largest classes of dyes used in textile industry, are released into the aquatic and terrestrial environments, through the effluents emerging from textile and dyestuff industries and are normally not removed by conventional waste-water treatment system. The characteristic chemical structure of azo dyes makes them recalcitrant to biological break down.

Dye degradation or decolorization is an important task to the researchers. Microorganisms have the ability to degrade such types of inorganic and organic compounds which are used in dye industries and generally persistent in nature. There are at least three groups of microorganisms, namely, fungi, bacteria and actinomycetes whose role in microbial decolorization of dyes is well established.

Fungi

The use of different fungi in biosorption has been studied more extensively because of the availability of large amount of waste fungal biomass from fermentation industry and the amenability of these microorganisms to genetic and morphological manipulations. Few reports are available on the biological treatment of industrial dye effluents (Bonnarme *et al*, 1991; Buswell *et al*, 1995). Attempts have been made to decolorize dye waste water from the textile and the dyestuff industries. Up to 99% color removal was obtained by adsorption of the dye to the cells of filamentous fungi. Heinfling *et al* (1997) reported that *Bjerkandera adusta* and *T. versicolor* remove 95% of HRB 8 dye within four days.

Traditionally, fungi have been classified as white-, brown-, and soft-rot fungi on the basis of technical decay descriptions (Nilsson, 1988), regardless of their taxonomic position because the enzyme systems and metabolic pathways involved in the breakdown of carbohydrates and lignins probably are truly distinct in these fungi, rather than just modified in one or a few specific enzyme activities, decay type is of significant taxonomic importance. One important physiological characteristic of decay fungi in culture is production of extracellular phenoloxidases and peroxidases. Certain fungi produce brown diffusion zones in agar plates supplemented with 0.5% (w/v) gallic or tannic acid, as a result of oxidation of the respective phenolic acid by extra- or intra-cellular phenoloxidases. Bavendamm (1928) suggested that the presence of phenoloxidases was correlated with fungi causing white-rot decay and that only these fungi were able to completely decompose lignin. Davidson *et al* (1938) were the first to extend this method to 210 species of wood-decaying fungi. Of all tested white-rot fungi, 96% were positive on gallic acid agar, tannic acid agar, or both. Kaarik (1965) tested 173 wood-decaying species on the plates supplemented with 28 substances and found wide variations in the reactions of given strains to specific phenolic compounds. Decolorization of dye is related to the presence of extracellular peroxidases, particularly manganese peroxidases (Gold *et al*, 1988). However, the evidence relating degradation of polymeric dyes to phenol-oxidizing enzymes and to lignin degradation is largely circumstantial.

The basidiomycete *Phanerochaete chrysosporium* is a fungus with unusual degradative capabilities. This organism is termed a "white-rot fungus" because of its ability to degrade lignin, a randomly linked phenylpropane-based polymeric component of wood. *P. chrysosporium* possesses great potential for its commercial use in bioremediation of dyes and ligninocellulosic materials present in textile effluents, and now this fungus has become a model example for its biotechnological use in bioremediation. In fact, the rates of pollutant degradation are proportional to the concentration of the chemical. Several of the mechanisms used by white-rot fungi to degrade chemicals have recently been elucidated. The first step in the degradation of many chemicals by white-rot fungi often involves the formation of highly reactive free radical intermediates. Free radicals are formed any time; one electron is removed or added to the ground state of a chemical. Such free radicals are very reactive and will rapidly give up or abstract an electron from another chemical. Free radical reaction often occurs as chains in which many different radicals are generated subsequent to the formation of the initial radical species. This free radical process provides some basis for the non-specific nature of white-rot fungi. The free radical reactions catalyzed by the peroxidases from white-rot fungi also appear to be involved in the degradation of many pollutants.

The biodegradation of other commercially important classes of textile dyes have also being reported. Bumpus & Aust (1986) reported that *P. chrysosporium,* a ligninolytic white-rot fungus degrades a wide variety of structurally diverse organo-pollutants. This fungus has been shown to degrade polymeric dyes (Glenn & Gold, 1983), Crystal Violet (Bumpus & Brock, 1988), azo and heterocyclic dyes (Cripps & Bumpus, 1990). Another white-rot fungus, *Thelephora* sp. was also used for decolorization of azo dyes, such as Orange G, Congo Red, and Amido Black 10B (Selvam *et al*, 2003). Decolorization of various kinds of azo dyes by various fungi has been reported (Rafii *et al*, 1990). Cripps *et al* (1990) reported the aerobic degradation of three azo dyes (Congo Red, Orange II and Tropaeolin) by the fungus *P. chrysosporium*. The degradation of azo, anthraquinone, heterocyclic, triphenylmethane and polymeric dyes by *P. chrysosporium* has been intensively studied (Bumpus, 1989; Cripps *et al*, 1990; Ollikka *et al*, 1993). *P. chrysosporium* could remove 87-93% of Orange, Tropaeolin O and Congo Red within five days. Spadaro *et al* (1992) established that *P. chrysosporium* was capable of mineralizing a variety of toxic azo dyes and the mineralization of aromatic rings of azo dyes that was dependent on the nature of ring substituents.

The fungal lignin-degrading system was implicated in the decolorization process, since crude lignin peroxidase was required for the initial step of Orange II and Tropaeolin O decolorization. Paszczynski & Crowford (1991) showed that while ligninase recognized Acid Yellow 9 as a substrate, Mn(II) peroxidase was responsible for decolorization of other azo dyes. Further, they reported a new approach to enhancing aerobic azo dye transformation by *P. chrysosporium* and achieving azo dye decolorization by selected *Streptomyces* spp. The extracellular ligninolytic enzyme systems of these fungi have been directly linked to the degradation of these compounds.

The other major group of extracellular oxidative enzymes involved in the white-rot fungal lignin degradative process is laccases. The laccases from *Trametes. versicolor* can catalyze the initial oxidation step in the biotransformation of anthracene and benzo[a]pyrene. This catalysis appears to involve either a direct laccase oxidation mechanism or an indirect mechanism involving the participation of an oxidation mediator, such as ABTS, or a putative mediator present in the ultrafiltrate fraction. Under *in vivo* studies in *P. chrysosporium*, manganese peroxidases have been shown to play a major role in the initial break down and decolorization of high-molecular-weight chlorolignin in bleach plant effluents. *P. chrysosporium* may also be able to transform other naturally occurring polymers such as lignite and subbituminous coals (Paszczynski *et al*, 1992).

Freitage and Morell (1992) screened 170 strains of white-, brown-, and soft-rot decay fungi and non-decaying xylophilous fungi for phenoloxidase activity using polymeric dye — Poly-478. Their primary objectives were (i) to relate dye decolorization to know ligninolytic activity and the presence of phenoloxidase and peroxidases and (ii) to find out whether dye decolorization may serve as a simple additional method for identifying genera of basidiomycetes in culture. On the basis of their growth, these fungi were classified into fast-growing, slow-growing and little-growing fungi (Table 20.1).

Table 20.1: Three types of fungi based on technical decay description–white-rot fungi, brown-rot fungi and soft-rot fungi

	WHITE-ROT FUNGI
Fast-growing fungi	*Bjerkandera adusta, Ceriporiopsis rivulosa, C. subvermispora, Diplomitoporus lindbladii, Hyphoderma puberum, P. chrysosporium, P. sordida, P. gigantea, Phlebia concentrica, Ph. subserialis, Ph. tremellosa, Ph. albida, Ph. merismoides, Ph. rufa, Pycnoporus sanguineus, Trametes versicolor, T. villosa, T. hirsuta, Irpex lacteus.*
Slow-growing fungi	*Acanthophysium lividocoerulem, Ganoderma oregonense, Gymnopilus* sp., *Heterobasidion annosum, Lentinula edodes, Mycoacia uda, Peniophora* sp., *Peniophora pseudo-pini, Perenniporia subacida, Phellinus weirii, Phlebia rufa, Pleurotus ostreatus, Schizophyllum commune, Sistotrema brinkmannii, Stereum hirsutum, Stereum sanguinolentum, Trametes versicolor, Trichaptum abietinum, Trichaptum biforme, Tyromyces* sp.
Little-growing fungi	*Acanthophysium lividocoeruleum, Armillaria mellea, Bondarzewia montana, Cystostereum pini-canadense, Dichostereum pallescens, Heterobasidion annosum, Pholiota adiposa, Pleuroflammula puberula, Polyporus varius, Schizophyllum commune, Sistotrema octosporum, S. brinkmannii, Stereum gausapatum, Trichaptum biforme.*
	BROWN-ROT FUNGI
Fast-growing fungi	*Fomitopsis palustris, Gloeophyllum subferrugineum, Postia placenta, Crustoderma dryinum, Gloeophyllum trabeum, Laetiporus sulfureus, Neolentinus lepideus.*
Slow-growing fungi	*Antrodia carbonica, A. serialis, A. vaillantii, A. xantha, Coniophora puteana, Crustoderma dryinum, Fomitopsis cajanderi, F. officinalis, F. pinicola, F. rosea, Gloeophyllum sepiarium, G. striatum, Neolentinus lepideus, Phaeolus schweinitzii, Postia placenta.*
Little-growing fungi	*Antrodia radiculosa, A. serialis, Dacryobolus sudans, Postia sericeomollis.*

SOFT-ROT FUNGI	
Fast-growing fungi	*Gliocladium virense, Trichoderma polysporum, Aspergillus niger, Henningsomyces* sp.
Slow-growing fungi	*Alternaria alternata, Cladosporium elatum, Cunninghamella* sp, *Hormoconis resinae, Periconiella* sp., *Phialophora fastigiata, Phialocephala dimorphospora, Scytalidium* sp., *Ulocladium* sp., *U. botrytis, U. chartarum.*
Little-growing fungi	*Bispora betulina, Exophiala mansonii, E. jeanselmei, Pachnocybe* sp., *Paecilomyces variotii, Penicillium glabrum, Phialophora* sp.

The majority of the experiments on the degradation of dyestuffs by fungi have been carried out with either whole cultures or crude enzyme preparations of extracellular enzymes of the ligninolytic fungi. For example, many studies have demonstrated that the white-rot fungi, such as *P. chrysosporium, Bjerkandera adusta, Trametes versicolor* and *Phlebia radiata*, are able to degrade a broad spectrum of structurally diverse dyes. Decolorization of azo, anthraquinonic, heterocyclic, triphenylmethane, and polymeric dyes and their partial mineralization by enzymatic and non-enzymatic systems of these fungi have also been reported (Swamy & Ramsay, 1999a).

Bumpus *et al* (1985) reported that *Lentinus edodes* (Shiitake), a most popular edible mushroom, produces high amounts of manganes peroxidase (MnP, EC 1.11.1.14). The production of the extracellular ligninolytic enzymes is strongly affected by the nature and amount of the nutrients, especially nitrogen (N) and microelements, in the growth substrate (Buswell *et al*, 1995). The extracellular production of ligninolytic enzymes by *L. edodes* mycelium growing on a solid medium decolorized several dyes such as Poly-478 and Remazol Brilliant Blue R (RBBR). Yesilada (1995) reported decoloration of crystal violet by *Coriolus versicolor* and *Funalia trogii*. Bakshi *et al* (1999) found enhanced biodecoloration of synthetic commercial textile dyes by *P. chrysosporium* by improving Kirk's medium in respect to buffer, C:N ratio, Mg^{2+} and Zn^{2+}, temperature shifts, agitation, and sunflower oil addition. Bioremediation of dyes, containing toxic chlorinated phenols used in Kraft bleach dyeing, has been observed by several workers.

Use of ligninolytic enzymes were investigated but their yields were too low for industrial applications (Bajpai, 1999). Commercial enzymes, such as xylanases were produced in large quantities in *Trichoderma reesei* and *Streptomyces lividans* to be used for pulp bleaching (Bajpai, 1999). Application of xylanases was shown to decrease chlorine consumption, leading to a more environmentally friendly process, and an increase in the final brightness of the pulp. Other enzymes, such as laccases, have been studied for their *in vivo* capacity to degrade lignin (Mayer & Staples, 2002). Laccases are commercially available and produced in fungal strains such as *Aspergillus* (Berka *et al*, 1997; Yaver *et al*, 1991). *A. niger* has been used to produce feruloyl esterase FAEA by homologus expression for pulp bleaching application (Record *et al*, 2003).

Bacteria

Apart from fungi, bacteria can also degrade dyes. Various bacterial strains reduce azo dyes under anaerobic conditions to the corresponding amines (Mayer, 1981). The latter may be mineralized in a subsequent aerobic process. The unspecific anaerobic reduction of azo compounds in this system usually proceeds rather slowly and an acceleration of these reactions would facilitate possible technical applications. It has been demonstrated recently that the rate of anaerobic reduction of azo dyes by *Sphingomonas xenophaga* BN6 could be significantly increased by the addition of different quinines, such as anthraquinone-2-sulfonate or 2-hydroxy-1,4-naphthoquinone. Thus, it was demonstrated that the addition of naphthaquinone and "natural organic matter" could significantly enhance the reduction rate of nitro aromatic compounds and hexachloroethane in the presence of "bulk reductant"(e.g., H_2S). Furthermore, it has been shown that strictly anaerobic Fe(III)-reducing bacteria use the reduction of quinine moieties of humic substances (and also sulfonated anthraquinones) to transfer the reduction equivalents released during the anaerobic oxidation of organic substances. Thus, it becomes evident that in the presence of redox mediators, many heterotrophic aerobic bacteria decolorize azo dyes under anaerobic conditions. Quinones may undergo one-electron reduction processes to the corresponding hydroquinone radicals or two-electron reduction to the corresponding hydroquinones. Therefore, one-electron potentials can be used to compare the relative tendency of different quinines to take up reduction equivalents.

The bacteria, which reduce azo dye, Amaranth under anaerobic conditions, include Gram-negative bacteria (*Sphingomonas xenophaga* BN 6, *Agrobacterium tumefaciens, Ralstonia eutropha* 335, *Hydrogenophaga palleronii, Escherichia coli* K12 and *Flexibacter filiformis)*, Gram-positive bacteria (*Bacillus subtilis, Rhodococcus erythropolis* and *Lactobacillus plantarum*) and Archea (*Halobacterium salinarum*). The toxicity potential and decolorization of three acid azo dyes (Acid Orange 6, Acid Orange 7, and Acid Orange 52) by methanogenic granular sludge from an anaerobic expanded granular sludge bed reactor was assayed. Complete bioreduction was found for all three azo dyes. Sulfanilic acid and 4-aminoresorcinol were detected from the decolorization of Acid Orange 6, sulfanilic acid and 1-amino-2-naphtol were detected from the reduction of Acid Orange 7, and sulfanilic acid and N,N-dimethyl-1,4-phenylenediamine (DMP) were found to be intermediates of Acid Orange 52 degradation. Sulfanilic acid and 1-amino-2-naphtol were persistent in the anaerobic conditions, whereas 4-aminoresorcinol was completely mineralized by anaerobic sludge and DMP was transformed into 1,4-phenylenediamine. Enrichment cultures obtained via consecutive passages on basal medium with only azo dye as a carbon and an energy source seemed to be morphologically heterogeneous. Baculiform and coccus cells were found when viewed under a light microscope. Cocci were joined in chains. Because anaerobic sludge contains sulfate-reducing bacteria and, therefore, may generate sulfide, azo dyes were tested for chemical decolorization by sulfide to compare rates of chemical and biologic reduction (Yemashova *et al*, 2004).

Actinomycetes

The ability of actinomycetes to decolorize several anthraquinone and azo dyes was reported more then a decade ago (Rafii *et al*, 1990). The anthraquinone dyes Remazol Brilliant Blue R, Poly B-411 and Poly R-478 were decolorized by strains of *Streptomyces* and *Thermomonospora*, while *Streptomycetes* also decolorized the azo dyes 4-(3-methoxy-4-hydroxy-phenylazo)-azobenzene-3,4′-disulfonic acid, 3-methoxy-4-hydroxy-azobenzene-4′-sulfonic acid and Orange I. Actinomycetes can catalyze hydroxylations, O, N and S oxidations and O- and N-dealkylation reactions against various xenobiotic compounds. *Streptomyces chromofuscus*, a lignocellulolytic microorganism belonging to actinomyces group, is capable of degrading azo dye in presence of veratryl alcohol (Paszczynski & Crowford, 1991). The ability of *S. chromofuscus* to mineralize certain of the dyes was significant, since azo dyes are resistant to aerobic degradation by most other soil bacteria. The mineralization of a guaiacol-substituted azo dye by *Streptomyces chromofuscus* has also been shown.

The genus *Streptomyces* has been reported to degrade benzene derivatives via classic aromatic catabolic pathways. Little research has been devoted to determining whether actinomycetes efficiently degrade condensed polycyclic aromatics, although some strains are able to metabolize naphthalene derivatives. *Streptomyces* spp. degrade some recalcitrant compounds such as carbamates, diazinon and bromoxylin. Actinomycetes have been shown to degrade organochlorine compounds in spent sulfite bleach plant effluent and a correlation between the abilities of *Streptomycetes* to decolorize anthron-type dyes and the degradation of lignocellulose has been examined. Some evidence indicates that *Streptomyces* can act synergistically with other soil microorganisms to degrade recalcitrant compounds (Gunner & Zuckman, 1968). *Streptomyces* spp. degradability depends on the substitution pattern of the aromatic ring. Pasti & Crowford (1991) showed that lignocellulolytic *Streptomyces* spp. are also able to partly mineralize azo dyes. Bacterial cytochrome P-450 is believed to catalyze most of these reactions.

III. MECHANISM OF DYE DEGRADATION

Our understanding about mechanism of microbial degradation and decolorization of azo and reactive dyes is still incomplete due to limited investigation. However, results indicate that maximum microorganisms adopt reductive process of degradation. In this endeavor, Kulla *et al* (1983) employed strains of *Pseudomonas* in chemostat culture for removal of dyes. Some anaerobic bacteria and *Streptomyces* have been characterized for decolorization of chromogenic dyes. A basidiomycetous fungus *Ganodermaa lucidum* has been found as a suitable organism for removal of Rhodamine-B and Sandolan Rhodine. The decolorization of dyes, containing toxic chlorinated phenols used in Kraft bleach dyeing, has been observed by several workers. Bennette (1971) reported that brownish color of the effluent in the textile industry is due to chlorolignin.The ability of microfungi to degrade this component of the effluent,which is an enzymatic process, has been studied by Cammarota & Santa-Anna Jr. (1992).

Adsorption of dyes to the microbial cell surface is the primary mechanism of decolorization (Knapp *et al*, 1995). Absorption was found to be important mechanism that contributes to the decolorization process, although there is a possibility that initial transformation of the dye, hence degradation, was also involved. Microscopic examinations showed that the fungal spores, instead of the fungal hyphae, were actually absorbing the dye. A hydrophobic-hydrophillic interaction between the fungus and the dye might contribute to the absorption phenomenon. The dye color decreased, when either the enzyme or the cell mass concentrations were increased, indicating that the dye degradation occurred both by enzymatic action as well as adsorption to the cell mass. The concerted action of both led to greater degradation than when they occurred singly.

Wong & Yu (1999) reported the adsorption of Acid Green 27, Acid Violet 7 and Indigo Carmine dyes on living and dead mycelia of *Trametes versicolor*. Also, enzymes such as lignin peroxidase (LiP), manganese-dependent peroxidase (MnP) and laccase, all of which are involved in lignin degradation, have been reported to decolorize dyes (Vyas & Molitoris, 1995). Dyes with different structures are decolorized at different intrinsic enzymatic rates. Kim *et al*

(1996) demonstrated the presence of H_2O_2-dependent enzyme activity declourizing Remozol Brilliant Blue R in the culture filtrate of *Pleurotus ostreatus* in a chemically defined medium. Yong & Yu (1997) suggested the binding of dyes to the fungal hyphae, physical adsorption and enzymatic degradation by extracellular and intracellular enzymes as major mechanisms for the color removal. Kim & Shoda (1999) have purified and characterized the novel peroxidase (Dyp) that is responsible for the dye decolorizing activity of *Geotrichum candidum* Dec 1. Nine of the 21 types of dyes were decolorized by Dec 1, and in particular, anthroquinone dyes were highly decolorized. Swamy & Ramsay (1999b) reported that in fungus, *Trametes versicolor*, lignin peroxidase (LiP) was not detected during decolorization of the azo dye of Amaranth, while laccase and manganese peroxidase (MnP) were detected in the decolorizing cultures. A white-rot fungus, *Thelephora* sp. was isolated from the Western Ghats of South India and characterized for its lignolytic enzymes (Selvam, 2000).

Several amino-substituted azo dyes including 4-phenylazoaniline and N-methyl- and N,N-dimethyl-4-phenylazo anilines are mutagenic as well as carcinogenic. In mammals, azo dyes are reduced to aryl amines by cytochrome p-450 and a flavin-dependent cytosolic reductase (Fujita & Peisach, 1977).

Secretion of azoreductases, which catalyse azo-linkage, has been discovered in *Pseudomonas* strains. The degradation mechanism described for *Pseudomonas* spp. involves an oxygen-insensitive azoreductase, which catalyzes the reductive cleavage of the azo group, using NAD(P)H as electron donor. Various anaerobic bacteria have been reported to degrade azo dyes. However, under aerobic condition these dyes have been considered to be essentially non-biodegradable. Cripps & Bumpus (1990), however, reported that *P. chrysosporium* degrades polycyclic hydrocarbons, containing azo and sulphur groups. Paszczynski *et al* (1992) adopted a new approach to increase susceptibility of azo dyes to degradation by aerobic microorganisms, especially by *Streptomyces* spp. and *P. chrysosporium*. They made azo dyes more degradable by introduction of a metabolizable substituent into the dye's chemical structure. It was found that lignase and manganese peroxidase preparations from *P. chrysosporium* were involved in dye degradation. Ferrer *et al* (1991) reported that *P. chrysosporium* produces multiple lignin peroxidase isoenzymes. They purified and characterized these heme-containing peroxidases. Leisola *et al* (1987) used analytical isoenzyme focusing to distinguish lignin peroxidase (LiP) enzymes in the extracelluar growth medium. The enzyme secretion by the fungus has been reported to increase by immobilizing the fungus on polyurethane foam (PUF). Buswell *et al* (1995) have demonstrated that *P. chrysosporium* INA-12 produces high LiP activity under nitrogen sufficiency, when glycerol is used as a carbon source. Supplementing the medium with oleic acid and soybean phospholipids as well as temperature shifts can also enhance LiP production. In a controlled experiment, Ferrer *et al* (1991) compared the color removal efficiency of lyophilized fungal culture with those of lignin peroxidase and horse-radish peroxidase. A well-defined lignin-degrading enzyme system, consisting of lignin peroxidase (LiP), Mn (II)-dependent peroxidase (MnP) and glyoxal oxidase, was found to be involved in the dye degradation. The LiP and MnP are glycosylated heme-proteins containing photoporphyrin IX. Enhanced secretion of LiP and MnP in immobilized cell cultures of *P. chrysosporium* has been reported by Bonnarme *et al* (1991). These authors discovered that soybean phospholipid fractions in the culture of *P. chrysosporium* increase succinate dehydrogenase, a mitochondrial marker and cytochrome-oxidoreductase, an endoplasmic marker.

Production of LiP and MnP in a low shear environment has been successfully achieved in airlift bioreactors. *P. chrysosporium* degrades a wide variety of structurally diverse organo-pollutants (Bumpus & Aust, 1986), through its non-specific H_2O_2-dependent extracellular lignin-degrading enzyme system was developed during secondary metabolism in nitrogen-limited medium.

Seshadri *et al* (1994) proposed that biotransformation of azo dyes is a two-step process, in which the azo bond is reduced under anaerobic conditions, producing two aromatic amines, which are then mineralized by aerobic microorganisms. In contrast to this proposed process, subsequent research showed that in the presence of readily utilizable cosubstrates, two azo dyes were reduced and decolorized under methanogenic conditions (Razo-Flores *et al*, 1997), and break down products from one azo dye were further mineralized. Plumb *et al* (2001) hypothesized that coenzyme-reducing equivalents involved in normal electron transport through oxidation of organic substrates may act as electron donors for reducing of azo dyes. This could explain the observation that the azo dye reduction or degradation occurs more readily as a cometabolic events, when additional readily degradable substrates are provided. When a readily utilizable synthetic feed was provided together with the dye waste, it is hoped that metabolism of the dye compound in the waste would be enhanced because of the greater metabolic activity due to degradation of synthetic feed components. Recent research has shown not only that the sulfide is able to chemically decolor azo dyes, but also that the azo reduction in living sulfidogenic anaerobic sludge is three times more rapid than the chemical reaction due to sulfide alone (Van der Zee *et al*, 2000).

IV. CONCLUSION

Microorganisms can remove organic pollutants including dyes from waste-water, leading to decrease in their toxicity. Different microbes degrade xenobiotics of health and environmental concern due either to adsorption mechanism or enzymatic degradation mechanism or both. This chapter highlights the role of major groups of microbes i.e., fungi, bacteria and actinomycetes, which are involved in degradation of dyes. However, essential works on this subject are still confined at laboratory level and are of less industrial value.

ACKNOWLEDGEMENTS

The fellowship granted to one of us (RK) by the University Grant Commission, New Delhi is thankfully acknowledged.

REFERENCES

Bajpai, P. 1999. Application of enzymes in the pulp and paper industry. Biotechnol. Prog. 57: 147-157.

Bakshi, D.K., K.G. Gupta & P. Sharma 1999. Enhanced biodecolorisation of synthetic textile dye effulent by *Phenerochaete chrysosporium* under improved condition. World. J. Microbiol. Biotechnol. 15: 507-509.

Bavendamm, W. 1928. Uber das vorkommen und den nachweis von oxydasen bei holzzerstorenden pilzen. Z. Pflanzenkr. Pflanzenschutz. 38: 257-276.

Bennett, D.J. 1971. Degradation of polycholorinated biphenyles by extracellular enzymes of *Phenerochaete chrysosporium* produced in a perforated plate bioreactor. Tappi. J.12: 2019-2026.

Berka, R.M., P. Schneider.; E.J. Golightly.; S.H. Brown.; M. Madden.; K.M. Brown.; T. Halkier.; K. Mondorf & F. Xu. 1997. Characterization of the gene encoding an extracellular laccase of *Myceliophthora thermophila* and analysis of the recombinant enzyme expressed in *Aspergillus oryzae.* Appl. Environ. Microbiol. 63: 3151-3157.

Bonnarme, P., M. Delatter, G. Corriev & M. Asthar. 1991. Peroxidase secreation by pellets or immobilized cells of *Phenerochaete chrysosprorium* BKM-F-1767 and INA-12 in relation to organelle content. Enzyme Microbial. Technol. 13: 727-733.

Bumpus, J.A., M. Tien.; D. Wrigh & S.D. Aust. 1985. Oxidation of persistent environmental pollutants by a white-rot fungus. Science 228: 434-1436.

Bumpus, J.A. & S.D. Aust. 1986. Biodegradation of environmental pollutants by the white-rot fungus *Phanerocheate chrysosporium.* Involvement of lignin-degrading system. Bioassay 6: 166-170.

Bumpus, J.A. & B.J. Brock. 1988. Biodegradation of crystal violet by the white-rot fungus *Phanerocheate chrysosporium.* Appl. Environ. Microbiol. 54: 1143-1150.

Bumpus, J.A. 1989. Biodegradation of polycyclic aromatic hydrocarbons by *Phanerocheate chrysosporium.* Appl. Envioron. Microbiol. 55: 154-158.

Buswell, J.A., Y.J. Cai & S.T. Chang. 1995. Effect of nutrient N and manganese on manganese peroxidase and laccasse production by *Lentinula (Lentinus) edodes.* FEMS Microbiol. Lett. 128: 81-88.

Cammarota, M.C. & G.L. Santa-Anna Jr. 1992. Decolorization of Kraft bleach plant E 1 stages effluents in fungal Bioreactor. Environ. Technol. 13: 65-71.

Cripps, C. & J.A. Bumpus. 1990. Biodegradation of azo and heterocyclic dyes by *Phanerochaete chrysosporium.* Appl. Environ. Microbiol. 56: 1114-1118.

Davidson, R.W., W.A.Campbell & D.J. Blaisdell. 1938. Differentiation of wood-decaying fungi by their reactions on gallic or tannic acid medium. J. Agric. Res. 57: 683-695.

Ferrer, I., M. Dezotti & N. Duran. 1991. Decolorization of Kraft effluent by free and immobilized ligin peroxidases and horseradish peroxidase. Biotechnol. Lett. 13(8): 577-582.

Freitag, M & J.J. Morrell. 1992. Decolorization of the polymeric dye poly R-478 by wood-inhabiting fungi. Can. J. Microbiol. 38: 811-822.

Fujita, S. & J. Peisach. 1977. Electron transfer between liver microsomal cytochrome b_5 and cytochrome p-450 in the azo reductase reaction. Biochem. Biophys. Res. Commun. 78: 328-325.

Glenn, J.K. & M.H. Gold. 1983. Decolorization of several polymeric dyes by the lignin-degrading basidiomycete *Phanerochaete chrysosporium.* Appl. Environ. Microbiol. 45: 1741-1747.

Gold, M.H., J.K. Glenn & M. Alic. 1988. Use of polymeric dyes in lignin biodegration assays. Methods Enzymemol. 161: 74-78.

Gunner, H.B. & B.M. Zuckman. 1968. Degradation of diazinon by synergistic microbial action. Nature (London) 217: 1183-1184.

Heinfling, A., M. Bergbauer & Szewyk. 1997. Biodegradation of azo and phthalocyanine dyes by *Trametes versicolor* and *Bjerkandera adusta.* Appl. Microbiol. Biotechnol. 48: 261-266.

Kaarik, A. 1965. The identification of the mycelia of wood-decaying fungi by their oxidation reactions with phenolic compounds. Stud. For. Suec. 31: 1-80.

Kim, S.J., K. Ishikawa.; M. Hirai & M. Shoda. 1996. Characteristic of a newly isolated fungus, *Geotrichum candidum* Dec 1, which decolorizes various dyes. J. Ferment. Bioeng. 79: 601-607.

Kim, S.J. & M. Shoda. 1999. Purification and characterization of a novel peroxidase from *Geotrichum candidum* Dec.1 involved in decolorization of dyes. Appl. Environ. Microbiol. 65: 1029-1035.

Knapp, J.S., P.S. Newby & L.P. Reece. 1995. Decolorization of dyes by wood-roting basidiomycete fungi. Enzyme Microbial. Technol.17: 664-668.

Kulla, H.G., F. Klausener., U. Meyer., B. Ludeke & T. Leisinger. 1983. Interference of aromatic sulfo groups in microbial degradation of azo dyes orange I and orange II. Arch. Microbiol. 135: 1-7.

Leisola, M.S.A., B. Kozulic., F. Meussdoeffer & A. Eiechter. 1987. Homology among multiple extracellular peroxidase from *Phanerocheate chrysosorium.* J. Biol. Chem. 262: 419-424.

Mayer, A.M. & R.C. Staples. 2002. Laccase: new function for an old enzyme. Phytochemistry. 60: 551-565.

Meyer, U. 1981. Biodegradation of synthetic organic colorants, In Microbiol Degradation of Xenobiotics and Recalcitrant Compounds, eds. Leisinger T, Nilsson, T. 1988. Defining fungal decay types–final proposal. International research group on wood preservation, Stockholm, Sweden. Document IRG/WP/1355.

Ollikka, P., K. Alhonmaki., V.M. Leppanen., T. Glumoff., T. Raijola & L. Suominen. 1993. Decolorization of azo, triphenylmethane, heterocyclic, and polymeric dyes by lignin peroxidase isoenzymes from *Phanerochaete chrysosporium*. Appl. Environ. Microbiol. 59: 4010-4016.

Pasti, M.B. & D.L. Crowford. 1991. Relationship between the ability of Streptomycetes to decolourize three anthron-type dyes and to degrade lignocellulose. Can. J. Microbiol. 37: 902-907.

Paszczynski, A. & R.L Crowford. 1991. Degradation of azo compounds by ligninase from *Phanerocheate chrysosporium*: involment of veratryl alcohol. Biochem. Biophys. Res. Commun. 178: 1056-1063.

Paszczynski, A., M.B. Pasty-Grigsby., S. Goszczynske., R.L. Crawford & D.L Crawford. 1992. Mineralization of sulfonated azo dyes and sulfanilic acid by *P. chrysosporium* and *Streptomyces chromofuscus.* Appl. Environ. Microbiol. 58: 3598-3604.

Plumb, J.J.; J. Bell & D.C. Stuckey. 2001. Microbial populations associated with treatment of an industrial dye effluents in an anaerobic baffled reactor. Appl. Environ. Microbiol. 67: 3226-3235.

Rafii, F., W. Franklin & C.E. Cerniglia. 1990. Azo reductase activity of anaerobie bacteria isolated from human intestinal microflora. Appl. Environ. Microbiol. 56: 2146-2151.

Razo-Flores, E., M. Luijten., B. Donlon., B. Lettinga & J. Fiel. 1997. Biodegradation of selected azo dyes under methanogenic conditions. Water. Sci. Technol. 36: 65-72.

Record, E., M. Asther., C. Sigoillot & S. Pages. 2003. Overproduction of the *Aspergillus niger* feruloyl esterase for pulp bleaching application. Appl. Microbiol. Biotechnol. 62: 349-355.

Selvam, K. 2000. Biotechnical application of some white-rot fungi *Fomes lividus, Thelephora* sp. and *Trametes versicolor*. Ph.D. Thesis. Bharathiar University, India.

Selvam, K., K. Swaminathan & K.S. Chae. 2003. Decolourization of azo dyes and a dye industry effluent by a white-rot fungus *Thelephora* sp. Bioresource. Tech. 88: 115-119.

Seshadri, S., P.L. Bishop & A.M. Agha. 1994. Anaerobic/aerobic treatment of selected azo dyes in wastewater. Waste. Manag. 15: 27-137.

Spadaro, J.T., M.H. Gold & V. Renganathan. 1992. Degradation of azo dyes by the lignin-degrading fungus *Phanerochaete chrysosporium.* Appl. Environ. Microbiol. 58: 2397-2401.

Swamy, J. & J.A. Ramsay. 1999a. The evaluation of white-rot fungi in the decolouration of textile dyes. Enzyme. Microbial. Technol. 24: 130-37.

Swamy, J. & J.A Ramsay. 1999b. Effect of glucose and NH_4^+ concentrations on sequential dye decolouration by *Trametes versicolor*. Enzyme. Microbial. Technol. 25: 278-284.

Van der Zee, F.P., G. Lettinga & J.A. Field. 2000. The role of (auto) catalysis in the mechanism of an anaerobic azo reduction. Water. Sci. Technol. 42: 301-308.

Vyas, B.R.M. & H.P. Molitoris. 1995. Involvement of an extracellular H_2O_2-dependent ligninolytic activity of the white-rot fungus *Pleurotus ostrus* in the decolourization of remazol brilliant blue R. Appl. Environ. Microbiol. 61: 3919-3927.

Wong, Y.X. & J.Yu. 1999. Laccase-catalysed decolorization of synthetic dyes. Water Res. 33: 3512-3520.

Yemashova, N., A. Telegina., I. Kotava., A. Netrusov & S. Kalyuzhnyi. 2004. Decolorization and Partial Degradation of Selected Azo dyes by Metanogenic Sludge. Appl. Biochem. Biotechnol. 119: 31-40.

Yever, D.S., M.D. Carmen Overjero., F. Xu., B.A. Nelson., K.M. Brown., T. Halkier., S. Bernauer., S.H. Brown & S. Kauppinen. 1991. Molecular characterization of laccase genes from basidiomycete *Coprinus cinereus* and heterologous expression of the laccase Lec 1. Appl. Environ. Microbiol. 65: 4943-4948.

Yesilada, O. 1995. Decolourization of crystal violet by fungi. World J. Microbiol. Biotechnol. 11: 601-602.

Young, L. & J.T. Yu. 1997. Ligninase-catalysed decolourization of synthetie dyes. Water Res. 31: 1187-193.

21

Composting Technology: Significance and Potential

D.J. BAGYARAJ, S. SHARATH and A.N. BALAKRISHNA

Department of Agricultural Microbiology,
University of Agricultural Sciences,
GKVK Campus, Bangalore, India

ABSTRACT

Composting though followed through generations is being refined over time. Humans have directly or indirectly involved in composting with or without technical knowledge. However, post-green revolution, people have found the grave need for organic amendments to soil for sustainable agriculture. With the present increased yearn for compost makes it imperative for hastened production to fill abyss requirement.

It is critical to understand the technicalities required in compost production, right from the substrate availability to the final evaluation of the compost. In this review chapter, a panoramic view is given about the realm of compost. A successful compost technology can be standardized through sound knowledge of substrate availability, its composition, micro- and macroorganisms involved and their action, methods available for composting, factors influencing and methods available to evaluate compost for maturity.

Some of the industries like distillery, paper mills etc. find it difficult to dispose the waste materials due to environmental concerns. These industries are now depending on composting technology to convert their waste products (distillery spent wash, press mud, wood wastes etc.) to value added products in a short period of time for faster efficiency to handle bulk of the waste produced. Plethora of alternative substrates has been evaluated to serve as inputs for compost preparation. All the necessary minuscules are standardized for better efficiency of compost production keeping in mind the concern for Mother Nature.

I. INTRODUCTION

Two decades after independence we were proud of green revolution, which was necessary to feed the growing population in the country. With the green revolution high yielding varieties of crop plants were introduced. For cultivating high yielding varieties, which demand higher levels of nutrients and plant protection measures, it became essential to use higher levels of chemical fertilizers and pesticides. Now, after a few decades, we have realized that use of higher level of chemicals is hazardous not only to the soil, animal and human health but also deleterious to the environment. Hence, sustainable agriculture or organic farming wherein biological inputs are maximum and chemical inputs are minimal, is drawing the attention of farmers across the world including our country. For replacing chemical fertilizers, which supply plant nutrients by organic manure, the basic requirement is the availability of raw material i.e. the organic matter or biomass that can be decomposed to organic manure. The question that is often asked is what are the different sources of available biomass and how much of it is available? Different sources of biomass can be categorized as

1. Livestock and human wastes: Cattle shed wastes, other livestock and human excreta, slaughter house waste, piggery and poultry wastes.
2. Crop residues, tree wastes and aquatic weeds: cereal residues, sugarcane trash, forest/ tree litter, aquatic weeds, groundnut shells etc.
3. Green manures: *Gliricidia maculata, Pongamia glabra, Calotropis gigantea, Tephrosia purpurea, T. candida, Indigofera teysmanni, Cassia tora, Sesbania speciosa, Ipomoea carnea* etc.
4. Urban and rural wastes: Solid wastes, sewage and sullage, biogas slurry, kitchen wastes etc.
5. Agro-industrial wastes: Press mud, bagasse, paper and pulp wastes, oil cakes, paddy bran and husk, coir pith, sawdust etc.
6. Marine wastes: Seaweeds, fishmeal etc.

Estimates of available biomass (Gaur, 1997) available in the country for composting is given below

Animal wastes	2018 mt
Crop residues	407 mt
Municipal wastes	29 mt
Sewage	292,000 million gallons
Blood meal	55000 t
Bone meal	450000 t
Forest litter	30 mt
Non-edible cakes	300000 t
Rice husk	15 mt
Rice bran	2.5 mt
Bagasse	5.3 mt
Press mud	2.0 mt
Sawdust	2.2 mt
Vegetable and fruit processing industry	25000 t
Cotton stubbles	12 mt
Cotton dust	33000 t
Tea wastes	10000 t
Tobacco wastes	62000 t
Jute sticks	2.5 mt
Marine algae and sea weeds	15000 t

A careful cataloguing of wastes accumulating in India every year reveals plentiful of substrate. Crop residues from rice, maize, sugarcane, sorghum, pearl millet, pulses, and biomass from aquatic weeds and from wasteland weeds if properly processed can supply 2.20 million tonnes of fertilizers equivalent as NPK. With all these resources there is still a constraint in the availability of organic matter for composting. Cow dung is burnt as fuel and most of the crop residues are used as fodder for animals or for making chip-board etc. Under these circumstances sugar factory and distillery wastes like press mud and spentwash can be profitably used for producing organic manure.

Composting is one of the oldest methods of preparation of organic manure and has developed into a key technology for disposal and recycling of rural and urban organic wastes. References to compost in Renaissance literature like Rigveda and Kautilya's Arthashasthra are numerous and composting of animal wastes was practised in 18th century. The scientific aspects of composting was initiated at Rothamsted by Hutchinson & Richards in 1921 and was adopted on farm scale by Howard & Wad (1931). Composting has been basis for organic methods of farming and gardening. Howard (1940) also systematized the concept of compost making in India and showed that best compost was prepared when crop residues and cattle dung are mixed together in the ratio of 3:1.

Various organic compounds which reach the soil by way of animal and plant residues are made up of complex carbohydrates, simple sugars, starch, cellulose, hemicelluloses, pectin's, gums, mucilage, proteins, fats, oils, waxes, resins, alcohols, aldehydes, ketones, organic acids, lignin, phenols, tannins, hydrocarbons, alkaloids, pigments and other products. The organic sources used for composting, in general, contain 15-60% cellulose, 10-30% hemicellulose, 5-30% lignin, 5-40% proteins, 5-13% minerals (ash) beside 2-30% of cold water soluble fraction (sugars, starch, amino acids, urea, ammonium salts) and 1-15% of ether and alcohol soluble fats, oils and waxes. Cellulose consists of unbranched polymers of β-linked glucose residues arranged in linear chains, where every other glucose residue is rotated approximately 180°. As a result cellobiose is the structural repeating unit of the glucan chains in cellulose. The structure of cellulose appears to be controlled by the biosynthetic machinery. At least six different crystalline forms of cellulose exist, but higher plants synthesize normally only two. They are closely related forms of cellulose-I, namely, Iα and Iβ. Hemicelluloses has been widely used to refer to those cell wall polysaccharides normally found in association with cellulose but soluble in dilute alkali. The hemicelluloses are second only in quantity to cellulose and represent a significant source of energy and nutrients to the microflora. Most of the hemicelluloses are found in close physical proximity to cellulose in primary and secondary cell walls of higher plants. Straw and bark consist up to 30% hemicellulose, sugarcane bagasse has 30%, conifer wood has 7-12% hemicellulose and deciduous wood has 20-25% hemicellulose. They function as storage and supporting substance in plants. Xylans are the predominant constituents of hemicelluloses but mannans, glucomannoglycans, galactans, arabinogalactoglycans, arabinoxyloglycans and other polysaccharides are also included in hemicelluloses. Pectin is an example of hemicelluloses and is an important constituent of the middle lamella of cell walls. Pectin is degraded by enzyme pectinase, which is a complex of several enzymes. Lignin is much more complex than celluloses and is formed by chemical reaction involving phenols and free radicals without any specific order. Lignin gets encrusted on the cellulose and hemicellulose matrix. The lignin molecule has only three elements, carbon, hydrogen and oxygen. The molecule is a polymer of aromatic nuclei with either a single repeating unit or several similar units as building blocks. The repeating units range from about 200 to 1000 depending upon the origin of lignin and the methods used to determine the molecular weight. Microorganisms act on each of these components and degrade them to a form that can be taken up by plants as nutrients (Table 21.1). Hence, the Nobel laureate Selman A. Waksman said, "Without microorganisms, life on this planet would stand still".

Table 21.1: Predominant microorganisms that degrade the organic constituents

Substrate	*Microorganisms*	*Genera of Microorganisms*
Cellulose	Bacteria	*Bacillus, Cellulomonas, Cytophaga, Clostridium, Corynebacterium, Pseudomonas, Vibrio, Achromobacter, Angiococcus, Cellfalcicula, Polyangium, Sporocytophaga*
	Fungi	Aspergillus, Alternaria, Chaetomium, Coprinus, Fomes, Fusarium, Myrothecium, Penicillium, Polyporus, Rhizoctonia, Rhizopus, Trichoderma, Trametes, Trichothecium, Verticellium, Agaricus, Zygorynchus, Acremonium, Irpex, Phanerochaete, Schizophyllum, Sporotrichum, Talaromyces, Thielavia
	Actinomycetes	Streptomyces, Nocardia, Micromonospora, Streptosporangium
Lignin	Bacteria	Pseudomonas, Flavobacterium, Arthrobacter, Corynebacterium, Micrococcus
	Fungi	Agaricus, Armillaria, Clavaria, Clitocybe, Coprinus, Fomes, Ganoderma, Marasmium, Panus, Pleurotus, Polyporus, Pori, Collybia, Flammula, Hypholoma, Lepiota, Mycena, Pholiota, Arthrobotrys, Cephalosporium, Humicola, Phanerochaete, Volvariella, Pycnoporus, Aspergillus, Polystictus, Stereum, Trametes, Ustulina
	Actinomycetes	Nocardia, Thermomonospora
Hemicellulose	Bacteria	Bacillus, Cytophaga, Erwinia, Pseudomonas, Achromobacter, Sporocytophaga, Lactobacillus, Vibrio
	Fungi	Aspergillus, Alternaria, Penicellium, Fusarium, Trichoderma, Chaetomium, Trichothecium, Rhizopus, Zygorynchus, Helminthosporium, Coriolus, Fomes, Polyporus, Glomerella
	Actinomycetes	Streptomyces
Starch	Bacteria	Bacillus, Clostridium, Cytophaga, Flavobacterium, Micrococcus, Pseudomonas, Achromobacter, Chromobacterium,
	Fungi	Aspergillus, Fomes, Fusarium, Polyporus, Rhizopus,
	Actinomycetes	Micromonospora, Streptomyces, Nocardia
Pectin	Bacteria	Bacillus, Clostridium, Pseudomonas
	Fungi	Fusarium, Verticillium
Chitin	Bacteria	Pseudomonas, Chromobacterium, Cytophaga, Flavobacterium, Clostridium, Micrococcus, Achromobacter, Beneckea,
	Fungi	Mortierella, Verticillium, Paeciliomyces, Gleomastix, Fusarium, Mucor, Trichoderma, Aspergillus, Gliocladium, Penicilluim, Thamnidium, Absidia
	Actinomycetes	Actinoplanes, Micromonospora, Nocardia, Streptomyces, Streptosporangium,
Inulin	Bacteria	Beneckea, Micrococcus, Clostridium, Cytophaga, Pseudomonas, Flavobacterium

Contd.

Table 21.1: Contd.

Substrate	*Microorganisms*	*Genera of Microorganisms*
	Fungi	Aspegillus, Penicillium, Fusarium
Proteins and nucleic acids	Bacteria	Pseudomonas, Bacillus, Clostridium, Serratia, Micrococcus
Cutin	Bacteria	Bacillus
	Fungi	Rhodotorula, Mortierella, Penicillium
	Actinomycetes	Streptomyces
Tannin	Fungi	Penicillium, Aspergillus
Humic acid	Fungi	Penicillium, Polystictus
Fulvic acid	Fungi	*Poria*

II. COMPOSTING

Composting can be defined as a process of bioconversion of complex organic substances by the activities of a mixed microbial population into simpler, readily available nutrient substances under optimum conditions of moisture and aeration. Composting is a microbiological, non-polluting and safe method for disposal and recycling of organic wastes by bioconversion to manures. It is a solid waste management method where in the organic component of the wastes is biologically decomposed and stabilized under controlled conditions to a state where it can be handled, stored and applied to supply essential plant nutrients without affecting the environment (Radhakrishna & Bagyaraj, 1998). Staggering volumes of organic wastes are generated everyday by the human/industrial activity. They are nuisance at the point of generation. Also they cannot be used due to their phyto-toxicity, nitrogen immobilization and perhaps structural incompatibility. Composting reduces the odours, reduces difficulties in handling and inactivates pathogens and weed seeds, thus, providing uniform organic material suitable for soil application. The benefits of compost/composting, thus, range from recycling of locally available agro-industrial wastes for hygienic disposal, reduction in the bulkiness of the material, conservation of plant nutrients, elimination of human pathogens and mitigation of environmental pollution.

Bioconversion of organic materials is carried out by different groups of heterotrophic microorganisms such as bacteria, fungi, actinomycetes and protozoa. During the course of composting, both qualitative and quantitative changes occur in the active flora, some species multiply rapidly at the first stage, change the environment and then disappear to allow other population to succeed them. Microorganisms utilize organic matter (which comprises living or dead bodies of plants, animals and microorganisms) mainly to derive their carbon and energy sources in addition to obtaining other nutrients to synthesize their own cellular components for their growth and reproduction (Bagyaraj & Radhakrishna, 2000).

Composting process can be basically divided into four phases, namely:

(a) Mesophilic or initial low active phase during which the readily available constituents like simple sugars and some organic components are used up by mesophilic microorganisms leading to rise in temperature; fungi and acid-producing bacteria appear during this stage. This phase is characterized with a temperature range of 20° to $40^{o}C$. Yeasts show profound activity directly proportional to the amount of sugars present in the composting mix. It is extremely difficult to assign the percentage composting done by bacteria or fungi. However, bacterial population plays a major role, at least in the initial phase and the fungal population predominates the later phase.

(b) Thermophilic phase in which the temperature is high encouraging thermophilic microorganisms which degrade more complex/recalcitrant constituents of the organic matter. This phase starts with an increase in temperature from 40°C and goes upto 70°C, which establishes predominance of thermophiles, eradicates pathogens and expedites composting. Microscopic examination revealed that hydrolysis of hemicellulose and lignin apparently leaves a porous structure and thin film of lignin may remain present on a large surface area, amenable to enzymatic action/composting. Scanning electron microscopy and transmission electron microscopy have shown the presence of bacteria, fungi and actinomycetes on the surface of bark before and during composting. Their number increased substantially by the breakup of bark surface at later stages of composing and progressive simplification of bacterial population as a function of composting (Davis *et al*, 1992). Davis *et al* (1992) also found tremendous heterogeneity of microbial population during pine bark composting. Microbial numbers increased concurrently with the increase in the temperature and the bacterial colonies were more as compared to fungal colonies. It is also found that in this phase the herbicides if present in the substrate are broken down to non-toxic forms.

(c) Cooling phase is wherein the mesophilic phase with mesophilic microorganisms are restored; this is characterized by the decrease in temperature below 40°C. During subsequent process of cooling and maturing as in Table 21.2, there is abundance of nitrogen fixers, sulfur oxidizers, hydrogen oxidizers and nitrifiers, thus, products of extracellular polysaccharides are comparatively more in this phase.

Table 21.2: Predominant organisms in the different phases of composting

Microorganism	*Mesophilic phase*	*Thermophilic phase*	*Cooling and maturation phase*
Bacteria	*Cellulomonas* sp., *Corynebacterium* sp. *Arthrobacter* sp. *Pseudomonas* sp. *Bacillus* sp. *Myxococcus* sp.	*Bacillus subtilis* *B. stearothermophilus* *Thermus thermophilus* *Clostridium thermocellum*	*Thiobacillus* sp. *Azotobacter* sp. *Azospirillum* sp. *Beijerinckia* sp. *Derxia* sp. *Acetobacter* sp. *Klebsiella* sp. *Beggiatoa* sp.
Fungi	*Penicillium* sp. *Aspergillus* sp. *Trichoderma viride* *Trichurus spiralis* *Paeciliomyces* sp.	*Absidia sp.* *Humicola januginose* *Thermoascus aurantiacus* *Humicola insolens* *Chaetomium thermophile* *Geoitrichum candidum* *Aspergillus fumigatus* *Mucor pusillus* *Torula thermophila* *Humicola griseus*	*Penicillium* sp. *Aspergillus* sp.
Actinomycetes	*Streptomyces* sp. *Nocardia* sp. *Micromonospora* sp.	*Thermoactinomyces* sp. *Thermomonospora* sp. *Micromonospora vulgaris*	*Streptomyces* sp. *Nocardia* sp.

Contd.

Table 21.2: Contd.

Microorganism	Mesophilic phase	Thermophilic phase	Cooling and maturation phase
		Nocardia brasiliensis *Pseudonocardia thermophila* *Thermopolyspora sp.* *Streptomyces thermofuscus* *Streptomyces thermoviaceus* *Streptomyces thermophilus*	

(d) Maturity/stabilization phase with least microbial activity as most of the microbially degradable substances have already been utilized by microorganisms. This phase is initiated with decrease in temperature; the quality of compost increases due to continuous alteration in the nature of humus, as there is increase in humic acid content and decrease in fulvic acid content. Deep reddish brown and easy fragility of the compost and enhanced water holding capacity are the first visual criteria of maturation/stabilization of compost. The temperature fluxes during composting are depicted in Fig. 21.1.

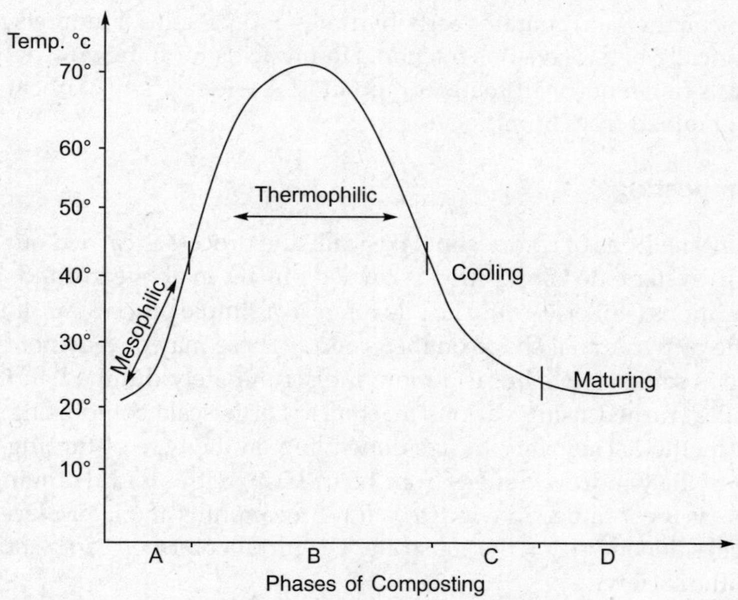

Fig. 21.1: Temperature fluxes during different stages of composting

Mechanism of Action

Approximately, 50% lignin and some hemicellulose are partially degraded during composting. Three-dimensional structure of lignin (phenylpropanoids) is considerably resistant to microbial degradation; the degradation of lignin is done by fungi, which hastens solubilization and detoxification. The byproducts of hemicellulose degradation provide nutrition to the microorganisms. Lignin degradation is a prerequisite for cellulose degradation, which in turns removes barriers for composting. Amorphous forms of cellulose are degraded at a faster rate as compared to crystalline form of cellulose.

The final product of decomposition of organic matter is CO_2, water and humus. Humus is the dark amorphous material, which consists of undecomposed organic matter, which is resistant for degradation along with dead microbial cells plus plant nutrients in water-soluble form. This is the material, which is incorporated to soil as organic manure to provide plant nutrients and also to improve physical conditions of the soil. Humus can be defined as a ligno-protein complex or an amino acid-lignin complex containing approximately 45% lignin compounds, 35% amino acids, 11% carbohydrates, 4% cellulose, 7% hemicellulose, 3% fats, waxes and resins and 6% other miscellaneous substances including plant growth-promoting substances and inhibitors. However, the age and composition of the humus is dependant upon its origin and environment. By radio carbon dating, the age of humus in podzols have been estimated to be in the range of 1580 to 2860 years, while that of chernozemic soils as 1000 years old. Among the fractions of humus, humic acids have been regarded as the abundant and oldest. Further, biodegradation products of lignin are closely linked with those of humus substances. Fulvic acid, humin and humic acid have been recognized as the three major fractions by subjecting humus complexes to solvent extraction procedures. Fulvic acid is alkali as well as acid soluble part of soil organic matter and contains carbohydrates and proteins. Humin is resistant to cold alkali and is chemically heterogeneous fraction. Humic acids form the bulk of humus complex and are regarded as polymers of aromatic compounds. A wide variety of phenolic degradation products can be obtained from humic acids.

Methods of Composting

There are different methods of composting. Basically the process is carried out either in pits or heaps. In the heap system, decomposition is carried out 1-2 m above ground. This consists of different configurations (Miller, 1993). The pile refers to a simple process caused by dumping of wastes into a cone or pyramidal shape on the ground. These may or may not be mechanically turned. The heap is same as pile but is a more indiscriminately dumped and of larger mass. Windrows are piles, turned using various machines. Large scale composting at outdoors involves windrow method. They vary in size depending on the type of turning equipment utilized. The height of the windrows ranges from 1.5 to 3.0 m, with a width ranging from 2 to 3 m. Composting process is continued in windrows for three months and later placed in curing pile for four to six weeks depending on the substrate. The product formed can be utilized by selling or distributed in the society.

Different methods of composting have been proposed and used, altering the methodology for convenience and for faster degradation.

Bangalore method

This method of composting was developed by Acharya (1939). This method is suitable in places where night soil and refuse are used for preparing the compost and in areas having scanty rain fall. Here composting is carried out in trenches; their size vary based on population; a population of more than 50 can have trench dimension of 10 m × 2.5 m × 1m, length, breadth and depth respectively. The trenches should preferably have sloping walls and a floor of 90 cm slope to prevent waterlogging.

A layer of refuse about 15 cm depth is spread at the bottom of trench, over this night soil is added at 5 cm, similarly like this alternative layers are made till about 30 cm above the ground level. Finally the pit is covered with a 15-20 cm thick layer of refuse. The materials are allowed to remain in the pit without turning and watering for 90 days. During this duration, there is reduction in the volume of biomass and we can add additional layer of night soil and refuse in alternative layers on the top, and this is plastered with mud to prevent loss of moisture and breeding of flies. Initially, there is aerobic decomposition and the temperature rises to 60-70°C, then this is taken over by anaerobic digestion occurring at a slower rate for a period of 180-240 days to obtain end product which is odourless and has high manurial value.

Chinese method

A) Another method of composting night soil is using heaps called as high temperature composting. Here the base material consists of stalks and roughages of crops and this is followed by layering with night soil, urine, sewage, animal dung and chopped plant residues. Here water is added at optimum levels. The heap is finally shaped and covered with mud plaster. Poles of bamboo are pierced vertically into the mud plaster and all the way to the bottom of the heap; this is kept for 24 hours and later removed, which leaves holes for ventilation. The temperature of the heap rises to about 60-70°C and at this stage the holes of aeration are closed by sealing with mud plaster. After a period of two weeks, the mud plaster is dismantled and the contents are mixed followed by resealing with mud plaster. This is left for a period of two months and the compost is ready and free from pathogens.

B) Generally, composting is carried out in a corner of a field and in a circular or rectangular pit. Rice straw, animal dung (usually pig), aquatic weeds or green manure crops are used and often silt pumped from river beds is mixed with the crop residues. The pits are filled layer by layer, each layer being 15 cm thick. Usually, the first layer is of a green manure crop or water hyacinth, the second layer is a straw mixture and the third layer is of animal dung. These layers are alternated until the pit is full, when a top layer of mud is added; a water layer of about 4 cm depth is maintained on the surface to create anaerobic conditions which help to reduce losses of nitrogen. Approximate quantities of the different residues in tonnes per pit are: river silt 7.5, rice straw 0.15, animal dung 1.0, aquatic plants or green manure 0.75 and superphosphate 0.02. Three turnings are given in all, the first one month after filling the pit and, at this time, the superphosphate is added and thoroughly mixed in. Water is added as necessary. The second turning is done after another month and the third, two weeks later. The material is allowed to decompose for three months and produces about eight tonnes of compost per pit.

Indore method

This method is suitable when there is mellifluous plant residues, animal dung, urine, wood ash etc. The plant wastes include vegetable wastes, weeds, stalks, stems, fallen prunings, chaff, fodder remnants, green matter etc. Here the total composting period taken is about three months.

Pit method: The site selected for pit method should be near cattleshed and at a high level so that no rainwater gets in during the monsoon season. Pits are dug at about 1 m deep and 1.5-2.0 m wide and of any suitable length. The material got from cattleshed like bedding material and other wastes are mixed and spread in the pit, each layer of the substrate is covered with slurry of dung made with 4.5 kg urine – earth and 4.5 kg of inoculum taken from a 15-day-old compost pit. The pit is sufficiently moistened and care should be taken to avoid compaction of the substrate. The substrate in the pit is turned after 15 days, 30 days and 60 days from the date of filling the pit. During each turning, the substrate is properly mixed and sufficiently moistened.

Heap method: Heap method is preferred in places where there is low rainfall. The site for heap preparation is selected at an elevation. The Indore pile should have a dimension of 2 m long, 1.5 m high, 2 m wide and the top of 0.5 m narrowed by tapering. A layer of carbonaceous materials such as leaves, hay straw, saw dust, wood chips, corn stalks etc. are spread to about 20 cm height, this is then covered by nitrogenous materials like grass, weeds, garbage, other plant residues, manure, digested sewage sludge. This layering is repeated similarly till the pile is 1.5 m high and moistened. The heaps are turned at 6 and 12-week interval.

NADEP method

This method is in parallelism with heap method of composting. The NADEP method employs a brick lined tank of size about 1×2×4 m. Aeration is achieved by holes on all sides of the brick tank, helping air circulation.

Common substrates employed in this method are agro wastes *viz.*, cattle feed refuse, weeds, grasses, leaves, sugarcane trash, stalks, roots, plant clippings etc. Methodically the substrates consisting of organic wastes, cow dung slurry and soil are layered; at the bottom, a layer of 150 kg organic wastes are spread, on this cow dung slurry (6 kg cow dung in 100 l water) is spread, a third layer consisting of 60 to 80 kg cleaned sieved dry soil is spread and over it water is sprinkled. The layering is repeated similarly up to about 1 m. Finally a layer of cow dung soil paste is added. Aerobic decomposition process takes about 90 to 120 days.

Japanese method of composting

In this method instead of pits, vats of stone slabs are made which measures 5.5 - 9.0 m in length and 0.9 - 1.2 m in width and have a height of 0.75 - 1.5 m. The stonewalls are laid in such a way that aeration can be made. The stones can also have holes. At the bottom of the vat, the stone slabs are laid and plastered with cement to prevent leaching of the nutrients. This method employs five different substrates viz.,

1. Cow dung and poultry wastes.
2. Grass, crop residues and weeds that include fresh and dry green leaves.
3. Wastes like agro-based industries like press mud, sericulture wastes, wastes from vegetable and fruit processing.
4. Naturally occurring nutrient source like rock phosphate.
5. Organic wastes from towns and cities like sewage and sludge.

In the bottom of the vats, a layer of highly lignaceous substrates like coconut leaves, coconut shells and other fibrous materials are layered up to 10-15 cm height. The second layer is spread with dried leaves, grass, groundnut shells etc. up to 10-15 cm height. These layers absorb the moisture and nutrients leached from the top. The layers after being laid are sprinkled with cow dung slurry or with bio-gas spent slurry. The third layer is made up of green leaves of pongamia, green grass, weeds and other crop residues rich in nitrogen. The fourth layer contains organic waste rich in phosphorous like roots of legume crop, green manures and rock phosphate, along with these potassium rich substrates like calotropis, other weeds, tobacco residues, tomato crop residues, ash, poultry wastes are used. Over these 2-3 buckets of cattle dung slurry is sprinkled and covered with straw. This is finally covered with cattle dung and some old composted material or tank silt or ash, which acts as good casing material. The substrate is turned once in 15 days.

Synthetic compost

In the preparation of synthetic compost, nitrogen in the form of dung required by microorganisms can be completely substituted with inorganic nitrogenous compounds like ammonium sulphate or urea, which are utilized equally effectively for decomposition of carbonaceous materials into compost. This facilitates the utilization of large quantities of various organic waste materials, where supplies of dung are either short of the requirement or not available at all as on mechanized farms. In this method nitrogenous fertilizers are added to bring down the C/N ratio to favour the decomposition. The material to be composted is sprayed with fertilizer solution and then with lime. Super phosphate may be added to fortify the phosphorous content of the manure. The treatment is continued layer-wise until the heap or pit is filled to the size and allowed to ferment. The manure becomes ready for application in about 4 to 6 months.

Windrow method

Turned windrows: The windrow method is a widely practised method of composting in USA. In this process, the waste materials like garbage received are dumped and is sorted for plastic, glass and rubber. The waste materials are piled in long rows of 2-3 m width and 1.5-3 m height on a hard surface and usually in the open area. Aeration of windrow is by periodic turning using equipment such as front-end loader or specially designed machinery like aero tillers. Occasionally forced aeration in conjunction with turning has been applied to the windrow process. The windrow method is also exploited by distillery and sugar industry for composting of press mud with distillery-spent wash.

Passively aerated windrows: This method passively aerates by supplying air to composting materials through perforated pipes embedded in each windrow, thus, eliminates the need for turning. The plastic pipe is laid down in two rows, and 1.25 cm diameter holes are present in the pipe, the holes are oriented downward to minimize plugging and to allow condensate to drain.

This method is commonly seen in Canada for composting seafood wastes with peat moss, manure slurries with peat moss, and solid manure with straw or wood shavings.

Static pile method

A) In this method a mixture of solid wastes is stock piled in the open and turned occasionally for aeration. This method is considered to waste both ammonia and energy. These disadvantages have been substantially mitigated by forcing air through perforated pipes at the bottom of the static pile, to keep the temperature below 60°C.

B) Aerated static pile method: The aerated static pile makes use of pipes embedded in the piles for aeration. The raw material mixture is piled over a base of wood chips, chopped straw, or other porous material. This porous material contains a perforated pipe. The pipe is connected to a blower, which either pulls or pushes air through the pile.

Sheet method

Sheet composting is carried out by spreading organic materials on the surface of the soil or untilled ground and allowing it to decompose naturally. Over time, the material will decompose and filter into the soil. This method is ideally suited for forage land, no-till applications, erosion control, roadside landscaping etc. The process does not favour the destruction of weed seeds, fly larvae, pathogens etc. and composting materials should be limited to plant residue and manure. Decomposition time is governed by environmental conditions and can be quite lengthy.

Factors Influencing Composting

There are numerous factors determining the fate of composting. It can be acceded that the time of composting and quality of the final product is affected by the factors. Hence, it is imperative to follow composting by employing optimized parameters; some of the grave important factors being:

C/N Ratio

The nitrogen and carbon content of the biomass substrate determine the rate of biodegradability of the substrate. The microbes in compost use carbon for energy and nitrogen for protein synthesis. Of the carbon present in the raw, uncomposted material, about 2/3 of it will be burned away as it supplies energy to the microbes. This fraction is lost to the atmosphere as carbon dioxide. The remaining 1/3 is incorporated into their cellular structure: amino acids, nucleic acids, lipids and polysaccharides. For every kilogram of usable carbon, approximately 320 grams of it goes directly into microbial cells. If analyzed, the average compost microbe cell would have a C:N ratio of between 5:1 and 8:1. The ideal C/N ratio of 30:1 is suitable for microbial growth. The decomposition is slow in the environment, if the C/N ratio is greater than 60, since, microorganisms require 1 unit of nitrogen to degrade every 10 units of carbon. Most organic materials do not have this ratio and, to accelerate the composting process, it may be necessary to balance the numbers. Hence, for composting of crop residues and leaf litter, which have a wider C/N ratio, the process can be hastened by mixing with blending agents like cattle dung, slurry or nitrogen rich plant materials. For optimization of C/N ratio for composting, proper blending material is required to hasten composting and to get good quality compost; approximation of ratio of C/N can be done using the formula

$$\text{C/N of the mixed substrate to be composted} = \frac{\text{Sum of C/N ratios of substrates used}}{\text{Number of substrates used}}$$

Example: If you have one bag of vegetable produce (C:N = 19:1) and one bag of barley straw (C:N = 100:1) then combined you have a C:N ration of (19:1 + 100:1)/2 = (119:1)/2 = 59.5:1

The C:N ratios of materials can be calculated by using Table 21.3.

Table 21.3: The list of C/N ratios of materials used for composting

Materials	*C/N content*
None	0:1
Humus	10:1
Crop residues like straw, hay and silage	
Straw general	80:1
Hay general	15-32:1
Hay legume	16:1
Hay non-legume	32:1
Rice straw	70-80:1
Rice hulls	121:1
Wheat straw	80-100:1
Barley straw	100-120:1
Oat straw	60:1
Maize stalks and leaves	50-60:1
Cotton stalks	70:1
Sugarcane trash	110-120:1
Alfalfa hay	13:1
Lucerne residues	19:1
Groundnut foliage	20:1
Corn silage	38-43:1
Corn cobs	98:1
Corn stalks	60-73:1
Cottonseed meal	7:1
Olive husks	30-35:1
Coffee grounds	18:1
Ferns	25:1
Flax	44-80:1
Jute leaves	25:1
Dairy wastes	60:1
Tobacco	13:1
Soybean meal	4-6:1
Vegetable residues	
Vegetable waste	11:1
Vegetable produce	19:1
Cull potatoes	18:1
Potato	27:1
Potato tops	25:1
Amaranthus sp.	11:1

Contd.

Table 21.3: Contd.

Materials	*C/N content*
Cabbage	12:1
Onion	15:1
Pepper	15:1
Tomato	12:1
Tomato-processing waste	11:1
Carrot	27:1
Turnip	19:1
Potato-processing sludge	28:1
Vegetable trimmings	12-20:1
Manures	
Turkey litter	16:1
Sheep	14:1
Horse-race track	41:1
Horse-general	27:1
Dairy free stall	13:1
Dairy tie stall	18:1
Broiler litter	14:1
Laying hen manure	6:1
Poultry manure	10:1
Dairy cow manure	13:1
Swine manure	13:1
Cattle manure	17:1
Municipal wastes	
Refuse (mixed food, paper, and so on)	34-80:1
Paper from domestic refuse	127-178:1
Pharmaceutical wastes	19:1
Night soil	6-10:1
Garbage (food waste)	15:1
Table scraps	15:1
Activated sludge	6:1
Digested sludge	16:1
Sewage sludge	5-16:1
Fruit wastes	
Fruit waste	32:1
Cranberry plant (stems, leaves)	61:1
Cranberry filter cake	31:1
Apple filter cake	13:1
Apple pomace	48:1
Apple-processing sludge	7:1
Cocoa shells	22:1
Fish and meat processing	
Blood wastes (slaughterhouse waste and dried blood)	3-3.5:1
Crab and lobster wastes	4-5.4:1
Fish-breading crumbs	28:1
Fish-processing sludge	5.2:1

Contd.

Table 21.3: Contd.

Materials	*C/N content*
Fish wastes (gurry, racks, and so on)	3.6:1
Mixed slaughter house waste	2-4:1
Mussel wastes	2.2:1
Poultry carcasses	5:1
Paunch manure	20-30:1
Shrimp wastes	3.4:1
Wood and paper	
Bark-hardwoods	223:1
Bark-softwoods	496:1
Corrugated cardboard	563:1
Lumbermill waste	170:1
Newsprint	398-852:1
Paper fiber sludge	250:1
Paper mill sludge	54:1
Paper pulp	90:1
Sawdust	442:1
Telephone books	772:1
Wood-hardwoods (chips, shavings, and so on)	560:1
Wood-softwoods (chips, shavings, and so on)	641:1
Office paper	129:1
Card board	378:1
Yard wastes and other vegetation	
Grass (compacted)	15:1
Grass (loose)	15:1
Leaves loose – dry	47:1
Leaves compact – wet	47:1
Leaves fresh	37:1
Seaweed	17:1
Green weeds	13:1
Shrub trimmings	53:1
Tree trimmings	16:1
Water hyacinth	18:1
Pine needles	60-110:1
Peat moss	58:1
Oak leaves	26:1
Gliricidia clippings	14:1
Cassia sp.	18:1
Sesbania sp.	18:1

Sharath and Jagadeesh, (2004b) has reported biocompost prepared from wood waste and press mud in 1:1 had least C/N in comparison with bio-compost prepared from other blending materials with wood waste. This C/N optimization is required, since, if there is wider C/N ratio, then the time for composting is increased and if there is narrow C/N ratio, then there is loss of valuable nitrogen to atmosphere in the form of ammonia. Addition of urea for enrichment, results in loss of nitrogen as ammonia volatilization (Balakrishna *et al*, 1991). Solid wastes generally contain sufficient P for microbial growth and activity. The optimum carbon to phosphorus (C/P) ratio is around 100:1 (Hagerty *et al*, 1973).

Now the concept of adjuvants are used, these are raw materials or waste products or by products or agro-residues or chemicals which provide easily assimilable nutrients to promote microbial growth, sustain their activity and complement in the composting process, by providing energy, thus, helping speedy conversion and maturation of compost. Though, nitrogen is lost as ammonia when urea enrichment is done, but urea can be added at low concentration in absence of organic nitrogen amendment to increase the number of microorganisms, which in turn will hasten composting. A circumstantial evidence is put forth by Rahman and Richards (1988) that urea solulbilizes condensed tannins, presumably to expedite the rate of composting and they have reported that the bark from older trees required more nitrogen for composting than the dust or bark from younger trees. The recommended urea amendment for soft wood bark is 0.5-0.8% and for hard wood bark is 1.0-1.2%. The addition of lime is recommended, wherein lignocelluloses along with alkali leads to swelling structure of lignocellulose and weakens the bonds between them, and once it becomes soft, it is easy for microorganisms to break and degrade.

Moisture

For proper decomposition, the moisture content of the organic wastes has to be maintained at 60 to 70 percent; excess moisture brings in anaerobic condition and delays the process of decomposition (Gowda, 1997). The level of hydration of the compost mix significantly affects fungal physiology and in turn, compost efficiency. A study by Gervais *et al* (1996) on the thermodynamic and dynamic properties of water (viz., water potential and water diffusion) and their correlation with its activity level on the survival, growth and functioning of fungi revealed the following:

- If the quantity of water is insufficient, it does not allow diffusion of solutes and gases thereby slowing or stopping the microbial growth/metabolism. This limits permeation of enzymes necessary for composting.
- If the intracellular or extracellular water level does not maintain osmotic equilibrium, it creates disequilibrium in the metabolic chain of microbes and retards composting rate.
- Excess water or water stress leads to mechanical rupture of plasma membrane, thereby affecting the permeability and kinetics of composting.

Moisture is essential to all living organisms, and most microorganisms, lacking mechanisms for moisture retention (like our skin) are particularly sensitive in this regard. At moisture content between 35 to 40%, decomposition rates are greatly reduced and below 30% they virtually stop. For most compost mixtures, moisture content of 55 to 60% is the recommended upper limit. In a composting mix consisting of wood waste and press mud in the ratio of 1:1, a moisture level of 60% was found to be optimum for faster degradation (Sharath & Jagadeesh, 2004b).

The moisture content suitable for composting various wastes is related to the structural strength of the material. For example, fibrous material like straw or wood chips can absorb relatively large quantities of water and still maintain their structural integrity and porosity.

Studies have been done with straw and vegetable trimmings using an initial moisture content as high as 85%, but 76% was too high when paper was used in place of straw.

Composting materials such as municipal waste and agricultural residues begin the process in a somewhat dry form. Systems have been developed to process the material and allow for moisture addition as composting proceeds. Due to the bulky nature of these materials, they allow for the absorption of relatively large quantities of water. Wet organics differ because they are not fibrous or bulky and are not able to support high moisture contents. If dried cakes were composted, constant mechanical agitation would be necessary to provide aeration due to the lack of porosity to store oxygen. Wet organic wastes include materials such as wastewater treatment material, animal manures, and some industrial wastes. There are five techniques used to lower the moisture content of wet organic materials and to increase the amount of solids. These techniques are as follows:

1. Recycling of previously composted material
2. Addition of bulking agents such as wood chips
3. Using constant agitation to aerate the compost
4. Addition of dry amendments
5. Drying the material before composting.

Maintaining proper moisture levels is a matter of balancing numerous competing forces. The level of moisture must be high enough to ensure good biological stabilization rates yet low enough not to compromise void ratio. If the void ratio is compromised the rate of oxygen transfer is reduced, thus, reducing the rate of biological activity. Generally there are tradeoffs that take place, because it may not be possible to optimize all the factors at the same time.

Aeration

Composting is an aerobic process. Oxygen demand is very high during the initial decomposition stage that is characterized by a rapid rise of the microbial population, temperature and high rate of biochemical activity. Oxygen content is mainly maintained at a level above 10% whether it is done by turning or by forced aeration. Temperature raise is achieved only by the oxidation of the organic substances. Under anaerobic conditions many intermediary compounds formed such as volatile organic acids, volatile S and N compounds, cause foul odour (Miller *et al*, 1989). If oxygen is unavailable, the organic matter produces methane, hydrogen sulphide, skatoles, indoles, mercaptans etc., accounting for the characteristic strong odours. In horticulture, organic acids in such compost are toxic to the plants and remain phytotoxic over long periods (Biddlestone & Gray, 1985).

There are three factors contributing to the maintenance of aerobic conditions within a composting material: bulk weight, porosity, and free air space. Moisture content is related to free air space and porosity. When void spaces within a compost pile are filled with water, the material has little strength and acts like a plastic. This stops aerobic composting due to the restriction of oxygen transfer. Only when the voids are filled with air, aerobic composting occurs.

The free air space of compost can be increased with the addition of conditioning agents, such as recycled compost product, and bulking agents like wood chips. There is evidence that the minimum optimum moisture content for a given material is related to a certain minimum free air space percentage. Aeration is most difficult to achieve during the early stages of composting before evaporation dries the mixture and increases the void volume. The rate at which greater void volume is achieved is increased with the use of conditioning agents. The additions of bulking agents, add structural support to the drying compost allowing for easier movement of oxygen throughout the mixture.

Oxygen is required for all composting material to meet the demand created by organic decomposition. The oxygen is heated from within the compost, picks up moisture, and thus dries the remaining material. The drying occurs generally because of oxygenation, which is an important benefit during composting. Both aeration and drying depend on supplying air to the pile but achieving both simultaneously is difficult to accomplish.

Another purpose of aeration is to remove heat from the system. Without the removal of heat from the system the temperature can climb to levels where it impedes biological activity. Aeration is used to control the rate of heat removal and thereby adjust the system's temperature.

Providing the optimum aeration is an important cardinal factor for obtaining high temperatures and for producing good quality compost. Frequent aeration in form of turning may lead to drying of the material (Gaur, 1996). And, delayed turning may accumulate excess moisture, leading to anaerobic condition with foul smell development. Turning the compost material once in two weeks was found optimum for decomposing a mixture of wood waste and press mud blended in 1:1 proportion (Sharath & Jagadeesh, 2004b).

Particle size

Smaller the particle, faster is the decomposition rate as smaller materials have more surface area available for microbes to attack. Therefore, reducing the particle size of raw materials will increase the speed of the composting process (Table 21.4). Size reduction also reduces the volume of the compost pile, thereby saving space. Particle size cannot be too small, for example, sawdust and wet materials can decrease aeration, reduce the rate of composting and perhaps cause anaerobic conditions leading to odour problems. Chopping of the substrate to 10 mm is recommended when windrows are formed with forced aeration (Gaur, 1999).

Table 21.4: Influence of particle size and enrichment on decomposition

	Organic C%	*N%*	*C/N ratio*	*Available P (ppm)*
Unchopped paddy straw (UPS)	31.8	1.26	25.20	140
Chopped paddy straw (CPS)	36.0	1.30	20.40	152
UPS + 1% RP + PSM + *A. chroococcum*	30.2	1.58	19.10	234
CPS + 1% RP + PSM + *A. chroococcum*	25.0	1.72	14.50	240

Chopping reduced decomposition period by 30 days.

Temperature

Temperature is one of the important parameter, as any change from the normal range is considered a faulty gesture of composting. Increase in temperature beyond optimum is not desirable as it is stressful for the survival of the microorganisms and reduces moisture and oxygen, which retards the rate of composting. This sets in partial decomposition of the composting mix, its partial desiccation and presumably death of certain mesophiles. The temperature indicates the decomposition rate of the substrate. Increase in temperature suggests rapid growth and activity of the microorganisms and this should be followed by optimum maintenance of the temperature for healthy compost. The harmonious temperature fluxes is very important for expediting composting; initial mesophilic temperature below 40°C followed by thermophilic activity from 40°C to 70°C, finally cooling phase with temperature dipping below 40°C have subsequent impact on the biodegradation. The external temperature is also detrimental for hastening composting. It is seen that composting takes a long time in temperate regions as compared to composting in tropical climate. Temperature generated with in a compost mass is important for two reasons: a) to maximize decomposition rates; b) to produce a material which is microbiologically safe for use (Gowda, 1996). The increase in temperature is desirable as it eliminates most of the animal and plant pathogenic microorganisms, depending on the source of the substrate employed, and kills most of the weed seeds.

The increased temperature created by the energy released from the organic decomposition drives the composting process and produces the drying that is desirable for composting. A temperature of 55 to 60°C for a couple of days was found optimum for maximum degradation (Gowda, 1996; Hajra and Kole, 1998; Sharath and Jagadeesh, 2004a).

Nutrient availability

Microorganisms need nutrients to grow and reproduce. These nutrients occur in the raw materials used in the compost mix. Additional fertilizer from any source (organic or inorganic) usually is not needed. Nitrogen fertilizer may be beneficial for mixes consisting mainly of bulking agents. The best way to add fertilizer is to dissolve it in water and wet the pile with a dilute fertilizer solution. Compost additives such as blood meal and bone meal are simply organic fertilizers. When microorganisms grow and multiply on organic debris, carbon is utilized for building the cellular material of microbial cells with release of carbon dioxide, methane and other volatile substances. In this process microorganisms also assimilate nitrogen, phosphorous, potassium and sulfur, which get bound in the cell protoplasm. The carbon to phosphorous ratio (C/P) should be around 100:1 for proper microbial growth and digestion. Solid wastes usually contain sufficient amount of phosphorous for microbial growth.

pH

The compost process will be more effective if the pH is in the range of 6.5 to 7.5. For initial mixtures in the range of 5.5 to 9.0, composting may take longer to establish. In particular, bacterial activity may be slow when pH is below 6.0. pH is likely to vary during composting process, but the final product will usually have a value near neutral.

Micro- and macroorganisms

The biochemical actions of microbes result in the transformation of the raw organic material of the compost into humus-like material. It does not produce humus. There are numerous organisms that work together in the process of degrading the raw material including slime molds, insects, arthropods, earthworms and bacteria. These organisms are primarily ordered into three categories. Primary decomposers are the true decomposers of organic matter. The second level of organisms feed on the primary decomposers. The third group prey on the second group as well as on each other. All of these organisms, along with the organic matter form the soil biomass. The decomposition of this biomass is what eventually results in humus.

Actinomycetes produce greyish cobweb like growth throughout the compost giving a pleasing, earthy smell, like the forest floor. Actinomycetes usually colonize the compost after the thermophilic bacteria have been active. Some common invertebrates found in the compost include: sowbugs, spiders, earthworms, millipedes and centipedes. When we see these "critters" in the compost, it is a good sign that the bacterial activity and temperatures in the pile are low. This may mean that the compost is ready for use. Sowbugs and pillbugs, sometimes called "rolly-pollies" because they roll up in a ball when touched, come to the pile after temperatures have lowered to the mesophilic range. They feed on woody materials that were not decomposed by others in the compost community. Millipedes have multiple body segments, each having two pairs of leg segments (except the front few segments which have one leg pair per segment). Millipedes help the compost process by feeding on dead plant matter. Centipedes are multi-segmented invertebrates, with each body segment having one pair of legs. Centipedes feed on other invertebrates of many sizes, therefore, centipedes do not directly assist in decomposing the plant material in compost. However, centipedes are part of the compost community and can coexist with the decomposers. Spiders may be present in compost, where they feed on invertebrates. Spiders help control garden pests. Earthworms are gardener's dream workers because they perform many functions that produce excellent compost. Their primary activity is eating through organic matter, decomposing materials into rich castings that help plant growth. Earthworms also move up and down in the soil and compost creating passageways for air and water. Furthermore, material that has passed through the earthworm, the castings, is higher in available nutrients. Hot temperature composting will kill earthworms. Therefore, earthworms enter the composting system near the end of the process when the pile temperatures have stabilized.

Some of the Amendments made to Substrates for Composting Keeping in Mind about the Factors Influencing Composting

1. Sludge: Sewage sludge has a C/N ratio on low end of the acceptable range; it is too wet and lacks porosity. It is usually conditioned by mixing with wood chips, saw dust, bark, dry compost and fly ash (to reduce moisture content).
2. MSW: Most plans for composting municipal solid wastes (MSW) are part of an overall resource recovery system, which removes a variety of recyclable materials from the MSW. Thus, it is difficult to generalize about its need for conditioning. However, fine grinding prior to composting should be avoided, if possible, to maintain porosity. The MSW and most of its fractions will need added water and will benefit from additional nitrogen.
3. Yard waste: Leaves have a very high C/N ratio and tend to be dry. They benefit from added water and nitrogen. Shredding improves its porosity and exposes the interior of the leaf, which is more susceptible for decomposition than the outer surface. Lawn clippings lack the structure to maintain their porosity for aeration, but have a favourable C/N and moisture content for decomposition. Thus, the lawn clippings are likely to deplete the oxygen in a windrow and become anaerobic and generate odours. Bush and

tree trimmings need shredding for size reduction. A combination of these yard wastes can make a good composting mix.

4. Food wastes: Wastes from food processing plants can vary widely in their characteristics. Many will be too wet to have the necessary porosity and so will need a bulking material. Dewatering could help reduce the amount of bulking material required, others may need grinding to reduce particle size. The C/N ratio may be adjusted by judicious blending with other wastes, for effective decomposition.

Rapid Composting using Microorganisms to Hasten/Activate the Process

Basically, composting is catalyzed by a consortium of heterogeneous population of bacteria and fungi. These are contributed by the raw material, adjuvants, soil where pits/heaps for composting are created and periodic water addition for retention of desired amount of moisture and environment. Besides, substrate-specific microbes for expediting composting are added.

It generally takes about 4-6 months to get finished compost under normal conditions. Considerable research has gone to reduce the time of composting and to upgrade nutrient content (Gaur, 1999). Rapid method will be more acceptable because there would be greater turnover in a lesser area of land with less effort. Fungi with ligno-cellulolytic activity such as *Aspergillus niger*, *Penicillium* sp., *Trichoderma viride*, *T. ressei*, *Trichurus spiralis*, *Chaetomium* sp., *Phanerochaete chrysosporium* and *Paeciliomyces fusisporus* are used to hasten the decomposition process (Bagyaraj & Radhakrishna, 2000). These cultures are added at the rate of 2 kg per ton. Several experiments have been conducted to study the role of efficient lignocellulolytic fungi in composting of crop residues richer in cellulose, hemicelluloses and lignin (Table 21.5).

Table 21.5: Enrichment and hastening of compost with lignocellulolytic fungi

Location	*Substrates*	*Total N%*		*C/N ratio*	
		Without inoculum	*With inoculum*	*Without inoculum*	*With inoculum*
New Delhi	Sorghum + Wheat straw	1.38	1.82	20.6	12.0
New Delhi	Paddy straw	1.52	1.76	15.5	12.3
Hisar	Mixed crop residue	1.38	1.52	24.0	20.0
Pune	Sugarcane trash	0.98	1.54	35.0	24.0

Phanerochaete chrysosporium has been found to secrete efficient lignin degrading enzymes, selection of its over producer mutant has been suggested to facilitate rapid composting (Farrell, 1987). Strategies have been developed for increasing the rate, efficiency and selectivity of the degradation of lignocelluloses by utilizing *Trichoderma viride* L-333, *Coriolus versicolor* 082 and *C. maxima* 0275 (Viesturs *et al*, 1992). Gomes *et al* (1993) reported that *Thermomyces lanuginosus*, a fungus isolated from preserved jute stacks, secreted high level, broad pH range (3.0-9.0), thermostable xylanase (hemicellulase); its ability for efficient saccharification may complement composting. Xylanases produced by *Cellulomonas* sp. are constitutive in nature and are potential for composting hemicellulose waste (Khanna and Gauri, 1993). The application of brown rot fungus, *Lenzites striata*, found to be superior, has been strongly recommended for composting of saw dust from different eucalyptus species (World bank, 1995). Apart from traditionally used substrates or wastes, more and more new materials are used for composting. Crop residues like coir pith are composted with the help of *Pleurotus sajor-caju*. Many of the organic wastes are converted to compost by a consortium consisting of *Trichoderma viride* and *Pleurotus sajor-caju*.

Singh *et al* (1992) also obtained good quality compost using sugarcane trash within four months with the inoculation of two cellulose degrading fungal cultures. Murkute *et al* (1992) opined that the good quality compost could be obtained from agro-industrial wastes like sugarcane trash, press mud, bagasse etc. with microbial inoculation. They observed that C/N ratio of different composts ranged from 13 to 33.76 with the use of fungal culture, while it ranged from 17.5 to 43.25:1 without the use of culture.

Enrichment of Compost with Nutrient Enhancing Microorganisms

Phosphocompost

India is producing 0.4% of world rock phosphate, which is of inferior quality and not used for fertilizer production. Nearly 95% of P fertilizer in the country is from imported rock phosphate. India has 260 million tonnes of low-grade rock phosphate, which can save a minimum of Rs. 3000 foreign exchange for every tonne of phosphatic fertilizer. The need for effective usage of low grade Indian rock phosphate as phosphatic fertilizer through composting has led to the process called phosphocompost. Rock phosphate can be added at the beginning of composting to get phosphorus enriched compost. The level of rock phosphate may be 12.5 - 25% depending on the required P enrichment. There is also a practice of addition of gypsum and some oil cakes, which are not used as cattle feed, during the last turning of the compost. Along with addition of rock phosphate, phosphate-solubilizing microorganisms like *Bacillus megaterium, Aspergillus awamori, Pseudomonas, Micrococcus, Flavobacterium, Penicillium, Fusarium* etc. are inoculated at the rate of 1-2 kg per tonne of the substrate. The phosphocompost is used as the phosphatic manure, which is superior to super phosphate as it serves as a good source of P to crop plants and is less chemically fixed in soils.

Addition of rock phosphate to wood waste-glyricidia mixture with subsequent inoculation with a phosphate solubilizer *Pseudomonas striata* resulted in a biocompost rich in phosphorus. *Pseudomonas striata* solubilized rock phosphate and released higher amounts of available P, hastening the composting process, besides increasing the P content in the biocompost (Sharath and Jagadeesh, 2004a).

Enrichment with plant growth promoting organisms

Nutrient status of the compost can be enhanced through inoculation with plant growth promoting organisms. Bacterial inoculants such as *Azotobacter*, *Azospirillum* are added to the decomposing materials at the rate of 2 kg/tonne after the thermophilic phase i.e. after a period of 30-45 days of initial decomposition. These nitrogen-fixing bacteria increase in population and enhance the nitrogen content. The fungal cultures such as *Trichoderma* spp. or bacteria like *Bacillus* spp. are also inoculated after 30-45 days initial decomposition, which on increasing in population improve the quality of the final product. These organisms can also help in suppressing root pathogens (Bagyaraj & Radhakrishna, 2000). The lignolytic and cellulolytic fungi degraded polymers and produced monomers (glucose), which serve as carbon and energy sources to *Azotobacter*. Increased population and nitrogen-fixing activity of *Azotobacter* increased the nitrogen content in the biocompost (Sharath & Jagadeesh, 2004a).

Period of Decomposition

The rate of composting depends upon chemical composition of the substrate, the percentage of amorphous and crystalline regions in the fibre structure, the three dimensional arrangement of lignin and cellulose and finally the kinetic parameters.

The period of decomposition varies for the different types of organic wastes. For example, the urban solid wastes with high quantity of biodegradable substrate take 45 days to degrade. Coir pith composting takes about 150-180 days. Crop residues take about 90-120 days for composting. Pressmud takes about 60-90 days under optimum conditions of moisture, temperature and aeration with suitable microbial inoculation.

Final Product

The physical and chemical properties of the final compost depend on the type of raw materials used viz., agricultural, municipal and industrial wastes, along with the extent of decomposition by soil microorganisms. The compost may have a C/N ratio of 15 - 20 and organic matter content of 8 - 10 percent. Compost is considered superior if the C/N ratio is 20 or less and organic matter content is 30 - 60 percent.

After a period of decomposition, the composted materials are removed from the yard, sieved in 4 mm sieves. The fine compost obtained, containing moisture of 30-40%, is stored in shade for further use. It is recommended that the packets must indicate specifications regarding the raw materials used and their origin and the chemical composition of the product, as follows:

1. Origin
2. Composition
3. Moisture content
4. pH
5. Electrical conductivity
6. C/N ratio
7. Cation exchange capacity
8. Organic matter content
9. Mineral nutrients content
10. Heavy metals concentration
11. Inert materials.

III. BY-PRODUCTS OF SUGAR INDUSTRY AND DISTILLERIES

The by-products of sugar industry and distillery are molasses, bagasse, press mud and spentwash (Gaur, 2002). Molasses serve as raw material for production of alcohol and rectified spirit. With production of 1 litre alcohol nearly 7-15 litres of spentwash is being discharged into the environment, depending on the process adopted.

Bagasse

About 5 million tons (mt) of dry bagasse is annually produced in the country. Bagasse is a fibrous residue left after extraction of sugarcane juice. The average composition of dry bagasse is cellulose 45%, lignin 20%, pentosans 28%, ash 2% and sugar 2%. Bagasse ash is rich in silica, potassium and iron. Bagasse produced in India is almost entirely used as fuel in sugar factories.

Press Mud

Press mud is a by-product of cane sugar mills, which is of the tune of 3 percent of the total cane crushed. About 5.6 mt of press mud is produced annually. As the constituents of press mud have been derived only from sugarcane, press mud does not contain any harmful or toxic substances deleterious to soil or plants (Table 21.6). Press mud is soft, spongy, light weight, dark brown to brownish white in colour containing sugar, fibre, coagulated colloids, cane wax, inorganic salts etc. This press mud is a finely pulverized organic material with higher contents of nutrients when compared to other bulky organic manures. In addition to this, it also contains important growth promoting substances (Padmanaban *et al*, 1993).

Table 21.6: The typical composition of press mud is as follows

Composition	*Concentration*
Organic carbon	22 - 35 %
Nitrogen	1.2 - 2.5 %
Phosphorous	3.83 %
Potassium	1.46 %
Calcium	3.20%
Magnesium	1.65 %
Sulfur	0.23 %
Copper	52 ppm
Zinc	69 ppm
Iron	24 ppm

Spentwash

It contains all the mineral nutrients whatever the sugarcane crop has taken up during its growth from soil. The effluent termed, "Spentwash" is dark in colour and is essentially containing residual nutrients from sugarcane and yeast cells. Spentwash is non-toxic and does not contain any heavy metals or other toxic residues as in Table 21.7. But it is highly polluting because it has a high BOD (45,000 to 50,000 ppm) and COD (100,000 to 200,000 ppm), high dissolved solids and organic matter concentration. It is highly acidic with pH around 4. Organic matter present in the spentwash is the cause of high BOD and COD. Most of the monosaccharides and disaccharides during alcoholic fermentation is used up. The residues are pentoses and their derivatives, peptides, proteins, phenolic compounds etc. and by-products of fermentation along with viable autolysed yeast cells. Further breakdown of spentwash is normally required before its use and disposal. One of the methods is to convert it to methane under anaerobic condition by passing through biogas plants with long detention period.

Table 21.7: The composition of typical spent wash is as follows:

Parameters	Values
pH	4.0 - 4.5
Total solids	2,20,000 - 2,30,000 mg/l
BOD	45,000 - 50,000 ppm
COD	1,00,000 - 2,00,000 ppm
Dissolved solids	1,80,000 - 2,00,000 mg/l
Suspended solids	15,000 - 25,000 mg/l
Acidity ($CaCO_3$)	16,000 - 18,000 mg/l
Sulphates	11,000 - 15,000 mg/l
Chlorides	11,000 - 14,000 mg/l
Sodium	3,000 - 5,000 mg/l
Ammonical nitrogen	500 - 700 mg/l

Yeast Sludge

Yeast is added to fermentor for production of alcohol from molasses. After fermentation is over, yeast sludge is separated as waste product. Yeast sludge contains proteins, vitamins and minerals. It can be composted with press mud and spentwash. Yeast sludge is also used as feed for cattle, poultry and fishes.

IV. SALIENT POINTS OF COMPOSTING PRESS MUD AND SPENTWASH

1. Composting should be carried out on impervious compost yard as prescribed by the Central Pollution Control Board, to avoid seepage of leachates.
2. Aerotillers can be used for aeration.
3. Press mud : spentwash in the ratio of 1:2.5 to 3 is usually used (Table 21.8).
4. Well-lined storm water collections (garland canals) to be provided to collect main water run off and leachates. The leachates collected can be reused in the composting process to prevent groundwater pollution.
5. Use of microbial inoculation (i) to hasten the process and (ii) to improve the quality is recommended.
6. Composting, thus, is an eco-friendly technology to check the pollution of surface and groundwater with spentwash.
7. Findings show that the manurial value of spentwash compost is higher than farmyard manure.

Press mud is spread in the compost yard to form windrows of 1.5 m height × 3.5 m width × 100-300 m length. Activator culture from an authentic source @ 2 kg/ton of press mud is

Table 21.8: Chemical analysis of press mud and spentwash

Organic material	Organic C%	Total N%	C/N ratio	pH	BOD (ppm)
Spentwash	25.0	1.87	15.5	4.3	45000
Press mud	33.0	1.50	22.0	6.5	–

diluted in water to form 100-200 litres and applied to the windrow. After 3 days, distillery effluent is sprayed on heaps to a moisture level of 60% and allowed for 4-5 hours to absorb the effluent. The heaps are then mixed by aerotiller, this machine agitates, aerates and shreds the composting mass, and, thus, lumps of press mud are shredded to uniform fine particles. This increases the surface area of the press mud and, hence, increases the absorption of the spentwash. The mixing also helps in spreading of the microbial inoculum uniformly to the entire length of windrows and also helps in dissipating the heat that is liberated due to the metabolic activities of the organisms. When moisture level drops to 40-50%, again effluent is sprayed and mixed. Effluent can be sprayed once or twice in a week depending on the moisture content of the heap. Press mud can be mixed with alternate substrate like coir pith, sugarcane trash, crop residues, yeast sludge, boiler ash etc. In Table 21.9 we can see the composition of compost prepared from press mud and distillery spentwash.

Table 21.9: Typical composition of biocompost prepared using press mud *

Moisture content	30-35%
Organic carbon	23-26%
Total nitrogen	1.5-2.5%
Total phosphorus	1.5-2.5%
Total potassium	2.5-3.5%
Calcium	30-50%
Magnesium	1.5-2.0%
Sulphur (as SO_4)	2.0-3.0%
Iron	1000-2000 μg/g
Zinc	150-200 μg/g
Manganese	150-175 μg/g
Copper	35-40 μg/g
Boron	3 μg/g
Molybdenum	<2 μg/g
pH	7.0-7.3

* The composition may vary within limited ranges in the biocompost obtained from different sugar factories/ distilleries.

V. STABILITY AND MATURITY OF THE COMPOST

The ultimate goal in composting is to produce a humus-like product that can be used for soil improvement and plant growth, and to reduce environmental pollution (Miller, 1993; Bagyaraj & Radhakrishna, 2000). The compost, thus, prepared must be allowed to stabilize and mature. Stability is a stage in the decomposition of organic matter and represents the state of microbial activity. Maturity is an organochemical condition of the compost, which indicates the presence or lack of phytotoxic organic acids. Thus, the two terms "stability" and "maturity" are not synonymous, though often used interchangeably. As stability represents the state of microbial activity, measurement of respiration through CO_2 evolution is often done. If the compost is stabilized it will have less microbial activity (as it would have utilized all degradable components in the substrate) and, thus, less CO_2 evolved through respiration.

Immature composts induce high microbial activity in soil potentially causing oxygen deficiency in soil and biological blockage of available soil nitrogen (Zucconi *et al*, 1981). They also leave odoriferous phytotoxic compounds into the soil viz., ammonia, ethylene oxide and low molecular weight fatty acids (Saviozzi *et al*, 1988). Plant health and disease suppressive properties of compost are affected by maturity, hence, the compost maturity and stability is of great importance. The recommended compost stability index is given in Table 21.10.

Table 21.10: Compost stability index based on carbon dioxide evolution

Respiration rate (mg CO_2-C/g compost C-day)	*Rating*	*Characteristics*
< 2	Very stable	Well cured; no malodours; earthy odour
2-5	Stable	Cured compost; minimal impact on soil dynamics
5-10	Moderately stable	Uncured compost; some malodour; addition to soil may immobilize N; high phytotoxicity potential; not recommended for growing plants from seed
10-20	Unstable compost	Very immature compost; high malodour and phytotoxicity potential; not recommended for growing plants from seed.
>20	Very unstable compost	Extremely unstable material; very high malodour and phytotoxicity potential; not recommended for use

Alternatively higher levels of ammonical N in the compost indicates that it is not cured, whereas presence of nitrate N indicates it is properly cured (ripened).

A plant assay is used to evaluate maturity of the compost. Zucconi *et al* (1981) studied the cress (*Lepidium sativum*) seed germination tests and established a germination index obtained by multiplying the % germination by the % root growth as related to control. The presence of phytotoxic compounds indicate that the compost is not matured. The general compost quality standard is given in Table 21.11.

Table 21.11: Standards for evaluating the quality of compost

Parameters	*Good quality*	*Poor quality*
Color	Brown to black	Varies
Smell	Pleasant earthy smell	Bad odor
pH	6.5-8.0	<6, >8
C/N ratio	10-20	<10, >20
Temperature	30-45°C	>45
Moisture	25-30%	>30
Humus percentage	>4%	<4%
Plant assay	Good growth	Growth inhibition
Nitrogen	>1.25%	<1%

VI. METHODS COMMONLY USED FOR TESTING THE QUALITY OF THE COMPOST

Though there are several methods for bio-assaying the quality of the compost, some of the commonly used methods are

1. Seed germination in compost tea
2. Seed germination in compost
3. Plant growth in compost.

Seed Germination in Composit Tea

This test will help to determine the compost quality and check whether it is finished. This method employs the use of fast germinating seeds like radish, cowpea, soybean etc. It is found that immature compost may contain phytotoxins that will often kill seed embryos. Seeds grown in immature compost will not sprout or may die immediately after sprouting.

Materials required

Compost sample, plastic cups, pure water, cheesecloth, petriplates, germination paper, dropping bottle.

Procedure

1. Take the cup and fill ¼ of the cup with the compost to be tested.
2. To this compost filled cup, add ½ cup of water.
3. Allow this set up to soak for 2 hours.
4. After sufficient soaking, filter the wet mixture through cheesecloth and collect the filtrate, called compost tea.
5. Take periplates, and put pieces of germinating paper at the bottom.
6. To these petriplates add 10 seeds, evenly distributed along the surface of the germinating paper.
7. With the help of dropping bottle add sufficient amount of water to moisten the petriplates containing germinating paper and seeds, this will serve as control.
8. To another pertiplate containing germination paper and seeds, add sufficient compost tea to moisten the treatment. Three replications of each of the control and treatment are kept.
9. Keep the experimental set up in a dark place at room temperature (27°C) under laboratory conditions. Observe and record of germination after 24, 48, 72, and 96 hours. Calculate average germination and percentage of average germination.

$$\text{Average \% germination} = \frac{\text{Average number of seeds germinated} \times 100}{10}$$

	Number of seeds germinated					
	Time	Plate 1	Plate 2	Plate 3	Average	% Average
Control	24 Hours 48 Hours 72 Hours 96 Hours					
Compost tea	24 Hours 48 Hours 72 Hours 96 Hours					

Compare percent average germination rate in compost tea treatment to germination rate in control. 80% germination rate is considered satisfactory.

Seed Germination in Compost

This method helps us to test the compost for maturity, based on seed germination rates.

Materials required

Compost sample, soil, pots, seeds.

Procedure

1. Fill the pots with soil, which serves as control.
2. Fill the pots with compost, which serves as compost treatment.
3. Sow 10 seeds in each of soil and compost pots. Keep 6 replications of soil and compost treatments.
4. Water the pots regularly to keep the pots moist.
5. After a week, count and record number of seeds germinated in each container and calculate germination rate.

	Number of seeds germinated							
	Pot 1	Pot 2	Pot 3	Pot 4	Pot 5	Pot 6	Average	% Average
Control								
Compost								

$$\text{Average \% germination} = \frac{\text{Average number of seeds germinated} \times 100}{10}$$

If the germination rate in the compost pots is less than the germination rate in soil control pots, then the compost is considered to be immature and needs further curing.

Plant Growth in Compost

This test helps to measure the quality of the compost and to see if it is providing nutrients to plants or, potentially robbing nutrients from plants.

Materials required

One week old seedlings, compost samples, soil, pots

Procedure

1. Fill the pots with soil and another set of pots with compost sample, which serves as control.
2. For mixed treatment mix the pots with soil – compost mix (75% soil + 25% compost). Six replications are kept for each of the treatments.

3. Plant 3 seedlings in each of the three treatments and keep the pots in green house.
4. Water the pots regularly to provide optimum condition. This experimental set up is kept for 30 days.
5. After 30 days, harvest the plants and record root weight, shoot weight and calculate average weight of the plant.

The plants in the soil-compost mix and 100% compost pots should be healthy and weigh equal or better than plants in 100% soil control. Poor plant growth indicates that the compost is unfinished. Microbes are using nutrients in the compost and soil to decompose the organic matter and, therefore, nutrients are not yet available to the plants.

VII. SOME OF THE COMMON PROBLEMS FACED DURING COMPOSTING AND ITS REMEDIES

Problem	*Cause/clues*	*Solution*
Nothing rots	• Not enough moisture	• Add enough moisture
	• Material too woody and not enough nitrogen	• Shred the woody substrate till optimum particle size
	• Pile too small	• Add nitrogen rich materials
Rotten and Sulfurous odor	• Too many food scraps or food wastes	• Avoid food wastes, or mix food wastes evenly throughout the pile
	• Too many grass clippings	• Mix grass clippings with carbon rich dry substrate
	• Anaerobic conditions	• Open pile apart, rebuild the pile loosely with different materials
	• Very small particles	• Turn pile while adding material of different sizes and coarseness. Layer pile with plenty of loose and bulky material
	• Too wet	• Cover the pile during monsoon, turn the pile while adding some dry, high carbon bulking agent, which absorbs excess moisture
	• Pile too large	• Decrease the pile size
Ammonia smell, mostly when turning the pile	• High nitrogen level	• Add high-carbon amendments
	• High pH	• Lower pH with acidic ingredients and/or avoid alkaline ingredients
	• Slowly available carbon source, like Large woody particles; C:N ratio less than 30:1	• Use another carbon amendment or increase the carbon proportion
	• Anaerobic condition	• Turning the pile helps circulating oxygen
Site related odors (piles not odoros)	• Raw materials, odour is characteristic of the raw material	• Handle raw materials promptly with minimal storage
	• Nutrient rich puddles because of poor drainage, standing puddles of water	• Divert runoff away
Pile has low temperature	• Gradually falling temperatures; pile does not reheat after turning or aeration as composting nearing completion	• None required
	• Low moisture, cannot squeeze water from materials	• Add water and remix

Contd.

Problem	Cause/clues	Solution
	• Pile too small, or if the surroundings are cool	• Rebuild the pile with optimized parameters, if cold surroundings increase the pile size.
	• High C/N	• Add sufficient nitrogen rich materials
	• Too dry or too wet	• If too moist, the pile will have an odour, then rebuild the pile and add dry carbonaceous material. Cover when raining. A dry pile having no odour, rebuild the pile while wetting the materials, a cover may be needed to retain the moisture
	• Material too coarse or too fine	• Add materials having sufficient particle size to counteract the effect of small or coarse sized particles
	• Anaerobic conditions	• Aerate the pile
	• pH low, pH measures less than 5.5; garbage-like odour	• Add lime or wood ash and remix
Pile too hot	• Insufficient aeration for heat removal, pile is moist	• Turn pile to increase airflow
	• Moderate to low moisture; limited evaporative cooling, pile feels damp but not excessively wet or dry	• Add water; continue turning and aeration to control temperature
	• Pile is too large	• Decrease the pile size to optimum
	• Too much nitrogen	• Turn pile while incorporating more carbon-rich materials
	• Pyrolysis or spontaneous combustion, low moisture content; pile interior looks or smells charred	• Decrease pile size; maintain proper moisture content; add water to charred or smoldering sections; break down pile, combine with other piles
	• Compost is not stable, High temperatures or odours in curing or storage pile	• Manage pile for temperature and odour control, turn piles as necessary; limit pile size
Uneven temperatures or varying odours in pile	• Materials at different stages of maturity, temperature varies along the pile length	• None required
	• Poorly mixed materials, visible differences in the pile moisture and materials	• Turn or remix pile
	• Uneven airflow or air short-circuiting, visible differences in the pile moisture and materials	• Remix pile
Compost contains clumps of materials and large particles; texture is not uniform	• Poor mixing of materials or insufficient turning, original raw materials discernible in compost	• Screen compost; improve initial mixing
	• Airflow uneven or short-circuiting, Wet clumps of compost	• Screen or shred compost; improve air distribution
	• Raw materials contain large particles and non-degradable or slowly degradable materials	• Screen compost; grind and/or sort raw materials

Contd.

Problem	Cause/clues	Solution
	• Active composting not complete	• Lengthen composting time or improve compost conditions.
Flies and insects in and around the pile	• Flies breed in compost piles, fresh manure or food material at pile surface; flies hover around piles	• Turn piles every four to seven days; cover static piles with a 6-inch layer of compost, or add lime and soil to top of the compost material and turn it the next day.
	• Flies breeding in raw materials, wet raw materials stored on site more than four days	• Handle raw materials promptly
	• Too much food wastes	• Add the food wastes only to the middle of the pile, if problem persists discontinue the use of food wastes
Mice and rats in pile	• Food wastes	• Try avoiding meat, dairy, animal feces, vegetable and fish wastes
	• Presence of burrows	• Aerate the piles as per the optimized parameters

VIII. CONCLUSION

(i) Compost can supply all the 14 essential nutrients for plant growth obtained from soil.

(ii) Compost acts as a soil conditioner as it improves physical condition of the soil such as soil aggregation, soil structure, water holding capacity and cation exchange capacity. Thus, it provides a plethora of goodies like efficient seed germination, easy root penetration, adequate root aeration etc.

(iii) It is an effective material providing resistance of soil to slacking, surface sealing, wind erosion etc.

(iv) It improves biological activity of soil, especially the enriched compost enhances the numbers of plant growth promoting organisms and those biologically controlling root pathogens.

(v) Most of the human and animal pathogens and weeds are killed during the thermophilic phase.

(vi) It reduces the application of chemical fertilizers deleterious to the environment.

(vii) It has better residual effect, hence helps subsequent crops.

(viii) Humic and fulvic acids present in compost play an important role in promoting plant growth.

(ix) Presence of plant growth promoting substances like gibberellins, IAA and triacontenol etc. in compost needs investigation.

(x) Composting technology has added value to press mud which was once considered a waste from sugar industry.

(xi) It helps to increase the efficiency of minerals and biofertilizers

(xii) More than anything it mitigates environmental pollution by spentwash coming out of distilleries. Thus, it is a boon to distilleries to transform spentwash which is considered to be a pollutant to eco-friendly product which helps the farming community.

REFERENCES

Acharya, C. N. 1939. Comparison of different methods of composting waste materials. Ind. J. Agril. Sci. 4: 135-149.

Bagyaraj, D. J. & Radhakrishna, D. 2000. Recent advances in composting. *in* L. V. Gangawane, Ganju, Y. Thomas S. & Aggarwal S. L. (eds.), Vistas in Mycology and Plant Pathology. Commonwealth Pub., New Delhi. pp. 88-100.

Balakrishna, A. N., Srinivas, K. N. Radhakrishna D. & Gowda, T. K. S. 1991. Enrichment of municipal waste compost (MPC) with urea. Curr. Res. 20: 160-162.

Biddlestone, A. J. and Gray, K. R. 1985. Composting. Vol 4. *in* C. W. Robinson, & J.A. Howel (eds.), Comprehensive Biotechnology. Pergamon Press, Oxford, U.K.

Davis, C. L., Hinch, S. A. Donkin C. J. & Germishuizen, P. J. 1992. Changes in microbial population numbers during the composting of pine bark. Biores.Tech. 39: 85-92.

Farrell, R. I. 1987. Industrial application of lignin transforming enzymes. Philosophical transactions, Royal Society of London. A 321: 549-553.

Gaur, A. C. 1996. Biomass potential and composting. *in* Proceedings of National Seminar on Organic Farming and Sustainable Agriculture. UAS, Bangalore. pp. 27-32.

_____. 1997. Biomass potential and composting. *in* G. K. Veersh, K. Shivashankar & M. A. Singlachar (eds.), Organic Farming and Sustainable Agriculture. APOF Pub., Bangalore. pp. 27-31.

_____. 1999. Microbial Technology for composting of Agricultural Residues by Improved Methods. ICAR Pub., New Delhi.

_____. 2002. Organic manure from distillery wastes. *in* Towards Sustainable Distillery Spentwash Management with Recycle-Reuse-Reduce-Value Add Concepts. SIMSA Pub., Chennai. pp. 50.64.

Gervais, P., Marechal P. A. & Molin, P. 1996. Water relations of solid state fermentation. J. Sci. Ind. Res. 55: 343-357.

Gowda, T. K. S. 1996. Compost making and enrichment techniques. Pages 46-51 *in* Proceedings of National Seminar on Organic Farming and Sustainable Agriculture. UAS, Bangalore. pp. 46-51.

_____. 1997. Compost making with enrichment techniques. *in* G. K. Veersh, K. Shivashankar & M. A. Singlachar (eds.), Organic Farming and Sustainable Agriculture. APOF Pub., Bangalore. pp. 46-51.

Hagerty, D. J., Pavoni J. I. & Jr. Heer, J. E. 1973. Solid Waste Management. Von Nostrand Reinhold, New York.

Hajra, J. N. & Kole, S. C. 1998. Enrichment of compost with inoculants. Biofert. Newsletter. 2-4.

Howard, A. 1940. An Agricultural Testament. The Other India Press, Goa.

_____. & Wad, Y. D. 1931. The Waste Products of Agriculture and Their Utilization as Humus. Oxford University Press, London.

Hutchinson, H. B. & Richards, E. H. 1921. Artificial farmyard manure. J. Min. Agril. 28: 398.

Khanna, S. & Gauri, V. 1993. Regulation purification and properties of xylanase from *Cellulomonas fimi*. Enzyme Microb. Technol. 15: 990-995.

Miller, F. C. 1993. Composting as a process based on the control of ecologically selective factors. *in* F. D. Metting (ed.), Soil Microbial Ecology. pp. 514-544.

_____, E. R. Harper & Macauley, B. J. 1989. Field examination of temperature and oxygen relationship in mushroom composting stacks, consideration of stack oxygenation based on utilization and supply. Aust. J. Exp. Agri. 29: 741-750.

Murkute, S. B., Thopate, A. M. More B. B. & Hapase, D. G. 1992. Biodegradation of agro-industrial wastes. Proceedings of National Seminar on Organic Farming. pp. 67-68.

Padmanaban, P., Rakkiyappan P. & Alexander, K. C. 1993. Sugarcane Breeding Institute Extension Publication No. 39.

Radhakrishna, D. & Bagyaraj, D. J. 1998. Microbial degradation of solid wastes as a source of plant nutrients. Pages 108-117 *in* B. Rai & I. S. Dkhar (ed.), New Trends in Microbial Ecology. N.E. Hill Univ. Pub., Shillong. pp. 108-117.

Rahman, M. D. & Richards, G. N. 1988. Interactions of starch and other polysaccharides with condensed tannins in hot water extracts of ponderosa pine bark. J. Wood Chem. Technol. 8: 111-120.

Saviozzi, A., Leviminzi, R. & Riffaldi, R. 1988. Maturity evaluation of organic wastes. Biocycl. 29: 54-56.

Sharath, S. & Jagadeesh, K. S. 2004a. Field scale composting of wood waste generated in a paper mill. Bull. Agri. Sci. 1: 9-18.

_____ & Jagadeesh, K. S. 2004b. Composting of wood waste of a paper mill: optimization of parameters. Bull. Agri. Sci. 1: 19-26.

Singh, S., Mishra, M. M. Goyal S. & Kapoor, K. K. 1992. Preparation of nitrogen and phosphorous enriched compost and its effects on wheat (*Triticum aestivum*). Ind. J. Agri. Sci. 62: 810-814.

Viesturs, U., Leite, M. Shvinka, J. Katkevics, K. J. Zeltina M. & Apine, A. 1992. Biotechnology for the conversion of photosynthetic products. Lignocellulose (Cellucon). 90: 13-24.

World Bank Technical Report on Bark and Sawdust Composting. 1995. Vol. 264: 63-65.

Zucconi, F., Pera, A. Forte A. & De Bertoldi, M. 1981. Evaluating toxicity of immature compost. Biocycle. 22: 54-57.

22

Seed Mycoflora and Aflatoxin Contamination in Mustard During Storage

MANJU TOMAR-BALHARA, RUPAM KAPOOR and A.K. BHATNAGAR

Department of Botany,
University of Delhi,
Delhi

ABSTRACT

Mustard, *Brassica juncea* (Linn.) Czern. & Coss. is an important oilseed crop of India. The harvesting season of mustard is followed by hot summer and then monsoon rains. Storage of seed under warm and humid conditions leads to moistening, favouring fungal growth. In recent years, the problem has been compounded by reduced demand, due to excessive import of various edible oils, which compels the farmers and the procurers to store the seeds for a year or more. As many as 25 species of fungi have been isolated from 150 samples of one year old and fresh seed collected during 2001-2002 from different districts of Rajasthan and Haryana. Heavy infestation results in the poor quality of stored seed and the oil extracted is not of a high quality.

Mustard seeds are prone to aflatoxin contamination during storage. Mustard seeds form good substrate for growth of toxigenic strains of *Aspergillus flavus*. In one year old stored seed samples collected from Rajasthan and Haryana, it was observed that 80% samples were contaminated with aflatoxins. Aflatoxin contamination in mustard seeds may be ascribed to several environmental, agronomic and biological factors among which the temperature and humidity of the storage site are most significant.

Much emphasis needs to be laid on improvement of storage environment to combat this problem through effective pre- and post-harvest operations. The losses can be minimized by adopting the necessary cultural operations, careful handling and effective storage management practices.

I. INTRODUCTION

India is the second largest rapeseed-mustard growing country in the world. Rapeseed and mustard are together the second most important edible oilseeds after groundnut. Rapeseed-mustard oilseed crops are important sources of edible oil in the Indian diet. Mustard oilseeds are grown in about 5 mha in India and the annual production is 5.8 million tonnes (Economy Database, 2004). Oilseed sector as a whole, and rapeseed and mustard in particular, has witnessed a significant increase in production since 1986 (Kumar, 1999). Increased production of rapeseed and mustard is because of development and widespread adoption of high-yielding varieties, region-specific agro-production and protection practices and an attractive support price. After launch of the Technology Mission on Oilseeds in 1986, irrigated area under these crops increased steadily from 51.8 percent to 68.8% in 1996-97. The production of rapeseed-mustard in India during 1985-86 increased from 2.68 mt from hectrage of 4.83 mha and yield of 650 kg/ha to 6.66 mt from 6.55 mha with a successive increase in yield to a level of 1017 kg/ha in 1996-97.

In India *Brassica juncea* is the dominant species covering around 75% of the area under brassicas. Seven important oilseeds of family Brassicaceae are cultivated worldwide:

- Brown Indian mustard (*Brassica juncea*);
- Black mustard (*Brassica nigra*);
- White mustard (*Brassica alba*);
- Brown sarson (*Brassica campestris* var. brown sarson);
- Yellow sarson (*Brassica campestris* var. yellow sarson);
- Toria (*Brassica campestris* var. toria);
- Rape or rapeseed (*Brassica napus*).

II. USES OF RAPESEED-MUSTARD

Rapeseed-mustard is widely and extensively grown as a vegetable. It is also grown for its seeds, which yield essential oil and serve as condiment. Mustard oil is primarily used as cooking medium but it has many applications for other purposes. Young and tender leaves of mustard are used in salads. Green leaves of mustard are often cooked and used in soups and stews. Mustard seed has been known as a spice from early times. The oil extracted from mustard is used in the preparation of pickles, cooking, frying and for flavouring curries and vegetables. It is a folk remedy for arthritis, footache, lumbago and rheumatism (Duke & Wain, 1981). Rapeseed-mustard meal is a good source of protein for livestock and poultry. Rapeseed oil is used for the production of rubber additives in USA, England, Germany and Sweden. Rapeseed oil has good lubricating properties and is traditionally used as lubricant for metal surface.

III. CULTIVATION AND HARVESTING OF RAPESEED-MUSTARD

The crop is grown in India as a rabi (winter) crop. The seeds are sown usually in October-November. The crop requires a good sandy loamy soil with about 50-75 kg N, 100-150 kg acid phosphate and 70-75 kg potash per hectare. The crop requires cool (18-25°C) climate and low humidity for satisfactory growth. The crop is generally harvested in March-April. The time of harvesting depends upon the sowing time, kind of variety grown, management practices adopted and environmental conditions to which the crop is exposed. Harvesting of the crop is carried out when the majority of siliquae turn from green to yellow. Harvesting is done with hand-sickels and threshing is carried out by beating the fruits with wooden sticks or by trampling the plants by bullocks.

IV. POST-HARVEST HANDLING AND STORAGE OF RAPESEED-MUSTARD

Seeds of all species can be classified in two categories: (i) Orthodox seeds: these seeds can be dried to a low moisture level, of 5% or less, and can be frozen without damage. (ii) Recalcitrant seed: these seeds cannot be dried to low moisture levels and cannot be frozen without loss of viability. Rapeseed-mustard seeds fall in the first category of orthodox seeds. However, mustard oilseeds are difficult to handle because of their small size and high mobility. Seeds are stored through summer and monsoon and much of the oil extraction is carried out in the months of September and October. In India, the traditional methods of storing oilseeds are not scientific. Traditionally farmers store the mustard in unthreshed state. This is because they often do not have time to thresh the seeds after harvesting. After threshing, the seeds are kept in a variety of traditional storage containers or gunny bags.

It has been estimated that one-fourth of the mustard produce is lost due to poor storage conditions. In recent years, the problem has been compounded by reduced demand, due to excessive import of various edible oils, which has forced the farmers and procurers to store the seeds for a year or more.

Most often the seeds are stored in warehouses in gunny bags. Before storage, the seeds are not subjected to any protective treatment. As moisture-testing facilities are not available in the fields, it is not feasible to optimize the storage conditions of seeds. This results in the poor quality of stored seeds, and the oil extracted too is not of a high quality. Some of the hazards faced by the seeds during storage are moisture, high temperature, poor ventilation, pests and pathogens.

V. FACTORS AFFECTING QUALITY OF MUSTARD SEEDS DURING STORAGE

Good quality seeds act as a catalyst for realizing the potential of their inputs. Availability of viable and vigorous seeds at planting time is important for achieving targets of agricultural production. Seeds lose their vigour and viability on storage. The loss of vigour precedes loss in germination. Neither the loss of viability during storage can be stopped nor the viability of seeds during storage can be increased (Agarwal, 1988). However, the loss of viability during storage can be minimized. The factors affecting quality of mustard seeds during storage are given below:

Seed Maturity

In *Brassica* species the inflorescence is a raceme and flowering as well as seed formation proceed acropetally i.e., from base toward the tip. Seeds mature in the lower pods first. If the crop is harvested pre-maturely, the terminal pods (siliquae) may still have immature seeds with high moisture content. Harvesting at green pod stage results in decline in oil content by 3-4%. In immature seeds, the chlorophyll concentration is high which gets extracted into the oil upon processing. The separation of chlorophyll from the oil is a difficult process involving cumbersome solvent extraction procedure. The pigment, in addition to imparting an undesirable colour, makes the oil less stable and rancid (LÖÖf, 1972).

Dockage

Dockage consists mainly of weed seeds, broken seeds, broken siliquae, dust and sand. The dockage normally has a moisture level 3-4% higher than that of mustard seeds. Large amounts of broken and foreign seeds influence the rate of respiration and provide a substrate for colonization by fungi. The presence of alien matter may be detrimental for the processing equipment and it may further lower the feed value of the meal.

Drying of Seeds

In India, there is generally little need for artificial drying, since the air is warm and dry during the time of crop harvest. However, due to unseasonal rainfall, seeds may be subjected to high moisture content, which may spoil the seeds during storage. Artificial drying of seeds can be recommended in such conditions. Sun-drying is considered the most effective way to reduce the moisture content. This does not involve high cost of equipment or labour. However, if the moisture is actually much greater than the permissible limits at a particular temperature, then drying machines should be used. Delay in drying of seeds causes observable damage to the seeds. Drying temperature plays an important role in maintaining the quality of the seeds. Often the seeds are over dried, resulting in cracking and loss of viability. Overheating damages the seeds and the heat-damaged seeds give an oil that is more susceptible to oxidation and rancidity (Appelqvist & LÖÖf, 1972). Moreover, non-viable seeds are more prone to infection.

Moisture Content

Moisture content of seeds prior to storage is the most important factor that determines the establishment, development and growth of fungi during storage. A dry seed is in a state of metabolic inertia. Barton (1961) reported that variation in the chemical composition of the oilseeds is responsible, at least partly, for the amount of moisture absorbed even under identical storage conditions. The moisture level of the mustard seeds plays an active role in determining its quality under storage. It is evident that higher moisture in seeds provides favourable conditions for growth of fungi, thereby escalating deterioration and finally leading to rapid decline in germinability. Various storage fungi have different moisture content requirements below which they fail to develop or may remain dormant within seeds without causing any damage. At higher moisture content fungi grow faster and the types of fungi and damage caused can change. *Aspergillus* spp. are reported to grow even at 65% moisture content corresponding to 7% moisture in the seed (Appelqvist & Loof, 1972).

The role of moisture content in grain infestation has been reviewed (Agarwal, 1988). Generally, storage fungi grow at moisture contents in equilibrium with relative humidity of 65 to 100%, which reduces seed germination and causes various biochemical changes. Christensen

(1973) reported that the major ecological character for storage fungi such as species of *Aspergillus* and *Penicillium* is their ability to grow in seeds with moisture content in equilibrium with 70 to 90% relative humidity. Greer (1953) studied the germination and growth conditions of various fungi. His observations revealed that 80 to 90% relative humidity was necessary for the members of mucorales. For *Penicillium* spp. 75% relative humidity, and for *Aspergillus* spp. 75% relative humidity was required but some species develop even at 65% relative humidity.

Fungi can begin to grow in pockets of moist oilseeds and produce more heat and moisture. The resulting increases in temperature and moisture cause spurt in the growth and respiration of fungi.

Temperature

High temperature kills the seeds, rendering them non-viable. Seeds stored at low temperature and low relative humidity can be preserved well for a long period. In oilseeds when biological respiration ceases because of the high temperature, chemical oxidation of the oil in the seed may continue. Temperature and moisture influence enzymatic and biological activities, which may lead to spoilage and self-heating (auto-oxidation). At high moisture levels the seed respiration rate becomes high, resulting in rise in temperature and a further increase in moisture content due to respired water.

Some fungi can grow and spread at temperatures as low as –10°C but usually little damage occurs until the grain temperature is above 15°C. Maximum growth occurs at 30 to 35°C and growth ceases at about 60°C. The range of minimum, optimum and maximum temperatures required for the growth of most storage fungi is 0 to 5°C, 30° to 35°C and 50 to 55°C respectively. Temperature below 20°C tend to favour cold tolerant fungi such as species of *Cladosporium* and *Penicillium,* while higher storage temperatures favor *Aspergillus* spp. Under tropical conditions seeds are more susceptible to *Aspergillus* than other fungi.

Christensen (1973) found that at temperature below 10°C storage fungi that invade seeds at moisture contents in equilibrium with relative humidity up to 85% would grow very slowly. Low temperature can be used as a substitute for control of high moisture content. If the seeds are already infected by fungi, low temperature or humidity is ineffective in inhibiting their subsequent growth.

Infestation of Fungi

Deterioration of seeds by fungi is a serious problem in warm and humid climate. Infestation of seeds by fungi may occur while the crop is maturing in the field. Two types of fungi invade seeds: field fungi and storage fungi.

Field fungi

The field fungi infect seeds during their development on maturing/drying plants in the field or later when the harvest is lying in the field. Species of *Alternaria, Cercospora, Colletotrichum, Drechslera, Phoma, Curvularia, Epicoccum, Nigrospora* and *Cladosporium* are included under field fungi.

Field fungi need high moisture content for their growth and, hence, damage caused by field fungi causes little or no damage under ordinary conditions. However, heavy rainfall at the time of harvesting often results in extensive seed deterioration. Field fungi require moisture content in equilibrium with a relative humidity of more than 90% and, hence, are not likely to continue development during storage (Christensen & Kaufman, 1965).

Storage fungi

Storage fungi are those which develop on and within the seeds at high moisture content. Storage species of *Aspergillus* and *Penicillium* are the important storage fungi. Other genera of storage fungi are *Absidia, Chaetomium, Mucor* and *Rhizopus.*

Invasion of seeds by storage fungi leads to an increase in respiration, water content and temperature, resulting in damage to the embryo and eventually loss of seed viability (Neergard, 1977). Seeds with a high rate of respiration, their own and of the fungi, attain high temperature and moisture. Due to invasion of these fungal species the stored seeds are likely to clump together. Fungi produce mycotoxins in the stored seeds.

The major deleterious effects of storage fungi are:

- Reduced viability
- Discolouration
- Heating of seeds
- Increase in level of hydrolytic enzymes
- Production of mycotoxins
- Seed caking and decay in the final stages of seed development.

VI. MECHANISM OF SEED DETERIORATION

The maintenance of high quality in stored oilseeds has always been a problem in conditions that permit the invasion of the seeds by fungi. The mechanism of seed deterioration is not fully understood. Even under ideal conditions of storage, loss in seed viability cannot be checked completely. High moisture and high temperature during storage are highly detrimental factors for the survival of most seeds. The primary causes of seed deterioration can be grouped into two categories:

(i) Seed tissues may deteriorate due to natural aging.
(ii) Deterioration also may be caused by invasion of seed and damage to tissues by pests and pathogens.

Seed deterioration can begin in the field after the seeds have reached physiological maturity, particularly if the harvest is delayed by wet weather. Physiological changes, which occur during storage in association with deterioration, are: delayed germination, reduced seedling growth rate, decreased tolerance to adverse germination conditions, and loss of germinability (Abdul-Baki & Anderson, 1972). Biochemical changes associated with seed deteriaoration include alterations changes in metabolic activity (Abdul-Baki & Anderson, 1972) and membrane degradation, low O_2 uptake and high CO_2 output (Kharluki & Agarwal, 1984) .

Membrane Disintegration

Cell membranes are constituted of a certain proportion of unsaturated lipids. In presence of O_2, the lipids may react to form free radical intermediates and unstable peroxides (Agarwal, 1988). This results in destruction of lipid itself, the formation of insoluble protein complexes by cross-linkage and damage to cell membrane (Dadlani & Agarwal, 1983a).

Seeds leach out a number of water-soluble compounds when imbibed in water. Leaching increases during seed deterioration. Analysis of seed lechate for soluble sugars, amino acid and electrolytes has been used to access seed quality. Negative correlation has been reported between the leaching of sugars and electrolytes and seed viability (Dadlani & Agarwal, 1983b) and seedling vigour (Mullett & Wilkinson, 1979).

Membrane deterioration, in addition to the loss of solute control, results in dispersal of highly ordered system of membrane associated enzymes (Stewart & Bewley, 1980). This may cause, among other effects, a shift in respiratory pathway in hexose oxidation during germination.

Enzyme Activity

In oilseeds, seed coat is made up of cellulose and pectin. Hence, the ability of storage fungi to secrete cellulolytic and pectinolytic enzymes is important for invasion and colonization of seed tissues. Lipolytic enzymes liberated by fungi act upon triglycerides of oil thereby releasing free fatty acids and causing hydrolytic rancidity (Lalita Kumari *et al*, 1971). Fungal infestation causes biochemical changes in nutritional value of the oilseed. Protein molecules are altered by microorganism's enzymatic action and new proteins of microbial origins are synthesized. In addition, depletion of some enzymes and intensification of others and/or production of new multiple molecular forms of enzymes takes place.

VII. AFLATOXINS

Early harvesting, damage during threshing, and improper storage in hot and humid conditions lead to fungal growth in rapeseed-mustard seed. As a result, mycotoxin contamination is common in mustard seeds and oil samples. Among the mycotoxins aflatoxins are considered at present to be one of the most dangerous contaminats in food and feed. Aflatoxins are toxic secondary metabolites produced by certain strains of *Aspergillus flavus* and *Aspergillus parasiticus*. Toxicity studies have demonstrated their carcinogenic, hepatotoxic, teratogenic and immunosuppressive activity (Betina, 1984).

Occurrence

The aflatoxins are a family of potent mycotoxins. These mycotoxins are produced by at least four species of genus *Aspergillus* viz. *A. flavus, A. parasiticus, A. nomius* and *A. tamarii* (Goto *et al*, 1996; Kurtzman *et al*, 1987, Magan *et al*, 1969; Malini *et al* 1983). Goto *et al* (1996) reported that out of four aflatoxigenic species *A. tamarii* produces colonies of dark olive colour while the rest produce bright greenish yellow colonies on Czapek-Dox agar medium. The most well investigated species for aflatoxin production are *A. flavus* and *A. parasiticus.* These fungi occur in warm climate and contaminate various agricultural commodities. Aflatoxins are often isolated in cereals, cereal products, groundnut, cotton seed, copra and beans. These toxic metabolites have also been detected in eggs, milk and cheese. The highest levels and incidence of contamination are found in tropical and subtropical regions. Crops in tropical areas are more susceptible to contamination than those in temperate regions, since optimal conditions for toxin formation are prevalent in areas with high humidity and temperature.

Seventeen types of aflatoxins have been identified (Bhatnagar *et al*, 1987). Among these aflatoxins, B_1, B_2, G_1, and G_2 occur most frequently and in larger amounts, while B_1 is the most toxic of the group. Aflatoxin M, a major metabolic product of aflatoxin B_1 in animals, is and usually excreted in the milk and urine of dairy cattle and other mammalian species that have consumed aflatoxin contaminated food and feed.

Aspergillus flavus and *A. parasiticus* have worldwide distribution and are prevalent in tropical and subtropical regions. *A. flavus* is also known as *A. flavus-oryzae.* These fungi contaminate seeds and can grow as unspecialized saprophytes on crop debris left on and in the soil (Wicklow & Horn, 1984). These are pathogenic to insects (Zacharunk, 1981). These two species are different and can be distinguished by their colony appearance and microscopic features. *Aspergillus flavus* and *A. parasiticus* are closely related in their DNA homologies. However, these are not vegetatively compatible and are, therefore, genetically isolated from one another (Neal *et al*, 1998). *Aspergillus flavus* produces only B_1 and B_2 but *A. parasiticus* produces same metabolites along with G_1 and G_2 aflatoxin (Mclean & Dutton, 1995).

Structure and Chemical Properties of Aflatoxins

The term aflatoxins refer to the group of bisfurnaocoumarin metabolites. Aflatoxins contain a coumarin nucleus fused to bifuran and either a pentanon (AFB_1 and B_2) or a six-membered lacton (AFG_1 and G_2). Aflatoxins B_2 and G_2 are dihydroxy derivatives of the parent AFB_1 and AFG_1 (Asao *et al*, 1963). Aflatoxin B_2 is produced by partial hydrogenation of AFB_1 (Van Drop *et al*, 1963), and further hydrogenation produces tetrahydroxy aflatoxin (Asao *et al*, 1963, Van Drop *et al*, 1963). Allocraft and Carnaugham (1963) reported hydroxylated aflatoxin derivatives, AFM_1 and AFM_2 from the milk of cow fed toxin. The hydroxylated product of G_1 has been isolated and identified as aflatoxin GM_1 (Heathcote and Dutton, 1969). Dutton and Heathcote (1966) reported the other derivatives produced by *A. flavus*, the 2 hydroxyaflatoxins known as AFB2a and AFG_2a. The aflatoxins are intensely florescent when exposed to long wave ultraviolet (UV) light. Therefore these compounds can be detected at extremely low levels on thin layer chromatogram. A concentration of aflatoxin M_1 of 0.02 µg/litre can be detected in liquid milk (Schuller *et al*, 1976).

These four compounds are separated by the colour of their fluorescence under long wave ultraviolet illumination. Aflatoxins B_1 and B_2 produce a blue fluorescence whereas G_1 and G_2 produce a green fluorescence. Hence, B designates blue and G designates green (Hartley *et al*, 1963). Aflatoxins are freely soluble in polar solvents (e.g., Chloroform and methanol), the solubility of aflatoxins in water ranges from 10-20 mg/litre. Aflatoxins in the dry state are very stable at high temperatures. In presence of moisture there is destruction of aflatoxins over a period of time. Mann *et al*, (1967) observed such type of destruction in oilseed meal. Aflatoxins are relatively unstable when exposed to light and particularly to UV radiation.

Symptoms of Aflatoxicosis

Mycotoxins in general and aflatoxins in particular are being studied as health hazard for humans. The ingestion of food containing aflatoxin may have serious adverse health effects in human. Diseases in animals and human beings resulting from consumption of aflatoxins are called aflatoxicosis. Aflatoxicosis causes acute liver damage, liver cirrhosis, induction of tumours, attack on central nervous system, skin disorders and hormonal effects (Pitt, 1989). Epidemiological studies have indicated a relationship between aflatoxin intake and incidence of liver cancer in several developing countries. Aflatoxicosis and liver cancer in human have been reported in India (Bilgrami and Sinha, 1986).

Indian childhood cirrhosis (ICC) is the most common cause of death of children under the age of 5 years. The characteristic features of the disease involve low grade fever, mild abdominal distension followed by enlarged liver with characteristic leafy borders, the disease may progress to jaundice, ascites, fibrosis, cirrhosis and hepatic comma (Yadgiri *et al*, 1970, Amla *et al*, 1971). Childern exposed to aflatoxins through mother's milk and food such as para boiled rice and unrefined peanut oil may acquire ICC (Amla *et al*, 1970).

Linsell and Peers (1977) observed the possibility of infection by hepatitis B virus prior to the incidence of primary liver cancer in aflatoxin-exposed persons. The studies revealed that aflatoxin could be a major disease determinant in primary liver cancer and hepatitis B virus served as cofactor in the etiology.

Acute toxicity

The general symptoms of aflatoxicosis are edema of the lower extremities, abdominal pain and vomiting. The principal target organ for aflatoxins is the liver. After the invasion of aflatoxins in to the liver, lipids infiltrate hipatocytes and leads to necrosis or liver cell death. The aflatoxin metabolites react negatively with different cell proteins and inhibited protein synthesis and carbohydrate and lipid metabolism.

Chronic Toxicity: Aflatoxin B_1 is mainly related to chronic toxicity. Chronic aflatoxicosis results in congested liver with haemorrhagic and necrotic zones, proliferation of hepatic parenchyma and epithelial cells of the duct and congestion of kidneys that show occasional haemorrhagic enteritis. Liver damage is apparent due to yellow colour that is characteristic of jaundice and the gall bladder will become swollen. The chronic aflatoxicosis reduces growth rate and reproductive efficiency.

Cellular effects

Aflatoxins have high affinity for nucleic acids and polynucleotides and inhibit the synthesis of nucleic acids. Aflatoxin B_1 is genotoxic, producing adducts in humans *in vivo*. In human cell culture, it produces DNA damage, gene mutation and chromosomal abnormality (IARC 56).

Immune suppression

In low range aflatoxins would be immunomodulatory (Pestka and Bondy, 1994) Aflatoxins reacts with T-cell and resulted in immune suppression and decrease the activities of vitamin K and phagocytic activity in macrophages.

VIII. AFLATOXINS CONTAMINATION IN MUSTARD

Aflatoxin Contamination in Preharvest Mustard Crop

Natural incidence of aflatoxins in various oilseeds and their derivatives has been reported in different parts of India. Studies have shown mustard to be a susceptible substrate for aflatoxin contamination (Llwellyn *et al*, 1981; Prasad & Sahay, 1987; Sharma *et al*, 1988; Sinha *et al*, 1993). Pre-harvest mustard seeds also form a good base for the growth of toxigenic strains of *Aspergillus flavus* and the natural incidence of aflatoxin. Bilgrami *et al* (1991) observed significant effect of planting date and cropping pattern as well as combined effect of planting date and cropping pattern on mycotoxin build-up. Delayed planting resulted in a very high level of aflatoxin contamination.

Both monocropping and mixed cropping are practised in mustard cultivation. Sinha *et al* (1988) observed that the number of fungi obtained from mixed cropping samples was invariably higher than from monocropping samples. Bilgrami *et al* (1991) observed that mustard variety Varuna was more susceptible to aflatoxin than variety BR-40. Differences between the two varieties in respect of aflatoxin contamination can be attributed to the variation in their maturity period as well as ability to resist aflatoxin elaboration. Thus, aflatoxin contamination in pre-harvested mustard can be minimized through suitable cultural practices viz. selection of hybrid varieties, sowing date and cropping pattern.

Aflatoxin Contamination in Mustard Seeds During Storage

Natural contamination of agricultural produce with aflatoxins is comparatively higher in under developed and developing countries because of socio-economic backwardness and faulty, unhygienic and outdated agricultural and storage conditions. In India, fairly high relative humidity and temperature, unseasonal rainfall and floods prevail in different eco-regions of the country at the time of harvesting of major crops (Bilgrami, 1991). Chauhan & Bhargava (1996) reported aflatoxin in 30 percent of the seed samples purchased from major retail outlets of Allahabad district. Quantity of aflatoxin ranged from 5 to 61 μg/kg whereas the permissible level of aflatoxin for human consumption is 20 μg/kg as determined by World Health Organization. Poddar & Purohit (1994) reported aflatoxin contamination of mustard and taramira seeds under storage in Rajasthan. Aflatoxin B_1 was invariably present in all the contaminated samples, with the amount of aflatoxin B_1 ranging from traces to 729 μg/kg in mustard and it varied from traces to 489 μg/kg in taramira. This high incidence was due to heavy pre- and post-monsoon rains as well as cloudy and foggy nights alternating with moderate hot days in this area. The type of storage structure used also influences aflatoxin formation.

Since mustard seeds are hygroscopic, a significant disadvantage of open storage is the risk of growth of fungi during period of high environmental moisture levels. Ranjan *et al* (1992) evaluated the extent of aflatoxin contamination in mustard stored in traditional system such as gunny bags, mora (made from paddy hay ropes), kothi (cylindrical structure, made from mud and paddy straw mixture) and iron bins. A total of 17 fungal species belonging to 11 genera were isolated from gunny bags and kothi. The highest percentage of toxigenic isolates was obtained from gunny bags. Iron bin samples were the least affected. The high levels of *Aspergillus flavus* and aflatoxin in gunny bag samples may be due to the porous nature of the bags, usually kept in semi-ventilated godowns or rooms. The other storage structures do not provide such an easy exchange of moisture with seeds. However, kothi and mora being made from mud and paddy straw have the capacity to absorb atmospheric moisture, which provides the favourable conditions for fungal infection.

IX. SEED MYCOFLORA AND AFLATOXIN CONTAMINATION IN MUSTARD COLLECTED FROM RAJASTHAN AND HARYANA: A CASE STUDY

In India mustard is cultivated in the Himachal Pradesh, Jammu, Madhya Pradesh, Maharashtra, North Eastern states, Orissa, Punjab, Rajasthan, Uttar Pradesh and West Bengal. In non-traditional areas of southern states the cultivation has been extended in recent years. The major rapeseed-mustard growing states are Rajasthan, Haryana, Madhya Pradesh and Uttar Pradesh.

Studies on seed mycoflora of mustard samples collected from Rajasthan and Haryana were undertaken. Seed samples (freshly harvested and one year old stored seeds) of mustard, *Brassica juncea* (Linn.) Czern & Coss. were collected from different places of Rajasthan and Haryana during the period from Apri-May in the years 2000 and 2001. The area surveyed for the mycoflora and aflatoxin analysis in Rajasthan and Haryana is shown in Fig. 22.1a and 22.1b respectively. Total 150 seed samples were collected from farmers, wholesellers, stockists and oil mills. Finally, 150 seed samples collected were reduced to 40 working samples.

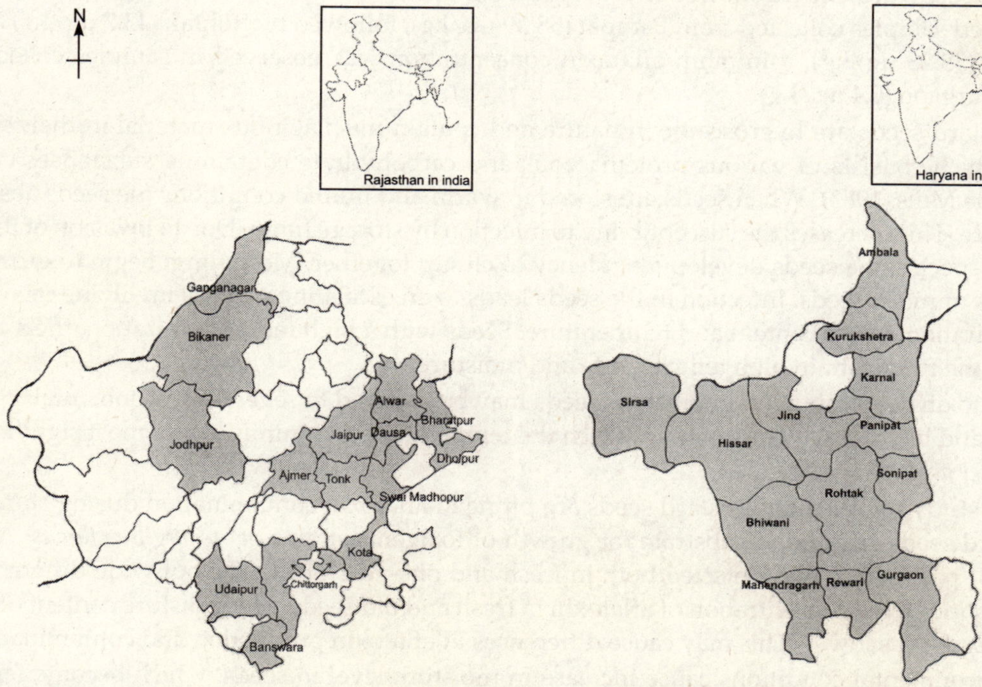

Fig. 22.1: Districts of Rajasthan and Haryana from where seeds of *Brassica* spp. were collected.

Seed samples (fresh and one year old) from different districts of Rajasthan were analysed for percent incidence of fungal infection. Among the fungi isolated, *Alternaria alternata*, *A. tenuissima*, *Aspergillus candidus*, *A. flavus*, *A. fumigatus*, *A. nidulans*, *A. niger*, *Fusarium moniliforme* and *F. oxysporum* were dominant (Fig. 22.2).

Twenty-five species of various fungi were isolated from old and fresh mustard seed samples collected from Haryana. The dominant species were *Alternaria alternata, Aspergillus candidu. A. flavus, A. fumigatus, A. nidulans, A. niger, A. parasiticus, A. sulphurus, Fusarium moniliforme* and *F. oxysporum.* Tables 22.1 and 22.2 list the mycoflora of seeds cultured on Czapek's Dox agar medium collected from Rajasthan and Haryana respectively.

Isolates of *Aspergillus flavus* from all the seed samples were pure cultured on SMKY medium and tested for their toxigenic potential. Out of the twelve strains isolated from seed samples of Rajasthan, ten strains produced aflatoxin. Sixteen strains of *A. flavus* were isolated from seed samples collected from Haryana. Of these thirteen showed toxin producing capacity.

Aflatoxin contamination was detected in seed samples using TLC. Forty percent fresh seed samples collected from different districts of Rajasthan showed mycotoxin contamination. The concentration of aflatoxin ranged from 0.23 μg/kg to 22.7 μg/kg. All one-year-old seed samples from Rajasthan except those collected from Alwar, Bharatpur, Dausa and Tonk districts, all showed aflatoxin contamination. The concentration of aflatoxin ranged from 0.32 μg/kg to 38.5 μg/kg of seeds.

In case of Haryana, freshly harvested seed samples from six districts i.e., Ambala, Karnal, Mahendragarh, Panipat, Riwari and Rohtak showed aflatoxin contamination. The concentration varied from 0.12 μg/kg to 29.7 μg/kg of seeds. One-year-old seeds samples from all districts except Kurukshetra showed mycotoxins on TLC. Maximum concentration was obtained in oilseed samples collected from Panipat (58.39 μg/kg), followed by Rohtak (43.7 μg/kg) and Riwari (33.28 μg/kg). Minimum aflatoxin concentration was observed in samples collected from Gurgaon (0.4 μg/kg).

Mustard seeds are hygroscopic in nature and contain mucilaginous material in their seed coat which consists of various proteinaceous and carbohydrate containing substances (Van Caesecle Mills, 1983). When seeds are stored in warm and humid conditions the seeds absorb moisture. This increases the susceptibility to infection by storage fungi. Due to invasion of these fungal species, the seeds develop a tendency to clump together. More fungi begin to grow in pockets of moist seeds. Infection in the seeds leads to an escalating process involving increase in respiration, water content and temperature. Seeds with a high rate of respiration, their own and of the fungi, attain high temperature and moisture.

Aflatoxin contamination in mustard seeds may be ascribed to several environmental, agronomic and biological factors among which the temperature and humidity are most significant (Sinha *et al*, 1988).

The study reveals that mustard seeds are prone to aflatoxin contamination during storage. Mustard seeds form good substrate for growth of toxigenic strains of *Aspergillus flavus.* Aflatoxin B_1 production was detected both in fresh and one-year-old seeds, but wide differences were found in the concentration of aflatoxin in fresh and old seeds. The moisture content of the fresh seeds was lower. This may cause differences in aflatoxin production and contamination.

Environmental conditions cause increase in moisture level in seeds, which become fertile base for growth and sporulation of *Aspergillus flavus* isolates and consequently aflatoxin production in seeds. Ranjan *et al* (1992) reported that aflatoxin synthesis in mustard is greatly influenced by storage structures. The range of aflatoxin B_1 varied from 1040 to 1820 μg/kg in mustard seeds stored in gunny bags.

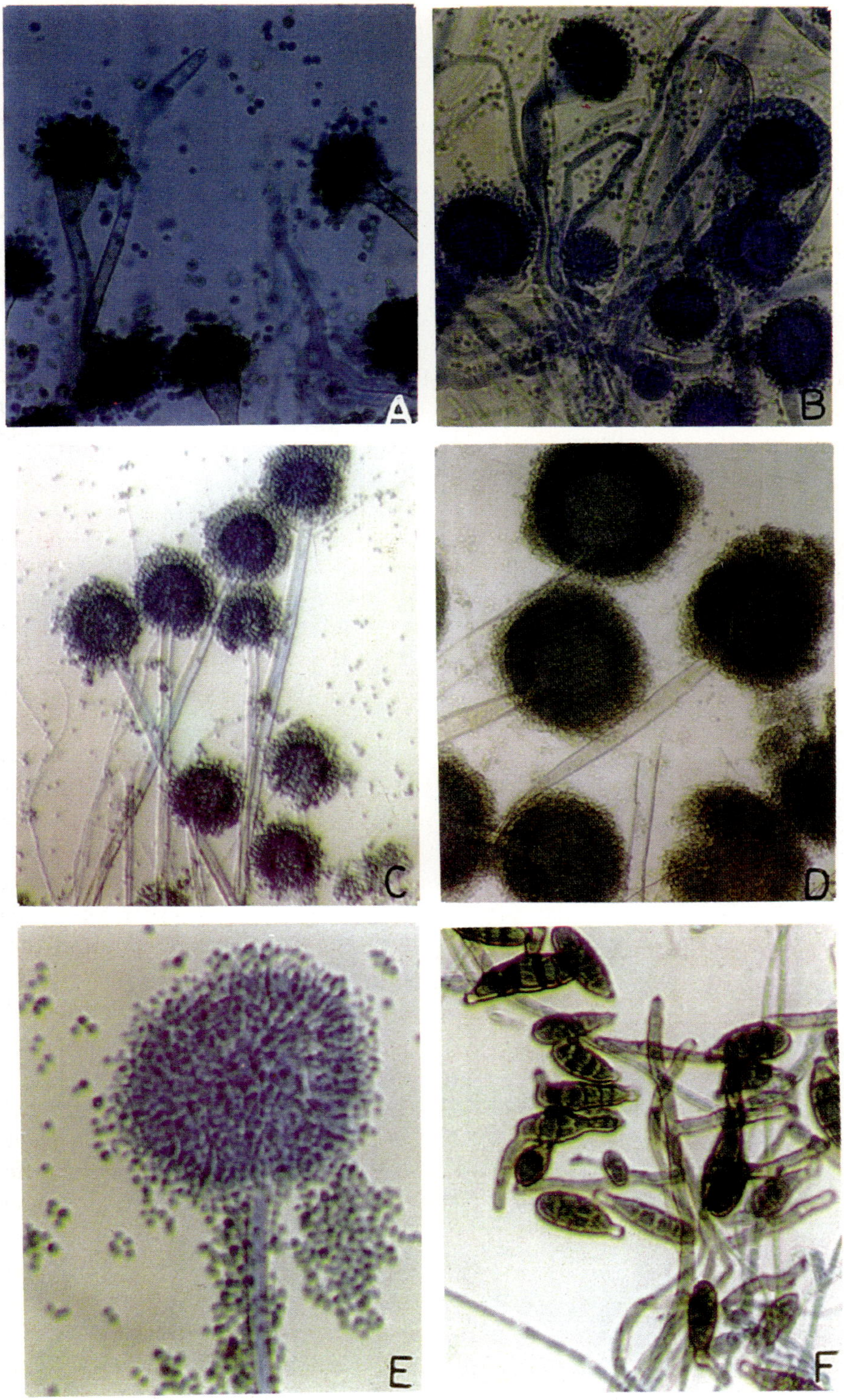

Fig. 22.2: Seed mycoflora of *Brassica* spp.: Conidiophores of (A) *Aspergillus nidulans*; (B) *Aspergillus sydowir*; (C) *Aspergillus niveus;* (D) *Aspergillus niger;* (E) *Aspergillus flavus* and (F) phragmospores of *Alternaria alternata.*

Table 22.1: Percent incidence of mycoflora associated with seed samples of various districts of Rajasthan in Czapek's Dox agar method

Districts	*Ajmer*		*Alwar*		*Banswara*		*Bharatpur*		*Bikaner*		*Dausa*		*Dholpur*		*Ganga Nagar*		*Jaipur*		*Kota*		*Swai Madhopur*		*Chittorgarh*		*Tonk*		*Udaipur*		*Jodhpur*	
Fungi	*F*	*O*	*F*	*O*	*F*	*O*	*F*	*O*	*F*	*O*	*F*	*O*	*F*	*O*	*F*	*O*	*F*	*O*	*F*	*O*	*F*	*O*	*F*	*O*	*F*	*O*	*F*	*O*	*F*	*O*
Alterneria alternata	-	-	4	22	-	-	-	3	6	11	-	-	-	-	-	-	8	14	-	-	-	-	-	-	-	-	-	-	-	-
A. tenuissima	-	-	-	-	-	4	-	-	3	14	-	13	-	-	-	-	-	-	-	-	-	-	-	-	-	-	-	-	-	-
Aspergillus candidus	2	18	-	5	-	3	-	-	-	-	-	-	-	-	8	-	-	-	-	-	12	36	-	-	-	-	-	-	-	4
A. flavus	11	32	3	8	9	15	-	6	43	86	-	18	32	46	-	14	-	13	4	18	14	42	-	13	-	6	4	15	26	45
A. fumigatus	-	4	-	-	-	-	-	-	6	10	-	8	-	-	4	33	-	-	-	-	-	-	-	-	-	-	-	7	-	-
A. nidulans	-	2	-	-	-	4	-	-	-	-	-	3	-	-	-	-	-	-	-	-	-	-	-	32	-	-	-	-	-	-
A. niger	4	13	-	28	6	18	-	13	18	26	-	-	10	-	-	-	6	13	6	23	-	14	4	18	-	16	-	3	-	6
Chaetomium globosum	-	-	-	3	-	-	-	-	-	-	-	-	-	-	-	-	-	-	-	6	-	-	-	-	-	-	-	4	-	-
Claadorporium cladosporoides	-	-	6	9	-	-	-	-	-	-	-	-	-	-	-	-	-	-	-	2	-	-	-	-	6	-	-	8	-	-
C. herbarum	-	3	-	4	2	-	-	-	-	-	-	2	-	-	-	-	-	-	-	-	-	-	-	2	-	-	-	-	-	-
Curvularia lunata	-	-	-	-	-	-	-	3	-	-	-	-	2	4	-	-	-	-	-	-	-	-	-	-	-	2	-	-	-	-
Botrytis cineraria																														
Dreschlera rostrata	-	3	-	-	-	-	-	3	-	-	-	-	-	6	-	-	3	9	-	-	-	-	-	-	-	-	-	-	-	-
D. tetramera	-	-	-	-	-	2	-	-	-	-	-	-	-	-	-	-	-	4	-	-	-	-	-	-	-	2	-	-	6	13
Fusarium moniliforme	3	20	-	16	-	13	-	4	16	39	-	11	-	-	2	14	6	18	-	3	-	-	-	-	-	-	-	-	4	22
F. oxysporum	2	33	-	-	-	-	-	-	-	-	3	14	-	-	-	16	-	-	2	18	12	18	-	28	-	6	6	16	3	16
F. solani	-	-	-	-	-	-	18	23	-	-	-	-	-	-	-	-	-	-	-	-	-	-	-	-	-	-	3	-	-	-
Penicillium *citrinus*	-	-	-	-	-	-	-	-	-	6	-	-	-	16	-	-	-	-	-	-	-	6	-	-	-	-	-	-	-	-
Penicillium funiculosum	-	-	-	4	-	-	-	-	-	-	-	-	-	-	-	-	-	18	-	-	-	-	-	-	-	-	-	-	-	-
Rhizopus nigricans	-	4	-	-	-	-	-	6	-	-	-	-	-	3	-	-	-	23	-	-	-	-	-	-	-	-	-	-	4	8
Stachybotrys atra	-	-	-	-	-	-	-	-	-	-	-	3	-	-	-	-	-	-	-	-	-	-	-	-	-	4	-	-	-	-

F- Fresh seeds O- One year old seeds

Table 22.2: Percent incidence of mycoflora associated with seed samples of various districts of Haryana in Czapek's Dox agar method

Districts	*Ambala*		*Bhiwani*		*Gurgaon*		*Hissar*		*Jind*		*Karnal*		*Sirsa*		*Sonipat*		*Kru-keshtra*		*Mahen-dragarh*		*Panipat*		*Riwari*		*Rohtak*	
Fungi	*F*	*O*	*F*	*O*	*F*	*O*	*F*	*O*	*F*	*O*	*F*	*O*	*F*	*O*	*F*	*O*	*F*	*O*	*F*	*O*	*F*	*O*	*F*	*O*	*F*	*O*
Absidia repens	-	-	-	-	-	-	*3*	*8*	-	-	-	-	-	-	-	-	-	-	-	-	-	*6*	-	-	-	*3*
A. ramosa	-	-	-	-	-	-	*4*	*10*	-	*6*	-	-	-	*3*	-	-	-	-	-	-	-	-	-	-	-	-
Alterneria alternata	43	64	12	35	-	28	-	-	-	-	32	54	-	-	-	18	-	-	16	42	4	6	18	35	6	17
A. longissima	2	6	-	-	-	-	9	-	-	-	-	-	-	-	-	3	-	-	-	-	-	-	-	-	-	-
A. solani	-	-	-	-	-	-	-	2	-	-	-	6	-	-	-	-	-	-	-	-	-	5	-	-	-	-
Aspergillus anneus	6	17	-	-	-	3	-	-	-	8	-	-	-	4	-	-	-	-	-	-	-	-	-	-	-	-
A. candidus	-	-	-	-	-	-	-	3	-	-	-	-	-	-	-	-	-	-	-	-	-	-	-	-	-	42
A. fumigatus	25	62	-	6	-	18	-	-	4	30	2	4	-	6	3	18	-	3	13	48	11	39	-	28	6	18
A. flavus	23	48	3	16	6	12	8	19	6	24	33	68	-	16	12	23	4	16	23	84	53	98	23	64	33	69
A. nidulans	6	10	-	3	-	-	-	-	-	-	-	-	-	3	-	-	-	-	-	-	-	-	-	-	-	13
A. niger	-	23	-	14	-	-	4	16	-	4	-	-	-	-	-	11	-	4	-	-	-	-	6	15	2	16
A sulphurus	8	28	6	33	-	-	-	-	16	22	11	36	-	-	12	44	-	-	-	-	-	-	-	-	23	48
A. terreus	-	3	-	-	-	-	-	-	-	-	-	2	-	-	-	-	-	-	-	-	-	-	-	-	-	-
A. proliferatus	-	*4*	-	*6*	-	-	-	*3*	*4*	*18*	-	-	-	-	-	-	-	-	-	-	*2*	*18*	-	-	-	-
A. parasiticus	2	18	3	39	-	3	2	16	-	8	26	49	-	6	-	-	-	-	-	-	11	23	13	43	33	69
Claadorporium cladosporoides	-	6	-	8	4	-	-	-	-	-	-	-	-	-	-	-	-	-	-	-	-	-	-	-	-	-
Curvuleria lunata	-	-	-	2	-	-	-	-	-	-	-	6	-	-	-	-	2	8	-	4	-	-	-	-	-	-
Dreschlera tetramera	-	-	-	3	4	-	18	23		-	-	16	-	-	2	6	-	-	-	-	-	-	18	93	12	16
Fusarium moniliforme	2	6	-	-	-	-	4	8	4	17	-	-	-	-	6	18	-	-	-	-	12	23	-	-	3	19
Rhizopus stolonifer	-	-	-	13	-	-	-	3	-	-	-	-	-	-	-	-	-	-	-	-	-	-	-	-	-	-
Rhizopus nigricans	-	-	-				-	28	-	-	-	-	-	-	-	-	-	-	-	-	-	-	-	-	-	-
Glyocladium roseum	-	-	-	-	-	5	-	-	-	-	-	-	-	05	-	-	-	-	-	-	-	-	-	-	-	-
Mucor basiformis	-	-	-	-	-	2	-	-	-	-	-	-	-	-	-	-	-	-	-	-	-	-	-	6	-	-
Penicillium	-	4	-	-	-	-	-	-	-	16	-	-	-	-	-	-	-	6	-	-	-	-	17	26	-	-
Stachypotrys atra	-	-	-	-	2	6	-	-	-	-	-	-	-	-	-	-	3	-	-	-	-	-	3	18	-	-

The presence of variable amount of aflatoxins in different types of samples could be due to environmental factors, toxigenic potential of the fungal strains and composition of substrate (Bilgrami, 1986).

REFERENCES

Abdul-Baki, A.A. & Anderson, J.D. 1972. Physiological and biochemical deterioration of seeds. Pages 283-316 in T.T. Kozlowski (ed.), Seed Biology. Academic Press, New York.

Agarwal, P.K. 1988. Seed deterioration during storage. Pages 1271-1278. Proc Int. Cong Pl Physiol New Delhi.

Allocraft, R. & R.B.A. Carnaugham. 1963. Groundnut toxicity: An examination for toxin in human food products from animal fed toxic groundnut meal. Vet. Record. 75: 259-263.

Amla, I., S.Y. Kumari, V. Sreenivasmurthy, A.P. Jayaraj & H.A.B. Parpia. 1970. Role of Aflatoxin in Indian childhood cirrhosis. *Indian Pediatr.* 262-270.

Amla, I., C.S. Kamala, G.S. Gopalakrishna, A.P. Jayaraj, V. Sreehivasamurthy & H.A.B. Parpia, 1971. Cirrhosis in children from peanut meal contaminated by aflatoxin. Am. J. Clin. Nutr. 24: 609.

Appelqvist, L.Å & B. LÖÖf. 1972. Postharvest handling and storage of rapeseed. Pages: 60-99. in L. Å. Appelqvist R. Ohlson (eds.), Rapeseed: Cultivation, Composition, Processing and Utilization Elsevier Publishing Company, London, New York.

Asao T., G. Buchi, M.M. Abdel-Kader, S.B. Chang, E.L. Wick and G.N. Wogan. 1963. Aflatoxin B and G.J. Am. Chm. Soc. 85: 1706-1707.

Barton, L.V. 1961. Seed preservation and longevity. Pages 216. Leonard Hill (Books) Ltd. Inter science Publishers Inc. New York.

Betina, V. 1984. Mycotoxins, production, separation and purification. Developments in Food Science, 8: 89-130.

Bhatnagar, R.K., R. Malini, K.G. Mukerji & T.A. Venkita Subramanian. 1987. Secondary metabolism in *Arpergillus* with special reference to aflatoxin production by *Arpergillus parasiticus* front. Appl. Microbiol. 2: 127-150.

Bilgrami, K.S., K.S. Ranjan, & A.K. Choudhary. 1991. Aflatoxin contamination in monocropping and mixed cropping mustard crops of Bihar. J. Sci. Food Agri. 44: 529-531.

Bilgrami, K.S. & K.K. Sinha. 1986. Aflatoxin in India: I In. Aflatoxin in Maize. A Proceedings of the Workshop of CIMMYT. EL Batan, Mexico. pp. 349357.

Chauhan, S. & V. Bhargava. 1996. Quantification of aflatoxin in oilseeds. Indian Phytopath. 49: 281-282.

Christensen, C.M. & H.H. Kaufman. 1965. Deterioration of stored grains by fungi Ann. Rev. Phylopath. 3: 69-84.

Christensen, C.M. 1973. Loss of seed viability by storage microflora. Seed, Sci. Tech. 1: 547-562.

Dadlani, M. & P.K. Agarwal. 1983a. Mechanism of soybean seed deterioration. Plant Physoil. & Biochem. 10: 23-30.

Dadlani, M. & P.K. Agarwal, 1983b. Factors influencing leaching of sugars and electrolytes from carrot and okra seeds. Scientia Hort. 19: 39-44.

Duke, J.A. &, K.K. Wain. 1981. Medicinal plants of the world. Computer index with more than 85,000 entries. 3 vols.

Economy Database, 2004. http://www.indiainfoline.com/econ/andb/agri/agri0.html

Ellis, W.U., J.P. Smith & B.K. Simpson. 1994. Aflatoxins in food occurrence and effects on organisms, detection and methods of control. Crit. Rev. Food Sci. Nutr. 30: 403-439.

Goto, T., D.T. Wicklow & Y. Ito. 1996. Aflatoxin and cyclopiazonic acid production by a sclerotium-producing *A. tamarii* strain. Appl. Environ. Microbiol. 62: 4036-4038.

Greer, F.N. 1953. Risk in the storage of grains. Agric. Merch. 33: 369-371.

Hartley, R.D., B.F. Nesbitt & J. O'kelly. 1963. Toxic metabolites of *Aspergillus flavus*. Nature (Load.), 198: 1056-1058.

Heatchote, J.G. & M.F. Dutton. 1969. New metabolites of *A. flavus*. Tetrrahedron. 25: 1497-1500.

Hoffman, C. & A.C. Evans. 1911. The use of species as preservatives. J. Ind. Eng. Chem. 3: 835-838.

IARC (International Agency for Reaearch on Cancer). 1993. Monograph 56. IARC Monograph series on the evaluation of carcinogenic risks to humans.

Kharluki, L. & P.K. Agarwal. 1984. Evidence for participation of pentose phosphate.

Kumar, P.R. 1999. Scenario of rapeseed-mustard oilseed in India. In: Mustard oil crisis. Why how and what next. Published by MRPC, New Delhi, Eco-Transformation Centre, New Delhi.

Kurtzman, C.P., B.W. Horn & C.W. Hesseltine. 1987. *Aspergillus nomius* a new aflatoxin-producing species related to *A. flavus* and *A. tamarii*. Antonievon Leeuwenhoek. 53: 147-158.

Lalitha Kumari, D., C.V. Govindswany & P. Vidyasekaran. 1971. Effect of seed-borne fungi on the physico-chemical properties of groundnut oil. Indian Phytopath. 24: 283-289.

Linsell, C.A. & F.G. Peers. 1977. Aflatoxin and liver cell cancer. Trans. Roy. Soc. Trop. Med. Hyg. 71: 471-473.

Llwellyn, G.C., M.L. Burkett & T. Eadie. 1981. Potential Mold growth, aflatoxin production, and antimycotic activity of selected natural spicies and herbs. J. Associ. Off. Anal. Chem. 64: 955-960.

LÖÖf, B. 1972. Postharvest handling and storage of rapeseed. Pages 49-59 in L.Å. Appelqvist and R. Ohlson (eds.) Rapeseed: Cultivation, Composition, Processing and Utilization. Elsevier Publishing Company, London, New York.

Magan, G., L. Vishwanathan, T.A. Venkitasubramanian & K.G. Mukerji 1969. Studies on Aflatoxin production by Indian strains of *Arpergillus flavus* Link. ex Fries. J. Gen. Microbiol. 59: 119-129.

Malini, R., K.G. Mukerji T.A. Venkitasubramanian 1983. Microbial interactions and Aflatoxin production by *Arpergillus flavus*. Biblotheca Mycol. 91: 285-290.

Mann, G.E., L.P. Codifer & F.G. Dollear. 1967. Effect of heat on aflatoxins in oil seeds meals. J. Agric. Food Chem. 15: 1090-1092.

Mclean, M. & M.F. Dutton, 1995. Cellular interactions and metabolism of aflatoxin: An update. Pharmacology and Therapeutics. 65: 163-192.

Mullett, J.M. & R.I. Wilkinson. 1979. The relationship between amounts of electrolyte lost on leaching seeds of *Pisum sativum* and some parameters of plant growth. Seed Sci. & Tech. 7: 393-398.

Neal, G.E., D.L. Eaton & D.J. Judah. 1998. Metabolism and toxicity of aflatoxins M sub (1) B sub (1) in human derived *in vitro* system. *Toxicology and Applied Pharmacology*. 151: 152-158.

Neergard, P. 1977 Seed pathology. Page 839 John Wiley. New York.

Perry, M. 1980. An ATP-synthesizing system in seeds. Planta 149: 497-509.

Pestka, J.J. & G.S. Bondy. 1994. Immunotoxic Effects of mycotoxins. Pages 339-358 in J.D. Miller (ed.). Mycotoxins in Grain, Compounds other than Aflatoxin Eagan Press, St. Paul Mn.

Pitt, J.T. 1989 Toxigenic Aspergillus and Penicillium Species. A paper presented on the International Training Course on Mycotoxin Prevention and Control. Dept. of Agriculture, Bangkok, Thailand.

Poddar, S.R. & D.K. Purohit. 1994. Aflatoxin contamination of mustard and taramira seeds under storage in Rajasthan. Ann. Arid Zone 33: 67-68.

Prasad, T. & S.S. Sahay. 1987. Aflatoxin production in mustard seeds. Biol. Bull. India 67-69.

Ranjan, K.S., S.S. Sahay, & A.K. Sinha. 1992. The influence of storage structure on aflatoxin contamination in wheat and mustard. J. Stored Prod Res. 28: 221-224.

Schuller, P.L., W. Horowitz & L. Stoloff. 1976. A review of sampling plans and collaboratively studied methods of analysis for aflatoxins. J. Assoc. Off. Analyt. Chemists 59: 1315-1343.

Sharma, N., L. Chand & G.K. Garg. 1988. Aflatoxin production by *Aspergillus parasiticus* on seed and cake of *Brassica* cultivars under varied moisture and temperature regime. Curr. Sci. 57: 846-847.

Sinha, A.K., K.S., Bilgrami, K.S. Ranjan & T. Prasad. 1988. Incidence of aflatoxins in mustard crop in Bihar. Indian Phytopath. 41: 434-437.

Sinha, K.K., N. Kumar & G. Prasad, 1993. The use of mustard (*Brassica juncea* L.) and gram (*Cicer arietinum* L.) seedling germination inhibition assay for aflatoxin B1. *Mycopathologia* 121: 175-178.

Stewart, R.R. C. & J.D. Bewely, 1980. Lipid peroxidation associated with accelerated ageing of soyabean axes-Plant Physiol. 65: 245-248.

Van Caeseele, C. & J.T. Mills. 1983. Mucilage in canola seeds: rapid detection and interaction with storage fungi. Pages 170-173 in E.E. McGregor, (ed.) Seventh Progress Report: Research on Canola Seed oil, Mael and Meal Fractions. Canola Council of Canada Publication No. 61, Woinnipeg, Canada.

Van Drop, D.A., A.S.M. Vande Zijden, R.K. Beerthuis, S. Spurreboom, W.O. Ord, K. de Jong & R. Keuning. 1963. Dihydroaflatoxin B, a metabolite of *A. flavus*: Remarks on the structure of aflatoxin. B. Rec. Trav. Chim. 82: 587-592.

Wicklow, D.T., & B.W. Horn. 1984. *Aspergillus flavus* sclerotia form in Wound-inoculated preharvest corn. Mycologia 76: 503-505.

Yadgiri, B., V. Reddy, P.G. Tupule, S.G. Srikantia, & C. Copalan. 1970. Aflatoxin and Indian childhood cirrhosis. Am. J. Clin. Nutr. 23: 94.

Zacharuk, R.Y. 1981. Fungal diseases of terres trial insects. Pages 385-387 in (eds.) E.W. Davidson & Totowa, N.J Pathogenesis of Invertebrate Microbial Diseases, Allanheld.

23

Air- and Soil-transmitted Fungal Pathogens of Onion: Diseases Caused and Control

NURAY ÖZER
Department of Plant Protection
Faculty of Agriculture Trakya University 59030 Tekirdağ/TURKEY
E-mail: nurayozer@tu.tzf.edu.tr; nurayozer@hotmail.com

ABSTRACT

Alternaria porri, *Botrytis squamosa*, *Cladosporium-allii cepae*, *Peronospora destructor*, *Puccinia allii* and *Stemphylium vesicarium* are air-transmitted pathogens of onion (*Allium cepa* L.) and are known as causal agents of purple blotch, leaf blight, leaf blotch, downy mildew, rust, leaf blight and stalk rot diseases, respectively. They generally infect aerial parts of the plant. The spores of *P. destructor*, which persist on the infected plant residue in soil, may be transmitted from soil to the roots and develop systemically through the bulb. *C. allii-cepae*, *P. allii* and *S. vesicarium* seldom cause serious damage in onion. *Pyrenochaeta terrestris*, *Sclerotium cepivorum*, *Urocystis cepulae* are important soil-transmitted pathogens of onion and cause pink root, white rot and smut diseases, respectively. They form resting spores, may persist in soil for many years. These species infect the roots or bulbs, or both. *U. cepulae* can occur on cotyledon leaves and has different transmission ways. Different control practices are suggested for the important air- and soil-transmitted pathogens. *Colletotrichum circinans*, *Glomeralla cingulata*, *Lasiodiplodia bromae*, *Rhizopus microsporus*, *R. stolonifer* and *Sclerotium rolfsii* are other soil-transmitted fungi of onion, but they appear infrequently on the bulbs in the field and store. *G. cingulata* is also air-transmitted and cause the symptoms on the leaves.

1. INTRODUCTION

Onion (*Allium cepa* L.) is an excellent food source for us. It is also used in traditional medicine, including the treatment of chickenpox, the common cold, influenza, measles and rheumatism. The extract of onion decreases sugars, lipids, and platelet aggregation and enhance fibrinolysis in the blood, indicating that the onion may help to prevent arteriosclerosis and other cardiovascular diseases. The distinctive flavour or odour of onion bulbs is produced when plant tissues are bruised or cut and the enzyme alliinase hydrolyzes S-alk(en)yl cycteine sulfoxide precursors to form volatile sulfur compounds.

Today onions are important crop worldwide and China has the highest production of dry onion with 18 035 000 tonnes and, India, the United States and Turkey follow it with 5 500 000 tonnes, 3 162 750 tonnes and 1 800 000 tonnes, respectively (FAO, 2004).

Onions require fertile, well-drained, non-crusting soils, and are often produced in muck soils. A slightly acidic soil in the 6.0-6.8 pH range is optimum. Onions require substantial amounts of nutrients. Excess nitrogen causes onions to be more susceptible to storage pathogens. Adequate potassium levels are especially important in improving bulb quality and storage life. Well-composted manure is preferred because it is usually free of weed seeds and has a better nutrient balance. Many farmers irrigate their onion fields. Irrigation often produces larger onions, which buyers prefer.

Many fungal pathogens attack the onion plants during growing season and cause economically important yield loss. Until now, numerous workers have described the fungi that cause different diseases on onion, and their control possibilities. Among these fungi, which have economic importance, purple blotch caused by *Alternaria porri* and basal rot caused by *Fusarium oxysporum* f. sp. *cepae* have been reviewed by Aveling (1998) and Cramer (2000), respectively. However, the knowledge on the transmission mechanism of plant pathogenic fungi is necessary to decide their control management. The transmission ways of onion pathogens can be grouped into three categories, *viz.* seed-, air-, and soil-transmitted fungi. Some plant pathogenic fungi can use the three transmission ways in its life cycle. Seed-tranmitted fungi of onion, *Aspergillus niger*, *Botrytis allii* and *Fusarium oxysporum* f. sp. *cepae* are summarized in a previous review (Özer and Köycü, 2004). These fungi are also transmitted from soil to the plant. Furthermore, *A. niger* can be transmitted to onion seed stalks and flowers by air. This present review does not include these species because of the presence of the information in previous review in detail, although it provides up to date information about air- and soil-transmitted fungi of onion, diseases caused by them and epidemiology, some interactions between these pathogens and onion diseases, and their control possibilities.

II. AIR-TRANSMITTED FUNGI

Alternaria Porri (Ellis) Cif.

The pathogen is the causal agents of purple blotch in onion. The fungus is distributed through Australia (Suheri and Price, 2001), India (Ajrekar, 1921; Gupta *et al*, 1991; Padule and Utikar, 1981), Kenya (Bock, 1964), South Africa (Aveling and Naude 1992) and United States (Skiles, 1953). It causes severe disease with 62-92% foliar infection leading to economic losses under humid climates (Utikar and Padule, 1981). Incidence of the pathogen can reach 43.4% causing seed losses in plots of 1-5 ha for transplantation sowing and of 0.55-2.63 ha for direct sowing (Castellanos-Linares *et al*, 1986). The disease has also been reported on Egyptian onion (*A. proliferatum*), Welsh onion (*A. fistulosum*) and leek (*A. ampeloprasum*) (Miller and Lacy, 1995).

Description

Conidiophores of pathogen arise singly or in groups and are straight or flexuous, sometimes geniculate, septate and pale to midbrown. Conidia usually occur singly and are straight or curved and obclavate (the body of the conidium may be ellipsoidal) and taper to a beak that is commonly about the same length as or slightly larger than the body of conidium. The sexual stage of fungus is not known (Miller and Lacy, 1995).

Disease caused and epidemiology

First symptoms on onion leaves or seed stalks are numerous, tiny, white circular or irregular spots, less than 1 mm in diameter. As the spots increase in size, they become oval-shaped or irregular and the white colour eventually changes to violet. Later stages of development show the central portion of the spots changing to purple, immediately surrounded by a pale yellow orange to salmon band beyond which is pale green zone. The dark purple colour is the most distinctive symptom of the disease. Yellowing extends from both ends of the spots often reaching the tips and bases of the leaves. Similar symptom is observed on seed stalks and floral parts of seed onion. In this case, seeds do not develop and are shriveled, stalks distort or break before the seeds mature (Aveling *et al*, 1993; Aveling, 1998). Breaking of floral stalks reduces the yield and production of onion seed (Munoz de Con *et al*, 1985). Severe infection reduce the number of seed stalks/plant, the number of umblets/umbels, the number of seeds/umblet, 1000 grain weight, the number of seeds/plant and seed yield/plant (Lakra, 1999).

A. porri is a seed-borne pathogen of in onion (Neergaard, 1979; Boff *et al*, 1995; Tylkowska and Dorna, 2001). However, it has not yet been determined how the onion seeds become infected with the pathogen (Aveling, 1998).

Typically, the lesions are purple or brown. However, there are also few to many white lesions which remain small. The two types of lesions vary depending on the weather at the time of infection and the climatic conditions afterwards. Extensive typical lesions take place only at comparatively high relative humidity and the lesions described as small white flecks occur at the low relative humidity (Aveling, 1998). The lesions may be invaded by *Stemphylium vesicarium* and turn black with fruiting structures of this fungus (Miller and Lacy, 1995). It was suggested that conidia of *A. porri* and *S. vesicarium* in mixtures caused similar number of lesions, and purple blotch symptoms were considered a complex caused by pathogen (Suheri and Price 2000).

The fungus also invades through the neck of the bulbs and sets, may enter through wounds on the fleshy scales. The disease develops rather quickly producing a semi-watery decay, which is especially conspicuous because of the yellow to red colour. The decay of bulbs gradually turns to wine red, then to black with age. As the tissue becomes desiccated, affected scales eventually dry to a papery texture. Often only one or two outer scales are affected (Miller and Lacy, 1995; Aveling, 1998).

Penetration of leaves by *A. porri* is via the epidermis and stomata, but the frequency of stomatal penetration exceed that of epidermal penetration (Aveling *et al*, 1994; Suheri and Price, 2001). When conidia reach on susceptible tissues, germinate and each conidium cell develops germ tube. Each germ tube usually terminates in a bulbous appressorium formed directly on the epidermal cell or on stoma. Following direct penetration of the outer epidermal cell wall or the stoma bulbous primary hyphae develop below appressoria. Secondary hyphae developed from the primary hyphae grow within.

In Australia, disease incidence was found to be higher during spring and autumn than during summer, because of leaf wetness period occurred during autumn (Suheri and Price 2001).

The disease occurs on the older leaves and scapes, and young leaves are relatively more susceptible when they emerge close to bulb maturity (Lakra, 1999). Young leaves of plants infested by onion thrips (*T. tabaci*) are most susceptible to infection (Thind and Jhooty, 1982; McKenzie *et al*, 1993; Memane *et al*, 2002).

Host-Parasite interaction

A. porri produces the metabolic pigments macrosporin, alterporriol A, B, C, D and E, altersolanol A and B, and the phytotoxins tentoxin, zinniol and porritoxin. The role of these toxins during the infection of onion leaves by *A. porri*, has not yet been determined. However, it was found that; *A. porri* declined dry matter, total chlorophyll, chlorophyll "b", total carbohydrates, free amino acids and proteins in onion leaves (Aveling, 1998).

Control

Cultivation under conditions favourable for rapid host growth, removal of dead foliage and rotten bulb from field, practices that reduce hours of leaf wetness (i.e. good field drainage and reduced plant density) and crop rotation are important cultural practices for control of purple blotch (Miller and Lacy, 1995; Aveling, 1998).

Spore germination on the leaves decreases with increased nitrogen supplied to the host (Khare and Nema, 1982). Nitrogen application (100 kg/ha) reduces disease incidence (Srivastava *et al*, 1999).

The cultivar Taliani Red is resistant to disease. This cultivar contains phenolic compounds and high oxidase activity, which inhibit macerating enzymes of pathogen (Michail and Salem, 1981). Additionally some cultivars (Yellow Globe Danvers, Red Creole), that have a good covering of wax on the foliage and seed stalks, are more resistant to *A. porri* than those having less wax (Aveling, 1998).

Hot water soak (50°C for 20 min) of seeds reduce the percentage infection of seedling. However, none of the fungicides including anilazine, benomyl, a carbendazim/flusilazole mixture, procymidone, tebuconazole and thiram, eradicate the pathogen from seeds (Aveling *et al*, 1993).

An integrated use of different chemicals, organic and biological treatments controlled the disease (Table 23.1).

Botrytis Squamosa Walker

B. squamosa is known as the causal agent of onion leaf blight. The disease is especially severe under prolonged moist conditions at temperatures of 15-24°C (Sutton *et al*, 1978; Lacy and Pontius, 1983) and is most important on *A. fistulosum* (a Japanese bunching onion), *A. bouddhae* (another Japanese bunching onion) and *A. schoenoprasum* (chives). It is present wherever onion is grown (Boff, 1996). Yield loss caused by *Botrytis squamosa* ranges from 7-30% (Shoemaker and Lorbeer, 1977a).

Table 23.1: Effective treatments to control purple blotch disease, their mode of action and cited references

Treatment	*Mode of action*	*Reference*
Chemicals		
Mancozeb (0.2%) sprays for 6 times from the onset of infection	Reduction in disease severity in the field	Sharma, 1986
Fentin acetate/maneb mixture sprays at 7-14 day intervals after the first signs of the disease	Reduction in the number of infected plant in the field	Naude, 1988
Iprodione sprays (0.25%) to leaves of plants at the 3-5 leaf stage, at 14-21 day intervals	Reduction in the number of infected plant in the field	Naude, 1988
Mancozeb (0.25) sprays starting 30th days after transplanting and 4 sprays at fortnightly intervals	Reduction in the number of infected plant, increase of yield	Srivastava *et al*, 1991; Srivastava *et al*, 1995
Chlorothalonil (0.2%) or Iprodione (0.25% for 3 times	Reduction in the number of infected plant in the field	Gupta *et al*, 1996
Pendimethalin (3.5 l/ha)	Reduction in the number of infected plant in the field	Srivastava *et al*, 1999
Copper oxychloride sprays to leaf	Reduction in the number of infected plant in the field	Raja *et al*, 2000
Manzozeb (0.3) + monocrotophos (0.05%) or Cypermethrin (0.01%) spray	Reduction in the number of infected plant and thrips population in the field	Memane *et al*, 2002
Organic		
Leaf extract of Polyalthia longifolia	Reduction in conidia germination *in vitro*	Datar, 1994)
Palmarosa oil (from Pithecelobium dulce and Propis juliflora) (1%) sprays at first appearance of disease symptoms and 15 days after	Reduction in the number of infected plant in the field	Mohan *et al*, 2001
Biological		
Myrothecium verrucaria	Inhibition of conidia germination in vitro	Aveling, 1998

Description

B. squamosa (telemorph *Botryotinia squamosa* Vien.–Bourg.) produces a septate, branching, hyaline mycelium. Conidia are formed on senescent or dead leaves after 60-72 hour of moderate (average of 12-24°C) temperatures and high (greater than 75%) relative humidity. Conidia are spheroidal, hyline and nonseptate. Conidiophores arise as branches of the mycelium with side branches at the tips, each of which has many ampullae that swell gradually at the tips to form conidia on fine denticles. Masses of conidia and conidiophores can be seen with the unaided eye on senescent or dead onion leaves when conditions are favourable. Sclerotia may form on leaf debris in cull piles or on the necks of onion bulbs. Germinating sclerotia form stipes on which conidiophores and conidia are produced; less commonly, apothecia and ascospores are formed (Lacy and Lorbeer, 1995).

Disease caused and epidemiology

The development of lesions of the green leaves (leaf spots) followed by blighting of the leaves are characteristic for this disease. The pathogen does not colonize bulb onion tissue (Presly,

1985). The lesions have a white, necrotic centre about 2 mm in diameter surrounded by a light green halo about 1-1.5 mm in width. When the lesions may expand slightly with age and take on an elliptical shape, the halo may disappear. The centre of the lesions become sunken, straw colored and may develop a characteristic slit.

The symptoms of the disease can be confused with other problems. These are *Botrytis* leaf fleck (*B. cinerea*), ozone injury typified by irregular areas with tiny necrotic flecks, leaf tip die back caused by several factors such as hot and dry weather and water stress, rain and hail injury resulting in pale green to silvery spots, in irregular size and shape and injuries from herbicides.

However, the halo around the lesions can be used to distinguish lesions from other symptoms in early stages of lesion development.

B. squamosa also causes blight of all floral structures and immature seed capsules. Ellerbrock and Lorbeer (1977a) suggested that the pathogen reduced seed yield by 93% after artificial inoculations of onion umbels under field conditions. Wet weather during the periods from anthesis to seed harvest appears to favour disease development (Lacy and Lorbeer, 1995). Lesions caused by *B. squamosa* on onion seed stalks are very important, expand sufficiently to girdle the stalk, abort the umbel and blight the florets (Ellerbrock and Lorbeer, 1977a; Ramsey and Lorbeer, 1986 a). Open florets are more susceptible to the disease than to unopen florets or immature seed capsules (Ramsey and Lorbeer, 1986b). Free moisture duration of 12-24, 6-12, and 6-12 hours are necessary for infection and 36-48, 12-24, and 24-36 hours are for blighting of unopen florets, open florets, and immature seed capsules, respectively by pathogen at 21°C (Ramsey and Lorbeer, 1986c). The fungus is unable to demonstrate disease transmission to seedlings grown from infected seeds (Ellerbrock and Lorbeer, 1977b). It was detected from infected seeds less frequently (Stewart and Franicevic, 1994; du Toit and Derie, 2004).

The pathogen over winters as sclerotia in cull piles, onion debris, and bulbs or on volunteers. Sclerotia in the soil result from desintegration and decay of infected leaves on which sclerotia formed. Sclerotia may survive in the soil and in cull piles to produce conidia.

Primary inoculum is predominantly conidia arising from germinating sclerotia or from leaf or bulb debris. Sclerotia on leaf debris also produce ascospores in perithecia that may serve as the source of new strains of the pathogen that are tolerant to fungicides used to control *Botrytis* leaf blight. Conidia carried by wind currents on leaf surface germinate, then form appressoria and infection hyphae, and the lesions at 20°C. Hyphal development within lesions increase with time during 6 days of continuous dew at 20°C, but is restricted under conditions of no dew and moderate (60±10%) relative humidity (Alderman and Lacy, 1983). Increasing lengths of dry periods result in decreasing number of lesions (Alderman *et al*, 1985).

Leaves of onion plants grown for commercial production are repeatedly infected and these serve as the source of inoculum for infecting new leaves throughout the growing season. Older leaves are more susceptible to blighting than younger leaves (Shoemaker and Lorbeer, 1977b; Alderman and Lacy, 1984). The more blighting and senescence of leaves that occur, the more conidia will be produced to cause further disease development. Growth of infection hyphae within senescent tissues is about three times faster than in non-senescent tissue on the same leaf (Lacy and Lorbeer, 1995).

Host-Parasite interactions

B. squamosa is necrotrophic in its rapid growth through epidermis killing cells in advance of colonization. Clark and Lorbeer (1976) determined that the pathogen produced a cavity within each lesion as a result of collapse and separation of mesophll and epidermal cells and that pectinase C, produced by this pathogen, caused swelling of the outer periclinal epidermal wall, collapse of epidermal cells and in some cases collapse of palisade parenchyma which mimicked histological changes associated disease. Stewart and Mansfield (1984) explained the infection process and the changes in the cells after infection in detail. During germ tube growth by pathogen on tissue surface and during the development of appressoria, no changes are apparent in the underlying plant cell wall or cytoplasm. Following penetration of the cuticle, expansion of infection hyphae within epidermal cell walls is associated with marked wall swelling. Vacuolation of cells with very swollen walls is followed by tonoplast rupture, contraction of the plasmalemma and disorganization of organels. The cytoplasm of cells at the edge of expanding lesions change rapidly from normal appearance to a disorganized state as their cell walls become swollen in advance of fungal invasion. Production of distinct infection hyphae is associated with the death of epidermal cells at the penetration site.

Control

Plants should be sown in single rows spaced at least 30 cm apart, which allow free air movement and more rapid leaf dryings (Boff *et al*, 1998). Cull piles or leaf debris should be eliminated and volunteer onions should be rogued out of the field. To reduce the incidence and severity of this disease, sprinkler irrigation should be timed so leaves have a change to dry quickly after irrigation ceases. This practice prevents extending normal nightly dew periods into longer, disease-conducive wetness periods. Regular rotations with non-susceptible crops may prevent to build-up of sclerotia in soil (Lacy and Lorbeer, 1995).

Allium fistulosum was found more resistant than *A. cepa* to *B. squamosa* (Currah and Maude, 1984).

Thermophilic compost increases the survival of onion seedlings and reduces the disease incidence (Boff *et al*, 2001).

Fungicide sprays with active ingredients chlorothalonil, iprodione, mancozeb, maneb, metenoxam and vinclozolin have been using on a weekly basis from the moment that onion plants touch each other between rows until the leaves fall over. Depending on the development, rate of the onion crop this amounts to 6-8 sprays per year. The spray interval depends on persistence of the fungicide used and weather conditions. Presly and Maude (1982) suggested that iprodione-insensitive isolates from treatment with low and high concentrations of the fungicides sporulated less than the sensitive parent forms although they readily infected onion leaves. They also found that insensitive isolates were cross-tolerant to quintozene, tecnazene, and dicloran.

Fungicide spray forecasting systems have been devised in different countries. SIV model, Sporulation Index Value, is based on hourly measurement of temperature and relative humidity (Lacy and Pontius, 1983). The SIV represents the extent to which the weather has been favourable for sporulation of *B. squamosa*. It is difficult to consider the infection using by this model. BOTCAST (*Botrytis* casting) model, based on the data of temperature, relative humidity, rainfall and leaf wetness, is used to time the first spray (Sutton *et al.*, 1986). BOTCAST calculates a Daily Inoculum Value and a Daily Infection Value. Daily Inoculum Value indicates whether the circumstances are suitable for spore production or not. Daily Infection Value represents the extent to which the circumstances were suitable for infection. Vincelli and Lorbeer (1988) found the BOTCAST sporulation submodel DINOV forecasted minor spray episodes better than the SIV model at a threshold of 50%. The authors developed an alternative sporulation model (IPI) that was superior to SIV and DINOV in forecasting spore episodes. de Visser (1996) found that application of both BOTCAST and SIV could reduce the number of spray by 54% compared to weekly spraying programme without any yield loss or higher disease severity.

Other *Botrytis* Species

Botrytis cinerea causes brown stain of the bulbs from the neck onto outer scales of storage onion. Generally, discoloured scales slough off during storage and handling, and the remainder of stored bulb is marketable (Crowe *et al*, 1995). In naturally infected dry scales, the pathogen grows intercellularly with little visible damage to tissue structure. Hypha are far more abundant in the mesophyll parenchyma than in other tissues of the bulb scale. Sclerotia are produced in the abaxial epidermis (Clark and Lorbeer, 1973). This pathogen also causes blight on seed stalks, flowers and capsules.

B. byssoidea causes mycelial neck rot (Walker, 1925), but it is not pathogenic to umbels (Ramsey and Lorbeer, 1986a). However, it can cause severe infection in seedling leaves (Presly, 1985).

Timely application of fungicide sprays may help reduce losses from these diseases.

Cladosporium allii-cepae (Ranojevic) Ellis (syn: *Heterosporium allii-cepae* Ranojevic)

The disease caused by *C. allii-cepae* is known as leaf blight or leaf spot. The disease causes significant yield loss when severe blighting occurs after bulbing and more commonly after leaves to senesce. This fungus is infrequent pathogen of *Allium* host throughout the world. The severe outbreaks of the disease have been reported in Britain and Ireland (Ryan, 1978; Ryan *et al*, 1981). The pathogen can attack a wide range of cultivated and wild *Allium* spp. such as *Allium altaicum*, *A. fistulosum* (Japanese bunching onion), *A. cepa* var. *ascalonicum* (shallot), *A. galanthum*, *A. pskemense* and *A. rarilorii* (Jordan *et al*, 1987).

Description

Conidiophores form solitary or fasciculate in groups of 2-7, straight or slightly flexuous, unbranched, somewhat sinuous in pure culture, smooth, septate, pale brown to dark brown or sometimes olivaceous brown in pure culture. Conidia form solitary or rarely in short unbranched chains (Kirk and Crompton, 1984). Dark pseudothecia readily form on mature lesions in nature (Hill, 1995a).

Disease caused and epidemiology

The pathogen produces small, yellow or grayish lesions which later become elliptical, olivaceous-brown and coalesce on severely infected plants leading to withering and death of the leaves (Jordan *et al*, 1990c). Olivaceous-brown colour is a result of the production of conidia and conidiophores.

The pathogen can occur on onion seed at very low level and possibly be a source of initial inoculum (Boff *et al*, 1995). It has been demonstrated that the incidence of contaminated seeds was low despite severe symptoms in the flowering inflorescence of onion dusted with conidia of the pathogen. In addition no infected onion seeds were obtained from naturally infected plants in the field (Jordan *et al*, 1990c). It occurs on the surface of the testa and is, therefore, unlikely to survive for long periods (Hill, 1995a).

Infected debris is a possible mode of transmission of disease between successive crops (Jordan *et al*, 1990c).

The pathogen sporulates in the light in culture and grows more rapidly in darkness (8 hours) with the relative humidity at 90% at 5-8.5°C (Hill, 1995a)). Maximum sporulation on leaf spots (in England) occurs in late autumn (November-December) with a secondary peak in early spring (March-April). The air-borne conidia are found in air over diseased onion crops (Jordan *et al*, 1990a). Conidia require 18-20 h at 100% relative humidity to germinate and infect the leaves (Jordan *et al*, 1990., a,b). Older leaves, leaf tips and wounds are more susceptible to infection than young healthy leaves (Hill, 1995a).

Control

To ensure vigorous growth and plowing under debris are important cultural practices. Application of maneb and chlorothalonil or fentin acetate at late in the season for interval of 14 days or less reduces the number of infected plants (Hill, 1995a).

Peronospora Destructor (Berk.) Casp.

P. destructor causes downy mildew on onion in many parts of the world including India (Mir, 1977), Giorgia (Langston and Sumner, 2000), New York (Smith *et al*, 1985), New Zealand (Fullerton *et al*, 1986), Pakistan (Ahmad and Karimullah, 1998). It can become severe on leaves of commercially grown onion plants and on leaves and seed stalks of onions grown for seed, especially when relatively cool, moist weather prevails. Actual yield losses in bulb onion of 60-70% have been recorded (Develash and Sugha, 1977a). Other reported hosts are shallot (*A. cepa* var. *ascalonicum*), leek (*A. porrum*), garlic (*Allium sativum*) and chive (*A. schoenosprasum*) (Schwartz, 1995).

Description

Sporangiophores of the pathogen are non-septate and various shades of violet, and emerge from stomate. Sporangia are pyriform to fusiform and attached to the sporangiophore by a pointed end. The mycelium is non-septate and located intercellularly and is with filamentous haustoria that coil within plant cells. Oospore is produced to overwinter (Schwartz, 1995).

Disease caused and epidemiology

The symptoms start in spot in the field and spread to surrounding areas. On the leaves, small light coloured spots appear which turn yellow then white. The spots enlarge during wet weather and affected areas become water soaked with a white streaked appearance. Greyish-violet spore masses develop on infected tissue (Fullerton *et al*, 1986). The affected leaves gradually become pale green and later yellowish, and the diseased parts collapse. In seed onion, the pathogen attacks the seed stalks and the stalks are broken over from the weight of seed umbel. Lesions on the seed stalks are circular or elongate often affecting only one side of the stalk. The fungus may infect flower parts, but does not infect onion seeds (Cheremushkina *et al*, 1990; Schwartz, 1995).

Late infection of onion seed crops with the pathogen causes only yield reduction, but early infection adversely affects both seed quality and yield (Rudolph and Wolf, 1986). Pathogen produces sporangia on the surface of the leaves and seed stalks (Schwartz, 1995).

Pathogen requires cool temperatures. High day temperatures exceeding 24 to 25°C or exceeding 27, 26, 29 or 30°C for more than 8, 6, 4, or 2, respectively inhibit sporulation during subsequent night (Hildebrand and Sutton, 1982; de Visser, 1998). A combination of night temperature, time of onset of high humidity (95%) and duration of high humidity affect the quantity of sporangia produced (Hildebrand and Sutton, 1982; Hildebrand and Sutton, 1984b). Spores survive and infect the leaves when dew was deposited rapidly. The spores are killed rapidly by alternating wetness and dryness, a phenomenon associated with low rates of dew position (Hildebrand and Sutton, 1984a). Then, mycelium in the leaves produces sporangia. As the upper portions of a leaf are killed, the fungus infects the next lower parts of the leaf in each successive cycle of spore formation. Such cycles can be repeated several times until the leaf may be completely killed. These repeated cycles of spore formation can result in severe and continued epidemics of downy mildew, if disease-favourable conditions persist (Lorbeer and Andaloro, 1984). The fungus may destroy the onion foliage in a field during the course of four infection cycles (Schwartz, 1995).

Control

Cultural control measured outlined are the use of mildew-free sets, destruction of initial inoculum sources such as cull and volunteer onions, the use of well drained land, appropriate choice of site, avoidance dense population of plants, the use of surface irrigation and crop rotation with the plants excluded that *Alliums* spp.

Late planting of the crop may increase disease severity. Similarly, application of higher doses of fertilizers or more number of irrigation may encourage higher severity of disease (Ahmad and Karimullah, 1998).

Line IC-48045 (Sharma, 1997) and *Allium roylei* X *A. cepa* hybrid (Kofoet and Zinkernagel, 1990) are resistant, however, IC-32149, IC-33617, IC-49371 and DOPII (Sharma, 1997) show moderate resistance reaction to the disease.

Effective fungicide sprays on the disease control in the onion field are given in Table 23.2

Most growers employ protective spray programme as frequently as 10 days. Chemical control is affected by the waxy nature of the onion leaf and by continual emergence of new leaves from the neck. The use of whether-based disease forecasting may offer a means to reduce the fungicides without increasing the risk of disease control failure (Whiteman and Beresford, 1998). Forecasting system of onion downy mildew referred to as DOWNCAST (Jesperson and Sutton, 1987) summarizes weather conditions that allow the infection of *P. destructor*. The crite-

Table 23.2: Effective treatments to control downy mildew disease, their mode of action and cited references

Treatment	*Mode of action*	*Reference*
Chlorothalonil (1.07 l/h) and Mancozeb (3.54 kg/ha)	Reduction in the number of infected plants	Smith *et al*, 1985
Metalaxyl+Mancozeb (0.3%)	Reduction in the number of infected plants, increase of bulb and seed yield	Georgy *et al*, 1986; Khokhar and Jaffrey, 2000
Integration of heat treatment (35°C for 12 h), bulb dip in metalaxyl (3.5 g/kg bulb) for 1 h and 3 sprays of metalaxyl+mancozeb (0.2%)	Reduction in the number of infected plants	Develash and Sugha, 1997b
Copper oxychloride	Reduction in the number of infected plants	Hamid *et al*, 2002
Pre and post infection sprays of oil based adjuvant fungicide (90 g dimethomorph/kg+600 g mancozeb/kg; + Synertrol oil-832 g emulsifiable vegetable oil/l to young onion plant	Reduction in the lesions and sporulation	MacManus *et al*, 2002b
Strobilurin group	Reduction in the number of infected plants in the field	Robak and Smolinska, 2002
Pencycuron + imidaclorprid (100 ml/kg seed)	Reduction in the number of infected plants, increase of yield Shynkorenko, 2003	

ria are based on quantitative and temporal relationships of temperature, rain, high humidity, rate of dew deposition and dew duration with the infection cycle of the pathogen. Sequence of weather conditions is used to predict "sporulation-infection periods" when the pathogen sporulated, survived and infected the onions. ONIMIL model developed by Battilani *et al* (1996) gave a quantitative prediction for sporulation. Then de Visser (1998) found that DOWNCAST (forecaster of onion downy mildew) gave positive predictions of sporulation for only 11 out of 24 nights when sporulation was observed and that this model failed to predict sporulation events in which the weather conditions were highly variable. In following year, de Visser (2001) evaluated a new model based on calculated leaf wetness in forecasting downy mildew and the influence of this method on DOWNCAST. Results from this study showed that calculated leaf wetness could potentially lead to some unnecessary sprays advices, but less probably to unpredicted outbreaks of downy mildew. Write *et al* (2002) determined that the number of fungicide application was reduced in New Zealand when they used the fungicides based on calendar timing, presence of disease and weather patterns using a modified version of DOWNCAST. Gilles *et al* (2004) suggested that DOWNCAST model could not predict the quantity of sporangia produced, developed MILIONCAST model and compared the accuracy of prediction of sporulation by DOWNCAST; de Visser 's modified version of DOWNAST, and ONIMIL with new mathematical empirical models. MILIONCAST model (acronym for MILdew on onION foreCAST) was based on the data from controlled environmental studies (8-12°C, at 100% relative humidity) to predict the rate of sporulation in relation to temperature and relative humidity. This model generally corresponded well to the pattern of sporulation observed on the outdoor potted plants.

P. allii Rudolphi (syn.: *Puccinia porri* Wint.)

It is the causal agent of rust disease in onion. The disease occurs throughout most of the north temperate zones but seldom causes serious damage. Disease in enhanced on stressed plants. The circular to elongate uredial pustules usually occur between leaf veins. Later in season, dark brown teliospores may form in pustules. Pycnial and aecial stages are rare in nature (Hill, 1995b).

In recent year, Metcalf (2002) determined that the Tasmanian strains of the pathogen infected onions, spring onions, shallots, chives and garlic.

Stemphylium Vesicarium (Wallr.) Simmons

The pathogen causes leaf blight and stalk rot in onion. The disease occurs sporadically on onion in India (Rao and Pavgi, 1975), New York (Shishkoff and Lorbeer, 1989), United States (Miller, 1995), Portugal South Africa (Verwoerd and du Plessis, 1931) and Spain (Basalllote-Ureba *et al*, 1993), and probably in other onion growing regions of the world. Foliar losses of 80 to 90% have been reported, leading to poor bulb yield (Miller, 1995).

Description

Aveling (1994) explained the development and the morphology of conidia using scanning electron microscopy. Conidiophores are straight or flexuous, simple, smooth or verrucose and cylindrical but enlarge apically at the site of conidium production. Smooth, round, but like conidial initials are produced singly at the apex of the verrucose conidiophores. As conidia matured, they became oblong to ovoid then densely verrucose conidiophores proliferate per currently at the distal region, forming secondary conidiophores and conidia. The perfect state may be *Pleospora allii* (Miller, 1995).

Disease caused and epidemiology

This pathogen, in conjunction with *A. porri*, is destructive foliar and seed stalk pathogens of onion under warm and moist conditions. The disease is characterized by elongated spindle-like lesions on the leaves, initially small and white, which later become sunken with a purple colour surrounded by a whitish margin (Miller, 1995; Basallotte-Ureba *et al*, 1999). Infection often starts at the leaf tips. Similar symptoms may also occur on the inflorescence stalks of onion (Miller, 1995).

The pathogen associated with *A. porri* is usually a weak parasite on onion, and can invade downy mildew (*P. destructor*) lesions (Brar *et al*, 1991; Basallote-Ureba *et al*, 1999).

Penetration of the leaf surface by conidia, whether directly through the epidermis or via stomata occur after the formation appressoria (Aveling and Snyman, 1993).

Perithecia may appear in the blighted areas of leaves and inflorescence stalks as small, black, pinhead-like raised bodies. Pseudothecia of the telemorph stage are found on leaf debris from affected plants (Basallote-Ureba *et al*, 1999).

Infection usually remains confined to leaves. The pathogen normally invades dead and dying onion tissue, i.e. necrotic leaf tips, purple blotch and downy mildew lesions, injured tissue and senescent tissue. Rainy weather lasting more than 24h induces the disease. The pathogen can also infect seeds (Aveling *et al*, 1993; Jakhar *et al*, 1996).

Control

Welsh onion (A. *fistulosum*) lines are resistant or moderately resistant. These lines can be used in breeding programme to develop *Stemphylium* leaf blight-resistant onion lines (Pathak *et al*, 2001).

Cultural and fungicide practices recommended for the control of purple blotch also control *Stemphylium* leaf blight and stalk rot (Miller, 1995).

III. SOIL-TRANSMITTED FUNGI

The Fungi Causing Seed Rot and Damping-off

Fusarium spp., *Pythium* spp. and *Rhizoctonia solani* cause seed rot, pre-emergence or post-emergence damping-off in onion seedling. They are common in agricultural soil, and the diseases they cause can occur wherever onions are grown under conditions of poor soil drainage (Sumner, 1995a,b; Vincelli, 1995).

Details on *Fusarium* spp. are given in previous review (Özer and Köycü, 2004). *Pythium irregulare* (Robertson, 1976; Bruckart and Lorbeer, 1982; Vincelli and Lorbeer, 1990; Sumner *et al*, 1997), *P. coloratum* (Robertson, 1976; Bruckart and Lorbeer, 1982) and *Rhizoctonia solani* (Abd-ElRazik *et al*, 1990b; Sumner *et al*, 1997) are strongly pathogenic to onion seedlings.

Germinating seeds may directly rot or seedling may rot and die before emergence. The seedlings lack vigour and grow slowly, then gradually turn yellow, wilt and collapse as the root systems deteriorate when the fungi attack them.

The fungi are commonly all soil-borne, however, some species are also known to be seed-borne, (Özer and Köycü, 2004).

The disease caused by *Pythium* spp. and *R. solani* is favoured by cool wet soil conditions (Fullerton *et al*, 1986). The disease caused by *Pythium* is most severe under conditions of high soil moisture and soil temperature below 18°C. Spores can be dispersed with anything that moves soil, such as equipment and boots. Active zoospores can also be dispersed in irrigation water, runoff from rain or splashing water (Vincelli, 1995). Infection by *R. solani* occurs at diurnal temperature ranges of 10-21° and 21-32°C. The fungus can be disseminated on tillage equipment and irrigation water (Sumner, 1995a).

Cultural practices preventing *Pythium* spp. include breaking up compacted soil horizons, land levelling and installing tile drains or planting on raised bed (Vincelli, 1995).

Turning soil 20-30 cm deep with a moldboard plough is useful cultural practice for the control of *R. solani*. This procedure buries colonized plant debris and sclerotia (Sumner, 1995a).

The use of clean seeds reduces the disease caused by both fungi. Crop rotation is of limited value for *R. solani*, since most crops, except grasses and cereals are susceptible to the pathogen. In addition, soil solarization for at least 1 month in the warm, dry climate of Israel during periods of intense solar radiation has been recommended for the reduction of inoculum potential of *R. solani* in soil (Sumner, 1995a).

In recent years, ARAD-H60 (White and Sivan H-202 (red) was found as resistant to *R. solani* in Cuba (Melendrez-Rodriguez *et al*, 2002).

Gupta *et al* (1987) determined that seed treatment with thiram (2 g/kg seed) combined with thiram soil treatment at 5 g/m^2 and another drenching at 0.2% 30 days after sowing were effective on the infection of *Pythium butleri* in the field.

Drenching the soil with carbendazim or quintozene followed by a second drench with captofol or thiram reduces the number of infected seedling with both pathogens in the greenhouse (Suhag and Rana, 1984). Furthermore, effective protection against both pathogens is provided by captan applied to seed as a dust, slurry, or in pellet in combination with iprodione and vinclozolin (Fullerton *et al*, 1986).

Pyrenochaeta Terrestris (Hans.) Gorenz-Walker & Larson

The pathogen causes the pink root disease in onion. The fungus occurs in most parts of the world and survives for many years in the soil. The disease may be severe on organic soil during periods of relatively high soil temperatures (Gorenz *et al*, 1949). Some isolates of fungus are pathogenic on pepper, eggplant, tomato, soybean, oats, barley, wheat, squash, cantaloupe, muskmelon, cucumber, cauliflower, sweet and field corn, sorghum, millet, ryegrass, spinach and carrot (Sumner, 1995c).

Description

The pathogen fungus, initially is identified as Phoma terrestris but later as *Pyrenochaeta terrestris* (Gorenz et al, 1948). It produces dark brown to black picnidia with setae. The picnidia may be formed in infected roots and bases of the leaf scales of onion seedling and older plants (Sumner, 1995c).

Diseases caused and epidemiology

Pathogenesis is described as the presence of the characteristic pink discolouration of the root tissue followed by loss of root turgor, disintegration of root cortical tissues, intensification of pigmentation and ultimately results in complete impairment of root function and root death (Sparace *et al*, 1986). Reduction in root volume can decrease the plant's ability to obtain nutrients and water from soil (Thornton and Mohan, 1996). Severely infected plants appear to be suffering from nutrient deficiency or drought and generally include etiolation wilting of leaves and leaf tip dieback. Leaf number and size are reduced and the development of bulbs starts earlier in infected plants than in uninfected ones. Plants infected with the pathogen are generally stunted, but the bulbs remain healthy and are still potentially marketable, although often too small and poor storage quality (Gasim-Ahmed and Harrington, 1974; Sneh *et al*, 1974).

When transplants or bulbs of white cultivars are grown in infested soil, pink to purple red blemishes may be observed on the outer scales 1-3 weeks after transplanting. In yellow or red cultivars there may be water-soaked areas on the outer scales (Sumner, 1995c).

The pathogen survives in soil (as deep as 45 cm), probably as chlamydospores, pycnidia or pycnidiospores; or in colonized roots or plant debris of susceptible crops. Seeds are not infested (Sumner, 1995c). The pathogen infects only the roots of the host. Since the optimum temperature for pathogen activity ranges between 24°C and 28°C, marked reductions in bulb field may occur in subtropical and tropical regions when onions are grown in infested soils (Levy and Gornik, 1981). Infection is reduced at 20°C and very little disease occurs at temperatures less than 16°C.

Host-Parasite interactions

The disease symptoms are accompanied by the biosynthesis of characteristic red pigments (Gunesekaran and Weber, 1981). The pathogen produces toxic metabolites described by Sparace *et al*, (1984) as pyrenocines. In the following year, Sparace *et al* (1986) determined three pyrones (pyrenocines A, B and C) showing varying degrees of phytotoxicity produced by this pathogen. Among them, Pyrenocine A had the most inhibitor effect to onion seedling elongation. It was also toxic to onion protoplasts, causing reduction in protoplast viability.

Control

Crop rotation of 3 to 6 with plants that are not susceptible to pathogen, reduction of irrigation interval from 16 or 12 days to 8 or 4 days are important cultural practices.

Different onion cultivars or lines were tested for resistance to disease. 'Granex', 'Dessex' and 'Laredo' were tolerant cultivars (Levy and Gornik, 1981); 'Empire', 'Epoch', 'Yellow Sweet', 'Spanish', 'Inca', 'XPH25', XPH70', 'XPH419', 'Sportan Banner', 'MSU8155 X 826' (Lacy and Roberts, 1982), 'Granex 429' (Gonzalez *et al*, 1994), 'SR2316-3', 'Sun Re 1463', 'Ax1846', Ax2246' and 'PSR72788' were determined as resistant to disease under field conditions. Additionally, Zappacosta *et al* (2003) suggested that *A. fistulosum* (cv. Nagiwa Negi) was resistant to disease and exhibited high activity of peroxidases and glucanases.

In heavily infested field, control measures are required for continuously profitable onion production. Soil solarization and fungicide treatments have been suggesting to the control the disease (Table 23.3).

Table 23.3: Effective treatments to control pink rot disease, their mode of action and cited references

Treatment	*Mode of action*	*Reference*
Soil solarization	Reduction in the number of infected plants in the field	Katan *et al*, 1980;Porter *et al*, 1989; Satour *et al*, 1989
Dazomet at 750 kg/ha either alone or in combination with solarization	Reduction in the number of infected plants and disease severity, increase in yield, in the field	Porter *et al*, 1989
Metham-sodium fumigation and solarization	Reduction in the number of infected plants in the field	Hartz *et al*, 1989
Metham sodium fumigation-solarization-rape biomass incorporation to soil	Reduction in the number of infected plants and disease severity, increase in yield, in the field	Cho *et al*, 2001

Sclerotium Cepivorum Berk

S. cepivorum is the causal agent of white rot, which is one of the most serious diseases of onion (*Allium cepa* L.). The disease causes major crop losses in many onion growing regions throughout Europe, Asia, Africa, the Middle East, Central and South America, Australia and New Zealand (Crowe, 1995). It affects plants during growing season, but predominantly at the first 3-4 months and can continue as a bulb rot during storage under adequate humidity and temperature conditions.

Description

Perfect stage of the fungus is not known. The pathogen has not functional spores, although it produces small spermatia like phialospores on water agar. The only reproductive structures are sclerotia, which normally are uniformly round and have diameters commonly between 0.35 and 0.50 mm (Crowe, 1995).

Diseases caused and epidemiology

Pathogen initially infects the root systems. Foliage symptoms result only after the pathogen grows into the stem plate or bulb of the plants. Infected leaves turn yellow and stunting of plants are followed by death of all foliage. Coincident with the above ground symptoms, white fluffy fungal growth can be seen on the roots and stem plate that extends around the base of the bulb and moves up to bulbs, and inward from storage leaf to storage leaf. The roots are rotten completely and diseased plants are lifted from the soil. Small, poppy, seed-sized sclerotia form in decaying tissues. Root and stem contact can spread the disease from plant to plant during the season (Fullerton *et al*, 1986).

Mycelium may grow 1-2 cm through soil to infect host roots. Appresoria are formed. A few sclerotia may form in roots. Mycelium also radiates 1-2 cm away from infested roots and may infect nearby roots. In this manner, the pathogen may grow on the roots of neighbouring host plants, moving along the planted row. Lateral movement is fastest 2-7 cm below the stem plate. The pathogen may successfully incite bulb decay. The white rot fungus reaches the stem plate, the plant rapidly rots and sclerotia are produced in abundance. Sclerotia become free in soil after host tissues completely decay (Crowe, 1995). The disease also continues as a bulb rot during storage. Hag *et al* (2003) detected the pathogen DNA only in the two outermost scales of the bulb.

Sclerotia of *S. cepivorum* are constitutively dormant for several weeks to months after farming on decayed host plants. There is a positive relationship between the density of the sclerotia in infested soil at planting time and white rot severity and incidence on onion (Crowe *et al*, 1980; Adams, 1981; El-Razik *et al*, 1986a). The pathogen persists in the soil in the absence of host plant as sclerotia and can survive in this form for more than 20 years (Coley-Smith and Parfitt, 1986; Coley-Smith *et al*, 1990). Following their production, sclerotia are constitutively dormant for 1-3 months and then only germinate in the presence of host plant (Coley-Smith *et al*, 1987). In following year, it was determined that sclerotia produced by most isolates of *S. cepivorum* under sterile and non-sterile conditions showed a constitutive dormancy immediately after maturing, in contrast to this the sclerotia of some isolates germinated quite well as soon as they were mature (Brix and Zinkernagel, 1992 a). Volatile and water-soluble breakdown products of propyl and allyl cysteine amino acids are responsible for triggering the mass germination of sclerotia under non-sterile conditions (Coley-Smith, 1979; Coley-Smith and Parfitt, 1986). These volatile compounds elicit germination of sclerotia as far as 10 cm away. Stimulated germination is optimum at 14-18°C and terminates near 24°C. Germination proceeds most abundantly at soil moisture levels optimum for host root growth (-300 mbar). Nearly all sclerotia may be stimulated to germinate during periods of 9-24°C, if the host plants are present in the field. Mycelial growth occurs at 5-27°C and stops if the temperature exceeds 27°C for several hours per day (Crowe, 1995).

The disease can be transmitted by infected onion sets and transplants, the movement transport of infested soil on equipment, harvest containers, shoes, animals which may feed diseased bulbs and defecate viable sclerotia some distance away.

Host-Parasite interactions

The fungus penetrates the root epidermis then invades cortical parenchyma (Abd El-Razik *et al*, 1973). Infection of onion tissue by the pathogen occurs by germ tubes penetrating between adjacent epidermal cell walls or directly, via penetration pegs produced from slightly swollen hyphal tips. After passing through the cuticle, the infection peg enlarges to form an infection hypha within the primary cell wall. Extensive degradation of epidermal cell wall occurs at a distance of 2-3 cells from the advancing hyphae. As infection advanced, hyphae spread rapidly from the epidermis to the cortex growing between and within dead/dying host cells. Extensive host cell death results in localized collapse of the tissue around infection points (Stewart *et al*, 1989). *Sclerotium cepivorum* produces the phytotoxin oxalic acid (Stone and Armentrout, 1985) and a range of extracellular enzymes that degrade the cell wall, including polyglacturonases (Mankarios and Friend, 1980; Favaron *et al*, 1993), pectinesterases (Abd-ElRazik *et al*, 1973), cellulases, arabinases and xylanases (Mancarios and Friend, 1980). The pathogen also produces two polygalacturonases and two pectin esterases during infection. These enzymes associate with cell death and cell wall degradation (Metcalf and Wilson, 1999).

Control

Cultural practices include; restriction of movement of equipment, plant materials and grazing animals from infected field; avoidance of *Allium* plantings in infested field; the removal and burning infected plant debris, the reduction of host plant density (Littley and Rahe, 1987). It is difficult to control the white rot disease using crop rotation because of the longevity of the sclerotia in the soil (Coley-Smith, 1983; Bank and Edgington, 1989). However, rotation of *Trifolium sorghum*, *Trifolium*/maize, wheat/sorghum, faba bean/sorghum, faba bean/maize and lentil/cotton decrease sclerotia number in soil in the presence of *Trichoderma harzianum*; in the absence of this antagonist fungus rotations of lentil/cotton and wheat/soybean decrease the pathogen population (Abdel-Rahman *et al*, 1990). Additionally, rotation between spring-planted and fall-planted *Allium* crops reduces sclerotial inoculum where the soil temperature was too high (Crowe, 1995).

Sclerotia rapidly decay under the conditions of prolonged periods of winter soil saturation and/or flooding, but this is not sufficient for commercial disease control (Leggett and Rahe, 1985; Banks and Edgington, 1989). However, periodic warm season flooding may allow regular recropping at host crop without excess white rot (Crowe, 1995).

The cultivars Beth Alpha and Pukekohe Longkeeper (Van Der Meer *et al*, 1983), and Sweet Sandwich (Brix and Zinkernagel, 1992b), were found as resistant to the disease. Somkuwar *et al* (1997) suggested that Hibrid 3 and IIHR Yellow showed moderate resistant reaction.

Numerous soil treatments with chemicals, organic compounds and microorganisms were tested for the control of *S. cepivorum*. Effective treatment under different experimental conditions and their modes of action are listed in Table 23.4.

Table 23.4: Effective treatments to control white rot disease, their mode of action and cited references

Treatment	*Mode of action*	*Reference*
Chemicals		
Single injection of 0.2 ml Diallyl disulphide (DADS) at 0.156% into soil	Reduction in sclerotial germination and the number of sclerotia	Enstwistle *et al*, 1982
Metham application through springler irrigation at 234 l/ha to soil	Reduction in the number of infected plants	Adams and Johnston, 1983
Integration of crop rotation flooding and fungicide application with vinclozolin as seed dressing (50 g a.i./kg seed, followed by three stem-base sprays (4 kg a.i. in 1200 l water/ha)	Reduction in the viability of sclerotia in the field	Banks and Edgington, 1989
Seedling dipping with Procymidone (40 g a.i./l)	Reduction in the number of infected plants in the greenhouse	Fahmy and Omar, 1989
Seedling dipping with Procymidone (20-30 g a.i./l and Triton B-60) or vinclozolin (15 g a.i./l) for 5 min before planting	Reduction in the number of infected plants in the field	Abd-ElRazik *et al*, 1990a
Procymidone as combination of seed treatment (5 g a.i./kg seed) and foliar sprays (0.75 kg a.i./ha monthly X4)	Reduction in the number of infected plants in the field	Fullerton and Stewart, 1991
Procymidone as dispersible granule (2.5 kg a.i./ha) to soil surface after planting	Reduction in the number of infected plants in the field	Stewart and Fullerton, 1991
Tebuconazole (0.5 kg a.i./ha) and Triadimenol (0.375 a.i. g/ha) as foliar spray, or Tebuconazole as soil spray, or seed treatment with procymidone (40 g a.i./kg seed) and foliar sprays of either procymidone or Triadimenol	Reduction in the number of infected plants in the field	Fullerton *et al*, 1995
Diallyl disulphide (DADS)-Scleroti germination stimulant-under 200 kPa pressure to a maximum soil dept at 30 cm	Reduction in the number of infected plants in the field	Mac Manus *et al*, 2002a
DADS injection into soil at 10 l/ha in 500 l/ha of water	Reduction in the number of infected plants and the viability of sclerotia in the field	Hovius and McDonald, 2002
Physical		
Gamma irradiation of seeds	Reduction in the number of infected plants	Utkhede and Rahe, 1982
Organic		
Dipping onion seedling in 1% and/or 2.5% natural bee (propolis) alcoholic extract for 4 h	Reduction in the number of infected plants and sclerotia number in greenhouse	Fahmy and Omar, 1989
Soil solarization	Reduction in the number of infected plants in the field	Satour *et al*, 1989
Soil solarization combined with organic soil amendment	Reduction in the number of infected plants in the field	Forleo, 2002
Addition of residues from air-dried and crushed mustard (*Brassica juncea*) to soil (1% w/v)	Reduction in the viability of sclerotia in the pot	Smolinska, 2000
Addition of rapeseed (*Brassica napus* cv. Balko, *B. napus* cv. Gorczanski)	Reduction in the number of sclerotia; increase of the total number of	Smolinska, 2000

Contd.

Table 23.4: Contd.

Treatment	*Mode of action*	*Reference*
to soil (1% w/v)	bacteria, spore forming bacteria, fluorescent pseudomonads in the pot	
Saponin from seeds of oldman salt bush (*Atriplex nummularia*)	Reduction in linear growth of pathogen *in vitro*	Abdel-Momen *et al*, 2000
Saponin from Lucerne roots	Reduction in the number of infected plants in the pot and field	Abdel-Momen *et al*, 2000
50% incorporation of composted onion wastes to soil	Reduction in viability of sclerotia in greenhouse	Coventry *et al*, 2002
Microorganisms		
Application of suspension of *Penicillium godlewskii* and *Aspergillus candidus* and *Bacillus subtilis* to soil 15 days before transplanting	Reduction in the number of infected plants in the greenhouse	El-Razik *et al*, 1986b
Mutants of *Trichoderma harzianum* induced by ethanol amine and diethyl-sulphate	Reduction in the number of infected plants in the pot	Mohamed and Fahmy, 1990
Addition of Mycorrhiza (*Glomus* sp) isolated from bean plants around the roots of seedling	Reduction in the number of infected plants in the field	Torres-Barrágan *et al*, 1996.
Trichoderma koningii	Colonization on *S. cepivorum*, infected roots *in vitro*	Metcalf and Wilson, 2001
T. viride as fluid drilled in guar gum with seeds	Reduction in disease symptoms in the field	Clarkson *et al*, 2002
Native microbial isolates such as Actinomycete, *Pichia onychis* and *Serratia marcescens*	Reduction in the number of infected plants in the pot	Paris and Cotes, 2002
Trichoderma harzianum (Trichodex)	Reduction in the number of infected plants in the pot	Paris and Cotes, 2002
Mutants of *T. harzianum*, *T. viride* and *T. koningii* generated by exposing their sporulated cultures to 20 and 50 kilo rad doses of gamma irradiation	Reduction in the number of infected plants and the pathogen growth in soil	Haggag and Mohamed, 2002
Conyothyrium minitans	Inhibition of fungal growth *in vitro*	McQuilken *et al*, 2003
T. koningii	Reduction in the number of infected plants; increase colonization of antagonist on the roots in the pot and field	Metcalf *et al*, 2004
Trichoderma pseudokoningii in silty clay soil	Scleroti degradation and reduction in the number of infected plants in the pot	Clarkson *et al*, 2004
T. viride	Scleroti degradation and reduction in the number of infected plants in the pot	Clarkson *et al*, 2004

Urocystis Magica Pass.

U. magica synonym: *U. cepulae* Frost and *U. colchici* (Schlechtend.) Rabenh. is the causal agent of smut disease in onion. The disease is problem for seed-grown onions wherever favorable moisture and temperature conditions exist. *Allium* species; onion, leek and shallot are the most susceptible. Garlic, golden garlic, wild garlic, chives, certain ornamental species of *Allium* are resistant to the disease (Hill, 1995c).

Description

Chlamydospores develope from terminal cells of sporogenous hyphal branches which are curved back upon themselves and surrounding hyphal cells became appendage cells of the chlamydospores. Central spore walls differentiate into three distinct layers. Appendage cell walls have only two distinguishable layers, and are attached to the central spore walls by an amorphous matrix. (Grayson and Lacy, 1975). Chlamydospores are single celled spherical to ellipsoid, reddish brown, smooth, and thick walled. (Hill, 1995c).

Diseases caused and epidemiology

First symptoms appear on cotyledon leaves as dark, lead-coloured, slightly thickened spots and streaks. Similar symptoms occur on the scales or leaves as they develop. Affected leaves are often twisted and distorted. The streaks develop into silvery blisters (sori) which rupture to release black masses of spore balls in the outer scales of the developing bulb (Fullerton *et al*, 1986). Diseased plants are usually severely stunted and died within 3-4 weeks after emergence. Smut in slightly affected bulbs appears as somewhat raised, dark brown to black pustules in the outer fleshy scales. Some pustules may occur as deeply as the third or fourth scale. Infected bulbs do not root in storage from smut, but are more susceptible to infection by other pathogen (Hill, 1995c).

The causal fungi may remain dormant but alive for 15 years or possible larger as multicelled chlamydospores (Fullerton *et al*, 1986).

The causal fungi are transmitted from one field to another, on infected bulbs and sets, tool or farm equipment, the feet of humans, domestic animals and bird, and by surface drainage water and wind-blown soil.

Smut spores are highly resistant to environmental changes and once incorporated in the soil may remain alive for up to 15 years even in the absence of a susceptible host crop (Fullerton *et al*, 1986). The pathogen can only penetrate and infect the onion plant the young seedling (10-21 days after seed germination). If penetration is delayed, further movement into the bulb is prevented even in heavily infested soil. Optimum temperature for spore germination and growth are 13-22°C. Spore germination and fungal growth and reduced above 25°C; however, infection can occur at temperatures as low as 10-12°C (Hill, 1995c). Soil moisture has no direct effect on spore germination or the disease, but an excess of water in soil generally means low temperature there and slow growth, thus, favouring the chances of attack. The smut fungi grow slowly in the soil and contaminated areas do not rapidly enlarge.

Control

Healthy onion sets or transplants may be planted in smut-infested soil without danger of becoming infected. All diseased seedling and plants should be collected and burned. Cultivation under conditions favourable for rapid host growth may reduce the time that seedlings can be infected.

No resistant varieties are available. Crop rotation can reduce the smut inoculum, but is not always practical (Hill, 1995c).

Seed treatments with chemicals protect seedling through the susceptible stage (Table 23.5).

Table 23.5: Effective treatments to control smut disease, their mode of action and cited references

Treatment	*Mode of action*	*Reference*
Carboxin+Thiram (25 kg/seed)	Reduction in the number of infected plants in the field	Tartier *et al*, 1976
Carbathiin+thiram to seed+granular in furrow application of mancozeb and chlorpyrifos	Reduction in the number of infected plants in the field	Hoepting *et al*, 2000
Seed treatment with Tebuconazole (100 or 200 mg/100 g seed)+metalaxyl (28%, 30 mg) ±1,3 –dichloropropen (5 or 50 mg/100 g seed)	Reduction in the number of infected plants in the field	McDonald *et al*, 2002
Seed treatment with Carboxin+Thiram (20g/kg seed), mancozeb (50 g/kg seed) and granule application of Mancozeb in furrow (1.2, 4.8 or 9.6 kg/ha)	Reduction in the number of infected plants in the field	Hoepting *et al*, 2003

Other Soil-borne Fungi

Colletotrichum circinans (Berk.)Voglino (syns. *Vermicularia circinans* Berk. and C. *dematium* (Pers.) Grove f. *circinans* (Berk.) Arx) causes smudge occurred primarily on the dried wrapper scales and lower portions of white onions. Minute stromata forms just beneath the cuticle in dark color. Stromata frequently form in circular, concentric rings approximately 1 cm in diameter. Mycelium proliferates between the cuticle and outer epidermal wall, and acervuli form on stromata. The pathogen requires warm and wet weather conditions (Sumner, 1995g).

Glomerella cingulata (Stoneman) Spauld.&H. Schrenk (anamorph *C. gloeosporioides* (Penz.) Penz.&Sacc. in Penz.) is the causal agent of twister disease in onion. Curling, twisting and chlorosis of the leaves, stunting of the roots are typical symptoms of the disease. The disease is more important in the tropics and subtropics area. The pathogen may also infect the plant from above ground inoculum (Hill, 1995d; Weeraratne, 2002).

Lasiodiplodia theobromae (Pat.) Griffon&Maubl. (syns. *Diplodia natalensis* Pole-Evans and *Botryodiplodia theobromae* Pat.) is the cause of silvery gray to black discolouration of the outer dry scales on the upper half of harvested bulbs. The disease reduces market value of onion (Sumner, 1995e).

Rhizopus microsporus Tiegh. and *R. stolonifer* (Ehrenb.:Fr) Vuill. (syn. *R. nigricans* Ehrenb.) is observed infrequently. Bulbs may have soft, mushy spots, particularly in the neck region. *Rhizopus* spp. are good saprophytes and colonize necrotic tissues and organic materials whenever there is a moist environment. The pathogens require wounds to penetrate the bulbs (Sumner, 1995f).

Sclerotium rolfsii Sacc. causes dirty white spots on the outer scales and lesions on the neck area and fleshy scales appeared water-soaked. It is known as southern blight. Characteristics spherical brown sclerotia may form on the bulb and surrounding soil and debris. The disease appears rarely in temperate regions (Sumner, 1995d).

REFERENCES

Abd-ElRazik, A.A., M.N. Shatla, & M. Rushudi. 1973. Studies on the infection of onion plants by *Sclerotium cepivorum* Berk. Phytopothologische Zeitschrift, 76: 108-116.

Abd-ElRazik, A.A., A.M. Amein, A.M. El-Shabrawy, & M.H. Rushdi. 1990a. Effect of certain cultural practices and fungicides on control of *Sclerotium cepivorum* on winter onion. Egyptian Journal of Phytopathology, 20: 87-97.

Abd-ElRazik, A.A., F.G. Fahmy, A.M. Amein, & A.I. El-Amein. 1990b. Role of onion seeds in transmission of damping-off causal fungi and chemical control of the disease. Review of Plant Pathology, 1992, 71: 3456.

Abdel-Momen, S.M., S.A. Omar, A.A. Hanafi, & T.M. Abdel-Rahman. 2000. Different sources of saponin affecting white rot disease in onion (*Allium cepa* L.). Bulletin of Faculty of Agriculture, University of Cairo, 51: 365-377.

Abdel-Rahman, T.M., M.S. Mohamed, A.M. El-Shabrawy, & M.H. Rushdi. 1990. Effect of different crop sequences on biological control of onion white rot disease by *Trichoderma harzianum* Rifai. Assiut Journal of Agricultural Sciences, 21: 149-164.

Adams, P.P. 1981. Forecasting onion white rot disease. Phytopathology, 71: 1178-1181.

Adams, P.B. & S.A. Johnston. 1983. Factor affecting efficacy of metham applied through sprinkler irrigation for control of *Allium* white rot. Plant Disease, 67: 978-980.

Ahmad, S. & Karimullah, 1998. Relevance of management practices to downy mildew in onion. Sarhad Journal of Agriculture, 14: 161-162.

Ajrekar, S.L. 1921. Annual report of the work done under plant pathologist. Department of Agriculture, Pune. Government of Bombay: 121-122.

Alderman, S.C. & M.L. Lacy. 1983. Influence of dew period and temperature on infection of onion leaves by dry conidia of *Botrytis squamosa*. Phytopathology, 73: 1020-1023.

Alderman, S.C. & M.L. Lacy. 1984. Influence of leaf position and maturity on development of *Botrytis squamosa* in onion leaves. Phytopathology, 74: 1461-1463.

Alderman, S.C., M.L. Lacy, & K.L. Everts. 1985. Influence of interruptions of dew periods on number of lesions produced by *Botrytis squamosa*. Phytopathology, 75: 808-610.

Aveling, T.A.S. 1998. Purple blotch (*Alternaria porri*) of onion. Recent Research and Development in Plant Pathology, 2: 63-76.

Aveling, T.A.S. & S.P. Naude. 1992. First report of *Alternaria porri* on garlic in South Africa. Plant Disease, 76: 643.

Aveling, T.A.S. & I.H. Rong. 1994. Scanning electron microscopy of conidium formation of *Stemphylium vesicarium* on onion leaves. Journal of Phytopathology, 140: 77-81.

Aveling, T.A.S. & H.G. Snyman. 1993. Infection studies of *Stemphylium* vesicarium on onion leaves. Mycological Research, 97: 984-988.

Aveling, T.A.S., H.G. Snyman, and S.P. Naude. 1993. Evaluation of seed treatments for reducing *Alternaria porri* and *Stemphylium vesicarium* on onion seed. Plant Disease, 77: 1009-1011.

Aveling, T.A.S., H.G. Snyman, & F.H. Rijkenberg. 1994. Morphology of infection of onion leaves by *Alternaria porri*. Canadian Journal of Botany 72: 1164-1170.

Bock, K.R. 1964. Purple blotch (*Alternaria porri*) of onion in Kenya. Annals of Applied Biology, 54: 303-311.

Bank, E. & L.V. Edgington. 1989. Effect of integrated control practices on the onion white rot pathogen in organic soil. Canadian Journal of Plant Pathology, 11: 268-272.

Basallote-Ureba, M.J., A.M. Prados-Ligero, & J.M. Melero-Vara. 1993. The first report in Spain of two leaf spots of garlic caused by *Stemphylium vesicarium*. Plant Disease, 77: 376.

Basallote-Ureba, M.J., A.M. Prados-Ligero, & J.M. Melero-Vara. 1999. Aetiology of leaf spot of garlic and onion caused by *Stemphylium vesicarium* in Spain. Plant Pathology, 48: 139-145.

Battilani, P., V. Rossi, P. Racca, & S. Giosue. 1996. ONIMIL., a forecaster for primary infection of downy mildew of onion. European Mediterranean Plant Protection Organisation Bulletin, 26: 567-576.

Boff, P. 1996. Survey of onion pathogens in Santa Catarina State, Brazil. Fitopatologia Brasileira, 21: 110-114.

Boff, P., J.F. Debarba, E. Silva, & H. Werner. 2001. Thermophilic compost to increase onion plant health. Bulletin OILB/SROP, 24 : 15-17.

Boff, P., M.J. Stadnik, R. Ferrari, & T.O. Da Silva. 1995. Assessment of seed health of the onion seeds commercialized in the state of Santa Catarina. Revista Brasileira de Sementes, 17:165-170.

Boff, P., H. Stuker, & P.A.S. Gonçalves. 1998. Effect of plant density on leaf diseases and bulb production in onion. Fitopatologia Brasileira, 23: 448-452

Brar, S.S. H.S. Rewal, D. Singh, R.C. Sharma, & H. Sinh. 1991. Efficacy of fungicides in controlling *Stemphylium botryosum* on downy mildew infected onion seed crop. Onion Newsletter for the Tropics, 3: 50-51.

Brix, H.D. & V. Zinkernagel. 1992a. Effect of cultivation, conditioning and isolate on sclerotium germination in *Sclerotium cepivorum*. Plant Pathology, 41: 13-19.

Brix, H.D. & V. Zinkernagel. 1992b. Screening for resistance of *Allium* species to *Sclerotium cepivorum* with special reference to non-stimulatory resistance. Plant Pathology, 41:308-316.

Bruckart, W.L. & J.W. Lorbeer. 1982. *Pythium* species pathogenic to onion seedlings grown on organic soils in New York. Phytopathology, 72: 469-475.

Castellanos Linares, J.J., F. Auchet-Jenckens, I. Garcia Correoso, & A. Fraga Aguiai, N. 1986. Effect of *Alternaria porri* (Ell.) Cif. on onion seed production under experimental conditions in Cuba. Review of Plant Pathology, 1988, 67: 2730.

Cheremushkina, N.P., M. Orekhovskaya, & T. Koptyaeva. 1990. Is the causal agent of onion perenosporose transmitted by seed? Review of Plant Pathology, 1992: 7060.

Cho, J.J., R.S. Shimabuku, H.R. Valenzuela, J. Uchida, & R.F. Mau. 2001. The effect of solarization, metham sodium, biological soil treatments and cover crop amendments on pink root incidence and yields of sweet onion in Maui, Hawaii. In "Proceedings of 113rd Annual Meeting, the Florida State Horticultural Society, USA" July 23-25, 2000, pp. 218-221.

Clark, C.A. & J.W. Lorbeer. 1973. Comparative histopathology of *Botrytis squamosa* and *B. cinerea* on onion leaves. Phytopathology, 66: 1279-1289.

Clark, C.A. & J.W. Lorbeer. 1976. Symptomatology, etiology and histopathology of *Botrytis* brown stain of onion. Phytopathology, 63: 1231-1235.

Clarkson, J.P., A. Mead, T. Payne, & J.M. Whipps. 2002. Selection of biocontrol agents of *Sclerotium cepivorum* for control white rot by sclerotial degradation in a UK soil. Plant Pathology, 51: 735-745.

Clarkson, J.P., Mead, A., Payne, T. and Whipps, J.M. 2004. Effect of environmental factors and *Sclerotium cepivorum* isolate on sclerotial degradation and biological control of white rot by *Trichoderma*. Plant Pathology, 53: 353-362.

Coley-Smith, J.R. 1979. Germination of sclerotia. In "Proceedings of the First International Workshop on White Rot of onion". Wageningen, The Netherlands.

Coley-Smith, J.R. 1983. Integrated control of *Sclerotium cepivorum*. In "Proceedings of the Second International Workshop on White Rot of Onions". Beltsville, Maryland.

Coley-Smith, J.R., C.M. Mitchell, & C.E. Sansford. 1990. Long term survival of sclerotia of *Sclerotium cepivorum* and *Stromatinia gladioli*. Plant Pathology, 39: 58-69.

Coley-Smith, J.R. & D. Parfitt. 1986. Some effects of diallyl disulphide on sclerotia of *Sclerotium cepivorum*: possible novel control method for white rot disease of onion. Pesticide Science, 37: 587-594.

Coley-Smith, J.R., D. Parfitt, I.M. Taylor, & R.A. Reese. 1987. Studies of dormancy in sclerotia of *Sclerotium cepivorum*. Plant Pathology, 36: 594-599.

Coventry, E., R. Noble, A. Mead, & J.M. Whipps. 2002. Control of *Allium* white rot (*Sclerotium cepivorum*) with composted onion waste. Soil Biology and Biochemistry, 34: 1037-1045.

Cramer, C.S. 2000. Breeding and genetics of *Fusarium* basal rot resistance in onion. Euphytica, 115 :159-166.

Crowe, F. 1995. White rot. In, "Compendium of onion and garlic diseases", (eds. Schwartz, H.F. and Mohan, S.K.), APS Press, St. Paul, Minnesota, U.S.A., pp. 14-16.

Crowe, F.J., D.H. Hall, A.S. Greathead, & K.G. Baghott. 1980. Inoculum density of *Sclerotium cepivorum* Berk. and the incidence of white rot of onion and garlic. Phytopathology, 70: 64-69.

Crowe, F., S.K. Mohan, & H.F. Schwartz. 1995. Other *Botrytis* Diseases. In, "Compendium of onion and garlic diseases", (eds. Schwartz, H.F. and Mohan, S.K.), APS Press, St. Paul, Minnesota, U.S.A., pp. 19-20.

Currah, L. & B. Maude. 1984. Laboratory tests for leaf resistance to *Botrytis squamosa* in onion. Annals of Applied Biology, 105: 277-283.

Datar, V.V. 1994. Investigation on purple blotch of onion in India. Acta Horticulturae, 358: 259-263.

Develash, R.K. & S.K. Sugha. 1997a. Incidence of downy mildew and its impact on yield. Indian Phytopathology, 50: 127-129.

Develash, R.K. & S.K. Sugha. 1997b. Management of downy mildew (*Peronospora destructor*) of onion (*Allium cepa*). Crop Protection, 16: 63-67.

de Visser, C.L.M. 1996. Field evaluation of supervised control system for *Botrytis* leaf blight in spring sown onions in the Netherlands. European Journal of Plant Pathology, 102: 795-805.

de Visser, C.L.M. 1998. Development of a downy mildew advisory model based on downcast. European Journal of Plant Pathology, 104: 933-943.

de Visser, C.L.M. 2001. Evaluation and use of a simulation model for leaf wetness to forecast downy mildew (*Peronospora destructor*) in onions. Acta Horticulturae, 555: 133-136.

du Toit, L.J. & M.L. Derie. 2004. Prevalence of *Botrytis* spp. In onion seed crops in Columbia Basin of Washington. Plant Disease, 88: 1061-1068.

Ellerbrock, L.A. & J.W. Lorbeer. 1977a. Etiology and control of onion flower blight. Phytopathology, 67: 155-159.

Ellerbrock, L.A. & J.W. Lorbeer. 1977b. Sources primary inoculum of *Botrytis squamosa*. Phytopathology, 67: 363-372.

El-Razik, A.A.A, A.M. El-Shabrawy, M.A. Sellam, & M.H.A. El-Rehim. 1986a. Distribution of sclerotia of *Sclerotium cepivorum* in soil and their relationship with severity of white rot of onions. Egyptian Journal of Phytopathology, 17: 95-105.

El-Razik, A.A.A, A.M. El-Shabrawy, M.A. Sellam, & M.H.A. El-Rehim. 1986b. Effectiveness of certain fungi and bacteria associated with sclerotia of *Sclerotium cepivorum* in upper Egypt soil on controlling white rot of onion. Egyptian Journal of Phytopathology, 17: 107-114.

Entwistle, A.R., P.R. Merriman, H.L. Munasinghe, & P. Mitchell. 1982. Diallyl-disulphide to reduce the number of sclerotia of *Sclerotium cepivorum* in soil. Soil Biology and Biochemistry, 14: 229-232.

Everts, K.L. & M.L. Lacy. 1990. Influence of environment on conidial concentration of *Alternaria porri* in air and on purple blotch incidence on onion. Phytopathology, 80: 1387-1391.

FAO, 2004. Food and Agricultural Organization. http://www.fao.org/ page/collections? *subset=Agriculture.*

Favaron, F., C. Castiglioni, & P. Di Lenna. 1993. Inhibiton of some rot fungi polygalacturonases by *Allium cepa* L. and *Allium porrum* L. extracts. Journal of Phytopathology, 139: 201-206.

Fahmy, F.G. & M.O.M. Omar. 1989. Potential use of 'propolis' to control white rot disease of onion. Assiut Journal of Agricultural Sciences, 20: 265-275.

Forleo, L., 2002. What alternatives are there to methyl bromide? Colture Protette, 31: 29-35.

Fullerton, R.A. & A. Stewart. 1991. Chemical control of onion white rot (*Sclerotium cepivorum* Berk.) in the Pukekohe district of New Zealand. New Zealand Journal of Crop and Horticultural Science, 19: 121-127.

Fullerton, R.A., A. Stewart, C.N. Hale, & R.J. Wood. 1986. Diseases of onions in New Zealand. Proceedings Agronomy Society of New Zealand, 16: 111-114.

Fullerton, R.A. A. Stewart, & E.A. Slade. 1995. Use of demethylation inhibiting fungicides (DMIs) for the control of onion white rot (*Sclerotium cepivorum* Berk.) in New Zealand. New Zealand Journal of Crop and Horticultural Science, 23: 121-125.

Gasim-Ahmed, A.A. & J.F. Harrington. 1974. Onion seed yield as affected by pink root disease, soil fumigation, mother bulb fertilization and bulb size. Horticultural Science, 9: 394-396.

Georgy N.I. I.A. Radwan, H.A. Mohamed, & A.E. Shahabi. 1986. Chemical control of downy mildew and purple leaf blotch of onion in Egypt. Agricultural Research Review, 61: 25-41.

Gilles, T., K. Phelps, J.P. Clarkson, & R. Kennedy. 2004. Development of MILIONCAST, an improved model for predicted downy mildew sporulation on onion. Plant Disease, 88: 695-702.

Gonzalez, H., G. Garcia, & Luque, M. 1994. Response of three onion varieties, *Allium cepa* L. in soil inoculated *Pyrenochaeta terrestris* (Hansen) and *Fusarium oxysporum* Schlecht. Agronomia Tropical Maracay, 44: 699: 706.

Gorenz, A.M., R.H. Larson, & J.C. Walker. 1949. Factors affecting pathogenicity of pink root fungus of onions. Journal of Agricultural Research (Washington, DC), 78: 1-18.

Gorenz, A.M., J.C. Walker, & R.H. Larson. 1948. Morphology and taxonomy of the onion pink-root fungus. Phytopathology, 38: 831-840.

Grayson, R.L. & M.L. Lacy. 1975. Development and nuclear history of the teliospores of *Urocystis colchici*. Phytopathology, 65: 994-999.

Gunasekaran, M. & D.J. Weber. 1981. Influence of physicochemical factors on growth and pigment synthesis by *Pyrenochaeta terrestris*. Mycologia, 73: 844-852.

Gupta, R.P., P.K. Srivastava, & W.B. Pandey. 1987. Control of damping-off in kharif onion nursery. Review of Plant Pathology, 67: 5298.

Gupta, R.P., P.K. Srivastava, & W.B. Pandey. 1991. Management of onion diseases and insect pests in India. Onion Newsletter for the Tropics, 3: 15-17.

Gupta, R.P., P.K. Srivastava, & R.C. Sharma. 1996. Chemical control of purple blotch and *Stemphylium* blight diseases of onion. News Letter National Horticultural Research and Development Foundation, 16: 14-16.

Haggag, W.M. & H.A.A. Mohamed. 2002. Enhancement of antifungal metabolite production from gamma-ray induced mutants of some *Trichoderma* species for control onion white rot disease. Plant Pathology Bulletin, 11: 45-46.

Hamid, K., M.A. Gill, & J. Ihsan. 2002. Chemical control of downy mildew in onion. Pakistan Journal of Phytopathology, 14: 91-94.

Haq, M.A., H.A. Collin, A.B. Tomsett, & M.G. Jones. 2003. Detection of *Sclerotium cepivorum* within onion plants using PCR primers. Physiological and Molecular Plant Pathology, 62: 185-189.

Hartz, T.K., C.R. Bogle, D.A. Bender, & F. Avila. 1989. Control of pink root disease in onion using solarization and fumigation. Journal of the American Society for Horticultural Science, 114: 587-590.

Hildebrand, P.H. & J.C. Sutton. 1982. Weather variables in relation to an epidemic of onion downy mildew. Phytopathology, 72: 219-224.

Hildebrand, P.H. & J.C. Sutton. 1984a. Effects of weather variables on spore survival and infection of onion leaves by *Peronospora destructor*. Canadian Journal of Plant Pathology, 6: 119-126.

Hildebrand, P.H. & J.C. Sutton. 1984b. Interactive effects of the dark period, humid period, temperature, and light on porulation of *Peronospora destructor*. Phytopathology, 74: 1444-1449.

Hill, J.P. 1995a. *Cladosporium* Leaf Blotch. In, "Compendium of onion and garlic diseases", (eds, Schwartz, H.F. and Mohan, S.K.), APS Press, St. Paul, Minnesota, U.S.A., pp. 21-22.

Hill, J.P. 1995b. Rust. In, "Compendium of onion and garlic diseases", (eds, Schwartz, H.F. & S.K. Mohan), APS Press, St. Paul, Minnesota, U.S.A., pp. 24-25.

Hill, J.P. 1995c. Smut. In, "Compendium of onion and garlic diseases", (eds, Schwartz, H.F. & S.K. Mohan), APS Press, St. Paul, Minnesota, U.S.A., pp. 25.

Hill, J.P. 1995d. Twister. In, "Compendium of onion and garlic diseases", (eds, Schwartz, H.F. and Mohan, S.K.), APS Press, St., Paul, Minnesota, U.S.A., pp. 26.

Hoepting, C.A., C.D. Scott Dupree, C.R. Harris, & G. Ritcey. 2000. Evaluation of insecticide and fungicide combinations for the control of onion maggot (*Delia antiqua*) and onion smut (*Urocystis cepulae*) in Ontario. In "Proceedings of the British Crop Protection Council Conference, Pest and Disease, UK", November 13-16, pp. 279-284.

Hoepting, C.A., C.D. Scott Dupree, C.R. Harris, & M.R. McDonald. 2003. Insecticide and fungicide combinations to optimize control of onion maggot (*Delia antiqua*) and onion smut (*Urocystis cepulae*) in Ontario. Journal of Vegetable Crop Production, 9: 49-63.

Hovius, M.H.Y. & M.R. McDonald. 2002. Management of *Allium* white rot (*Sclerotium cepivorum*) in onion on organic soil with soil-applied diallyl disulfide and di-N-propyl disulfide. Canadian Journal of Plant Pathology, 24: 281-286.

Jakhar, S.S., J.C. Duhan, & L.S. Suhag. 1996. Studies of the epidemiology and survival of *Stemphylium vesicarium* (Wallr.) Simmons in debris and seeds of onion. Seed Research, 24: 135-140.

Jesperson, G.D. & J.C. Sutton. 1987. Evaluation of a forecaster for downy mildew of onion (*Allium cepa* L.). Crop Protection, 6: 95-103.

Jordan, M.M., R.T. Burchill, & R.B. Maude. 1990a. Epidemiology of *Cladosporium allii* and *Cladosporium allii-cepae*, leaf blotch pathogens of leek and onion. I. Production and release of conidia. Annals Applied Biology, 117: 313-326.

Jordan, M.M., R.T. Burchill, & R.B. Maude. 1990b. Epidemiology of *Cladosporium allii* and *Cladosporium allii-cepae*, leaf blotch pathogens of leek and onion. II. Infection of host plants. Annals of Applied Biology, 117: 327-336.

Jordan, M.M., R.B. Maude, & R.T. Burchill. 1990c. Sources, survival and transmission of *Cladosporium allii* and *C. allii-cepae*, leaf blotch pathogens of leek and onion. Plant Pathology, 39: 237-241.

Jordan, M.M., R.B. Maude, & R.T. Burchill. 1987. Investigation of the host ranges of the leaf blotch pathogens of onion (*Cladosporium allii-cepae*) and leek *C. allii-*. Plant Pathology, 36: 394-397.

Katan, J., I. Rotem, Y. Finkel, & J. Daniel. 1980. Solar heating of the soil fort he control of pink root and other soil-borne diseases in onions. Phytoparasitica, 8: 39-50.

Khare, U.K. & K.G. Nema. 1982. Factors affecting germination of spores of *Alternaria porri in vitro* and *in vivo*. Indian Phytopathology, 35: 100-103.

Khokhar, L.K. & A.H. Jaffrey. 2000. Eficacy of fungicides against downy mildew and yield of onion. Pakistan Journal of Agricultural Research, 16: 43-44.

Kirk, P.M. & J.G. Crompton. 1984. Pathology and taxonomy of cladosporium leaf blotch of onion (*Allium cepa*) and leek (*A. porrum*). Plant Pathology, 33: 317-324.

Kofaet, A. & V. Zinkernagel. 1990. Resistance to downy mildew (*Peronospora destructor* (Berk.) Cap.) in *Allium* species. Zietschrift für Pflanzenkrankheiten und Pflanzenschutz, 97: 13-23.

Lacy, M.L. & J.W. Lorbeer. 1995. *Botrytis* leaf blight. In, "Compendium of onion and garlic diseases", (eds, Schwartz, H.F. and Mohan, S.K.), APS Press, St. Paul, Minnesota, U.S.A., pp. 16-18.

Lacy, M.L. & G.A. Pontius. 1983. Prediction of weather mediated release of conidia of *Botrytis squamosa* from onion leaves in the field. Phytopathology, 73: 670-676.

Lacy, M.L. & D.L. Roberts. 1982. Yields of onion cultivars in Midwestern organic soils infested with *Fusarium oxysporum* f. spp. *cepae* and *Pyrenochaeta terrestris*. Plant Disease, 66: 1003-1006.

Lakra, B.S. 1999. Development of purple blotch incited by *Alternaria porri* and its losses in seed crop of onion (*Allium cepa*). Indian Journal of Agricultural Sciences, 69: 144-146.

Langstron, D.B. Jr., & D.R. Sumner. 2000. First report of downy mildew (*Peronospora destructor*) of onion in Georgia. Plant Disease, 84: 489.

Levy, D. & A. Gornik. 1981. Tolerance of onions (*Allium cepa* L.) to the pink root disease caused by *Pyrenochaeta terrestris*. Phytoparasitica, 9: 51-57.

Leggett, M.E. & J.E. Rahe. 1985. Factors effecting the survival of sclerotia of *Sclerotium cepivorum* in the Fraser Valley of British Columbia. Annals of Applied Biology, 106: 255-263.

Littley, E.R. & J.E. Rahe. 1987. Effect of host plant density on white rot of onion caused by *Sclerotium cepivorum*. Canadian Journal of Plant Pathology, 9: 146-151.

Lorbeer, J. & J. Andaloro. 1984. Onions-Downy mildew (*Peronospora destructor* (Berk.) Caspary). Vegetable MD. Online. Cornell University. http//vegetablemdonline.ppath.cornell.edu/onions_downy.htm.

Mac Manus, G.P.V., R.D. Davis, K.L. Bell, & R.A., Kopittke. 2002a. Sclerotial germination stimulant suppresses onion white rot. In "Proceedings Onion Conference, National Vegetable Industry Centre. Australia" (ed, Napier, T) June 3-7, pp. 50-52.

Mac Manus, G.P.V., V.J. Galea, & R.G. O'Brien. 2002b. Oil-based adjuvants improve fungicide activity against downy mildew in onion. In "Proceedings Onion Conference, National Vegetable Industry Centre. Australia" (ed, Napier, T) June 3-7, pp. 53-57.

Mankarios, A.T. & J. Friend. 1980. Polysaccharide-degrading enzymes of *Botrytis allii* and *Sclerotium cepivorum*: Enzyme production in culture and the effect of the enzymes on isolated onion cell walls. Physiological Plant Pathology, 93: 104.

McDonald, M.R., A.G. Taylor, J.W. Lorbeer, J.J. Heide, ED. van der, S. Nicola, J. Nowak, & C.S. Vavrina. 2004. Efficacy testing of onion seed treatments in the greenhouse and field. Acta Horticulturae, 631: 87-93.

McKenzie, C.L., B. Cartwright, M.E. Miller, & J.V. Edelson. 1993. Injury to onions by *Thrips tabaci* Zindeman (Thysanoptera: Thripidae and its role in the development of purple blotch. Environmental Entomology, 22: 1266-1277.

McQuilken, M., Gemmell, Jacqueline, R.A. Hill, & J.M. Whipps. 2003. Production of macrosphelide A by the mycoparasite *Coniothyrium minitans*. FEMS Microbiology Letters, 219: 27-31.

Melendrez-Rodriguez, J.F., M. Santana-Sotolongo, & L. Herrera-Isla. 2002. Study on the incidence of *Rhizoctonia solani* Kuhn in onion fields in Banao region. Centro Agricola, 29: 51-54.

Memane, S.A., M.B. Khetmalas, D.B. Pawar, and Warade. S.D. 2002. Management of leaf blight and thirips of onion during rainy season. Journal of Maharashtra Agricultural Universities, 26: 347-349.

Metcalf, D.A. 2002. The host range of Tasmanian strains of onion rust. In "Proceedings Onion Conference, National Vegetable Industry Centre. Australia" (ed, Napier, T) June 3-7, pp. 69-72.

Metcalf, D.A. & C.R. Wilson. 1999. Histology of *Sclerotium cepivorum* infection of onion roots and the spetial relationships of pectinases in the infection process. Plant Pathology, 48: 445-452.

Metcalf, D.A. & C.R. Wilson. 2001. The process of antagonism of *Sclerotium cepivorum* in white rot affected onion roots by *Trichoderma koningii*. Plant Pathology, 50: 249-257.

Metcalf, D.A., J.J.C. Dennis, & C.R. Wilson. 2004. Effect of inoculum density of *Sclerotium cepivorum* on the ability of *Trichoderma koningii* to suppress white rot of onion. Plant Disease, 88: 287-291.

Michail, S.H. & M.A. Salem. 1981. Reaction of onion cultivars to scald disease incited by *Alternaria porri*. Acta Phytopathologica Academiae Scientiarum Hungaricae, 16: 299-305.

Miller, M.E. 1995. *Stemphylium* leaf blight and stalk rot. In, "Compendium of onion and garlic diseases", (eds. Schwartz, H.F. and Mohan, S.K.), APS Press, St. Paul, Minnesota, U.S.A., pp. 25-26.

Miller, M.E. & M.L. Lacy. 1995. Purple blotch. In, "Compendium of onion and garlic diseases", (eds. Schwartz, H.F. and Mohan, S.K.), APS Press, St. Paul, Minnesota, U.S.A., pp. 23-24.

Mir, N.M. 1977. *Peronospora destructor* (Berk.) Casp., a new record from India. Current Science, 46: 206.

Mohan, K., E.G. Ebenezer, & K. Seetharaman. 2001. Management of leaf blight disease of onion caused by *Alternaria porri* by plant extracts, plant oils and biocontrol agents. News Letter National Horticultural Research and Development Foundation, 21/22: 11-14.

Mohamed, M.S. & F.G. Fahmy. 1990. Improvement of the biological control of *Trichoderma harzianum* Rifai to onion white rot. Egyptian Journal of Phytopathology, 20: 117-125.

Munoz de Con, L., J.J. Pérez Martínez, & A. Prats Prez. 1985. Onion seed production under tropical conditions. Review of Plant Pathology, 1986, 65: 1662

Naude, S.P. 1988. Purple blotch of onions. South Africa Vegetable Grower, 2: 19.

Neergaard, P. 1979. Seed Pathology. Vol 1. The Macmillan Press, London, 519 p.

Özer, N. & N.D. Köycü. 2004. Seed-borne fungal diseases of onion and their control. In, "Disease Management of Fruits and Vegetables: Fruit and Vegetable Diseases", (ed. Mukerji, K.G.), Kluwer Academic Publishers, The Netherlands, pp: 281-306.

Padule, D.N. & P.G. Utikar. 1981. Field reaction of onion varieties to *Alternaria* blight. Agricultural Science Digest, 1: 19-20.

Paris, M.A. & A.M. Cotes. 2002. Evaluation of microbial isolates for control of *Sclerotium cepivorum* in onion. Bulletin OILB/SROP, 25: 311-314.

Pathak, C.S. L.L. Black, S.J. Cherng, T.C. Wang, & S.S. Ko. 2001. Breeding onions for Stemphylium leaf blight resistance. Acta Horticulturae, 555: 77-81.

Porter, I.J., P.R. Merriman, & P.J. Keane. 1989. Integrated control of pink root (*Pyrenochaeta terrestris*) of onions by dazomet and soil solarization. Australian Journal of Agricultural Research, 40: 861-869.

Presly, A.H. 1985. Studies on *Botrytis* spp. Occurring on onions (*Allium cepa*) and leeks (*Allium porrum*). Plant Pathology, 34: 422-427.

Presly, A.H. & R.B. Maude. 1982. Tolerance in *Botrytis squamosa* to iprodione. Annals of Applied Biology, 100: 111-127.

Raja, J., B. Rajendran, K. Sachithanandam, & CM. Pappiah. 2000. Management of thrips and leaf spot disease in onion. Vegetable Science, 27: 80-81.

Ramsey, G.R. & J.W. Lorbeer. 1986a. Flower blight and scape girdling of onion grown for seed production in New York. Phytopathology, 76: 599-603.

Ramsey, G.R. & J.W. Lorbeer. 1986b. Pathogenicity of *Botrytis* species on onion umbels and scapes under controlled conditions. Phytopathology, 76: 604-612.

Ramsey, G.R. & J.W. Lorbeer. 1986c. The role of temperature and free moisture in onion flower blight. Phytopathology, 76: 612-616.

Rao, R.R. & M.S. Pavgi. 1975. *Stemphylium* leaf blight of onion. Mycopathologia, 56: 113-118.

Robak, J. & Smolinska, U. 2002. A new endargerment of vegetable diseases and possibility of their control. Progress in Plant Protection, 42: 59-66.

Robertson, G.I. 1976. *Pythium* species in market gardens and their pathogenicity to fourteen vegetable crops. New Zealand Journal Agricultural Research, 19: 97-102.

Rudolph, M. & P. Wolf. 1986. Possibilities of downy mildew control in onion seed growing. Nachrichtenblatt für den Pflanzenschutz in der DDR, 40: 190-193.

Ryan, E.W. 1978. Leaf spot of onions caused by *Cladosporium allii-cepae*. Plant Pathology, 27: 200.

Ryan, E.W. J.A. Doyle, & T.B. O Donnell. 1981. Control of *Cladosporium leaf* spot in spring-sown and overwintering bulb onions. In "Proceedings of the British Crop Protection Conference-Pests and Diseases" pp: 349-354.

Satour, M.M., M.F. Abdel-Rahim, T. El-Yamani, A. Radwan, A. Grinstein, H.D. Rabinowitch, & J. Katan. 1989. Soil solarization in onion fields in Egypt and Israel: short- and long-term effects. Acta Horticulturae, 255: 151-159.

Schwartz, H.F. 1995. Downy mildew. In, "Compendium of onion and garlic diseases", (eds, Schwartz, H.F. and Mohan, S.K.), APS Press, St. Paul, Minnesota, U.S.A., pp. 20-21.

Sharma, S.R. 1986. Effect of fungicidal sprays on purple blotch and bulb yield of onion. Indian Phytopathology, 39: 78-82.

Sharma, I.M. 1997. Reaction of onion lines to downy mildew caused by *Peronospora destructor*. Journal of Mycology and Plant Patholoy, 27: 324-325.

Shishkoff, N. & J.W. Lorbeer. 1989. Etiology of *Stemphylium* leaf blight of onion. Phytopathology, 79: 301-304.

Shoemaker, P.B. & J.W. Lorbeer. 1977a. Timing initial fungicide application to control *Botrytis* leaf blight epidemics on onion. Phytopathology, 67: 409-414.

Shoemaker, P.B. & J.W. Lorbeer. 1977b. The role of dew and temperature in the epidemiology of *Botrytis* leaf blight of onion. Phytopathology, 67: 1267-1272.

Shynkorenko, E., 2003. Preparation for onion crop seed treatment in one year culture. Zashichita Rastenii, 27: 269-277.

Skiles, R.L. 1953. Purple and brown blotch of onion. Phytopathology, 43: 409-412.

Smith, R.W., J.W. Lorbeer, & A.A. Abd-ElRazik. 1985. Reappearance and control of onion downy mildew epidemics in New York. Plant Disease, 69: 703-706.

Smolińska, U. 2000. Survival of *Sclerotium cepivorum* sclerotia and *Fusarium* chlamydospores in soil amended with Cruciferous Residues. Journal of Phytopathology, 148: 343-349.

Sneh, B., D. Netzer, & J. Krikunj. 1974. Isolation and identification of *Pyrenochaeta terrestris* from soil on dilution plates. Phytopathology, 64: 275-276,

Somkuwar, R.G., R.V. Gowda, & T.H. Singh. 1997. Varietal response to white rot in onion. Madras Agricultural Journal, 84:343-344.

Sparace, S.A., J.B. Mudd, B.B. Burke, & A.J. Aasen. 1984. Pyrenocine C, a phytotoxin-related metabolite produced by the onion pink root pathogen, *Pyrenochaeta terrestris*. Phytochemistry, 23: 2693-2694.

Sparace, S.A., E.A. Shanin, & J.B. Mudd. 1986. Accumulation of pyrenocines in cultures of the onion pink root pathogen *Pyrenochaeta terrestris* and some observations on their phytotoxicities. Physiological and Molecular Plant Pathology, 28: 381-392.

Srivastava, K.J., S.M.H. Qadri, B.K. Tiwari, S.R. Bhonde, & U.B. Pandey. 1991. Chemical control of purple blotch of onion bulb crop in kharif season. Indian Phytopathology, 44: 251-253.

Srivastava, P.K., R.C. Sharma, & P.P. Fupta. 1995. Effect of different fungicides on the control of purple blotch and *Stemphylium* blight diseases in onion seed crop. National Horticultural Research and Development Foundation, 15: 6-9.

Srivastava, P.K. B.K. Tiwari, & K.J. Srivastava. 1999. Studies on integrated disease management of onion. News Letter National Horticultural Research and Development Foundation, 19: 7-9.

Stewart, A., D. Bachouse, P.W. Sutherland, & R.A. Fullerton. 1989. The development of infection structures of *Sclerotium cepivorum* on onion. Journal of Phytopathology, 126: 22-32.

Stewart, A. & S. C. Franicevic. 1994. Infected seed as a source of inoculum for *Botrytis* infection of onion bulbs in store. Australasian Plant Pathology, 23:36-40.

Stewart, A. & R.A. Fullerton. 1991. Additional studies on the chemical control of onion white rot (*Sclerotium cepivorum* Berk.) in New Zealand. New Zealand Journal of Crop and Horticultural Science, 19: 129-134.

Stewart, A., R.A. Fullerton, & P.W. Sutherland. 1989. Infection of onion by the white rot pathogen, *Sclerotium cepivorum.* Journal of Phytopathology, 126: 33-42.

Stewart, A. & J.W. Mansfield. 1984. Fungal development and plant response in detached onion, onion bulb scales and leaves inoculated with *Botrytis allii, B. cinerea, B. fabae* and *B. squamosa*. Plant Pathology, 33: 401-409.

Stone, H.E & V.N. Armentrout. 1985. Production of oxalic acid by *Sclerotium cepivorum* during infection of onion. Mycologia, 77: 526-530.

Suhag, L.S. & R.S. Rana. 1984. Studies on the chemical control of seedling mortality of kharif onions in nursery. Indian Journal of Plant Pathology, 2: 13-15.

Suheri, H. & T.V. Price. 2000. Infection of onion leaves by *Alternaria porri* and *Stemphylium vesicarium* and disease development in controlled environments. Plant Pathology, 29: 375-382.

Suheri, H. & T.V. Price. 2001. Purple leaf blotch disease of *Alliums* spp. in Australia. Acta Horticulturae, 555: 171-173.

Sumner, D.R., 1995a. *Rhizoctonia* seed rot and seedling diseases. In, "Compendium of onion and garlic diseases", (eds, Schwartz, H.F. and Mohan, S.K.), APS Press, St. Paul, Minnesota, U.S.A., pp. 8-9.

Sumner, D.R., 1995b. *Fusarium* damping-off. In, "Compendium of onion and garlic diseases", (eds, Schwartz, H.F. and Mohan, S.K.), APS Press, St. Paul, Minnesota, U.S.A., pp. 9-10.

Sumner, D.R. 1995c. Pink root. In, "Compendium of onion and garlic diseases", (eds, Schwartz, H.F. and Mohan, S.K.), APS Press, St. Paul, Minnesota, U.S.A., pp. 12-13.

Sumner, D.R., 1995d. Southern blight. In, "Compendium of onion and garlic diseases", (eds, Schwartz, H.F. and Mohan, S.K.), APS Press, St. Paul, Minnesota, U.S.A., pp. 13-14.

Sumner, D.R., 1995e. *Diplodia* stain. In, "Compendium of onion and garlic diseases", (eds, Schwartz, H.F. and Mohan, S.K.), APS Press, St. Paul, Minnesota, U.S.A., pp. 28.

Sumner, D.R., 1995f. Mushy Rot. In, "Compendium of onion and garlic diseases", (eds, Schwartz, H.F. and Mohan, S.K.), APS Press, St. Paul, Minnesota, U.S.A., pp. 29.

Sumner, D.R., 1995g. Smudge. In, "Compendium of onion and garlic diseases", (eds, Schwartz, H.F. and Mohan, S.K.), APS Press, St. Paul, Minnesota, U.S.A., pp. 30.

Sumner, D.R., R.D. Gitatis, J.P. Gay, D.A. Smittle, B.W. Maw, E.W. Tollner, & Y.C. Hung. 1997. Control of seed-borne pathogenic fungi in fields of sweet onion. Plant Disease, 81: 885-891.

Sutton, J.C., T.D.W. James, & P.M. Rowel. 1986. BOTCAST: A forecasting system to time the initial fungicide spray for managing *Botrytis* leaf blight of onions. Agriculture, Ecosystems and Environment, 18: 123-143.

Sutton, J.C., C.J. Swanton, & T.S. Gllespie. 1978. Relation of weather variables and host factors to incidence of air-borne spores of *Botrytis squamosa*. Canadian Journal of Botany, 56: 2460-2469.

Tartier, L., R. Crete, & E. Hogue. 1976. Chemical control of onion smut, *Urocystis magica*, in organic soil. Phytoprotection, 57: 116-123.

Thind, T.S. & J.S. Jhooty. 1982. Association of thrips with purple blotch infection on onion plant caused by *Alternaria porri*. Indian Phytopathology, 35: 696-698.

Thornton, M.K. & J.K. Mohan. 1996. Response of Sweet Spanish onion cultivars and numbered hybrids to basal rot and pink root. Plant Disease, 80: 660-663.

Torres-Barragán, A., E., Zavaleta-Mejía, C. González-Chávez, & R. Ferrera-Cerrato. 1996. The use of arbuscular mycorrhizae to control onion white rot (*Sclerotium cepivorum* Berk.) under field conditions. Mycorrhiza, 6: 253-257.

Tylkowska, K. & H. Dorna. 2001. Onion (*Allium cepa*) seed and plant health with special reference to *Botrytis allii*. Phytopathologia Polonica, 21: 55-68.

Utikar, P.G. & D.N. Padule. 1981. A virulent species of *Alternaria* causing leaf blight of onion. Indian Phytopathology, 33: 335-336.

Utkhede, R.S. & J.E. Rahe. 1982. Reduction in white rot incidence by seed irradiation in *Allium cepa*. Plant Disease, 66: 723-725.

Van Der Meer, Q.P., J.L. Van Bennekom, & A.C. Van Der Giessen. 1983. Screening for resistance to white rot caused by *Sclerotium cepivorum* Berk. in onions (*Allium cepa* L.) and leek (*Allium porrum*). Euphytica, 32: 697-701.

Verwoerd, L. & S.J. du Plessis. 1931. Descriptions of some new species of South African fungi and of species not previously recorded in South Africa. III. South African Journal of Science, 28: 290-297.

Vincelli, P.C. 1995. *Pythium* Seed Rot and Damping-Off. In, "Compendium of onion and garlic diseases", (eds, Schwartz, H.F. and Mohan, S.K.), APS Press, St. Paul, Minnesota, U.S.A., pp. 7.

Vincelli, P.C. & J.W. Lorbeer. 1988. Forecasting spore episodes of *Botrytis squamosa* in commercial onion fields in New York. Phytopathology, 78: 966-970.

Vincelli, P.C. & J.W. Lorbeer. 1990. *Pythium irregulare* and *P. coloratum* causing root rot of onion. Mycopathologia, 11: 67-72.

Walker, J.C. 1925. Two undescribed species of *Botrytis* associated with neck rot diseases of onion bulbs. Phytopathology, 15: 708-713.

Weeraratne, W.A.P.G. 2002. Leaf twister disease of onion (*Allium cepa* L.). Tropical Agriculturist, 151: 25-33.

Whiteman, S.A. & R.M. Beresford. 1998. Evaluation of onion downy mildew disease risk in New Zealand using meteorological forecasting criteria. In "Proceedings of the 51st New Zealand Plant Protection Conference. New Zealand". August 11-13, 1998, pp. 117-122.

Write, P.J., R.W. Chynoweth, R.M. Beresford, & W.R. Henshall. 2002. Comparison of strategies for timing protective and curative fungicides for control of onion downy mildew (*Peronospora destructor*) in New Zealand. In " Proceedings BCPC Conference: Pest and diseases, UK" November 18-21, 2002, pp. 207-212.

Zappacosta, D., R. Delhey, & N. Curvetto. 2003. Growth and enzyme activity in roots and calli of resistant and susceptiple *Allium* line as influenced by sterile culture filtrates of *Phoma terrestris*. Biologia Plantarum, 46: 101-105.

24

Allelochemicals: The Possibility of Management of Plant Diseases

KAORI TOMITA-YOKOTANI[1*], TAKAKO KATO[1], TOSHISADA SUZUKI[2]

[1]*Graduate School of Life and Environmental Sciences, University of Tsukuba*

[2]*Faculty of Agriculture, Kagawa University, (*email; kaboka@sakura.cc.tsukuba.ac.jp)*

ABSTRACT

We show a case of the effect of allelochemicals released from sunflower seed (achene) on the fungal germination. The allelopathic activity during seed germination process dramatically changes. To manage of plant diseases using allelochemicals, we would need deep considerations about individual plant growth. To use allelochemicals for the management of plant diseases is thought very useful usage after the consideration of each plant growth.

1. INTRODUCTION

Functional Ways for Defense of Plant to their Living Environment

Plant diseases often are caused by bacteria, microbes, insects and any other organisms. Responses to the disease in plants are different from that in animals. Plant cannot move to any places where they want to live, although animals can move to anywhere, but plants have many ways for defenses to environmental organisms because they cannot avoid from attacking the other organisms (Yoshikawa *et al*, 1993; Shibuya, 2000). In many cases, a plant disease is caused by microorganisms (Ohki, 1994). Plants release some chemicals which can affect the other growth of them. Fig. 24.1 shows the guideline about allelopathy and allelochemicals in the case of plant-plant system. Allelochemicals may include many organic substances which are the functional substances for the defense in plant disease (Rice, 1984, Stuz and Christie, 2003).

Can Allelochemicals Work as Functional Substances in Management for Plant Diseases?

Although a plant does not have immunity system like an animal, it has various living body defense reactions to the disease by microorganisms (Schmelzer, 2002). Fig. 24.2 shows a general defense function in case a plant is attacked by the microorganisms (Shibuya, 2000). Phytoalexin is a general term of the antimicrobial function. These are the low molecule substances and a

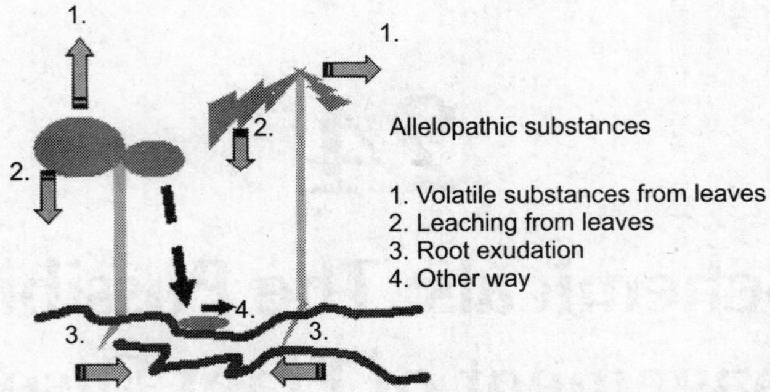

Fig. 24.1: How are the allelochemicals released from plants by ?

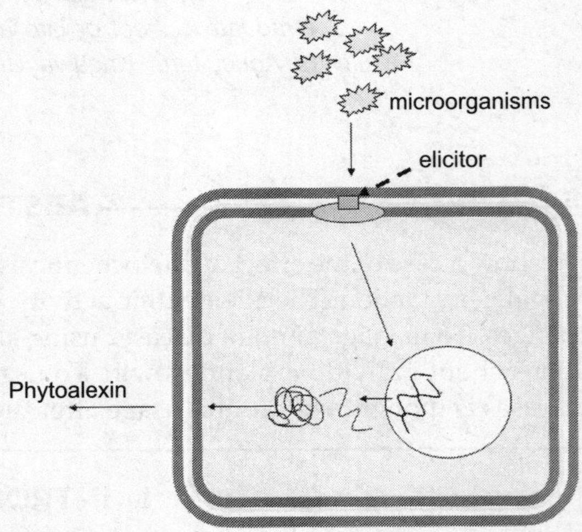

Fig. 24.2: A general defense function in case a plant is attacked by the microorganisms. (This figure shows the case where the receptor of elicitor exists on a plasma membrane).

plant compounds inductively, and is one of the living plants defensive functions. In these substances, some substances also act as allelochemicals. Some substances of flavonoids and terpenoids, which are known as groups of allelochemicals, are contained in phytoalexins (Iwashina 2003, Macias *et al*, 2003). Elicitor is the general term, the low molecule substance, which produces substances of phytoalexin etc. An elicitor is recognized by the receptor which exists in plant, and activates a gene which is related to defense function. In elicitor, there are a substance of the non-living thing origin like heavy metal ion and a substances of the living thing origin which is of disease germ origin, such as oligosaccharide. Some oligogalacturonides are substances known to act as elicitor. On the other hand, some oligosaccharides act as allelopathy (Yokotani-Tomita *et al.*, 1998; Shibuya, 2000; Nabetani *et al*, 2000). By taking into consideration the relation of the substance and allelochemicals which participate in a vegetative living plant body defense reaction, a possibility that disease management of the plant using allelochemical can be performed is suggested. Here, we indicate the applied method to use the exudates of sunflower (*Helianthus annuus* L.) seeds about the anti-microbe activity.

II. EFFECT OF EXUDATES FROM SUNFLOWER SEEDS ON FUNGAL GERMINATION

It is well known that sunflower has a strong allelopathic ability (Rice 1984; Macias *et al*, 2002; 2003). The portion which is equivalent to the seed of a sunflower is histologically called fruits. Here, it will be called a seed. It has found that the exudates from sunflower seeds have allelopathic functions (Ohno *et al*, 2001; Kato *et al*, 2004). The exudates from sunflower seeds have an inhibitory activity on the fungal germination (Kato *et al*, 2002). The exudates of sunflower seeds were collected by the method described in Fig. 24.3. The exudates were fractionated into water-soluble fraction and ethylacetate-soluble fraction.

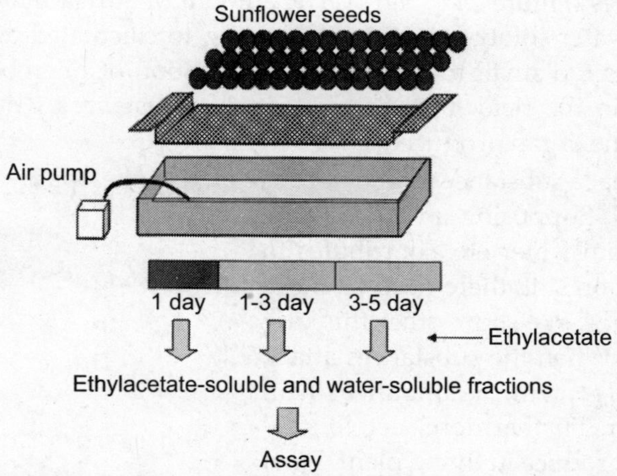

Fig. 24.3: Method for the correction of exudates from sunflower seeds

It is reported that plant disease is caused by fungi in most cases (Ohki, 1994; Brimner and Boland, 2003). We used *Neurospora crassa* which is a model fungus for the fungal assay experiment to the microbe of allelochemicals. Filamentous fungi have high homology among their species although the fungus which cause plant disease is often species-selective function. With the use of model organisms, we will be able to analyze the complex mechanisms of several organisms with multiple environmental responses. To investigate the inhibitory function of allelopathic substance(s), the fungus assay is useful (Tomita-Yokotani *et al*, 2003, Parvez *et al*, 2004). The inhibitory activity was recognized in the ethylacetate-soluble fraction. Figure 24.4 shows the results of the effects of exudates from sunflower seeds on the germination of *Neurospora crassa*. As shown in Figure 24.4, left, the ethylacetate-soluble fraction of exudates inhibited

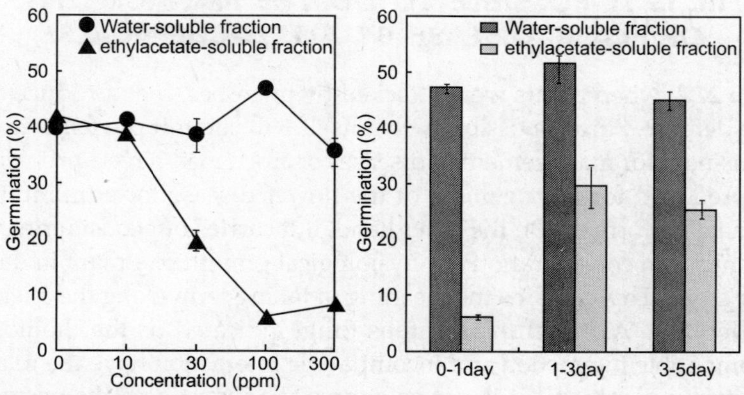

Fig. 24.4: Effects of water-soluble and ethylacetate-soluble fraction on the conidial germination in *Neurospora crassa* during sunflower seed germination process.

the conidial germination, although the water-soluble fraction of that did not inhibit them, in the exudates of 1 day incubated sunflower seeds. This inhibitory activity in the etylacete-soluble fraction was reduced during seed germination process although that in the water-soluble fraction showed no changes (Figure 24.4, right). The amount of ethylacetate-soluble fraction was smaller than that of water-soluble fraction. According to calculated activity, the amount of inhibitory fraction was too small to inhibit the germination of microbe after the substances rinsed with rain etc. in the field. It indicates that the substances would directly affect the growth of fungi near the organ produced.

It is often thought that a substance would not be related to allelopathy because amount of the substance is too small. Supposing an allelopathic substance rinsed by rain water etc. contributes to the role of allelopathy on soil, there need be more amount of the substance. We show other thinking as shown in Fig. 24.5. When the substances attack directly other microbial growth, the amount of them would not need a lot. Furthermore, active substances subsequently produce in living plant. Final amount of the substances that affect the growth of microbe would be high. It suggests that the allelochemicals can be used by a method which is applied directly partially.

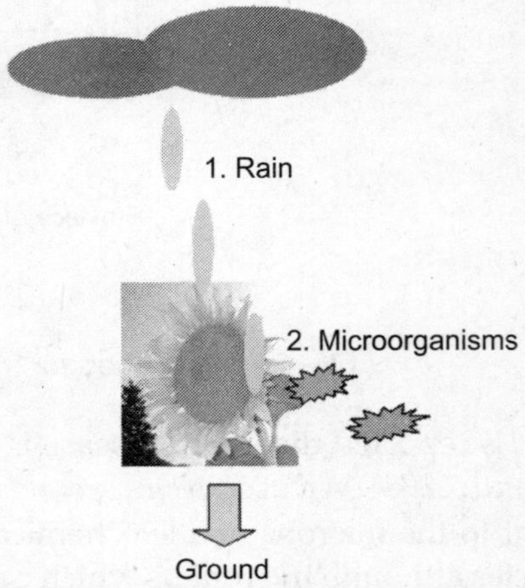

Fig. 24.5: The fate of allelochemicals. 1; Allelochemicals is rinsed by rain. 2; Allelochemicals directly attack on the growth of microorganisms at their productive region.

Here, we described the inhibitory ability of exudates of sunflower seed. At least in sunflower plant seed, first day of exudates of the seed have high ability of anti-microbial activity, and it has a possibility that the exudates would be useful usage for the defense about plant disease to filamentous fungi.

III. IS IT POSSIBLE TO DO THE MANAGEMENT OF PLANT DISEASE BY ALLELOCHEMICALS?

Return to Figure 24.2, when plants were attacked by microbes, they produce any substances, phytoalexins, to defense. Since a possibility that itself will commit phytoalexin as poison to an animal is high, using it for management of diseases of crops may have a problem (Ohki, 1994). However, it is predicted for management of the flower disease for admiration that effective practical use can be performed. On the other hand, if it carries out considering the method for use of allelochemicals in consideration of physiological growth of a plant in detail, use will be possible also for crops. This contains the meaning of taking term using the allelochemicals and time into consideration. Although filamentous fungi indicated us that homogeny is high in them, in fact, compatible interaction and incompatible interaction have the relation between a microorganism and a plant, and it is not so simple. To investigate the occurrences of these allelochemicals and its species-selective activity, to do management of plant disease by allelochemicals would be good way. Recently, the research on the recognition mechanism of a resistance gene and a pathogenic organ has been solved further in the several species of plants. But research of how to use allelochemicals directly relevant to field agriculture is also important. It is very difficult to know the actual rule in the field because of many parameters (Fuerst, and Putnam, 1983). The experiment in the fields is important because soil is out in the fields (Schafer, *et al*, 2003, Inderjit and Bhowmik, 2004). In the case of the tomato (*Lycopersicon esculentum*), it is reported that tomato bacterial wilt is controlled with the root exudates of Chinese chive (*Allium tuberosum*) (Yu, 1999). On the other hand, we showed when sunflower seed exude the high biological active property in this chapter. The method for use of effective interesting allelochemicals is thought from the experimental result of a sunflower. Recently, sunflower seedlings are used as a kind of food, and that is called "sprout" (ISS, website). Sprout is produced using a lot of water in a food plant. In the water after the production of sprout, it has a possibility of the existence a lot of useful allelochemicals such as shown in Fig. 4. It might become a good anti-microbial medicine. Integration of research of many fields related to allelopahty would become important for management of plant disease in the future.

REFERENCES

Brimner, Theresa A. & Greg J. Boland. 2003. A review of the non-target effects of fungi used to biologically control plant diseases Agriculture Ecosystem & Environment 100, 3-16.

Fuerst E.P., & A.R. Putnam. 1983. Separating the competitive and allelopathic components of interference: Theoretical principles. *J. Chem. Ecol.* 9, 937-944.

Inderjit & P.C. Bhowmik. 2004. Sorption of benzoic acid onto soil colloids and its implicaitions for allelopathy studies. *Biol. Fertil. Soils* 40, 345-348.

Iwashina, T., 2003. Flavonoid function and activity to plant and other organisms. *Biol. Sci. Space*, 17, 18-23.

Kato, T., K, Tomita-Yokotani, T. Suzuki, S. Kosemura, & K. Hasegawa. 2002. Allelopathic substances from exudate of germinating sunflower (*Helianthus annuus*) seeds. *Thied World Congress on Allelopathy*, 208.

Kato, T., K. Tomita-Yokotani, S. Kosemura, & K. Hasegawa. 2004. Bioactive substance, sundiversifolide, in sunflower (*Helianthus annuus* L.) plants. *Plant Cell Physiol.* 45, supplement 181.

Macias, F.A., A. Torres, J.L.G. Galindo, R.M. Varela, J.A. Alvarez, & M.G. Molinillo. 2002. Bioactive terpenoids from sunflower leaves cv. Peredovick. *Phytochemistry* 61, 687-692.

Macias, F.A., D. Marin, A. Oliveros-Bastidas, R.M. Varela, M. Simonet, C. Carrera, & M.G. Molinillo. 2003 Allelopathy as a new strategy for sustainable ecosystems development. *Biol. Sci. Space*, 17, 18-23.

Nabetani H., N., Goto, K., Tomita-Yokotani, K. Hasegawa & K. Hirose. 2000. Occurrence of lepidimoide-like substance in growing plant tissues and its physiological activity in *Arabidopsis thaliana*. *J. Weed Sci. Tech.* 45, 235-243.

Ohki, S., 1994. Shokubutu to byouki. SCIENCE IN ACTION SERIES 18, Tokyo-Kagaku Dojin, in Japanese. pp. 201.

Ohno S.K. Tomota-Yokotani, T. Suzuki, M. Node, S. Kosemura, S. Yamamura & K. Hasegawa. 2001. A new species-selective allelopathic substance from germinating sunflower (*Helianthus annuus* L.) seeds. *Phytochemistry* 56, 577-581.

ISS, www. Curlsprout.com/index.htm

Parvez, M. M., K., Tomita-Yokotani, T. Fujii, T. Konishi, & T. Iwashina. 2004. Effects of quercetin and its derivatives on the growth of *Arabidopsis thaliana* and *Neurospora crassa*. *Biochem. Syst. Ecol.* 32, 631-635.

Rice E.L. 1984. Allelopathy. 2nd ed. Academic Press, New York.

Schafer, M., & Peter M. Kotanen. 2003. The influence of soil moisture on losses of buried seeds to fungi. *Acta Oecologia*, 24, 225-263.

Schmezer, E. 2002. Cell polarization, a crucial process in fungal defence. TRENDS in Plant Science, 17, 411-415.

Shibuya, N. 2000. Biochemical studies on the receptor for elicitor. Edy. By Yamada, T., Shimamoto, K., Watanabe, Y. Published by SHUJUNSHA Co., Ltd., Tokyo.

Stuz, A.V. & B.R. Christie. 2003. Benefical microbial allelopathies in the root zone: the management of soil quality and plant disease with rhizobacteria. Soil & Tillage Research, 72, 107-123.

Tomita-Yokotani, K., T., Kato, M. M., Parvez, Y., Mori, N. Goto, & K. Hasegawa. 2003. Aproach of allelopathy study with *Arabidopsis thaliana* (L.) Heynh. and *Neurospora crassa*. *Weed Biology and Management* 3, 93-97.

Yokotani-Tomita K., N., Goto S., Kosemura, S. Yamamura & K. Hasegawa. 1998. Growth-promoting allelopathic substance exuded from germinating *Arabidopsis thaliana* seeds. *Phytochemistry* 47, 1-2.

Yoshikawa, M., N. Yamaoka, & Y. Takeuchi. 1993. Elicitors – Their significance and primary – modes of action in the induction of plant defense reactions. *Plant and Cell Physiol*. 34. 1163-1173.

Yu, J.Q. 1999. Allelopathic suppression of Pseudomonas solanacearum infection of tomato (Lycopersicon esculentum) in a tomato-chinese chive (Allium tuberoum) intercropping system. *J. Chem. Ecol*. 25, 2409-2417.

25

Developments in Production, Characteristics and Potential Applications of Yeast Phytases

PARVINDER KAUR and T. SATYANARAYANA*

Department of Microbiology,
University of Delhi South Campus,
Benito Juarez Road, New Delhi-110 021, India
E-mail address tsnarayana@vsnl.net, Running Title Yeast phytases

ABSTRACT

Phytases are phosphohydrolases, which catalyze the sequential release of phosphate from phytic acid, the main storage form of phosphorous in plants and plant products like cereal grains, legumes and oilseeds. The enzyme is fast-gaining acceptance as an animal feed supplement worldwide and its potential in improving human nutrition and other areas is being explored. Yeasts are among the simplest eukaryotes, which are ideal candidates for phytase research due to their mostly non-pathogenic and GRAS status. They have not, however, been utilized to their full potential. This review focuses attention on the present state of knowledge on yeast phytases.

I. INTRODUCTION

Phosphorus, an essential mineral for all life forms, is a basic component of nucleic acids, ATP, and numerous other biological compounds. Phytic acid is the organic form of phosphorus that is largely unavailable to monogastrics such as poultry, pigs, fish and humans due to the lack of adequate levels of phytases (phytate hydrolyzing enzymes). Furthermore, phytates act as antinutritional factor in more than one ways and their excess in the environment also leads to phosphorus pollution problems (Wodzinski and Ullah *et al*, 1996; Liu *et al*, 1998; Kaur *et al*, 2003).

Recent market trends clearly point toward the increasing importance of phytase as a feed supplement. Throughout the world, the ever-increasing urgency to protect the environment has led to implementation of strict regulatory measures by the authorities. Indiscriminate release of high-phosphate manure into the environment tends to result in phosphate accumulation in water bodies, culminating in serious problems like eutrophication. Therefore, reduction of phytic acid content in foods and feeds by enzymatic hydrolysis using phytase is not only desirable but also the need of the hour. Apart from the immense commercial value in feed and food industries, phytase has potential applications in other fields too. In this chapter, we have made an attempt to review the present status of yeast phytases.

Phytic Acid

Phytic acid, the organic form of P, is an abundant plant constituent comprising 1 to 5% by weight of edible legumes, cereals, oilseeds and nuts. It is chemically myo-inositol 1,2,3,4,5,6 hexakis dihydrogen phosphate ($C_6H_{18}O_{24}P_6$) and has a molecular weight of 659.86. Although Pfeffer was the first to discover phytic acid in 1872 in subcellular particles of wheat endosperm, it was described by Posternak in 1903. The structure of phytic acid was proposed by Anderson in 1914 and finally confirmed by NMR spectroscopy in 1969 (Johnson & Tate, 1969).

The term phytates refers to the salts of phytic acid, whereas phytin is the calcium/magnesium salt of phytic acid. Phytates are the primary source of inositol and storage form of phosphorous in the mature seeds of monocot as well as dicot plants, and most foods of plant origin contain 50-80% of their total P as phytate (Harland & Morris, 1995). The total P and phytate-P levels of some cereals are presented in Table 25.1. Earlier, the role of phytin-P was speculated to be mainly as a storage product, where it was believed that large amounts of phosphorus were stored in the seed, and the phosphorus was liberated on germination and incorporated into ATP. The investigations carried out later highlighted the importance of inositol phosphate intermediates in several cellular functions such as in transport as secondary messengers and in signal transduction in plant and animal cells (Sasakawa *et al*, 1995), in ATP metabolism (Safrany *et al*, 1999), controlling cancer and preventing the effects of heart disease and diabetes (Shamsuddin & Vucenik, 1999); and they also form important constituents of phospholipids in plant and animal tissues.

Table 25.1: The total P and Phytate-P of some cereals*

Cereals	*Total P(%)*	*Phytate P(%)*	*Phytate P/Total P(%)*
Maize	0.39	0.25	64
Rice	0.15	0.09	60
Wheat	0.44	0.27	61
Sorghum	0.33	0.22	73
Barley	0.30	0.20	61
Bajra	0.31	0.23	74

* The data is from Tyagi *et al* (1998)

In monogastric animals and humans, the dietary phytate is mostly unutilizable due to lack/low levels of phytase, thus, necessitating supplementation with external utilizable P sources. Moreover, phytate tends to act as an antinutrient in more than one ways. The molecular structure of phytic acid is the basis for its antinutritive effect as the six reactive groups make it a strong chelating agent that binds cations such as Ca^{2+}, Mg^{2+}, Fe^{2+} and Zn^{2+}. At gastrointestinal pH, stable and insoluble metal phytate complexes are formed, thus making the divalent cations unavailable for absorption in the intestinal tract of animals (Maga, 1982). Under acidic conditions, a negative influence of phytic acid on protein solubility and inhibition of digestive enzymes like trypsin, pepsin, amylase and others have also been observed (Harland & Morris, 1995).

The combo effect of phytate as an antinutrient and unavailable source of P makes it the target for investigations to overcome these hurdles. Several methods such as soaking, fermentation, cooking and germination of seeds have been suggested to eliminate phytates from foods. Harsh treatments like autoclaving, chemical treatment and ion-exchange techniques have also been attempted. These methods reduce the nutritional properties of the foodstuffs. Hence, the need for a mild but effective alternative led to the discovery of phytate-hydrolyzing enzymes called phytases.

Phytase

Phytase catalyses the hydrolysis of phytic acid to inorganic phosphate and myo-inositol phosphate derivatives. Although, phytase was discovered in 1907 by Suzuki *et al* (1907), the catalytic sequence of the enzyme was first established in 1962 (Tomlinson & Ballou, 1962). Two classes of phytases, 3-phytases (EC 3.1.3.8) and 6-phytases (EC 3.1.3.26) were classified by IUPAC-IUB in 1975 based on the position specificity of initial hydrolysis of phytates. The 3-phytase (myo-inositol hexakisphosphate 3-phosphohydrolase) first attacks 3-position phosphate of the phytate (Johnson & Tate, 1969) and is mainly of microbial origin. On the other hand, 6-phytase (myo-inositol hexakisphosphate 6-phosphohydrolase) attacks the 6-position phosphate of the phytate first (Cosgrove, 1969) and is generally observed in plants. Both these groups belong to histidine acid phosphatases (HAP).

In recent years, the number of enzymes described as phytases have increased (Konietzny & Greiner, 2002). These enzymes, however, exhibit variations in structure and catalytic mechanism. Consequently, the acid phytases have been categorized into HAPs, β propeller phytases (BPP) and purple acid phosphatases (PAP) based on their active site motifs. Most widely studied phytases are the HAPs, which share a unique and conserved active site motif, RXGXRXP (Van Etten *et al*, 1991). The recently discovered BPP class includes the phytases from *Bacillus subtilis* (Phy C) [Kerovuo *et al*, 1998] and *B. amyloliquefaciens* (TS-Phy) [Kim *et al*, 1998]. Their molecules display a basic structure similar to a BPP, and hence, the name of the group. Another recently isolated phytase, GmPhy (EC 3.1.3.2) from the cotyledons of germinating soybeans (Hegeman & Grabau, 2001) has the active site motif of a PAP. Apart from the acidic phytases, alkaline phytases (with optimal activity in alkaline range) are also widely distributed in nature (Liu *et al*, 1998; Oh *et al*, 2004). More focus has, however, been on acidic phytases due to their applicability in animal feeds and broader substrate specificity than those of alkaline phytases.

II. SOURCES OF PHYTASES

Phytases are widespread in nature, occurring in plants, microorganisms as well as in some animals.

Plant Sources

Among plants, it has been found in cereals and legumes and various kinds of seeds and pollen. Suzuki *et al* (1907) were the first investigators to describe the enzymatic activity of rice bran phytase and to prepare an extract that retained its activity. It has been reported in many plant species such as wheat, rye, barley, pea, bean, soybean, maize, rice, potato, and other legumes or oilseeds (Gibson and Ullah, 1990; Liu *et al*, 1998), where it is induced during germination to provide the growing seedling with inorganic phosphorous (P_i) and the plant with concomitant free myo-inositol, an important growth factor. Seeds are known to contain both constitutive and germination-inducible phytases (Meyer *et al*, 1971; Eastwood *et al*, 1969). The enzyme is also associated with the root cell wall and mucilage in apical root zones. Phytases from some plant sources have been isolated and extensively studied including those in soybeans (Morgan *et al*, 1998) and oats (Greiner & Alminger, 1999).

Animal Sources

The role of phytases in animal cells is more obscure. Phytase has been detected only in the blood of lower vertebrates like birds, reptiles, fishes, seaturtles (Rapaport *et al*, 1941). A phytase like enzyme, multiple inositol polyphosphate phosphatase (MIPP) with the capacity to regulate cellular activities of phytic acid and IP_5 has been found in the erythrocytes and plasma of various species of vertebrates and in the mammalian small intestine. The first report on animal phytase in calf liver and blood was that by Mc Collum and Hart (1908).

The ruminants probably digest phytate through the action of phytase produced by microbial flora in the rumen, and the inorganic phosphates produced by the splitting of phytate are utilized by both the microbial flora and ruminant host (Dvorakova, 1998). Phytase was reported in the intestinal mucosa of rat, chicken, calves and hens (Bitar & Reinhold, 1972). On investigation of the presence of phytase in the gut of various animals, it was confirmed that phytase activity in human intestines was 30 times lower than that of rat, and hence, dietary phytases were considered to be an important factor for phytate hydrolysis in higher animals (Frolich, 1990).

Microbial Sources

Since phytic acid is a very common compound in nature, many microbes have the ability to degrade it and utilize the hydrolyzed phosphorous. Although phytases have been reported in a number of bacteria, not many have been used for their production due to low enzyme yields and further, their pH optimum is neutral to alkaline and thus precluding their use as feed additives (Wodzinki & Ullah, 1996). Nonetheless there are reports on phytase producing *Bacillus* spp. (Kim *et al*, 1999), *E. coli* (Greiner *et al*, 1993) and *Treponema* (Yanke *et al*, 1998). Bacterial phytases are mostly cell-associated.

Among fungi, incidence of phytase production is highest in *Aspergillus* spp., but species of *Mucor*, *Rhizopus* and *Penicillium* are also good producers (Wodzinki & Ullah, 1996). For production of phytases, fungal cultures are the most widely used (Pandey *et al*, 2001; Liu *et al*, 1998) due to high yields and their acid tolerance, the prerequisites for feed supplementation. Phytases have also been reported from thermophilic fungi like *Thermomyces lanuginosus* (Berka *et al*, 1998) and *Sporotrichum thermophile* (Ghosh, 1997) and from cultivated edible mushrooms (Callopy & Royse, 2004). Inspite of several reports on yeast phytases (Table 25.2), very few have been extensively studied.

Table 25.2: A list of phytase producing yeasts

Yeast	Location of the enzyme	Reference
Arxula adeninivorans	Extracellular	Sano *et al*, 1999
Candida spp.	Extracellular	Nakamura *et al*, 2000
Candida krusei	Cell bound	Quan *et al*, 2001
Cryptococcus sp.	Extracellular	Greenwood & Lewis, 1977
Kluyveromyces lactis	Extracellular	Nakamura *et al*, 2000
Pichia anomala	Extracellular	Nakamura *et al*, 2000
P. anomala	Cell bound	Vohra & Satyanarayana, 2001
P. rhodanensis	Extracellular	Nakamura *et al*, 2000
P. spartinae	Extracellular	Nakamura *et al*, 2000
Rhodotorula gracilis	Cell bound	Bindu *et al*, 1998
Saccharomyces cerevisiae	Cell bound	Howson and Davis, 1983
S. cerevisiae	Extracellular	Nayini & Markakis, 1984
S. kluyveri	Extracellular	Nakamura *et al*, 2000
Schwanniomyces occidentalis	Extracellular	Segueilha *et al*, 1992
S. castellii	Cell bound	Lambrecht *et al*, 1992
Torulaspora delbrueckii	Extracellular	Nakamura *et al*, 2000

III. YEAST PHYTASES

Yeasts and their metabolic products have been exploited by human beings in numerous ways for a long time. The products of modern yeast biotechnologies form the backbone of many commercially important sectors, including food, beverages, pharmaceuticals, industrial enzymes and others. Although *Saccharomyces cerevisiae* and *Schizosaccharomyces pombe* are the most widely studied yeasts, various non-conventional yeasts like *Pichia, Arxula, Candida* and several others are fast-gaining popularity due to their beneficial biotechnological roles. Several researchers have also investigated the role of yeasts in phytate degradation.

Lambrechts *et al* (1992) screened 21 yeast strains for their ability to hydrolyze phytates and among all these, *Schwanniomyces castellii* CBS 2863 exhibited highest phytase activity. The presence of a translucent zone around the colony in the solid medium containing calcium phytate as the insoluble phosphorus source was observed. After 5 days of incubation at 28°C, 9 strains (*Candida brumptii* CBS6145, *C. tropicalis* CBS5696, *Debaryomyces castellii* CBS2923, *Kluyveromyces fragilis* 111, *K. fragilis* CBS1555, *K. fragilis* CBS 5795, *Saccharomyces cerevisiae* CBS1253, *Schwanniomyces castellii* CBS2863 and *Torulopsis candida* CBS940) displayed a translucent zone, the diameter of which was at least twice that of the diameter of the colonies. These strains were further tested in the liquid medium containing sodium phytate. Phytase activity was determined in the culture supernatant and in the crushed pellet. Based on phytase activity, five strains (*C. tropicalis* CBS5696, *T. candida* CBS940, *D. castellii* CBS2923, *K. fragilis* 111 and *S. castellii* CBS 2863) were selected. *S. castellii* CBS 2863 exhibited highest total specific activity which was 2.8 times higher than total specific activities measured for all strains at the end of exponential growth phase (Segueilha *et al*, 1992).

Nakamura *et al* (2000) screened several hundred yeast species for extracellular phytase. *Pichia spartinae* and *P. rhodanensis* exhibited highest levels of phytase with the optimal reaction temperature at 75-80°C and 70-75°C, at pH 3.6-5.5 and 4.5-5.0, respectively. Although Nakamura *et al* (2000) reported yeast to be a rich genetic resource for heat-resistant phytases, the possibility of using these phytases in industry was not investigated.

The yeast *Arxula adeninivorans* was reported to be one of the rare yeasts capable of assimilating phytate as a sole source of carbon and phosphate (Sano *et al*, 1999). This capacity was associated with a high phytase activity secreted into the medium. A high phytase titre was attained when the inoculum was raised on phytate minimal agar plate. The reason was thought that clonal selection could occur on PhyM agar plates because a fraction of *A. adeninivorans* cells was seen in filamentous form under the microscope. Use of galactose in place of glucose further enhanced the phytase titre several folds. The enzyme activity associated with the washed cells represented only 5.6% of the total activity secreted into the medium. A greater phytase production was recorded when the organism was grown at 44°C.

The red yeast *Rhodotorula gracilis* was also shown to produce phytase (Bindu *et al*, 1998). Several permeabilization methods were employed to enhance phytase activity of the yeast cells, but only the freeze-thaw method led to an increase up to 77.26 Ug^{-1} cells.

A strain of *Pichia anomala* was reported to produce a cell-bound phytase (Vohra & Satyanarayana, 2001). The enzyme was suitable for supplementing animal feeds due to the desired properties of thermostability, acid-stability and broad substrate specificity (Vohra & Satyanarayana, 2002a).

Quan *et al* (2001) demonstrated the production of a cell-bound phytase by *Candida krusei* WZ-001. The enzyme synthesis was regulated by phosphate concentration in the medium with a high titre at 0.5 mg phosphate per 100 ml medium.

IV. PRODUCTION OF YEAST PHYTASES

Several investigators have optimized nutritional and physical parameters for maximizing the production of phytases. In *Schwanniomyces castellii*, phytase production was carried out continuously in a fermenter aerated at 1 vvm and agitated at 600 rev/min. Galactose (1%) was found to be the preferred carbon source and sodium phytate (0.06%) was also required for phytase production (Segueilha *et al*, 1992). *Rhodotorula gracilis* produced phytase constitutively, but maximum production was achieved in a medium lacking phytate in 18 h (Bindu *et al*, 1998).

A. adeninivorans secreted high levels of phytase during its active growth phase at 44°C. Galactose was found to be a better carbon source than glucose, while yeast extract (1%) and peptone (1%) served as good nitrogen sources. Phytate did not show an inducible effect on phytase production (Sano *et al*, 1999).

Maximum phytase yield by *C. krusei* WZ-001 was attained in 48 h (late exponential phase) at 30°C and pH 7.0. Glucose (5%) and polypeptone (0.7%) were found to be suitable carbon and nitrogen sources, respectively; and phytase synthesis was repressed by phosphate in the medium (Quan *et al*, 2001).

Phytase production from *Pichia anomala* has been extensively studied. A high phytase titre was attained in the synthetic medium that contained glucose (4%) and beef extract (1%), supplemented with Fe^{2+} (0.15 mM), at 20°C in 24 h (Vohra & Satyanarayana, 2001). Further enhancement in the titres was achieved using the statistical approach, response surface methodology (RSM), whereby the interaction of 3 variables (glucose, beef extract and inoculum density) was studied. The concentrations of glucose (2%) and beef extract (0.5%) were just half of those actually required (Vohra & Stayanarayana, 2002b). A medium consisting of cane molasses, a by-product of sugarcane industry, was formulated that supported an improved phytase yield with reduced production cost as compared to that of the synthetic glucose-beef extract medium (Vohra & Satyanarayana, 2004). The economical cane-molasses medium was further optimized using statistical approaches Plackett-Burman and RSM for attaining a higher phytase production by *P. anomala* (Kaur & Satyanarayana, 2005).

V. PURIFICATION OF YEAST PHYTASES

A combination of various methods has been usually employed for purifying phytases. A typical salt or solvent precipitation for concentrating the enzyme followed by chromatography such as ion exchange or gel filtration is the general methodology employed.

A three-stage purification method including anion and gel filtration chromatography was used to purify the phytase from *S. castellii* (Segueilha *et al*, 1992). The phytase from *A. adeninivorans* was partially purified using Sephadex G50 filtration and DEAE chromatography (Sano *et al*, 1999). A phytase from *Candida krusei* WZ-001 was purified to homogeneity by ion-exchange chromatography, hydrophobic interaction chromatography and gel filtration (Quan *et al*, 2002), while the phytase of *Pichia anomala* was purified to homogeneity by a two-step process of acetone precipitation followed by anion exchange chromatography using DEAE-Sephadex (Vohra & Satyanarayana, 2002a).

VI. CHARACTERIZATION OF YEAST PHYTASES

The properties of enzymes are important in determining their potential applications in various industries. The physico-chemical properties of some yeast phytases are shown in Table 25.3.

Table 25.3: Kinetic properties of purified yeast phytases

Yeast	*Optimal temperature (°C)*	*Optimal pH*	*Molecular weight (kDa)*	*K_m (mM)*	*Reference*
Arxula adeninivorans	75	4.5	-	0.25	Sano *et al*, 1999
Candida intermedia	65	4.5	-	-	Nakamura *et al*, 2000
C. krusei WZ-001	40	4.6	330	0.03	Quan *et al*, 2002
C. tropicalis	65	4.5	-	-	Nakamura *et al*, 2000
Kluyveromyces thermotolerans	60-65	4-5	-	-	Nakamura *et al*, 2000
Pichia anomala	60	4.0	64	0.20	Vohra and Satyanarayana, 2002
P. rhodanensis	70-75	4.0-4.5	-	0.25	Nakamura *et al*
P. spartinae	75-80	4.5-5.5	-	0.33	Nakamura *et al*
Schwanniomyces castellii	77	4.4	490	0.038	Segueilha *et al*, 1992

Temperature

Phytases with high temperature optima are desirable in the animal feed industry because feed pelleting involves a step of 80-85°C for few seconds. Phytase of *Schwanniomyces castellii* was found to be optimally active at 77°C (Segueilha *et al*, 1992), while that of *A. adeninivorans* at 75 °C (Sano *et al*, 1999). In *Pichia rhodanensis, P. spartinae* and *P. anomala*, the phytases exhibited optimal activity at 70-75°C, 75-80°C (Nakamura *et al*, 2000) and 60°C (Vohra & Satyanarayana, 2002a), respectively. The phytase of *Candida krusei* WZ-001 was optimally active at 40°C (Quan *et al*, 2002).

Thermostability of phytase is considered to be an important and useful criterion for its industrial application due to the requirement of withstanding pelleting temperatures. *S. castellii* phytase was stable at 74°C for 1 h (Segueilha *et al*, 1992), whereas that of *C. krusei* was easily inactivated at temperatures above 50°C (Quan *et al*, 2002). The phytase of *P. anomala* was also found to be thermostable with a half-life of 5 min at 80 °C (Vohra & Satyanarayana, 2002a).

pH

Most phytases are active within the pH range 4.5-6.0 and the enzyme activity decreases rapidly at pH values lower than 3.0 or higher than 7.5. The optimum pH values for phytases of *S. castellii* (Segueilha *et al*, 1992) and *A. adeninivorans* (Sano *et al*, 1999) were 4.4 and 4.5, respectively. Nakamura *et al* (2000) studied a number of yeast strains for extracellular phytase. All the yeast phytases were optimally active in the pH range between 4.0 and 5.0 when assayed at 50-60°C. Many strains produced another phytase at 37°C with pH optimum of 3.0-4.0. This suggests that either there are more than one phytase or the same protein exhibits different pH optima at different temperatures. Phytases from *C. krusei* (Quan *et al*, 2002) and *P. anomala* (Vohra & Satyanarayana, 2002a) exhibited pH optima at 4.6 and 4.0, respectively.

Effect of Metal Ions

The requirement for metal ions for phytase activity varies with phytases. Phytase of *S. castellii* was slightly inhibited in the presence of 5 mM Ca^{2+} and Mg^{2+}, Mn^{2+} and Fe^{2+}. The cations Zn^{2+} and Cu^{2+} (0.5 mM) caused around 50% inhibition of enzyme activity, while 5mM Zn^{2+} and Cu^{2+} strongly inhibited (Segueilha *et al*, 1992). The phytase of *C. krusei* WZ-001 was strongly inhibited by Zn^{2+} and Mg^{2+}(1 and 5 mM); it was not inhibited to any significant extent in the presence of Ba^{2+} and Pb^{2+} (Quan *et al*, 2002). *P. anomala* phytase did not exhibit any requirement of metal ions for activity and it was strongly inhibited by Fe^{3+}, Cu^{2+}, Zn^{2+} and Hg^{2+} (Vohra & Satyanarayana, 2002a).

Substrate Specificity

Phytases usually show broad substrate specificity, with a high affinity for phytates. A broad substrate specificity was reported for phytases of *S. castellii*. The K_m value for phytate was the lowest, while for glucose-1-phosphate, glucose-6-phosphate, p-nitrophenyl phosphate and ATP, the values were high (Segueilha *et al*, 1992). Similarly, phytase of *P. anomala* exhibited a broad substrate specificity, with comparable efficiency on phytate, ATP and *p*-nitrophenyl phosphate (Vohra & Satyanarayana, 2002a). The phytase of *C. krusei* had a broad substrate specificity, but with maximum activity on phytate (Quan *et al*, 2002).

The rate of enzymatic hydrolysis is based on Michaelis-Menten kinetics and generally expressed as the Michaelis constant (K_m). The K_m values of some yeast phytases are given in Table 25.3.

Molecular Characteristics

Phytases are high molecular weight proteins ranging from 40-500 kDa. The phytase of *P. anomala* had a molecular mass of 64 kDa (Vohra & Satyanarayana, 2002a), while phytase of *C. krusei* had a molecular mass of 330 kDa, and was composed of 2 different subunits (Quan *et al*, 2002). The phytase from *S. castellii* had a molecular weight of 490 kDa with a glycosylation of around 31%. The glycosylated protein was tetrameric, with one large subunit (M.W. 125,000) and three identical small subunits (M.W. 70,000) (Segueilha *et al*, 1992).

There are very few reports on the active sites of yeast phytases. In *P. anomala* phytase, arginine residues were found to be involved in catalysis (Vohra & Satyanarayana, 2002a). The amino acid sequence analysis of some yeast phytases revealed the presence of a conserved eight-cysteine motif. These were not directly associated with catalysis, but appeared to have a role in the formation of disulfide bridges, and thus contribute to stability and heat tolerance in phytases (Mullaney & Ullah, 2005).

Detection of Dephosphorylation Pathway of Phytic Acid

Phytases of different origins produce specific isomers of myo-inositol phosphates with varying degree of phosphorylation and have different dephosphorylation pathways. The knowledge of the positional isomers formed by the enzymatic degradation of phytate is very limited. Among yeasts, the pathway of dephosphorylation of phytic acid by purified phytase of *C. krusei* WZ-001 was described by Quan and co-workers (2003a) and the structure of the isomers were analyzed by 1D- and 2D-NMR techniques. The catalytic reaction sequence is depicted in Fig. 25.1.

The route of enzymatic dephosphorylation by phytases of *S. cerevisiae* and *P. rhodanensis* was investigated upto the IP_3 level by Adelt *et al* (2003) and is shown in Fig. 25.1. The sequential hydrolysis of myo-inositolhexakisphosphate (IP_6) by phytases releases various lower inositolphosphates (IPs), which are required for studying their physiological effects. An efficient immobilized bioreactor comprising phytase would be useful for producing various lower isomers of phytic acid. Quan *et al* (2003b) immobilized cells of *C. krusei* in Ca-alginate gel beads for the preparation of various IPs and the pure isomers were isolated by ion-exchange chromatography.

VII. HETEROLOGOUS GENE EXPRESSION OF PHYTASE

In Yeasts

Several attempts have been made for cost-effective enhanced production of phytase using methods like mutagenesis and recombinant DNA technology. The development of heterologous

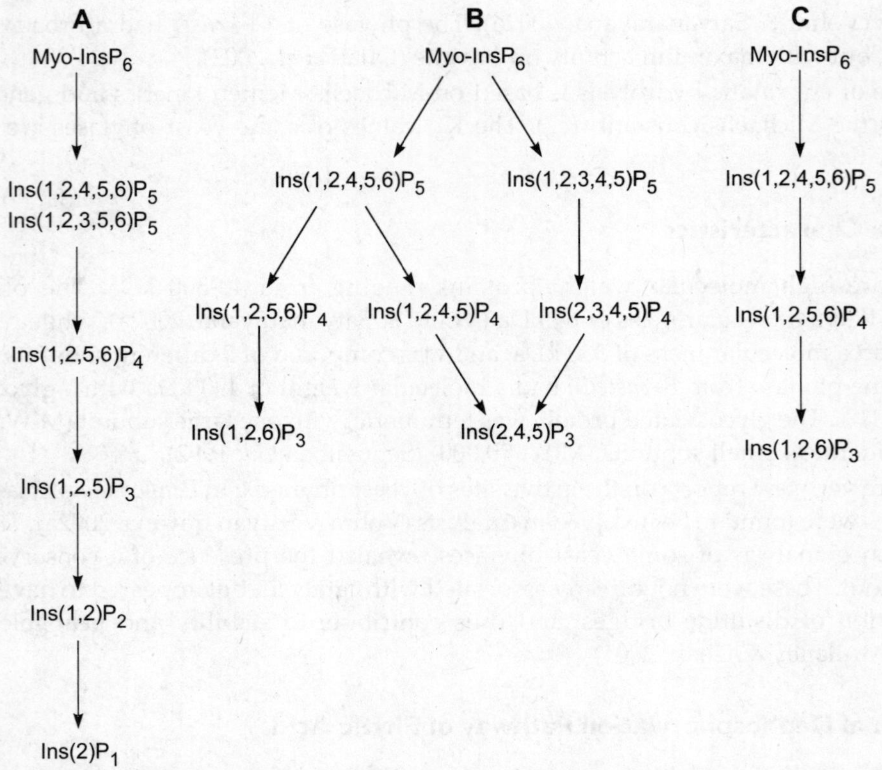

Fig. 25.1: Pathways of dephosphorylation of phytic acid catalyzed by the phytases of (A) *C. krusei*, (B) *P. rhodanensis* and (C) *S. cerevisiae*.

microbial expression systems has made possible high phytase production at low costs, thus, enhancing the commercial prospects of the enzyme. A summary of heterologous expression of phytase in yeast systems is shown in Table 25.4.

Table 25.4: Heterologous expression of phytases in yeast systems

Phytase gene source	*Expression host*	*Reference*
Aspergillus niger	*Saccharomyces cerevisiae*	Han *et al*, 1999
A. niger	*Pichia pastoris*	Han & Lei, 1999; Xiong *et al*, 2004
A. fumigatus	*P. pastoris*	Rodriguez *et al*, 2000
Escherichia coli	*P. pastoris S. cerevisiae* *Schizosaccharomyces pombe*	Lee *et al*, 2005
E. coli	*P. pastoris*	Stahl *et al*, 2003; Chan *et al*, 2004

As *S. cerevisiae* and its by-products have long been known to be safe for human and animal consumption, it was examined as a potential phytase expression host by Han *et al* (1999). They expressed the *phyA* gene from *A. niger* in *S. cerevisiae* using the expression vector pYES2 to construct the plasmid, pYPP1 for transforming *S. cerevisiae* cells.

The facultative methylotrophs, *Hansenula polymorpha* and *P. pastoris* have also been investigated as potential high-yielding production systems for phytase. Han and Lei (1999) expressed

A. niger NRRL 3135 *phyA* in *P. pastoris* and the recombinant strain of *H. polymorpha* was developed for cost-effective and efficient process of production of consensus (engineered) phytase (Mayer *et al*, 1999).

E. coli appA phytase was expressed in *Pichia pastoris* by Stahl *et al* (2003) and later by Chen *et al* (2004). Recently, four different yeast systems were used for expression of *E. coli* appA2 gene to develop an effective fermentation system for phytase production. Three inducible systems [*S. cerevisiae* (pYES2), *Schizosaccharomyces pombe* (pDS472a) and *P. pastoris* (pPICZαA)] and one constitutive system [*P. pastoris* (pGAPZαA)] were employed. The inducible *P. pastoris* system was found to produce the highest enzyme titres (Lee *et al*, 2005).

In Plants

Although there are several reports of generation of transgenic plants (Li *et al*, 1997; Ullah *et al*, 2002) and animals (Mullaney *et al*, 2000; Golovan *et al*, 2001) expressing phytases from bacterial and fungal sources, there are very few reports from yeasts. Very recently Hamada *et al* (2005) reported high-level production of *Schwanniomyces occidentalis* phytase in transgenic rice plants. In this investigation, codon modification combined with the use of a secretory signal sequence was used to produce substantial amounts of the yeast phytase in active and stable form. The aim of this investigation was phytase production in rice with a view to apply it in the animal feed industry.

VIII. COMMERCIAL PROSPECTS

There are basically two areas where the use of phytase application has been emphasized. The first one is phytate elimination in feed and food industries, and the second is preparation of myo-inositol phosphates (degradation products of phytic acid), as tools for biochemical investigation. Besides this, emphasis has also been laid on its use in aquaculture, soil amendment and in synthesis of peroxidase (due to structural similarity of the active site of peroxidase and fungal phytase).

Applications in Feed

Plant feedstuffs form a major constituent of animal feed, mainly swine and poultry diets. About two-thirds of phosphorus of feedstuffs of plant origin is in the form of phytate P. Under most dietary conditions, phytate P is unavailable to these animals, and the phytin-P is excreted in the manure. The phytic acid present in the manure of these animals is enzymatically cleaved by soil and water-borne microorganisms, and the phosphorus, thus, released is transported into the water bodies that leads to eutrophication. This causes oxygen depletion due to excessive algal growth, and the low dissolved oxygen in water bodies in turn results in fish kills. In addition, the phytate P chelates several important minerals, and thereby, reduces their availability to the animals feeding on the diet. The ruminant animals, however, sustain the microflora that enzymatically releases the inorganic phosphorus from phytic acid, thus, avoiding the need for inorganic phosphorus supplementation.

Incorporation of phytases into feeds reduces the need for phosphorus supplementation. Degradation of myo-inositol phosphates yield myo-inositol, an important growth factor, and trace minerals also become available. This improves the digestibility, and therefore, the nutritive value of the feed, thus the growth rate of monogastric animals is enhanced, and it also lowers the level of phosphorus excretion, thus reducing chances of environmental pollution.

There are two basic ways to use phytase in feeds. The first possibility is replacement of Pi supplementation with phytase. As the reaction conditions (pH, temperature, moisture, incubation time) in the animal stomach or intestine are not optimal, the second method of phytase use i.e., feed pretreatment, becomes more feasible.

Feed trial experiments with poultry birds as well as pigs confirmed a possibility of replacing Pi supplementation by using microbial phytases in a phytate-rich diet for monogastric animals. Besides increased phytate-P availability, it also enhances digestibility of amino acids and minerals. The use of phytase as a feed additive is limited by cost, by inactivation at the feed pelleting temperatures (60-90°C), and by the loss of activity during storage. These problems can be overcome by endogenous production of phytase by the monogastric animals, which would increase the bioavailability of plant phytate and in turn lead to reduced phosphorus excretion. Another approach is to use thermostable phytases obtained by cloning of phytase from thermophilic fungi, or protein engineering; or supplementation of phytase with salts, polyols, starch etc. to increase enzyme thermostability.

The availability of phosphorus can be improved by adding microbial phytase to the feed or by using phytase-rich cereal diet. The enzyme minimizes the need for supplementation of Pi due to improvement in the utilization of organic phosphorus in poultry, and thus markedly reducing the excretion of phosphorus in manure. Segueilha *et al* (1993) were able to remove phytic acid in wheat bran and glandless cotton flour using phytase from *S. castellii.* An *E. coli* derived phytase expressed in *P. pastoris* system was compared with 2 commercial feed phytases with respect to the phosphorus-releasing efficacy in young chicks and pigs, and the former was found to be more advantageous than the latter in chicks (Augspurger *et al*, 2003). The safety evaluation of an *E. coli* phytase expressed in *Schizosaccharomyces pombe* was done using various *in vitro* and *in vivo* toxicity tests, with the aim of its intended use in animal feed (Ciofalo *et al*, 2003).

Following first commercialization attempts by Gist-Brocades Europe (1993-1994), several companies like Alko Co. (Finland), Altech (USA) and BASF (USA) started the industrial scale production of phytase marketed under the names Finase, Allzyme and Natuphos, respectively, and successfully utilized it in feed applications. At the close of twentieth century, annual sales of phytase as an animal feed additive were estimated to be $500 million, and the sales are still growing (Abelson, 1999).

Applications in Food

The presence of phytates in plant foodstuffs is well known. Dietary phytase is inactivated during cooking, and therefore, phytate digestion is very poor, and thus affects mineral absorption in the small intestines. Addition of *A. niger* phytase to the flour containing wheat bran has been shown to increase iron absorption in humans. It has also been reported that microbial phytase could accelerate the process of steeping required in the wet milling of corn, thereby improving the properties of corn steep liquor (Caransa *et al*, 1988).

Another application is in bread making, wherein the use of phytase was suggested for producing low phytin bread. In commercial whole wheat bread, phytic acid is present at levels of 0.29 to 1.05% (w/w). By adding mould phytases during bread making, dough phytates could be almost completely eliminated. Degradation of phytate by lactic acid bacteria and yeasts (*S. cerevisiae*) during wholemeal dough fermentation has been studied using NMR technique (Reale *et al*, 2004).

Phytic acid, however, also has positive effects. It exerts an antineoplastic effect in animal models of breast carcinomas and the presence of undigested phytate in the colon may protect it against the development of colonic carcinoma (Igbal *et al*, 1994).

Potential in Aquaculture

In aquaculture, feed costs constitute up to 70% of total fish production costs. Several studies have been conducted on the use of soybean meal or other plant meals in aquaculture by substituting low cost plant protein for a more expensive protein source such as menhaden fish meal, a significant cost reduction could be achieved. As in poultry and swine, fish lack an adequate digestive enzyme to effectively utilize the phytin-P in feed thereby excreting it in the water. Therefore, phytase has been evaluated as a means to both promote the use of low-cost plant meals in the aquaculture industry, and to maintain acceptable phosphorus levels in water. Several fish feeding trials have supported the potential value of phytase in diets containing high levels of plant feedstuffs (Robinson *et al*, 1996; Mwachireya *et al*, 1999).

As a Soil Amendment

Phosphorus deficiency in soil is a major constraint for agricultural production worldwide. P-based fertilizers are, therefore, applied to avoid this deficiency and promote high yields. Only 10-20% of fertilizer P is available to the plants, as a major portion becomes fixed into organic and inorganic fractions that are only poorly available to the plant roots. Of particular significance is the occurrence of soil organic P as phytate, and in certain locations, phytic acid and its derivatives may represent up to 50% of the total organic phosphorus in the soil. Plants are known to produce phytases, but display only low activity in roots and other plant organs, and occurrence of plant-secreted phytase within the rhizosphere is not known. This suggests that plant roots may not possess an innate ability to acquire phosphorus directly from soil phytates. Recent studies have shown that phytase from *Schwanniomyces occidentalis* can be expressed in rice plants (Hamada *et al*, 2005). The phytase activity of certain plant root-colonizing bacteria has also been shown to contribute toward its plant growth promoting activity. Phytase plays a dual role in plant nutrition, making the phytate phosphorus available to the plant roots and at the same time elimination of chelate-forming phytate (Idriss *et al*, 2002).

Semisynthesis of Peroxidase

A semisynthetic peroxidase was designed by taking advantage of the structural similarity of the active site of vanadium-dependent haloperoxidases and fungal phytases and acid phosphatases. Vanadium ion was incorporated into the active site of *A. niger (ficuum)* NRRL 3135 phytase, and this resulted in transformation of native phosphohydrolase activity of phytase into semisynthetic peroxidase that catalyzes enantio-selective oxidation of prochiral sulfides. It was found that only histidine acid phosphatase with the active site sequence RHGXRXP could function as a peroxidase when vanadate ion was incorporated into the active site. The vanadium chloroperoxidase shares structural similarities with some membrane bound acid phosphatases (Hemrika *et al*, 1997).

IX. CONCLUSIONS

There has been significant progress on phytase research during the recent years. Phytases are now being increasingly recognized for their beneficial environmental role in reducing the phosphorus levels in manure, and thereby minimizing the need to supplement phosphorus in diets. Their usage as an animal feed additive is fast-growing since it is ecofriendly. There is an immense potential application of phytase in human nutrition as well. The important roles played by lower inositol phosphates and myo-inositol derivatives in cells and tissues draws attention on the investigations to understand the potential role of phytases. Another area that offers tremendous opportunity is the use of phytase in aquaculture so as to allow the use of low cost plant meals. Research programmes aiming at improving the industrial production of microbial phytase at a cost-effective level are underway in laboratories all over the world.

Fungal phytases, especially from *Aspergillus* spp. are being extensively investigated due to their use in animal feeds to improve phosphorus and mineral availability. Corresponding food grade phytase is, however, not available for human consumption. Due to the GRAS (Generally Regarded As Safe) status, yeasts are a preferable phytase sources for exploring their usability in human foods. Neither the capability to produce phytases nor the genes encoding yeast phytases have been investigated in depth. The potential applications of yeast phytases are immense, and therefore, further research efforts are needed.

ACKNOWLEDGEMENTS

We are grateful to DBT and ICAR for financial assistance during the course of writing this article. Parvinder Kaur gratefully acknowledges the financial assistance from CSIR in the form of JRF/SRF.

REFERENCES

Abelson, P.H. 1999. A potential phosphate crisis. Science 283: 2015.

Adelt, S., M. Podeschwa, G. Dallmann, H.J. Altenbach, & G. Vogel. 2003. Stereo- and regiospecificity of yeast phytases-chemical synthesis and enzymatic conversion of the substrate analogues *neo*- and L-*chiro*-inositol hexakisphosphate. Bioorganic Chem. 31: 44-67.

Augspurger, N.R., D.M. Webel, X.G. Lei, & D.H. Baker. 2003. Efficacy of an *E. coli* phytase expressed in yeast for releasing phytate-bound phosphorus in young chicks and pigs. J. Anim. Sci. 81: 474-483.

Berka, R.M., M.W. Rey, K.M. Brown, T. Byun, & A.V. Klotz. 1998. Molecular characterization and expression of a phytase gene from the thermophilic fungus *Thermomyces lanuginosus.* Appl. Environ. Microbiol. 64: 4423-4427.

Bindu, S., D. Somashekar, & R. Joseph. 1998. A comparative study of permeabilization treatments for in situ determination of phytase of *Rhodotorula gracilis.* Lett. Appl. Microbiol. 27: 336-340.

Bitar, K. & J.G. Reinhold. 1972. Phytase and alkaline phosphatase activities in intestinal mucosae of rat, chicken, calf and man. Biochim. Biophys. Acta. 268: 442-452.

Callopy, P.D. & D.J. Royse. 2004. Characterization of phytase activity from cultivated edible mushrooms and their production substrates. J. Agric. Food Chem. 52:7518-7524.

Caransa, A., M. Simell, M. Lehmussari, M. Vaara, and T. Vaara. 1988. A novel enzyme application in corn wet milling. Starch 40: 409-411.

Chen, C.C., P.H. Wu, C.T. Huang, & K.J. Cheng. 2004. A *Pichia pastoris* fermentation strategy for enhancing the heterologous expression of an *Escherichia coli* phytase. Enz. Microb. Technol. 35: 315-320.

Ciofalo, V., N. Barton, K. Kretz, J. Baird, M. Cook, & D. Shanahan. 2003. Safety evaluation of a phytase, expressed in *Schizosaccharomyces pombe*, intended for use in animal feed. Regulatory Toxicol. Pharmacol. 37: 286-292.

Cosgrove, D.J. 1969. Ion exchange chromatography of inositol polyphosphates. Ann. N.Y. Acad. Sci. 165: 677-686.

Dvorakova, J. 1998. Phytase: Sources, preparation and exploitation. Folia Microbiol. 43: 323-338.

Eastwood, D., R.J.A. Tavener, & D.L. Laidman. 1969. Induction of lipase and phytase activities in the aleurone tissue of germinating wheat grains. Biochem. J. 113 (Proc. Biochem. Soc): 32.

Frolich, W. 1990. Chelating properties of dietary fibre and phytate. The role for mineral availability. Adv. Exp. Med. Biol. 270: 83-93.

Ghosh, S. 1997. Phytase of a thermophilic mould *Sporotrichum thermophile apinis*. M.Sc dissertation, Department of Microbiology, University of Delhi, Delhi.

Gibson, D.M. & A.H.J. Ullah. 1990. Phytases and their action in phytic acid. Pages 77-92 in Inositol metabolism in plants. Wiley-Liss, Chichester.

Golovan, S.P., M.A., Hayes, J.P., Phillips, & C.W. Forsberg. 2001. Transgenic mice expressing bacterial phytase as a model for phosphorus pollution control. Nature Biotechnol. 19: 429-433.

Greenwood, A.J. & D.H. Lewis. 1977. Phosphatases and the utilization of inositol hexaphosphate by soil yeasts of the genus *Cryptococcus*. Soil Biol. Biochem. 9: 161-166

Greiner, R. & M.L. Alminger. 1999. Purification and characterization of a phytate-degrading enzyme from germinated oat (*Avena sativa*). J. Sci. Food Agric. 79: 1453-1460.

Greiner, R., U. Konietzny, & K.D. Jany. 1993. Purification and characterization of a phytase from *Klebsiella terrigena*. Archives Biochem. Biophy. 341: 201-206.

Hamada, A., Yamaguchi, K., Ohnishi, N, Harada, M., Nikumaru, S. & H. HondaH. 2005. High-level production of yeast (*Schwanniomyces occidentalis*) phytase in transgenic rice plants by a combination of signal sequence and codon modification of the phytase gene. Plant Biotechnol. J. 3: 43-56.

Han, Y.M. & X.G. Lei. 1999. Role of glycosylation in the functional expression of an *Aspergillus niger* phytase (*phyA*) in *Pichia pastoris*. Arch. Biochem. Biophys. 364: 83-90.

Han, Y.M., D.B. Wilson, and X.G. Lei. 1999. Expression of an *Aspergillus niger* phytase gene (*phy A*) in *Saccharomyces cerevisiae*. Appl. Environ. Microbiol. 65: 1915-1918.

Harland, B.F. & E.R. Morris. 1995. Phytate: A good or a bad food component. Nutr. Res. 15: 733-754.

Hegeman, C.E. & E.A. Grabau. 2001. A novel phytase with sequence similarity to purple acid phosphatases is expressed in cotyledons of germinating soybean seedlings. Plant physiol. 120: 1598-1608.

Hemrika, W., R. Renirie, H.L. Dekker, P. Barnett, & R.S. Weaver. 1997. From phosphatases to vanadium peroxidases: A similar architechture of the active site. Proc. Natl. Acad. Sci., U.S.A. 94: 2145-2149.

Howson, S.J. & R.P. Davis. 1983. Production of phytate hydrolysing enzymes by some fungi. Enz. Microb. Technol. 5: 377-389.

Idriss, E.E., O. Makarewicz, A. Farouk, K. Rosner, R. Greiner, H. Bochow, T. Richter, & R. Borriss. 2002. Extracellular phytase activity of bacillus amyloliquefaciens FZB45 contributes to its plant-growth-promoting effect. Microbiology 148: 2097-2109.

Igbal, T.H., K.O. Lewis, & B.T. Cooper. 1994. Phytase activity in the human and rat small intestine. Gut 35: 1233-236.

Johnson, L.F. & M.E. Tate. 1969. The structure of myo-inositol pentaphosphates. Ann. A.N. Acad. Sci. 165: 526-532.

Kaur, P. & T. Satyanarayana. 2005. Production of cell-bound phytase by *Pichia anomala* in an economical cane molasses medium: optimization using statistical tools. Process Biochem. (In press).

Kaur, P., B. Singh, A. Vohra, & T. Satyanarayana. 2003. Fabulous phytases: Diverse functions in the living world and commercial prospects. The Botanica 53: 35-42.

Kerovuo, J., M. Lauraeus, P. Nurminen, N. Kalkkinen, & J. Apajalahti. 1998. Isolation, characterization, molecular gene cloning and sequencing of a novel phytase from *Bacillus subtilis*. Appl. Environ. Microbiol. 64: 2079-2085.

Kim, D.H., B.C. Oh, W.C. Choi, J.K. Lee, & T.K. Oh. 1999. Enzymatic evaluation of *Bacillus amyloliquefaciens* phytase as a feed additive. Biotechnol. Lett. 21: 925-927.

Kim, Y.O., H.K. Kim, K.S. Bae, J.H. Yu, & T.K. Oh. 1998. Purification and properties of a thermostable phytase from *Bacillus* sp. DS11. Enz. Microb. Technol. 22: 2-7.

Konietzny, U. & R. Greiner. 2002. Molecular and catalytic properties of phytate-degrading enzymes (phytases). Int. J. Food Sci. Technol. 37: 791-812.

Lambrechts, C., H. Boze, G. Moulin, & P. Galzy. 1992. Utilization of phytate by some yeasts. Biotechnol. Lett. 14: 63-66.

Lee, S., T. Kim, C.H. Stahl, & X.G. Lei. 2005. Expression of *Escherichia coli* AppA2 phytase in four yeast systems. Biotechnol. Lett. 27: 327-334.

Li, J., C.E. Hegeman, R.W. Hanlon, G.H. Lacy, D.M. Denbow, and E.A. Grabau. 1997. Secretion of active recombinant phytase from soybean cell-suspension cultures. Plant Physiol. 114: 1-9.

Liu, B.L., A. Rafig, Y.M. Tzeng, & A. Rob. 1998. The induction and characterization of phytase and beyond: A review. Enz. Microb. Technol. 22: 415-424.

Maga, J.A. 1982. Phytate: Its chemistry, occurrence, food interactions, nutritional significance, and methods of analysis. J. Agric. Food Chem. 30: 1-9.

Mayer, A.F., K. Hellmuth, H. Schlieker, R. Lopez – Ulibarri, S. Oertel, U. Dahlems, A.W. Strasser, & A.P. van Loon. 1999. An expression system matures: a highly efficient and cost effective process for phytase production by recombinant strains of *Hansenula polymorpha.* Biotechnol. Bioengn. 63: 373-381.

Mc Collum, E.V. & E.B. Hart. 1908. On the occurrence of a phytin splitting enzyme in animal tissue. J. Biol. Chem. 4: 497-500.

Meyer, H., A.M. Meyer, & E. Harel. 1971. Acid phosphatases in germinating lettuce-evidence for partial activation. Physiol. Plant. 24: 95-101.

Morgan, A.J., M. Hessing, & H.E. Sleijster-Selis. 1998. "Phytase from Germinated Soybeans." International Patent Application. No. WO98/20139.

Mullaney, E.J., C.B., Daly, & A.H.J. Ullah. 2000. Advances in phytase research. Adv. Appl. Microbiol. 47: 157-199.

Mullaney, E.J. & A.H.J. Ullah. 2005. Conservation of cysteine residues in fungal histidine acid phytases. Biochem. Biophys. Res. Commun. 328: 404-408.

Mwachireya, S.A., R.M. Beames, D.A. Higgs, & B.S. Dosanjh. 1999. Digestibility of canola protein products derived from the physical, enzymatic and chemical processing of commercial canola meal in rainbow trout *Oncorhynchus mykiss* (Walbaum) held in freshwater. Aquacul. Nutr. 5: 73-82.

Nakamura, Y., H. Fukuhara, & K. Sano. 2000. Secreted phytase activities of yeasts. Biosci. Biotechnol. Biochem. 64: 841-844.

Nayini, N.R. & P. Markakis. 1984. The phytase of yeast. Lebensm. Wiss. Technol. 17: 24-26.

Oh, B.C., W.C. Choi, S. Park, Y.O. Kim, & T.K. Oh. 2004. Biochemical properties and substrate specificities of alkaline and histidine acid phytases. Appl. Microbiol. Biotechnol. 63: 362-372.

Pandey, A., G. Szakacs, C.R. Soccol, A. Jose, L. Rodriguez, & V.T. Soccol. 2001. Production, purification and properties of microbial phytase. Biores. Technol. 77: 203-214.

Posternak 1903. Compt. Rend. 137, 202. Phytase. In: Advances in Applied Microbiology, 1996. 42:263-302.

Quan, C.S., S. Fan, & Y. Ohta. 2003a. Pathway of dephosphorylation of myo-inositol hexakisphosphate by a novel phytase from *Candida krusei* WZ-001. J Biosci. Bioengn. 95: 530-533.

Quan C.S., S.D. Fan & Y. Ohta. 2003b. Immobilization of *Candida krusei* cells producing phytase in alginate gel beads: an application of the preparation of myo-inositol phosphates. Appl Microbiol Biotechnol. 62: 41-47.

Quan, C.S., L. Zhang, Y. Wang, & Y. Ohta. 2001. Production of phytase in a low phosphate medium by a novel yeast *Candida krusei*. J. Biosci. Bioengn. 92: 154-160.

Quan, C.S., S. Difan, L.H. Zhang, Y.J. Wang, & Y. Ohta. 2002. Purification and properties of a phytase from *Candida krusei* WZ-001. J. Biosci. Bioengn, 94: 419-425.

Rapoport, S., E. Leva, & G.M. Guest. 1941. Phytase in plasma and erythrocytes of vertebrates. J. Biol. Chem. 139: 621-632.

Reale, A., L. Mannina, P. Tremonte, A.P. Sobolev, M. Succi, E. Sorrentino, & R. Coppola. 2004. Phytate degradation by lactic acid bacteria and yeasts during the wholemeal dough fermentation: a ^{31}P NMR study. J. Agric. Food Chem. 52: 6300-6305.

Robinson, E.H., S. Jackson, & M.H. Li. 1996. Supplemental phytase in catfish diets. Aquacul. Mag. 22: 80-82.

Rodriguez, E., E.J. Mullaney, & X.G. Lei. 2000. Expression of *Aspergillus fumigatus* phytase gene in *Pichia pastoris* and characterization of recombinant enzyme. Biochem. Biophys. Res. Commun. 268: 373-378.

Safrany, S.T., T.T. Caffrey, X. Yang, & S.B. Shears. 1999. Diphosphoinositol polyphosphates: the final frontier for inositide research. J. Biol. Chem. 380: 945-951.

Sano, K., H. Fukuhara, & Y. Nakamura. 1999. Phytase of the yeast *Arxula adeninivorans*. Biotechnol. Lett. 21: 33-38.

Sasakawa, N., M. Sharif, & M.R. Hanley. 1995. Metabolism and biological activities of inositol pentakisphosphate and inositol hexakisphosphate. Biochem. Pharmacol. 50: 137-146.

Segueilha, L., C. Lambrechts, H. Boze, G. Moulin, & P. Galzy. 1992. Purification and properties of the phytase from *Schwanniomyces castellii*. J. Ferment. Bioeng. 74: 7-11.

Segueilha, L., G. Moulin, & P. Galzy. 1993. Reduction of phytate content in wheat bran and glandless cotton flour by *Schwanniomyces castellii*. J. Agric. Food Chem. 41: 2451-2454.

Shamsuddin, A.M. & I. Vucenik. 1999. Mammary tumor inhibition by IP6: a review. Anticancer Res. 19, 36-71.

Stahl, C.H., D.B. Wilson, & X.G. Lei. 2003. Comparison of extracellular *Escherichia coli* AppA phytases expressed in *Streptomyces lividans* and *Pichia pastoris*. Biotechnol. Lett. 25, 827-831.

Suzuki, U., K. Yoshimura, & M. Takaishi. 1907. Ueber ein Enzym "Phytase" das "Anhydro-oxy-methylen diphosphorsaure" Spaltet. Tokyo Imper. Univ. Coll. Agric. Bull. 7: 503-512.

Tomlinson, R.V. & C.E. Ballou. 1962. Myo-inositol polyphosphate intermediates in the dephosphorylation of phytic acid by phytases. Biochemistry 1: 166-177.

Tyagi, P.K., Tyagi, P.K., & S.V.S. Verma. 1998. Phytate phosphorus content of some common poultry feed stuffs. Indian J. Poult. Sci. 33: 86-88.

Ullah, A.H., K. Sethumadhavan, E. J. Mullaney, T. Zeigelhoffer, & S. Austin-Philips. 2002. Cloned and expressed fungal phyA gene in alfalfa produces a stable phytase. Biochem. Biophys. Res. Commun. 290: 1343-1348.

Van Etten, R.L., R. Davidson, P.E. Stevis, H. MacArthur, & D.L. Moore. 1991. Covalent structure, disulfide bonding and identification of reactive surface and active site residues of human prostatic acid phosphatase. J. Biol. Chem. 266: 2313-2319.

Vohra, A. & T. Satyanarayana. 2001. Phytase production by the yeast *Pichia anomala*. Biotechnol. Lett. 23: 551-554.

Vohra, A. & T. Satyanarayana. 2002a. Purification and characterization of the thermostable and acid stable phytase from *Pichia anomala*. World J. Microbiol Biotechnol. 18: 687-691.

Vohra, A. & T. Satyanarayana. 2002b. Statistical optimization of the medium components by response surface methodology to enhance phytase production by *Pichia anomala*. Process Biochem. 37: 999-1004.

Vohra, A. & T. Satyanarayana. 2004. A cost-effective cane molasses medium for enhanced cell-bound phytase production by *Pichia anomala*. J. Appl. Microbiol. 97: 471-476.

Wodzinski, R.J. & A.H.J. Ullah. 1996. Phytase. Adv. Appl. Microbiol. 42: 263-302.

Xiong, A.S., Q.H. Yao, R.H. Peng, X. Li, H.Q. Fan, M.J. Guo, & S.L. Zhang. 2004. isolation, characterization, and molecular cloning of the cDNA encoding a novel phytase from *Aspergillus niger* 113 and high expression in *Pichia pastoris*. J. Biochem. Mol. Bio. 37: 282-291.

Yanke, L.J., H.D. Bae, L.B. Selinger, & K.L. Cheng. 1998. Phytase activity of anaerobic ruminal bacteria. Microbiology, UK. 144: 1565-1573.

26

Insights into Mode of Action of Insecticidal Proteins of *Bacillus thuringiensis*

NEEMA AGRAWAL, NARESH ARORA and RAJ K. BHATNAGAR
International Center for Genetic Engineering and Biotechnology (ICGEB) P.O. Box 10504, Aruna Asaf Ali Marg, New Delhi-110067,
E-mail: raj@icgeb.res.in

I. INTRODUCTION

In recent years, *i*nsecticidal *c*rystal *p*roteins (ICPs) produced by *Bacillus thuringiensis* (B.t.) have been successfully used all over the world to control insects that affect economically important plants or are vectors of human diseases. B.t. is not only established as an ecofriendly alternative or supplement to chemical insecticides but it also serves as a key source of genes for developing insect resistant transgenic crops. Like chemical pesticides, the major concern in the deployment of B.t.-based biopesticides and B.t.-crops in the field is the development of resistance by insects, a problem that can undermine the sustainability of this highly specific and ecofriendly biopesticide. The judicious use of Bt proteins and a detailed study of their molecular mechanism of action will lead to a better understanding of the mechanism of insect resistance to these toxins and at the same time ascertain their long-term application in the fields. Moreover, pesticides with enhanced activity and broader host range can also be developed by understanding the mode of action of these toxins. In this review, the mechanism of action of Bt toxins is discussed in detail.

II. STRUCTURE OF CRY TOXINS

Bacillus thuringiensis comprises a group of aerobic, Gram-positive spore-forming bacteria characterized by their ability to produce insecticidal parasporal crystalline inclusions during stationary phase of the growth cycle (Carlson *et al*, 1994). ICPs also called as Cry toxins or δ-endotoxins exhibit specific toxicities to different orders of insect (Schnepf *et al*, 1998). Cry toxins were first grouped into six classes, CryI, CryII, CryIII, CryIV, CryV and CryVI based on their insecticidal activity (Hofte and Whiteley, 1989). To accommodate the ever-increasing number of *cry* genes, a novel nomenclature has been proposed based exclusively on amino acid identity (Crickmore *et al*, 1998). The complete list of currently recognized B.t. genes and proteins is available on the following website:

http://www.biols.susx.ac.uk/Home/Neil_Crickmore/Bt/index.html.

The *cry* genes code for Cry proteins with a range of molecular masses from 50 to 140 kDa. Some of the Cry proteins like Cry1, Cry4A and Cry4B are synthesized as inactive 130-140 kDa protoxins, which are proteolytically processed in the insect gut upon ingestion by susceptible larvae to 60-70 kDa active toxin by removal of C-terminal half of the protein (Choma *et al*, 1990). However, the carboxyl terminal fragment is not found in some toxins (Cry3A, Cry3Ba, Cry3Bb, Cry3Ca, Cry2Aa Cry2Ab, Cry2Ac, and Cry11Aa) or is very small in other protoxins, like Cry1Ia and Cry1Ib and Cry13A (Widner and Whiteley, 1989, Carroll *et al*, 1989, Bravo, 1997). Amino acid alignment of several Cry toxins with different specificity revealed five highly conserved sequence blocks in the active N-terminal half of the proteins, separated by hypervariable sequences (Geiser *et al*, 1986; Hofte and Whiteley, 1989). The presence of three additional conserved blocks lying outside the active toxic core was identified in the carboxyl-terminal halves of 130-140 kDa protoxin sequences (Schnepf *et al*, 1998). The putative function of this long carboxyl terminal segment is to aid in the formation of an ordered crystalline array. Since most of the cysteine residues are located in this fraction of the protoxin, it has been suggested that the alkaline and reducing conditions required for the solubility of crystal proteins is due to disulfide bridge formation within the protoxin fragment (Choma and Kaplan, 1992).

To date, the structures of three crystal proteins-Cry3A (Li *et al*, 1991), Cry1Aa (Grochulski *et al*, 1995), and Cyt2A (Li *et al*, 1996) have been solved by X-ray crystallography. The coleopteran-specific Cry3A protoxin has a molecular mass of 70 kDa and does not contain the large carboxyl-terminal region present in the 130 kDa lepidopteran-specific Cry1Aa protoxin. The crystal structure of Cry1Aa toxin was determined from the activated toxin fragment. Though these toxins share only 25% amino acid sequence identity, the 2 structures display a similar three-domain topology (Grochulski *et al*, 1995).

Domain I is a seven α-helix bundle in which a central helix (helix α-5) is completely surrounded by six outer helices. The six α-helices are amphipathic and span the hydrophobic region of a membrane bilayer. This domain has been implicated in the channel formation in the membrane. Domain II consists of three antiparallel β-sheets joined in a "Greek key" topology, arranged in a β-prism fold (Sankaranarayanan *et al*, 1996). The three surface-exposed loops of domain II which connect the three β-pleated sheets are of different lengths and are located between β1 and β2 (loop 1), β6 and β7 (loop 2) and β10 and β11 (loop 3). There is another exposed loop (α 8 loop) located between α-helix 8 and β1 sheet in Cry3A (Li *et al*, 1991) or between α8a and α8b in Cry1Aa (Grochulski *et al*, 1995). Extensive mutagenesis has been carried out in the residues constituting the loops of domain II specifically in Cry3A, Cry1A and Cry1C toxins to evaluate their role in binding and toxicity. Loop 2 (β6-β7) of Cry3A consists only of residues Pro-412 and Ser-413 whereas loop 2 of Cry1A type toxins is large and is the most divergent in the three-dimensional structures of Cry3A and Cry1Aa (Li *et al*, 1991; Grochulski *et al*, 1995). Domain II has been implicated in receptor binding and determining toxin specificity toward insect. Domain III consists of two twisted, antiparallel β-sheets forming a β-sandwich with a "jelly roll" topology. Domain III has been speculated to play multiple functions. It may be involved in ion channel formation as a voltage sensor, since conservative mutations in R521K and R527K reduced toxicity without reducing binding (Chen *et al*, 1993). The salt bridges and hydrogen bonds formed between arginine residues present in the β-17 sheet and other residues in the vicinity play an important role in stabilizing the toxin structure. Both β-sheets (β-17 and β-23) are important determinants for the proper folding of the toxin (Grochulski *et al*, 1995). Domain III has also been implicated in receptor binding.

The five highly conserved amino acid blocks identified in Cry toxin sequences by Hofte and Whiteley (1989) were either located at the centre of domains or were involved in interdomain contacts. Li *et al* (1991) proposed that the high degree of conservation of these blocks and their important structural location would imply that the Cry toxins, which possess these blocks, would share similar structure that of globular toxins composed of three structural domains. Block 1 encompasses helix 5 of domain I which has been implicated in pore formation. Being centrally located, it might play a role in maintaining the structural integrity of the helical bundle. Block 2 includes helix 7 of domain I and the first β–strand of domain II; these two form the region of contact between the two domains. The residues involved in forming salt bridges between domain I and domain II also lie in block 2 (Grochulski *et al*, 1995). Block 3 contains the last β–strand of domain II, a structure involved in interactions between domain I and domain III. The arginines present in block 4 may be involved in ion channel formation (Li *et al*, 1991; Grochulski *et al*, 1995).

III. THE INSECT GUT: TARGET ORGANELLE OF BT δ-ENDOTOXINS

δ-endotoxins are feeding poisons and primarily act on the midgut epithelial cells of insects. To understand their mode of action it would be important to describe the nature of this target. (For comprehensive review on structure and function of the insect midgut, see Dow, 1986 and Knowles, 1994). B.t. toxins are active against insects belonging to several orders including lepidoptera, coleotera and diptera. Here, we are describing the gut of only lepidopterans in details as most of the research on the mode of action of B.t. toxins is centered on them. The Cry1A and Cry1C class of ICPs are toxic to lepidopteran larvae. Lepidopteran larvae are characterized by a high blood ratio of K^+:Na^+ (Sutcliffe, 1963) and a very high ambient alkalinity in the gut (> pH 12 in some insects) (Dow, 1984 and 1986). The larval midgut is a simple, one layered, tubular epithelium composed of two major cell types; a columnar cell with a microvillate apical border, which is believed to play a role in nutrient absorption and a unique goblet cell, containing a large vacuolar cavity which is joined to the apical surface by a valve (Cioffi, 1979; Dow, 1986). The gap junctions present between the goblet and columnar cells provide electrical and chemical coupling between the two (Moffett and Koch, 1988). The "K^+ pump" located on the apical membrane of the goblet cell consists of a proton pump (of V-ATPase family) in parallel with H^+/K^+ exchanger. It actively transports K^+ from the blood side to the lumen side, thus, generating a transepithelial potential difference of over 150 mV (Harvey and Nedergaard, 1964; Dow, 1986; Moffett and Koch, 1988). The highly electric nature of the epithelium due to the activity of K^+ pump probably makes the gut susceptible to small dosage of toxin which forms pores in the membrane and, thus, disrupts the transepithelial potential.

Like lepidopteran larvae, dipteran larvae have a very high pH in their gut (Dadd, 1975; Lacey and Federici, 1979) but their epithelium lacks goblet cells. Unlike the above two orders, coleopterans have a gut pH that is neutral or slightly acidic (Weltens *et al*, 1992) and they are killed by Cry3 class of B.t. toxins.

IV. MOLECULAR MODE OF ACTION OF δ-ENDOTOXINS

The mechanism of action of the *B. thuringiensis* toxins is a multistage process. In general, it consists of three major steps: (1) solubilization of crystals and activation of protoxins in the insect midgut, (2) binding of the activated toxin to midgut receptors, and (3) insertion of the toxin into the midgut epithelial membrane to create ion channels or pores.

Solubilization and Activation

When ingested by a susceptible larva, the crystalline inclusions are solubilized in the midgut of the insect releasing inactive protoxins. The mixture of protoxins within an inclusion determines the solubility of these protoxins and, thus, the effectiveness for a particular insect (Aronson *et al*, 1991). The inclusions composed of Cry1, Cry2 and Cry4 protoxins are solubilized at high pH (9-10) and under reducing conditions, found in the midgut of lepidopteran and dipteran larvae. Inclusions containing the Cry3A protoxin are soluble at pH 7-8, close to that found in susceptible coleopteran larval guts. Differences in the extent of solubilization can sometimes account for the differences in the degree of toxicity of Cry proteins among target insects (Aronson *et al*, 1991; Du *et al*, 1994) suggesting that solubilization is an important factor in susceptibility. A reduction in the solubility is speculated to be one potential mechanism for insect resistance (McGaughey and Whalon, 1992; Oppert *et al*, 1994).

Following solubilization, the inactive protoxins are processed to active form by insect midgut proteases (Lecadet and Dedonder, 1967; Tojo and Aizawa, 1983). The major proteases of the lepidopteran insect midgut are trypsin-like (Milne and Kaplan, 1993) or chymotrypsin-like (Johnston *et al*, 1995; Peterson *et al*, 1995; Novillo *et al*, 1997). In resistant population of *Ostrinia nubilalis* (Europeon corn borer), reduced protoxin processing due to decreased activities of protoxin activating proteinases has been associated with resistance to Cry1Ab toxin (Li *et al*, 2004).

Receptor Binding

Elucidation of the receptor binding mechanism is one of the most extensively studied steps in the mode of action. After activation, binding of the toxin to the receptor in the brush border membrane presumably initiates the cascade of events leading to insect mortality. The interaction with binding sites in the midgut epithelium of susceptible insect species is a major determinant for the selectivity of the ICPs. In view of the important contribution of these high-affinity binding sites in the mode of action of *B. thuringiensis* ICPs, much effort is put on the cloning and characterization of ICP-binding proteins. An *in vitro* system using insect midgut brush border membrane vesicles (BBMV) and radiolabeled (^{125}I) toxins was developed to study binding at the molecular level (Wolfersberger *et al*, 1987). Binding is a two-stage process involving reversible (Hofmann *et al*, 1988) and irreversible steps (Van Rie *et al*, 1989; Ihara *et al*, 1993; Rajamohan *et al*, 1995). The later step may involve a tight binding between the toxin and receptor, insertion of the toxin into the apical membrane or both. Hofmann *et al* (1988) and Van Rie *et al* (1989) demonstrated a direct correlation between toxin binding affinity, binding site concentration, and toxicity i.e., the highly active toxin constituted more binding sites and bound with higher affinity to the BBMVs than did the less active toxins.

However, Wolfersberger (1990) observed an inverse relationship between binding affinity and toxicity in the gypsy moth (*Lymantria dispar*). A highly active toxin, Cry1Ab exhibited weaker binding affinity compared to a less active toxin, Cry1Ac. Garczynski *et al* (1991) demonstrated the presence of high-affinity binding sites for Cry1Ac toxin in BBMVs of both sensitive (*Manduca sexta* and *Heliothis virescens*) and naturally resistant insects (*Spodoptera frugiperda*). Ihara *et al* (1993) observed that Cry1Aa and Cry1Ab toxins bind to *Bombyx mori* BBMVs with similar affinity but Cry1Aa was found to be 100-fold more toxic than Cry1Ab. They demonstrated that the differences in the rate of irreversible association with BBMVs could account for variations in toxicity. These results support a hypothesis in which toxin receptors are necessary, but insufficient for insecticidal activity.

The toxin-binding proteins (receptors) from several insects have been identified by ligand blotting studies (Garczynski *et al*, 1991; Oddou *et al*, 1991; Oddou *et al*, 1993; Vadlamudi *et al*, 1993; Cowles *et al*, 1995) using BBMVs and iodinated or biotinylated toxins. The candidate membrane proteins that have been identified to serve as Cry protein receptors in different insects include the enzyme aminopeptidase N (APN), a cadherin-like glycoprotein and an anionic brush border membrane glycoconjugate, BTR-270.

Aminopeptidases

The majority of ICP-binding proteins characterized belong to aminopeptidase N family. Aminopeptidase N (APN; EC 3.4.11.2) is a zinc-dependent metallopeptidase having a well-conserved HEXXHXW motif, the marker for zinc binding site (Taylor, 1993; Hooper, 1994) that catalyzes the removal of N-terminal, preferentially neutral residues from peptides.

In mammals, aminopeptidase N is a type II membrane protein, anchored to epithelial cells of kidney and intestine by an uncleavable N-terminal signal peptide sequence of approximately 60 amino acids (Kenny *et al*, 1987). In insects, however, APN is a type I membrane protein, anchored to the midgut-pithelium by a glycosyl-phosphatidylinositol (GPI) moiety located at the C terminus (Takesue *et al*, 1989 and 1992; Garczynski and Adang, 1995). The differences in the mode of attachment and, thus, the orientation of the membrane anchored aminopeptidases in insects and mammals might account for the inability of B.t. toxins to interact with mammalian aminopeptidases. Due to the lack of interaction, the B.t. toxins do not exhibit toxicity to mammals and have been used safely for the control of agricultural pests. Cry toxins bind at a site distinct from the catalytic site as APN enzyme activity was not affected by their binding (Luo *et al*, 1996).

Isoforms of APN ranging in size from 105-170 kDa have been isolated from several lepidopteran insects including *Lymantria dispar*, *Plutella xylostella*, *H. virescens*, *B. mori*, *Plodia interpunctella*, *H. armigera* and *S. litura* (Oltean *et al* 1999; Luo *et al* 1996; Hua *et al* 1998; Knight *et al* 1994; Rajagopal *et al* 2003; Agrawal *et al* 2002). The presence of a 120 kDa Cry1Ac receptor was first observed on ligand blots of *M. sexta* BBMVs (Garczynski *et al*, 1991; Knowles *et al*, 1991). It was demonstrated that Cry1Ac- binding to this band was abolished by N-acetylgalactosamine (GalNAc) (Knowles *et al*, 1991), suggesting the presence of competing GalNAc moieties on the receptor itself (Sangadala *et al*, 1994). The protein was identified as the metalloprotease aminopeptidase N (Knight *et al*, 1994). Since the discovery of the *M. sexta* APN receptor (MsAPN-1), two more *M. sexta* APNs have been reported (Luo *et al*, 1996; Denolf *et al*, 1997).

Two Cry1Ac toxin binding APNs were isolated from *H. virescens* BBMVs that had the same N-terminal sequence but showed different molecular masses (130- and 170 kDa) due to the difference in their extent of glycosylation (Oltean *et al*, 1999). Like MsAPN-1, GalNAc competed with Cry1Ac binding to the 170 kDa protein but not with Cry1Aa and Cry1Ab binding in ligand blotting and surface plasmon resonance (SPR) experiments. Recently, in *M. sexta* a chemical deglycosylation technique was used to demonstrate that carbohydrates attached to 120 kDa APN are binding epitopes for Cry1Ac toxin. They also analyzed the glycan structure of APN by using carbohydrate compositional analysis and lectin-binding. In the average APN molecule, 2 of 4 possible N-glycosylation sites are occupied with fucosylated mannose type N-glycans and 5-6 of the 13 probable O-glycosylation sites are occupied with simple [GalNAc-peptide] type O-glycans. The O-glycosylated C-terminal stalk is the most likely binding site for Cry1Ac (Knight *et al*, 2004).

Luo *et al* (1996) purified and characterized a Cry1C-binding 106 kDa protein from *M. sexta* BBMVs that was found to be distinct from a previously described 115 kDa Cry1Ac-binding APN (Knight *et al*, 1994; Sangadala *et al*, 1994; Masson *et al*, 1995). Their results showed that Cry1Ac and Cry1C toxins recognize functionally related, but structurally distinct isoforms of APN in *M. sexta* midgut. Lee *et al* (1996c) demonstrated that Cry1Ac recognizes 120 kDa APN while Cry1Aa and Cry1Ab recognize 210 kDa molecule in *L. dispar* BBMV by ligand blotting experiments. Wang *et al* (2005) have identified four APN isoforms (TnAPN1, TnAPN2, TnAPN3 and TnAPN4) from the cabbage looper, *Trichoplusia ni*, by cDNA cloning. For the first time, they demonstrated that APN genes were expressed in Malphigian tubules in addition to the midgut. Multiple aminopeptidases have been isolated from different insects. The presence of multiple related aminopeptidases is important for hydrolysis of amino acids from peptides and their subsequent transport by amino acid transporter across the brush border membrane. The existence of multiple aminopeptidases has been explained by gene duplication mechanism (Chang *et al*, 1999; Garner *et al*, 1999).

A number of aminopeptidase-encoding genes have been cloned and expressed in *E. coli* and insect cell cultures. Yaoi *et al* (1999) successfully expressed a 110 kDa *B. mori* aminopeptidase in *E. coli* and demonstrated its binding to Cry1Aa toxin. A 105 kDa APN from *P. xylostella* and a 120 kDa APN from *M. sexta* were successfully expressed in *S. frugiperda* insect cells (Denolf *et al*, 1997; Luo *et al*, 1999). But, both the expressed APNs did not exhibit binding toward the Cry1A toxins, inspite of being glycosylated. Simpson and Newcomb (2000) could demonstrate the binding of the surface expressed *E. postvittana* APN to Cry toxins by ligand blot assays but failed to show toxin-binding to the intact cells. A 110 kDa APN-encoding gene was transiently expressed in *Drosophila* S2 cells. Although the recombinant APN bound Cry1Ac toxin but it could not mediate cytotoxicity (Banks *et al*, 2003). The plausible reason for the lack of binding of expressed APN molecules to the toxins could be the differences in the posttranslational modifications that occur in the epithelial cells of insect midgut and Sf insect cell culture. In another report, Agrawal *et al* (2002) isolated an aminopeptiadse-encoding gene from the midgut of polyphagous pest, *Spodoptera litura*, and expressed it in Sf21 cells using baculovirus expression system. The 108 kDa recombinant APN was catalytically active and was anchored to the membrane by GPI moiety. They successfully demonstrated the specific interaction of recombinant APN with Cry1C toxin by immunofluorescence toxin-binding studies on intact Sf21 cells and ligand blot and immunoprecipitation analyses (Agrawal *et al*, 2002). The protein showed no interaction with biologically inactive Cry1Ac toxin. Rajagopal *et al* (2003) reported the cloning and expression of two APNs from *Heliothis armigera* (American cotton bollworm). Both the genes were expressed in *Trichoplusia ni* cells using baculovirus expression vector system. The expressed APNs were membrane-associated, catalytically active and glycosylated. Ligand blot analysis of the 2 HaAPNs with bioactive Cry1A toxins displayed differential interaction. HaAPN1 interacted with all the 3 toxins, Cry1Aa, Cry1Ab and Cry1Ac whereas HaAPN2 interacted with only Cry1Ac. Heterogeneity within aminopeptidases has led to the speculation about multiple APNs being involved in initial recognition event of toxin-binding. The recombinant APN protein produced in large amounts can be used for structural and functional characterization, thus, facilitating further analysis of B.t. toxin-APN interactions.

The identification of toxin-binding molecules and the calculations of their affinities for different Cry toxins reveal the complexity of toxin-receptor interactions. Dissociation constants (K_d) values have been determined by toxin overlay assays, binding assays performed with BBMV and surface plasmon resonance. The consensus is that BBMV preparations have higher affinities for Cry1A toxins while purified proteins exhibit lower affinities (Masson *et al*, 1995; Lee *et al*, 1996c). This difference in affinity could partly be due to loss of the irreversible step, which involves toxin insertion into the BBMV membrane.

Site-directed mutagenesis of putative interacting residues on the toxin together with SPR analysis revealed the role of several residues involved in Cry1Ac binding to *L. dispar* APN and suggested a model for the toxin-receptor interaction. A bivalent sequential-binding model (Jenkins *et al*, 2000) was proposed where domain III residues of Cry1Ac (509-513, 544-547) initiate APN contact via a GalNAc moiety, followed by receptor adherence which is facilitated by domain II loop residues (α-8 loop, loop 2, loop3). Sequential-binding may represent domain II binding to a second APN site, or the occurrence of a conformational change. Binding to membrane-bound receptor is followed by insertion (ki). The following equation describes the specific interaction between Cry1Ac and APN in *L. dispar*:

Domains involved

$$\begin{array}{ccccccc} & & \text{III} & & \text{II} & & \text{I (II, III?)} \\ R + T & \underset{kd1}{\overset{ka1}{\leftrightarrow}} & RT & \underset{kd2}{\overset{ka2}{\leftrightarrow}} & RT & \overset{k_i}{\rightarrow} & RTM \end{array}$$

In this equation, the following steps occur: 1) Reversible APN-binding is initiated by domain III (at a GalNAc site). Both ka1 and kd1 are affected by the domain III mutations. 2) Secondary-binding occurs through domain II. On-rate, kd2, is affected by domain II, especially loop. 3) Finally, membrane (M) insertion (ki) occurs through domain I, although domains II and III may also be involved. Receptor may be released by toxin after membrane insertion.

The differences in receptor-binding properties between Cry1A toxins make it difficult to create a general binding model. These differences may be attributable to the dramatically dissimilar sequence of Cry1Ac domain III with respect to other Cry toxins. Thompson *et al* (1995) pointed out that Cry1Ac domain III shows only 50% similarity to domain III segments from other lepidoptera-active toxins. It has been proposed that Cry1Ac domain III evolved from an origin different from that of other Cry1A toxins (Bravo, 1997). This explains the unique specificity of Cry1Ac for a sugar moiety.

Cry toxins form ion channels in membranes (Schwartz *et al*, 1993). At higher concentrations, approx. 20 μg/ml, Cry1Ac toxin inserts into phospholipid membranes and forms pores permeable to $^{86}Rb^+$ (English *et al*, 1991). Brush border membrane components decrease the concentration of toxin necessary to alter membrane permeability (English *et al*, 1991, Carroll and Ellar, 1993). Multiple toxin-binding proteins have been identified in BBMVs of several insects, but it is important to identify the ones that serve as functional receptors in which binding leads to pore formation. The functional evidence that APN is a toxin receptor comes from Sangadala *et al* (1994) who reconstituted a mixture of 120 kDa APN and a 65 kDa protein into membrane vesicles and measured a 30% increase in Cry1Ac-binding and a 1000-fold decrease in the amount of Cry1Ac toxin required to induce $^{86}Rb^+$ efflux. Luo *et al* (1997) reconstituted the 170 kDa APN purified from *H. virescens* BBMVs into phospholipid vesicles and demonstrated that the APN promoted toxin-induced $^{86}Rb^+$ release for Cry1A toxins, but not Cry1C toxin. Furthermore, Cry1Ac being the most active toxin against *H. virescens* caused $^{86}Rb^+$ release at lower concentrations and to a greater extent than less active Cry1Aa and Cry1Ab toxins. The correlation between toxin-binding specificity and $^{86}Rb^+$ release strongly suggests that the purified 170 kDa APN is the functional receptor in the *H. virescens* midgut epithelial cell brush border membranes. Lee *et al* (1996c) investigated the effect of PIPLC on pore-forming activity of Cry1Ac with *L. dispar* by voltage clamping assays. After PIPLC treatment, the inhibition of I_{SC} (short circuit current inhibition) by Cry1Ac was significantly slower than inhibition by Cry1Ac alone. In contrast, PIPLC treatment had only a slight effect on the inhibition of I_{SC} by Cry1Aa. These voltage clamping assay data clearly demonstrated that a GPI-anchored protein, possibly APN, is the major functional receptor for Cry1Ac and suggest that Cry1Aa recognizes a different receptor.

Although the role of APN as a functional receptor for Cry toxins was well supported by the above mentioned *in vitro* experiments, its only recently that its role as a B.t. toxin receptor has been established by *in vivo* evidences. In one of the studies (Rajagopal *et al*, 2002), APN was demonstrated to be the receptor of Cry1C toxin in *Spodoptera litura* by silencing the *apn* employing the technique of RNA interference (RNAi). RNAi is a posttranscriptional gene silencing phenomenon in which the introduced dsRNA specifically degrades the cognate mRNA. The introduction of dsRNA corresponding to *S. litura* APN (SlAPN) into the hemolymph of fifth instar larvae, specifically silenced *apn* gene which in turn resulted in increased resistance of the larvae to Cry1C protein. Gill and Ellar (2002) also explored the *in vivo* role of APN by constructing transgenic *Drosophila*. The *apn* from *M. sexta* was expressed in midgut and mesodermal tissues of Cry1Ac1-insensitive *Drosophila* larvae by utilizing the Gal4 enhancer trap technique. Feeding Cry1Ac1 to transgenic larvae resulted in high mortality, thus, demonstrating the functionality of APN as a receptor for Cry1Ac *in vivo*.

The possibility of existence of alternative receptors in the gut of susceptible larvae was put forth by Carroll *et al* (1997) who demonstrated that APN in *M. sexta* midgut is predominantly located in the posterior midgut while binding of Cry toxins can take place throughout the midgut. This indicates that receptors other than APN are functional in binding Cry toxins in the anterior region of the gut.

Cadherin-like Receptors

Proteins belonging to cadherin superfamily are receptors for Cry1A toxins in many lepidopteran insects. Bt-R_1, a 210 kDa protein showing 30-60% similarity to proteins in the cadherin superfamily was purified and cloned from *M. sexta* (Vadlamudi *et al*, 1995). Though the estimated molecular mass of Bt-R_1 from its cDNA sequence is 172 kDa but due to N-glycosylation the mobility of the protein corresponds to molecular mass of 210 kDa. The binding affinities of Cry1Ab toxin for *M. sexta* 210 kDa receptor, estimated from ligand blotting and BBMV binding assays did not differ significantly (780 and 1000 pM respectively) (Vadlamudi *et al*, 1993). The cloned 210 kDa receptor expressed in human embryonic 293 cells showed binding affinities (1015 pM) similar to those of BBMV and ligand blots (Vadlamudi *et al*, 1995). These experiments imply that in case of cadherin-like receptors the insertion rate does not influence the binding affinity. When expressed in Sf21 insect cells, Bt-R1 was released in soluble form into culture media. The recombinant Bt-R1 interacted with Cry1Ab toxin with high affinity (Kd = 1.1 nM), which was similar to native Bt-R1 (Meng *et al*, 2001). Bt-R1 is a classical cadherin protein with predicted adhesive function and Ca^{2+} binding sites. This protein contains tandem repeats of homologous extracellular domains, called cadherin repeats (CR), which are numbered from the outermost 1 to the innermost 12, a membrane-proximal extracellular domain (MPED), a membrane-spanning domain and a cytoplasmic domain. Cry1A toxins bind Bt-R1 and Bt-R1a with high affinity (Vadlamudi *et al*, 1995; Francis and Bulla, 1997) and when expressed on the surface of COS-7 and *Drosophila* S2 cells induce Cry1A cytotoxicity (Dorsch *et al*, 2002; Hua *et al*, 2004b). Bt-R1a differs from Bt-R1 by two amino acid substitutions (Hua *et al*, 2004b). The expression of developmentally important BT-R(1) cadherin was studied during larval growth. The protein was differentially cleaved at each developmental stage and the cleavage was calcium dependent (Candas *et al*, 2002). BT-R(1) cadherin contains a number of motifs potentially involved in interactions with cadherin and integrins. It is believed that BT-R1 is required for structural integrity of midgut and to support differentiation of epithelial cells during larval development in *M. sexta*. Moreover, the binding of Cry1A toxins to BT-R1 disrupts such organization and functionality causing death of the insect (Midboe *et al*, 2003). In *M. sexta*, the effect of calcium and Cry1Ab toxin on adhesion and clustering of BBMVs was studied. The adhesion and aggregation of the vesicles, which involve protein-protein interactions were enhanced by calcium. Interestingly, a 50% decrease in calcium-induced vesicle aggregation was observed when BBMVs were preincubated with Cry1Ab toxin, which specifically binds to BT-R1 with high affinity and disrupts the midgut epithelium. This shows that the adhesive function of BT-R1 is disrupted by Cry1Ab toxin (Griko *et al*, 2004).

A Cry1Aa toxin-binding 175 kDa glycoprotein (Bt-R175) was purified from midgut of *B. mori* and its cDNA was cloned. The cDNA encoded a 193 kDa protein that included 9 extracellular cadherin repeats, a 23.5 kDa membrane-proximal domain, a membrane-spanning region and a 13.6 kDa cytoplasmic domain (Nagamatsu *et al*, 1998). The recombinant Bt-R175, expressed in Sf9 cells, was demonstrated to be a functional receptor as it induced swelling and lysis of cells after exposure to Cry1Aa toxin because of formation of ion channels in cell membrane (Nagamatsu *et al*, 1999). When expressed in mammalian COS-7 cells, Bt-R175 induced susceptibility to Cry1Aa (Tsuda *et al*, 2003). A Cry1Aa receptor of similar size has also been reported by analyzing ligand blots of *L. dispar* BBMV proteins (Lee *et al*, 1996c), although the protein has not been purified. With the use of Cry1Aa-Cry1Ac hybrids, it was demonstrated that domain III of Cry1Aa is required for binding to 210 kDa protein (Lee *et al*, 1995b). Very recently, a 220 kDa cadherin-like protein, OnBt-R1, was identified as receptor for Cry1Ab toxin in *O. nubilalis*. The recombinant OnBt-R1, expressed in Sf9 cells induced functional susceptibility to Cry1Ab toxin. Thus, it was suggested to be involved with Cry1Ab toxicity in the pest (Flannagan *et al*, 2005).

Elimination or alteration of cadherin-like proteins from the midgut correlates with insect resistance to *B. thuringiensis* Cry proteins in *H. viresecns* and *Helicoverpa armigera* (Gahan *et al*, 2001; Jurat-Fuentes *et al*, 2004; Xu *et al*, 2005) and *Pectinophora gossypiella* (Morin *et al*, 2003).

Novel Receptors

In *L. dispar*, besides APN and cadherin-like protein, a high molecular weight anionic glycocojugate BTR-270 was identified. SPR analyses revealed that it had high affinity for Cry1Aa and Cry1Ab toxins and weaker affinity for Cry1Ac toxin (Valaitis *et al*, 2001).

To identify the protein components of BBMV prepared from midgut of *M. sexta*, a midgut brush border proteome was established using one- and two-dimensional gel electrophoresis. Subproteomes for Cry1Ac binding proteins and GPI-anchored proteins led to the identification of two novel proteins, actin and alkaline phosphatase, that bind Cry1Ac in addition to APN (McNall and Adang, 2003). Later, Chen *et al* (2005) also localized membrane-type alkaline phosphatase (m-ALP) in sections of midgut as Cry1A–binding protein (apart from APN and cadherin-like protein, Bt-R1) in *M. sexta*. Predominantly, both APN and m-ALP were distributed along microvilli in the posterior region and Bt-R1 in the middle region of midgut.

Role of Domain II Loop Regions in Receptor Binding

This domain has been described as the specificity-determining domain. The reciprocal hybrid genes between closely related toxins (Cry1Aa and Cry1Ac) resulted in chimeric toxins with altered specificity (Lee *et al*, 1992). Site-directed mutagenesis studies of Cry1A toxin revealed that loop a-8, located in the junction of domains I and II and the two protruding loops oriented parallel with the helical bundle of domain I (loop 1 [between β-2 and β-3] and loop 2 [between β-6 and β-7]) and loop 3 regions of domain II are involved in receptor recognition and toxicity (Rajamohan *et al*, 1995; Rajamohan *et al*, 1996a, 1996b, 1996c; Jenkins *et al*, 2000; Lee *et al*, 2000 and 2001). Smith and Ellar (1994) showed that mutations in loops 1 and 2 of Cry1C toxin were able to modulate toxicity and specificity. Site-directed mutagenesis analysis of Cry3A toxin showed that in addition to loop 1, loop 3 (between β-10 and β-11) is involved in irreversible binding to *Tenebrio molitor* midgut membranes (Wu and Dean, 1996).

The deletion and substitution mutations in the loop 2 residues in Cry1A toxins affected the binding and toxicity of the toxins (Lu *et al*, 1994; Rajamohan *et al*, 1996a). Substitution mutants F371A and G374A showed a 400-fold decrease in toxicity for *M. sexta* while only a minor change was observed in the binding affinity (two-fold reduction in K_d) of these mutant toxins for BBMVs when compared to the wild-type toxin (Rajamohan *et al*, 1995). It was suggested that these mutations affected the irreversible binding. In contrast, in *H. virescens* mutant F371A had no effect on toxicity or binding (Rajamohan *et al*, 1996a). Cry1Ac toxin has exactly the same loop 2 residues as Cry1Ab but binds a different receptor in *M. sexta*, *H. virescens* and *L. dispar* (Lee *et al*, 1995a, 1996a and 1996b). These residues were found to be functionally different with regard to different receptors in the same insect. A model for Cry1 toxin-binding to midgut membrane vesicles has been constructed which supports a determinant role for domain II loops in Cry toxin specificity in *H. virescens* (Jurat-Fuentes *et al*, 2001).

In summary, the binding picture for domain II is complex. Results of mutagenesis studies suggest that all of the loops of domain II can participate in receptor binding, but the loop residues perform functionally distinct roles in different insects or receptors. Mutations in domain II were also shown to improve the binding affinity and toxin potency in some insects (Rajamohan *et al*, 1996a). These experiments illustrate that the understanding of receptor-toxin interactions could lead to the construction of second-generation potent toxins.

Role of Domain III in Receptor Binding

Domain III has also been implicated in receptor binding. Aronson *et al* (1995) mutated a hypervariable region of domain III (residues 500-509) of Cry1Ac. Only changes in Ser-503 and Ser-504 resulted in lower toxicity in *M. sexta* and were correlated to a decrease in binding affinity to BBMV proteins by ligand blots. Cry1Aa and Cry1Ac toxins recognize 210 and 120 kDa binding proteins, respectively in *L. dispar* BBMVs. Lee *et al* (1995b) analyzed homologue scanning mutants that exchanged domain III between Cry1Aa and Cry1Ac and demonstrated that hybrid proteins containing Cry1Aa domain III bind a 210 kDa receptor in *L. dispar*, whereas hybrid proteins containing Cry1Ac domain III bind a 120 kDa receptor in *L. dispar*. These experiments suggest a role (direct or indirect) for domain III in receptor recognition. In another case, a Cry1Ab-Cry1C hybrid toxin (domains I and II from Cry1Ab and domain III from Cry1C toxin) significantly improved the potency of Cry1Ab toxin in *S. exigua* without affecting the binding site specificity (deMaagd *et al*, 1996). This result suggests that domain III of Cry1C plays an important role in toxicity in *S. exigua*. Finally, mutations in the domain III residues of Cry1Ac toxin resulted in proteins that bind poorly to the *M. sexta* and *H. virescens* toxin-binding proteins and showed extensive loss to toxicity for both insects (Lee *et al*, 1999).

Understanding of the toxin-receptor interaction is limited but it is apparent that residues in different loops may be involved in binding and that the residues employed differ from toxin to toxin, depending on the sequence of receptor molecule. Availability of cDNA of aminopeptidases and cadherin–like proteins will facilitate domain mapping of the receptor ligand and provide useful insight into the mode of action of insecticidal crystal proteins.

Mapping of Insecticidal Protein Interacting Domains of the Receptor

The role of different domains, domain II and domain III of Cry proteins in receptor binding has been extensively studied by mutagenesis and domain swapping experiments. Recently, there is a lot thrust on the identification/mapping of epitopes involved in Cry-toxin interaction on receptors. Among several approaches used, phage display technique has proven to be highly successful to identify protein epitopes involved in protein-protein interactions (Deleo *et al*, 1995; Szardenings *et al*, 1997; Demangel *et al* 2000). Using phage display, stable expression of different fragments of Cry1Aa protein was achieved by Marzari *et al* (1997). The interaction of *M. sexta* Cry1A cadherin-like receptor Bt-R(1) with Cry1Ab toxin is worked out in detail. Using phage display, Gomez *et al* (2001) identified a key epitope on *M. sexta* cadherin-like receptor, Bt-R1 that is involved in toxin-receptor interaction. An scFv antibody (scFv73) was characterized that inhibited binding of Cry1A toxins to Bt-R1, but not to APN on midgut membrane vesicle. The CDR3 region of scFv showed homology with an eight amino acid epitope (^{869}HITDTNNK^{876}) of Bt-R1. This epitope was also conserved in Bt-R175, a cadherin-like receptor isolated from *B. mori*, supporting the notion that cadherin-like proteins are important receptors for the toxins. The scFv molecule bound to domain II of Cry1Ab toxin. Synthetic peptides corresponding to the scFv73 CDR3 and Bt-R1 epitope not only competed with the binding of Cry1A toxins to their receptors but also decreased the *in vivo* toxicity of Cry1Ab to *M. sexta* larvae. In order to study the mechanism of receptor interaction, it is important to identify the Cry1A cognate binding partner for Bt-R1 epitope. Till date, the molecular basis of protein-protein interaction is not clear. Recently, a concept of hydropathic complementarity is emerging, which proposes that proteins can interact through amino acid sequences having inverse hydropathic profiles (Blalock, 1995). Among the three exposed loop regions of domain II of Cry1Aa and Cry1Ab toxins, loop 2 was identified as the cognate-binding partner of Bt-R1 epitope (^{869}HITDTNNK^{876}). Analysis of hydropathic profiles of loop 2 regions in the toxins and of the Bt-R1 binding epitope revealed a stretch of 10-amino acid (^{363}SSTLYRRPFNI^{373}) from β-6-loop2 region of Cry1Ab toxin that showed hydropathic pattern complementarity to a 10-amino acid region (^{865}NITIHITDTNN^{875}) of Bt-R1, suggesting that the binding of Cry1A toxins to Bt-R1 is determined by hydropathic complementarity (Gomez *et al* 2002a). In another report, truncation analysis of Bt-R1 demonstrated a 66 amino acid region adjacent to the membrane-proximal extracellular domain as the only fragment capable of binding the Cry1A toxins of *B. thuringiensis*. Later, the Cry1A binding 66 amino acid epitope of Bt-R1 was narrowed to 12 amino acids. Gomez *et al* (2003) identified two Cry1Ab toxin-binding regions (TBRs) on Bt-R(1), an 11-residue long TBR1 (^{865}NITIHITDTNN^{875}) that interacted with loop 2 of domain II in the toxin and a 12-residue long TBR2 (^{1331}IPLPASILTVTV^{1342}) was demonstrated to interact with loop alpha-8 of Cry1Ab domain II. These interactions were determined by carrying out western blots and competition ELISA experiments using synthetic peptides as competitors for Cry1Ab binding to Bt-R1. TBR1 was located in CR7 and TBR2 in CR11. Noticeably, the specificity of Cry1A toxin involves at least 2 structural determinants on both molecules. Recently, Hua *et al* (2004) expressed the ectodomain units (CR7 to CR12) of BT-R1a on the surface of S2 cells and examined/evaluated their interaction with Cry1Ab and their role in mediating toxicity using fluorescent-based *Drosophila* S2 assay. Their results suggested that both CR11 and CR12 bind Cry1Ab but only CR12 was essential for mediating cytotoxicity in S2 cells.

Nagamatsu *et al* (1999) observed that a mutant of *B. mori* cadherin, Bt-R175, lacking CR9, failed to mediate Cry1Aa toxicity in Sf9 cells. Together, the results of Bt-R1 and Bt-R175 indicate that the CR unit adjacent to the membrane-proximal region is critical to Cry1 toxin binding and pore formation. For Bt-R175, the critical unit is CR9 and for Bt-R1a the critical unit is CR12.

The Cry1Ac toxin-binding region in *H. virescens* cadherin was mapped to a 40-amino acid fragment (1422 to 1440) and loop 3 of domain II was identified as its cognate-binding epitope. The substitution of Leu (1425) and Phe (1429) with charged amino acid resulted in the loss of toxin-binding indicating their importance in toxin interaction. A L1425R mutant which can result from a single point mutation may lead to Cry1A resistance in insects as it shows decreased toxin binding. However, a 10-fold increase in affinity was observed on mutating Gln (1430) to an Alanine (Xie *et al*, 2005).

In regard to receptor-binding epitopes, in *B. mori* APN, a 64 residue long epitope (135 Ile to 198Pro) at the N terminus was identified as Cry1Aa toxin-binding region. This site was specific for Cry1Aa toxin since it was not involved in Cry1Ac binding (Yaoi *et al*, 1999).

The identification of epitopes involved in Cry toxin-receptor interactions could provide insights into the mechanism of insect specificity and the mode of action of these toxins. This information will also be useful in designing more potent B.t. toxins by improving their specificity and toxicity. Knowledge of the functional receptor region would also facilitate detection of changes resulting in toxin-binding alteration, the most reported mechanism of resistance to B.t. toxins. Furthermore, receptor epitope mapping will be important for the characterization of insect populations that have developed resistance and to design toxins that could overcome receptor point mutations leading to Cry toxin resistance.

Membrane Insertion and Ion Channel Activity

Since domain I of Cry toxin consists of amphipathic helices of sufficient length to span a membrane, it has been implicated in the formation of lytic pores in the intestinal epithelium of the insect. The alpha helical structure of domain I is analogous to the membrane-translocating or pore-forming domains of other bacterial toxins including colicin A and diphtheria toxin.

Numerous mutations have been introduced into domain I in order to assess the effect of amino acid substitutions on toxicity and ion-channel activity. Mutations in domain I have been shown to affect the ability of the toxin to dissociate from the binding complex. The introduction of a negatively charged residue at the base of $\alpha3$ or in the loop joining $\alpha4$ and $\alpha5$ helices resulted in complete loss of activity due to the disruption of irreversible binding (membrane insertion) (Wu and Aronson, 1992; Chen *et al*, 1995). This revealed that a positively charged surface is important for toxicity. The mutations in the conserved stretch of amino acid residues of the α-helix 5 showed that only the conserved mutations retained the ion-channel activity. Interestingly, one of the mutants H168R showed enhanced activity toward several insects in bioassays and voltage clamping due to an increase in charge at this position. Studies of mutations in the helices of domain I have revealed that residues in alpha-helices 3, 4, and 5 are important in toxicity (Wu and Aronson, 1992; Chen *et al*, 1995; Kumar and Aronson, 1999). By PCR-based mutagenesis, Tyr-202 was identified as a critical residue within the alpha 4-alpha-5 loop of the 130 kDa Cry4A mosquito-larvicidal protein. Tyr202Ala mutant completely lost the larvicidal activity when bioassayed against *Aedes aegypti* larvae but a conserved mutation, Tyr202Phe did not affect toxicity. Thus, revealing a crucial role for the conserved aromatic residue at the 202 position in toxin activity (Pornwirron *et al*, 2004).

Pore formation constitutes a key step in the mode of action of B.t. d-endotoxins. Various Cry toxins have been shown to form ionic channels in receptor-free planar lipid bilayers at high concs. The radius of the pores formed by Cry1C in planar lipid bilayer was estimated to be between 1.0 and 1.3 nm by using a non-electrolyte exclusion technique (Peyronnet *et al*, 2002). It has been proposed that the pore could be formed by an oligomer containing 4 to 6 Cry toxin monomers. Soberon *et al* provided a direct evidence that the intermolecular interaction between Cry1Ab monomer is necessary step for pore formation and toxicity. They showed that 2 Cry1Ab mutant proteins affected in different steps of their mode of action (F371A in receptor-binding and H168F in pore formation) were able to form functional hetero-oligomers when mixed and both restored toxicity against *M. sexta* (Soberon *et al*, 2000).

The Cry proteins undergo several conformational changes from crystal inclusions to membrane inserted channels in the midgut epithelial cells of the target insect. Rausell *et al* (2004b) analyzed the stability of the different forms of Cry1Ab toxin and showed that the pre-pore complex and membrane inserted pore of Cry1Ab have a more flexible conformation than the monomeric toxin. Moreover, alkaline pH found in the midgut of lepidopteran insects further increases the flexibility of these structures. In the case of cadherin receptor BT-R1, toxin-binding initiates a conformational change that probably results in assembling of a pre-pore toxin oligomer (Gomez *et al*, 2002b). Binding of Cry1Aa toxins to the cadherin receptor promotes the formation of a 250 kDa oligomer. Trp spectroscopy studies and black lipid bilayer analysis revealed that the oligomeric structure of Cry1Ab toxin, as compared to monomer, interacts more efficiently with phospholipid membranes and forms stable pores (Rausell *et al*, 2004a). Aminopeptidases also bind Cry1 toxins and catalyze toxin-induced pore formation (Knight *et al*, 1994; Sangadala *et al*, 1994 and 2001).

The pore forming activity of Cry toxin has been studied with complete proteins, with domain I in isolation, with synthetic peptides mimicking α-helices, and with mutants that disrupt ion channel function. Channels formed by the α5 helix and domain I differed from those formed by whole toxins. Although domain I contains ample information to generate ion channels, such channels are significantly different than those formed by native toxin. Thus, the other domains may also play an indirect role in channel formation either by assisting in structural conformation, in oligomerization, or in regulation of channel activity. In summary, these results show qualitative support for the model that domain I constitutes or participates in the ion channel.

Domain III has also been reported to play a role in ion channel activity. On analyzing the alternating arginine region in β-sheet 17 (conserved block 4), Chen *et al* (1995) observed that mutations in central arginines caused structural instability in Cry1Aa whereas conservative substitutions of the outermost Arg caused reduction of toxicity due to alteration in the ion channel activity.

Several techniques employed to study the ion channel activity of the Cry toxins include voltage clamping using whole insect midgut (Harvey and Wolfersberger, 1979; Liebig *et al*, 1995), BBMV swelling assays (Carroll and Ellar, 1993; Wolfersberger *et al*, 1996), insect tissue culture cells (Knowles and Ellar, 1987; Schwartz *et al*, 1991), phospholipid vesicles (Haider and Ellar, 1989; English *et al*, 1991; Gazit and Shai, 1993b) and lipid bilayers (Schwartz *et al*, 1993; Gazit and Shai, 1995; Lorence *et al*, 1995; Martin and Wolfersberger, 1995). Knowles and Ellar (1987) used CF-1 (*Choristoneura fumiferana*) cells and proposed a "colloidal osmotic lysis" theory as a model for the cytolytic toxicity of Cry toxins. According to their proposal, the Cry toxins make nonspecific pores in the midgut membrane, leading to the influx of water and ions, resulting in cell swelling and eventual lysis. Despite the fact that a number of differences exist between insect cell culture and lepidopteran insect midgut, the experiments on cultured cells demonstrate the general capacity of the Cry toxins to insert into membranes and form large, nonspecific pores under certain conditions.

Harvey and Wolfersberger (1979) used electrophysiological analysis of sections of whole midgut of *M. sexta* to measure short circuit current inhibition (I_{SC}). The loss or enhancement of toxicity of mutant toxins correlated well with the changes in I_{SC}, thus, validating the use of voltage clamping technique to assess the specificity and potency of Cry toxins. The Na^+ and K^+ gradient-dependent amino acid uptake in BBMV of *M. sexta* was reportedly inhibited by bioactive Cry1A toxins in a dose-dependent fashion (Hendrickx *et al*, 1989; Wolfersberger, 1989), but the biologically inactive Cry1B protein had no effect on the amino acid uptake (Hendrickx *et al*, 1989).

Several groups have examined Cry toxin ion channel activity in planar lipid bilayer (PLB) systems and reported the effect of pH on the type and size of ion channels formed (Slatin *et al*, 1990; Schwartz *et al*, 1993). As described earlier in this chapter, reconstitution systems involving BBMV fused with lipid bilayers have shown the active involvement of BBMV proteins in the pore formation (English, 1991; Carroll and Ellar, 1993; Sangadala, 1994; Luo *et al*, 1997).

Models for the Mechanism of Pore Formation

The hydrophobic surfaces of the six outer amphipathic helices surrounding a central α-helix 5 face inward in the toxin structure. Since the helices have an opposite arrangement, a conformational change in the toxin is required for the insertion of the domain I into the membrane and pore formation. Association of the toxin molecule with the epithelial membrane is thought to cause large conformational change, possibly via the disruption of interactions between domain I and domain II that results in the insertion of a region of domain I into the membrane (Schwartz *et al*, 1997).

The exact mechanism of pore formation is not clear at present. Based on several independent experimental results, two putative models for pore formation have been proposed. These two models assume that binding of the toxin to its receptor triggers the conformational change needed for insertion but that the receptor does not form part of the pore (Knowles and Dow, 1993; Knowles, 1994).

The penknife model

This model was proposed by Hodgman and Ellar (1990) and is based on the model proposed for colicin A insertion. The strongly hydrophobic helices α-5 and α-6 (highly conserved amongst different Cry toxins) (Li *et al*, 1991) are joined by a loop and are furthest away from the membrane. They flip out of domain I in a penknife fashion and insert into the membrane. The remainder of the molecule would remain at the membrane surface or on the receptor.

Hodgman and Ellar (1990) proposed the formation of a hexameric toxin pore, which is lined by six helical hairpins, each donated by a toxin molecule. A hexamer was proposed on the grounds of geometrical considerations of the estimated area of the hydrophobic faces and internal pore radius of 0.6 nm (estimated by Knowles and Ellar, 1987), but has no direct experimental support.

The umbrella model

Li *et al* (1991) proposed an umbrella model in which the pore formation is initiated by insertion of a pair of hydrophobic helices α-4 and α-5 or α-6 and α-7 as helical hairpin and then the rest of domain I is rearranged. These helix pairs are expected to be closest to the plasma membrane and, thus, could drop down into the membrane while the remaining helices of domain I flatten out on the membrane surface like the ribs of an umbrella. The pore may be formed by oligomerization or by further insertion of other helices from domain I.

Both the models are proposed on theoretical grounds by comparison with membrane channels or channel forming toxins. The role of domain III, however, was largely ignored in the above proposed models. The results of site-directed mutagenesis studies of the loops between α-helices of domain I and the insensitivity of the inserted toxin to a variety of proteases and monoclonal antibodies (Chen *et al*, 1995; Hussain *et al*, 1996) argue against both the models.

Dean *et al* (1996) presented an alternative model that suggests penetration of the whole of domain I into the membrane. These models suggest the involvement of residues from at least α-helices 5 and 7 of domain I (Gazit and Shai, 1993a), β-sheet 17 of domain III (Chen *et al*, 1993; Wolfersberger *et al*, 1996) and a potential role of oligomers in toxin channel formation (Grochulski *et al*, 1995; Lee *et al*, 1996c).

Although the mechanism of action of insecticidal Cry toxins has been studied extensively, the understanding of the process by which Cry toxin-receptor binding leads to membrane pore formation remains fragmentary. The aggregation or oligomerization of toxin on the midgut epithelial membrane is required for pore formation but the mechanism of toxin aggregation is unknown (Aronson *et al*, 1999 and 2001). Recently, the role of lipid rafts in the mechanism of Cry toxins was suggested. Lipid rafts are detergent resistant microdomains (DRM) enriched in cholesterol, sphingolipids and GPI-anchored proteins. In mammals, lipid rafts play critical roles in signal transduction, sorting and trafficking of membrane proteins and also function as entry points for viruses, bacteria and toxins (Brown and London, 1998 and 2000; Hooper, 1999). Zhuang *et al* (2002) isolated and characterized lipid rafts from the midgut epithelium of 2 lepidopteran insects, *H. virescens* and *M. sexta*. Interestingly, they demonstrated the localization of Cry1A toxin-binding molecules, including APN and the toxin molecules themselves in lipid rafts isolated from these insects. For the first time they showed that lipid rafts play an important role in Cry1A toxin aggregation and pore formation.

Very recently, Bravo *et al* (2004) have proposed an updated model of mode of action of Cry1A toxins. In *M. sexta*, they have shown that the sequential binding of Cry1Ab toxin to its receptors, APN and cadherin-like protein (Bt-R1) depends on the oligomeric state of the toxin. The binding of the monomeric form of the toxin to Bt-R1 leads to a conformational change in the toxin such that a membrane-bound protease cleaves the helix a-1 of Domain I (Gomez *et al*, 2002b). The proteolytic processing triggers the formation of a tetrameric pre-pore structure that is insertion competent and has more affinity for the GPI-anchored APN. The oligomeric toxin then binds APN, which drives the pre-pore toxin complex into lipid rafts where it inserts and forms pore in the membrane that leads to ion leakage, cell lysis and insect death.

V. CONCLUSIONS

A plethora of information about biological activity and molecular mechanism of mode of action of insecticidal protein of *B. thuringiensis* has accumulated. Based on the safety of these proteins toward mammals and specifically humans, these proteins have been incorporated into several crops. Understanding their precise mode of action will benefit further application in protecting insect predation of crop and also help in designing novel insecticidal proteins.

REFERENCES

Agrawal, N., P. Malhotra, & R.K. Bhatnagar. 2002. Appl. Environ. Microbiol. 68: 4583-4592.

Aronson, A.I., D. Wu, & C. Zhang. 1995. J. Bacteriol. 177: 4059-4065.

Aronson, A.I. & Y. Shai. 2001. FEMS Microbiol. Lett. 195: 1-8.

Aronson, A.I., E.-S. Han, W.H. McGaughey, & D. Johnson. 1991. Appl. Environ. Microbiol. 57: 981-986.
Aronson, A.I., C. Geng, & L. Wu. 1999. Appl. Environ. Microbiol. 65: 2503-2507.
Banks, D.J., G. Hua, & M.J. Adang. 2003. Insect Biochem. Mol. Biol. 33: 499-508.
Blalock, J.E. 1995. Nat. Med. 1: 876-878.
Bravo, A. 1997. J. Bacteriol. 179: 2793-2801.
Bravo, A., I. Gomez, J. Conde, C. Munoz-Garay, J. Sanchez, R. Miranda, M. Zhuang, S.S. Gill, & M. Soberon. 2004. Biochim. Biophys. Acta 1667: 38-46.
Brown, D.A. & E. London. 1998. Annu. Rev. Cell dev. Biol. 14: 111-136.
Brown, D.A. & E. London. 2000. J. Biol. Chem. 275: 17221-17224.
Candas, M., B.R. Francis, N.B. Griko, E.G. Midboe, & L. A. Bulla, Jr. 2002. Biochemistry 41: 13717-13724.
Carlson, C.R., D. Caugant, A. & A.-B. Kolsto. 1994. Appl. Environ. Microbiol. 60: 1719-1725.
Carroll. J., M.G. Wolfersberger, & D.J. Ellar. 1997. J. Cell Sci. 110: 3099-3104.
Carroll, J. & D.J. Ellar. 1993. Eur. J. Biochem. 214: 771-778.
Carroll, J., J. Li, & D.J. Ellar. 1989. Biochem. J. 261: 99-105.
Chang, W.X.Z., L.J., Gahan, B.E. Tabashnik, & D.G. Heckel. 1999. Insect Mol. Biol. 8: 171-177.
Chen, J., M.R. Brown, G. Hua, & M.J. Adang. 2005. Cell Tissue Res. May18 (PMID: 15902495)
Chen, X.J., A. Curtiss, E. Alcantara, & D.H. Dean. 1995. J. Biol. Chem. 270: 6412-6419.
Chen, X.J., M.K. Lee, & D.H. Dean. 1993. Proc. Natl. Acad. Sci. USA 90: 9041-9045.
Choma, C.T. & H. Kaplan. 1990. Biochemistry 29: 10971-10977.
Choma, C.T. & H. Kaplan. 1992. J. Invertebr. Pathol. 59: 75-80.
Cioffi, M. 1979. Tissue Cell 11: 467-479.
Cowles, E.A., H. Yunowitz, J.-F. Charles, & S.S. Gill. 1995. Appl. Environ. Microbiol. 61: 2738-2744.
Crickmore, N., D.R. Zeigler, J. Feitelson, E. Schnepf, J. Van Rie, D. Lereclus, J. Baum, & D.H. Dean. 1998. Microbiol. Mol. Biol. Rev. 62: 807-813.
Dadd, R.H. 1975. J. Insect Physiol. 21: 1847-1853.
de Maagd, R.A., M.S.G. Kwa, H. van der Klei, T. Yamamoto, B. Schipper, J.M. Vlak, W.J. Stiekema, & D. Bosch. 1996. Appl. Environ. Microbiol. 62: 1537-1543.
Dean, D.H., F. Rajamohan, M.K. Lee, S.-J. Wu, X.-J. Chen, E. Alcantara, & S.R. Hussain. 1996. Gene 179: 111-117.
DeLeo, F., L. Yu, J.B. Burritt, L.R. Loetterle, C.B. Bond, & A.J. Jesaitis. 1995. Proc. Natl. Acad. Sci. USA 92: 7110-7114.
Demangel, C., R.C. Maroun, S. Rouyre, C. Bon, J. –C. Mazie, & V. Choumet. 2000. Eur. J. Biochem. 267: 2345-2353.
Denolf, P., K. Hendrickx, J. Van Damme, S. Jansens, M. Peferoen, D. Degheele, & J. Van Rie. 1997. Eur. J. Biochem. 248: 748-761.
Dorsch, J.A., M. Candas, N.B. Griko, W.S. Maaty, E.G. Midboe, R.K. Vadlamudi, & L.A. Bulla Jr. 2002. Insect Biochem. Mol. Biol. 32: 1025-1036.
Dow, J.A.T. 1984. Am. J. Physiol. 246: R633-R635.
Dow, J.A.T. 1986. Adv. Insect Physiol. 19: 187-238.
Du, C., P.A.W. Martin, & K.W. Nickerson. 1994. Appl. Environ. Microbiol. 60: 3847-3853.
English, L.H., T.L. Readdy, & A.E. Bastian. 1991. Insect Biochem. Mol. Biol. 21: 177-184.
Flannagan, R.D., C.G. Yu, J.P. Mathis, T.E. Meyer, X. Shi, H.A. Siqueira, & B.D. Siegfried. 2005. Insect Biochem. Mol. Biol. 35: 33-40.
Francis, B.R. & L.A., Bulla, Jr. 1997. Insect Biochem. Mol. Biol. 27: 541-550.
Gahan, L.J., F. Gould, & D.G. Heckel. 2001. Science 293: 857-860.
Garczynski, S.F. & M.J. Adang. 1995. Insect Biochem. Mol. Biol. 25: 409-415.
Garczynski, S.F., J.W. Crim, & M.J. Adang. 1991. Appl. Environ. Microbiol. 57: 2816-2820.

Garner, K.J., S. Hiremath, K. Lehtoma, & A.P. Valaitis. 1999. Insect Biochem. Mol. Biol. 29: 527-535.

Gazit, E. & Y. Shai. 1993a. Biochemistry 32: 3429-3436.

Gazit, E. & Y. Shai. 1993b. Biochemistry 32: 12363-12371.

Gazit, E. & Y. Shai. 1995. J. Biol. Chem. 270: 2571-2578.

Geiser, M., S. Schweitzer, & C. Grimm. 1986. Gene 48: 109-118.

Gill, M. & D. Ellar. 2002. Insect Mol. Biol. 11: 619-625.

Gomez, I., D.H. Dean, A. Bravo, and M. Soberon. 2003. Biochemistry 42: 10482-10489.

Gomez, I., J. Miranda-Rios, E. Rudino-Pinera, D.I. Oltean, S.S. Gill, A. Bravo, & M. Soberon. 2002a. J. Biol. Chem. 277: 30137-30143.

Gomez, I., J. Sanchez, R. Miranda, A. Bravo, & M. Soberon. 2002b. FEBS Lett. 513: 242-246.

Gomez, I., D.I. Oltean, S.S. Gill, A. Bravo, & M. Soberon. 2001. J. Biol. Chem. 276: 28906-28912.

Griko, N., M. Candas, X. Zhang, M. Junker, & L.A. Bulla, Jr. 2004. Biochemistry 43: 1393-1400.

Grochulski, P., L. Masson, S., Borisova, M.-C. Pusztai, J.-L. Schwartz, R. Brousseau, & M. Cygler. 1995. J. Mol. Biol. 254: 447-464.

Haider, M.Z. & D.J. Ellar. 1989. Biochim. Biophys. Acta 978: 216-222.

Harvey, W.R. & S. Nedergaard. 1964. Proc. Natl. Acad. Sci. USA 51: 757-765.

Harvey, W.R. & M.G. Wolfersberger. 1979. J. Exp. Biol. 83: 293-304.

Hendrickx, K., A. de Loof, & H. Mellaert. 1989. Comp. Biochem. Physiol. 95C: 241-245.

Hodgman, T.C. & D.J. Ellar. 1990. DNA sequence 1: 97-106.

Hofmann, C., P. Luthy, R. Hutter, & V. Pliska. 1988. Eur. J. Biochem. 173: 85-91.

Hofte, H. & H.R. Whiteley. 1989. Microbiol. Rev. 53: 242-255.

Hooper, N.M. 1999. Mol. Membr. Biol. 16: 145-156.

Hooper, N.M. 1994. FEBS Lett. 354: 1-6.

Hua, G., J.L. Jurat-Fuentes, & M.J. Adang. 2004a. J. Biol. Chem. 279: 28051-28056.

Hua, G., J.L. Jurat-Fuentes, & M.J. Adang. 2004b. Insect Biochem. Mol. Biol. 34: 193-202.

Hua, G., K. Tsukamoto, & H. Ikezawa. 1998. Comp. Biochem. Physiol. B Biochem. Mol. Biol. 121: 213-222.

Hussain, S.-R., A.I. Aronson, & D.H. Dean. 1996. Biochem. Biophys. Res. Commun. 226: 8-14.

Ihara, H., E. Kuroda, A. Wadano, & M. Himeno. 1993. Biosci. Biotechnol. Biochem. 57: 200-204.

Jenkins, J.L., M.K. Lee, A.P. Valaitis, A. Curtiss, & D.H. Dean. 2000. J. Biol. Chem. 275: 14423-14431.

Johnston, K.A., M.J. Lee, C. Brough, V.A. Hilder, A.M.R. Gatehouse, & J.A. Gatehouse. 1995. Insect Biochem. Mol. Biol. 25: 375-383.

Jurat-Fuentes, J.L., L.J. Gahan, F.L. Gould, D.G. Heckel, & M.J. Adang. 2004. Biochemistry 43: 14299-142305.

Jurat-Fuentes, J.L. & M.J. Adang. 2001. Appl. Environ. Microbiol. 67: 323-329.

Kenny, A.J., S.T. Stephenson, & A.J. Turner. 1987. Cell surface peptidases. In Mammalian Ectoenzymes (Kenny, A.J. and Turner, A.J. eds.) Vol. 14, pp. 169-210, Elsevier Science Publishing Co., inc., New York.

Knight, P.J., J. Carroll, & D.J. Ellar. 2004. Insect Biochem. Mol. Biol. 34: 101-112.

Knight, P.J.K., N. Crickmore, & D.J. Ellar. 1994. Mol. Microbiol. 11: 429-436.

Knowles, B.H. 1994. Adv. Insect Physiol. 24: 275-308.

Knowles, B.H. & J.A.T. Dow. 1993. BioEssays 15: 469-476.

Knowles, B.H. & D.J. Ellar. 1987. Biochim. Biophys. Acta 924: 509-518.

Knowles, B.H., P.J.K. Knight, & D.J. Ellar. 1991. Proc. Roy. Soc. Lond. B 245: 31-35.

Kumar, A.S.M. & A.I. Aronson. 1999. J. Bacteriol. 181: 6103-6107.

Lacey, L.A. & B.A. Federici. 1979. J. Invertebr. Pathol. 33: 171-182.

Lecadet, M.-M. & R. Dedonder. 1967. J. Invertebr. Pathol. 9: 310-321.

Lee, M.K., J.L. Jenkins, T.H. You, A. Curtiss, J.J. Son, M.J. Adang, & D.H. Dean. 2001. FEBS Lett. 497: 108-112.

Lee, M.K., F. Rajamohan, J.L. Jenkins, A. Curtiss, & D.H. Dean. 2000. Mol. Microbiol. 38: 289-298.

Lee, M.K., T.H. You, F.L. Gould, & D.H. Dean. 1999. Appl. Environ. Microbiol. 65: 4513-4520.

Lee, M.K., A. Curtiss, E. Alcantara, & D.H. Dean. 1996a. Appl. Environ. Microbiol. 62: 583-586.

Lee, M.K. & D.H. Dean. 1996b. Biochem. Biophys. Res. Commun. 220: 575-580.

Lee, M., T. You, B. Young, J. Cotrill, A. Valaitis, & D.H. Dean. 1996c. Appl. Environ. Microbiol. 62: 2845-2849.

Lee, M.K., F., Rajamohan, F. Gould, & D.H. Dean. 1995a. Appl. Environ. Microbiol. 61: 3836-3842.

Lee, M.K., B.A. Young, & D.H. Dean. 1995b. Biochem. Biophys. Res. Commun. 216: 306-312.

Lee, M.K., R.E. Milne, A.Z. Ge, & D.H. Dean. 1992. J. Biol. Chem. 267: 3115-3121.

Li, H., B. Oppert, R.A. Higgins, F. Huang, K.Y. Zhu, & L.L. Buschman. 2004. Insect Biochem. Mol. Biol. 34: 753-762.

Li, J., J. Carroll, & D.J. Ellar. 1991. Nature 353: 815-821.

Li, J., P.A. Koni, & D.J. Ellar. 1996. J. Mol. Biol. 257: 129-152.

Liebig, B., D.L. Stetson, & D.H. Dean. 1995. J. Insect Physiol. 41: 17-22.

Lorence, A., A., Darszon, C. Diaz, A. Lievano, R. Quintero, & A. Bravo. 1995. FEBS Lett. 360: 217-222.

Lu, H., F. Rajamohan, & D.H. Dean. 1994. J. Bacteriol. 176: 5554-5559.

Luo, K., Lu, Y.J. & M.J. Adang. 1996. Insect Biochem. Mol. Biol. 26: 783-791.

Luo, K., J.R. McLachlin, M.R. Brown, & M.J. Adang. 1999. Prot. Exp. Purif. 17: 113-122.

Luo, K., S. Sangadala, L. Masson, A. Mazza, R. Brousseau, & M.J. Adang. 1997. Insect Biochem. Mol. Biol. 27: 735-743.

Martin, G.G. & M.G. Wolfersberger. 1995. J. Exp. Biol. 198: 91-96.

Marzari, R., P. Edomi, R.K. Bhatnagar, S. Ahmad, A. Selvapandiyan, & A. Bradbury. 1997. FEBS Lett. 411: 27-31.

Masson, L., Y.-J, Lu, A. Mazza, R. Brousseau, & M.J. Adang. 1995. J. Biol. Chem. 270: 20309-20315.

McGaughey, W.H. & M.E. Whalon. 1992. Science 258: 1451-1455.

McNall, R.J. & M.J. Adang. 2003. Insect Biochem. Mol. Biol. 33: 999-1010.

Meng, J., M. Candas, T.P. Keeton, & L.A. Bulla, Jr. 2001. Protein Expr. Purif. 22: 141-147.

Midboe, E.G., M. Candas, L.A. Bulla, Jr. 2003. Comp. Biochem. Physiol. B Biochem. Mol. Biol. 135: 125-137.

Milne, R. & H. Kaplan. 1993. Insect Biochem. Mol. Biol. 23: 663-673.

Moffett, D.F. & Koch, A.R. 1988. J. Exp. Biol. 135: 39-49.

Morin, S., R.W. Biggs, M.S. Sisteron, L. Shriver, C. Ellers-Kirk, D. Higginson, *et al*, 2003. Proc. Natl. Acad. Sci. USA 100: 5004-5009.

Nagamatsu, Y., T. Koike, K. Sasaki, A. Yoshimoto, & Y. Furukawa. 1999. FEBS Lett. 460: 385-390.

Nagamatsu, Y., S. Toda, F. Yamaguchi, M. Ogo, M. Kogure, M. Nakamura, Y. Shibata, & T. Katsumoto. 1998. Biosci. Biotechnol. Biochem. 62: 718-726.

Novillo, C., P. Castanera, & F. Ortego. 1997. Arch. Insect Biochem. Physiol. 36: 181-201.

Oddou, P., H. Hartmann, & M. Geiser. 1991. Eur. J. Biochem. 202: 673-680.

Oddou, P., H. Hartmann, R. Radecke, & M. Geiser. 1993. Eur. J. Biochem. 212: 145-150.

Oltean, D.I., A.K. Pullikuth, H.-K. Lee, & S.S. Gill 1999. Appl. Environ. Microbiol. 65: 4760-4766.

Oppert, B., K.J. Kramer, D.E. Johnson, S.C. MacIntosh, & W.H. McGaughey. 1994. Biochem. Biophys. Res. Commun. 198: 940-947.

Peterson, A.M., G.J.P. Fernando, & M.A. Wells. 1995. Insect Biochem. Mol. Biol. 25: 765-774.

Peyronnet, O., B. Nieman, F. Genereux, V. Vachon, R. Laprade, & Schwartz, J.L. 2002. Biochim. Biophys. Acta. 1567: 113-122.

Pornwiroon, W., G. Katzenmeier, S. Panyim, & C. Angsuthanasombat. 2004. J. Biochem. Mol. Biol. 37: 292-297.

Rajagopal, R., N. Agrawal, A. Selvapandiyan, S. Sivakumar, S. Ahmad, & R.K. Bhatnagar. 2003. Biochem. J. 370: 971-978.

Rajagopal, R., S. Sivakumar, N. Agrawal, P. Malhotra, & R.K. Bhatnagar. 2002. J. Biol. Chem. 277: 46849-46851.

Rajamohan, F., O. Alzate, J.A. Cotrill, A. Curtiss, & D.H. Dean. 1996a. Proc. Natl. Acad. Sci. USA 93: 14338-14343.

Rajamohan, F., E. Alcantara, M.K. Lee, X.J. Chen, A. Curtiss, & D.H. Dean. 1995. J. Bacteriol. 177: 2276-2282.

Rajamohan, F., J.A. Cortill, F. Gould, & D.H. Dean. 1996b. J. Biol. Chem. 271: 2390-2396.

Rajamohan, F., S-R.A. Hussain, J.A. Cortill, F. Gould, & D.H. Dean. 1996c. J. Biol. Chem. 271: 25220-25226.

Rausell, C., C. Munoz-Garay, I. Gomez, Miranda-CassoLuengo, E. Rudino-Pinera, M. Soberon, & A. Bravo. 2004a. Biochemistry 43: 166-174.

Rausell, C., L. Pardo-Lopez, J. Sanchez, C. Munoz-Garay, C. Morera, M. Soberon, & A. Bravo. 2004b. J. Biol. Chem. 279: 55168-55175.

Sangadala, S., P. Azadi, R. Carlson, & M.J. Adang. 2001. Insect Biochem. Mol. Biol. 32: 97-107.

Sangadala, S., F.S. Walters, L.H. English, & M.J. Adang. 1994. J. Biol. Chem. 269: 10088-10092.

Sankaranarayanan, R.K., K. Sekar, R. Banerjee, V. Sharma, A. Surolia, & M. Vijayan. 1996. Nat. Struct. Biol. 3: 596-603.

Schnepf, E., N. Crickmore, J. Van Rie, D. Lereclus, J. Baum, J. Feitelson, D.R. Zeigler, & D.H. Dean. 1998. Microbiol. Mol. Biol. Rev. 62: 775-806.

Schwartz, J.-L., Y.J. Lu, P. Soehnlein, R. Brousseau, L. Masson, R. Laprade, & M.J. Adang. 1997. FEBS Lett. 410: 397-402.

Schwartz, J.-L., L. Garneau, D. Savaria, L. Masson, R. Brousseau, & E. Rousseau. 1993. J. Membr. Biol. 132: 53-62.

Schwartz, J.-L., L. Garneau, L. Masson, & R. Brousseau. 1991. Biochim. Biophys. Acta 1065: 250-260.

Simpson, R.M. & R.D. Newcomb. 2000. Insect Biochem. Mol. Biol. 30: 1069-1078.

Slatin, S.L., C.K. Abrams, & L. English. 1990. Biochem. Biophys. Res. Commun. 169: 765-772.

Smith, G.P. & D.J. Ellar. 1994. Biochem. J. 302: 611-616.

Soberon, M., R.V. Perez, M.E. Nunez-Valdez, A. Lorence, I. Gomez, J. Sanchez, & A. Bravo. 2000. FEMS Microbiol. Lett. 191: 221-225.

Sutcliffe, D.W. 1963. Comp. Biochem. Physiol. 9: 121-135.

Szardenings, M., S. Tornoth, F. Mutulis, R. Muceniece, K. Keinanen, A. Kuusinen, & J.E.S. Wikberg. 1997. J. Biol. Chem. 272: 27943-27948.

Takesue, Y., K. Yokota, K. Miyajima, R. Taguchi, & H. Ikezawa. 1989. J. Biochem. 105: 998-1001.

Takesue, S., K. Yokota, S. Miyajima, R. Taguchi, H. Ikezawa, & Y. Takesue. 1992. Comp. Biochem. Physiol. 102B: 7-11.

Taylor, A. 1993. FASEB J. 7: 290-298.

Thompson, M.A., H.E. Schnepf, & J.S. Feitelson. 1995. In genetic engineering principles and methods, (Setlow, J.K., ed.) 17 pp. 99-117 Plenum press, New York, NY.

Tigue, N.J., J. Jacoby, & D.J. Ellar. 2001. Appl. Environ Microbiol. 67: 5715-5720.

Tojo, A. & K. Aizawa. 1983. Appl. Environ. Microbiol. 45: 576-580.

Tsuda, Y., F. Nakatani, K. Hashimoto, S. Ikawa, C. Matsuura, T. Fukada, K. Sugimoto, & M. Himeno. 2003. Biochem. J. 369: 697-703.

Vadlamudi, R.K., B. Weber, I. Ji, T.H. Ji, & L.A. Bulla. 1995. J Biol. Chem. 270: 5490-5494.

Vadlamudi, R.K., T.H. Ji, & L.A. Bulla. 1993. J. Biol. Chem. 268: 12334-12340.

Valaitis, A.P., J.L. Jenkins, M.K. Lee, D.H. Dean, & K.J. Garner. 2001. Arch. Insect Biochem. Physiol. 46: 186-200.

Van Rie, J., S. Jansens, H. Hofte, D. Degheele, & H. Van Mellaert. 1989. Eur. J. Biochem. 186: 239-247.

Wang, P., X. Zhang, & J. Zhang. 2005. Insect Biochem. Mol. Biol. 35: 611-620.

Weltens, R., M. Peferoen, P. Steels, & E. Van Kerkhove. 1992. Arch. Int. Physiol. Biochim. 100: 6.

Widner, W.R. & H.R. Whiteley. 1989. J. Bacteriol. 171: 965-974.

Wolfersberger, M.G., X.J. Chen, & D.H. Dean. 1996. Appl. Environ. Microbiol. 62: 279-282.

Wolfersberger, M.G. 1990. Experimentia 46: 475-477.

Wolfersberger, M.G. 1989. Arch. Insect Biochem. Phys. 12: 267-277.

Wolfersberger, M.G., P. Luthy, A. Maurer, P. Parenti, V.F. Saachi, B. Giordana, & G.M. Hanozet. 1987. Comp. Biochem. Physiol. A86: 301-308.

Wu, D. & A.I. Aronson. 1992. J. Biol. Chem. 267: 2311-2317.

Wu, S.J. & D.H. Dean. 1996. J. Mol. Biol. 225: 628-640.

Xie, R., M. Zhuang, L.S. Ross, I. Gomez, D.I. Oltean, A. Bravo, M. Soberon, & S.S. Gill. 2005. J. Biol. Chem. 280: 8416-8425.

Xu, X., L. Yu, & Y. Wu. 2005. Appl. Environ. Microbiol. 71: 948-954.

Yaoi, K., K. Nakanishi, T. Kadotani, M. Imamura, N. Koizumi, H. Iwahana, & R. Sato. 1999. Biochim. Biophys. Acta 1444: 131-137.

Zhuang, M., D.I. Oltean, I. Gomez, A.K. Pullikuth, M. Soberon, A. Bravo, & Gill, S. S. 2002. J. Biol. Chem. 277: 13863-13872.

27

Enhancement of Wheat Performance with Plant Growth Promoting Bacteria in Different Soils

DILFUZA EGAMBERDIYEVA
Center of Agro ecology, University of Agriculture,
University str. 1, 700140 Tashkent, Uzbekistan
e-mail: dilfuza_egamberdiyeva@yahoo.com

ABSTRACT

Plant-growth-promoting bacteria isolated from the soil in different climatic regions of Uzbekistan, Germany and Finland were analysed for plant-growth-promoting effects on wheat on different soils. The investigations were carried out in pot experiments with typical nutrient rich loamy sand and sandy loam soils from Müncheberg, Germany, nutrient poor, calcareous Calcisol soil from Tashkent, Uzbekistan and sandy clay soil in Helsinki, Finland. The soil types were found to influence growth-promoting effects. Inoculation with bacterial strains was found to significantly increase the root and shoot growth of wheat in nutrient-poor calcisol than in nutrient-rich loamy sand. Bacterial inoculation also resulted in significantly higher values for N, P, and K contents of plant components. Many of the bacterial strains producing the phytohormone auxin, were able to fix N_2 and reacting antagonistically to soil-borne plant pathogen *Fusarium culmorum.*

Key Words: plant-growth-promoting bacteria, soil type, wheat

I. INTRODUCTION

Beneficial effects of microorganisms have most often been based on increased plant growth, faster seed germination, and better seedling emergence (Lazarovits and Nowak, 1996). Beneficial free-living soil bacteria are usually referred to as plant-growth-promoting rhizobacteria or PGPR (Kloepper, 1980). Different plant-growth-promoting rhizosphere bacteria, *Azospirillum, Bacillus, Pseudomonas, Enterobacter* have been used for their beneficial effects on plant growth (Chanway; 1991b; Kloepper *et al,* 1992; Höflich *et al,* 1994; Egamberdiyeva and Höflich, 2001). The mechanisms of plant growth stimulation by associative bacteria are mobilization of nutrients, stimulation of root growth by production of phytohormones and antagonism against soil borne plant pathogens (Bothe *et al*, 1992; Sarwar *et al*, 1992; Haahtela, 1990; Kloepper *et al*, 1988; Höflich *et al*, 1994). Plant-growth-promoting effects are influenced by soil type. Several studies clearly showed the effect of plant growth promoting bacteria on plant growth of different crops at different climates, soils and temperatures (Ruppel, 1987; Sarig *et al*, 1990; Döbereiner, 1992; Okon, 1991; Bilal *et al*, 1993; Höflich and Kühn, 1996; Javed and Arshad, 1997; Bashan and Holguin, 1997).

Knowledge of the plant-growth-promoting effects at different soils may be important for successful root inoculation. However, there are few reports to compare the effects of plant-growth-promoting bacteria on wheat in different climatic regions and soils. The major objectives of our research were to investigate the effect of plant-growth-promoting bacteria on the growth of wheat at different soils.

II. MATERIALS AND METHODS

Sites and Soil

Sites, used in this study are located in different climatic regions i.e., Finland (cold), Germany (temperate) and Uzbekistan (semi-arid). The soils were collected for pot experiments from nutrient poor calcareous Calcisol soil, near Tashkent, Uzbekistan; nutrient richer loamy sand and sandy loam near Muncheberg, Dedelow, Germany and sandy clay near Helsinki, Finland Table 27.1). Soils were characterised according to World Base soil Resources (Decker *et al*, 1998). The chemical and physical properties of soil are presented in Table 27.1. From loamy sand, sandy loam and Calcisol the total carbon content, C, was identified by elementary analysis while total nitrogen, N, content was determined by the Kjeldahl method (Egner, 1932). Soil pH-value was measured by means of electrometer. Soil particle distribution was determined by using sodium phosphate. From sandy clay soil the total C and N concentrations were determined by using an autoanalyzer (leco CHN-900). Soil texture was determined by Elonen (1971) pipette method.

Table 27.1: Soil chemical properties, and soil particle distribution at 0-30 cm soil layer

Site	*Type*	C_{tot}	N_{tot}	*pH*	*soil particle size, mm*		
		mg	*(100g)* $^{-1}$		*2–0.2%*	*0.2–0.02%*	*<0.02%*
Muencheberg	Loamy sand	700	60	6.9	7.6	79.8	12.6
Dedelow	Sandy loam	800	80	6.1	4.3	70.2	25.5
Helsinki	Sandy clay	410	23	5.7	39.0	24.0	37.0
Tashkent	Calcisol	200	6	7.8	2.2	54.5	43.3

Winter wheat (*Triticum aestivum*) cv. Bussard, (Germany), cv. Ziklon, (Uzbekistan) and spring wheat cv. Reno (Finland) were used as the tests plants for the inoculation experiments. Wheat seeds were obtained from the Centre for Agricultural Landscape and Land Use Research, Müncheberg, Germany; Tashkent University of Agriculture and Agricultural Department of Helsinki University, Finland.

Microorganisms

The soil bacterial strains from following plants: *Pseudomonas alcaligenes* PsA15 and *Mycobacterium phlei* MbP18 were isolated from of the root zone soil of wheat, *Bacillus polymyxa* BcP26 and

Arthrobacter simplex ArS50 from the root zone soil of cotton grown in Calcisol soil (Tashkent, Uzbekistan). Rhizosphere bacteria *Rahnella aguatilis* 6 was isolated from wheat, *Agrobacterium rhizogenes* A1A4 and *P. putida* A1A2 from raps and phyllosphere bacteria *Pantoea agglomerans* P294/2 from grass grown in loamy sand (Germany, Muencheberg). Humus bacteria *Paenibacillus borealis* KK19, *B. megaterium* KMNLA1, *P. fluorescens* KKCK11 and *P. fluorescens* HNMK14 were isolated from Norway spruce stands and identified (Elo *et al*, 2000).

For isolation of soil bacteria, 1 g soil was removed from cotton and wheat roots grown in field site and were shaken with 9 ml sterile water with 40% tetramethylthiuramdisulfide (TMTD) solution for 30 minutes. Rhizosphere and phyllosphere bacteria were isolated from wheat, raps, grass plants grown in pots in greenhouse. After four weeks, adhering soil was removed from plant roots and shoots were separated. For isolation of rhizosphere bacteria, 1 g of washed roots were macerated and shaken with 9 ml sterile water. For isolation of phyllosphere bacteria 1 g leaves were macerated and shaken with 9 ml sterile water and 10^{-1} to 10^{-7} serial dilutions were made. From each dilution 0.1 ml of suspension was spread on Hirte agar plates (Hirte, 1961). After an incubation time of four days at 28°C the bacterial strains were isolated from the plates and identified.

Microbiological Analysis

The identification of bacteria was done by using a combination of conventional methods like morphological, physiological and chemotaxonomic characterisation (Holt *et al*, 1994) and the use of commercial identification systems like Biolog (GN, GP) from the Biolog Inc. and API 20E or API 20NE from bioMérieux, (Behrendt *et al*, 1997). The auxin production was tested using Salkowsky´s reagent (Sarwar *et al*, 1992) and by rapid in situ assay method (Bric *et al*, 1991). The strains were tested for antagonistic activity. *Fusarium culmorum* was used as an indicator strain for antagonistic bacteria. Bacterial isolates were tested on Hirte agar plates (Hirte, 1961) and peptone dextrose agar (PDA). Small blocks of peptone dextrose agar culture containing the fungus were placed on the test plate. Bacteria isolates were streaked on the test plates perpendicular to the fungus. Plates were incubated at 28°C until the fungi had grown over the control plates without bacteria. Antifungal activity was recorded by taking the width of the zone of growth inhibition between the fungus and the organism tested.

Plant Growth and Inoculation in Pots

The study of the effect of isolated strains on plant growth and nutrient uptake was carried out in pot experiments using a nutrient-rich loamy sand and a sandy loam soil originating from Muencheberg and Dedelov (Germany) and a nutrient-poor calcareous calcisol soil from Tashkent (Uzbekistan) and sandy clay soil from Helsinki (Finland).

The inoculation treatments were set-up in a randomised design with six replicates. The day before sowing, pots (10 cm diameter and 13 cm deep) were filled with 350 g soil. Four seeds of wheat were sown per pot. After the emergence of the seeds, plants were thinned to two per pot. The bacteria were grown in glycerol-peptone, Lurea, nutrient medium and pseudomonads in King B medium. Tubes were placed on a rotary shaker (120 rpm; 23°C) and agitated for three days. Seedlings of these plants were inoculated with 1 ml of the bacterial suspension with an inoculum density of ca. $10^6 - 10^7$ cfu/ml. Control seeds were given 1 ml medium. Plants were grown in pots for four weeks under open natural weather conditions at a temperature of 34°C to 36°C during the day and 22°C to 26°C at night in summer, (July, Uzbekistan), and 20°C to 24°C during the day and 14°C to 16°C at night (June, Germany) and 20-24°C during the day and 12-15°C at night (July, Finland). The soil was moistened with water and maintained at 60% of its moisture-holding capacity (MHC). Four weeks after germination, shoots and roots were separated and dried overnight at 105°C before determining the root and shoot dry weight. The growth was determined by recording the root and shoot dry matter in 6-leaves-stage and N, P, K content of plants.

Statistical Analysis

The data were analysed with a two-way ANOVA and Student-Newman-Keuls test for testing the significant differences ($p<0.05$) of main effects.

III. RESULTS

Growth Promotion of Wheat by Bacterial Inoculants

Bacterial inoculation affected the early plant growth and the nutrient content of wheat grown in different soils. The inoculation of winter wheat grown in the loamy sand with bacterial strains *Rahnella aguatilis* 6, *A. rhizogenes* A1A4, *P. putida* A1A2 and *P. agglomerans* P294/2 significantly ($p<0.05$) increased the shoot and root dry weights from 13 to 22% as compared to the control. This growth-promoting effect was higher in the loamy sand than in the sandy loam (Table 27.2).

Table 27.2: The Influence of bacterial strains on root and shoot dry matter of winter wheat in different soils (pot experiment, control=100)

Bacterial Strains	*Loamy sand*		*Sandy loam*	
	Shoot	*Root*	*Shoot*	*Root*
Control	100 (0,1266)[1]	100 (0,0734)[1]	100 (0,1263)[1]	100 (0,0864)[1]
Rahnella aguatilis 6	122*	109	114*	108
Agrobacterium rhizogenes A1A4	113*	104	112*	112
Pseudomonas putida A1A2	117*	113*	113*	123
Pantoea agglomerans P294/2	115*	111	104	100
LSD α = 0.05	13	20	12	25

[1] g/pot

*Significantly different from the control for $P<0.05$

The bacterial strains *Penibacillus borealis* KK19, *B. megaterium* KMNL1A, *P. fluorescens* KKCK 11 and *P. fluorescens* HNMK14 (originating from Finland) showed more higher effect on wheat in sandy clay soil. The most effective shoot and root dry weight-promoting isolate was *P. fluorescens* KKCK11, which generated 58% increase over the control (Fig. 27.1).

The inoculation of wheat with bacterial strains *P. alcaligenes* PsA15, *M. phlei* MbP18 *and* *B. polymyxa* BcP26 from semi-arid region Uzbekistan significantly ($p<0.05$) increased the shoot

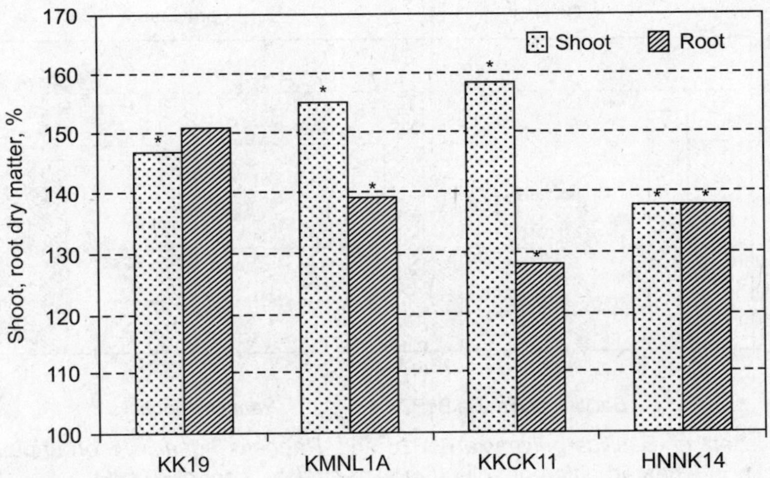

Fig. 27.1: The effect of bacterial strains on shoot, root dry matter of wheat in sandy clay (pot experiments, control=100%)

and root dry weights in Calcisol soil from 15 to 38% as compared to the control (Fig. 27.2). The increase in biomass translated into significantly higher N, P and K content. The bacterial strains significantly increased the total N, P and K contents 34–58% of wheat in calcisol (Fig. 27.3).

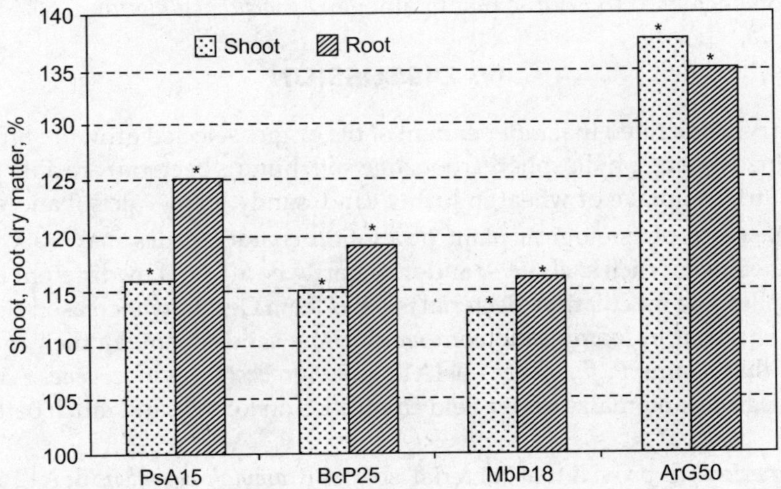

Fig. 27.2: The effect of bacterial strains on shoot, root dry matter of wheat in Calcisol soil (pot experiments, control=100%)

Rahnella agauatilis 6 (originating from Germany) significantly increased the P- and K-uptake of winter wheat grown at the loamy sand, but there was no effect on the percentage N of shoot material (Fig. 27.3).

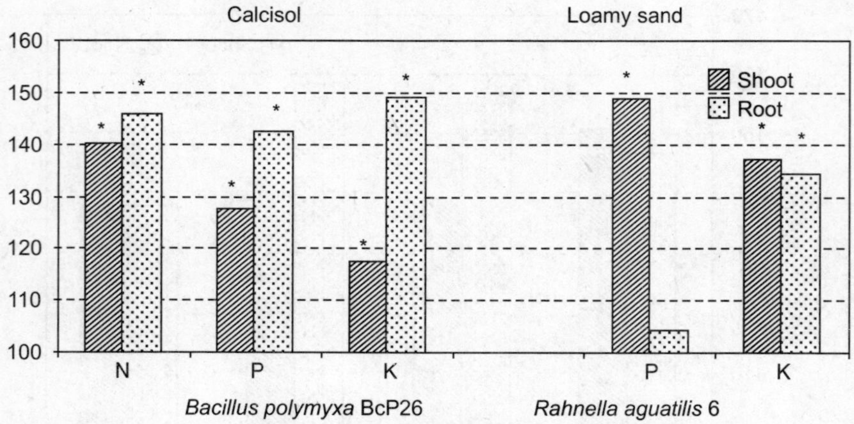

Fig. 27.3: The effect of *Bacillus polymyxa* BcP26 and *Rahnella aguatilis* 6 on shoot, root nutient uptake of wheat in different soils (pot experiments, control=100%)

Inoculation experiments with wheat at different soils showed that plant-growth-promoting bacterial isolates were more effective in sandy clay and calcisol soil than on loamy sand and sandy loam.

Plant-growth-promoting potential of bacterial strains from different soils were analysed by testing their capability to produce phytohormone auxin, cell wall hydrolysed enzymes, inhibit plant pathogenic fungi. Almost all bacterial isolates produced phytohormone auxin, which reacted antagonistically to soil-borne plant pathogen *Fusarium culmorum*.

IV. DISCUSSION

This study has demonstrated that independent of the origin, selected growth-stimulating bacterial isolates (rhizosphere, phyllosphere, root zone soil, humus bacteria) are able to increase the growth and nutrient uptake of wheat in loamy sand, sandy loam, calcisol and sandy clay.

The importance of physiological plant promotion characteristics may vary with soil and weather parameters (Höflich *et al*, 1994) and may partly be affected by different growth-stimulation effects following inoculation. Bacterial isolates from Germany increased significantly the plant growth of wheat in loamy sand compared to the sandy loam. Höflich *et al*, (1997) also demonstrated that strains *P. fluorescens* PsIA12 and *Agrobacterium rhizogenes* A1A4 promoted the growth of legume and maize under field conditions on loamy sand much better than sandy loam.

Also our experiments proved that bacterial isolates *B. amyloliquefaciens* BcA12 and *Mycoplana bullata* MpB46 from semi-arid region had a greater effect on wheat in nutrient-poor calcisol soil than in nutrient-rich loamy sand (Egamberdiyeva and Höflich, 2001).

Defreitas and Germida (1992) also demonstrated that in low fertility Asquith soil, *Pseudomonas* strains significantly enhanced early plant growth. Paula *et al*, (1992) reported that the magnitude of the plant response to any microbial inoculation can be greatly affected by the nutrient content of soil. Inoculation of plants with bacteria only marginally increased yields when tested under ideal climatic conditions. The greatest benefits occurred when crops encountered stressful conditions for prolonged periods (Lazarovits and Norwak, 1997). Non-treated plants by comparison performed poorly under such conditions. In our study, statistically enhancement in wheat growth promotion and nutrient uptake was observed at nutrient poorer soil calcisol and sandy clay rather than loamy sand and sandy loam.

Increased nutrient uptake by plants inoculated with bacteria has been attributed to the production of plant growth regulators, like phytohormone auxin by the bacteria at the root interface, which stimulated root development and resulted in better absorption of water and nutrients from the soil (Azc'on *et al*, 1975; Haahtela *et al*, 1990; Höflich *et al*, 1996; Zimmer *et al*, 1995). In our experiments auxin was detected in all bacterial suspensions. The bacterial strains reacted antagonistically to plant pathogenic fungi.

In summary, the final results of the plant growth promotion in our experiments show that plant growth-promoting bacteria can play an essential role in helping the plant establish and grow. From the results, we conclude that growth-promoting bacteria from different climates (cold, temperate, and semi-arid), the rhizosphere, phyllosphere, from the soil of different crop root zones as well as from humus may be isolated. Effective bacteria-plant partners must be selected through vegetation trials with consideration to the specific ecological conditions like soil types.

REFERENCES

Azcon, R., J.M. Barea & D.S. Heiman. 1975: Utilization of rock phosphate in alkaline soils by plants inoculated with mycorrhizal fungi and phosphate solubilizing bacteria. Soil Biology and Biochemistry 8: 64.

Bashan, Y., & G. Holguin. 1997. *Azospirillum*-plant relationships: environmental and physiological advances (1990-1996). Canadian Journal of Microbiology 43: 103-121.

Behrendt, U., Th. Müller, & W. Seyfarth. 1997: The influence of extensification in grassland management on the populations of microorganisms in the phyllosphere of grasses. – Microbiological Research 152: 75-85.

Bothe, H., H. Körsgen, T. Lehmacher, & L. Hundeshagen. 1992: Differential effects of *Azospirillum*, auxin and combined nitrogen on growth of the roots of wheat. Symbiosis 13: 167-179.

Bilal, R., G. Rasul, & K.A. Malik. 1993: Attachment, colonization and proliferation of *Azospirllium brasiliense* and *Enterobacter* spp. on root surface of grasses. World Journal of Microbiology and Biotechnology 9: 63-69.

Bric, J.M., R.M. Bostock, & S.E. Silverstone. 1991: Rapid in situ assay for indoleacetic acid production by bacteria immobilized on a nitrocellulose membrane. Appl. Environ. Microbiology, 57: 535-538.

Defreitas, J.R. & J.J. Germida. 1992A: Growth promotion of winter wheat fluorescent *Pseudomonas* under growth chamber conditions. Soil Biology and Biochemistry 24: 1127-1135.

Döbereiner, J. 1992: Recent changes in concepts of Plant Bacteria interactions, endophytic N_2-fixing bacteria.-Cien. Cult. 44: 310-313.

Egner, H. 1932: Meddelande fran Centralanstalten för Försöksvasndet pa Jordburksomradt: Nr. 425; Avdelningen för lantbrukskemie: Nr.51 (Stockholm).

Elonen, P. 1971: Particle-size analysis of soil. Acta Agric. Fennica 122: 1-122.

Elo S., L. Mannuksela, M. Salkinoja-Salonen, & A. Smolander, K. Haahtela. 2000. Humus bacteria of Norway spruce stands: plant growth-promoting properties and birch, red fescue and alder colonizing capacity. Fems Microbiology Ecology, 31: 143-152.

Haahtela, K., R. Rönkko, T. Laakso, P.M. Williams, & T.K. Korhonen. 1990: Root associated *Enterobacter* and *Klebsiella* in *Poa pratensis*. Characterization of an ion-scavenging system and a substance stimulating root hair production. Molecular Plant Microbial Interactions 3: 358-363.

Holt G.J., N.R. Krieg, P.H, Sneath, J.T. Staley & S.T. Williams. 1994. Bergey's Manual of Determinative Bacteriology, 9th edition, The Williams and Wilkins Co. Baltimore.

Höflich, G., W. Wiehe, & G. Kühn. 1994: Plant growth stimulation with symbiotic and associative rhizosphere microorganisms. Experientia 50: 897-905.

Höflich, G., & Kühn, G. 1996: Förderung das Wachstums und der Nährstoffaufnahme bei kurziferen Öl- und Zwischenfruhten durch inokulierte Rhizospherenmikroorganismen. Zetschrift für Pflanzenernährung und Bodenkunde 159: 575-578.

Höflich, G., E. Tappe, G. Kühn, & W. Wiehe. 1997: Einfluß associativer Rhizosphärenbakterien auf die Nährstoffaufnahme und den Ertrag von Mais. Archive Acker und Pflanzen Boden. 41: 323-333.

Hirte, W.F. 1961: Glyzerin-Pepton-Agar, ein vorteilhafter Nährboden für bodenbakteriologische Arbeiten. Zentralblatt Bakteriology II.114: 141-146.

Javed, M., & M. Arshad. 1997: Growth promotion of two wheat cultivars by plant growth promoting rhizobacteria. Pakistan Journal of Botany 29: 243-248.

Kloepper, J.W., J. Leong, M. Teintze, & M.N. Schroth. 1980: Enhanced plant growth by siderophore produced by plant growth-promoting rhizobacteria. Nature (London) 286: 885-886.

Kloepper, J.W., D.J. Hume, F.M. Schner, C. Singleton, B. Tipping, M. Laliberte, K. Frauley, T. Kutchaw, C. Simonson, R. Lifshitz, I. Zaleska, & L. Lee., 1988: Plant growth promoting rhizobacteria on canola. Plant Diseases 72: 42-46.

Kloepper, J.W., & C.J. Beauchamp. 1992: A review of issues related to measuring of plant roots by bacteria. Canadian Journal of Microbiology 38: 1219-1232.

Lazarovits, G., & J. Norwak. 1997. Rhizobacteria for Improvement of Plant growth and Establishment. Horticultural Science 32: 188-192.

Lifshitz, R., J.W. Kloepper, M. Kozlowski, C. Simonson, J. Carlso, E.M. Tipping, & I. Zaleska. 1987. Growth promotion of canola (rapeseed) seedlings by a strain of *Pseudomonas putida* under gnotobiotic conditions. Canadian Journal of Microbiology 33: 390.

Okon, Y. 1991: Associative symbioses. International symposium Congress, Jerusalem, Israel, 17-22 November, Abstracts, 59.

Paula M.A., S. Urquiaga, I.O. Siqueira, & J. Döbereiner. 1992: Synergistic effects of vesicular-arbuscular mycorrhizal fungi and diazotrofic bacteria on nutrition and growth of sweet potato (*Ipomoea batatas*). Biology and Fertility of Soils 14: 61-66.

Ruppel S. 1987. Isolation diazotropher Bakterien aus der Rhizosphäre von Winterweizen und Charakterisierung ihrer Leistungsfähigkeit. Diss. Akademie der Landwirtschafts-wissenschaften der DDR, Müncheberg.

Sarig, S., Y. Okon, & A. Blum. 1990: Promoting of leaf area development and yield in Sorghumbicolor inoculated with *Azospirillum brasilense*. Symbiosis 9: 235-245.

Sarwar, M., D.A. Arshad, W.T. Martens, & J.R. Frankenberger. 1992: Tryptophane dependent biosynthesis of auxins in soil. Plant and Soil 147: 207-215.

Zimmer, W., K. Kloos, B. Hundeshagen, E. Neiderau, & H. Bothe. 1995: Auxin biosynthesis and denitrification in plant growth promotion bacteria. - In: Fendrik, J., De Gallo, J. Vandeleyden, & D. De Zamoroczy, (Eds.). *Azospirillum* VI and related microorganisms. Series G: Ecological Science 37: 120-141.

28

Phytoremediation: A Cost-Effective Approach for the Removal of Soil Contaminants

I.S. DUA*, DAVOOD ERADATMAND ASLI, RITU SARSOHA and VIKAS KUMAR
Department of Botany Punjab University, Chandigarh-160 014

ABSTRACT

Remediation of sites contaminated with toxic metals is particularly challenging. Unlike organic compounds, metals cannot be degraded and the clean up usually require their removal. However, this energy-intensive approach can be prohibitively expensive. In addition, the metal-removing process often employs stringent physicochemical agents which can dramatically inhibit soil fertility with subsequent negative impacts on the ecosystem. Phytoremediation has been proposed as a cost-effective, environmental-friendly alternative technology. A great deal of research indicates that plants have the genetic potential to remove many toxic metals from the soil. Despite this potential, phytoremediation is yet to become a commercially available technology. Progress in the field is hindered by a lack of understanding of complex interactions in the rhizosphere and plant-based mechanisms which allow metal translocation and accumulation in plants.

I. INTRODUCTION

At many hazardous waste sites requiring clean up, the contaminated soil, groundwater, and/or wastewater contain a mixture of contaminant types, often at widely varying concentrations. These may include salts, organics, heavy metals, trace elements and radioactive compounds. The simultaneous clean-up of multiple, mixed contaminants using conventional, chemical and thermal methods is technically difficult and expensive, and further these methods also destroy the biotic component of soils.

*Corresponding author: sd47@yahoo.com

Phytoremediation, an emerging clean up technology for contaminated soils, groundwater and wastewater is both low-tech and low-cost. We define phytoremediation as the engineered use of green plants, including grasses, forbs and woody species to remove, contain or render harmless such environmental contaminants as heavy metals, trace elements, organic compounds and radioactive compounds in soil and water. This definition includes all plant-influenced biological, chemical and physical processes that aid in the uptake, sequestration, degradation, and metabolism of contaminants, either by plants or by free-living organisms that constitutes the plant rhizosphere. Phytoremediation takes advantage of the unique and selective uptake capabilities of plant root systems, together with the translocation, bio-accumulation, and contaminant storage abilities of the entire plant body. Plant based soil remediation systems can be viewed as biological solar driven pump and treat systems with an extensive self-extending uptake network that enhances the below ground ecosystem for subsequent productive use.

Examples of simpler phytoremediation systems that have been used for years are constructed or engineered wetlands. The phytoremediation of a site contaminated with heavy metals or radionuclides involves "farming" the soil with selective plants to "biomine" the inorganic contaminants, which are concentrated in the plant biomass (Salt *et al*, 1995a). For soils contaminated with toxic organics, the approach is similar, but the plant may take up or assist in the degradation of the organic. Several sequential crops of hyper accumulating plants could possibly reduce soil concentrations of toxic inorganics or organics to the extent that residual concentrations would be environmentally acceptable and no longer considered hazardous. The potential also exists for degrading the hazardous organic component of mixed contamination, thus reducing the waste (which may be sequestered in plant biomass) to a more manageable radioactive one.

Though it is not a panacea, phytoremediation is well suited for applications in low-permeability soils, where most currently used technologies have a low degree of feasibility or success, as well as in combination with more conventional clean up technologies (electromigration, foam migration etc.). In appropriate situations, phytoremediation can be an alternative to the much harsher remediation technologies of incineration, thermal vaporization, solvent washing, or other soil washing techniques, which essentially destroy the biological component of the soil and can drastically alter its chemical and physical characteristics as well, creating a relatively non-viable solid waste. Phytoremediation actually benefits the soil, leaving an improved, functional, soil ecosystem at costs estimated at approximately one-tenth of those currently adopted technologies.

To exploit plant potential to remediate soil and water contaminated with a variety of compounds, several technological subsets have been proposed. Phytoextraction is the use of higher plants to remove inorganic contaminants, primarily metals, from polluted soil. In this approach, plants capable of accumulating high levels of metals are grown in contaminated soil. At maturity metal-enriched above ground biomass is harvested and a fraction of soil contaminated with metal is removed. Plants have a natural propensity to take up metals. Like Cu, Co, Fe, Mo, Mn, Ni and Zn which are essential mineral nutrients. Others, however, such as Cd and Pb have known physiological activity. Perhaps not surprisingly phytoremediation as an environmental clean up technology was initially proposed for the remediation of metal-contaminated soil (Utsunamyia, 1980). The identification of metal hyperaccumulators, plants capable of accumulating extraordinarily high metal level, demonstrated that plants have the genetic potential to clean up contaminated soil. As early as the 19th century Baumann (1885), identified plants capable of accumulating uncommonly high Zn levels. One decade later, Minguzzi and Vergnano (1948) identified plants capable of hyperaccumulating up to 1% Ni in shoots.

Following the identification of these and other hyperaccumulator species great deal of research has been conducted to elucidate the physiology and biochemistry of metal hyperaccumulation in plants. Significant results have been obtained and the understanding of metal accumulating mechanisms substantially advanced. However, a better understanding of the biological processes is needed if phytoremediation is to become a reliable, commercially available technology.

The success of phytoremediation, as an environmental clean up technology depends on several factors including the extent of soil contamination, metal availability for uptake into roots (bioavailability) and plant ability to intercept, absorb and accumulate metals in shoots (Ernst, 1996). Ultimately, the potential for phytoremediation depends on the interaction between soils metal and plant. The complexity of this interaction, controlled by climatic conditions argues against generic and in favour of a site-specific phytoremediating approach. This underlines the importance of understanding the mechanisms and processes that govern metal uptake and accumulation in plants.

II. BACKGROUND AND APPLICATIONS

Phytoremediation is the name given to a set of technologies that use plants to clean contaminated sites. Many techniques and applications have been called phytoremediation, possibly leading to confusion .This document uses the term phytoremediation to refer to a set of plant-contaminant interactions, and not to any specific application. Many of the phytoremediation techniques involve applying information that has been known for years in agriculture, silviculture and horticulture to environmental problems.

The term phytoremeditation (phyto= plant and remediation= correct evil) is relatively new. Basic information for what is now called phytoremediation comes from a variety of research areas including constructed wetlands, oil spills, and agricultural plant accumulation of heavy metals. The term has been used widely since its inception, with a variety of specific meanings. In this document, phytoremediation is used to mean the overall idea of using plant-based environmental technologies, not any specific application.

Research efforts into remediation can be roughly categorized into two sets: exploration of mechanisms and evaluation of claims. Mechanism work has centered on finding theoretical limits and explanations for results observed in the field. Pilot-scale field work has both preceded and followed explanatory laboratory research, and early successes have piqued interest. Long-term, objective field evaluation is critical to understanding how well phytoremediation may work, what the real cost of application will be, and how to build models to predict the interaction between plants and contaminants. Most of the projects are ongoing and, thus, provide only preliminary data.

Phytoremediation classification may be based on the contaminant fate i.e., its degradation, extraction, containment or a combination of these. Phytoremediation applications can also be classified based on the mechanisms involved in the extraction of contaminants from soil or groundwater; concentration of contaminants in plant tissue; degradation of contaminants by various biotic or abiotic processes; volatilization or transpiration of volatile contaminants from plants to the air; immobilization of contaminants in the root zone; hydraulic control of contaminated groundwater (plume control); and control of run-off, erosion, and infiltration by vegetative covers. A brief explanation of these application categories is as follows:

Degradation

Plants may enhance degradation or organic entitus in the rhizosphere (root zone of influence). Microbial counts in rhizosphere soils can be 1 or 2 orders of magnitude greater than in non-rhizosphere soils. It is not known whether this is due to microbial or fungal symbiosis with the plant, plant exudates including enzymes, or other physical/chemical effects in the root zone. There are, however, measurable effects on certain contaminants in the root zone of planted areas. Several projects examine the interaction between plants and such contaminants as trinitrotoluene (TNT), total petroleum hydrocarbons (TPH), pentachlorophenol (PCP), and polynuclear aromatic hydrocarbons (PAH).

Another possible mechanism for contaminant's degradation is metabolism within the plant. Some plants may be able to take in toxic compounds and in the process of metabolizing the available nutrients, detoxify them. Trichloroethylene (TCE) is possibly degraded in poplar trees and the carbon used for tissue growth while the chloride is expelled through the roots.

Extraction

Phytoextraction, or phytomining, is the process of planting a crop of a species that is known to accumulate contaminants in the shoots and leaves of the plants and then harvesting the crop and removing the contaminant from the site. Unlike the destructive degradation mechanisms, this technique yields a mass of plant and contaminant (typically metals) that must be transported for disposal or recycling. This is a concentration technology that leaves a much smaller mass to be disposed of when compared to excavation and landfilling. This technology is being evaluated in a Superfund Innovative Technology Evaluation (SITE) demonstration and may also be a technology amenable to contaminant recovery and recycling.

Rhizofiltration is similar to phytoextraction in that it is also a concentration technology. It differs from phytoextraction in that the mechanism is root accumulation and harvest using hydroponic-growing techniques. This is useful for separating metal contaminants from water.

Volatilization or transpiration through plants into the atmosphere is another possible mechanism for removing a contaminant from the soil or water or a site. It is often raised as a concern in response to a proposed phytoremediation project, but has not been shown to be an actual pathway for many contaminants. Mercury (Hg) has been shown to move through a plant and into the air, in a plant that was genetically altered to allow it to do so. The thought behind this media switching is that elemental Hg in the air poses less risk than other Hg forms in the soil. However, the technology or the associated risk has not been evaluated.

Containment and Immobilization

Containment using plants binds the contaminants to the soil, renders them non-bioavailable, or immobilizes them by removing the means of transport. Physical containment of contaminants by plants can take the form of binding the contaminants within a humic molecule (humification), physical sequestration of metals as occurs in some wetlands, or by root accumulation in non-harvestable plants. Certain trees sequester large concentrations of metals in their roots, and although harvesting and removal are difficult or impractical, the contaminants present a reduced human or environmental risk while they are bound in the roots.

Risk reduction may also be achieved by transforming the contaminant into a form that is not hazardous, or by rendering the contaminant non-bioavailable. Hydraulic control is another form of containment. Groundwater contaminant plume control may be achieved by water consumption, using plants to increase the evaporation and transpiration from a site. Some species of plants use tremendous quantities of water, and can extend roots to draw from the saturated zone.

III. SOIL MICROORGANISMS AND METAL PHYTOREMEDIATION

Several lines of evidence suggest that soil microorganisms possess mechanisms capable of altering enviromental mobility of metal contaminants with subsequent effects on the potential for root uptake. For example, a strain of *Xanthomonas maltophyla* was shown to catalyze the reduction and precipitation of highly mobile Cr^{6+} to Cr^{3+}, a significant less mobile and enviromentally less hazardous compound (Blake *et al* 1993). The same strain was also found to induce the transformation of other toxic metal ions including Pb^{2+}, Hg^{2+}, Au^{3+}, Te^{4+}, Ag^{+} and oxyanions, such as SeO_4^-. Strains of *Escherichia coli* and *Psuedomonas putida* with similar characteristics were subsequently identified (Shen and Wang, 1994, 1995; Wang and Shen, 1995,1997; Wang and Chiwra, 1998). These properties prompted some scientists to propose the use of selected microorganisms to remove toxic metals from wastewater (Blake *et al*, 1993). A recent study revealed that As mobility can be enhanced by the activity of dissimilatory non-reducing bacterium *Shewanella alga* (Cummings *et al*, 1999). Strain BrY of this bacterium was shown to induce As mobilitization from sorption sites within whole sediments. Ionic mercury (Hg^{2+}) which is susceptible to methylation and subsequent generation of methly1 – mercury, a highly neurotoxic compound, was shown to be reduced by some bacteria to Hg^0, a volatile element which posses less enviromental risk. Recently, Crusberg (1998) proposed the investigation of the fungus *Penicillium ochrochloron* for the ability to precipitate a variety of heavy metals including Cu, Ni, Pb and Cd. In addition, soil microorganisms have been shown to exude organic compounds which stimulate bio-availability and facilitate root absorption of a variety of metal ions including Fe^{2+} (Crowley *et al*, 1991; Bural *et al*, 2000), Mn^{2+} (Barber and Lee, 1974), and possibly Cd^{2+}(Salt *et al*, 1995a). Microbes are also known to alter chemical properties of the rhizospheric soil with subsequent effects on the environmental mobility of metal contaminants. For example, chemolithotrophic bacteria have been shown to enhance environmental mobility of metal contaminants via soil acidification, or in contrast, to decrease their solubility due to precipitation as sulfides (Kelley and Tuovinen, 1988). It is generally accepted that under natural conditions a majority of plants have mycorrhizae (Smith and Reed 1997). Fungal symbiotic associations have the potential to enhance root absorption area, and stimulate the acquisition of plant nutrients including metal ions (Khan *et al*, 2000). For example, the length of mycorrhizae associated with pine and willow seedlings was estimated to be two orders of magnitude greater than the length of host root system (Jones *et al*, 1990; Rousseau *et al*, 1994). In addition, fungal associations were shown to enhance root absorption area up to 47-fold (Smith and Reed, 1997). Mycorrhiza has been reported in plants growing on soil contaminated with heavy metals (Shetty *et al*, 1994; Chaudry *et al*, 1998,1999). In addition, arbuscular mycorrhizal fungi are known to colonize ferns (Sharma, 1998), suggesting a possible role of mycorrhizal associations in the recently reported arsenic hyperaccumulation (Ma *et al* 2001). Some of the mycorrhizal fungal species identified in metal-rich soils are *Glomus* spp. and *Gigaspora* spp. as well as *Eutrophosphora* spp. (Raman and Sambandan, 1998; Chaudry *et al*, 1999; Pawlowska *et al*, 1996). There are contradictory reports as to the effect of mycorrhizae on metal uptake. Weissenhorn and Leyval (1995); Joner and Leyval(1997), indicated that Cd is sequestered in the hyphae of *Glomus mosseae* with subsequent restriction of metal transfer from the fungus to the root of subterranean clover plants. An inhibition of Zn and Cu accumulation in mycorrhizal plants was also reported by Heggo *et al*, (1990). Binding to the cell wall of the hyphae has been proposed as a mechanism by which ectomycorrhizal fungi protect the plant from metal toxicity (Galli *et al*, 1994). It was reported that Zn was deposited as crystaloids within fungal mycelium and cortical cells of *Euphorbia cyparissias* mycorrhizal roots. However, other studies could not confirm the inhibition of metal uptake and accumulation in mycorrhizal plants. On the contrary, Zn concentration in pine tree was stimulated by *Thelephora terrestris* (Colpaert and Vanassche, 1992). Furthermore, tolerance of *Picea abies* to Cd conferred by *Paxillus involutus* could not be related to an inhibition of Cd uptake (Goldbold *et al*, 1998). In addition to the effect on metal root uptake, mycorrhiza has also been reported to affect metal transport within plant. For example, mycorrhiza has been shown to alter the pattern of Zn translocation from root to shoot in the grass *Andropogon gerardii* (Shetty *et al*, 1994). An inhibition of Zn translocation to shoots was also reported in mycorrhizal maize seedlings (Khan *et al*, 2000). However, this effect appeared to be species specific since mycorrhizae did not affect Zn translocation pattern in *Festuca arrundinaceae* (Shetty *et al*, 1994). In support of this species specific effect, Dhillion (1992) has indicated that "host preference" is an important characteristic of the symbiotic association.

In the greenhouse study, Zn accumulation in roots was lower in plants inoculated with mycorrhizal fungi although the effect of mycorrhizae on the rate of Zn uptake into root was not clear (Banks *et al*, 1994). These results suggest that one way of increasing the rate of metal uptake is to minimize root colonization by mycorrhizae. Recently, Burke *et al* (2000) showed that soil application of the fungicide Benomyl significantly decreased maize root colonization by arbuscular mycorrhizae. This treatment was associated with an increase in Pb concentration in shoots. However, total Pb removal was inhibited due to a reduction in shoot biomass caused by fungicide application.

A great deal of data indicates that soil microorganisms have the potential to alter the rate of metal uptake and accumulation in plants. Thus, microbes have been shown to catalyze redox transformations leading to changes in soil metal bioavailability and potential for phytoextraction. In addition, the exudation of organic compounds from roots has been shown to affect root absorption of certain metals. The effect of mycorrhizal associations on metal root uptake is not clear and appears to be specific to metal and plant. The bulk of evidence tends to indicate an inhibition of metal uptake by mycorrhizae. However, it is possible that adapted mycorrhizal fungi may play an important role in hyperaccumulation of some metals and metalloids.

IV. METAL BIOAVAILABILITY FOR UPTAKE INTO ROOTS

In soil, metals exist as a variety of chemical species in a dynamic equilibrium governed by soil physical, chemical, and biological properties (Chaney, 1988). The kinetics of this interaction, which ultimately dictates the extent of metal bioavailability, is poorly understood. In general, only a fraction of soil metal is readily available (bioavailable) for plant uptake. The bulk of soil metal is commonly found as insoluble compounds unavailable for transport into roots. The investigation of metal bioavailability routinely requires soil extraction studies. In general, aqueous extraction provides an estimate of the amount of bioavailable metal in the soil solution. Estimates of total bioavailable metal (which also includes metal ions bound to soil exchange sites) are obtained by extracting the soil with organic compounds such as ammonium acetate (Ernst, 1974) or DTPA-TEA (diethylentriamine-pentaacetic acid-triethanolamine).

With the exception of mercury, metal uptake into roots occurs from the aqueous phase. In soil, some metals such as Zn and Cd, occur primarily as soluble or exchangeable, readily bioavailable form. Others, such as Pb occurs as insoluble precipitates (phosphates, carbonates, and hydroxy-oxides) which are largely unavailable for plant uptake (Turnau, 1998). Binding and immobilization within the soil matrix can significantly restrict the potential for metal phytoextraction. Despite the adverse effect on metal root uptake, soil inactivation with chemical amendments has been proposed as a temporary solution for the remediation of Pb-contaminated soil (Cunningham and Berti, 1993). It should be added that although in-place inactivation has the potential to temporarily reduce the risk of spreading metal contamination (Berti and Cunningham, 1993, 1997). Soil-sequestered Pb may act as a long-term pollution source posing a risk to humans and animals due to accidental ingestion of contaminated soil or inhalation of air-borne contaminated dust. Pierzynski and Schwab (1993) investigated the effect of chemical amendments on the potential for phytoextraction of several toxic metals including Cd, Pb and Zn. They showed that addition of limestone, cattle manure, and poultry litter to soil significantly reduced Zn bioavailable fraction. This result correlated well with a reduction of soybean tissue Zn concentration in limestone treatment compared with control. However, Cd and Pb tissue concentrations were not significantly affected by limestone or manure application, suggesting that the effect of soil amendment on bioavailabilty is metal specific. Plants possess highly specialized mechanisms to stimulate metal bioavailability in the rhizosphere, and to enhance uptake into roots (Romeheld and Marschner, 1986). Thus, grass species have been documented to exude a class of organic compounds termed siderophores (mugineic and avenic acids) capable of enhancing the availability of soil iron for uptake into roots (Fushiya *et al* 1982, Takagi *et al* 1984). Dicotyledonous species facilitate iron uptake by acidifying the rhizosphere via H^+ extrusion from roots. Acidic environment stimulates the reduction of ferric to ferrous iron which is readily taken up by plants (Chaney *et al*, 1972; Beinfait *et al*, 1982). Fan *et al* (1997, 1999), investigated the effect of humic substances on root exudation of organic compounds and metal uptake. Exudation of major ligands including deoxymugineic acid, malate and acetate from wheat roots was inhibited by the addition of humic substances.

Lead is a major metal contaminant notorious for posing a significant risk to humans, especially children. For example, it has been estimated that in the USA alone lead poisoning affects more than 800,000 children between the ages of one to five (Pirkle *et al* 1998). The potential for Pb phytoextraction is limited primarily due to low soil mobility and little propensity for lead uptake into roots. Recent research has focused on artificially inducing Pb desorption from soil solids. Several chelating agents have been tested for this purpose. Addition of EDTA (ethylene-diamine-tetraacetic-acid), at a rate of 10 mmol/kg soil, stimulated Pb accumulation in maize to levels as high as 1.6 percent of shoot dry weight (Blaylock *et al*, 1997 and Vassil *et al*, 1998). A surge in pea Pb accumulation was also associated with the addition of synthetic chelates (Huang and Cunningham, 1996). Recently, Kirkham (2000) employed a sequential extraction to investigate the effect of EDTA on the solubility of Pb associated with different soil fractions. The results indicated that soil EDTA addition mobilized Pb associated with the ion-exchangeable and carbonate soil fractions. Lead retained in the oxide and the organic fraction was less affected by EDTA addition indicated that a variety of synthetic chelates have a potential to induce Pb desorption from the soil matrix (Elless and Blaylock, 2000; Huang *et al*, 1997). Their effectiveness, in decreasing order, was EDTA > HEDTA > DTPA > EGTA > EDDHA. Unfortunately, highly soluble Pb-EDTA easily percolates through soil profile, posing a high risk of groundwater contamination (Wu *et al*, 1999). The effect of HBED application on Pb translocation from root to shoot was less clear. In a recent field study, application of nitriloacetate (NTA) and elemental sulfur increased the solubility of Zn, Cd and Cu by as much as 21-, 58-, and 9-fold, respectively (Kayser *et al*, 2000). However, plant accumulation of these metals increased by a factor of 2 to 3 only. An interesting effect was reported by Robinson *et al* (1997) who indicated that EDTA addition increased soil soluble Ni but decreased metal uptake by *Berkheya codii*. Lasat *et al* (1998a) investigated the effect of ammonium ion addition on radiocesium bioavailability and uptake in three plant species. Addition of NH_4^+ released more ^{137}Cs from soil solids into solution. Mobilized ^{137}Cs, however, quickly percolated through soil before being absorbed by roots. These results clearly illustrate the potential risk of enhancing the solubility of soil contaminants with chemical additives and the need to better manage and contain the spread of pollution. Research has been conducted to maximize metal bioavailability while minimizing the environmental risk posed by chelate application. Elless and Blaylock (2000) suggested that for Pb removal, an amendment formulation combining lower EDTA doses and surfactants may be an attractive alternative to higher rates of soil EDTA application. In support of this, Mulligan *et al* (2001) have recently pointed out that biosurfactants have the potential to enhanced metal bioavailability in contaminated soil and sediments. Wu *et al* (1999) showed that the risk of pollution spread can be significantly reduced by applying EDTA to plants transplanted into contaminated soil. These results suggest that with innovative agronomic practices notable improvements over current chelate-assisted phytoextraction may be possible. Blaylock *et al* (1997) indicated that in addition to Pb, bioavailability of other metals including Cd, Cu, Ni and Zn can be enhanced by the application of synthetic chelates. For example, EGTA (ethylenbis[oxyethylenetrinitrilo] tetraacetic acid) has been shown to have a high affinity for Cd^{2+}, but not for Zn^{2+}. EDTA, HEDTA, and DTPA are selective for Zn^{2+}. The addition of EDTA, citric and oxalic acid to a contaminated soil dramatically increased (>200-fold) Cr^{3+} uptake and accumulation in roots and shoots (Shahandeh and Hossner, 2000). Chen and Hong (1995); Hong *et al*(1995) indicated that acetomido-iminodiacetic acid stimulated Pb bioavailability while S-carboxymethylcysteine was effective for Cu. Phytoextraction of precious metals was investigated by Anderson *et al* (1999). They indicated that up to 57 mg/kg Au was accumulated by Indian mustard grown in soil treated with ammonium thiocyanate. The same authors have reported unusual high accumulations of thallium; 0.4 percent and 1.5 percent, on a dry weight basis, in *Iberis intermedia* and *Biscutella laevigata*, respectively.

The potential for phytoextraction of several major metal contaminants including Pb is adversely affected by metal adsorption to soil solids and/or precipitation as insoluble compounds. Addition of synthetic chelates has been shown to stimulate the release of metals into soil solution and enhance the potential for uptake into roots. In the case of Pb, however, EDTA addition to the soil has been documented to increase the risk of spreading contamination due to high solubility of Pb-chelate complex. Recent research aims at eliminating this risk by implementing alternative chelate formulations and innovative agronomic practices.

V. PLANT-BASED REMEDIAL MECHANISMS

Interest in phytoremediation has grown significantly following the identification of metal hyperaccumulator plants. Hyperaccumulators are species capable of accumulating metals at level 100- fold greater than those typically measured in shoots of the common non-accumulator plants. Thus, a hyperaccumulator will concentrate more than 10 ppm Hg, 100 ppm Cd, 1000 ppm Co, Cr, Cu and Pb, 10,000 ppm Zn and Ni. To date, approximately 400 plant species from at least 45 plant families have been reported to hyperaccumulate metals (Baker *et al*, 2000). Most of these species accumulate Ni, about 30 accumulate either Co, Cu, and Zn, even fewer accumulate Mn and Cd, and there are no known Pb hyperaccumulators. Recently, a fern (*Pteris vitatta*), collected from an abandoned wood preservation site has been shown to accumulate about 14,500 ppm arsenic in fronds without showing symptoms of toxicity (Ma *et al*, 2001). These results suggest that phytoremediation of As contaminated site is feasible.

Possibly, the best-known metal hyperaccumulator is *Thlaspi caerulescens* (Alpine pennycress). While most plants exhibit toxicity symptoms at Zn concentration of about 100 ppm, *T. caerulescens* was shown to accumulate up to 26,000 ppm without showing any injury (Brown *et al*, 1995a). In addition, this species extracted up to 22 percent of soil exchangeable Cd from a contaminated site (Turnau , 1998) and showed remarkable Cd tolerance (Schat *et al*, 2000; Escarre *et al*, 2000). Because of these properties, *T. caerulescens* represents an excellent experimental system for studying the mechanisms of metal uptake, accumulation, and tolerance as they relate to metal phytoextraction (Lasat *et al*, 1996, 1998b, 2000; McGrath *et al*, 1997; Shen *et al*, 1997). In addition other *Thlaspi* spp. and a *Brassica juncea* have been used as a model system to investigate the physiology and biochemistry of metal accumulation in plants (Salt *et al*, 1995b, 1997,1999; Kramer *et al*, 1996, 1997).

Transport across root cellular membrane is an important process which initiates metal absorption into plant tissues. The electrical charge prevents metal ions from diffusing freely across the lipophilic cellular membranes into the cytosol. Therefore, ion transport into cell must be mediated by membrane proteins with transport functions. Root uptake kinetics have been investigated for a variety of metal ions including Cd^{2+} (Cohen *et al*, 1998; Hart *et al*, 1998), Cu^{2+} (Thornton, 1991) and Zn^{2+}(Santa Maria and Cogliatti, 1988; Vazquez *et al*, 1994). A comprehensive characterization of Zn^{2+} transport and accumulation in related hyperaccumulator and nonaccumulator plant species was reported by Lasat *et al* (1996, 1998b). These authors indicated that in *T. caerulescens*, Zn hyperaccumulation was correlated with a stimulation of Zn influx into root cells. In addition, an alteration of Zn intracellular transport was also documented in *T. caerulescens*. Lasat *et al* (1998b) showed that sequestration in the root vacuole prevented Zn translocation from root to shoot in *T. arvense* (a related non-accumulator), whereas in *T. caerulescens,* the mechanism of vacuolar sequestration is disabled allowing Zn translocation and hyperaccumulation in leaves.

Accumulation of elevated levels of metal ions is phytotoxic. Plants have developed strategies to defend against this potential stress. Metal tolerance may result from two basic strategies i.e. metal exclusion and metal detoxification (Baker, 1981). The excluders prevent metal uptake into roots avoiding translocation and accumulation in shoots. Excluders have a low potential for metal extraction, but they can be used to stabilize the soil, and avoid further contamination spread due to erosion. Such a species is *Agrostis tenuis* (a pseudometallophyte) which avoids Cd, Cu, Pb and Zn uptake by precipitating the metal in the rhizosphere (Dahmani-Muller *et al*, 2000). In contrast, hyperaccumulators absorb high levels of metals in cells. Such elevated metal concentrations are toxic to common non-accumulator plants. A primary effect of metal toxicity is the inhibition of a number of cytoplasmic enzymes. Therefore, the homeostasis of metal ions in the cyoplasm is essential for avoiding toxicity (Meharg, 1994). Metal tolerance is a major characteristic of hyperaccumulator species. In fact, in some species, such as the Ni-hyperaccumulator *T. goesingense,* hyperaccumulating phenotype is exclusively conferred by exceptionally high metal tolerance (Kramer *et al*, 1997). Early studies suggested that following uptake intracellular nickel is detoxified via binding to low molecular weight organic compounds (<10 KD) (Lee *et al*, 1997). More recently, Kramer *et al*, (2000) showed that in the Ni-hyperaccumulator *T. goesingense,* intracellular Ni is predominantly localized in the vacuole as a Ni-organic acid complex, possibly citrate whereas cytoplasmic Ni is detoxified via binding to histidine (Persans *et al*, 1999). Zinc is another metal for which several hyperaccumulator species have been identified. Brune *et al* (1994) discussed the mechanism of Zn tolerance in plants and proposed five putative mechanisms for regulation of cytoplasmic Zn: i) low uptake across the plasma membrane, ii) sequestration in a subcellular organelle, iii) precipitation as insoluble salts, iv) complexation to low molecular weight organic ligands, and, v) active extrusion across the plasma membrane into the apoplast. A considerable body of evidence argues against hypotheses: 1, 3, and 5 as the mechanism of Zn tolerance in *T. caerulescens* (Lasat *et al*, 1996; Tolra *et al*, 1996; Shen *et al*, 1997; Zhao *et al*, 1998). In contrast, there is substantial evidence supporting the mechanism of cellular sequestration. Several transporters have been shown to mediate Zn fluxes across cellular membranes (Paulsen and Saier, 1997) including tonoplast (MacDiarmid *et al*, 2000; Lyons *et al*, 2000). In yeast, overexpression of membrane proteins MTP's (Metal Tolerance Proteins) from *T. goesingense* was shown to confer resistance to Cd, Co, Ni and Zn possibly due to transport into the vacuole (Persans *et al*, 2000). A family of transporters, similar to *Nramp* transporters identified in animals, bacteria and yeast, with a possible role in Cd transport was identified in *Arabidopsis* although the cellular localization of these transporters is not known (Thomine *et al*, 2000). Furthermore, metal compartmentalization in the vacuole has been previously documented (Vazquez *et al*, 1994; Kupper *et al*, 1999). Vazquez *et al*(1994) proposed that following transport across the tonoplast, Zn is precipitated in the vacuole of epidermal and subepidermal cells of the *T. caerulescens* leaf. Recent results from single-cell sap extraction studies confirmed this observation (Kupper *et al*, 1999). Other studies, however, suggested that most of Zn accumulated in *T. Caerulescens* leaves is water soluble (Tolra *et al*, 1996; Zhao *et al*, 1998). Ernst (1975), and Mathys (1975) suggested that a Zn-malate complex, formed in the cytoplasm, shuttles Zn into the vacuole where it dissociates. In the vacuole, zinc was proposed to be complexed and stored as oxalate (which is a much stronger Zn ligand at the acidic pH of the vacuole). Subsequent studies, however, have not confirmed this hypothesis. For example, Tolra *et al* (1996) found constitutively high levels of malate in shoots of *T. caerulescens*, although a clear correlation to zinc accumulation was not observed. Shen *et al* (1997) also found very high concentrations of malate in *T. caerulescens* shoots. However, high malate concentrations were also measured in *T. ochroleucum*, which the authors used as a reference non-accumulator species. This result argues against a role of malate in Zn tolerance as proposed by Mathys (1977). More recently, precipitation in the vacuole as Zn-phytate has been proposed to account for Zn inactivation in the vacuole (Van Steveninck *et al*, 1990).

It has been documented that in animals and certain fungi metal uptake induces the production of small cysteine-rich proteins known as metallothioneins (MTs) (Hamer, 1986). In plants, MT proteins have been identified only in wheat (Lane *et al*, 1987) and *Arabidopsis* (Murphy *et al*, 1997). Although detection of plant MTs has been problematic, evidence suggests that they have the ability to bind heavy metals. For example, Murphy and Taiz (1995) have shown that in *Arabidopsis* MT gene expression can be modulated in response to copper. In addition, expression of *Arabidopsis* MTs, restored copper tolerance in a copper-sensitive *S. cearevisiae* mutant lacking the MT encoding *cup 1 gene.* In a similar approach, modified yeast cells were employed to demonstrate that MTs are also involved in Zn tolerance (Robinson *et al*, 1996). Despite these significant results, *Arabidopsis* MTs could not be isolated in protein extracts obtained from the transformed yeast strains emphasizing the difficulty of detecting plant MTs. It should be added that the detection of MT proteins in plant tissues has relied mostly on the analysis of MT RNA levels. However, this approach fails to address the possible involvement of post-transcriptional processes in the function of MT genes.

Accumulation of heavy metals in plants has been shown to induce the production of phytochelatins (PCs), a family of thiol-rich peptides (Steffens, 1990; Rauser, 1995). The synthesis of phytocheltains has been documented to be induced by a variety of metals. However, PCs have been shown to be primarily involved in Cd and Cu tolerance (Rauser, 1990; Ow, 1996). In support of this, Chen and Goldsbrough (1994) found an increased activity of gamma-glutamylcysteine synthetase activity in tomato cells selected for cadmium tolerance. Howden *et al* (1995a) (1995b), have isolated *A. thaliana* mutants with increased sensitivity to Cd. These *cad1* mutants, deficient in PC sysnthesis, showed greatly reduced levels of PC synthase activity, a major enzyme of the PC biosynthetic pathway. By screening plant genes for metal tolerance, a wheat cDNA, *TaPCSI*, was identified whose expression in *Saccharomyces cerevisisiae* significantly stimulated cadmium tolerance (Clemes *et al*, 1999, Vatamaniuk *et al*, 1999). In addition, cells expressing PCS genes accumulated more Cd than untransformed cells. Possibly, PCs act as a cytoplasmic scavenger and carrier for Cd transport into the vacuole where it is stored as a stable compound. Oritz *et al* (1995) demonstrated that *htm1* is a gene encoding a transporter which mediates the transport of the Cd-PC complex into the vacuole of *Schizosaccharomyces pombe.* In support of this, yeast mutants lacking the ability to express *hmt1* were Cd sensitive. A tonoplast transporter capable of transporting both PCs and Cd-PC complexes from the cytoplasm into the vacuole has also been identified in plants. Salt and Rauser (1995), studied the metal-chelating characteristics of a synthetically constructed PC analogue peptide $(Glu\text{-}Cys)_2$ and found that $(Glu\text{-}Cys)_2$ exhibited a Cd-binding stoichiometry of 0.5 Cd atom per peptide molecule. The stoichiometry for Hg^{2+}, and Pb^{2+} was one metal atom for each $(Glu\text{-}Cys)_2$ molecule. Salt *et al* (1995b) investigated Cd accumulation in Indian mustard and found that high

Cd levels were associated with a rapid accumulation of phytochelatins in roots. A recent study suggested that phytochelatins may also be involved in arsenic detoxification (Schmoger *et al*, 2000).

To avoid toxicity, plants have also been documented to catalyze redox reactions and alter the chemistry of metals ions. For example, it was shown that from a solution supplemented with toxic Cr^{6+}, water hyacinth (*Eichornia crassipes*) accumulated nontoxic Cr^{3+} in roots and shoots. The Cr^{6+} to Cr^{3+} reduction apparently took place in the fine lateral roots. The reduced ion (Cr^{3+}) was subsequently translocated to leaf tissues. DeSouza *et al* (2000) investigated Se accumulation and showed that in non-accumulator plants, following uptake via a sulfate transporter, selenate is reduced to selenite, process catalyzed by the enzyme ATP sulfurilase. However, the evidence does not support the existence of a similar process in Se hyperaccumulator species. Pickering *et al* (2000) indicated that in *B. juncea,* As^{5+} is transported in roots as a phosphate analogue which is subsequently reduced to As^{3+} in shoots where it is primarily stored as As^{3+} -tris-thiolate.

There is evidence suggesting that biochemical mechanism that confers hyperaccumulating properties is both species and metal specific. For example, in *T. goesingense,* high Ni tolerance was shown to primarily account for the expression of the Ni-hyperaccumulator phenotype. However, in *T. caerulescens,* stimulated transport into the cells was also shown to play a role in Zn hyperaccumulation. In support of this, Lasat *et al* (2000) and Pence *et al* (2000), demonstrated that the expression of a gene, *ZNTI,* encoding a Zn transporter was stimulated in *T. careulescens* compared with a non-accumulator relative, *T. arvense. ZNTI* was shown to be a member of the *ZIP* family of plants micronutrient transporters (Eide *et al*, 1996; Grotz *et al*, 1998; Guerinot, 2000). Lasat *et al* (2000) and Pence *et al* (2000) further speculated that high expression of *ZNTI* in *T. careulescens* is related to a mechanism that allows detoxification of ionic zinc to a physiologically less active compounds which is subsequently stored in the leaf cells of the hyperaccumulator species.

The genetic control of metal hyperaccumulation in plants is not well understood. For example, analysis of the heritability of Zn hyperaccumulation in *T. careunesces* yielded inconclusive results (Pollard and Baker, 1996; Pollard *et al*, 2000). Early work suggested that the genetic control of metal tolerance is complex (Antonovics *et al*, 1971). However, more recent studies have indicated that metal tolerance is regulated by few major genes (Macnair 1993). Smith and Macnair (1998) conducted a genetic analysis of copper tolerance with Cu-tolerant and susceptible lines of *Mimulus guttatus,* and showed that the difference in tolerance was caused by a modifier gene which is active only when the gene for tolerance is present. However, the two genes acted in conjuction to define one independent mechanism. Recently, Macnair *et al* (2000) crossed the tolerant Zn hyperaccumulator *A. halleri* and the nontolerant non-accumulator relative *A. petrea,* and concluded that the two characters are genetically independent. Their data also suggested that Zn tolerance is controlled by a single major gene. The observation that tolerance to toxic metals is regulated by one or few major genes suggests that biotransformation of selected crop species for tolerance, and ultimately superior metal extraction potential is feasible (Macnair *et al*, 2000).

Understanding the physiology and biochemistry of metal accumulation in plants is important for several reasons. Firstly, this would allow the identification of agronomic practices capabale of optimizing the potential for phytoremediation. In addition, understanding the physiology and biochemistry of metal tolerance is needed in order to identify and isolate genes responsible for the expression of the hyperaccumulating phenotype. A significant body of evidence suggests that metal tolerance is an important feature of hyperaccumulator plants. Research data indicates that metal tolerance is regulated by few major genes. This provides hope that biotechnology may be used to engineer more efficient hyperaccumulator plants.

VI. MODIFICATION OF PLANTS FOR ENVIRONMENTAL REMEDIATION

To improve the potential for metal phytoextraction, Brown *et al* (1995b) proposed that the transfer of the hyperaccumulator phenotype from small and slow-growing hyperaccumulator species (Ebbs *et al*, 1997) to fast growing, high biomass-producing non-accumulator plants. To achieve this goal, Cunningham and Ow (1996) and Chaney *et al*(2000) suggested the use of conventional breeding to improve plants for metal extraction. In this approach, however, the potential for success can be limited because of anatomical constraints that can severely restrict sexual compatibility between species (Ow, 1996). As an alternative, biotechnology offers the opportunity for direct gene transfer, thus, circumventing limitation imposed by sexual incompatibility.

Biotechnology has been employed to identify Pb-tolerance mutants from mutagenized *Arabidopsis* populations (Chen *et al*, 1997) providing an opportunity to isolate genes for Pb tolerance. Screening of mutagenized populations of *Arabidopsis* has also resulted in the identification of mutants capable of accumulating elevated levels of Mn (Delhize *et al*, 1993) and P (Delhaize and Randall, 1995). Possibly, the most spectacular application of biotechnology for environmental restoration has been the bioengineering of plants capable of removing methylmercury from the contaminated soil (Rugh *et al*, 1996; Pilon-Smits and Pilon, 2000). Methylmercury, a strong neurotoxic agent, is biosynthesized in Hg-contaminated soils. To detoxify this compound, transgenic plants have been engineered to express modified bacterial genes *mer B* and *mer A*. In transformed plants, *mer B* catalyzes, the protonolysis of the carbon-mercury bond with the release of Hg^{2+}, a less mobile mercury species. Subsequently, *mer A* converts Hg^{2+}, taken up by roots to Hg^0, a less toxic volatile element which is released into the atmosphere. Rugh *et al* (1998) genetically modified yellow poplar trees. Genetically modified plantlets germinated and grew vigorously in media normally containing toxic levels of Hg^{2+} and released elemental mercury approximately 10 times the rate of untransformed plants. In a recent study, Bizily *et al* (2000), engineered transgenic *A. thaliana* plants to express bacterial genes, *mer A* and *mer B*. Modified plants were able to grow on media containing 50-fold higher methylmercury concentrations than wild type plants and up to 10-fold higher concentrations than plants that expressed *mer B* alone. To improve the potential for phytoremediation of Se-contaminated soil, Pilon-Smits *et al* (1999) overexpressed an ATP-sulfurylase (APS) gene, cloned from *A. thaliana* (Leustek *et al* 1994), in Indian mustard. The transgenic plants had fourfold higher APS enzymatic activity, and accumulated 3 times more Se than the wild type. In a recent study, tomato plants have been modified to express the bacterial gene 1-aminocyclopropane-1-carboxilic acid (ACC) deaminase. These transgenic plants showed enhanced metal tolerance and were able to acquire greater metal levels (Cd, Co, Cu, Mg, Ni, Pb and Zn) than untransformed plants (Grichko *et al*, 2000). Furthermore, when *PsMTA* gene encoding the metalothionein-like compounds in peas was expressed in *A. thaliana*, more Cu was accumulated in the roots of the transformed plants than the control plants (Evans *et al*, 1992). Enhanced accumulation was attributed primarily to a stimulation of Cu tolerance in transgenic plants since copper uptake and translocation was unaffected by the genetic manipulation. To investigate the importance of glutathione and PC for Cd tolerance, Indian mustard was genetically engineered to overexpress the *E. coli gshI* gene encoding glutamylcysteine synthetase (Zhu *et al*, 1999). Concentration of Cd in the shoots of transgenic plants was up to 90 percent greater than wild type plants. In a different study, plants engineered to overexpress the *A. thaliana gene (ZAT)* related to putative zinc-transport genes (ZnT) from animals exhibited enhanced Zn resistance and accumulation in roots (Vander Zaal *et al*, 1999).

To improve the potential for metal phytoextraction, Brewer *et al* (1999) generated somatic hybirds between the Zn hyperaccumulator *T. caerulescens* and *B. napus.* The hybirds accumulated high Zn levels that would have been toxic to *B. napus* indicating that the transfer of the metal hyperaccumulating phenotype is feasible. Somatic hybirds capable of removing significant amounts of Pb were also obtained from *B. juncea* and *T. caerulescens* (Gleba *et al*, 1999). Collectively, this research illustrates the potential of biotechnology to enhance the effectiveness of phytoremediation.

Metal hyperaccumulators are notorious for small size and slow growth. These properties have an adverse impact on the potential for metal phytoremediation and severely restrict the employment of effective agronomic practices such as, mechanical harvest. To overcome these disadvantages, conventional breeding approaches have been proposed to improve plants for metal extraction. Unfortunately, the success of this approach may be precluded due to sexual incompatibility caused by anatomical differences between parents. Biotechnology has the potential to overcome this limitation by allowing direct gene transfer. Research data indicate that manipulation of relevant plant features, including metal tolerance, via biotechnology is a realistic possibility.

VII. CONCLUSION

A great deal of evidence indicates that plants have the genetic potential to clean up soil contaminated with toxic metals. Identification of metal hyperaccumulator species has been an impetus for phytoremediation research. Despite significant research efforts, phytoremediation is still an emerging technology. Investigators as well as science administrators have strived to identify and address research needs in an effort to close knowledge gaps and make phytoremediation a reliable clean up technology. It is possible, however, that in addition to scientific issues this transition would also require a new research approach. Complex interactions that take place under site-specific conditions require that metal phytoextraction must be approached as a multi disciplinary research effort. Success will ultimately depend on the employment of a holistic approach to integrate the work of plant biologists, soil microbiologists, agronomists, and environmental engineers. These experts will have to work together focusing on identifying and solving numerous and diverse scientific issues posed by metal phytoextraction. For example, for Pb phytoextraction soil chemists will have to work to enhance Pb bioavailability for uptake while avoiding groundwater contamination. To ensure a research continuum, biologists will have to identify solutions to optimize the ability of plants to take up and store Pb. Agronomists will have to further provide solutions to applied issues such as, how to incorporate amendments including synthetic chelates in soil, and when and how to harvest. Such an integrated approach will allow the formulation of a comprehensive research plan, ensure research continuity, and avoid redundance.

REFERENCES

Anderson, C.W.N, R.R. Brooks, A. Chiarucci, C.J. LaCoste, M. Leblanc, B.H. Robinson, R. Simack, and R.B. Stewart. 1999. Phytomining for nickel, thalium and gold. J. Geochem. Explor.67: 407-415.

Antonovics, J., A.D. Bradsham, and R.G. Turner. 1971. Heavy metal tolerance in plants. Adv. Ecol. Res. 7: 1-85.

Baker, A.J.M. 1981. Accumulators and excluders – strategies in the response of plants to heavy metals. J. Plant Nutr. 3: 643-654.

Baker, A.J.M., S.P. McGrath, R.D. Reeves, and J.A.C. Smith. 2000.Metal hyperaccumulator plants: A review of the ecology and physiology of a biological resource for phytoremediation of metal polluted soils. *In:* N. Terry and G.Banuelos (ed.) Phytoremediation of contaminated soil and water. Lewis Publishers, Boaca Raton, Fl. pp. 85-107.

Banks, M.K., A.P. Schwab, G.R. Fleming, and B.A. Hetrick. 1994. Effects of plants and soil microfauna on leaching of zinc from mine tailings. Chemosphere. 29: 1691-1699.

Barber, S.A., and R.B. Lee. 1974. The effect of microorganisms on the absorption of manganese by plants. New Phyto. l73: 97-106.

Baumann, A. 1885. Das verhalten von zinksatzen gegen pflanzen und imboden. Landwirtsch. Verss. 3: 1-53. Variation of vegetation type occurring on seleniferous soils. J. Am. Pharm. Assoc. 26: 394-405.

Berti, W.R., and S.D. Cunningham. 1993. Remediating soil Pb with green plants. Intl. Conf. Soc. Environ. Geochem. Health, 25-27 July, New Orleans, LA.

Berti, W.R., and S.D. Cunningham. 1997. In place inactivation of Pb in Pb-contaminated soils. Eviron. Sci.Technol. 31: 1359-1364.

Beinfait, H.F., J Duivenvoorden, and W. Verkerke. 1982. Ferric reduction by roots of chlorotic bean plants: Indications for an enzymatic process. J. Plant Nutr. 5: 451-456.

Bizily, S.P., C. L. Rugh, R.B.Meagher. 2000.Phytodetoxification of hazardous organomercurials by genetically engineered plants. Nature Biotech. 18: 213-217.

Blake., R.C., D.M. Choate, S. Bardhan, N. Revis, L.L. Barton, and T.G. Zocco. 1993. Chemical transformation of toxic metals by a pseudomonas strain from a toxic waste site. Environ.toxicol. Chem. 12: 1365-1376.

Blaylock M. J., D.E. Salt, S. Dushenkov, O. Zakharova, and C. Gussman. 1997. Enhanced. accumulation of Pb in indian mustard by soil-applied chelating agentes. Environ. Sci. technol. 31: 860-865.

Brewer, E.P., J.A. Saunders, J.S. Angle, R.L. Chaney, and M.S. McIntosh. 1999. Somatic hybridization between the zinc accumulator *Thlaspi caerulescens* and *Brassica napus*. Theor. Appl. Genet. 99: 761-771.

Brown, S.L., R.L. Chaney, J.S. Angle, and A.J.M. Baker. 1995a. Zinc and cadmium uptake by hyper accumulator *Thlaspi caerulescens* grown in nutrients solution. Soil. Sci. Soc. Am. J. 59: 125-133.

Brown, S.L., R.L. Chaney, J.S. Angle, and A.J.M. Baker. 1995b. Zinc and cadmium uptake by hyper accumulator *Thlaspi caerulescens* and metal tolerant *Silene vulgaris* grown on sludgeamended soils. Environ. Sci. Technol. 29: 1581-1585.

Brune, A., W. Urbach, and K.J. Dietz. 1994. Compartmentation and transport of zinc in barley primary leave as basic mechanism involved in zinc tolerance. Plant Cell Environ. 17: 153-162.

Bural, G.I., D.G. Dixon, and B.R. Glick. 2000. Plant growth – promoting bacteria that decrease heavy metal toxicity in plants. Can.J. Microbiol. 46: 237-245.

Burke, S.C., J.S. Angle, R.L. Chaney, and S.D. Cunnigham. 2000. Arbuscular mycorrhizae effects on heavy metal uptake by corn. Int. J. Phytol. 2: 3-39.

Chaney, R.L. 1988. Metal speciation and interactions among elements affect trace elements transfer in agricultural and environmental food – change. *In* J.R. Kramer and H.E. Allen (ed.) Metal speciation: theory, analysis and applications . Lewis Publishers, Chelsea, MI. pp. 218-260.

Chaney, R.L., J.C. Brown, and L.O. Tiffin. 1972. Obligatory reduction of ferric chelates in iron uptake by soybeans. Plant Physiol. 50: 208-213.

Chaney, R.L., Y.M.Li, J.S. Angle, A.J.M. Baker, R.D. Reeves, S.L. Brown, F.A. Homer, M. Malik, and M. Chin. 2000. Improving matter hyper accumulator wild plants to develop commercial phytoextraction systems: Approaches and progress. *In* N. Terry and G. Banuelos (ed.) Phytoremediation of contaminated soil and water. Lewis Publishers, Boca Raton, Fl. pp. 129-158.

Chaudhry, T.M., W.J. Hayes, A.J. Khan, C.S. Khoo. 1998. Phytoremadiation Focusing on accumulator plants that remediate metal contaminated soils. Aust. J. Ecotoxicol. 4:37-51.

Chaudhry, T.M., L. Hill, A.G. Khan, and C. Keuk. 1999. Colonization of iron and zinc – contaminated dumped filter cake waste by microbes, plants and associated mycorrhizae. *In* M.H. Wong and A.J.M. Baker(ed.) Remediation and management of degraded land CRC Press, Boca Raton, FL. pp. 275-283.

Chen, J., and P.B. Goldsbrough. 1994. Increased activities of - glutamylcysteine synthetase in tomato cells selected for cadmium tolerance. Plant Physiol. 106: 223-239.

Chen, T.C., and A. Hong. 1995. Chelating extraction of lead and copper from an authentic contaminated soil using N- (2-acetamido) iminodiacetic acid and S-carboxymethyl-L-cysteine. J. Hazard. Mater. 41: 147-160.

Chen, J., J.W. Huang, T. Caspar, and S.D. Cunningham. 1997. Arabidopsis as a model system for studying lead accumulation and tolerance in plants. In E.L. Kruger *et al* (ed.) Phytoremediation of soil and water contaminants. Am. Chem. Soc., Washington, DC. pp. 264-273.

Clemens, S., E.J. Kim, D. Newmann, and J.I. Schroeder. 1999. Tolerance to toxic metals by a gene family phytochelatin synthases from plants and yeast. EMBO J. 18: 3325-3333.

Cohen, C.K., T.C. Fox, D.F. Garvin, and L.V. Kochian. 1998. The role of iron – deficiency stress responses in stimulating heavy-metal uptake transport in plants. Plant Physiol. 116: 1063-1072.

Colpaert, J.V., and J.A. Vanassche. 1992. Zinc toxicity in ectomycorrhizal *Pinus sylvestris* L. New Phytol. 123: 325-333.

Crowley, D.E., Y.C. Wang, C.P.P. Reid, and P.J. Szansiszla. 1991. Mechanism of iron acquisition from siderophores by microorganisms and plants. Plants Soil 130: 179-198.

Crusberg, T.C. 1998. Biomineralization of heavy metals within fungal mycelia. A new technology for bioremedation of hazardous wastes.U.S EPA STAR Grant R823341.

Cummings, D.E., F. Caccavo, Jr.S. Fendorf, and R.F. Rosenzeig. 1999. As mobilization by the dissimilatory Fe(iii)-reducing bacterium *Shewanella alge Br Y*. Eniron. Sci. Technol. 33: 723-729.

Cunningham, S.D., and W.R. Berti. 1993. Phytoremediation of contaminated soils. Progress and promise. Symposium on Bioremediation and Bioprocessing. 28 Mar.-2 Apr., Am. Chem. Soc., Denver., CO.

Cunningham, S.D., and D.W. Ow. 1996. Promises and prospects of phytoremediation. Plant physiol. 110: 715-719.

Dahmani-Muller, H.F. Van Oort, B. Gelie, and M. Balabane. 2000. Strategies of heavy metal uptake by three plant species growing near a metal smelter. Environ. Pollut. 109: 231-238.

Delhaize, E, P.J. Randall, P.A. Wallace, and A.Pinkerton. 1993. Screening Arabidotsis for mutants in mineral nutrition. Plant soil 155/156: 141-134.

Delhaize, E and P. J. Randall. 1995. Characterization of a phosphate-accumulator mutant of Arabidopsis thaliana. Plant Physion. 107: 207-213.

DeSouza, M.P., E.A.H. Pilon-Smiths, and N. Terry. 2000. The physiology and biochemistry of selenium volatilization by plants. *In* B.D.Ensley and I. Raskin(ed.) Phytoremediation of toxic metals: using plants to clean up the environment. John Wiley & Sons, NY. pp. 171-190.

Dhillion, S.S. 1992. Evidence for host-mycorhizal preference in native grassland species. Mycol. Res. 96: 359-362.

Ebbs D.S., M.M. Lasat, D.J. Brady, J. Cornish, R. Gordon, and L.V. kochian. 1997. Phytoextraction of cadmium and zinc from a contaminated site. J. Environ. Qual. 26: 1424-1430.

Eide, D., M. Broderius, J. Fett, M.L. Guerinot. 1996. A novel iron-regulated metal transporter from plants identified by functional expression in yeast. Proc. Natl. Acad. Sci. USA 93: 5624-5628.

Elless, M.P., and M.J. Blaylock. 2000. Amendment optimization to enhance lead extractability from contaminated soil for phytoremediation. Int. J. Phytol. 2: 75-89.

Ernst, W.H.O. 1974. Schwermetal vegetation der Erde. G. Fischer, Verlag, Stuttgart.

Ernst, W.H.O. 1975. Physiology of heavy metal resistance in plants. In T.C. Hutchinson *et al*(ed.) Proceeding of International Conference on Heavy Metals in the Environment. CEP Consultants, Edinburgh. pp. 121-136.

Ernst, W.H.O. 1996. Bioavailability of heavy metals and decontamination of soil by plants. Appl. Geochem. 11: 163-167.

Escarre, J., C. Lefebre, W. Grubber, M. LeBlanc, J. Lepart, Y. Riviere, and B. Delay. 2000. Zinc and cadmium hyperaccumulation by *Thalaspi caerulescens* from metalliferous and nonmetalliferous sites in Mediteranean area: Implications for PsMTA function. Plant Mol. Biol. 20: 1019-1028.

Evans, K.M., J.A. Gatehouse, W.P. Linsay. J. Shi, A.M. Tommey, and N.J. Robinson. 1992. Expression of pea metallothionein – like gene PsMTA in *Escherichia coli* and *Arabidopsis thaliana* and analysis of trace metal ion accumulation: Implications for PsMTA function. Plant Mol. Biol. 20: 1019-1028.

Fan. T.W.M., A.N. Lane, J. Pedler, D.E. Crowley, and R.M. Higashi. 1997. Comprehensive analysis of organic ligands in whole root exudates using nuclear magnetic resonance and gas chromatography. Mass spectroscopy. Anal. Biochem.251: 57-68.

Fan, T.W. M., M. Shenker, R.M. Higashi, D.E. Crowley, and A.N. Lane. 1999. Rhizosphere mobilization of heavy metals via plant root exudation. Abstr. Am. Chem. Soc. 217 BIOT 043.

Fushiya, S., K. Takahashi, S. Nakatsuyama, Y. Sato, S. Nozoe, and S. Takagi. 1982. Co-occurrence of nicotianamine and avenice acids in Avena sativa and Oryza sativa. Phytochem. 21: 1907-1908.

Galli, U., H. Scheupp, and C. Brunold. 1994. Heavy metal binding by mycorrhizal fungi. Physiol. Plantarum 92: 364-368.

Gleba, D., N.V. Borisjuk, L.G. Borisjuk, R. Kneer, A. Poulev, M. Skarzhinskaya, S. Dushenkov, S. Logendra, y.y. Gleba, and I. Raskin. 1999.Use of plant roots for phytoremediation and molecular farming. Proc. Natl. Acad. Sci. USA 96: 5973-5977.

Goldbold, D.L., G. Jentscke, S. Winter, and P. Marschner. 1998. Ectomycorrhizas and amelioration of metal stress in forest trees. Chemosphere 36: 757-762.

Grichko, V.P., B. Filby, B.R. Glick. 2000. Increased ability of transgenic plants expressing the bacterial enzyme ACC deaminase to accumulate Cd, Co, Cu, Ni, Pb and Zn. J. Biotech. 81: 45-53.

Grotz, N., T. Fox, E. Connolly, W. Park., M.L. Guerinot, and D. Eide.1998 identificationof a family of zinc transporter genes from *Arabidopsis* that respond to zinc deficiency. Proc. Natl.

Guerinot, M. L. 2000. The ZIP family of metal transporters. BBA 1465: 190-198.

Hamer, D.H. 1986. Metallothioneins. Ann. Rev. Biochem. 55: 913-951.

Hart, J.J., R.M. Welch, W.A. Norvell, L.A. Sullivan, and L.V. Kochian. 1998 Characterzation of cadmium binding, uptake and translocation in intact seedlings of bread and durum wheat cultivars. Plant Physiol. 116: 1413-1420.

Heggo, A., J.S. Angle, and R.L. Chaney. 1990. Effects of cesicular-arbuscular mycorrhizal fungi on heavy metal uptake by soybean. Soil Biol. Biochem. 22: 856-869

Hong, A., T.C. Chen, R.W. Okey. 1995. Chelating extraction of copper from soil using scarboxymethylcysteine. Water Environ. Res. 67: 971-978.

Howden, R., P.B. Goldsbrough, C.R. Anderson, and C.S. Cobbet. 1995a. Cadmium-sensitive, cad1 Mutants of Arabidopsis thaliana are phytochelatin deficient. Plant Physiol. 107: 1059-1066.

Howden, R., P.B. Goldsbrough C.R. Anderson, and C.S. Cobbet. 1995b. A cadmium-sensitive, Glutathione-deficient mutant of Arabidposis thaliana. Plant Physiol. 107: 1067-1073.

Huang, J.W., and S.D. Cunningham. 1996. Lead phytoextraction: Species variation in lead uptake and Translocation. New Phytol. 134: 75-84.

Huang. J.W.J. Chen, W.R. Berti and S.D. Cunningham. 1997. Phytoremediation of lead-contaminated soil. Role of synthetic chelates in lead phytoextraction. Environ. Sci. Technol. 31: 800-805.

Joner. E.J. and C. Leyval. 1997. Uptake of 109Cd by roots and hyphae of a *Glomus mosseae/Tripholium subterraneum* mycorrhiza from soil amended with high and low concentration of cadmium. New Phytol. 135: 353-360.

Jones, M.D., D.M. Durall, and P.B. Tinker. 1990. Phosphorus relationship and production of extramatrical hyphae by two types of willow ectomycorrihizas at different soil phosphorus levels. New Phytol . 115: 259-267.

Kayser, A., K. Wenger, A. Keller, W. Attinger, H.R. Felix, S.K. Gupta, R. Schulin. 2000. Enhancement of phytoextraction of Zn, Cd and Cu from calcareous soil: The use of NTA and sulfur amendments. Environ. Sci. Technol. 34: 1778-1783.

Khan, A.G., C.Keuk, T.M. Chaudhry, C.S. Khoo, and W.J. Hayes. 2000. Role of plants mycorrhizae and phytochelators in heavy metal contaminated land remediation. Chemosphere 41: 197-207.

Kelley, B.C., and O.H. Tuovinen. 1988. Micriobial oxidation of minerals in mine tailings. *In* W. Solomons and V. Foerstner (ed.) Chemistry and biology of soild waste. Springer Verlag, Berlin. pp. 33-53.

Kirkham, M.B. 2000. EDTA-Facilitateed phytoremediation of soil with heavy metals from sewag sludge. Int. J. Phytol. 2: 159-172.

Kramer, U., J.D. Cotter-Howells, J.M. Charnock, A.J.M. Baker and A.C. Smith. 1996. Free histidine as a metal chelator in plants that accumulate nickel. Nature. 379: 635-638.

Kramer, U., R.D. Smith, W. Wenzel, I. Raskin and D.E. Salt. 1997. The role of metal transport and tolerance in nickel hyperaccumulation by *Thlaspi goesingense* Halacsy. Plant Physiol. 115: 1641-1650.

Kramer, U., I.J. Pickering, R.C. Prince, I. Raskin, and D.E. Salt. 2000. Subcellular localization and speciation of nickel in hyperaccumulator and non-accumulator *Thlaspi* species. Plant Physiol. 122: 1343-1353.

Kupper, H., F.J. Zhao, and S.P. McGrath. 1999. Cellular compartmentation of Zinc in leaves of the hyperaccumulator Thalspi caerulescens. Plant Physiol. 119: 305-312.

Lane, B., R. Kajioka, and T. Kennedy. 1987. The wheat germ Ec protein is a Zinc-containing metallothionein. Biochem. Cell Bio. 65: 1001-1005.

Lasat, M.M., A.J.M. Baker, and L.V. Kochian. 1996. Physiological characterization of root Zn^{2+} absorption and translocation to shoots in Zinc hyperaccumulator and nonaccumulator species of Thlaspi. Plant Physiol. 112: 1715-1722.

Lasat, M.M., M. Fuhrmann, S.D. Ebbs, J.E. Cornish,and L.V. Kochian. 1998a. Phytoremediation of a radiocesium-contaminated soil: Evaluation of cesium-137 bioaccumulation in the shoots of three plant species. J. Environ. Qual. 27: 165-169.

Lasat, M.M., A.J.M. Baker, and L.V. Kochian. 1998b. Altered Zn compartmentation in the root symplasm and stimulated Zn absorption into the leaf as mechanisms involved in Zn hyperaccumulation in *Thlaspi caerulescens*. Plant Physiol. 118: 875-883.

Lasat, M.M., N.S. Pence, D.F. Garvin, S.D. Ebbs, and L.V. Kochian. 2000. Molecular physiology of Zinc transport in the Zn hyperaccumulator *Thlaspi caerulescens*. J. Exp. Bot. 51: 71-79.

Lee, J., R.D. Reeves, R.R., Brooks, and T. Jaffre. 1997. Isolation and identification of a citratocomplex of nickel from nickel-accumulating plant. Phytochem. 16: 1502-1505.

Leustek, T., M. Murillo, and M. Cervantes. 1994. Cloning of a cDNAencoding ATP sulferylase from Arabidopsis thaliana by funcational expression in Saccharomyces cerevisiae. Plant Physiol. 105: 897-902.

Lyons, T.J., A. Gasch, L.A. Gaither, D. Botstein, P.O. Brown, and D. Eide. 2000. Genomewide characterization of the Zap lp Zinc-responsive regulon in yeast. Proc. Natl. Acad. Sci. USA 97: 7957-7962.

Ma, L.Q., K.M. Komar, C.Tu, W. Zhang, Y. Cai, and E.D. Kendelley. 2001. A fern that hyperaccumulation.

MacDairmid, C.W., L.A. Gaither and D. Eide. 2000. Zinc transporter that regulate vocular zinc storage in *Saccharomyces cervisiae* EMBOJ. 19: 2845-2855.

Macnair M.R 1993. The genetic of metal tolerance in vascular plants New Phytol 124: 541-559.

Mac Nair M.R., G.H Tilstone S.S Smith. 2000. The genetic os metal tolerance and accumulation in higher plants In. N Terry and G Banuelos (ed.) Phytoremediation of contaminated soil and water Lewis Publisher Boca Raton pp. 235-250.

Mathys, W. 1975. Enzymes of heavy – metal- resistant and non-resistant Population of *silene cucubalus* and there interaction with some heavy metals *in vitro* and *in vivo*. Physiol. Plantarun 33: 161-165.

Mathys, W. 1977. The role of malate, oxalate, and mustared oil glucosides in the evolution of zinc – resistance in herbage plants. Physiol plant 40: 130-136.

McGrath , S.P., Z.G. Shen, F.J. Zhao. 1997. Heavy metal uptake and chemical changes in the rhizosphere of *Thlaspi careulesels* and *Thlaspi ochroleucul* grown in contaminated soil. Plant soil 188: 153-159.

Meharg, A.A. 1994. Integrated tolerance mechanisms: constitutive and adaptive plant responses to elevated metal concentrations in the enviroment. Plant Cell Environ. 17: 989-993.

Minguzzi, C., and O. Vergnano. 1948. II Contento di nichel nelli ceneri di Alyssum bertlonii Desv. Atti dalls secieta Toscana di Scince Naturali Mem Ser A 55: 49-77.

Mulligan, C.N., R.N. Yong, and B.F. Gibbs. 2001. Remediation technologies or metal contaminated soil and groundwater: An Revaluation Eng. Geol. 60: 193-207.

Murphy, A., and L. Taiz. 1995. Comparison of metall othionein gene expressiopn nonprotein thiols in ten *Arabidopsis* ecotypes. Plant physiol. 109: 945-954.

Murphy, A., J. Zhou, P. Dgoldsbrough, and L. Taiz. 1997. Purification and immunonogical identification of metallothioneins 1 and 2 from *Arabidopsis thaliana*. Plant Physiol. 113: 1293-1301.

Ow. D.W. 1996. Heavy metal tolerant genes: prospective tools for bioremedation. Res. Conserv. Recycling 18: 135-149.

Ortiz, D.F., T. Ruscitti, K.F. McCue, and D.W. Ow. 1995. Transport of metal-binding peptides by *HMT1*, a fission yeast ABC – type B vacuolar membrrane protein. J. Biol. Chem. 270: 4721-4728.

Paulsen, I.T., and M.H. Saier. 1997. A novel family of ubiquitous heavy metal ion transport proteins. J. Meme. Biol. 156: 99-103.

Pawlowska, T.E., J. Blaszkowski, and A. Ruhling. 1996. The mycorrhizal status of plants colonizing a calamine spoil mound in southern Poland Mycorrhiza 6: 499-505.

Pence, N.S., P.B. Larsen, S.D. Ebbs, D.L.D. Letham, M.M. Lasat, D.F. Garvin D. Eide, and L.V. Kochian. 2000. The molecular physiology of heavy metal rtransport in the Zn/Cd hyperaccumunator *Thlaspi caerulesecns*. Proc. Natl . Acad. Sci. USA 97: 4956-4960.

Persans, M.W., X.Yan, J.-MMl. Patnoer, U. Kramer, and D. Esalt. 1999. Moleculer dissection of the role of histidine in nickel hyperaccumulation in *Thlaspi goesinganse* (Halacsy). Plant Physion. 121: 1117-1126.

Persans, M.W., L. Huynh, and D.E. Salt. 2000. A novel family of putative vacuolar metal transport proteins involved in nickel tolrance in the nickel hyperaccumulator *Thlaspi goesingense*. Am. Soc. Plant Physiol., CA, abstract 747.

Pickering, I.J., R.C. Prince, M.J. George, R.D. Smith, G.N. Geogrge, and D.E. Salt. 2000. Reduction and corrdination of arsenic in Indian mustard. Plant Physiol. 122: 1171-1177.

Pierzynski, G.M., and A.P. Schwab. 1993. Bioavailability of zinc, cadmium and lead in a metal – contaminated alluvial soil.J.Environ. Qual. 22: 247-254.

Pilon- Smits, E.A.H., S. Hwang, C.M. Lytle,Y. Zhu. J.C. Tai, R.C. Bravo, Y. Chen, T. Leustek, and N. Terry. 1999. Overexpression of ATP sulfurylase in Indian mustard leads to incrased selenate uptake. reduction, and tolerance. Plant Physiol. 119: 123-32.

Pilon-Smits, E.A.H., and M. Pilon. 2000. Breeding mercury-breathing plants for environmental…cleanup\phytoextraction of Toxic Metals A Review of Biological Mechanisms—31(1) 109 – Jour 3/7/05.

Pirkle, J.L., R.B. Kaufmann, D.J. Brody, T. Hickman, E.W.Gunter, D.C. Paschal. 1998. Exposure of the U.S. population to lead, 1991-1994. Environ. Health Prespectives 106: 745-750.

Pollard, A.J., and A.J.M. Baker. 1996. Quantitative genetics of zinc hyperaccumulation in *Thlaspi caerulescens*. New phytol. 132: 113-118.

Pollard, A.J., K.L. Dandridge, and E.M. Jhee. 2000. Ecological genetics and the evolution of trace elements hyperaccumulation in plants. In N. Terry and G. Banuelos (ed.). Phytoremediation of contaminated soil and water. Lewis Publishers, Boca Raton, F.L. pp. 251-264.

Raman, N., and S. Sambandan. 1998. Distribution of VAM fungi in tannery effluent polluted soils of Tamilnadu. India Bull. Contamin. Toxicol. 60: 142-150.

Rauser, W.E. 1990. Phytochelatins. Ann. Rev. Biochem. 59: 61-86.

Rauser, W.E. 1995. Phytochelatins and related peptides. Plant Physiol. 109: 1141-1149.

Robinson, N.J., J.R. Wilson, and J.S. Turner. 1996. Expression of the type 2 metallotionein-like gene MT2 from *Arabidopsis thaliana* in Zn^{2+} -metabolism deeficient *Synechococcus* PCC 7942: Putative role for *MT2* in Zn^{2+} -metabolism. Plant Mol. Biol. 30: 1169-1179.

Robinson, B.H., R.R. Brooks, A.W. Howes, J.H. Kirkman, and P.E.H. Gregg. 1997. The potential of the high-biomass nickel hyperaccumulator *Berkheya codii* for phytoremediation and phytomining. J. Geochem. Explor. 60: 115-126.

Romheld, V., and H. Marschner. 1986. Mobolization of iron in the rizosphereof different plant species. Adv. Plant Nutr. 2: 155-204.

Rousseau, J.V.D., D.M. Sylvia, and A.J. Fox. 1994. Contribution of ectomycorrhiza to the potential nutrient absorbing surface of pine in pure culture. Agric. Ecosys. Environ. 28: 421-424.

Rugh, C.L., H.D. Wilde, N.M. Stack, D.M. Thompson, A.O. Summers, and R.B. Meagher. 1996. Mercuric ion reduction and resistance in transgenic *Arabidopsis thaliana* plants expressing a modified bacterial *mer* A gene. Proc. Natl. Acad. Sci. USA 93: 3182-3187.

Rugh, C.L., J.F. Senecoft, R.B. Meagher, and S.A. Merkle. 1998. Development of transgenic yellow-poplar for mercury phytoremediation. Nature Biotech. 33: 616-621.

Salt, D.E., and W.E. Rauser. 1995. MgATP-dependent transport of phytochelatins across the tonoplast of oat roots. Plant Physiol. 107: 1293-1301.

Salt, D.E., M. Blaylock, P.B.A.N. Kumar, V. Dushenkov, B.D. Ensley, I. Chet, and I. Raskin. 1995a. Phytoremediation: A novel strategy for the removal of toxic metals from the environment using plants. Biotechnol. 13: 468-475.

Salt, D.E., R.C. Prince, I.J. Pickering, and I. Raskin. 1995b. Mechanisms of cadmium mobility and accumulation in Indian mustard. Plant Physiol. 109: 1427-1433.

Salt, D.E., I.J. Pickering, R.C. Prince, D. Gleba, S. dushenkov, R.D. Smith, and I. Raskin. 1997. Metal accumulation by aquacultured seedlings of Indian mustard. Environ. Sci. Technol. 31: 1636-1644.

Salt, D.E., R.C. Prince, A.J.M. Baker, I. Raskin, and I.J. Pickering. 1999. Zinc ligands in the metal hyperaccumulator *Thlapsi caerulescens* as determined using X-ray absorption spectroscopy. Environ. Sci. Techonol. 33; 713-717.

Santa Maria, G.E., and D.H. Cogliatti. 1988. Bidirectional Zn-fluxes and compartmentation in wheat seedling roots. J. Plant Physiol. 132: 312-315.

Schat, H., M. Llugany, and R. Bernhard. 2000. Metal-specific patterns of tolerance, uptake and transport of heavy metals in hyperaccumulating and nonhyperaccumulating metallophytes. In N. Terry and G.Banuelos (ed.) Phytoremediation of contaminated soil and water. Lewis Publishers, Boca Raton, FL. pp. 171-188.

Schmoger, M.E.V., M. Oven, E. Grill. 2000. Detoxification of arsenic by phytochelatins in plants. Plant Physiol. 122: 793-801.

Shahandeh, H., and L.R. Hossner. 2000. Enhancement of Cr(III) phytoaccumulation. Int. J. Phytol. 2: 269-286.

Sharma, B.D. 1998. Fungal associations with Isoetes species. Am. Fern J. 88: 138-142.

Shen, H., and Y.T. Wang. 1993. Characterization of enzymatic reduction of hexavalent chromium by *Escherichia coli* ATCC 33456. Appl. Environ. Microbiol. 59: 3771-3777.

Shen, H., and Y.T. Wang. 1994. Modeling chromium reduction in *Escherichia coli* 33456. Biotechnol. Bioeng. 43: 293-300.

Shen, H., and Y.T. Wang. 1995. Modeling siumltaneous hexavalent chromium reduction and phenol degradation by a defined coculture of bacteria. Biotechnol. Bioeng. 48: 606-613.

Shen, Z.G., F.J. Zhao, and S.P. McGrath. 1997. Uptake and transport of zinc in the hyperaccumulator *Thlaspi caerulescens* and the non-hyperaccumulator *Thlaspi ochroleucum.* Plant Cell Environ. 20: 898-906.

Shettey, K.G., B.A.D. Hetrick, D.A.H. Figge, and A.P. Schwab. 1994. Effects of mycorrhizae and other soil microbes on revegetation of heavy metal contaminated mine spoil. Environ. Pollut. 86: 181-188.

Smith, S.E., and D. J. Reed. 1997. Mycorrhizal symbiosis, 2nd edition. Academic press, London. pp. 589.

Smith, S.E., and M.R. Macnair. 1998. Hypostatic modifiers cause variation in degree of copper tolerance in *Mimulus guttatus.* Heredity 80: 760-768.

Steffens, J.C. 1990. The heavy metal-binding peptides of plants. Ann. Rev. Plant Physiol. Plant Mol.Biol. 41: 553-575.

Takagi, S., K. Nomoto, and T. Takmoto. 1984. Physiolgical aspects of mugineic acid, a possible phytosiderophore of graminaceous plants. J. Plant Nutr. 7: 469-477.

Thomine, S., R. Wang, J.M. Ward, N.M. Crawford, J.I. Schroeder. 2000. Cadmium and iron transport by member of a plant metal transporter family in *Arabibopsis* with homology to *Nramp* genes. Proc. Natl. Acad. Sci. USA. 97: 4991-4996.

Thornton, B. 1991. Indirect compartmental analysis of copper in live ryegrass roots: Comparison with model system. J. Exp. Bot. 42: 183-188.

Tolra, R.P., C.H. Poschenrieber, and J. Barcelo. 1996. Zinc hyperaccumulation in *Thlaspi caerulescens*. II. Influence of organic acids. J. Plant Nutr. 19: 1541-1550.

Turnau, K. 1998. heavy metal content and localization in mycorrizal *Euphorbia cyparissias* from zinc wastes in Southern Poland. Acta Sov. Bot. Poloniae 67: 105-113.

Utsunamyia, T. 1980. Japanese Patent Application No. 55-72959.

van der Zaal, B.J., L.W. Neuteboom, J.E. Pinas, A.N. Chardonnens, H. Schat, J.A.C. Verkleij, P.H.I. Hoykass. 1999. Overexpression of a novel. *Arabiopsis* gene related to pulative zinc transporter genes from animals can lead to enhanced zinc resistance and accumulation. Plant Physiol. 119: 1047-1056.

Van Steveninck, R.F.M., M.E. Van Steveninck, A.J. Wells, and D.R. Fernando. 1990. Zinc tolerance and the binding of zinc as zinc phytate in *Lemna minor.* X-ray microanalytical evidence. J. Plant Physiol. 137: 140-146.

Vassil, A.D., Y. Kapulmik, I. Raskin, D.E. Sail. 1998. The role of EDTA in lead transport and accumulation by Indian mustard. Plant Physiol. 117: 447-453.

Vatamaniuk, O.K.S. Mari, Y.P. Lu, and P.A. Rea. 1999. At PCSI, a phytochelatin synthase from Arabidiopsis: Isolation and *in vitro* reconstruction. Proc. Natl Acad. Sci. USA 96: 7110-7115.

Vazquez, M.D., C. Poschenrieder, J, Barcelo, A.J.M. Baker, P. Hatton, and G.H. Cope. 1994. Compartmentation of zinc in roots and leaves of the zinc hyperaccumulator *Thlaspi caerulescens* J&C Presl. Bot. Acta 107: 243-250.

Wang, Y.T., and H. Shen. 1995. Bacterial reduction of hexavalent chromium. J. Ind. Microbiol. 14: 159-163.

Wang, Y.T. and H. Shen. 1997. Modeling Cr (VI) reduction by pure bacterial cultures. Water Res. 31: 727-732.

Wang, Y.T., and E.M. Chirwa. 1998. Simultaneous removal of Cr(VI) and phenol in chemostat culture of *E. coli* ATCC 33456 and *P. putida* DMP-1. Water Sci. Technol. 38: 113-119.

Weissenhorn, I., and C. Leyval. 1995. Root colonization of maize by a Cd-sensitive and a Cd-tolerant *Glomous mossae* and cadmium uptake in sand culture. Plant Soil 175: 223-238.

Wu, J., F.C. Hsu, and S.D. Cunningham. 1999. Chelate-assisted Pb phytoextraction Pb availability, uptake, and translocation constraints. Environ. Sci. Technol. 33: 1898-1904.

Zhao, F.J., Z.G. Shen, and S.P. McGrath. 1998. Solubility of zinc and interactions between zinc and phosphorus in the hyperaccumulator *Thaispi caerulescens*. Plant Cell Environ. 21: 108-114.

Zhu, Y.L., E.A.H. Pilon-Smiths, A.S. Tarun, S.U. Weber, L. Jouanin, and N. Terry. 1999. Cadmium tolerance and accumulation in Indian mustard is enhanced by overexpressing–glutamylcysteine synthetase. Plant Physiol. 121: 1169-1177.

29

Genetically Modified Cotton and its Biosafety Concerns: A Review

I.S. DUA*, VIKAS KUMAR, DAVOOD ERADATMAND ASLI and BHAVNEET

Department of Botany,
Punjab University, Chandigarh-160 014
E-mail: isd47@yahoo.com

ABSTRACT

This review represents some ethical dilemmas that have arisen as a result of the development and deployment of Bt cotton risks and benefits to the environment resulting from genetic engineering of crops for resistance against biotic and abiotic stresses, in addition to its effects on biodiversity. The socio-economic impacts and distribution of benefits from transgenic technologies are reviewed and almost unprecedented amount of discussion has been stimulated on the merits and demerits of genetic engineering of crops. The arguments for and against transgenics are invariably based on visions of new technology from widely different ethical perspectives.

I. INTRODUCTION

The subject of transgenic crops embraces many of the major issues in philosophy, the meaning of words, free will, the concepts of right and wrong and justice, and ultimately the meaning of life itself. It has become a battle ground involving scientists, commerce, politicians, journalists, lobby-groups and the public. The arguments are about values that are neither absolute nor universal, and the controversies surrounding transgenic crops have to a large extent polarized society into the proponents and opponents. Much of the argument is emotive, with talk of "Cashing in on Hunger", "Demon Seeds", "Terminator Technology" and "Frankenstein Foods", and it is difficult for anyone to appreciate any underlying truths, should they even exist. Yet farmers have been plant breeders for ten thousand years, altering the genetic integrity of most crops without any knowledge of heredity. These farmers were the first genetic engineers

Corresponding Author: isd47@yahoo.com, Department of Botany, Punjab University, Chandigarh-160 014

(Jones, 1994), although genetics as such did not come into existence until the work of Mendel was rediscovered by De Vries (1900) and others. Using knowledge of the inheritance of traits, plant breeding assumed scientific status and during the past 100 years has developed to allow very direct control over crop evolution for man's benefit. Until the advent of genetic engineering, plant breeding was confined to making crosses within and between crop species, which could occur naturally, albeit, that some crosses were forced (Forster *et al*, 1997). Developing transgenic crops have become routine only within the last few years and have changed the nature of plant breeding substantially, raising a whole host of ethical considerations as a consequence. These centre on the associated risk and benefits to man and the environment, the balance of the distribution of benefits, and on the technology being good or bad in itself. The moral and ethical concerns are important factors in influencing a risk-averse public (Callahan, 1996), and are a pivotal feature of the debate on transgenic crops and their products (Wadman, 1996; Geary, 1996; Newton *et al*, 1999).

The unprecedented rapid adoption of transgenic crops during, 1996 to 2000, when GM crops were first adopted, reflects the significant multiple benefits realized by large and small farmers in industrial and developing countries that have grown transgenic crops commercially. Between 1996 and 2000, a total of 15 countries—10 industrial and 5 developing, contributed to more than 25-fold increase in the global area of transgenic crops from 1.7 million hectares to 44.2 million hectares in 2000. Global population exceeded 6 billion in 2000 and is expected to reach approximately 9 billion by 2050, when approximately 90 per cent of the global population will reside in Asia, Africa and Latin America. Today, 815 million people in the developing countries suffer from malnutrition and 1.3 billion are afflicted by poverty. Transgenic crops, often referred to as genetically modified crops (GM), represent promising technologies that can make a vital contribution to global food, feed and fibre security. The principal aim of this publication is to provide a comprehensive global overview of the experience with Bt cotton over the last 8 years since its introduction in 1996; with regards to the agronomic, environmental, economic and socio-benefits that it has delivered and its global potential in the future.

This article is devoted to assessing the performance of Bt cotton and to assess the future global potential for Bt cotton or cotton with other novel genes, that confirm resistance to the major caterpillar/moths (*Lepidoptera*). These insect pests are of significant economic importance in cotton-producing countries. Bt cotton was first adopted commercially nine years ago in 1996. In 2002 it was deployed commercially in nine countries—seven of which were developing countries (China, India, Indonesia, Argentina, Mexico, South Africa and Columbia), and two industrial countries (USA and Australia). There is no considerable published experimental and survey data of commercial Bt crops generated from independent studies to assess the impact of Bt cotton to-date. These studies have documented the production, environment, health, and economic and social impact of Bt cotton in both large and small farms in developing and industrial countries.

The contents of this review are structured chronologically to provide the reader with a global overview of the cotton crop and the available data for assessing the performance of Bt cotton and project its future global potential. The principal aim is to present a consolidated set of data that will facilitate a knowledge-based discussion of the potential benefits that Bt cotton offers to the global society. The topics presented in this article are:

- The cotton crop;
- Global distribution of cotton in developing and industrial countries;
- Insect pests of cotton;
- The use of Bt genes in cotton;
- Assessment of Bt cotton experience with respect to—agronomic, environmental, social and economic benefits;
- Future global potential for Bt cotton—opportunities and challenges.

The Cotton Crop

Cotton is the world's principal fibre, used in almost half of all textiles (Rabobank, 1996). The cultivation of the cotton crop has impacted on the economic development of countries since it was first cultivated, 5,000 to 10,000 years ago (Stephens and Mosley, 1974). Cotton was not known in Europe until the late Middle Ages. However, during the Industrial Revolution in western Europe it played an important role; large scale processing made it the world's main clothing fibre by the end of the 19th century when cotton comprised 80 per cent of all textile material. Today, whereas cotton still retains its status as the most valuable and preferred natural fibre, manmade synthetic fibres have eroded cotton's share in textiles to around 40 per cent. About 20 million tonnes of raw cotton valued at about $ 2 billion is now produced annually, 30 per cent of which is traded as raw material. In 2000–01 cotton prices were at historical lows of $0.40 per pound and the level of direct assistance provided by the governments in 2000–01 was approximately $ 4 billion. Trade in cotton-related products represents almost half of the total $115 billion trade in clothing (Rabobank, 1996).

Cotton is a member of genus *Gossypium* and belongs to the *Malvaceae* family, which also includes the flowering shrub *Hibiscus* and *Okra*. More than 95 per cent of commercial cotton is upland cotton, *Gossypium hirsutum*, while long-staple cotton, *Gossypium barbadense* occupies a small area of less than 5 per cent. Both cultivated cotton species are self-pollinating allotetraploids, are incompatible with the diploid wild species of cotton, and there are no identified non-cotton wild relatives with which upland and long-staple cotton will outcross. Two species of Asiatic cotton, *G. arboretum* and *G. herbaceum* are grown in restricted areas of Asia and Africa; these are diploid and incompatible with upland and long-staple cotton.

Global Distribution of Cotton

Of the 33.5 million hectares of cotton grown in 2001–02 approximately 70 per cent were planted in developing countries and only 30 per cent in industrial countries. The top 10 cotton-growing countries, by area (Table 29.1), accounted for 80 per cent of the planted area of global cotton with the balance of 20 per cent grown in other 55 countries. Of the top 5 countries that planted more than 1 million hectares each (70% of global total), India has by far the largest area (8.7 million hectares), followed by the USA (5.6 million hectares), China (4.8 million hectares), Pakistan (3.1 million hectares) and Uzbekistan (1.5 million hectares). It is noteworthy that 70 per cent of global cotton planted in countries of the South, three Asian countries predominate, India, China and Pakistan that together plant 50 per cent of global cotton.

Table 29.1: Top ten cotton-growing countries by area during 2001-2002

Sr. No.	*Country*	*000 Ha*
1.	India	8,730
2.	U.S.A.	5,596
3.	China	4,824
4.	Pakistan	3,125
5.	Uzbekistan	1,453
6.	Brazil	750
7.	Turkey	654
8.	Turkmenistan	550
9.	Mali	516
10.	Benin	415
	Subtotal	26,613 (80%)
	Others	6,844
	World Total	33,457

Source: ICAC, 2002

Insect Pests of Cotton

Insect pests of cotton are a major constraint to production because of the significant yield losses and quality degradation they cause. However, infestation levels of specific pests vary enormously from year to year, from country to country, and from region to region (Benedict and Altman, 2001). Although up to 1,326 species of insects have been reported on cotton worldwide (Mathews, 1994), the number of insect pests that are economically important are few. Most of the major insect pests of cotton belong to the caterpillar species (*Lepidoptera*). For the purpose of this review, which is to assess the performance of Bt cotton and its future global potential, the following are the most important lepidopteran insect pests: pink bollworm (*Pectinophora gossypiella*), American bollworm (*Helicoverpa armigera*), spiny bollworm (*Earias* spp.), cutworms (*Argeotis* spp.), bollworm, (*Helicoverpa zea*), Egyptian cotton leaf worm (*Spodoptera liteoralis*), tobacco budworm (*Heliothis virescens*) and Australian budworm (*Helicoverpa punctigera*).

The Use of Bt Genes in Cotton

Bt cotton is a genetically engineered form of natural cotton. The main advantage of utilizing biotechnology in agriculture is to increase the productivity through the use of newer varieties that possess properties such as resistance to pests, diseases, and other stressful conditions like drought, salinity and water logging. Of these measures, imparting the property of insect (specific) resistance through the transfer of a gene from *Bacillus thuringiensis* (Bt) into target plants by modern biotech methods is presently considered to be one of the most advanced applications of biotechnology.

The toxin in Bt cotton exists in nature within the micro-organism *Bacillus thuringiensis*. The Japanese Bacteriologist Ishiwata Shigetane first discovered it in 1901. Subsequently, in 1915, a German scientist, named Ernst Berliner, isolated this toxin from a dead moth in Thuringen region of Germany and, thus, the name *Bacillus thuringiensis*.

Bacillus thuringiensis is a spore-forming bacterium species that is commonly found in soil. Bt contains a native crystal protein which when ingested by insect pests, causes paralysis in the digestive tract that is lethal. Bt foliar sprays have been used for 50 years to control insect pests and have a long history of safe use. Bt sprays are one of the few insecticides permitted for use in organic farming. The commercial Bt cotton available today contains genes from the isolate *B. thuringiensis,* spp. *kurstaki* that produces Cry1Aa, Cry1Ab, Cry1Ac, Cry2A (Benedict and Altman, 2001).

Bollgard® and the Chinese Bt Fusion Gene

The Bt genes that are currently deployed are from two sources. Monsanto developed and deployed the Cry1Ac gene in its Bollgard® varieties, which are the most widely used in all nine countries that grow Bt cotton. The second source is the Bt fused gene that was developed by the public sector Chinese Academy of Agricultural Sciences (CAAS) in Beijing, China. The commercial plantings of the CAAS Bt cotton feature a modified Bt fusion gene, Cry1Ab/Cry1Ac, planted in the four provinces of Anhui, Shangdong, Shanxi and Hubei (Jia 1998, James 1998). The cowpea trypsin gene, CpTi with a different mechanism of insect resistance to Bt, has also been incorporated by CAAS as a stacked gene with Bt in some varieties. By 1999, the CAAS single gene Bt cotton and the stacked Bt/CpTi cottons, designed to provide more durable resistance, was planted in nine provinces compared with four in 1998.

The most prevalent Bt gene on a global basis Cry1Ac, was incorporated into Coker 312 cotton designated MON 531 by Monsanto (Perlak *et al*, 2001) and was later named Bollgard® cotton; high transformation efficiency was achieved in Coker with *Agrobacterium tumefaciens.* The transformed Coker was then backcrossed with lines form Delta and Pine Land and other companies that had the necessary agronomic qualities for commercial acceptance.

The advantages of the Cry1Ac in Bollgard®, over the Bt cotton spray, summarized by Benedict and Altman (2001) are as follows:

- Active protein provides moderate to high dose control that allows fair to excellent control of selected important lepidopteran pests.
- Active protein expressed in all plant parts.
- Active protein expressed throughout the season, hence, timing of insecticide applications in relation to an infestation is not an issue.
- Wash off of insecticide during rain, and degradation in sunlight are not issues as they are with spray formulations.
- Less farmer exposure to insecticide.
- Labour saving technology, due to elimination or reduction of insecticide sprays.
- Decreases production risks and provides peace of mind and insurance to farmers at cost-effective control rates.
- Contributes to, and provides the foundation for an integrated pest management (IPM) strategy.

Bollgard® - II Cotton

Bollgard®, the first generation Bt cotton developed by Monsanto, with one Bt gene Cry1Ac, has been successfully grown on over 10 million hectares by millions of farmers in nine countries since its introduction in 1996. Producers have benefitted from reduced insecticide usage, higher yields and higher economic returns, whereas society has benefitted from a safer environment and more affordable cotton prices. Bollgard® has delivered substantial agronomic, environmental health and economic benefits to both small and large farmers in developing and industrial countries.

The Insect Resistance Management (IRM) strategy for Bt cotton that Monsanto in function with USDA and universities, developed prior to the introduction of Bollgard® had anticipated further developments of Bt cotton and planned for the development of a second generation of an improved Bt cotton with two Bt genes, now designated Bollgard® II. The new product, Bollgard® II, Event 1985 was developed using particle acceleration plant transformation procedures to add the Cry2Ab gene to the cotton line DP50B that already had the Cry1Ac (Carpenter *et al*, 2002; Rahn *et al*, 2001). The Bollgard®-II second generation of Bt cotton technology contains two different genes that encode proteins from *Bacillus thuringiensis* Cry2Ab and Cry1Ac; the latter is the protein in the first generation of Bt cotton products. The dual gene cultivars are expected to provide growers with a broader control over a wider variety of insects. In addition to enhancing efficacy, the dual Bt gene product can, most importantly, serve as a new tool to combat the potential development of insect resistance in cotton fields by providing a second mode of action to control these pests. The Cry2A proteins have characteristics distinct from the Cry1Ac protein (English and Slatin, 1992; English *et al*, 1994) and the amino acid sequences of the proteins are quite dissimilar with less than 30 per cent sequence identity (Crickmore *et al*, 1998). Paired genes are one tool used to delay the onset of resistance (Roush, 1994; Gould, 1998).

Thus, Bollgard®-II represents an important development from three perspectives. First, the two genes reduce the probability of resistance developing and this is a very important contribution to the durability of Bt resistance (Gould, 1998); second, it increases the efficacy of control for some of the major lepidopteran pests, and third, it increases the spectrum of pests that can be controlled to include several secondary pests, including armyworms and loopers.

II. AGRONOMIC, ENVIRONMENTAL, SOCIAL AND ECONOMIC BENEFITS

Bt cotton provides significant multiple benefits to producers and society (Qaim and de Janvry, 2002; Ismael *et al*, 2002a,b,c 2001; Carpenter *et al*, 2002; Gianessi *et al*, 2002; ISAAA, 2002a,b,c,d,e; Huang *et al*, 2002; Pray *et al*, 2002, 2001; Naik, 2001; Benedict and Altman, 2001; Edge *et al*, 2001; Perlak *et al*, 2001; Traxler *et al*, 2001). These benefits result in significant agronomic, environmental, social and economic advantages. The benefits are realized by growers through increased yields, reduced production costs, environmental benefits from reduced use of broad spectrum synthetic insecticides, health benefits from using a safer form of insect control and social benefits that arise from increasing incomes while saving valuable time for farmers and their families, particularly in developing countries. The principal agronomic, environmental, social and economic benefits are discussed.

Agronomic Benefits

The major agronomic attribute of Bt cotton over conventional cotton is its ability to produce significantly higher yields in the presence of infestations of the major lepidopteran pests. This reflects Bt cotton's higher level of resistance conferred by Bt genes, particularly to the bollworm complex. Higher yields, plus decreased insect control costs are usually the principal contributors to the increased profitability of Bt cotton. Given that pest infestations vary significantly from country to country and year to year, yield advantage of Bt cotton would also be expected to be highly variable. The data in Table 29.2 list the average increases in yield (expressed as a percentage of the corresponding non-Bt yield) in eight countries that have approved and now grow Bt cotton commercially. It is evident that the yield increases are highly variable and there are several features that deserve comment.

Naik (2001) has reported an average increase of 38 per cent; similar high increases in yield have been consistently reported for extensive multi-year, multi-locational field trials in India. In 2001 when cotton insect pest infestation was severe in India, yield increment in experiments were exceptionally high, up to 90 per cent (ICAR 2002).

Table 29.2: Global yield increases (%) in Bt Cotton in selected countries

Country	*% Yield Increase*
India	38
Argentina	35
Indonesia	30
South Africa	24
Mexico	11
USA	10+
China	5 to 10
Australia	Not Significant

Source: Compiled by James, 2002

Environmental Benefits

Bt cotton has resulted in a significant decrease in the number of insecticide sprays required for the control of the major lepidopteran insect pests. In turn, this reduction has had a major impact on the total number of insecticide sprays applied to Bt cotton. As expected, the reduction in the number of sprays is related to the degree of infestation which varies significantly in different seasons and countries.

Benedict and Altman (2001) estimated that in the US, a reduction of 2.2 insecticide sprays on 1.1 million hectares of Bt cotton resulted in elimination of 1.09 million kg (1,090 MT) of insecticide active ingredient (a.i.) in 1998; this calculation is based on 0.45 kg/hectare of a.i. per single spray/hectare. This quantity, 0.45 kg/hectare is comparable to the amount of active ingredient often applied per spray/hectare in developing countries, where the number of sprays is usually much higher than the United States because of significantly higher pest infestations, pest-conducive tropical or sub-tropical climates, and where inefficient control programmes often have to cope with severe problems related to insect resistance to insecticides.

The substantial decrease in insecticide use associated with the cultivation of Bt cotton lead to significant decreases in insecticide run off into watersheds, aquifers, soils and generally into the environment. Whereas the insecticides that are currently approved meet maximum toxicity requirements of regulatory bodies, many insecticides have lethal effects on non-target organisms, and aquatic animals in ponds and streams (Edge *et al*, 2001). This is of particular concern in developing countries where monitoring is not undertaken to detect pollution of natural resources.

The use of broad-spectrum insecticides, such as pyrethroids, on cotton has adversely affected and decreased the populations of non-target species including the arthropod natural enemies that can provide effective control of non-lepidopteran pests. Prior to the introduction of the Cry1Ac gene in Bt cotton in 1996, there was speculation that given insecticidal specificity and effectiveness of the Bt proteins (English and Slatin, 1992) used as topical sprays that arthropod natural enemies would be protected, increase in numbers, and have the potential to act as biological control agents and, thus, contribute to integrated pest management. Experimental studies confirmed that the arthropod natural enemy populations in Bt cotton are greater than in non-Bt cotton (Roof and DuRant, 1997). In addition to reducing the number of sprays for the bollworm/budworm complex in the United States, Bt cotton has also reduced the number of sprays for other insects such as thrips and aphids, by one or two sprays (Benedict and Altman, 2001). This effect has been attributed to higher populations of beneficial predators and parasitic insects that are depleted or eliminated by broad-spectrum insecticide sprays.

Social Benefits

Bt cotton significantly increases income and saves time for farmers, thus, providing an opportunity to impact the quality of life for farmers and their families. In China, the increased income allows farm families to increase food purchases and food consumption, thus, improving nutritional standards (Pray *et al*, 2001). The time savings for women in South Africa gives them more time to devote to high value activities such as caring for children and the sick and allows them to generate additional income by participating in non-farming activities (Ismael *et al*, 2002a). Children in South Africa who no longer have to participate in spraying activities can now devote more time to educational and other worthwhile pursuits (Ismael *et al*, 2002a). These are important examples of how Bt cotton can offer social benefits that extend beyond the farmers' fields and into their home and community. Small resource-poor cotton farmers, 50 per cent of whom are women in South Africa, spend much of their time carrying water for domestic use and for farm use. It is estimated that annually women and girls in Africa spend 40 billion hours carrying water for domestic and agricultural use including water for spraying pesticides on crops such as cotton (Johns Hopkins University, 2002).

Economic Benefits

Acknowledging that the economic advantage of Bt cotton is related to varying level of pest infestation, all countries growing Bt cotton have reported economic gains. Based on surveys of the commercial Bt cotton and extensive replicated multi-locational trials over several years, the largest economic gains per hectare have accrued to China. The economic gain per hectare for commercial Bt cotton in China has been up to $ 550 per hectare (Pray *et al*, 2002). Significant and consistent economic gains have been recorded in large scale national field trials in India, ranging from $ 75 to $ 200 or more per hectare, with most of the estimates on the higher end of the range. Extensive field trials and surveys in the United States have reported average gains per hectare of $ 50 to $ 80 and substantially more in some cases. Economic gains from Bt cotton in South Africa are up to $ 50 per hectare with those in Australia and Argentina ranging from $ 25 to $ 50 per hectare. In general, the overall economic benefit of Bt cotton results from yield increase and decrease in insect control costs, which are partially offset by the higher price of Bt cotton seed, including the technology fee where it is applicable. Taking all these factors into account, the highest national economic returns in 2001 accrued to China and the US, which grew 35 per cent and 60 per cent of world Bt cotton respectively.

Naik (2001) also explored the effect of a hypothetical 11 per cent and 17 per cent drop in the international price of cotton in the event that Bt cotton production would increase supply and reduce prices. Under the most pessimistic price scenario of a maximum 17 per cent decrease in cotton prices, the benefits of Bt cotton over farm practices were reduced in 1998/99 from ($ 255 to $ 278/hectare) to ($ 185 to $ 230 hectare), which provided handsome returns to Bt cotton farmers.

Based on the ICAR data set from large scale field trials in 2001, the economic advantages for three Bt hybrids were relatively high due to severe pest infestations. The overall economic advantages for the three Bt hybrids ranged from $ 96/hectare, (29% increase compared to conventional cotton) to $ 210/hectare (86% increase over conventional cotton). The magnitude of the economic advantage is of the same order as the 1998–99 data set analyzed by Naik, 2001 ($ 236/hectare). However, the benefits in terms of farming practice can be expected to reflect a higher contribution from insecticide savings, because of the high number of prophylactic sprays applied by farmers.

In summary, the results of field experiments in India indicate that, with typical high insect pest infestations, the overall economic benefits from commercial Bt cotton are likely to be high, ranging from $ 75 to $ 200/hectare or more. The first commercial plantings of approximately 45,000 hectares were planted in India in 2002.

III. GLOBAL POTENTIAL OF BT COTTON: OPPORTUNITIES AND CHALLENGES

A wealth of past experience and knowledge is usually invaluable in projecting future trends and developments. Accordingly, in attempting to assess the global potential for Bt cotton, the information on lepidopteran pests of cotton, their control, as well as the assessment of the impact of Bt cotton reviewed in this provides knowledge on which assumptions and future projections of potential benefits can be based. It is evident that the major lepidopteran pests, particularly the bollworm complex, are a major constraint to increased productivity, and are of economic importance in most cotton-growing countries. Ninety per cent of the cotton area in the 50 key countries have medium to high levels of lepidopteran pests (James, 2002). This conclusion is not only supported by the voluminous published and unpublished data on the incidence and severity of lepidopteran cotton pest infestations, but corroborated by the need to apply over 80,000 MT of cotton insecticide (a.i.) annually in the 50 key countries that grow cotton throughout the world. The cost of cotton insecticides at the farm level in 2001 was $ 1.7 billion, with more insecticide consumed by cotton than any other single crop 20 per cent of the insecticide applied globally to all crops in 2001 was applied to cotton. Damage to cotton from insect pests will vary and is dependent on the level of pest infestation, weather and variety.

Thus, the number of sprays applied to control lepidopteran pests in different countries varies from an average of about 2 sprays per season to 12, and the maximum can reach up to 30 sprays in countries like China where infestation levels can be extremely high.

From a global viewpoint, an international initiative to extend the adoption of Bt cotton must also anticipate and consider the implications of a significant expansion in the global area of Bt cotton. These considerations at the international level are similar to those at the national level and include necessary global strategies for responsibly managing and optimizing the durability of resistance, and the spatial and temporal deployment of different varieties carrying different sources of resistance. It is a prerequisite to establish an international mechanism to formulate, coordinate and oversee a global strategy for deploying Bt cotton responsibly and effectively without the onerous bureaucracy that usually paralyses such legitimate endeavours. Whereas globalization presents such new challenges in terms of international collaboration, it also presents new opportunities for developing countries to access new technologies that otherwise would be unavailable to them. The global deployment of Bt cotton presents new opportunities for public and private organizations to collaborate in a win-win mode to bring environmental, economic, health and social benefits to small and large farmers and society at large in developing countries. Failure to extend the adoption of Bt cotton to more developing countries that wish to benefit from it will deny their farmers superior technology and further disadvantage them relative to their counterparts in industrial and developing countries who are adopting Bt cotton. From a biosafety viewpoint the natural barriers to outcrossing in the cotton crop makes it a model candidate for adoption in developing countries. Similarly, its principal use as a fibre crop, rather than a food/feed-crop, facilitates its acceptance by the public at large, particularly in view of the multiple and significant environmental, economic and social benefits it delivers.

Bt cotton is a proven safe and effective technology that can deliver substantial and significant benefits to society at large and this makes Bt cotton a unique candidate for extended deployment in developing countries. With the adoption of any technology, there is always unintended or unforeseen effects that could present new challenges. However, with the significant and substantial proven benefits that Bt cotton offers developing countries, the greatest risk is not to explore the technology, and thus be certain to suffer the consequences of inferior technology that will disadvantage farmers in developing countries who have to compete in international markets.

IV. DISCUSSION

The application of genetically modified plants in agriculture is anticipated to be a major event in evolving more productivity and sustainable agriculture in the coming years. The concerns are the risks and the magnitude of the consequences to environment. Technology rarely does not have some kind of adverse effect, yet people continue innovating and accepting the new challenges that come with new freedoms. The environmental groups fighting Bt cotton most likely sit in an office in Delhi cooled by an air-conditioner that spews some kind of harmful matter into the atmosphere. If societies were to abandon all technologies that involved some kind of risk, everyone would have to move to forests and survive off nuts and berries. Nevertheless, society must be responsible for technologies that it develops.

There is a serious concern among scientists, policy makers, regulatory agencies and the public regarding the widespread release of GM food crops. Currently, the relationship between the scientific community and the general public is poor. Such negative public perception might prohibit advancement of this technology and prevent realization of its great potential. Emerging new technology especially environmentally friendly approaches to genetic engineering, hold the key to addressing these public concerns.

Bt crops are increasingly becoming a large part of modern agriculture. Bt cotton holds great promises for Indian farmers as well as globally but has yet to be commercially used. Members of the GEAC have delayed the introduction of something that could be very beneficial to farmers. The members of GEAC will receive a guaranteed paycheck but Indian farmers do not have this facility. Until the complexities of biotechnology regulations are addressed in India, farmers will go on facing the same hardships as before and continue waiting for committee approval.

Logic has been defied as Bt cotton has moved through the complicated machinery of approval of transgenic crops in India. Bureaucrats, environmentalists, and industrialists have debated the dangers of Bt cotton while the cotton farmers who are the ones most affected have been left out. The Indian Bt cotton story tells the tale of the complicated maze of approval for transgenic crops. The appropriate role of government in biotechnology regulation is highly contested. This experience sheds light on the controversial issue and can give direction for future policy.

REFERENCES

Benedict, J. and D.W. Altman. 2001. Commercialization of transgenic cotton expressing insecticidal crystal protein. In Jenkins, J. and S. Saha (ed.). Genetic Improvement of Cotton: Emerging Technologies. Science Publications, Enfield, New Hampshire, USA. 8: 137-201.

Callahan, D. 1996. Biotechnology and ethics: a blueprint for the future. Keynote: setting and communicating the limits in biotechnology. http:www.biotech.nwu.edu/nsf/Callahan.html.

Carpenter, J., A. Felsot, T. Goode, M. Hammig, D. Onstand and S. Sankula. 2002. Comparative environmental impacts of biotechnology -derived and traditional soyabean, corn, and cotton crops. Council for Agricultural Science and Technology (CAST): Ames, lowa, USA. p. 189.

Crickmore, N., D.R. Ziegler, J. Feitelson, E. Schnepf, J. Van Rie, D. Lereclus, J. Baum and D.H. Dean. 1998. Revision of the nomenclature for the *Bacillus thuringiensis* pesticidal crystal proteins. Microbiol. Mol. Biol. Rev. 62: 807-813.

De Vries, H. 1900. The law of separation of characters in crosses. Journal of the Royal Horticultural Society 25:243-248.

Edge, J.M., J.H. Benedict, J.P. Carroll and H.K. Reding. 2001. Bollgard® cotton: An assessment of global economic, environmental, and social benefits. The Journal of Cotton Science. The Cotton Foundation. 5: 121-136.

English, L., and S.L. Slatin. 1992. Mode of action of delta–endotoxin from *Bacillus thuringensis*: A comparison with other bacterial toxins. Insect Biochem. Mol. Biol. 22: 1-7.

English, L., H.L. Robbins, M.A. Vontersch, C.A. Kulescza, D.Avi, D. Coyle, C.S. Jany and S.L. Slatin. 1994. Mode of action of CryllA: A *Bacillus thuringiensis* endotoxin. Insect Biochem. Mol. Biol. 24: 1025-1035.

Forster, B.P., Lee, M.A., Lundqvist, U., Millam, S., Vamling, K. and T.M.A. Wilson. 1997. Genetic engineering of crop plants: from genome to gene. Experimental Agriculture 33: 15-33.

Geary, J. 1996. Battle of the bean genes. Time 28th October 1996. Pp. 46-47.

Gianessi, L.P., C.S. Silvers, S.Sankula and J.E. Carpenter. 2002. Plan biotechnology: Current and potential impact for improving pest management in U.S. agriculture. An analysis of 40 case studies. National Center for Food and Agricultural Policy, Washington DC, USA. p. 75.

Gould, F. 1998. Sustainability of transgenic insecticidal cultivars: Integrating pest genetics and ecology. Ann. Rev. Entomol. 43: 701-726.

Greenplate, J. T., J.W. Mullins, S.R. Penn, A. Dahm, B.J. Reich, J.A. Osborn, P.R. Rahn, L. Ruschke and Z. Shappley. In press. Partial characterization of cotton plants expressing two toxin proteins from *Bacillus thuringiensis:* Relative toxin contribution, toxin interaction, and resistance management. J. Appl. Entomol (In press).

Huang, J., S. Rozelle, C. Pray, and Q. Wang. 2002. Plant biotechnology in China. Science. 295: 674-677.

ICAC. 2002. Cotton: Review of the world situation. Vol. 55, No. 5. International Cotton Advisory Committee, Washington DC, USA.

ICAR. 2002. Personal communication. Report on 2001 IPM trial Cost Benefit Analysis. Indian Council of Agricultural Research, New Delhi, India.

ISAAA. 2002a. Bt Cotton in China. http://www.isaaa.org/kc.

ISAAA. 2002b. Bt Cotton in India. http://www.isaaa.org/kc.

ISAAA. 2002d.Bt Cotton in South Africa.http://www.isaaa.org/kc.

ISAAA. 2002c. Bt Cotton in Mexico. http://www.isaaa.org/kc.

ISAAA. 2002e. Bt Cotton in Indonesia.http://www.isaaa.org/kc.

Ismael, Y., R. Bennett and S. Morse. 2001. Can farmers in the developing countries benefit from the modern technologies? Experience from Makhathini Flats, Republic of South Africa. Crop Biotech Brief. Vol. 1. No., 5. 2001. ISAAA Publication.

Ismael, Y., R. Bennett, S. Morse and T.J. Buthelezi. 2002a. Bt cotton, pesticides, labour and health. Presentation at the 6th International ICABR Conference, Ravello, Itally, 11-14 July 2002.

Ismael, Y., R. Bennet and S. Morse. 2002b. Bt cotton, pesticides, labour and health: A case study of smallholder farmers in the Makhathini Flats, Republic of South America. Paper for the 6th International Conference on Agricultural Biotechnology: New Avenues for production, consumption and Technology Transfer, Ravello, Italy.

Ismael, Y., R. Bennet, and S. Morse. 2002c. Do small-scale Bt cotton adopters in South Africa gain an economic advantage? 6th International ICABR Conference, Ravello, Italy, 11-14 July 2002. pp. 1-16.

James, C. 1998. Global Review of Commercialized Transgenic crops: 1998. ISAAA Briefs No. 8. ISAAA: Ithaca, NY, USA. pp. 43.

James, C. 2001. Global Review of Commercialized Transgenic Crops: 2001. ISAAA Briefs No. 24: Preview. ISAAA: Ithaca, NY, USA. pp. 20.

James, C. 2002. Global Review of Commercialized Transgenic Crops: 2002. ISAAA Briefs No. 26: Preview. ISAAA: Ithaca, NY, USA. pp. 138.

Jia, S.R. 1998. Development of resistance management strategies for commercial cultivation of Bt cotton in China. In Proceedings of 5th International Symposium. The Biosafety Results of Field Tests of Genetically Modified Plants and Microorganisms, Braunschweig, Germany, 6-10 September 1998.

Johns Hopkins University. 2002. Population reports. http://www.jhuccp.org/pr/m14/m14chap2.

Jones, S. 1994. The language of the genes. Flamingo, London, pp. 347.

Matthews, G.A. 1994 Insect and mite pests: General introduction. In Matthews, G.A. and J.P. Tunstall (eds.). Insect Pests of Cotton. CAB International, Wallingford, UK. pp. 29-37.

Naik G. 2001. An analysis of socio-economic impact of Bt technology on Indian cotton farmers. Center for Management in Agriculture, Indain Institute of Management, Ahmedabad, India, April 2001.

Newton P., Brown, D. and Clover, C. 1999. Alarm over 'Frankenstein' foods. Electronic Telegraph, Issue 1358, 12th February 1999.

Perlak, F.J., M. Oppenhuize, K. Gustafson, R. Voth, S. Sivasupramaniam, D. Heering, B. Carey, R.A. Ihrig and J.K. Roberts, 2001. Development and commercial use of Bollgard® cotton in USA-Early promises versus today's reality. The Plant Journal. 27: 489-501.

Pray, C., D, Ma. J. Huang and F. Qiao. 2001. Impact of Bt cotton in china. World Development. 29: 1-34.

Pray, C., J. Huang, R. Hu and S. Rozelle. 2002. Five years of Bt cotton in China – The benefits continue. The Plant Journal. 31: 423-430.

Qaim, M. and A. de Janvry. 2002. Bt cotton in Argentina: Analyzing adoption and farmers' willingness to pay. Presented at the Annual Meeting of the American Agricultural Economics Association (AAEA), Long Beach, California, USA, 28-31 July 2002.

Rabobank. 1996. The International cotton Complex. Utrecht. Netherlands. Pp. 104.

Rahn, P.R., L. Ruschke, and Z.W. Shappley. 2001. Efficacy and agronomic performance of Bollgard® II. Proceedings of the Beltwide Cotton Conference. Memphis, TN, USA. National Cotton Council. 2: 832.

Roof, M.E. and J.A. DuRant. 1997. On–farm experiences with Bt cotton in South Carolina. Proceedings of the Beltwide Cotton Conference. National Cotton Council, Memphis, TN, USA. pp. 861.

Roush, R.T. 1994. Managing pests and their resistance to *Bacillus thuringiensis*: Can transgenic plants be better than crop sprays? Biocontrol Sci. Technol. 4: 501-516.

Stephens, S.G., and M.E. Mosley. 1974. Early domesticated cottons from archaeological sites in central coastal Peru. A. Antiquity. 39: 109-122.

Traxler, G., S. Godoy-Avila, J. Falck-Zepeda, and J. de Jesus Espinoza-Arellano. 2001. Transgenic cotton in Mexico: Economic and environmental impacts. Paper presented at the 5th International Conference, Biotechnology, Science and Modern Agriculture: A new Industry at the Dawn of the Century, Ravello, Italy, 15-18 June 2001.

Wadman, M. 1996. Genetic resistance spreads to consumers. Nature 383: 564.

30

Phase Change in Plants: Some Aspects

PRIYANKA KAPOOR* and I. USHA RAO**

**Department of Botany, University of Delhi, Delhi-110 007, India*
E-mail: smailpriyanka@yahoo.com
***Department of Botany, University of Delhi, Delhi-110 007, India*
E-mail: usharaolab@yahoo.com

ABSTRACT

All organisms progress through a series of distinct developmental phases during their growth. In higher plants, the shoot apex passes through three more or less distinct phases during its post-embryonic development: a juvenile vegetative phase, an adult vegetative phase, and a reproductive phase. Structures formed during a specific phase of shoot growth retain morphological and physiological features characteristic of that phase even after the shoot has entered a new phase. Consequently, phases of shoot development are permanently recorded as variations in the character of structures along the axis of the shoot. This phenomenon is known as heteroblasty and is the most obvious result of a change in the development phase of the shoot (Poethig, 1990). *Hedera helix* is a classical example of a species with a distinct juvenile and mature phenotypic characteristics. It has long been a subject to study physiological basis of the phase change (Hackett, 1992). GA_3 can activate the process of rejuvenation in *Hedera helix* (Rogler and Hackett, 1975a). *Zea mays* has excellent genetics and relatively short life cycle compared to woody plants (Poethig, 1990). Tp1, Tp2 and Tp3 mutations are significant because these reveal the existence of a genetically regulated juvenile developmental programme in maize and these mutations define components of this programme (Poethig, 1988). New Zealand has large number of heteroblastic plants. Metamorphic (abrupt) phase change like that reported in *E. hookerianus* (Day *et al*, 1997) results from a switch between developmental programmes, whereas sallometric (gradual) changes indicate that developmental programmes overlap (Poethig, 1990; Ray, 1990). But this hypothesis has to be tested for *E. hookerianus* (Day *et al*, 1997).

I. INTRODUCTION

Morphology has dealt with the comparative study of "external, visible tangible" plant structures (Goebel, 1900). It has achieved the ability to describe extremely heterogeneous organs as variations of rather simple schemes (Troll, 1935).

Plant structure is the product of continued embryonic development. This means that organs such as leaves have a time dimension and cannot be understood only on the basis of their mature state (Sachs, 2001). For example, the leaves of *Tropaeolum* have well-defined lobes when these are young and these are barely visible when the leaf matures. It can be explained that early stages of leaf development rather than their mature states, are the ones that determine topological relations with other organs and are, therefore, specially relevant to morphological studies (Sachs, 2001).

The perpetually embryonic apical meristems in many higher plants undergo a more or less abrupt switch in potential from a juvenile to an adult type of growth during development. This phenomenon is termed "phase change". Phase change is one of the manifold expressions of differentiation. It is a complex of orderly processes whereby a part is modified into a special form or acquires a special function during growth of the individual. In higher plants, the developmental phases are episodes in the life of a part of an organism which is called as shoot apex (Fig. 30.1). The shoot apex of higher plants passes through three more or less distinct phases during the post-embryonic development: a juvenile vegetative phase, an adult vegetative phase and reproductive phase. The juvenile phase of shoot development starts when the shoot meristem begins to initiate a stem, true leaves, and axillary buds. This phase may last for a few days to many years in different species and is distinguished by a variety of unique vegetative traits and by absence of reproductive structures. The adult phase that follows is

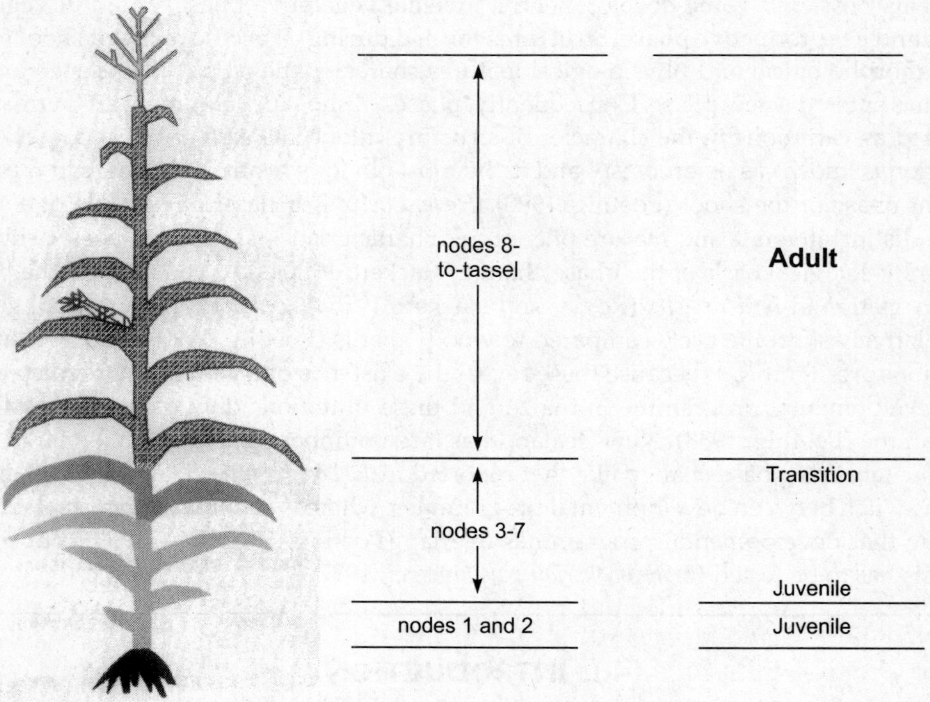

Fig. 30.1: Wild type *Zea mays* (W64A inbred) showing different regions with juvenile, transition and adult phases of soot apex development (after; Moose and Sisco, 1994)

characterized by a different set of vegetative traits and is usually also defined by the ability of the shoot to undergo sexual reproduction. The transformation of the shoot apex into a reproductive structure, such as inflorescence, flower or cone, marks the end of its growth and involves dramatic changes in the differentiation. In some plants, reproduction is the last phase in the life of the shoot; in other types of plants, the growth of the shoot is perpetuated by a lateral vegetative meristem after the terminal meristem becomes reproductive, whereas in some species the primary meristem remains permanently vegetative and only lateral shoots form reproductive structures. Even within the reproductive phase, position-related transitions can occur, such as a change in flower form (Poethig, 1990).

The primary axis of shoot elongates by the addition of new structure at one pole or end, structures formed early in development are located at the base of the shoot and structures formed later are located in more apical positions. Structures formed during a specific phase of shoot growth retain morphological and physiological features characteristic of that phase even after the shoot has entered a new phase. Consequently, phases of shoot development are permanently recorded as variation in the character of structures along the axis of the shoot. This phenomenon is known as "Heteroblasty" and is the most obvious result of change in the developmental phase of the shoot (Poethig, 1990).

Goebel (1900) first recognized that the ontogeny of a plant comprises orderly sequence of developmental stages each possessing some degree of autonomy. The entire life cycle of a seed plant is spanned by two broad phases, juvenile and adult that usually differ significantly in various ways, and are distinctive in that the first one is sterile i.e., juvenile phase is sterile or non-flowering throughout. The change from the juvenile to the adult type of growth in some species is abrupt and overt. Goebel (1900) referred to such development as heteroblastic, although he recognized that the term was meaningful only in a relative sense. Phase change also occurs in some cases of so-called homoblastic development also, but is merely discernible visually. Goebel (1900) also pointed that phase change in many instances is adaptive to external conditions i.e., it is a strategy for survival.

Some of the general characteristics of phase change are:

- Phase change occurs in somatic cells, and is a conspicuous feature of development in many species of higher plants.
- Alternative phases, in some cases, differ sharply in phenotype, but change from one phase to the other may involve transitional phenotypes.
- Phase change is reversible under certain physiological conditions.
- Change from one phase to the other is a directed response to stimuli originating outside the apical meristem.
- Once established, a given growth phase tends to be constant under continuous vegetative propagation, even in the absence of the environmental factor which induced it.
- The frequency of phase change, however, is far higher than that characteristic of germinally transmissible mutations or mutations that are transmitted through germlines.
- Phase change is associated with meristematic growth, not with aging of quiescent meristems.
- The reversal of phase, which in variably occurs during sexual reproduction, is not directly attributable to either meiosis or fertilization, as genetic processes.

- Chimerism is different from phase change. A chimera can be maintained as such only by vegetative propagation. Chimerism in a plant terminates with sexual reproduction. The change from the juvenile to the adult form of growth occurs in each ontogenetic cycle, and juvenile phase is regularly reinstated in each embryo.

II. FEATURES OF PHASE CHANGE IN PLANTS

Phase Change in Woody Plant Species

The process of phase change is widespread in woody plants and is normally associated with a change in morphology from the juvenile vegetative condition to the adult reproductive condition. Phase change in *Hedera helix* is of particular interest for several reasons. First, the juvenile and mature forms represent one of the most dramatic examples of dimorphism in higher plants. Second, both forms can be maintained in a stable condition and, therefore, may be used as reference states in physiological and biochemical experiments (Rogler & Hackett, 1975a,b). These forms can be propagated and maintained as separate individual plants for many years through asexual means of propagation such as rooting of the stem cutting (Hackett, 1992). Hackett (1992) has maintained and experimented with separate, individual juvenile and mature plants of a single clone for about 30 years. Juvenile and mature phase differ in flowering competence. The applications of GA_3 to the mature form induces the morphological reversion to the juvenile form of growth (Rogler & Hackett, 1975a,b; Miller & Goodin, 1976).

The juvenile and adult stages of English ivy (*Hedera helix*) exhibit a series of pronounced morphological differences (Miller & Goodin, 1976). The juvenile plant is a vigorously growing prostrate vine, vertically flattened, possessing aerial roots and 3- or 5- lobed distichous leaves on the anthocyanin pigmented stem. The adult form is a slower-growing form. Plants of the mature growth phase are upright shrubs or small trees with ovate leaves which are arranged spirally. Stems and leaves of this growth phase are low in anthocyanins (Girouard, 1967a,b).

Hedera helix displays such striking differences between juvenile and adult stages that these plants or phases were once considered as two different species (Stein & Fosket, 1969).

In the process of phase change in woody angiosperms, like *Hedera helix,* gibberellins are considered to function as a juvenile hormone (Frydman & Wareing, 1974; Rogler & Hackett, 1975a,b). Maturation is thought to be negatively influenced by gibberellins. The change from the juvenile to the mature phase is considered to be a gradual process in woody angiosperms (Zimmerman *et al*, 1985). There is an evidence that GA_3 can activate the process of rejuvenation in *Hedera helix* (Rogler & Hackett, 1975a). GA_3 has marked effects on the adult growth phase of English ivy bringing about its total reversion to the juvenile phase (Robbins, 1957).

Phase Change in Herbaceous Plant Species

Maize is an excellent species in which we can study the phenomenon of phase change because its developmental morphology, physiology and genetics are well characterized and the architecture of the shoot is based on the succession of a simple basic structural or segmental unit called the phytomer (Coe & Neuffer, 1978). Each phytomer, during the vegetative development has four components – leaf, internode (the section of the stem between the leaves), a bud located in the axil of the leaf and a leaf-like structure called as prophyll located between the bud and the internode (Poethig, 1988a,b). Several mutations affecting shoot morphology in maize have been identified and characterized (Coe & Poethig, 1982).

Genetic analysis of vegetative phase change is complicated by the fact that species with stable, distinctive vegetative phases are not readily amenable to genetic analysis because of their long life cycle. In these respects, maize has several advantages as an experimental system. It has excellent genetics and relatively short life cycle compared to woody plants. Maize shoot apex passes through three more or less distinct phases during the post-embryonic development – a "juvenile vegetative phase, a juvenile adult phase and a reproductive phase" (Poethig. 1990).

The vegetative development of the maize shoot can be divided into juvenile and adult phases based on the types of leaves produced at different times in shoot development (Orkwiszewski & Poethig, 2000). Plants produce different types of leaves or leaf-like organs at different stages in their development. This phenomenon is called as heteroblasty or phase change (Jones, 1955). Among these are cotyledons, the variety of leaf types produced during the vegetative growth of shoot, inflorescence bracts and highly modified leaves found in flowers and other reproductive structures.

Orkwiszewski and Poethig (2000) have described two models for the regulation of leaf formation during shoot development. These are called as meristem patterning model and apex patterning model. Meristem patterning model proposes that leaf identity is specified by dividing the shoot apical meristem into discrete identity domains. The identity of cells in the shoot apical meristem determines the identity of the leaves produced by these cells; thus, juvenile leaves arise from the juvenile region of the meristem, adult leaves arise from an adult region of the meristem and transition leaves arise from cells at the boundary between these domains. Apex patterning model proposes that leaf identity is regulated by factors that originate outside the shoot apical meristem and which act simultaneously on the shoot apical meristem and all existing undetermined organs. This model assumes that changes in leaf identity can occur without a prior change in the identity of cells in the shoot apical meristem (Sylvester *et al*, 1990). In juvenile leaves, epidermal cells produce epicuticular wax, stain purple with toluidine blue, have weakly crenulated anticlinal walls and do not produce bulliform cells or epidermal hairs except along the leaf margins. Whereas in adult leaves, epidermal cells do not produce epicuticular wax, stain aqua with toluidine blue, have highly crenulated anticlinal walls and produce both epidermal hairs and bulliform cells. Leaves that develop during the transition between these phases produce a combination of juvenile and adult traits, with juvenile traits located at the tip of the leaf blade and adult traits located at the base of the leaf blade (Bongard-Pierce *et al*, 1996).

Numerous mutations affecting shoot morphology in maize have been identified and characterized (Coe & Poethig, 1982). Several of these mutations have profound effect on shoot morphology. Some of these mutations may have regulatory functions. Among these are four semi-dominant mutations: "Corngrass (Cg)" (Whaley & Leech, 1950; Singleton, 1951; Galinat, 1954 a,b). "Teopod1 (Tp1), Teopod2 (Tp2) and Teopod3 (Tp3)" (Poethig, 1988; Bassiri & Irish, 1992).

Tp1, Tp2 and Tp3 mutations are significant because these reveal the existence of a genetically regulated juvenile developmental programme in maize and these mutations define the components of this programme (Poethig, 1988). These mutations represent genes that regulate juvenile growth because each mutation causes several traits, which are normally expressed in the basal part of maize (juvenile parts) to be expressed at higher nodes. All three mutations affect the same set of traits, none of these mutations has any obvious deleterious effect on cell growth and Tp1 and Tp2 do not affect the overall duration of shoot growth or rate of leaf initiation but do prolong the vegetative phase of shoot growth at the expense of reproductive development (Bassiri & Irish, 1992). The morphology of mutant phytomers (Tp1, Tp2 and Tp3) reflects the imposition of juvenile traits on otherwise normal pattern of shoot growth rather than the replacement of one type of structure with normal structure of a different type (Poethig, 1988). The phenotype of Tp1, Tp2 and Tp3 suggests that these genes have major regulatory functions.

The possibility that GA_3 may play a role in the vegetative maturation of maize was first suggested by the report that GA_3 is capable of phenotypically correcting the vegetative and reproductive phenotype of Tp1 and Cg mutants (Nickerson, 1960), the two dominant, gain of function mutants that prolong the expression of the juvenile vegetative phase (Poethig, 1988). The Tp and Cg mutants are highly tillered and cause phytomers in adult positions to express juvenile traits and change reproductive structures into vegetative structures. Evans and Poethig (1995) have investigated the effect of GA_3 on vegetative and reproductive maturation in maize by examining the phenotype of several GA-deficient andromonoecious dwarf mutants: dwarf1 (d1), dwarf3 (d3), dwarf5 (d5) and anther ear 1 (an 1). GA_3 promotes vegetative phase change but is not actually required for this process because the heterochronic phenotype of dwarf mutants is relatively minor compared to other heterochronic mutations such as Tp mutants and also GA_3 treatment of wild type plants has only a small effect on the timing of vegetative phase change. The transition from juvenile vegetative to adult vegetative development may depend on several factors (including GA_3), the combined activity of which determines the exact time of vegetative maturation (Bassiri & Irish, 1992). The vegetative phase change will only be blocked completely if all of these factors are absent. Although all the three phases juvenile vegetative, adult vegetative and reproductive development are affected by a reduction or an increase in GA_3, changes in the expression of these phases are not completely coupled to GA_3. GA_3 is involved in vegetative maturation as well as transition between different phases of development in maize. In maize, GA_3 affects multiple partially independent processes associated with reproductive and vegetative maturation.

Phase Change and Heteroblastic Species of New Zealand

New Zealand has a large number of heteroblastic plants. Cockayane (1911) estimated that 200 native species in 37 families have distinct juvenile and adult stages of vegetative development. Juvenile and adult plants differ markedly in leaf size, shape, inclination and growth habit (11 native tree species begin their ontogeny as divaricating shrubs). The juvenile form may persist for many years and some species produce flowers before they reach the adult stage of vegetative development (Gould, 1993).

Pseudopanax crassifolius, is a species endemic to New Zealand. *P. crassifolius* is profoundly heteroblastic, successive leaves are produced that differ in size, shape, angle of inclination and anatomy but the mechanism behind this, still remains unclear (Cockayne, 1967). A relative of *P. crassifolius, Hedera helix* (Araliaceae) also shows prominent juvenile and adult phase (Gould, 1993).

Elaeocarpus hookerianus is a heteroblastic tree native to New Zealand. It is a canopy tree. It shows shoot heteroblasty. It shows discrete juvenile, adolescent and adult phases. It has a divaricating juvenile stage i.e., small-leaved woody shrubs. It is the most extreme example of a plant that changes from a juvenile divaricating shrub to an arborescent adult. According to Day *et al* (1997) a three-phased strategy adopted by this plant is required for:

(a) Providing responsive, energy – efficient shoot systems under low-light regimes,
(b) Growing rapidly to the forest canopy, and
(c) Exploiting the forest canopy environment.

The most noteworthy examples of the juvenile – adult phase phenomenon are those where the persistent juvenile is of a different life form from adult and it may be so persistent that in some species a tree may be one life-form below and of another above. Sometimes both may flower, or the juvenile may bloom before the adult stage is reached. The following plants have been observed blooming when juvenile: *Dacrydium intermedium, D. laxifolium* and *Podocarpus dacrydioides* (Podocarpaceae), *Clematis indivisa* and *Ranunculus lyalli* (Ranunculaceae), *Pittosporum divaricatum* (Pittosporaceae), *Weinmannia racemosa, W. sylvicola* (Cunoriaceae), *Pennantia corymbosa* (Icacinaceae), *Aristotelia fruticosa* (Elaeocarpaceae), *Discaria taumatou* (Rhamnaceae), *Plagianthus betulinus* (Malvaceae), *Hoheria sexstylosa* (Malvaceae), *Edwardsia microphylla* (Papilioniodeae), *Nothopanax simplex, N. edgerleyi* (Araliaceae), *Anisotome filifolia* (Umbelliferae), *Dracophyllum arboretum* (on reversion shoots), *Parsonsia heterophylla, P. capsularis* (Apocynaceae) (Allan 1961; Cockzayne, 1958).

Maturation and phase change in woody plants are developmental processes which are poorly understood, especially from a physiological point of view. Maturation indices, such as phytohormones, can be used to understand more about the events that accompany the increase in the developmental state and to choose optimal conditions in order to improve forest programs and make them more profitable (Valdés *et al*, 2002).

ACKNOWLEDGEMENTS

The fellowship granted by UGC (2000-2001) to Ms Priyanka Kapoor is thankfully acknowledged.

REFERENCES

Allan, H.H. 1961. Flora of New Zealand. Wellington: Government Printer.

Bassiri, A. & E.E. Irish. 1992. Heterochronic effects of Teopod 2 on the growth and photosensitivity of the maize shoot. Plant Cell 4: 497-504.

Bongard-Pierce, D.K.; M.M.S. Evans & S. Poethig. 1996. Heteroblastic features of leaf anatomy in maize and genetic regulation. Intern. J. Pl. Sci. 157: 331-340.

Cockayne, L. 1911. Observation concerning evolution, derived from ecological studies in New Zealand. Trans. New Zealand Inst. 44: 1-50.

Cockayne, L. 1958. The Vegetation of New Zealand (ed. 3). Anton Hain: Weinheim/Bergstr. (Germany).

Cockayne, L. 1967. New Zealand Plant and their Story. ed. 2 Wellington: Government Printer.

Coe, E.H. (Jr.) & R.S. Poethig 1983. Evidence bearing on the orientation of the first division of the zygote. Maize Genet. Coop. Newslett. 57: 34.

Coe, E.H. (Jr.) & M.G. Neuffer. 1978. Embryo cells and their destinies in the corn plant. In Subtelny, S. & Sussex, I.M. (Eds.) The Clonal Basics for Development. Academic Press, New York. pp. 113-129.

Day, J.S.; K.S. Gould P.A. & Jameson. 1997. Vegetative architecture of *Elaeocarpus hookerianus.* Transition from juvenile to adult. Ann. Bot. 79: 617-624.

Evans, M.M.S. & R.S. Poethig. 1995. Gibberellins promote vegetative phase change and reproductive maturity in maize. Pl. Physiol. 108: 475-487.

Frydman, V.M. & P.F. Wareing. 1974. Phase change in *Hedera helix* L.J. exp. Bot. 25: 120-129.

Galinat, W.C. 1954a. Corn-grass. I. Corn-grass as a prototype or a false progenitor of maize. Am. Nat. 839: 101-104.

Galinat, W.C. 1954b. Corn grass. II. Effect of the corn grass gene on the development of the maize inflorescence. Am. J. Bot. 41: 803-806.

Girouard, R.M. 1967a. Initiation and development of adventitious roots in stem cuttings of *Hedera helix.* Anatomical studies of the juvenile growth phase. Can. J. Bot. 45: 1877-1882.

Girouard, R.M. 1967b. Initiation and development of adventitious roots in stem cuttings of *Hedera helix.* Anatomical studies of the mature growth phase. Can. J. Bot. 45: 1883-1886.

Goebel, K. 1900. Organography of Plants. Vol. I. I.B. Balfour, Trans Oxford Press, London.

Gould, K.S. 1993. Leaf heteroblasty in *Pseudopanax crassifolius:* Functional significance of leaf morphology and anatomy. Ann. Bot. 71: 61-70.

Hackett, W.P. 1992. Use of juvenile and mature phases of *Hedera helix* to study flowering competence. FNL 14: 21-24.

Jones, H. 1955. Further studies on heterophylly in *Callitriche intermedia:* Leaf development and experimental induction of ovate leaves. Ann. Bot. 75: 369-388.

Miller, D.R. & J.R. Goodin. 1976. Cellular growth rates of juvenile and adult *Hedera helix*. Pl. Sci. Letts. 7: 397-401.

Moose, P.S. & P.H. Sisco. 1994. Glossy 15 controls the epidermal juvenile-to-adult phase transition in maize. Plant Cell 6: 1343-1355.

Nickerson, N.H. 1960. Sustained treatment with gibberellic acid of maize plants carrying one of the dominant genes Teopod and Corn grass. Am. J. Bot. 47: 713-808.

Orkwiszewski, J.A.J. & R.S. Poethig. 2000. Phase identity of the maize leaf is determined after leaf initiation. PNAS. 97: 10631-10636.

Poethig, R.S. 1988a. Heterochronic mutations affecting shoot development in maize. Genetics 119: 959-973.

Poethig, R.S. 1988b. A non-cell-autonomous mutation regulating juvenility in maize. Nature 336: 82-83.

Poethig, R.S. 1990. Phase change and regulation of shoot morphogenesis in plants. Science 250: 923-930.

Poethig, R.S. 1994. The maize shoot: 11-17. In: Freeling, M. & Walbot, V. (ed.). The Maize Handbook. Springer, New York.

Ray, T.S. 1990. Metamorphosis in the Araceae, Am. J. Bot. 77: 1599-1602.

Robbins, W.J. 1957. Gibberellic acid and the reversal of adult *Hedera* to a juvenile state. Am. J. Bot. 44: 743-746.

Rogler, C.E. & W.P. Hackett. 1975a. Phase change in *Hedera helix:* Induction of the mature to juvenile phase change by gibberellin A_3. Physiologia Pl. 34: 141-147.

Rogler, C.E. & W.P. Hackett. 1975b. Phase, change in *Hedera helix:* stabilization of the mature form with abscisic acid and growth retardant. Physiologia Pl. 34: 148-152.

Sachs, T. 2001. The future of plant morphology. Phytomorphology. Golden Jubilee Issue: pp: 1-17.

Singleton, W.R. 1951. Inheritance of corn grass a macromutation in maize, and its possible significance as an ancestral type. Am. Nat. 821: 81-96.

Stein, O.L. & E.B. Fosket. 1969. Comparative developmental anatomy of shoots of juvenile and adult *Hedera helix.* Am. J. Bot. 56: 546-551.

Sylvester, A.W.; Z.W. Sande & M. Freeling. 1990. Division and differentiation during normal and liguleless-1 maize leaf Development. 110: 985-1000.

Valdés, A.E., M.L. Centeno, S. Espinel & B. Fernández. 2002. Could plant hormones be the basic of maturation indices in *Pinus radiata.* PlantPhyisol. Biochem. 40 : 211-216

Whaley, W.G. & J.H. Leech. 1950. The developmental morphology of the mutant "Corn-Grass". Bull. Torr. Bot. Club 77: 274-286.

31

Introduction of Glycinebetaine Pathway through Genetic Engineering Enhances Potential of Plants to withstand Abiotic Stress

P. SHARMILA, SHIPRA GUPTA, FIROZ ANWAR and P. PARDHA SARADHI

Department of Environmental Biology,
University of Delhi,
Delhi-110 007, India

I. INTRODUCTION

In nature, plants are often exposed to variety of abiotic stresses such as drought, salinity, low temperature, high temperature, heavy metal and high CO_2 (Aspinall and Paleg, 1981; Nikolopoulos and Manetas, 1991; Alia and Pardha Saradhi, 1993; Pardha Saradhi *et al*, 1995; Prasad *et al*, 1999; Chen and Murata, 2002; Jia *et al*, 2002; Xiong *et al*, 2002; Xiong and Zhu, 2002; Pardha Saradhi and Sharmila, 2003). Due to ever increasing population, the pressure on the land is continuously increasing. This is resulting in an extensive damage to the overall environment of the earth. Salinity and drought are the most widespread soil constraints that limit plant growth and crop productivity. These environmental stresses are, thus, serious threats to sustainable food production. However, the world is finite but has to provide food and energy to meet the needs of growing population. Hence, the significant reduction in potential crop yield due to abiotic stresses is being increasingly felt and has received serious attention of scientists throughout the globe (McCue and Hanson, 1990; Bohnert and Jensen, 1996; Prasad *et al*, 2000a,b).

Plants respond to various abiotic stresses by altering their metabolic events, in order to adapt maximally to the changed environmental conditions. Some of the common metabolic alterations observed in plants exposed to abiotic stresses are i) a decrease in net photosynthesis (Dubey, 1997; Giardi *et al*, 1997); ii) an increase (25-70%) in respiration (Livena and Levin, 1967; Rains, 1972); iii) decrease in the activities of some important enzymes of nitrogen metabolism such as nitrate reductase (Srivastava, 1980; Aryan *et al*, 1983; Campbell and Smarrelli, 1986; Rao and Gnanam, 1990); iv) synthesis and accumulation of low molecular weight organic solutes such as proline, glycinebetaine, trehalose, mannitol, spermine etc. (McCue and Hanson, 1990; Alia and Pardha Saradhi, 1991; Pardha Saradhi *et al*, 1995; Hayashi *et al*, 1997; Rajam *et al*, 1998); v) excessive generation of toxic oxygen species (Foyer, *et al*, 1994; Alscher *et al*, 1997) and vi) increase in membrane permeability (Leopold and Willing, 1984).

In this chapter attempts have been made to summarize the work carried out so far in enhancing tolerance of crop plants to withstand abiotic stress through genetic transformation/manipulation.

II. IMPROVEMENT OF PLANTS FOR ABIOTIC STRESS TOLERANCE

In order to obtain maximum output from the agricultural fields that are increasingly being exposed to the conditions that affect plant growth and productivity, efforts are being made to develop genotypes of crop plants that can tolerate/survive well under the adverse conditions prevalent in these areas. Therefore, identification of the genes for imparting abiotic stress tolerance from various sources and transferring them to crop plant species in order to develop drought, salt and low temperature tolerant genotypes has been the major goal of the plant breeders all over the globe. Although conventional plant breeding methods have been encouraging, overall progress has been very slow because of i) lengthy and laborious procedures; ii) difficulty in modifying a single trait and iii) its dependence on the availability of required gene(s) in a close relative/compatible genotype (Holmberg and Bulow, 1998). Identifying genes regulating abiotic stress tolerance and transferring them from closely related species to cultivated crop plants has so far been a tough job for the plant breeders. The second approach, which has been successful to a larger extent has been i) to identify the gene(s) that impart stress tolerance from plants/bacteria or some other organism growing well under the conditions that are otherwise adverse for crop plant and ii) transfer/introduction of the gene(s) into susceptible organisms using genetic engineering technology (Tarczynski *et al*, 1993).

In recent years genetic engineering for introduction of biosynthetic pathways associated with stress responses, has evolved as one of the most promising methods for improving stress tolerance in plants (McCue and Hanson, 1990; Bohnert and Shen, 1999). At cellular level, the most common type of adaptation to tackle abiotic stress, involves the accumulation of non toxic low molecular weight organic molecules widely known as compatible solutes (Alia and Pardha Saradhi, 1993; Hanson *et al*, 1994). These are termed compatible solutes, as they do not interfere with regular cellular metabolism even when present at higher concentrations (Alia and Pardha Saradhi, 1993; Bohnert and Shen, 1999).

Most common compatible solutes that get accumulated in plants under stress are proline, glycinebetaine, mannitol, trehalose, fructans and D-ononitol (Arora and Pardha Saradhi, 1995; Pardha Saradhi *et al*, 1995; Sivakumar *et al*, 1998, 2000, 2001; Bohnert and Shen, 1999). Accumulation of these compatible solutes results in enhancing the tolerance of plants to various types of abiotic stresses (Alia and Pardha Saradhi, 1993; Arora and Pardha Saradhi, 1995; Bohnert and Jensen, 1996; Bohnert and Shen, 1999). The compatible solutes have been shown to improve tolerance of plants by i) osmotic adjustment through mass action, which results in increased water retention and/or sodium exclusion; ii) acting like low molecular weight chaperons through the association with lipids or proteins prevent membrane disintegration or the dissociation of protein complexes/inactivation of enzymes (Arakawa and Timasheff, 1985;

Bohnert and Jensen, 1996; Sivakumar *et al*, 1998); iii) scavenging toxic oxygen species/free radicals (Smirnoff and Cumbes, 1989; Alia *et al*, 1991, 1993, 1995, 1997; Pardha Saradhi *et al*, 1995); iv) regulating the ratio of $NAD(P)^+$ to NAD(P)H (Pardha Saradhi *et al*, 1993; Alia and Pardha Saradhi, 1993) and v) regulating cytosolic pH (Venekamp *et al*, 1989; Alia and Pardha Saradhi, 1991, 1993). In fact, compatible solutes like proline and glycinebetaine have been demonstrated to protect photosynthetic machinery (PS II in particular) against photodamage/photoinhibition by scavenging toxic oxygen species and/or by preventing dissociation of 18 and 23 kDa polypeptides of water oxidation complex (Alia *et al*, 1991, 1993, 1997; Papageorgiou and Murata, 1995).

One of the most successful approach, for enhancing the tolerance of plant to abiotic stresses such as salt, drought and low temperature stresses, so far has been through identification and successful transfer of the gene(s) for various compatible solutes through genetic engineering technology. The genes for synthesis/over-production of compatible solutes viz. proline, glycinebetaine, mannitol, trehalose, D-ononitol and fructans (which are known to impart stress tolerance in plants) have so far been used successfully (Tarczynski *et al*, 1993; Holmstrom *et al*, 1994, 1996; Kishore *et al*, 1995; Pilon-Smits *et al*, 1995; Thomas *et al*, 1995; Lilius *et al*, 1996; Hayashi *et al*, 1997; Shen *et al*, 1997a,b; Sheveleva *et al*, 1997; Grover *et al*, 1999). Further, increased levels of these solutes often enhanced the tolerance of plants to many abiotic stresses viz. salt, drought and low temperature stress, implying that genetic engineering for biosynthesis/over-production of compatible solutes is a fruitful approach for obtaining genotypes of crop plants with enhanced tolerance to various kinds of abiotic stresses.

As described earlier, the plants exposed to abiotic stresses generate excess of toxic oxygen species. To detoxify these highly reactive oxygen species which can disrupt entire cellular metabolism, the cells of plants exposed to stress quickly switch on or over express the genes responsible for the synthesis of enzymes and other components that are associated with instantaneous removal/detoxification of toxic oxygen species and their toxic products. Activities of the enzymes viz. catalase, peroxidase, superoxide dismutase increase in plants exposed to abiotic stresses (Bowler *et al*, 1992; Scandalios, 1993; Foyer *et al*, 1994, 1997). Activities of enzymes associated with ascorbate glutathione cycle viz. ascorbate peroxidase, monodehydroascorbate reductase, dehydroascorbate reductase and glutathione reductase, which play a vital role in the detoxification of toxic oxygen species, also increase in plants under stress.

Owing to importance of these antioxidant enzymes in providing protection (to cellular metabolism) against the toxic oxygen species, several attempts have been successfully made to introduce the gene(s) for these enzymes in plants to enhance their tolerance to abiotic stresses which results in an oxidative stress. Earlier, over expression of the gene for plastidic Cu/ZnSOD has been shown to increase tolerance of transgenic tobacco plants to high light and chilling stress (Gupta *et al*, 1993a,b). In contrast, studies with transgenic tobacco plants expressing 30-50 fold higher SOD than controls failed to display increased protection against paraquat or photoinhibition (Tepperman and Dunsmuir, 1990).

Although genetic engineering approach has now become a very useful technique for transferring gene(s) for abiotic stress tolerance, the success has so far been mostly achieved in either *Nicotiana tabacum* or *Arabidopsis thaliana*.

III. GLYCINEBETAINE

General Properties of Glycinebetaine

Glycinebetaine is a zwitterionic, fully *N*-methyl-substituted derivative of glycine that is found in a large variety of microorganisms, higher plants, and animals (Rhodes and Hanson, 1993). It is structurally the simplest of the betaines. It is always present as a free dipole, but this is not the case with the other betaines. Unlike amino acids it has only one titratable group, the carboxyl, which has a pKa of 1.83 at 25°C. Thus over the physiological pH range, glycinebetaine will have no net charge and no buffering capacity. Glycinebetaine is extremely water soluble (157g 100 ml^{-1} at 19°C), but insoluble in ether (Dawson *et al*, 1969). Glycinebetaine does not inhibit a number of plant and animal enzymes. It is also found to be non-toxic. In contrast to proline, it is slowly metabolized in plants and is a semi- permanent end product of metabolism. Thus, glycinebetaine (*N*-trimethyl glycine) has been presented as a solute notable for its lack of toxicity and intracellular metabolic stability. Soil microorganisms however, rapidly degrade it extracellularly (Kortstee, 1970; Wyn Jones *et al*, 1973).

Being a zwitterion, glycinebetaine requires no counter ions for electrical neutrality in the cytoplasm. It is also an amphiphilic compound with both hydrophobic and hydrophilic ends. This property helps to neutralize anionic groups situated within the hydrophobic protein domains. As glycinebetaine lacks the lone electron pair, it cannot interact with the electrophilic centres (e.g. the Mn cluster) and, therefore, it does not alter electric properties (Papageorgiou and Murata, 1995). Nakaya *et al* (1991) reported that glycinebetaine does not react with water although it is strongly hydrated (the ratio of water to glycinebetaine at the rate of 8.6:1, mol/mol) in aqueous media.

Distribution

Glycinebetaine, is found in a wide variety of organisms viz. bacteria, cyanobacteria, algae, animals and higher plants (Hayashi *et al*, 1997; Prasad *et al*, 2000a,b; Sakamoto and Murata, 2001). In general, higher plant species adapted to dry and saline environments (e.g., members of Chenopodiaceae and Poaceae) accumulate glycinebetaine as an adaptive means (Rhodes and Hanson, 1993). The most extensively studied plants are spinach and barley (Papageorgiou *et al*, 1991).

Among plants, glycinebetaine exhibit sporadic distribution. Taxonomically distinct plants belonging to Chenopodiaceae, Amaranthaceae, Portulacaceae, Malvaceae, Convolvulaceae, Cuscutaceae, Solanaceae, Avicenniaceae and Asteraceae accumulate glycinebetaine (Weretilnyk *et al*, 1989). All the members of Chenopodiaceae show its accumulation and more critical evaluation is required for confirming about accumulation reports in other families. Highest levels of glycinebetaine content in plant kingdom were recorded in *Pachycornia tenuis* (533 µmol. g dry wt^{-1}) and *Atriplex canescens* (418 µmol. g dry wt^{-1}) (Wyn Jones and Storey, 1981).

There are indications that glycinebetaine accumulation is associated mainly in halophytes, xerophytes and the other plant species that grow under salty and arid habitats. However, not all the xerophytic or halophytic species accumulate glycinebetaine or other betaines under field conditions. So far no salt-sensitive or drought sensitive plant with low osmotic pressure has been found to accumulate glycinebetaine. As glycinebetaine was found in halophytes, much of the later works were confined to saline and dry habitats giving relatively less importance to glycophytes and mesophytes. In fact, keeping in mind probable adaptive role played by glycinebetaine, several scientific groups have successfully attempted to enhance the tolerance of non-glycinebetaine accumulating plants to salinity and/or drought (Hayashi *et al*, 1997; Alia *et al*, 1998a,b, 1999; Sakamoto *et al*, 1998; Nuccio *et al*, 1998; Prasad *et al*, 2000a,b).

Adaptive Significance of Glycinebetaine

In general, organisms (irrespective of whether plants, animals or microbes) adapted to dry and saline environment accumulate glycinebetaine as an adaptive strategy (Rhodes and Hanson, 1993). Glycinebetaine is known to protect cell from osmotic stress by a) maintaining an osmotic balance with the environment (Robinson and Jones, 1986, Deshnium *et al*, 1995; 1997); b) stabilizing several functional units such as oxygen evolving photosystem II complex; c) protecting the structure of proteins by facilitating hydration and thereby retaining its native conformational state (Anjum *et al*, 1999) including that of Rubisco; d) maintaining structural integrity of membranes (lncharoensakdi *et al*, 1986; Bernard *et al*, 1988; Santoro *et al*, 1992; Winzor *et al*, 1992; Papageorgiou and Murata, 1995; Sakamoto and Murata, 2000; Quan *et al*, 2004); e) protecting complex II of mitochondria against Na^+ toxicity (Hamilton and Heckathron, 2001) and f) protecting the transcriptional and translational machinery under stress (Allard *et al*, 1998; Houry *et al*, 1999; Bourot *et al*, 2000).

Biosynthetic Pathways

Four distinct biosynthetic pathways for glycinebetaine have so far been reported in living organisms (Fig. 31.1).

Pathway I: In Chenopodiaceae, the most extensively studied family of glycinebetaine-accumulating higher plants, glycinebetaine is synthesized from choline in a two-step oxidation catalyzed by a ferredoxin (Fd)-dependent choline monooxygenase (CMO) (Brouquisse *et al*, 1989) and a betaine aldehyde dehydrogenase (BADH) with a strong preference for NAD^+ (Weigel *et al*, 1986; Weretilnyk and Hanson, 1989; Rhodes and Hanson, 1993).

$^{18}O_2$ tracer studies confirm that CMO utilizes molecular oxygen as a substrate. The immediate product of the CMO reaction may be betaine aldehyde hydrate [$(CH_3)_3N^+$-CH_2-$CH(OH)_2$] which is in spontaneous equilibrium with betaine aldehyde [$(CH_3)_3N^+$-CH_2-C(OH)=O] (Lerma *et al*, 1988).

Both CMO and BADH enzymes are predominantly localized in the chloroplast stroma of chenopods (Weigel *et al*, 1986, 1988; Brouquisse *et al*, 1989), although in spinach leaves approximately 10% of the BADH may exist as a cytosolic isoform (Weretilnyk and Hanson, 1988). Glycinebetaine is also localized predominantly in the chloroplasts of salinized spinach leaves, where it provides osmotic adjustment (Robinson and Jones, 1986). The spinach CMO is not membrane bound, requires ferredoxin and Mg^{2+}, and unlike the bacterial choline oxidizing enzymes (CDH and COX) does not appear to catalyze the oxidation of betaine aldehyde to glycinebetaine (Brouquisse *et al*, 1989).

Both CMO (Burnet *et al*, 1995) and BADH (Weretilnyk and Hanson, 1989) have been purified to homogeneity from spinach. CMO appears to be a homodimer (or possibly homotrimer) of subunits of MW 42.9 kDa (Burnet *et al*, 1995; Rathinasabapathi *et al*, 1997), whereas BADH is a homodimer with a subunit MW of ~ 60 kDa (Weretilnyk and Hanson, 1990).

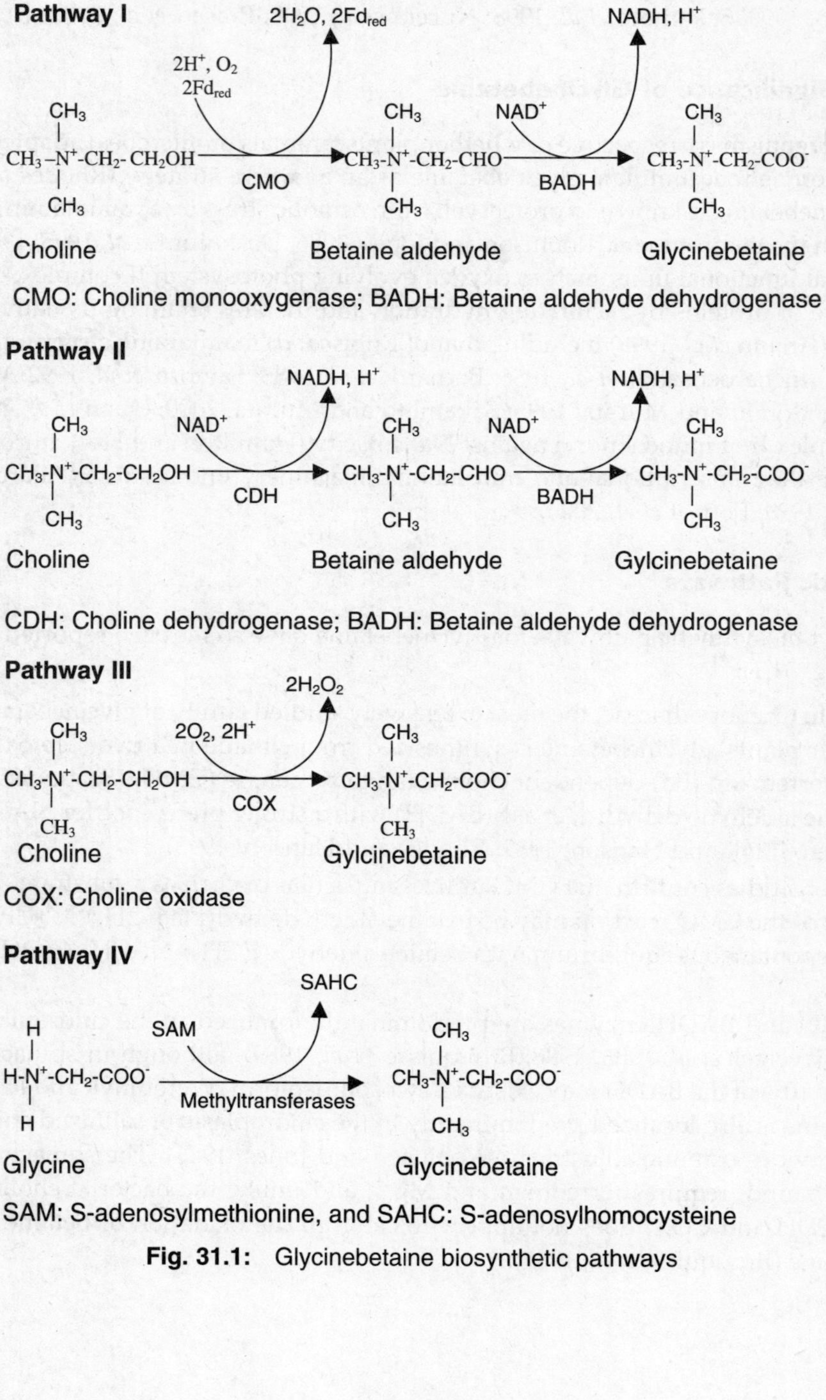

Fig. 31.1: Glycinebetaine biosynthetic pathways

cDNAs encoding BADH have been isolated from spinach (Weretilnyk and Hanson, 1990) and sugar beet (McCue and Hanson, 1992). The spinach and sugar beet BADHs show significant sequence identity to the human and bacterial BADHs described above (Chern and Pietruszko, 1995). The spinach BADH has an unusual chloroplast transit target sequence (Rathinasabapathi *et al*, 1994). The chenopod BADH cDNAs have facilitated cloning of BADHs from a number of other higher plant sources (reviewed in Rhodes and Hanson, 1993), including several monocotyledons (barley, rice and sorghum) (Ishitani *et al*, 1995; Wood *et al*, 1996; Nakamura *et al*, 1997). In leaves of both chenopods and grasses BADH transcripts and enzyme level increase in response to salinity stress (Arakawa *et al*, 1992; McCue and Hanson, 1992). A biochemical signal from the roots is implicated in this response in sugar beet (McCue and Hanson, 1992). Although transgenic tobacco plants over-expressing chloroplast localized BADH have been obtained they are unable to accumulate glycinebetaine in the absence of exogenously supplied betaine aldehyde as tobacco lacks CMO (Rathinasabapathi *et al*, 1994).

It should be cautioned, however, that BADH may be a general aldehyde dehydrogenase, acting on other aldehyde substrates in addition to betaine aldehyde (Trossat *et al*, 1997). BADH will utilize 3-dimethylsulfoniopropionaldehyde, an intermediate in DMSP synthesis, and certain aldehydes (3-aminopropionaldehyde and 4-aminobutyraldehyde) involved in polyamine metabolism (Trossat *et al*, 1997). Transgenic plants engineered for BADH expression (Holmstrom *et al*, 1994; Rathinasabapathi *et al*, 1994) may possibly exhibit phenotypes related to perturbed polyamine metabolism in addition to phenotypes attributable to alterations in the glycinebetaine synthesis pathway (Trossat *et al*, 1997). The multiple substrate specificities of BADH may explain the occurrence of BADH in plants that do not accumulate glycinebetaine (Weretilnyk *et al*, 1989; Ishitani *et al*, 1993), and in organs of glycinebetaine-accumulating plants which do not contain glycinebetaine (e.g. roots of cereals) (Ishitani *et al*, 1995).

Recent studies suggest that BADH is localized in peroxisomes in monocots (Nakamura *et al*, 1997). All monocotyledenous BADHs have a C-terminal tripeptide (SKL) that is known to be a signal for targeting preproteins to microbodies (Nakamura *et al*, 1997). This raises the question of whether the subcellular localization of the choline oxidation pathway is the same in grasses as in chenopods. If glycinebetaine synthesis occurs in the peroxisome in grasses, then CMO is unlikely to be the choline-oxidizing enzyme of grasses, because its Fd requirement would not be met within the peroxisome (Russell *et al*, 1998).

A cDNA encoding CMO has recently been cloned from spinach. The cDNA sequence confirms that CMO is a Fe-S protein containing one [2Fe-2S] cluster per subunit, and that CMO has a typical chloroplast transit peptide sequence (Rathinasabapathi *et al*, 1997). Spinach CMO does not show any homology with the other choline oxidizing enzymes (viz. CDH and COX) (Rathinasabapathi *et al*, 1997). The activity of CMO rises incrementally in response to salinity levels probably as a result of increased transcription (Brouquisse *et al*, 1989; Rathinasabapathi *et al*, 1997). The magnitude of salt induction of CMO mRNA is comparable to that reported for BADH (Rathinasabapathi *et al*, 1997). The spinach CMO has been used to clone the sugar beet CMO, and to detect salt-inducible CMO mRNA in leaves of *Amaranthus caudatus* (Amaranthaceae) (Russell *et al*, 1998).

Pathway II: In *E.coli* and certain animals, synthesis of glycinebetaine from choline is mediated by two dehydrogenases viz. membrane bound choline dehydrogenase and NAD^+-dependent betainealdehyde dehydrogenase.

In mammals, where glycinebetaine plays an important role as a compatible solute, particularly in the kidney (Bagnasco *et al*, 1986; Garcia-Perez and Burg, 1991), the pathway of glycinebetaine synthesis resembles that in *E. coli*, and is catalyzed by a particulate, mitochondrial CDH (Haubrich and Gerber, 1981) and a soluble BADH (Chern and Pietruszko, 1995).

Pathway III: The third pathway unlike the above two pathways, converts choline to glycinebetaine using a single enzyme and is operational only in certain soil bacteria such as *Arthrobacter globiformis*, *A. pascens* and *Alcaligenes* sp. (Ikuta *et al*, 1977; Rozwadowski *et al*, 1991).

A pathway of choline oxidation similar to *Arthrobacter* sp. has also been described for the fungus *Cylindrocarpon didymum* M-1. A choline oxidase (a dimer of two subunits of MW 64 kDa, containing two mol of FAD per mol of enzyme), utilizes O_2 as electron acceptor and generates H_2O_2 as product. Like the *Arthrobacter* enzyme, the fungal enzyme is also capable of oxidizing betaine aldehyde (K_m for betaine aldehyde = 5.8 mM, cf. K_m for choline = 1.3 mM), but occurs together with a NAD^+-dependent BADH (Tani *et al*, 1979 Yamada *et al*, 1979).

Pathway IV: In halophytic phototrophic bacterium *Ectothiorhodospira halochloris*, glycinebetaine is synthesized from glycine through direct *N*-methylation (Galinksi and Truper, 1994; Nyyssölä *et al*, 2000). Three *N*-methylations are catalyzed by two methyltransferases, glycine sarcosine methyltransferase and sarcosine dimethylglycine methyltransferase, with partially overlapping substrate specificity. The methyltransferases have been successfully expressed in *E. coli*, where they confer betaine accumulation and improved salt tolerance (Nyyssölä *et al*, 2000; Waditee *et al*, 2003).

Successful Engineering of Plants for Production of Glycinebetaine

Owing to the importance of glycinebetaine, there have been increased efforts to identify the genes involved in various biosynthetic pathways of glycinebetaine (Andresen *et al*, 1988; Weretilnyk and Hanson, 1990; Deshnium *et al*, 1995; Rathinasabapathi *et al*, 1997; Fan *et al*, 2004) and introduce the same in plant species, which lack the potential to synthesize glycinebetaine. Gene(s) associated with synthesis of glycinebetaine through the above pathways using choline as the precursor could be successfully introduced into higher plants. Plant species that have so far been transformed by the introduction of one of the above mentioned biosynthetic pathways for glycinebetaine is presented in Table 31.1.

Seed germination and early seedling growth have been reported to be the most vulnerable stages in the life cycle of plants that are sensitive to osmotic stress (Bewley and Black, 1982; Bliss *et al*, 1986). Prasad *et al*, (2000a,b) demonstrated successful insertion of *codA* gene into Indian mustard. Unlike wild type, the seed of transgenic lines with *codA* gene germinated well under salt (NaCl) (Fig. 31.2) in comparison to wild type. Seedling growth in terms of shoot length, fresh weight and dry weight of the transgenic line were significantly superior to that of wild type under salt (Fig. 31.3) as well as drought stress (unpublished).

Figure 31.4 depicts the photosynthetic activity (PSII activity) of the shoots of wild type and transgenic line of *Brassica juncea* grown on medium with various concentrations of choline. Irrespective of the choline concentration in the medium, the PSII activity in the shoots of the transgenic line was significantly higher than that of wild type.

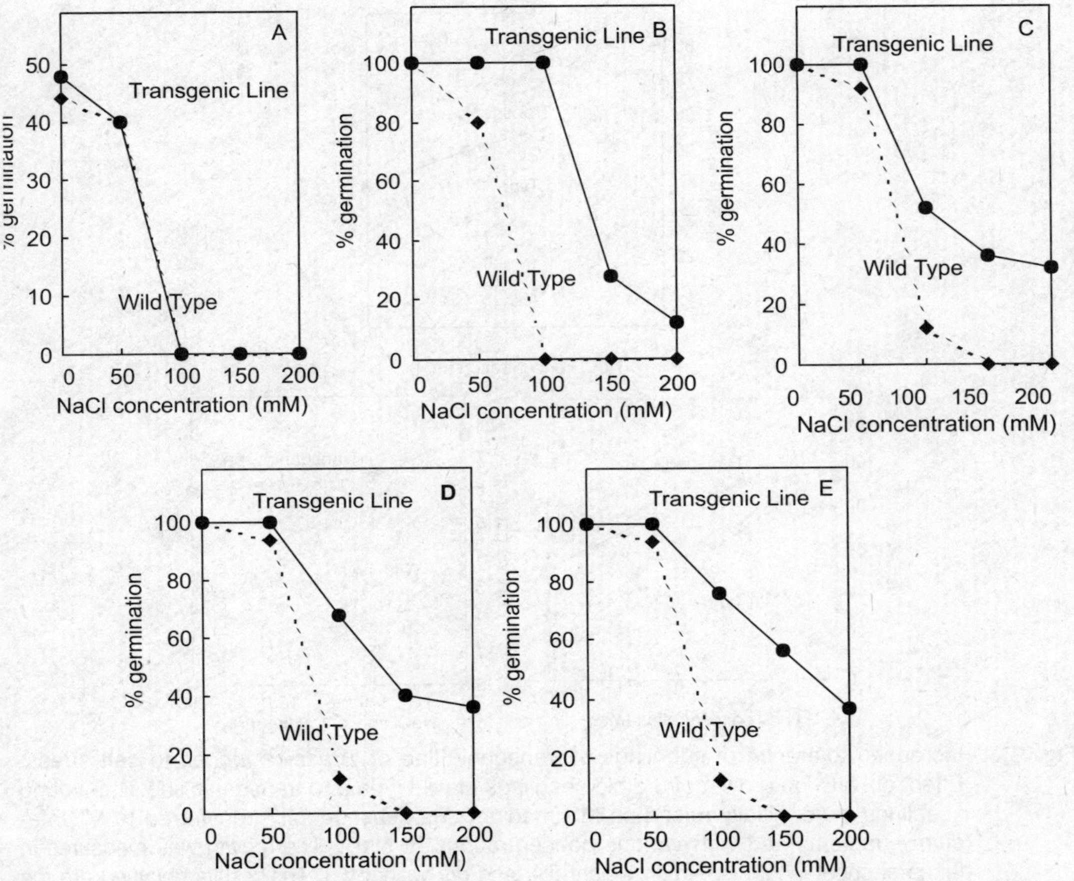

Fig. 31.2: Effect of various concentrations of NaCl on germination of seeds from wild type and transgenic line of *Brassica juncea* on 2 d (A), 4 d (B), 6 d (C), 8 d (D) and 10 d (E).

The tolerance of the transgenic line to NaCl stress was assessed in terms of the alteration in PSII activity in the leaf discs exposed to various concentrations of NaCl. A continous decline in PSII activity in the leaf discs (of both wild type and transgenic line) with time of their exposure to any levels of NaCl was observed. The extent of decline was more pronounced with an increase in the level of NaCl. However, the rate of decline was significantly lower in the leaf discs of transgenic line in comparison to that of wild type (Fig. 31.5).

Is Choline Really A Limiting Factor?

It is note worthy that the level of choline in transformed plants of *Arabidopsis* (Hayashi *et al*, 1997), Japonica rice (Sakamoto *et al*, 1998) and mustard (Prasad *et al*, 2000a,b) did not differ significantly from that of wild type plants suggesting that the transformed lines seem to have potential to maintain the required level of choline inspite of the introduction of a new pathway which uses choline for the synthesis of glycinebetaine.

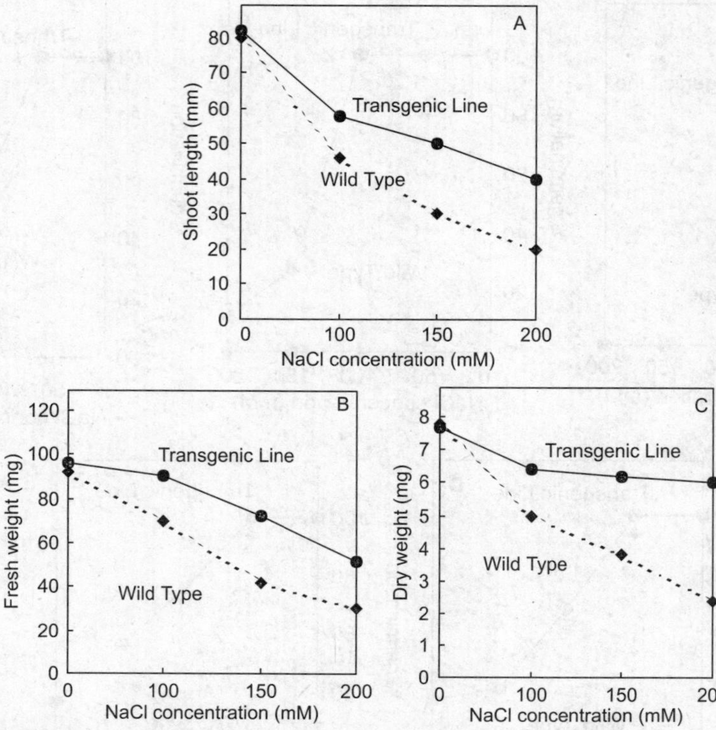

Fig. 31.3: Increased tolerance of seedlings of transgenic line of *Brassica juncea* to salt stress. Effect of NaCl on growth of 3 d old seedlings of wild type and transgenic line. The young seedlings were initially raised on MS medium and subsequently transferred to MS medium supplemented with various concentrations of NaCl. The growth was measured in terms of shoot length (A), fresh weight (B), and dry weight (C). The results obtained are the average from three independent experiments each with a minimum of four replicates.

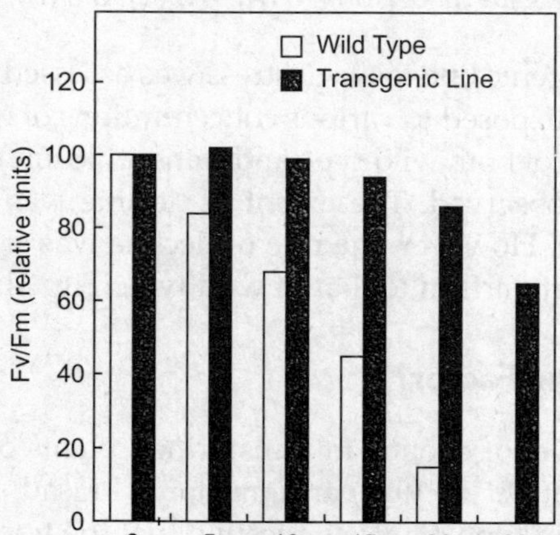

Fig. 31.4: Change in PSII activity in the leaf of wild type and transgenic line of *Brassica juncea* exposed to various choline concentrations. PSII activity was measured in terms of Fv/Fm ratio. The initial value of Fv/Fm was taken as 100%. The initial value of Fv/Fm was 0.83±0.015. For both wild type as well as transgenic line.

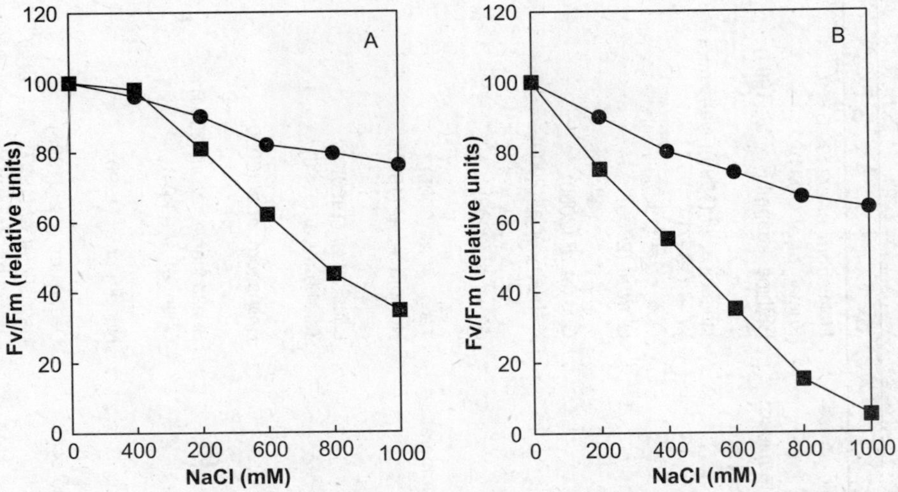

Fig. 31.5: Changes in PSII activity in the leaf discs of wild type and transgenic line of *Brassica juncea* exposed to various concentrations of NaCl at 12 h (A) and 24 h (B). PSII activity was measured in terms of Fv/Fm ratio. The initial value of Fv/Fm was taken as 100%. The initial value of Fv/Fm was 0.820 ± 0.02 for both wild type as well as transgenic line.

Earlier, Nuccio *et al* (1998) reported that low levels (than that observed in natural glycinebetaine accumulators) of glycinebetaine in the transgenic lines of tobacco might be due to limitation of endogenous choline. These investigators managed to enhance the level of glycinebetaine from 50 nmol g^{-1} fresh weight to a maximum of 800 nmol g^{-1} fresh weight by exogenous application of choline (5 mM). In contrast, even in the absence of any exogenously applied choline, the level of glycinebetaine in plants of *B. juncea* transformed with the *codA* gene ranged between 815 and 950 nmol g^{-1} fresh weight. Moreover, the level of choline in *B. juncea* was several times higher than that reported in tobacco (Nuccio *et al*, 1998; Prasad *et al*, 2000a,b). Hayashi *et al* (1997) and Sakamoto *et al* (1998) reported that the genotypes of *Arabidopsis* and *Oryza* bearing the *codA* gene also accumulate glycinebetaine to levels similar to that observed by us in *B. juncea*. Accordingly, Sakamoto and Murata (2000) also seem to believe that choline may not be a limiting factor.

Low glycinebetaine levels noted by Nuccio *et al* (1998) in tobacco plants could be due to (a) low level of choline; (b) lower potential to convert toxic betainealdehyde (Holmstrom *et al*, 1994; Rathinasabapathi *et al*, 1994) to non-toxic glycinebetaine due to low activity of betainealdehyde dehydrogenase (if present) or by some other dehydrogenase with broad specificity and/or (c) the site of production of betainealdehyde in transgenic tobacco is the chloroplast, whereas betainealdehyde dehydrogenase might be located in some other site such as peroxisomes as observed in rice (Nakamura *et al*, 1997).

Earlier, we have demonstrated that exogenously supplied choline suppresses growth and development of plants even at concentrations as low as 5 mM (Prasad *et al*, 2000b). As is clear from Fig. 31.6, besides growth and chlorophyll content choline severely affects photosynthetic machinery. Accordingly, there seems to be perfect homeostasis with regard to choline biosynthesis in plants. Therefore, any efforts that are made to promote over production of choline can disrupt the essential cellular metabolic events.

Table 31.1: Various plant species genetically engineered with the gene(s) for glycinebetaine synthesis

Enzyme and Source	*Gene*	*Plant/Cyanobacterial species transformed*	*Product synthesized and stress tolerant traits*	*References*
		Nuclear Transformation		
BADH, *Escherichia coli*	*betB*	*Nicotiana tabacum*	Glycinebetaine, osmotic stress	Holmstrom *et al* (1994)
BADH, *Escherichia coli*	*betB*	*Daucus carota*	Glycinebetaine, salt	Kumar *et al* (2004)
BADH, *Spinacia oleracea*, *Beta vulgaris*	*BADH*	*Nicotiana tabacum*	Glycinebetaine, not determined	Rathinasabapathi *et al* (1994)
BADH, *Hordeum vulgare*	*BADH*	*Nicotiana tabacum*	Glycinebetaine, salt	Ishitani *et al* (1995); Nakamura *et al* (1997)
BADH	*BADH*	*Oryza sativa*	salt	Guo *et al* (1997)
BADH, *Spinacia oleracea*	*BADH*	*Nasturtium officinale* (Watercress)	Not determined, salt	Li *et al* (2000)
BADH, *Atriplex hortensis*	*BADH*	*Triticum aestivum*	Not determined, salt	Guo *et al* (2000)
BADH, *Hordeum vulgare*	*BADH*	*Oryza sativa*	Glycinebetaine, salt, cold and heat	Kishitani *et al* (2000)
BADH, *Atriplex hortensis*	*BADH*	*Lycopersicon esculentum*	Glycinebetaine, salt	Jia *et al* (2002)
BADH, *Spinacia oleracea*	*BADH*	*Nicotiana tabacum*	Glycinebetaine, high temperature	Yang *et al* (2005)
CDH and BADH, *Escherichia coli*	*betA* and *betB*	*Synechococcus* sp.	Glycinebetaine, salt	Nomura *et al* (1995)
CDH, *Escherichia coli*	*betA*	*Nicotiana tabacum*	Betaine aldehyde, salt	Lilius *et al* (1996)
CDH, *Escherichia coli*	Modified *betA*	*Oryza sativa*	Betaine aldehyde, salt and drought	Takabe *et al* (1998)
CDH, *Escherichia coli*	Modified *betA*	*Nicotiana tabacum*	Betaine aldehyde, salt and cold	Holmstrom *et al* (2000)
CDH, *Escherichia coli*	*betA*	*Brassica oleracea*	Betaine aldehyde, salt	Bhattacharya *et al* (2004)
CDH, *Escherichia coli*	*betA*	Elite *Zea mays* inbred DH4866	Glycinebetaine, chilling	Quan *et al* (2004)
CMO, *Spinacea oleracea*	-	*Nicotiana tabacum*	Betaine aldehyde, not determined	Nuccio *et al* (1998)
COX, *Arthrobacter globiformis*	*codA*	*Synechococcus* sp.	Glycinebetaine, salt and cold	Deshnium *et al* (1995)
COX, *Arthrobacter globiformis*	*codA*	*Arabidopsis thaliana*	Glycinebetaine, salt, cold, heat and freezing	Hayashi *et al* (1997); Alia *et al* (1998a,b); Sakamoto *et al* (2000); Sulpice *et al* (2003)

Table 31.1: Contd.

COX, *Arthrobacter globiformis*	*codA*	*Oryza sativa*	Glycinebetaine, salt, cold and drought	Sakamoto *et al* (1998); Mohanty *et al* (2002); Sawahel (2003)
COX, *Arthrobacter globiformis*	*codA*	*Diospyros kaki*	Glycinebetaine, salt	Gao *et al* (2000)
COX, *Arthrobacter globiformis*	*codA*	*Brassica juncea*	Glycinebetaine, salt	Prasad *et al* (2000a,b); Prasad and Pardha Saradhi (2004)
COX, *Arthrobacter globiformis*	*codA*	*Lycopersicon esculentum*	Glycinebetaine, chilling	Park *et al* (2004)
COX, *Arthrobacter pascens*	*cox*	*Arabidopsis thaliana*	Glycinebetaine, salt, drought and freezing	Huang *et al* (2000)
COX, *Arthrobacter pascens*	*cox*	*Brassica napus*	Glycinebetaine, salt and drought	Huang *et al* (2000)
COX, *Arthrobacter pascens*	*cox*	*Nicotiana tabacum*	Glycinebetaine, salt	Huang *et al* (2000)
GSMT and SDMT, *Ectothiorhodospira halochloris*	*GSMT* and *SDMT*	*Escherichia coli*	Glycinebetaine, salt	Nyyssölä *et al* (2000)
GSMT and DMT, *Aphanothece halophytica*	*GSMT* and *DMT*	*Escherichia coli*	Glycinebetaine, salt	Waditee *et al* (2003)
GSMT and DMT, *Aphanothece halophytica*	*GSMT* and *DMT*	*Synechococcus* sp. PCC 7942	Glycinebetaine, salt and cold	Waditee *et al* (2005)
GSMT and DMT, *Aphanothece halophytica*	*GSMT* and *DMT*	*Arabidopsis thaliana*	Glycinebetaine, salt and cold	Waditee *et al* (2005)

BADH- Betaine aldehyde dehydrogenase; CDH- Choline dehydrogenase; CMO- Choline monooxygenase; COX- Choline oxidase; GSMT- Glycine sarcosine methyltransferase; DMT- Dimethyl sarcosine methyltransferase

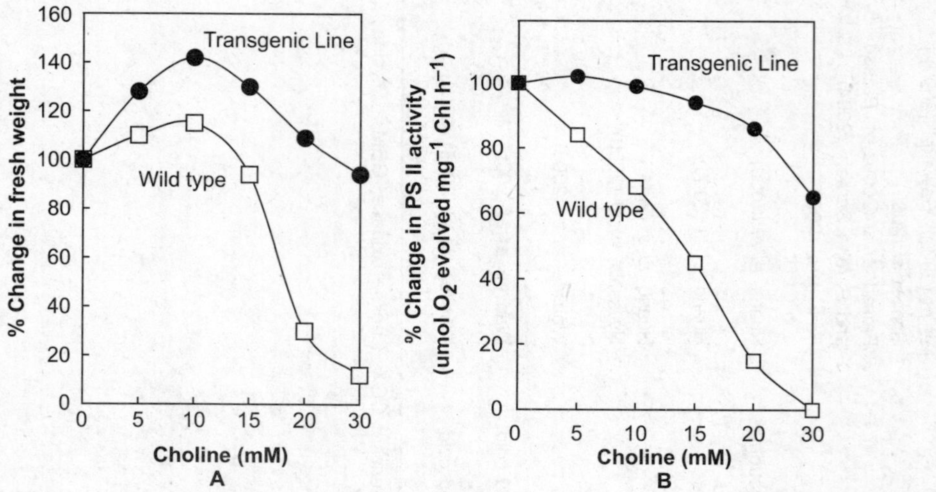

Fig. 31.6: Change in fresh weight (A) and PSII activity (B), of shoot of wild type and transgenic line of *Brassica juncea* grown on medium with various concentration of choline. The initial values of fresh weight, dry weight, chlorophyll content, PSII activity for wild type and transformed lines were taken as 100%. The initial value of fresh weight of the shoot was 380 ± 10 mg and oxygen evolution by thylakoids was 210.9 ± 18.6 μmol O_2 evolved mg^{-1} chl h^{-1}. The results are the average of data obtained from three independent experiments.

How are Glycinebetaine Synthesizing Transgenics Superior?

Earlier, glycinebetaine had been proposed to play a key role in osmoregulation (Robinson and Jones, 1986; McCue and Hanson, 1990). However, as the level of glycinebetaine in the leaves of transgenic genotypes of *B. juncea* was only 815-950 nmoles g^{-1} fresh weight (~0.04% of dry matter) and it did not alter in plants exposed to drought, it does not seem to contribute significantly toward osmoregulation. In contrast, the level of total sugars in the leaves of *B. juncea* exposed to drought stress go to the levels as high as ~25% of the dry matter (unpublished data). Therefore, we believe that glycinebetaine might be playing an important role in enhancing drought tolerance in *B. juncea* by protecting biological components such as cell membranes and proteins/enzymes (Hayashi *et al*, 1997). This facilitates protection and maintenance of major metabolic events in plant cells such as photosynthesis. Earlier, Papagerogiou and Murata (1995) had reported that glycinebetaine helps in protection of PSII activity by stabilizing its machinery against stress.

In summary, our findings along with the findings of Murata's research team clearly show that irrespective of the developmental stage at which the plants were exposed to salt, drought or low temperature stress, the transformed lines show better tolerance than those of wild type (in terms of maintenance of turgidity of leaves, level of chlorophyll and photosynthetic functions). Moreover, we believe that the introduction of this pathway into the crop plants would be beneficial than introduction of other pathways for glycinebetaine synthesis.

REFERENCES

Alia and P. Pardha Saradhi. 1991. Proline accumulation under heavy metal stress. *J. Plant Physiol.* 138: 554-558.

Alia and P. Pardha Saradhi. 1993. Suppression in mitochondrial electron transport is the prime cause behind stress induced proline accumulation. *Biochem. Biophy. Res. Commun.* 193: 54-58.

Alia, H. Hayashi, A. Sakamoto and N. Murata. 1998a. Enhancement of the tolerance of *Arabidopsis* to high temperatures by genetic engineering of the synthesis of glycinebetaine. *Plant J.* 16: 155-161.

Alia, H. Hayashi, T.H.H. Chen and N. Murata. 1998b. Transformation with gene for choline oxidase enhances the cold tolerance of *Arabidopsis* during germination and early growth. *Plant Cell Environ.* 21: 232-239.

Alia, P. Pardha Saradhi and P. Mohanty. 1991. Proline enhances primary photochemical activities in isolated thylakoid membranes of *Brassica juncea* by arresting photoinhibitory damage. *Biochem. Biophy. Res. Commun.* 181: 1238-1244.

Alia, P. Pardha Saradhi and P. Mohanty. 1993. Involvement of proline in protecting thylakoid membranes against free radical-induced photodamage. *J. Photochem. Photobiol.* 38: 253-257.

Alia, P. Pardha Saradhi and P. Mohanty. 1997. Involvement of proline in protecting thylakoid membranes against free radical-induced photodamage. *J. Photochem. Photobiol.* 38: 253-257.

Alia, K.V.S.K. Prasad and P. Pardha Saradhi. 1995. Effect of zinc on free radicals and proline in *Brassica juncea* and *Cajanus cajan*. *Phytochemistry* 39: 45-47.

F. Allard, M. Houde, M. Krol, A. Ivanov, N.P.A. Hunter and F. Sarhan. 1998. Betaine improves freezing tolerance in wheat. *Plant Cell. Physiol.* 39: 1194-1202.

R. G. Alscher, J. L. Donahue and C. L. Cramer. 1997. Reactive oxygen species and antioxidants: Relationships in green cells. *Physiol. Plant.* 100: 224-233.

P. A. Andresen, I. Kaasen, O. B. Styrvold, G. Boulnois, and A. R. Strøm. 1988. Molecular cloning, physical mapping and expression of the *bet* genes governing the osmoregulatory choline-glycine betaine pathway of *Escherichia coli*. *J. Gen. Microbiol.* 134: 1737-1746.

F. Anjum, V. Rishi and F. Ahmad. 1999. Compatibility of osmolytes with Gibbs energy of stabilization of proteins, *Biochim. Biophys. Acta.* 1476: 75-84.

K. Arakawa, K. Mizuno, S. Kishitaini and T. Takabe. 1992. Immunological studies of betaine aldehyde dehydrogenase in barley. *Plant Cell Physiol.* 108: 581-588.

T. Arakawa and S.N. Timasheff. 1985. The stabilization of proteins by osmolytes. *Biophys. J.* 47: 411-414.

S. Arora and P. Pardha Saradhi. 1995. Light-induced enhancement in proline levels in *Vigna radiata* exposed to environmental stresses. *Aust. J. Plant Physiol.* 22: 383-386.

A. P. Aryan, R.G. Batt and W. Wallace. 1983. Reversible inactivation of nitrate reductase NADH and the occurrence of partially inactive enzyme in wheat leaf. *Plant Physiol.* 71: 582-587.

D. Aspinall and L.G. Paleg. 1981. Proline accumulation: Physiological aspects. In: *The physiology and biochemistry of drought resistance in plants*. L.G. Paleg and D. Aspinall (eds.), pp. 215-228. Academic Press, Sydney.

S. Bagnasco, R. Balaban, H.M. Fales, Y.M. Yang and M. Burg. 1986. Predominant osmotically active organic solutes in rat and rabbit renal medullas. *J. Biol. Chem.* 261: 5872-5877.

T. Bernard, M. Ayache and D.L. Rudlier. 1988. Restoration of growth and enzymic activities of *Escherichia coli* Lac mutant by glycinebetaine. C. R. Acad. Sci. III 307: 99-104.

J.D. Bewley and N. Black. 1982. Physiology and Biochemistry of Seed in Relation to Germination. 2, Berlin: Spring- Verlag.

R.C. Bhattacharya, M. Maheshwari, V. Dineshkumar, P.B. Kirti, S.R. Bhat and V.L. Chopra. 2004. Transformation of *Brassica oleracea* cv. Capitata with bacterial *betA* gene enhances tolerance to salt stress. *Sci. Hortic.* 100: 215-227.

R.D. Bliss, K.A. Platt-Aloka and W.W. Thomson. 1986. The inhibitory effect of NaCl on barley germination. *Plant Cell Environ.* 9: 727-733.

H.J. Bohnert and R.G. Jensen. 1996. Strategies for engineering water stress tolerance in plants. *TIBTECH* 14: 89-97.

H.J. Bohnert and B. Shen. 1999. Transformation and compatible solutes. *Sci. Hortic.* 78: 237-260.

S. Bourot, O. Sire, A. Trautwetter, T. Touze, L.F. Wu, C. Blanco, and T. Bernard. 2000. Glycinebetaine assisted protein folding in a lysA mutant of *Escherichia coli*. *J. Biol. Chem.* 275: 1050-1056.

C. Bowler, M.V. Montagu and D. Inze. 1992. Superoxide dismutase and stress tolerance. *Ann. Rev. Plant. Physiol. Plant Mol. Biol.* 43: 83-116.

R. Brouquisse, P. Weigel, D. Rhodes, C.F. Yocum and A.D. Hanson. 1989. Evidence for a ferredoxin-dependent choline monoxygenase from spinach chloroplast stroma. *Plant Physiol.* 108: 581-588.

M. Burnet, P.J. Latontaine and A.D. Hanson. 1995. Assay, purification, and partial characterization of choline monooxygenase from spinach. *Plant Physiol.* 108: 581-588.

W.H. Campbell and J. Smarelli Jr. 1986. Nitrate reductase biochemistry and regulation In: Biochemical basis of plant breeding, Vol. II. C. A. Neyra (ed.) pp. 1-39. CRC Press Inc. Boca Raton, FL.

M.K. Chern and R. Pietruszko. 1995. Human aldehyde dehydrogenase E3 isoenzyme is a betaine aldehyde dehydrogenase (BADH): structural similarities to the mammalian ALDHs and a plant BADH. *Gene* 103: 45-52.

T.H.H. Chen and N. Murata. 2002. Enhancement of tolerance of abiotic stress by metabolic engineering of betaines and other compatible solutes. *Curr. Opin. Plant. Biol.* 5: 250-257.

R.W.C. Dawson, D.C. Elliot, W.H. Elliot and K.M. Jones. 1969. Data for Biochemical Research. 2nd edition, Clarendon Press, Oxford, p. 12.

P. Deshnium, D.A. Los, H. Hayashi, L. Mustardy and N. Murata. 1995. Transformation of *Synechococcus* with a gene for choline oxidase enhances tolerance to salt stress. *Plant Mol. Biol.* 29: 897-907.

P. Deshnium, Z. Gombos, Y. Nishiyama and N. Murata. 1997. The action of glycine betaine in enhancement of tolerence of *Synechococcus* sp. strain PCC 7942 to low temperature. *J. Bacterol.* 179: 339-344.

R.S. Dubey. 1997. Photosynthesis in plants under stressful conditions. In: *Handbook of Photosynthesis*. M. Pessarakli, (ed.) pp. 859-876 Marcel Dekker, Inc. New York.

F. Fan, M. Ghanem and G. Gadda. 2004. Cloning, sequence analysis, and purification of choline oxidase from *Arthrobacter globiformis*: a bacterial enzyme involved in osmotic stress tolerance. *Arch. Biochem. Biophys.* 421: 149-158.

C.H. Foyer, H. Lopez-Delgado, J.F. Dat and I.M. Scott. 1997. Hydrogen peroxide- and glutathione-associated mechanism of acclimatory stress tolerance and signalling. *Physiol. Plant.* 100: 241-254.

C.H. Foyer, M. Lelandais and K.J. Kunert. 1994. Photooxidative stress in plants. *Physiol. Plant.* 92: 696-717.

E.A. Galinski and H.G. Truper. 1994. Microbiol behaviur in salt-stressed ecosystems. *FEMS Microbiol. Rev.* 15: 95-108.

M. Gao, A. Sakamoto, K. Miura, N. Murata, A. Sugiura and R. Tao. 2000. Transformation of Japanese persimmon (*Diospyros kaki* Thunb.) with a bacterial gene for choline oxidase. *Mol. Breed.* 6: 501-510.

A. Garcia-Perez and M.B. Burg. 1991. Renal medullary organic osmolytes. *Physiol. Rev.* 71: 1081-1115.

M.T. Giardi, J. Masojidek and D. Godde. 1997. Effects of abiotic stresses on the turnover of the D_1 reaction centre II protein. *Physiol. Plant.* 101: 635-642.

A. Grover, C. Sahi, N. Sanan and A. Grover. 1999. Taming abiotic stress in plants through genetic engineering: current strategies and perspective. *Plant Sci.* 143: 101-111.

Y. Guo, L. Zhang, G. Xiao, S.Y. Cao, D.M. Gu, W.Z. Tian and S.Y. Chen. 1997. Expression of betaine aldehyde dehydrogenase gene and salinity tolerance in rice transgenic plants. *Sci. China* 40: 496-501.

B.H. Guo, Y.M. Zhang, H.J. Li, L.Q. Du, Y.X. Li, J.S. Zhang, S.Y. Chen and Z.Q. Zhu. 2000. Transformation of wheat with a gene encoding for the betaine aldehyde dehydrogenase (BADH). *Acta. Bot. Sinica.* 42: 279–283.

A.S. Gupta, J.L. Heinen, A.S. Holaday, J.J. Bruke and R.D. Allen. 1993a. Increased resistance to oxidative stress in transgenic plants that over expresses chloroplastic Cu/Zn superoxide dismutase. *Proc. Natl. Acad. Sci. USA* 90: 1629-1633.

A.S. Gupta, R.P. Webb, A.S. Holaday and R.D. Allen. 1993b. Overexpression of superoxide dismutase protects plants from oxidative stress. Induction of ascorbate peroxidase in superoxide dismutase-overexpressing plants. *Plant Physiol.* 103: 1067-1073.

E.W. Hamilton III and S.A. Heckathron. 2001. Mitochondrial adaptations to NaCl. Complex I is protected by anti-oxidants and small heat shock proteins, whereas complex II is protected by proline and betaine. *Plant Physiol.* 126: 1266-1274.

A.D. Hanson, B. Rathinasabapathi, J. Revoal, M. Burnet, M.O. Dillon and D.A. Gage. 1994. Osmoprotective compounds in plumbaginaceae: A natural experiment in metabolic engineering of stress tolerance. *Proc. Natl. Acad. Sci. USA* 91: 306-310.

D.R. Haubrich and N.H. Gerber. 1981. Choline dehydrogenase: assay, properties and inhibitors. *Biochem. Pharmacol.* 30: 2993-3000.

H. Hayashi, Alia, L. Mustardy, P. Deshnium, M. Ida and N. Murata. 1997. Transformation of *Arabidopsis thaliana* with the *codA* gene for choline oxidase; accumulation of glycinebetaine and enhanced tolerance to salt and cold stress. *Plant J.* 12: 133-142.

N. Holmberg and L. Bulow. 1998. Improving stress tolerance in plants by gene transfer. *Trends Plant Sci.* 3: 61-66.

K.O. Holmstrom, B. Weilin, A. Mandal, I. Kristiansdottir, T.H. Teeri, T. Lamark, A.R. Strom and E.T. Palva. 1994. Production of the *Escherichia coli* betaine-aldehyde dehydrogenase, an enzyme required for the synthesis of osmoprotectant glycinebetaine in transgenic plants. *Plant J.* 6: 749-758.

K.O. Holmstrom, E. Mantyla, B. Welin, A. Mandal, and E.T. Palva. 1996. Drought tolerance in tobacco. *Nature* 379: 683-684.

K.O. Holmstrom, S. Somersalo, A. Mandal, T.E. Palva and B. Welin. 2000. Improved tolerance to salinity and low temperature in transgenic tobacco producing glycine betaine. *J. Exp. Bot.* 51: 177-185.

W.A. Houry, D. Frishman, C. Eckerskorn, F. Lottspeicch and F. Hartl. 1999. Identification of in vivo substrates of the chaperonin GroEL. *Nature* 402: 147-154.

J. Huang, R. Hirji, L. Adam, K.L. Rozwadowski, J.K. Hammerlindi, W.A. Keller and G. Selvaraj. 2000. Genetic engineering of glycinebetaine production toward enhancing stress tolerance in plants: Metabolic limitations. *Plant Physiol.* 122: 747-756.

S. Ikuta, S. Imamura, H. Misaki and Y. Horiuchi. 1977. Purification and characterization of choline oxidase from *Arthrobacter globiformis*. *J. Biochem.* 82: 1741-1749.

A. Incharoensakdi, T. Takabe and T. Akazawa. 1986. Effect of betaine on enzyme activity and subunit interaction of ribulose 1-5-bisphosphate carboxylase/oxygenase from *Aphanothece halophytica*. *Plant Physiol.* 81: 1044-1049.

M. Ishitani, K. Arakawa, K. Mizuno, S. Kishitani and T. Takabe. 1993. Betaine aldehyde dehydrogenase in gramineae: levels in leaves of both betaine-accumulating and non-accumulating cereal plants. *Plant Cell Physiol.* 34: 493-495.

M. Ishitani, T. Nakamura, S.Y. Han and T. Takabe. 1995. Expression of the betaine aldehyde dehydrogenase gene in barley in response to osmotic stress and abscisic acid. *Plant Mol. Biol.* 27: 307-315.

W. Jia, Y. Wang, S. Zhang and J. Zhang. 2002. Salt-stress-induced ABA accumulation is more sensitively triggered in roots than in shoots. *J. Exp. Bot.* 53: 2201-2206.

S. Kishitani, T. Takanami, M. Suzuki, M. Oikawa, S. Yokoi, M. Ishitani, A.M. Alvarez-Nakase, T. Takabe and T. Takabe. 2000. Compatibility of glycinebetaine in rice plants: Evaluation using transgenic rice plants with a gene for peroxisomal betaine aldehyde dehydrogenase from barley. *Plant Cell Environ.* 23: 107-114.

P.B.K. Kishor, Z. Hong, G.H. Miao, C.A.A. Hu and D.P.S. Verma. 1995. Overexpression of Δ^1-pyrroline-5-carboxylate synthetase increases proline production and confers osmotolerance in transgenic plants. *Plant Physiol.* 108: 1387-1394.

G. J. J. Kortstee. 1970. The aerobic decomposition of choline by microorganisms. *Arch. Microbiol.* 71: 235-244.

S. Kumar, A. Dhingra and H. Daniell. 2004. Plastid-expressed *betaine aldehyde dehydrogenase* gene in carrot cultured cells, roots, and leaves confers enhanced salt tolerance. *Plant Physiol.* 136: 2843-2854.

A.C. Leopold and R.P. Willing. 1984. Evidence for toxicity effects of salt on membranes. In: *Salinity tolerance in plants*, R.C. Staples and G.H. Toenniessen (eds.), pp. 67-76 Wiley, New York.

C. Lerma, A.D. Hanson and D. Rhodes. 1988. Oxygen-18 and deuterium labeling studies of choline oxidation by spinach and sugar beet. *Plant Physiol.* 88: 695-702.

Y.X. Li, F.Q. Chang, L.Q. Du, B.H. Guo, H.J. Li, J.S. Zhang, S.Y. Chen and Z.Q. Zhu. 2000. Genetic transformation of watercress with a gene encoding for betaine aldehyde dehydrogenase (BADH). *Acta. Bot. Sinica.* 42: 480-484.

G. Lilius, N. Holmberg and L. Bûlow. 1996. Enhanced NaCl stress tolerance in transgenic tobacco expressing bacterial choline dehydrogenase. *Bio/Technology* 14: 177-180.

A. Livena and N. Levin. 1967. Tissue respiration and mitochondrial oxidative phosphorylation of NaCl treated pea seedlings. *Plant Physiol.* 42: 407-414.

K.F. McCue and A.D. Hanson. 1990. Drought and salt tolerance: towards understanding and application. *Trends Biotech.* 8: 358-362.

K.F. McCue and A.D. Hanson. 1992. Salt–inducible betaine aldehyde dehydrogenase from sugar beet: cDNA cloning and expression. *Plant Mol. Bio.* 18: 1-15.

A. Mohanty, H. Kathuria, A. Ferjani, A. Sakamoto, P. Mohanty, N. Murata and A.K. Tyagi. 2002. Transgenics of an elite *indica* rice variety Pusa Basmati 1 harbouring the *codA* gene are highly tolerant to salt stress. *Theor. Appl. Genet.* 106: 51-57.

T. Nakamura, S. Yokota, Y. Muramoto, K. Tsutsui, Y. Oguri, K. Fukui, and T. Takabe. 1997. Expression of betaine aldehyde dehydrogenase gene in rice, a glycinebetaine nonaccumulator, and possible localization of its protein in peroxisomes. *Plant J.* 11: 1115-1120.

S. Nakaya, Y. Tanaka, Y. Yoshioka and N. Yasuda. 1991. Interaction of water with nitrogenous compounds in human urine as studied by nuclear magnetic resonance spectroscopy, *J. Iwat. Med. Asc.* 43: 311-320.

D. Nikolopoulos and Y. Manetas. 1991. Compatible solutes and in vitro stability of Salsola soda enzymes: proline incompatibility. *Phytochemistry* 30: 411-413.

Nomura, M. Ishitani, T. Takabe, A.K. Rai and T. Takabe. 1995. *Synechococcus* sp. PCC7942 transformed with *Escherichia coli bet* genes produces glycine betaine from choline and acquires resistance to salt stress. *Plant Physiol.* 107: 703-708.

M.L. Nuccio, B.L. Russel, K.D. Nolte, B. Rathinasabapathi, D.A. Gag and A.D. Hanson. 1998. The endogenous choline supply limits glycinebetaine synthesis in transgenic tobacco expressing choline monooxygenase. *Plant J.* 16: 487-496.

A. Nyysola, J. Kerovuo, P. Kaukinen, N.V. Weymarn and T. Reinikainen. 2000. Extremes halophiles synthesis betaine from glycine by methylaton. *J. Biol. Chem.* Vol. 275: 22196-22201.

G.C. Papageorgiou and N. Murata. 1995. The unusually strong stabilizing effects of glycinebetaine on the structure and function in the oxygen-evolving photosystem-II complex. *Photosyn. Res.* 44: 243-252.

G.C. Papageorgiou, Y. Fujimura and N. Murata. 1991. Protection of oxygen-evolving Photosystem II complex by glycinebetaine. *Biochem. Biophys. Acta* 1057: 361-366.

P. Pardha Saradhi, Alia and B. Vani. 1993. Inhibition of mitochondrial electron transport is the prime cause behind proline accumulation during mineral deficiency in *Oryza sativa. Plant Soil* 155/156: 465-468.

P. Pardha Saradhi, Alia, S. Arora and K.V.S.K. Prasad. 1995. Proline accumulates in plants exposed to UV radiation and protects them against UV induced peroxidation. *Biochem. Biophy. Res. Commun.* 209: 1-5.

P. Pardha Saradhi and P. Sharmila. 2003. Improvement of crop plants for abiotic stress tolerance through introduction of glycinebetaine pathway. In: *Plant Genetic Engineering*. (R.P. Singh and P.K. Jaiwal (eds.), SCI Tech Publishing LLC, U.S.A., pp 147-177.

E.J. Park, Z. Jekniæ, A. Sakamoto, J. DeNoma, R. Yuwansiri, N. Murata and T.H.H. Chen. 2004. Genetic engineering of glycinebetaine synthesis in tomato protects seeds, plants, and flowers from chilling damage. *Plant J.* 40: 474-487.

E.A.H. Pilon-Smits, M.J.M. Ebskamp, M.J. Paul, M.J.W. Jeuken, P.J. Weisbeek and C.M. Smeekens. 1995. Improved performance of transgenic fructan accumulating tobacco under drought stres. *Plant Physiol.* 107: 125-130.

K.V.S.K. Prasad and P. Pardha Saradhi. 2004. Enhanced tolerance to photoinhibition in transgenic plants through targeting of glycinebetaine biosynthesis into the chloroplasts. *Plant Sci.* 166: 1197-1212.

K.V.S.K. Prasad, P. Pardha Saradhi and P. Sharmila. 1999. Concerted action of antioxidant enzymes and curtailed growth under zinc toxicity in *Brassica juncea. Environ. Exp. Bot.* 42: 1-10.

K.V.S.K. Prasad, P. Sharmila, P.A. Kumar and P. Pardha Saradhi. 2000a. Transformation of *Brassica juncea* (L.) Czern with bacterial *codA* gene enhances its tolerance to salt stress. *Mol. Breed.* 6: 489-499.

K.V.S.K. Prasad, P. Sharmila and P. Pardha Saradhi. 2000b. Enhanced tolerance of transgenic *Brassica juncea* to choline confirms successful expression of the bacterial *codA* gene. *Plant Sci.* 159: 233-242.

R. Quan, M. Shang, H. Zhang, Y. Zhao and J. Zhang. 2004. Improved chilling tolerance by transformation with *betA* gene for the enhancement of glycinebetaine synthesis in maize. *Plant Sci.* 166: 141-149.

D.W. Rains. 1972. Salt transport by plants in relation to salinity. *Ann. Rev. Plant Physiol.* 23: 516-525.

M.V. Rajam, S. Dagar, B. Waie, J.S. Yadav, P.A. Kumar, F. Shoeb and R. Kumaria. 1998. Genetic engineering of polyamine and carbohydrate metabolism for osmotic stress tolerance in higher plants. *J. Biosci.* 23: 473-482.

K.R. Rao and A. Gnanam. 1990. Inhibition of nitrate and nitrate reductase activities by salinity stress in *Sorghum vulgare. Phytochemistry* 29: 1047-1050.

B. Rathinasabapathi, K.F. McCue, D.A. Gage and A.D. Hanson. 1994. Metabolic engineering of glycinebetaine synthesis: plant betainealdehyde dehydrogenase lacking transit peptide are targeted to tobacco chloroplast where they confer betainealdehyde resistance. *Planta* 193: 155-162.

B. Rathinasabapati, M. Burnet, B.L. Russel, D.A. Gage and A.D. Hanson. 1997. Choline monooxygenase, an unusual iron-sulphur enzyme catalyzing the first step of glycine betaine synthesis in plant: prosthetic group characterization and cDNA cloning. *Proc. Natl. Acad. Sci. USA* 94: 3454-3458.

D. Rhodes and A.D. Hanson. 1993. Quaternary ammonium and tertiary sulfonium compounds in higher plants. *Ann. Rev. Plant. Physiol. Plant Mol. Biol.* 44: 357-384.

S.P. Robinson and G.P. Jones. 1986. Accumulation of glycinebetaine in chloroplast provides osmotic adjustments during salt stress. *Aust. J. Plant Physiol.* 13: 659-668.

K.L. Rozwadowski, G.G. Khachatourians and G. Selvaraj. 1991. Choline oxidase, a catabolic enzyme in *Arthrobacter pascens*, facilitates adaptation to osmotic stress in *Escherichia coli. J. Bacteriol.* 173: 472-478.

B.L. Russell, B. Rathinasabapati, A.D. Hanson. 1998. Osmotic stress induced expression of choline monooxygenase in sugar beet and *Amaranth. Plant Physiol.* 116: 859-865.

A. Sakamoto, Alia and N. Murata. 1998. Metabolic engineering of rice leading to biosynthesis of glycinebetaine and tolerance to salt and cold. *Plant Mol. Biol.* 38: 1011-1019.

A. Sakamoto and N. Murata. 2000. Genetic engineering of glycinebetaine synthesis in plants: current status and implications for enhancement of stress tolerance. *J. Exp. Bot.* 51: 81-88.

A. Sakamoto and N. Murata. 2001. The use of bacterial choline oxidase, a glycinebetaine-synthesizing enzyme, to create stress-resistant transgenic plants. *Plant Physiol.* 125: 180-188.

A. Sakamoto, R. Valverde, Alia, T.H.H. Chen and N. Murata. 2000. Transformation of *Arabidopsis* with the *codA* gene for choline oxidase enhances freezing tolerance of plants. *Plant J.* 22: 449-453.

M.M. Santoro, Y. Liu, S.M. Khan, L.X. Hou and D.W. Bolen. 1992. Increased thermal stability of proteins in the presence of naturally occurring osmolytes. *Biochemistry* 31: 5278-5283.

W. Sawahel. 2003. Improved performance of transgenic glycinebetaine-accumulating rice plants under drought stress. *Biol. Plant.* 47: 39-44.

J.G. Scandalios. 1993. Oxygen stress and superoxide dismutase. *Plant Physiol.* 101: 7-12.

B. Shen, R.G. Jensen and H.J. Bohnert. 1997a. Increased resisitance to oxidative stress in transgenic plants by targeting mannitol biosynthesis to chloroplasts. *Plant Physiol.* 113: 1177-1183.

B. Shen, R.G. Jensen and H.J. Bohnert. 1997b. Mannitol protects against oxidation by hydroxyl radicals. *Plant Physiol.* 115: 527-532.

E. Sheveleva, W. Chmara, H.J. Bohnert and R.G. Jensen. 1997. Increased salt and drought tolerance by D-ononitol production in transgenic *Nicotiana tabacum* L. *Plant Physiol.* 115: 1211-1219.

P. Sivakumar, P. Sharmila and P. Pardha Saradhi. 1998. Proline suppresses Rubisco activity in higher plants *Biochem. Biophys. Res. Commun.* 252: 428-432.

P. Sivakumar, P. Sharmila and P. Pardha Saradhi. 2000. Proline alleviates salt stress induced enhancement in ribulose 1,5-bisphosphate oxygenase activity. *Biochem. Biophys. Res. Commun.* 279: 512-515.

P. Sivakumar, P. Sharmila and P. Pardha Saradhi. 2001. Proline suppresses Rubisco activity by dissociating small subunits from holoenzyme. *Biochem. Biophys. Res. Commun.* 282: 236-241.

N. Smirnoff and Q. J. Cumbes. 1989. Hydroxyl radical scavenging activity of compatible solutes. *Phytochemistry* 28: 1057-1060.

H.S. Srivastava. 1980. Regulation of nitrate reductase activity in higher plants. *Phytochemistry* 19: 728-733.

R. Sulpice, H. Tsukaya, H. Nonaka, L. Mustardy, T.H.H. Chen and N. Murata. 2003. Enhanced formation of flowers in salt-stressed *Arabidopsis* after genetic engineering of the synthesis of glycine betaine. *Plant J.* 36: 165-176.

T. Takabe, Y. Hayashi, A. Tanaka, T. Takabe and S. Kishitani. 1998. Evaluation of glycinebetaine accumulation for stress tolerance in transgenic rice plants. In: *Proceedings of International Workshop on Breeding and Biotechnology for Environmental Stress in Rice*, Hokkaido Agricultural Experiment Station, Sapporo, Japan, pp. 63-68.

Y. Tani, N. Mori, K. Ogata, H. Yamada. 1979. Production and purification of choline oxidase from *Cylindrocarpon didymium* M-1. *Agric. Biol. Chem.* 43: 815-820.

M. Tarczynski, R. Jensen and H.J. Bohnert. 1993. Stress protection of transgenic tobacco by production of osmolytes mannitol. *Science* 259: 508-510.

J.M. Tepperman and P. Dunsmuir. 1990. Transformed plants with elevated levels of chloroplastic SOD are not more resistant to superoxide toxicity. *Plant Mol. Biol.* 14: 501-511.

J.C. Thomas, M. Sephai, B. Arendall and H.J. Bohnert. 1995. Enhancement of seed germination in high salinity by engineering mannitol expression in *Arabidopsis thaliana. Plant Cell Environ.* 18: 801-806.

C. Trossat, B. Rathinasabapathi and A.D. Hanson. 1997. Transgenically expressed betaine aldehyde dehydrogenase efficiently catalyzes oxidation of dimethylsulphoniopropionaldehyde and w-aminoaldehydes. *Plant Physiol.* 113: 1457-1461.

J.H. Venekamp, J.E. Lampe and J.T.M. Koot. 1989. Organic acids as a source of drought-induced proline synthesis in field bean plants, *Vicia faba* L. *J. Plant Physiol.* 133: 654-659.

R. Waditee, Y. Tanaka, K. Aoki, T. Hibino, H. Jikuy , J. Takano, T. Takabe and T. Takabe. 2003. Isolation and functional characterization of *N*-methyltransferases that catalyze betaine synthesis from glycine in a halotolerant photosynthetic organism *Aphanothece halophytica. J. Biol. Chem.* 278: 4932-4942.

R. Waditee, N.H. Md. Bhuiyan, V. Rai, K. Aoki, Y. Tanaka, T. Hibino, S. Suzuki, J. Takano, A.T. Jagendorf, T. Takabe and T. Takabe. 2005. Genes for direct methylation of glycine provide high levels of glycinebetaine and abiotic-stress tolerance in *Synechococcus* and *Arabidopsis*. *Proc. Natl. Acad. Sci. USA* 102: 1318-1323.

P. Weigel, C. Lerma and A.D. Hanson. 1988. Choline oxidation by intact spinach chloroplast. *Plant Physiol.* 86: 54-60.

P. Weigel, E.A. Weretilnyk and A.D. Hanson. 1986. Betaine aldehyde oxidation by spinach chloroplast. *Plant Physiol.* 82: 753-759.

E.A. Weretilnyk and A.D. Hanson. 1988. Betaine aldehyde dehydrogenase polymorphism in spinach: genetic and biochemical characterization. *Biochem. Genet.* 26: 143-151.

E.A. Weretilnyk, S. Bednarek, F. Kent, K.F. McCue, D. Rhodes and A.D. Hanson. 1989. Comparative biochemical and immunological studies of the glycine betaine synthesis pathway in diverse families of dicotyledons, *Planta* 178: 342-352.

E.A. Weretilynk and A.D. Hanson. 1990. Molecular cloning of a plant betaine aldehyde dehydrogenase in an enzyme implicated in adaptation to salinity and drought. *Proc. Natl. Acad. Sci. USA* 87: 2475-2749.

C.L. Winzor, D.J. Winzor, L.G. Paleg, G.P. Jones and B.P. Naidu. 1992. Rationalization of the effects of compatible solutes on protein stability in terms of thermodynamic non ideality. *Arch. Biochem. Biophys.* 296: 102-107.

A.J. Wood, H. Saneoka, D. Rhodes, R.J. Joly and P.B. Goldsbrough. 1996. Betaine aldehyde dehydrogenase in sorghum. Molecular cloning and expression of two related genes. *Plant Physiol.* 110: 1301-1308.

R.G. Wyn Jones and R. Storey. 1981. Betaines. In: *The Physiology and biochemistry of drought resistance in plants*, pp. 171-204, (L.G. Paleg and D. Aspinall (eds.), Academic Press, Sydney New York London Toronto San Francisco.

R.G. Wyn Jones, A.J. Rippin and R. Storey. 1973. Metabolism of choline in the rhizosphere and its possible influence on plant growth. *Pestic Sci.* 4: 375-383.

H. Yamada, N. Mori and Y. Tani. 1979. Properties of choline oxidase of *Cylindrocarpon didymium* M-1. *Agric. Biol. Chem.* 43: 2173-2177.

X. Yang, Z. Liang and C. Lu. 2005. Genetic engineering of the biosynthesis of glycinebetaine enhances photosynthesis against high temperature stress in transgenic tobacco plants. *Plant Physiol.* 138: 2299-2309.

Index